Sociology

IN A CHANGING WORLD

9e

William Kornblum

*City University of New York,
Graduate School and University Center*

In collaboration with
Carolyn D. Smith

WADSWORTH
CENGAGE Learning

Australia • Brazil • Japan • Korea • Mexico • Singapore • Spain • United Kingdom • United States

Sociology in a Changing World, Ninth Edition
William Kornblum

Acquiring Editor: Erin Mitchell

Developmental Editor: Robert Jucha

Assistant Editor: John Chell

Editorial Assistant: Mallory Ortberg

Media Editor: Melanie Cregger

Marketing Manager: Andrew Keay

Marketing Assistant: Dimitri Hagnere

Marketing Communications
   Manager: Tami Strang

Content Project Manager: Cheri Palmer

Design Director: Rob Hugel

Art Director: Caryl Gorska

Print Buyer: Becky Cross

Rights Acquisitions Specialist: Dean Dauphinais

Production Service: MPS Limited,
   a Macmillan Company

Text Designer: Jeanne Calabrese

Photo Researcher: Premedia Global

Text Researcher: Sue Brekka

Copy Editor: Heather McElwain

Illustrator: MPS Limited, a Macmillan Company

Cover Designer: Riezebos Holzbaur/
   Tae Hatayama

Cover Image: © Steve Schapiro/Corbis (top image)
   AFP/Getty Images (bottom image)

Compositor: MPS Limited,
   a Macmillan Company

For product information and technology assistance, contact us at **Cengage Learning Customer & Sales Support, 1-800-354-9706.**

For permission to use material from this text or product, submit all requests online at **www.cengage.com/permissions** Further permissions questions can be emailed to **permissionrequest@cengage.com**

Library of Congress Control Number: 2001012345

ISBN-13: 978-1-111-30157-6

ISBN-10: 1-111-30157-3

Wadsworth
20 Davis Drive
Belmont, CA 94002-3098
USA

Cengage Learning is a leading provider of customized learning solutions with office locations around the globe, including Singapore, the United Kingdom, Australia, Mexico, Brazil, and Japan. Locate your local office at **www.cengage.com/global**

Cengage Learning products are represented in Canada by Nelson Education, Ltd.

To learn more about Wadsworth, visit **www.cengage.com/wadsworth**

Purchase any of our products at your local college store or at our preferred online store **www.cengagebrain.com**

Printed in the United States of America
1 2 3 4 5 6 7 14 13 12 11

Brief Contents

Entries that appear in magenta refer to the optional Global Social Change chapter.

Contents

PART TWO : SOCIAL DYNAMICS

PART THREE : SOCIAL DIVISIONS

Features

Entries that appear in magenta refer to the optional Global Social Change chapter.

Preface

The more I study, teach, and conduct research in sociology, the more I love the subject. My primary goal in writing this book, therefore, is to share the power and pleasures that come with sociological knowledge. No area of intellectual work overlaps with so many other fields of knowledge; none straddles the sciences and the humanities so squarely or offers such varied insights into the troubling questions of our own time. As the new century has opened on unimaginable acts of terror and bravery, sociology is now more than ever an essential field of study.

This edition of *Sociology in a Changing World* was revised against a background of war and severe economic recession. Many of the key global trends that sociologists use to track social well-being, such as incomes of the poor or improvements in the rights of women, are on a downturn. Widening gaps in wealth and health persist in the United States as well. But there are also long-term trends that are positive, such as growing awareness of the dangers of global warming, or the increasing signs of greater tolerance in racial and ethnic diversity that the election of Barack Obama represents in the United States. Meaningful social change, however, almost never occurs without upsetting traditional cultures, or without raising many issues of social justice as some groups see their lives improved while others experience new forms of inequality. As the late Senator Ted Kennedy liked to say, people want progress but fear change, often for good reasons. To explore the social and political consequences of social change more systematically, we have added a more explicit social justice perspective in this edition of *Sociology in a Changing World*.

Social justice is about fairness in the distribution of opportunities and resources in societies. When we ask researchable questions about who gains and who loses when changes in societies occur, we begin to take a social justice stance on the issues. Although we can assert that sociologists and other social scientists have a primary obligation to address issues of social justice, they are not politicians or preachers who explicitly promote particular arguments about morality and justice. Though we have strong beliefs and opinions about inequalities and injustices, we are responsible for bringing facts to bear on thorny social issues. Thus, every chapter of this book includes a social justice feature that demonstrates how sociologists are addressing social justice issues through their research. For example, Chapter 10, which deals with global inequalities of wealth and power, shows how researchers take a "human needs" approach to developing solutions to address extreme poverty. It shows how sociologists and other scientists help fight global poverty by addressing its root causes: hunger and malnutrition, access to health care, water, sanitation, energy, trade barriers, gender equality, and access to education. In Chapter 6, which describes how we behave in groups and organizations, the social justice feature raises questions about the individual's rights, or lack of them, in different types of organizations. It shows, among other things, that the vast majority of workers in U.S. companies believe that they cannot be fired without just cause, but this is simply not true, as the research clearly demonstrates. This text demonstrates in every chapter how much sociological research has already contributed to improving our social institutions and enhancing our capabilities for acting wisely in difficult social situations. Throughout, it focuses on issues of social change at every level of social organization, because an emphasis on social changes in culture, social structures, inequalities, and social institutions gives students a clear understanding of the types of theoretical and empirical questions that motivate our field.

This book introduces sociology as the preeminent field for the study of social change. Students who use this text will come to understand that whereas social change may be wrenching for some and welcomed by others, it is never without major complications. The tragic events of September 11, 2001, which I witnessed firsthand, were a terrible reminder of this fact. Some of the terrorists viewed our society as threatening to their social order because of the relative freedom women enjoy here or because our movies and television programs show a society that tolerates many differing viewpoints in religion, sexual behavior, and dress.

Of course, there are groups in the United States who are also shocked by some aspects of Western culture. In our own society, we are experiencing major changes, and sociologists study them with all the concepts and methods at their command. These changes include globalization of trade and commerce, the export of industrial production and good jobs outside the developed nations, the rapid growth of metropolitan regions, the aging of our society, soaring energy consumption and the prospect of global warming, new challenges to science and technology because of infectious diseases, and many more changes discussed in this book. Among these is sexuality. Issues of sexuality and gender have assumed such great relevance in the world today that it became clear to me that these subjects deserve even more coverage in the text.

Social change does not mean lack of continuity. Human social institutions like the family, the college, and the corporation adapt to change, but they also resist change and display many similarities over time. A challenge of sociology is to understand and predict both change and its absence. Why, for example, do poor people tend to remain poor, even as new technologies unleash new potentials for wealth and well-being? Why do people of different religions often clash, even though their faiths preach love and brotherhood? Why are women still subordinate to men in so much of the world, even as women continue to press for full equality in their societies? The opportunity to address important questions like these is what makes sociology such a challenging and satisfying field.

I wanted to create a comprehensive text, one that covers all the basic methods and subjects in sociology but one that also has its own clear identity. That is why I chose to focus on the explanation and prediction of social change as a unifying theme throughout the book. After years of research as a working sociologist, and over decades of teaching sociology at the City University of New York, I also believed that I had some insights into what features of a sociology text best help students understand the concepts and their applications. As a result, I had some firm convictions about how to organize the text to assist instructors in their own approaches to the course, and I incorporated some of my own experiences in teaching the course into the book's pedagogical features.

ORGANIZATION OF THE TEXT

Some subjects in sociology, such as class, race, and gender, are so important that they cannot be segregated to a single chapter. They need to be brought up wherever they are salient, and this text tries to do that to the fullest degree possible. I also believe that students need a chance to learn more about how sociologists actually do their work than it is possible to convey in one chapter. That is why I offer Sociological Methods boxes and in-depth discussions of sociological research in each chapter. I believe that instructors can judge whether their students should be required to use this material or whether it can be optional, and this provides many opportunities to use the text with students at different levels of skill and interest.

The book is divided into the following five parts:

Part 1, Social Foundations, introduces the "human science" of sociology: Chapter 1 traces the history of sociology and introduces the major perspectives sociologists use, and Chapter 2 describes the methods sociologists use in conducting their research. The next two chapters focus on some of sociology's most fundamental concepts: culture (Chapter 3) and the elements of social structure (Chapter 4).

Part 2, Social Dynamics, covers many of the processes that seem to account for social stability and change: Chapter 5 discusses socialization, and Chapter 6 examines the structure of groups and how people behave in groups of different kinds. Chapters 7, 8, and 9 move from the structure and function of groups in a society to the processes that change societies. Chapter 7 deals with sexuality; Chapter 8 with deviance; and Chapter 9 looks at collective behavior and social movements.

The five chapters in **Part 3, Social Divisions,** examine social inequality, particularly in the United States. Chapter 10 introduces the concepts of social stratification and global inequality. Chapter 11 deals with inequalities of social class, and Chapters 12, 13, and 14 focus on social inequalities based on race, ethnicity, gender, and age.

Part 4, Social Institutions, applies the concepts and perspectives discussed earlier in the book to an analysis of several major institutions: Chapter 15 explores the changing nature of the family, Chapter 16 introduces religion, and Chapter 17 covers education. Chapter 18 deals with economic institutions, and Chapter 19 discusses political institutions. Chapter 20 analyzes the institutions of science, technology, and medicine.

Part 5, Social Continuity and Change, completes the text by focusing on questions of how and why societies change. Chapter 21 explores the relationships among population growth, urbanization, and community, and Chapter 22 summarizes the ways in which sociologists think about and conduct research on the causes and consequences of social change.

DISTINCTIVE FEATURES

Early in my career, I began using photography and film in research and teaching. I also studied filmmaking at the New York University film school and made some professional documentary films. So when I began working on this book, I wanted it to have a strong visual dimension. We are constantly bombarded with visual images, especially through television and other media, and I wanted students to have a chance to look at photos critically and incorporate direct observation of social phenomena into the introductory course. All textbooks use photos, of course, but I try to put a lot of thought and research into selecting all the visual elements in the book and suggesting their placement and pedagogical function.

Sociology in a Changing World also differs from its competitors in its emphasis on social change. Throughout, the book highlights the tension between attempts to modify social institutions and efforts to maintain traditional modes of behavior. The text also tries to emphasize contemporary social-scientific research. Following are the special features that incorporate this teaching philosophy:

Focus Questions draw attention to key issues discussed in the chapter.

Thinking Critically features appear at places where I feel that students could benefit from some probing questions about the material just presented. These short but provocative features are designed to get students to think critically about key aspects of the material and come up with answers based on sociological facts and personal values.

Opening vignettes begin each chapter to capture students' attention and introduce them to the subject matter. The "story" presented in each vignette is continued in a Visual Sociology box within the chapter.

Research Methods boxes show how sociologists go about conducting their research.

Global Social Change boxes apply sociological concepts to explain major social changes around the world.

Then and Now features contrast photos of a particular social condition in the past and today.

Mapping Social Change features use maps and photos to show the distribution of a condition or characteristic throughout the United States or the world.

For Review and *Perspectives* charts summarize material for ease in studying. They organize topics such as the major sociological perspectives, the basic research methods, effects of cultural contact, and elements of social structure. Several new charts have been added in this edition.

Visual Sociology, a short photo essay, shows how sociologists analyze photographs or other visual material as part of their research.

Chapter summaries are found at the end of each chapter. Each summary is a thorough but concise listing of the key concepts and relationships presented in the chapter.

NEW IN THIS EDITION

The ninth edition maintains the thematic and content strengths of the previous edition but adds new content and features designed to help students better understand the momentous changes that are shaking our peace and security, as well as those that are altering our social institutions—the family, education, religion, and others—in ways that may or may not benefit future generations and invariably generate a good deal of conflict in our own.

Social Continuity and Change

How can we evaluate social change? If change heightens inequalities or increases animosities among groups, especially over a significant period, we can say with confidence that it is not positive social change. But what criteria do we apply to determine whether change is positive? An answer that inspires this book is that to be positive, *social changes must bring about an expansion and increase in social justice throughout the world.* That means more people are safe from hunger, violence, and crime; fewer infants die at birth; laws are applied more equally to rich and poor, women and men; and workers who produce goods and services can afford to feed and clothe and provide health care for their families at a level considered adequate in their societies. Positive social change should improve the likelihood that, in a given society, opportunities are more equal for people regardless of gender, race, religion, national origins, and much more. When sociological research points to ways in which change is moving us away from greater social justice, as is unfortunately so often the case, a critical sociological viewpoint emphasizes why this is so and what lessons we can draw. In consequence, we have included examples of current and critical sociological research that offer constructive criticism of the status quo. Most of the popular chapter-opening vignettes and Visual Sociology boxes, on the other hand, have not been substantially changed because many of the sociologists and subjects dealt with in these sections also figure in the telecourse.

Indeed, throughout, the text emphasizes research by sociologists featured in the *Exploring Society* telecourse, produced by DALLAS TeleLearning in 2002. This insightful and captivating series of videos is keyed to *Sociology in a Changing World* and features sociologists like Arlie Hochschild, Troy Duster, Juan Battle, Victor Ayala, Claire Renzetti, and many others who conduct original research and are experienced, charismatic classroom teachers.

Exploring Society includes twenty-two thirty-minute videos, each of which features documentary footage of real social situations—including, for example, the drop-in center for the homeless in New York City that sociologist Terry Williams and I help manage. These real social situations either explore examples offered in the text (such as what parades reveal about social structure and social conflict) or develop additional examples that complement those in the text. This new version of the telecourse is organized by content modules in each video so that instructors have greater flexibility in presenting the videos during a class session if they so choose, whereas students working at home will have an easier time following the presentation of concepts and examples. A telecourse study guide prepared by my colleague Jane Penney provides the essential integration of videos and text.

New and Expanded Topics

This edition includes numerous discussions of the causes and consequences of the recent wave of terrorism, the impact of Hurricane Katrina, the AIDS pandemic, population growth and planetary atmospheric warming, and globalization. New material dealing with these topics appears throughout the text.

Each chapter of the book has been updated and revised to reflect contemporary trends in sociological research and teaching. In addition to including a new chapter on sexuality, I have added or expanded discussions of the following important topics:

Chapter 1 introduces the theme of social justice by emphasizing W. E. B. DuBois's description of how African Americans released from slavery were denied opportunities in work and education. The chapter also uses the metaphor of parades to illustrate how sociologists view and interpret human activities.

Chapter 2 discusses the differences between quantitative and qualitative research methods and shows how quantitative data on suicide rates among returning veterans point to an important social justice issue: the lack of adequate mental health care for veterans.

Chapter 3 shows how social Darwinists applied Darwin's theories to human inequality, and describes how the next generation of social scientists rejected social Darwinism and its proponents.

Chapter 4 discusses the controversy over Harvard University President Lawrence Summers's statement that genetic differences might explain why fewer women succeed in science and math careers. The text points out that some of the world's most important scientific discoveries have been made by women.

Chapter 5 uses hate crimes against Latino immigrants to illustrate the concept of failures of socialization.

Also featured is recent research on the effects of video games on children's behavior.

Chapter 6 describes William F. Whyte's classic study of street-corner groups in detail, showing how status in the group can have a strong influence on performance. Also covered are the costs of gang membership in terms of increased risk of violence and injury.

Chapter 7 includes a new section on sexuality and social justice, discussing the persecution of people whose sexual behavior or preferences differ from the norms of their cultures.

Chapter 8 discusses the Madoff Ponzi scheme and other white-collar crime. Also covered is the "don't ask, don't tell" policy and the likelihood of its eventual repeal.

Chapter 9 describes the protests for and against the proposed Arizona state law cracking down on illegal immigrants. Also new is a discussion of the hippie movement as an extension of earlier movements, a pattern that can be seen in other movements as well, such as the civil rights movement and the women's movement. The discussion of revolutionary movements is expanded, as is the influence of demagogues on public opinion.

Chapter 10 uses the example of the Haiti earthquake to illustrate the plight of the world's poorest people. The effects of colonialism and capitalism are discussed, along with the efforts of the United Nations to meet the World Millennium Development Goals.

Chapter 11 introduces the Gini index, an international measure of inequality. The importance of social justice in terms of human capital is explored. The discussion of the situation of farmers has been expanded.

Chapter 12 notes the impact of the Great Recession on the situation of minority workers. Also described is the Cherokee "Trail of Tears."

Chapter 13 discusses the feminization of poverty, not just in the United States but throughout the world, and includes a new section on the sociology of masculinities.

Chapter 14 describes recent research on different generations, including the "Millennials," teenagers and young adults, and the impact of the Great Recession on this generation.

Chapter 15 explores numerous issues related to families, such as unemployment in married-couple families, mothers in the labor force, families in poverty, the plight of children during the current recession, family poverty and food insecurity, and the "boomerangers"—young adults forced to move back in with their parents due to the recession.

Chapter 16 describes new findings on religious belief and practices in the United States. Issues related to the sexual abuse scandal in the Catholic Church are discussed.

Chapter 17 includes a new section on education and military service.

Chapter 18 describes state capitalism in China, Vietnam, Cuba, and other former socialist nations. The impact of the Great Depression in the United States is emphasized. Also discussed is the debt crisis experienced by consumers who have been induced to borrow beyond their ability to repay.

Chapter 19 refers to current conditions of political instability throughout the world, and places particular emphasis on human rights. A summary of the Universal Declaration of Human Rights is included.

Chapter 20 now includes an extensive section on medicine and health in global perspective, with discussions of investing in global health care systems, the importance of safe water, social justice and health in the developed nations, and the need for preventive care and control of rising medical costs. A Research Methods box presents a typology of health care systems.

Chapter 21 emphasizes life expectancy, infant mortality, and women's reproductive rights.

Chapter 22 discusses modernization and the rule of law. Also covered is the accelerating impact of climate change.

SUPPLEMENT PACKAGE

Sociology in a Changing World is accompanied by a wide array of supplements prepared for both instructors and students to create the best learning environment inside as well as outside the classroom. All the continuing supplements for *Sociology in a Changing World* have been thoroughly revised and updated, and several resources are new to this edition.

Supplements for the Instructor

Instructor's Manual Created by author William Kornblum and collaborator Carolyn D. Smith, this resource contains a detailed lecture outline; instructional goals; extensive teaching suggestions; topics for discussion; strategies for how to effectively use the text's rich selection of tables, charts, and other visual features in class; student exercises; InfoTrac® College Edition exercises; Internet activities; and a list of resources for each chapter of the text. The Instructor's Resource Manual contains extended Internet exercises with critical thinking questions, a separate list of Internet resources for class discussion and reference, and a variety of print, video, and Internet resources to supplement the new edition's "Sociology and Social Justice" feature.

Test Bank This extensive test bank includes 100 to 125 multiple-choice, 30 to 40 true-false, 8 to 10 short-answer, and 3 to 5 essay questions for each chapter of the text, all with page references and answer explanations. Each multiple-choice item has the question type (factual, applied, or conceptual) indicated. All questions are labeled as new, modified, or pickup so instructors know if the question is new to this edition of the test bank, picked up from the previous edition of the test bank but modified, or picked up straight from the previous edition of the test bank. The revised test items now include additional questions based on the main text's rich selection of box features and visuals. Further test questions specific to the "Exploring Society: An Introduction to Sociology" telecourse are provided as an appendix.

ExamView® for Windows/MAC Create, deliver, and customize tests and study guides (both print and online) in minutes with this easy-to-use assessment and tutorial system. ExamView includes a Quick Test Wizard and an Online Test Wizard to guide instructors step by step through the process of creating tests. The test appears on screen exactly as it will print or display online. Using ExamView's complete word-processing capabilities, instructors can enter an unlimited number of new questions or edit questions included with ExamView.

Optional Custom Chapter 22, "Global Social Change" The optional chapter on Global Social Change is available for inclusion in the main text via the custom group. Supporting material for this chapter is available in all of the print supplements and in the computerized testing program. Please consult your sales representative for details.

CourseReader: Sociology This is a fully customizable online reader that provides access to hundreds of readings and video selections from multiple disciplines. This easy-to-use solution allows you to select exactly the content you need for your courses and is loaded with convenient pedagogical features like highlighting, note taking, and digital video downloads. YOU have the freedom to assign individualized content at an affordable price. CourseReader: Sociology is the perfect complement to any class.

PowerPoint® Presentations Available for download from the instructor's companion website, these updated and redesigned PowerPoint Presentations outline each chapter to help guide classroom lectures and stimulate student discussion.

Instructor Companion Website To access additional course materials, please add this title to your Instructor Account at **www.login.cengage.com**. On the instructor site, you will find resources like the Instructor's Manual, Test Bank, and more!

Classroom Presentation Tools for the Instructor

Wadsworth's Lecture Launchers for Introductory Sociology An exclusive offering jointly created by Thomson Wadsworth and DALLAS TeleLearning, this video contains a collection of video highlights taken from the *Exploring Society: An Introduction to Sociology* telecourse (formerly *The Sociological Imagination*). Each three- to six-minute video segment has been specially chosen to enhance and enliven class lectures and discussions of twenty key topics covered in an introduction to sociology course. Accompanying the video is a brief written description of each clip, along with suggested discussion questions to help effectively incorporate the material into the classroom. Available on VHS or DVD.

Sociology: Core Concepts Video Another exclusive offering jointly created by Wadsworth and DALLAS TeleLearning, this video contains a collection of video highlights taken from the *Exploring Society: An Introduction to Sociology* telecourse (formerly *The Sociological Imagination*). Each fifteen- to twenty-minute video segment will enhance student learning of the essential concepts in the introductory course and can be used to initiate class lectures, discussion, and review. The video covers topics such as the sociological imagination, stratification, race and ethnic relations, social change, and more. Available on VHS or DVD.

The Wadsworth Sociology Video Library Vol. 1 The Wadsworth Sociology Video Library drives home the relevance of course topics through short, provocative clips of current and historical events. Perfect for enriching lectures and engaging students in discussion, many of the segments on this volume have been gathered from BBC Motion Gallery. Ask your Cengage Learning representative for a list of contents.

Supplements for the Student

Student Telecourse Guide The Telecourse Guide for Kornblum's *Sociology in a Changing World*, Ninth Edition, is designed to accompany the *Exploring Society: Introduction to Sociology* telecourse produced by DALLAS TeleLearning of the Dallas County Community College District. The Telecourse Guide provides the essential integration of videos and text, providing students with valuable resources designed to direct their daily study in the *Exploring Society* telecourse. Each chapter of the Telecourse Guide contains a lesson that corresponds to one of the twenty-two video segments in the *Exploring Society* telecourse. Each lesson includes the following components: Overview, Lesson Assignment, Lesson Goal, Lesson Learning Objectives, Review, Lesson Focus Points, Related Activities, Practice Tests, and an Answer Key.

Student Companion Website

To access additional course materials, please visit the student companion website at **www.cengagebrain.com**. At the CengageBrain.com home page, search for the ISBN of your title (from the back cover of your book) using the search box at the top of the page. This will take you to the product page where resources like tutorial quizzes, glossary, flash cards, and more can be found.

Internet-Based Supplements

CourseReader: Sociology This is a fully customizable online reader that provides access to hundreds of readings and video selections from multiple disciplines. This easy-to-use solution allows you to select exactly the content you need for your courses, and is loaded with convenient pedagogical features like highlighting, note taking, and digital video downloads. YOU have the freedom to assign individualized content at an affordable price. CourseReader: Sociology is the perfect complement to any class.

WebTutor™ ToolBox for WebCT® and Blackboard®

WebTutor ToolBox combines easy-to-use course management tools with content from this text's rich companion website. It is ready to use as soon as you log on—or customize WebTutor ToolBox with weblinks, images, and other resources.

ACKNOWLEDGMENTS

This new edition of *Sociology in a Changing World* is the culmination of two years of steady work—weekly writing and editing for a year, plus another year of production-related tasks. The value in a book like this is the work of many minds and hands. From the contributions of the social scientists who have reviewed the manuscript and the sociology students who have commented on previous editions, to the final touches of detail before the book is printed, there are scores of people whose contributions are essential at every step in the process. It is exciting to launch a new edition of the text, just as it is gratifying to thank those who made it possible. I have had the benefit of continuity in editors and professional friends who have advised me since the first edition. But each edition also brings in new publishing professionals whose influences help keep the book fresh and up to date.

First among the many people who continue to influence this project is my editor and collaborator Carolyn Smith. Carolyn first encouraged me to take on this project, and when my resolve wavered, her faith in our ability to complete it sustained me. Carolyn has untangled my prose, agonized with me over reviewers' comments, and kept track of the endless details that go into creating a scholarly text. There is not one word in this book that she has not thought about; every page bears the stamp of her expertise. Guy Smith also has my gratitude for his support during the long and demanding revision process.

The visual identity of this book, its use of photographic material for sociological example and method, has always been extremely important to me and to everyone involved in its creation. Our aim has been to create a textbook that would integrate sociological ideas into the design and layout of the book. In this edition, the design created by Jeanne Calabrese, along with the cover design created by Yvo Riezebos, RHDG Design, and the art and photo programs continue to emphasize creative treatments of data presentation and visual sociology.

My colleagues at the City University of New York have been unstinting with their suggestions and advice. My dear friend and colleague, Vernon Boggs, was always my greatest personal ally in the making of this book. His untimely death deprives me of someone whose deep commitment to teaching and research in sociology were a sustaining example. My longtime collaborator and friend, Terry Williams, continues to influence my thinking on key issues in sociology and life. Rolf Meyersohn has been heroic in his generosity with books and articles and words of encouragement. Paul Attewell's mastery of trends in sociological research and methods, and the profound knowledge of social movements and class theory shared by Stanley Aronowitz, William Helmreich, William DiFazio, and Dean Savage, have been invaluable on many occasions. If I have managed to avoid a natural predisposition toward male bias, much of the credit must go to the late Susan Kornblum, Edith Goldenhar, Cynthia Epstein, Judith Lorber, June Nash, Barbara Katz Rothman, Julia Wrigley, and many other colleagues who have shared their insights and knowledge. My colleagues Lily Hoffman, John Mollenkopf, and Mitchell Duneier are unstinting in sharing their sociological insights and knowledge of new literature. Special thanks are due to my dear colleague Juan Battle for his generous help.

The thoughtful comments we receive about the book and the encouragement of many students and colleagues are immensely rewarding. Their corrections and constructive suggestions, too, are most welcome. Many colleagues throughout the United States offer me wise counsel and scholarly advice, for which I continue to be deeply grateful. But I owe a special debt to my colleague and friend, Carol Chenault of Calhoun Community College, Decatur, Alabama, for her generosity and her abiding interest in the textbook. Professor Chenault's comparative visual sociology is featured in Chapter 13 of the text.

No single author can hope to adequately represent a field as large and complex as sociology. Over several years of textbook writing, I have learned that the comments and suggestions offered by reviewers and users of the book are essential to its success. Many dedicated sales representatives have taken the time to introduce me to instructors who use the book. In addition, the comments of numerous reviewers have helped me to correct mistakes of fact, interpretation, and emphasis. I might disagree with some of them, but I always find them helpful.

Each edition of the book has benefited from the insights of reviewers. I gratefully acknowledge the following instructors who contributed by reviewing or participating in a survey to the ninth edition of *Sociology in a Changing World*:

Charles Baker, Olympic College

Carol Chenault, Calhoun Community College

Mark G. Eckel, McHenry County College

Kathryn Feltey, University of Akron

Alan G. Hill, Delta College

Susan Holbrook, Southwestern Illinois College

Eileen Kaufman, Union County College

Annie Leslie, Bowie State University

Lori Maida, Westchester Community College

Sean Noonan, Harper College

Sharon Squires, California State University, Dominguez Hills

Isaac Sakyi-Addo, Calhoun Community College

Madeline Troche-Rodriguez, Harry S. Truman, City College of Chicago

Stacey Tucker, University of Tennessee

In preparing this edition of *Sociology in a Changing World*, I greatly benefited from the counsel and help of a team of professional editors and publishers who recognized the strengths of the book and worked with me to make improvements. In this regard, I am especially grateful to Erin Mitchell, Acquisitions Editor; Robert Jucha, Development Editor; Melanie Cregger, Media; John Chell, Assistant Editor; Mallory Ortberg, Editorial Assistant; Cheri Palmer and Jill Traut for their skillful production management and editorial coordination; Jaime Jankowski for her help with the photo program; and Andrew Keay for his marketing efforts. Thanks also go to my students at the City University of New York, both graduates and undergraduates, for keeping me "turned on" to sociology.

William Kornblum is a professor of sociology at the Graduate School of the City of New York, where he helps train future instructors and researchers in the social sciences. He also teaches undergraduates at various campuses of the City University, including Queens College, Hunter College, and City College.

A specialist in urban and community studies, Kornblum began his teaching career with the Peace Corps in the early 1960s, when he taught physics and chemistry in French-speaking West Africa. He received his doctorate in sociology from the University of Chicago in 1971. He has also taught at the University of Washington at Seattle and worked as a research sociologist for the U.S. Department of the Interior. At the CUNY Graduate School, he directs research on environmental issues and urban policy. Kornblum's latest book is *At Sea in the City: New York from the Water's Edge* (Algonquin Books, 2002). With his longtime research partner, Terry Williams, he co-authored *The Uptown Kids* (Putnam, 1998, translated into Japanese in 2010), a sociological portrait of teenagers and young adults growing up in high-rise public housing projects. He was also the principal investigator of Project TELL, a longitudinal study of the ways in which home computers can improve the life chances of young people at risk of dropping out of school. In 2005, the Russell Sage Foundation published his study of the impact of 9/11 on the region's airline workers in *Wounded City*, edited by Nancy Foner. That same year, he became a proud grandfather.

The author's other publications include *Blue Collar Community*, a study of the steel-making community of South Chicago; *Growing Up Poor* (with Terry Williams), a study of teenagers growing up in different low-income communities in the United States; and *Social Problems*, a comprehensive textbook about social problems and social policies in the United States.

Sociology: An Introduction

FOCUS QUESTIONS

What is the sociological imagination, and how does it generate the questions that make sociology a "human science"?

How do sociologists study social change at different levels of social reality?

Who are the key thinkers and researchers who have transformed social thought into social science?

How do the major sociological perspectives approach different social issues?

What is the Memorial Day parade like in your town? Who participates? What feelings do people have as they watch the parade? For the residents of Waterloo, New York, one of the earliest communities to hold a Memorial Day parade, the annual event is a source of local pride, and its parade is like those held in many other towns, villages, and cities throughout the United States each year at the end of May. At the opposite extreme, the largest and most elaborate Memorial Day parade is held in the nation's capital, where the ceremony includes decoration of the hundreds of thousands of graves of men and women who sacrificed their lives in the defense of the nation. With military personnel engaged in conflict in Iraq, Afghanistan, and elsewhere in a troubled and often violent world, each Memorial Day parade in recent years has served as a reminder of the costs of war in deaths, wounds, and loss of loved ones.

Throughout the world, significant social occasions are marked by parades. The people walking or marching in the parade enjoy dressing in the appropriate uniforms or costumes and showing off before an admiring public. Spectators enjoy the pageantry, the excitement, the bands, the costumes, and much more. To sociologists and other keen observers, most parades are an occasion to observe people displaying many of the attributes of their societies that they value and enjoy.

Although the parades on Memorial Day commemorate military sacrifice and service to the nation, these parades have many additional meanings for a sociologically observant spectator. In many communities, despite the underlying sadness of the occasion, a spirit of pride and joyfulness is evident in the marchers and the crowds. Children and teenagers may decorate their bicycles and ride alongside the marching bands and floats. Families wave and cheer for loved ones as they march by. The local school bands and organizations are often part of the march, as are many of the community's prominent citizens and elected officials. An observant sociologist can learn a great deal about a community or society by just watching the parade in Washington, D.C., or the more modest ones like that of Waterloo, New York. So much of the way a parade is organized is a purposeful display of what we value and how our society is changing. That a woman's marching band takes a prominent role in the Washington parade, for example, tells us that gender equality and the role of women in the military is being emphasized by the parade's organizers. Decisions as to which groups and individuals will lead the parade tell us a good deal about who has power or recognition in the community.

Parades may have religious significance, as when a town in Mexico or Central America gathers to parade its local saints through the joyful streets outside the church. The parade of worshippers who stream through the holy Muslim shrines of Mecca or Medina in Saudi Arabia attracts Muslims from throughout the world and presents a dazzling display of styles of dress and pious devotion. Parades may combine religious themes with celebrations of the people of the society themselves, as we see in the annual Mardi Gras parade in New Orleans. There, an insightful observer can watch how people from different social backgrounds—rich, poor, black, white, women, men, and so on—interact in the parade. And the ways in which the parade and other Mardi Gras celebrations, especially the famous masked balls, have changed over the last few decades tell us a great deal about how New Orleans society is changing. For example, as the world witnessed during and after Hurricane Katrina, the people of New Orleans are quite diverse racially and ethnically. African Americans and other people of color are the majority of the city's population, and many were too poor even to pay bus fare to escape from the city. Their neighborhoods were often below sea level and therefore were most likely to be destroyed. As New Orleans struggles to rebuild, the annual Mardi Gras parade is halted for a moment of silence in commemoration of Katrina's 1,300 victims and for the losses so many residents suffered (Robertson, 2009).

Parades may also be revealing to a sociological observer who notes what groups are missing or underrepresented in the passing contingents and in the watching crowds of spectators. Young people from poor families and men and women of racial and ethnic minority status are more likely to fight and die in wars in foreign lands. But are they adequately represented in the Memorial Day events? Are they represented among the parade's leaders? Where do they figure in the parade? Their presence or absence, and the places of honor they may be accorded, can tell us a good deal about the equity of sacrifice and recognition in the given town or city.

All these subjects will come up in later chapters. The point here is that sociologists love parades for what they can reveal about underlying social relations.

THE SOCIOLOGICAL IMAGINATION

It may take some thought to see why various kinds of parades tell us a great deal about our own and other societies. After all, one might argue that a parade is a special event. It is not meant to be "read" for its deeper social meanings. And no doubt most of the people who watch Mardi Gras parades, for example, are caught up in the fun of the event—by the creativity of the costumes, the flirtations that go on, the music and dance, and perhaps the drinking. They might be offended if someone asked them to look more analytically at the larger significance of their behavior. But a sophisticated understanding of social life requires some imagination. Although some participants in an event like a parade may be entirely caught up in the moment, others may also have fun while simultaneously thinking about what is going on at a deeper level.

One of the main goals of sociology courses is to help you develop the ability to both participate in social life and step back and analyze the broader meanings of what is going on. This ability is often called the **sociological imagination.** In this book, we hope to help you develop this special insight, which will equip you to use sociological knowledge in your daily life. Most of all, we hope to enable you to use your sociological imagination to gain wisdom about the society in which we all participate and for whose future we are all responsible.

Most people need some help in developing a sociological imagination. This is especially true when it comes to understanding their own place in what might be thought of as the "parade" of social life. People with a limited sociological imagination often fail to distinguish between social forces and personal troubles. If they are excluded from the "parade" because they are unemployed, they blame themselves for failing to do better; if they divorce, they blame each other. When they see crime, they blame "human nature"; when they see success, they praise individual achievement. But this tendency to think of life as a series of individual mistakes or successes blinds them to the fact that social conditions also shape their lives, often in ways for which they can hardly be held accountable. The habit of seeing events mainly in terms of how they affect individuals blinds people to the possibility of improving the way their society is organized.

Sociologists are concerned with how **social conditions** influence our lives as individuals. Social conditions are the realities of the life we create together as social beings. Conditions such as poverty, wealth, crime, and drug use, for example, differ from biological facts (facts concerning our behavior and needs as animals) and psychological facts (facts about our patterns of behavior as individuals). Sociologists do not deny that psychological facts are important. Differences among individuals help some people cope with stress better than others, seize opportunities that others

sociological imagination: According to C. Wright Mills, the ability to see how social conditions affect our lives.

social conditions: The realities of the life we create together as social beings.

allow to slip by, or fail where others succeed. But before saying that the success or failure of an individual or group is the result of psychological causes, sociologists try to look at how social conditions such as poverty or wealth, war, or changes in the availability of jobs affect an individual's chances of success.

According to sociologist C. Wright Mills, who made famous the term *sociological imagination,* people often believe that their private lives can be explained mainly in terms of their personal successes and failures. They are critical of themselves but not of their societies. They fail to see the links between their own individual biographies and the course of human history. Often, they blame themselves for their troubles without grasping the effects of social change on their lives. "The facts of contemporary history," Mills points out, "are also facts about the success and the failure of individual men and women." Mills further states:

> When a society is industrialized, a peasant becomes a worker; a feudal lord is liquidated or becomes a businessman. When classes rise or fall, a man is employed or unemployed; when the rate of investment goes up or down, a man takes new heart or goes broke. When wars happen, an insurance salesman becomes a rocket launcher; a store clerk, a radar man; a wife lives alone; a child grows up without a father. (1959, p. 3)

According to Mills, neither a person's biography nor the history of a society can be understood unless we consider the influence of each on the other:

> People do not usually define the troubles they endure in terms of historical change and institutional contradiction. The well-being they enjoy, they do not usually impute to the big ups and downs of the societies in which they live. Seldom aware of the intricate connection between the patterns of their own lives and the course of world history, ordinary people do not usually know what this connection means for the kinds of people they are becoming and for the kinds of history-making in which they might take part. (1959, p. 1)

The social forces of history—war, depression or recession, increases in population, changes in production and consumption, and many other social conditions—become the forces that influence individuals to behave in new ways. But those new ways of behavior themselves become social forces and, in turn, shape history.

To take just one example, the not-so-distant ancestors of many African Americans were brought to the Western Hemisphere in slave ships. African Americans have experienced slavery, war, emancipation, segregation, and rural and urban poverty. In reaction to the historical forces that deprived them of full citizenship in the United States, they developed a variety of behaviors, from the spirituals that expressed their deep feelings of religious faith and protest to boycotts and demonstrations against segregation. These protests and demonstrations, which often took the form of a special kind of parade, became powerful social forces that continue to shape the history of the American people.

By applying the sociological imagination to events such as the way Hurricane Katrina changed New Orleans and its people, or to many of the episodes of tragedy and heroism that have occurred during the war in Iraq, one can begin to "grasp history and biography and the relations between the two within society" (Mills, 1959, p. 6). Individual greed may explain why funds for maintaining levees were mismanaged, or why the federal government's response to the suffering of the city's most vulnerable residents was so negligent. But in addition to these more personal or psychological reasons, widespread failures of communication and organization severely hampered the response to the disaster. In other words, failures at the sociological level were as important as individual motivations.

The invasion of Iraq by the United States and its allies also demonstrates a host of sociological failures, the most important of which were lack of planning for a prolonged occupation of a proud nation and underestimation of the likelihood of insurgency and civil war (Ricks, 2009).

One of the main objectives of this book is to help you apply your sociological imagination to an understanding of the social forces that are shaping your own place in the parade—that is, the forces that are shaping your life and those of the people you care about. The sociological imagination can help us avoid blaming ourselves needlessly for the troubles we encounter in life. It can help us understand, for example, why some people are rich and powerful but many others are not; why the benefits of good health care or enriching education are available to some but not to others; or why women may find themselves resenting the men in their lives. The sociological imagination helps us sort out which facts about ourselves are explained by our place in society and which ones are a result of our own actions. Above all, the sociological imagination can suggest ways in which we can realistically effect change in our lives and in society itself.

SOCIOLOGY AND SOCIAL CHANGE

Sociology is the scientific study of human societies and human behavior in the many groups that make up a society. Sociologists must ask difficult, sometimes embarrassing questions about human life in order to explore the consequences of cataclysmic events such as those that shut down factories or enslave an entire people. To understand the possible futures of people who confront such drastic changes, sociologists are continually seeking knowledge about what holds societies together and what makes them bend under the impact of major forces such as war and migration.

Throughout this book, we will see that there are many ways in which sociologists can already predict the social changes that are likely to occur in coming years and decades. For example, as the proportion of elderly people in our society increases, we can predict how this important social change will bring about other changes in the need for health care or for improved retirement systems, or new definitions of the responsibilities of middle-aged children for the elderly. But there are also changes, such as the rise of terrorism directed against the United States, that might have been predicted and better anticipated but were not. Sociologists are trying to develop better knowledge about the causes of terrorism in order to help diminish the likelihood of terrorism in the future. To do so, we conduct research to better understand how different societies are changing or remaining stable, which groups are benefiting from social change, and which groups are being left behind.

The Social Environment

The knowledge sociologists gather covers a vast range. Sociologists study religious behavior; conduct in the military; the behavior of workers and managers in industry; the activities of voluntary associations such as parent–teacher groups and political parties; the changing relationships between men and women or between aging individuals and their elderly parents; the behavior of groups in cities and neighborhoods; the activities of gangs, criminals, and judges; differences in the behaviors of entire social classes—the rich, the middle classes, the poor, the down-and-out; the way cities grow and change; the fate of entire societies during and after revolutions; and a host of other subjects. But how to make sure the information gathered is reliable and precise, how to use it to build theories of social cohesion and social change—that is the challenge faced by the young science of sociology.

As in any science, there are many debates in sociology about the appropriate ways to study social life and about which theories or types of theories best explain social phenomena. Most sociologists, however, would agree with the following position:

> Human actions are limited or determined by "environment." Human beings become what they are at any given moment not by their own free decisions, taken rationally and in full knowledge of the conditions, but under the pressure of circumstances which delimit their range of choice and which also fix their objectives and the standards by which they make choices. (Shils, 1985, p. 805)

This statement expresses a core idea of sociology: Individual choice is never entirely free but is always determined to some extent by a person's environment. In sociology, *environment* refers to all the expectations and incentives established by other people in a person's social world. For sociologists, therefore, the environment within which an individual's biography unfolds is a set of people and groups and organizations, all with their own ways of thinking and acting. Certainly each individual has unique choices to make in life, but the social world into which that person was born—be it a Native American reservation, an urban ghetto, a comfortable suburb, or an immigrant enclave in a strange city—determines to varying degrees what those choices will be.

Levels of Social Reality

In their studies of social environments, sociologists look at behaviors ranging from the intimate glances of lovers to the complex coordination of a space shuttle launch. For purposes of analysis, however, they often speak of social behavior as occurring at three different levels of complexity: micro, middle, and macro.

The Micro Level Micro-level sociology is concerned with observation of the behaviors of individuals and their immediate others—that is, with patterns of interaction among a few people. One example is Erving Goffman's studies of the routine behaviors of everyday life.

sociology: The scientific study of human societies and human behavior in the groups that make up a society.

micro-level sociology: An approach to the study of society that focuses on patterns of social interaction at the individual level.

Each of these passengers in a railroad station waiting room has claimed a personal territory. Each has taken a separate bench and has chosen to sit at one end of the bench. If another person came to that bench, he or she would probably choose to sit as far away as possible.

Michael Dwyer/Stock, Boston Inc.

Goffman's research showed how seemingly insignificant ways of acting in public actually carry significant meanings. In a study titled "Territories of the Self" (1972), Goffman categorized some of the ways in which we use objects as "markers" to claim a personal space:

> Markers are of various kinds. There are "central markers," being objects that announce a territorial claim, the territory radiating outward from it, as when sunglasses and lotion claim a beach chair, or a purse a seat in an airliner, or a drink on a bar the stool in front of it. . . . There are "boundary markers," objects that mark the line between two adjacent territories. The bar used in supermarket checkout counters to separate one customer's batch of articles from the next is an example. (pp. 41–42)

The last time you placed your sweater or book on the empty seat next to you on a bus, you told yourself that when someone came for the seat you would take up your things. But you hoped that the stranger who was coming along the aisle would get your message and choose another seat; you would claim your extra space as long as possible. You communicated all this by the manner in which you placed your marker and the persistence with which, by your body language, you defended "your" space.

The Macro Level Some sociologists deal almost exclusively with analysis on a much larger scale, termed **macro-level sociology**. The macro level of social life refers to whole societies and how they are changing—that is, to revolutions, wars, major changes in the production of goods and services, and similar social phenomena that involve very large numbers of people. One example of macrosociological analysis is the study of how the shift from heavy manufacturing to high-tech industries has affected the way workers earn a living. Another example is the study of how the invasion and settlement of the American West in the nineteenth and early twentieth centuries gave rise to the beliefs and actions that drove Native Americans onto reservations.

macro-level sociology: An approach to the study of society that focuses on the major structures and institutions of society.

middle-level sociology: An approach to the study of society that focuses on relationships between social structures and the individual.

The Middle Level **Middle-level sociology** involves social phenomena that occur in communities or in organizations such as businesses and voluntary associations. Middle-level social forms are smaller than entire societies but are larger and involve more people than micro-level social forms such as the family or the peer group, in which everyone involved knows everyone else or is in close proximity to the others (as on a bus or in a classroom). The drama that surrounds the firing of a coach on a sports team, and the reorganization of personnel that often follows, is an example of social change at the middle level of social analysis.

These three levels of sociological analysis can be helpful in understanding the experience of the immigrants from Asia, Africa, or Latin America who may be appearing in your town or community or in one nearby. Macro-level social forces such as war or overpopulation may account for the influx of immigrants. Middle-level social forces such as the availability of work at low wage and skill levels, or the presence of earlier arrivals from other nations, may help explain why certain immigrant groups become concentrated in particular communities within the United States. And at the micro level of analysis, there will be important differences between the ways in which immigrant and native-born people interact on a daily basis, especially at first.

Note that the three levels of social reality are not defined according to fixed or standard measures. Instead, they are used in relative fashion. The social impact of a global corporation such as General Electric, for example, can be analyzed at the macro level of the entire corporation and its dealings on the world stage. In that case, its component factories, and the trade unions they often battle against, are middle-level forms, and specific groups within those factories operate at the micro level within the global corporation. How one uses and defines these levels of social reality often depends on the type of sociological analysis one is performing.

The For Review chart on page 7 presents the three basic levels of sociological analysis along with some examples of the types of studies conducted at each level. Throughout this book, we show how the sociological imagination can be applied at different levels of society. In this chapter, we set the stage by describing the basic perspectives from which modern sociologists approach the study of social conditions. In the next chapter, we outline the procedures and methods sociologists use in conducting their research. We begin here with a brief description of the origins and development of the science of sociology.

Levels of Sociological Analysis

Analytical Level	Social Behaviors Studied	Typical Questions
Macro	Revolutions; intercontinental migrations; emergence of new institutions.	How are entire societies or institutions changing?
Middle	Relations in bureaucracies; social movements; participation in communities, organizations, tribes.	How does bureaucracy affect personality? Do all social movements go through similar stages?
Micro	Interaction in small groups; self-image; enactment of roles.	How do people create and take roles in groups? How are group structures created?

FROM SOCIAL THOUGHT TO SOCIAL SCIENCE

Like all the sciences, sociology developed out of prescientific longings to understand and predict. The world's great thinkers have pondered the central questions of sociology since the earliest periods of recorded history. The ancient Greek philosophers believed that human societies inevitably arose, flourished, and declined. They tended to perceive the past as better than the present, looking back to a "golden age" in which social conditions were presumed to have been better than those of the degraded present. Before the scientific revolution of the seventeenth century, the theologians and philosophers of Medieval Europe and the Islamic world also believed that human misery and strife were inevitable. As the Bible put it, "The poor always ye have with you." Mere mortals could do little to correct social conditions, which were viewed as the work of divine Providence.

The Age of Enlightenment

The roots of modern sociology can be found in the work of the philosophers and scientists of the Great Enlightenment, which had its origins in the scientific discoveries of the seventeenth century. That pivotal century began with Galileo's "heretical" proof that the earth was not the center of the universe; it ended with the publication of Isaac Newton's *Principia Mathematica*. Newton is often credited with the founding of modern science. He not only discovered the laws of gravity and motion but, in developing the calculus, also provided later generations with the mathematical tools whereby further discoveries in all the sciences could be made.

Hard on the heels of this unprecedented progress in science and mathematics came a theory of human progress that paved the way for a "science of humanity." Francis Bacon in England, René Descartes and Blaise Pascal in France, and Gottfried Wilhelm Leibniz in Germany were among the philosophers who recognized the social importance of scientific discoveries. Their writings emphasized the idea of progress guided by human reason and opposed the dominant notion that the human condition was ordained by God and could not be improved through human actions (Bury, 1932; Nisbet, 1969).

Today, we are used to inventions crowding one upon another. Between the childhood of our grandparents and our own adulthood, society has undergone some major transformations: from agrarian to industrial production; from rural settlements and small towns to large cities and expanding metropolitan regions; from reliance on wood and coal as energy sources to dependence on electricity and nuclear power; from typewriters to computers. In the seventeenth century, however, people were used to far more stability. Ways of life that had existed since the Middle Ages were not expected to change in a generation.

The rise of science transformed the social order. As was often said at the time, science "broke the cake of custom." New methods of navigation made it possible to explore and chart the world's oceans and continents. Applied to warfare, scientific knowledge enabled Europeans to conquer the peoples of Africa, Asia, and the Western Hemisphere. In Europe, those conquests opened up new markets and stimulated new patterns of trade that hastened the growth of some regions and cities and the decline of others. The entire human world had entered a period of rapid social change that continues today and shows no signs of ending.

The Age of Revolution

The vehicle of social change was not science itself, because relatively few people at any level of society were practicing scientists. Rather, the modern era of rapid social change is a product of the many new ideas that captured people's imagination during the eighteenth century. The series of revolutions that took place in the American colonies, in France, and in England all resulted in part from social movements unleashed by the triumphs of science and reason. The ideas of human rights (that is, the rights of all humans, not just the elite), of democracy versus rule by an absolute monarch, of self-government for colonial peoples, and of applying reason and science to human affairs in general—all are currents of thought that arose during this period.

The revolutions of the eighteenth century loosed a torrent of questions that could not even have been imagined before. The old order of society was breaking down as secular (that is, nonreligious) knowledge replaced sacred traditions. The study of laws and lawmaking and debates about justice in society began to replace the idea that kings and other leaders had a "divine right" to rule. Communities were breaking apart; courts, palaces, and great estates were crumbling as people struggled to be free. What would replace them? Would the rule of the mob replace the rule of the monarch? Would greed and envy replace piety and faith? Would there be enough opportunities in the New World for all the people who were being driven off the land in the Old World? Would the factory system become the new order of society, and if so, what did that imply for the future of society?

Karl Marx

Émile Durkheim

No longer could the Scriptures or the classics of ancient Greece and Rome be consulted for easy answers to such questions. Rather, it was becoming evident that new answers could be discovered through the **scientific method**: repeated observation, careful description, the formulation of theories based on possible explanations, and the gathering of additional data about questions arising from those theories. Why not use the same methods to create a science of human society? This ambitious idea led to the birth of sociology. It is little wonder that the French philosopher Auguste Comte thought of sociology, even in its infancy, as the "queen of sciences," one that would soon take its rightful place beside the reigning science of physics. It was Comte who coined the term *sociology* to designate the scientific study of society. He believed that the study of social stability and social change was the most important subject for sociology to tackle. He made some of the earliest attempts to apply scientific methods to the study of social life.

The Founders of Sociology

In the nineteenth century, an increasing number of philosophers and historians began to see

themselves as specializing in the study of social conditions and social change. They attempted to develop global theories of social change based on the essential qualities of societies at different stages of human history, and they devoted much of their attention to comparing existing societies and civilizations, both past and present. As Comte put it, the age of discovery had revealed such an array of societies that, "from the wretched inhabitants of Tierra del Fuego to the most advanced nations of Western Europe," there is such a great diversity of societies that comparisons among them will yield much insight into why they differ and how they change (1971/1854, p. 48).

The early sociologists tended to think in macrosociological terms. Their writing dealt with whole societies and how their special characteristics influence human behavior and social change. Most sociologists would agree that the nineteenth-century social theorists who had the greatest and most lasting influence on the field were Karl Marx, Émile Durkheim, and Max Weber. All three applied the new concepts of sociology to gain an understanding of the immense changes occurring around them.

Karl Marx German-born Karl Marx (1818–1883) became a radical philosopher as a young man and was embroiled in numerous insurrections and attempts at revolution in Germany and France. Forced to flee Germany after the abortive European revolution of 1848, he lived and worked in England for the rest of his life. Often penniless, Marx worked for hours on end in the library of the British Museum, where he developed the social and economic theories that would have a major influence on sociological thought. His famous treatise *Capital* is a detailed study of the rise of capitalism as a dominant system of production. In this work and elsewhere, Marx set forth an extremely powerful theory to explain the transformations taking place as societies became more industrialized and urbanized. He argued that those transformations would inevitably end in a revolution in which the workers would overthrow capitalism, but he also believed that revolution could be hastened through political action.

Émile Durkheim Émile Durkheim (1858–1917) was the founder of scientific sociology in France. His books, among which the best known are *The Division of Labor in Society, Rules of the Sociological Method,* and *Suicide,* were pioneering examples of the use of comparative data to assess the directions and consequences of social change. The first university professor with a chair in the "social sciences,"

scientific method: The process by which theories and explanations are constructed through repeated observation and careful description.

Durkheim was soon surrounded by a brilliant group of academic disciples who were deeply interested in understanding the vast changes that occurred in societies as they became more populous, more urbanized, and more technologically complex. In 1898, Durkheim and his colleagues established the first scientific journal in sociology, *L'Année sociologique (The Sociological Year)*. This journal and much of Durkheim's own writing were among the first examples of the application of statistics to social issues. (See the discussion of suicide in Chapter 2.)

Max Weber Max Weber (1864–1920) was a German historian, economist, and sociologist. Weber's life, like Durkheim's, spanned much of the second half of the nineteenth century and the early decades of the twentieth. Like the other early sociologists, therefore, Weber witnessed the tumultuous changes that were bringing down the old order. He saw monarchies tottering in the face of demands for democratic rule. He observed new industries and markets spanning the globe and linking formerly isolated peoples. He saw and described the rise of modern science and jurisprudence and modern ways of doing business. The growing tendency to apply rational decision-making procedures, rather than merely relying on tradition, was for Weber a dramatic departure from the older ways of feudal societies and mercantile aristocracies. Weber compared many different societies to show how new forms of government and administration were evolving.

All three of these pioneers in sociology were scholars of great genius. They were also political activists. Marx, of course, was the most revolutionary of the three and devoted much of his energy to the international socialist movement. Durkheim was a lifelong socialist but was more moderate than Marx. Although he took stands on many political issues, Durkheim did not devote himself to political activities. As a young man, Weber had been involved in the movement to create a unified German nation, but as a mature scholar, he developed a belief in "value-free" social science. A social scientist might draw research questions from personal political beliefs, but the research itself must apply scientific methods. This view, which Durkheim also shared, did much

Max Weber

Max Weber (1864–920) c. 1896–97 (b/w photo), German Photographer, (19th century)/Private Collection/Archives Charmet/The Bridgeman Art Library

Women in Sociology In the late nineteenth century, as it emerged from philosophy to become an independent scholarly field, sociology was dominated by white male university professors. There were a few exceptions, such as W.E.B. DuBois and the brilliant Muslim social thinker Ibn Khaldun, but for the most part, the founders of sociology were European men.

A notable exception was Harriet Martineau (1802–1876), who is thought to have been the first woman to contribute to the emerging field of sociology.

Martineau wrote important sociological interpretations of the early phases of capitalism and modernity. She was also a social reformer whose writings on the plight of children and women in British factories were highly influential (Hoecker-Drysdale, 1992).

In the United States as well, early female sociologists were usually barred from university positions, and their work was never accorded the intellectual respect it merited. An example is Ida B. Wells-Barnett (1862–1931), an African American woman raised in the South after the Civil War. She wrote extensively about the new forms of discrimination and racism directed against her people and became a spokesperson for the civil rights of African Americans and women. Recent scholarship by feminist sociologists has drawn renewed attention to her work and that of other women founders of sociology (Lengermann & Niebrugge-Brantley, 1998).

Ida B. Wells-Barnett

R. Gates/Hulton Archive/Getty Images

Today, sociology is far more diverse in every aspect, and women throughout the world are well represented among the field's most innovative minds. The photo of Manjula Giri shows a modern sociologist from a developing nation, Nepal, who is both a university-trained scholar and a dedicated social activist. Giri has written about the failure of revolution to improve the conditions of rural women in Nepal. In her own village, she is using her sociological skills to help local women develop a women's farming cooperative. She has also started literacy classes and is planning to expand her work to other villages in the region. Giri is an example of the scholar-activist, a dual role that is increasingly common throughout the poorer regions of the world.

Harriet Martineau

Spencer Arnold/Hulton Archive/Getty Images

Manjula Giri

Courtesy of Manjula Giri

to advance sociology to the level of a social science rather than just a branch of philosophy.

To become a science, sociology had to build on the research of its founders. The twentieth century brought changes of such magnitude at every level of society that sociologists were in increasing demand. Their mission was to gain new information about the scope and meaning of social change.

The Rise of Modern Sociology

We credit the European social thinkers and philosophers with creating sociology, but nowhere did the new science find more fertile ground for development than in North America. By the beginning of the twentieth century, sociology was rapidly acquiring new adherents in the United States and Canada, partly because of the influence of European sociologists like Marx and Durkheim, but even more because of the rapid social changes occurring in North America at the time. Waves of immigrants to cities and towns, the explosive growth of population and industry in the cities, race riots, strikes and labor strife, moral crusades against crime and vice and alcohol, the demand for woman suffrage—these and many other changes caused American sociology to take a new turn. There was an increasing demand for knowledge about exactly what changes were occurring and whom they affected. In North America, therefore, sociologists began to emphasize the quest for facts about changing social conditions—that is, the empirical investigation of social issues. As Peter Berger writes in *Invitation to Sociology*, "The sociologist . . . is someone concerned with understanding society in a disciplined way. The nature of this discipline is scientific" (1963, p. 16). This means that the sociologist is bound by scientific rules of evidence (discussed in the next chapter) and must strive to remain objective and base his or her conclusions on empirical findings rather than on personal hopes or prejudices.

Empirical Investigation *Empirical* information refers to carefully gathered, unbiased data regarding social conditions and behavior. In general, modern sociology is distinguished by its relentless and systematic search for empirical data to answer questions about society. Journalists, for example, also seek the facts about social conditions, but they must cover many different events and situations and present them as "stories" that will attract their readers' interest. Because they usually cannot dwell on one subject very long, journalists often must content themselves with citing examples and quoting experts whose opinions may or may not be based on empirical evidence. In contrast, sociologists study a situation or phenomenon in more depth; when they do not have enough facts, they are likely to say, "That is an empirical question. Let's see what the research tells us. If the answers are inconclusive, we will do more research." Evidence based on measurable effects and outcomes is required before one can make an informed decision about an issue.

Among the chief goals of modern sociology are using the sociological imagination to ask relevant questions and seeking answers to those questions backed by evidence that others can verify. Anyone can make assertions about society or about why people behave the way they do. "I think it's human nature to act selfishly, no matter what kind of education people have" is a common assertion that is not backed up by any solid evidence. As you read this book, you will learn that, to strengthen your sociological imagination, you must learn how to apply evidence to your views and admit that you may modify your opinions upon careful consideration of that evidence.

The Social Surveys The empirical focus of American sociology began largely as an outgrowth of the reform movements of the late nineteenth and early twentieth centuries. During this period, the nation had not yet recovered from the havoc created by the Civil War. Southern blacks were migrating to the more industrialized North in ever-increasing numbers, and at the same time, millions of European immigrants were finding ill-paying jobs in larger cities, where cheap labor was in great demand. By the turn of the twentieth century, therefore, the nation's cities were crowded with poor families for whom the promise of "gold in America" had become a tarnished dream. In this time of rapid social change, Americans continually debated the merits of social reform and proposed new solutions for pressing social issues. Some called for socialism, others for a return to the free market, a ban on labor organizations, an end to immigration, or the removal of black Americans to Africa. But where would the facts to be used in judging those ideas come from?

To gain empirical information about social conditions, dedicated individuals undertook numerous "social surveys." Jacob Riis's (1890) account of life on New York's Lower East Side; W.E.B. DuBois's (1967/1899) survey of Philadelphia blacks; Emily Balch's (1910) depiction of living conditions among Slavic miners and steelworkers in the Pittsburgh area; and Jane Addams's famous *Hull-House Maps and Papers* (1895), which described the lives of her neighbors in Chicago's West Side slum area—these and other carefully documented surveys of the living conditions of people experiencing the effects of rapid industrialization and urbanization left an enduring mark on American sociology.

Sociology &
Social Justice

W.E.B. DUBOIS The pioneering sociologist W.E.B. DuBois used objective research to fight for social justice for

African Americans. Social justice, as we will see throughout this book, involves fairness in the distribution of opportunities and resources in societies. DuBois's book *The Philadelphia Negro* (1967/1899) showed how African Americans, who had technically been released from slavery after the Civil War, were denied opportunities in work and education that would allow them to escape conditions of dire poverty. DuBois helped direct sociological research to racial and social issues in minority communities, using empirical data to provide an objective account of the dismal social conditions of northern blacks at the turn of the twentieth century. By presenting the facts about the racism and social exclusion they experienced, DuBois created a model for using sociological research to press for greater social justice for African Americans and other minorities. The Research Methods box on page 13 presents an excerpt from the landmark survey conducted by DuBois, the first black sociologist to gain worldwide recognition.

The Chicago School and Human Ecology

By the late 1920s, the United States had become the world leader in sociology. The two great centers of American sociological research were the University of Chicago and Columbia University. At these universities and others influenced by them, two distinct approaches to the study of society evolved. The Chicago school emphasized the relationship between the individual and society, whereas the major East Coast universities, which were more strongly influenced by European sociology, tended toward macro-level analyses of social structure and change.

The sociology department at the University of Chicago (the oldest in the nation) extended its influence to many other universities, especially in the Midwest, South, and West. At that time, the department was under the leadership of Robert Park and his younger colleague Ernest Burgess. Park in particular is associated with the Chicago school. His main contribution was to develop an agenda for sociological research that used the city as a "social laboratory." Park favored an approach in which facts concerning what was occurring among people in their local communities (at the micro and middle levels) would be collected within a broader theoretical framework. That framework attempted to link macro-level changes in society, such as industrialization and the growth of urban populations, to patterns of settlement in cities and to how people actually lived in cities.

In one of his essays on this subject, Park began with the idea that industrialization causes the breakdown of traditional "primary-group"

The infamous strike of steelworkers in Homestead, Pennsylvania, in 1892 pitted workers and their families against the private security police hired by the Carnegie Corporation and later by the National Guard. The resulting violence and class conflict was the subject of a pathbreaking social survey of social conditions among the workers by Emily Balch, who, like W.E.B. DuBois and Jane Addams, was an early practitioner of the new field of sociology.

attachments (those of family members, age-mates, or clans). After stating the probable relationship between the effects of industrialization and high rates of crime, Park asked several specific questions:

> What is the effect of ownership of property . . . on truancy, on divorce, and on crime?

In what regions and classes are certain kinds of crime endemic? In what classes does divorce occur most frequently? What is the difference in this respect between farmers and, say, actors?

To what extent in any given [ethnic] group . . . do parents and children live in the same world, speak the same language, and share the same ideas, and how far do the conditions found account for juvenile delinquency in that particular group? (1967/1925, p. 22)

This set of research questions, of which those quoted here are a small sample, inspired and shaped the work of hundreds of sociologists who were influenced by the Chicago school. To this day, Chicago remains the most systematically studied city in the United States, although similar research has been carried out in other large cities throughout the nation. From studies of the linguistic diversity of African peoples to attempts to understand the subtle negotiations by which youth gangs divide up an urban "turf," the insights of the Chicago school remain a vital aspect of contemporary sociology (A. D. Abbott, 2001).

As you can see from the types of questions Park asked, the distinctive orientation of the Chicago school was its emphasis on the relationships among social order, social disorganization, and the distribution of populations in space and time. Park and Burgess called this approach **human ecology**. With many modifications, it remains an important, though not dominant, perspective in contemporary sociology.

Human ecology, as Park and others defined it, is the branch of sociology that is concerned with population growth and change. In particular, it seeks to discover how populations organize themselves to survive and prosper. Human ecologists are interested in how groups that are organized in different ways compete and cooperate. They also look for forms of social organization that may emerge as a group adjusts to life in new surroundings.

A key concept for human ecologists is *community*. There are many ways of defining this term, just as there are many ways of defining most of the central concepts of sociology. From an ecological perspective, however, the term *community* usually refers to a population that carries out major life functions (e.g., birth, marriage, death) within a particular territory. Human ecology does not assume that there will ever be a "steady state" or an end to the process of change in human communities. Instead, it attempts to trace the change and document its consequences for the social environment. What happens when newcomers "invade" a community? In what way are local gangs a response to recent changes in population or in the ability of members of a community to compete for jobs? Not only do populations change, but people's preferences and behaviors also continually change. So do the technologies for producing the goods and services we want. As a result, our ways of making a living, our modes of transportation, and our choices of leisure activities create constant change, not just in communities but in entire societies.

The Chicago school became known for this "ecological" approach, the idea that the study of human society should begin with empirical questions about population size, the distribution of populations over territories, and the like. The human ecologists recognized that many other processes shape society, but their most important contribution to the discipline of sociology was to include the processes by which populations change and communities are formed.

Modern ecological theories also consider the relations between humans and their natural environment. We will see that the way people earn a living, the resources they use, the energy they consume, and their efforts to control pollution all have far-reaching consequences, not only for their own lives but also for the society in which they live. These patterns of use and consumption also have an increasing impact on the entire planet—so much so that ecological problems are becoming an ever more important area of sociological research. (See the Mapping Social Change box on pages 14–15.)

MAJOR SOCIOLOGICAL PERSPECTIVES

Sociological *perspectives* are sets of ideas and theories that sociologists use in attempting to understand various problems of human society, such as the problems of population size, conflict between populations, how people become part of a society, and other issues that we will encounter throughout this book. Although human ecology remains an important sociological perspective, it is by no means the only one modern sociologists employ. Other perspectives, to which we now turn, guide empirical description and help explain social stability and social change.

Interactionism

Interactionism is the sociological perspective that views social order and social change as resulting from the immense variety of repeated interactions among individuals and groups. Families, committees, corporations, armies, entire societies—indeed, all the social forms we can think of—are a result of interpersonal behavior in which people communicate, give and take, share, compete, and so on. If there were no exchange of goods, information, love, and all the rest—that is, if there were no interaction among people—obviously there could be no social life at all.

---

human ecology: A sociological perspective that emphasizes the relationships among social order, social disorganization, and the distribution of populations in space and time.

interactionism: A sociological perspective that views social order and social change as resulting from all the repeated interactions among individuals and groups.

The Social Survey

W.E.B. DuBois was one of the first American sociologists to publish highly factual and objective descriptions of life in American cities. An African American, DuBois earned his doctorate in philosophy at Harvard before the turn of the twentieth century, when sociology was still regarded as a subfield of that discipline. But, as can be seen in this excerpt from his account of black life in Philadelphia about a century ago, DuBois was able to sharpen his arguments about the effects of racial discrimination with simple but telling statistics.

For a group of freedmen the question of economic survival is the most pressing of all questions; the problem as to how, under the circumstances of modern life, any group of people can earn a decent living, so as to maintain their standard of life, is not always easy to answer. But when the question is complicated by the fact that the group has a low degree of efficiency on account of previous training; is in competition with well-trained, eager and often ruthless competitors; is more or less handicapped by . . . discrimination; and finally, is seeking not merely to maintain a standard of living but steadily to raise it to a higher plane—such a situation presents baffling problems to the sociologist. . . .

And yet this is the situation of the Negro in Philadelphia; he is trying to better his condition; is seeking to rise; for this end his first need is work of a character to engage his best talents, and remunerative enough for him to support a home and train up his children well. The competition in a large city is fierce, and it is difficult for any poor people to succeed. The Negro, however, has two especial difficulties: his training as a slave and freedman has not been such as make the average of the race as efficient and reliable workmen as the average [native-born] American or as many foreign immigrants. The Negro is, as a rule, willing, honest and good-natured; but he is also, as a rule, careless, unreliable and unsteady. This is without doubt to be expected in a people who for generations have been trained to shirk work; but an historical excuse counts for little in the whirl and battle of breadwinning. Of course, there are large exceptions to this average rule; there are many Negroes who are as bright, talented and reliable as any class of workmen, and who in untrammeled competition would soon rise high in the economic scale, and thus by the law of the survival of the fittest we should soon have left at the bottom those inefficient and lazy drones who did not deserve a better fate. However, in the realm of social phenomena the law of survival is greatly modified by human choice, wish, whim and prejudice. And consequently one never knows when one sees a social outcast how far this failure to survive is due to the deficiencies of the individual, and how far to the accidents or injustice of his environment. This is

especially the case with the Negro. Every one knows that in a city like Philadelphia a Negro does not have the same chance to exercise his ability or secure work according to his talents as a white man. Just how far this is so we shall discuss later; now it is sufficient to say in general that the sorts of work open to Negroes are not only restricted by their own lack of training but also by discrimination against them on account of their race; that their economic rise is not only hindered by their present poverty, but also by a widespread inclination to shut against them many doors of advancement. . . .

What has thus far been the result of this complicated situation? What do the mass of the Negroes of the city at present do for a living, and how successful are they in those lines? And in so far as they are successful, what have they accomplished, and where they are inefficient in their present sphere of work, what is the cause and remedy? These are the questions before us, and we proceed to answer the first in this chapter, taking the occupations of the Negroes of the Seventh Ward first . . .

Of the 257 boys between the ages of ten and twenty, who were regularly at work in 1896, 39 percent were porters and errand boys; 25.5 percent were servants; 16 percent were common laborers, and 19.5 percent had miscellaneous employment. The occupations in detail are as follows:

W.E.B. DuBois

Total population, males 10 to 20		651	
Engaged in gainful occupations		257	
Porters and errand boys		100	39.0%
Servants		66	25.5%
Common laborers		40	16.0%
Miscellaneous employment			
Teamsters	7		
Apprentices	6		
Bootblacks	6		
Drivers	5		
Newsboys	5		
Peddlers	4		
Typesetters	3		
Actors	2		
Bricklayers	2		
Hostlers	2	51	19.5%
Typists	2		
Barber	1		
Bartender	1		
Bookbinder	1		
Factory hand	1		
Rubber worker	1		
Sailor	1		
Shoemaker	1		
		257	100.0%

Note: This simple table includes much useful information, but modern tables present data more efficiently. See Chapter 2 for a detailed discussion of the presentation of data in tables.

Source: DuBois, 1967/1899, pp. 97–99.

The interactionist perspective usually generates analyses of social life at the level of interpersonal relationships, but it does not limit itself to the micro level of social reality. It also looks at how middle- and macro-level phenomena result from micro-level behaviors or, conversely, how middle- and macro-level influences shape the interactions among individuals. From the interactionist perspective, for example, a family is a product of interactions among a set

Worldwide Calorie Consumption and Areas of Famine

From a biological perspective, the success of any species is measured by how well it meets the broad requirements of population growth and maintenance. Every day the world's 6 billion people seek and obtain enough food to convert into bodily energy, a minimum of perhaps 1,500 calories a day for survival at starvation levels (although this varies greatly with climate and other factors). Over 70 percent of the world's population is inadequately nourished (that is, obtains fewer than 2,500 calories a day), often while engaging in hard physical labor. At the same time, a smaller proportion, including most (but by no means all) of the North American population, live so comfortably and well above the daily minimum that overweight and obesity are a major health problem.

Ecological theories help explain malnutrition and starvation in some areas of the world. In Somalia, for example, the combination of a desert environment and warring political factions produces social instability that leads to persistent famine. The problem is not so much that food is unavailable as that supplies cannot be delivered or distributed effectively.

Because children are the most vulnerable segment of any society's population, they are usually affected most by adverse social conditions. This table shows that overall the percentages and numbers of underweight, malnourished children in the developing countries have declined from 37.4 percent (175.7 million) in 1980, to 26.7 percent (149.6 million) in 2000. The greatest improvements have occurred in Asia, where positive economic and social changes have brought higher levels of living to many. But in Africa, where civil strife in many regions blocks economic growth, the proportion of malnourished children is actually growing.

Global and Regional Trends in the Estimated Prevalence of Protein-Energy Malnutrition in Underweight Children Under 5, Since 1980

Region	1980 %	1980 Million	1990 %	1990 Million	1995 %	1995 Million	2000 %	2000 Million
Africa	26.2	22.5	27.3	30.1	27.9	34.0	28.5	38.3
Asia	43.9	146.0	36.5	141.3	32.8	121.0	29.0	108.0
Latin America	14.2	7.3	10.2	5.6	8.3	4.5	6.3	3.4
Developing countries	37.4	175.7	32.1	177.0	29.2	159.5	26.7	149.6

Source: WHO, 2000.

of individuals who define themselves as family members. But each person's understanding of how a family ought to behave is a product of middle- and macro-level forces: religious teachings about family life, laws dealing with education or child support, and so on. And these are always changing. You may have experienced the consequences of changing values that cause older and younger family members to feel differently about such issues as whether a couple should live together before marrying. In sum, the interactionist perspective insists that we look carefully at how individuals interact, how they interpret their own and other people's actions, and the consequences of those actions for the larger social group (Blumer, 1969b; Frank, 1988).

The general framework of interactionism contains at least two major and quite different sets of issues. One set concerns the problems of exchange and choice: How can social order exist and groups or societies maintain stability when people have selfish motives for being in groups—that is, when they are seeking to gain as much personal advantage as they can? The second set of issues involves how people actually manage to communicate their values and how they arrive at mutual understandings. Research and explanations of the first problem fall under the heading of "rational choice" (or exchange theory); the second issue is addressed by the study of "symbolic interaction." In recent years, these two areas of inquiry have emerged as quite different yet increasingly related aspects of the study of interaction.

Rational Choice: The Sociological View

Adam Smith, whose famous work *The Wealth of Nations* (1965/1776) became the basis for most subsequent economic thought, believed that individuals always seek to maximize their pleasure and minimize their pain. If they are allowed to make the best possible choices for themselves over time, they will also produce an affluent and just society. They will serve others, even when they are unaware that they are doing so, in order to increase their own benefit. They will choose a constitution and a government that protect their property and their right to engage in trade. They will seek the government's protection against those who would infringe on their rights or attempt to dominate them, but the government need do little more than protect them and allow them to make choices based on their own reasoning.

You may have already encountered this theory, known as *utilitarianism,* in an economics or political science course. In sociology, it is applied

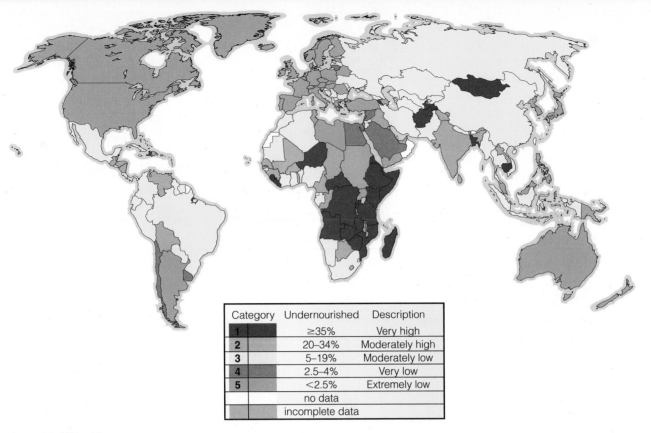

Category	Undernourished	Description
1	≥35%	Very high
2	20–34%	Moderately high
3	5–19%	Moderately low
4	2.5–4%	Very low
5	<2.5%	Extremely low
	no data	
	incomplete data	

Source: World Food Programme.

to a variety of issues. Often this rational-choice view of interaction is referred to as *exchange theory* because it focuses on what people seem to be getting out of their interactions and what they in turn are contributing to the relationship or to the larger group. In every interaction, something is being exchanged: It may be time or attention, friendship, material values (e.g., wages or possessions), or less easily calculated values such as esteem or allegiance. The larger the number of interacting members, the more complex the types of exchanges that occur among them. When people perceive an interaction as being one-way, they begin to feel that they are being exploited or treated unfairly and will usually leave the relationship or quit the group (Homans, 1961). In industry, for example, if workers feel that they are not being paid enough for their work, they may form a union, bargain collectively with the bosses, or even go on strike. But in doing so, each worker will weigh potential benefits against potential losses—losses in pay, in esteem, in friendship, and so forth. The choices are not always easy, nor are the motivations always obvious. When many values are involved, the rational calculation of benefits and costs becomes even more difficult.

Rational-choice models of behavior prompt us to look at *patterns* of behavior to see how

they conform to and depart from normal expectations of personal profit and loss. But those models do not always identify the underlying values. How we learn what to value in the first place, how we communicate our choices and intentions, how we learn new values through interaction—all are subjects that require other concepts besides those found in rational-choice theories of behavior. Such questions lead us toward research about

The players on opposing teams are paying their respects to one another after a game of soccer. Here is an example of social interaction that is highly structured—the players form opposing lines that walk past each other and touch hands—but the players are also free to add their own comments and perform the ritual with a bit of personal style.

how people actually carry out and understand human interaction in their daily lives.

Symbolic Interactionism When we make choices about our interactions with other people, we may be said to be acting rationally. Other forces are likely to shape our behavior as well. For example, you may select a particular course because the instructor is rumored to be good. But what does "good" mean—clear and well organized? An easy grader? Friendly? Humorous? The dimensions of a choice can be complicated, and we may not be aware of everything that goes into our decisions. You may, without realizing it, choose a course as much to be with certain other people as to be in that particular course. Our choices tell other people about us: what we like, what we want to become, and so on. Indeed, the way people dress, the way they carry themselves (body language), the way they speak to each other, and the gestures they make convey a great deal of information that is not always intentional or expressed in speech. Some forms of communication give information without speaking it, or speak one thing and mean another. But words are of great importance too, and the content of communication is made explicit in words and sentences. Sociologists refer to all these aspects of behavior as symbolic interaction. From the symbolic-interactionist perspective, "society itself tends to be seen as a mosaic of little scenes and dramas in which people make indications to themselves and others, respond to those indications, align their actions, and so build identities and social structures" (Rock, 1985, p. 844).

Symbolic interactionists call attention to how social life is "constructed" through the mundane acts of social communication. For example, in all the choices students make—joining friendship groups, learning the school's informal rules of school, challenging and breaking those rules—the social order of student society, or "college culture," is actually "constructed." Erving Goffman, whose work was mentioned earlier, is known for his research on these processes. Goffman applied the symbolic-interactionist perspective to the study of everyday interactions such as rituals of greeting and departure, of daily life in asylums and gambling houses, and of behavior in streets and public places. His work examines how people behave in social situations and how others rate their "performances."

The power of symbolic interactionism lies in its ability to generate theories about how people learn to play certain roles and how those roles are used in the social construction of groups and organizations. However, if we want to think sociologically about more complex phenomena, such as the rise of bureaucratic organizations or the reasons why some societies experience revolutions, we also need the concepts developed by two other perspectives: functionalism and conflict theory.

Functionalism

Is society simply the sum total of countless micro-level exchanges and communicative interactions, or do the organizations within a society have properties independent of the actions of individuals? When we

The cubicles of an open office space—whose occupants perform the specific duties required by their jobs—have become a visual metaphor for the functionalist arrangement of modern work life.

speak of the family, the army, the corporation, or the laboratory, we generally have in mind an entity marked by certain specific functions, tasks, and types of behavior. The army requires that its members learn to engage in armed combat, even if that is not what they will be doing most of the time. The family requires that its members behave in nurturant ways toward one another. The farm requires that those who run it know how to plant and harvest. Individual interactions may determine how well a given person performs these various tasks; but the larger organization—the army, the family, the farm—establishes specific ways of behaving and doing the work of that organization. In this sense the organization, which exists longer than any of its members, has its own existence.

The **functionalist** perspective asks how society manages to carry out the functions it must perform in order to maintain social order, feed large masses of people each day, defend itself against attackers, produce the next generation, and so on. From this perspective, the many groups and organizations that make up a society form the structure of human society. This social structure is a complex system designed to carry out the essential functions of human life. The function of the family, for example, is to raise and train a new generation to replace the old; the function of the military is to defend the society; the function of schools is to teach the next generation the beliefs and skills they will need to maintain the society in the future; and a major function of religion is to develop shared ideas of morality.

When a society is functioning well, all its major parts are said to be well integrated and in equilibrium. But periods of rapid social change

functionalism: A sociological perspective that focuses on the ways in which a complex pattern of social structures and arrangements contributes to social order.

can throw social structures out of equilibrium. Entire ways of life can lose their purpose or function. When that happens, the various structures of society can become poorly integrated, and formerly useful functions can become dysfunctional.

Consider an example: In agrarian societies, in which most people work the land, families typically include three generations, with many members in each generation living close to one another. Labor is in great demand; many hands are needed where there are no machines to perform work in fields and barnyards and granaries. Early marriage and large numbers of children in the agrarian family help such a society function optimally. When the society industrializes and its agriculture becomes mechanized, however, families may continue to produce large numbers of children even though the demand for farmhands has decreased. When they grow up, those children may migrate to towns and cities. The migrants are likely to continue to value large families and to have numerous children, but if they are unable to find jobs, they may join the ranks of the unemployed and their children may grow up in poverty. In this situation, the family can be said to be poorly integrated with the needs of the society; the value of large family size has become dysfunctional—it no longer contributes to the well-being of groups or individuals.

Conflict Theory

A major flaw in the functionalist perspective is that we have rarely seen anything approaching equilibrium in human societies. Inequalities of wealth and power, racism, religious intolerance, ethnic hatreds, grinding economic competition, and criminality constantly threaten peace and the smooth functioning of even the most affluent and stable societies (Barash, 2002). In the twentieth century alone, two world wars and many civil wars disrupted the lives of millions of people. Almost as devastating was the Great Depression of the 1930s, the most severe economic slump in modern history. Worst of all were the nightmares of the Nazi Holocaust and the purges of Stalinist Russia, in which more than 20 million people were exterminated.

The world wars, the depression, and the Holocaust shocked and demoralized the entire world. They also called into question the optimism of the nineteenth- and early twentieth-century social philosophers, many of whom believed in the promise of progress through modern science and technology. Between 1914 and the end of World War II, modern ideas and technologies were used for horrible purposes often enough to disillusion all but the most

These Iraqi refugees living on the streets of Syrian and Lebanese cities are unable to return to destroyed and dangerous neighborhoods in Iraq. They and millions of others in similar situations are a reminder that the physical and emotional scars of war last long after the cessation of hostilities. As wars become more destructive, the study of social conflict and its solutions takes on even greater urgency.

ardent optimists. Bewildered intellectuals and political leaders turned to sociology to find some explanation for those horrors.

Marxian Theory One explanation was provided by Marxian theory. According to Marx, the rise of capitalism was the cause of conflict in modern times. Under capitalism, forms of exploitation and domination spread. For example, in the early period of industrial capitalism, workers were forced to work twelve hours a day, six days a week; in less developed areas of the world, large populations were virtually enslaved by the new colonial powers.

At the heart of capitalism, for Marx, is conflict among people in different economic classes, especially between those who control wealth and power and those who do not. Marx argued that the division of people in a society into different classes, defined by how they make a living, always produces conflict. Under capitalism, this conflict occurs between the owners of factories and the workers. Marx believed that class conflict would eventually destroy or at least vastly modify capitalism. His theory is at the heart of what has come to be known as conflict theory or the **conflict perspective.**

In the 1960s, when protests against racism and segregation, the Vietnam War, pollution of the environment, and discrimination against women each became the focus of a major social movement, the conflict perspective became more prominent. It clearly was not possible to explain the rapid appearance of major social movements with theories that emphasized how the social system would function if it were in a state of equilibrium. Even Marxian theory did not do a very good job of predicting the protest movements of the 1960s, or their effects on American society. The environmental movement and the women's movement, for example, were not based on economic inequalities alone, nor were the people who joined them necessarily exploited workers. Sociologists studying the role of conflict in social change therefore had to go beyond

conflict perspective: A sociological perspective that emphasizes the role of conflict and power in society.

Society on Parade Everybody loves a parade, and no one more than sociologists. We see parades as events that intentionally display the people, the values, and the social structures we cherish. Parades also yield endless sociological insights into other people's social worlds. The way particular sociologists analyze a parade may tell us something about their concerns and values as well. Take the annual Rose Bowl parade in Pasadena, California. A functionalist sociologist would argue

© Crandall/The Image Works

© Zbigniew Bzdak/The Image Works

the Marxian view. Many turned to the writings of the German sociologist Georg Simmel (1904), who argued that conflict is necessary as a basis for the formation of alliances. According to Simmel, conflict is one means whereby a "web of group affiliations" is constructed. The continual shifting of alliances within this web of social groups can help explain who becomes involved in social movements and how much power those movements are able to acquire.

Power The concept of *power* holds a central place in conflict theory. From the functionalist perspective, society holds together because its members share the same basic beliefs about how people should behave. Conflict theorists point out that the role of power is just as important as the influence of shared beliefs in explaining why society does not disintegrate into chaos. **Power** is the ability of an individual or group to change the behavior of others. A nation's government, as we will see in Chapter 4, usually controls the use of force (a form of power) to maintain social order. For sociologists who study conflict and power, the

important questions are who benefits from the exercise of power and who loses. For example, when the government intervenes in a strike and obliges workers to return to their jobs, does the public at large or the corporation against which the workers were striking benefit most? And what about the workers themselves? What do they gain or lose? Such questions are central to conflict theory today (Foucault, 2001; Gramsci, 1971).

The Multidimensional View of Society

Each of the sociological perspectives leads to different questions and different kinds of observations. These are extremely powerful analytical tools. Anyone who masters them and learns to apply them in appropriate ways will have an immense advantage over more naive observers of the ever-changing parade of social life (Merton, 1998). From the ecological perspective come questions about how populations can exist and flourish in various natural and social environments. From the interactionist perspective come questions about how people get along and behave in groups and organizations of all kinds. Functionalism asks

power: The ability to control the behavior of others, even against their will.

that the parade affirms the society's values of competition and success and, in the process, helps the merchants of Pasadena. A sociologist with a more critical eye might call attention to the way the parade affirms values that perpetuate inequality and social injustices. The Rose Bowl queen may be beautiful and talented, but a critical interpretation might emphasize how she is set on a pedestal and brought along to adorn the male field of combat. From this perspective, the queen symbolizes women as beautiful objects to be possessed by men with power and money. An interactionist analysis might see the parade and its viewers as a form of public theater in which the symbols of American culture are displayed in endless variations as the different floats pass by.

A saint's day parade in a devout Roman Catholic village of Central or South America offers another glimpse into the interactions that mark a special event in a village society. Who carries the village's patron saint? By what rules of social standing are people arranged in the parade? Simply by describing who's who in the parade, a local informant could offer many insights into the way the village is organized and how its people think and feel.

A military parade may be a joyous or ominous event, depending on whether one is on good terms with the generals or not. From a functionalist perspective, this is a display of force. Official communications may portray it as a way of demonstrating that the nation has

©ISSOOF/SANOGO/Getty Images

the capacity to defend itself against other nations. In many parts of the world, however, military parades are also meant to let citizens know in no uncertain terms who controls the use of deadly force. From a conflict perspective, these parades reveal who has power in the society and how that power is used to favor some groups, who are usually featured in the parade, over others, who often are not included among the marchers.

PERSPECTIVES
Major Sociological Perspectives

Perspective	Description	Generates Questions About...	Applications
Interactionism	Studies how social structures are created in the course of human interaction	How people behave in intimate groups; how symbols and communication shape perceptions; how social roles are learned and society is "constructed" through interaction	Education practice, courtroom procedure, therapy
Functionalism	Asks how societies carry out the functions they must perform; views the structures of society as a system designed to carry out those functions	How society is structured and how social structures work together as a system to perform the major functions of society	Study of formal organizations, development of social policies, management science
Conflict Theory	Holds that power is just as important as shared values in holding society together; conflict is also responsible for social change	How power affects the distribution of scarce resources and how conflict changes society	Study of politics, social movements, corporate power structures

questions about how society is structured and how it works as a social system. And the conflict perspective asks how power is used to maintain order and how conflict changes society.

These different perspectives developed as sociologists asked different questions about

society. In contemporary sociology, each continues to stimulate relatively distinct research based on the types of questions being asked. Yet a great deal of research combines the insights of different perspectives in ways that vastly increase the power of the resulting analysis. The Perspectives Chart above summarizes the major sociological perspectives and the kinds of questions they ask.

SUMMARY

What is the sociological imagination, and how does it generate the questions that make sociology a "human science"?

- *Sociology* is the scientific study of human societies and of human behavior in the groups that make up a society. It is concerned with how *social conditions* influence our lives as individuals. The ability to see the world from this point of view has been described as the *sociological imagination.*

How do sociologists study social change at different levels of social reality?

- Sociologists study social behavior at three levels of complexity. *Micro-level* sociology deals with behaviors that occur at the level of the individual and immediate others. The *middle level* of sociological observation is concerned with how the social structures in which people participate actually shape their lives. *Macro-level* studies attempt to explain the social processes that influence populations, social classes, and entire societies.

Who are the key thinkers and researchers who have transformed social thought into social science?

- The scientific discoveries of the seventeenth century led to the rise of the idea of progress, as opposed to the notion of human helplessness in the face of divine Providence.

- In the eighteenth century, revolutions in Europe and North America completely changed the social order and gave rise to new perspectives on human social life.

- Out of this period of social and intellectual ferment came the idea of creating a science of human society. Sociology, as the new science was called, developed in Europe in the nineteenth century.

- During that formative period, several outstanding sociologists shaped and refined the new discipline. Among them were Karl Marx, Émile Durkheim, and Max Weber.

- In the twentieth century, sociology developed most rapidly in North America, spurred by the need for empirical information concerning social conditions. Numerous "social surveys" were conducted around the turn of the century; by the late 1920s, two distinct approaches to the study of society had evolved at American universities. The "Chicago school" focused on the relationship between the individual and society, whereas the major East Coast universities leaned toward macro-level analysis.

- Under the leadership of Robert Park and Ernest Burgess, the Chicago school developed the approach known as *human ecology.* This perspective emphasizes the relationships among social order, social disorganization, and the distribution of populations in space and time. Fundamental to this approach is the concept of *community,* meaning a population that carries out major functions within a particular territory.

How do the major sociological perspectives approach different social issues?

- Modern sociologists employ other basic perspectives in addition to human ecology. *Interactionism* is a perspective that views social order and social change as resulting from all the repeated interactions among individuals and groups.

- One version of this approach is rational-choice or exchange theory, which focuses on what people seem to be getting out of their interactions and what they contribute to them. Another is the symbolic-interactionist perspective, which studies how social structures are actually created in the course of human interaction.

- *Functionalism,* in contrast, is concerned primarily with the large-scale structures of society; it asks how those structures enable society to carry out its basic functions.

- In the decades since World War II, functionalism has been strongly challenged by the *conflict perspective,* which emphasizes the role of conflict and *power* in explaining not only why societies change but also why they hold together.

- None of the major sociological perspectives is fully independent of the others. Each emphasizes different questions and different observations about social life. Used in combination, they greatly increase our ability to understand and explain almost any aspect of human society.

The Kornblum Companion Website

www.cengagebrain.com

Supplement your review of this chapter by going to the companion website to take one of the tutorial quizzes, use the flash cards to master key terms, and check out the many other study aids you'll find there. You'll also find special features, such as GSS data and Web links that will put data and resources at your fingertips to help you with that special project or to do some research on your own.

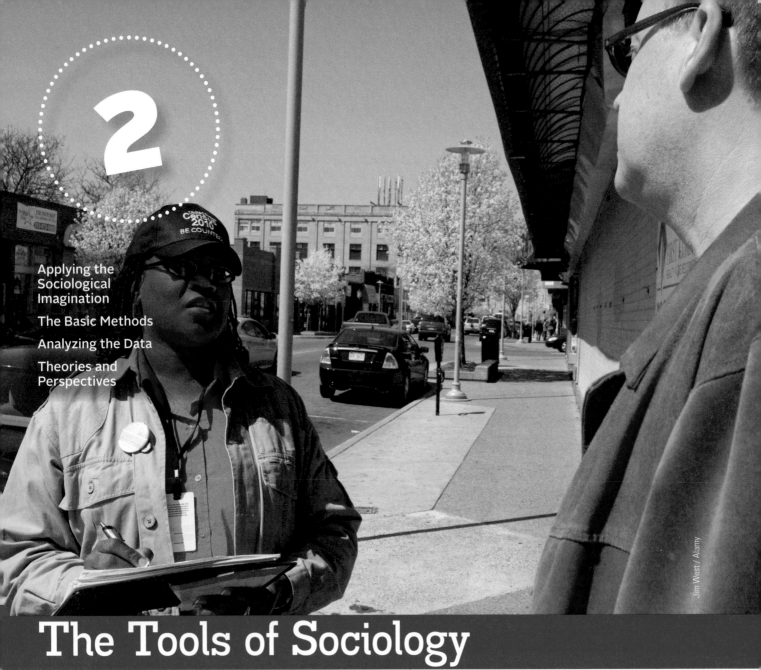

2

Applying the
Sociological
Imagination

The Basic Methods

Analyzing the Data

Theories and
Perspectives

Jim West / Alamy

The Tools of Sociology

FOCUS QUESTIONS

How are the questions sociologists ask about social life the same as, and yet different from, the questions people from all walks of life normally ask?

What basic scientific methods do sociologists employ in their research?

How do sociologists analyze the data they collect using the various research methods?

What is the difference between theories and perspectives?

Have you or any of your friends ever seriously considered committing suicide? This may not be the kind of question most people want to discuss openly, yet suicide is a serious social issue in the United States and many other nations. Perhaps the most tragic aspect of the phenomenon is the number of suicides committed by teenagers and young adults who have their entire adult lives ahead of them. What can a person do to counteract persistent thoughts about suicide? To whom can one turn when the urge to kill oneself becomes overwhelming? And what are the causes of suicide? These are all important sociological questions.

Most people who have not studied sociology assume that suicide is closely related to an individual's mental state. They will cite cases of people who ended their lives because of intense feelings of depression, or hopelessness, or shame due to crime, debt, loss of a career, or similar misfortunes. In other words, people who lack a sociological imagination will generally attribute suicide to personal problems that overwhelm the individual and make life seem intolerable. In some cases, they may also cite examples of groups of people who have committed suicide together, as members of the Heaven's Gate and People's Temple cults did. But these examples of mass suicide tend to be viewed as exceptions to the notion that suicide is an individual act based on an intense desire to "end it all."

Sociologists do not deny that people who kill themselves are suffering from intensely negative feelings, but they do not believe that all suicide is caused by intense mental states. Too much evidence suggests that other forces operate on individuals who commit suicide. For example, in recent research on suicide in different nations, sociologists have found that the highest numbers of suicides occur in Russia, China, and the nations of the former Soviet Union. Among the Western democracies, numbers of suicides vary considerably. Finland, Austria, Denmark, and France have the highest levels of suicide. The United States is in the middle of the range, and Italy and Greece are at the bottom.

What explains these differences in how many people in a given society kill themselves? When sociologists ask a question like that, they are asking not why an individual commits suicide but why an entire society has a high level of suicides. This is an important question, because if you live in a society where suicide is more common, it may be more likely that you will consider this fatal option at some point in your life.

The study of suicide, like the study of any form of human behavior, raises many issues of scientific observation and interpretation. Consider this question: If in a given year 1,000 people kill themselves in a small nation and 1,000 people kill themselves in another nation with a far larger population, can we say that both nations have the same amount of

AP Photos

suicide? Clearly, we need to consider the difference in population size when comparing the levels of suicide in these nations. Measurement of social phenomena like suicide is a vital aspect of understanding the behavior and possibly preventing it. This chapter, therefore, is devoted to the basic scientific methods that sociologists employ when trying to understand complex human behaviors.

APPLYING THE SOCIOLOGICAL IMAGINATION

As citizens of a democratic nation, we are encouraged to form opinions about the social issues of our time. Among the most troubling issues we face are many that involve life and death. Capital punishment, living wills, abortion, fertility drugs, suicide, assisted suicide—all pose extremely difficult ethical problems. These problems also indicate the rapid pace of social change in our lives. For the most part, our grandparents did not have to face many of these issues.

As in all democracies, candidates seeking our votes are eager to tell us what to believe about social issues and what society should do about them. But often the claims and recommendations we hear are contradictory. The explanations given for such problems as suicide, for example, may suggest very different solutions. How is an informed citizen to make decisions about the causes of social conditions or what should be done to remedy them? This is where sociology comes in. Sociology brings scientific methods to bear on these debates.

Most people associate social-scientific research with questionnaires, opinion surveys, and statistical reports. It is true that sociologists throughout the world use these research techniques, but they also use many other methods to explore social conditions. In this chapter, we review many of the most common research tools that sociologists use. We will show how sociological research conforms to the rules of the scientific method and how the findings lead to changes in theories and to new research. We will also show how sociological knowledge differs from "common sense."

Steps in the Research Process

In all research projects, certain basic steps must be completed, though not always in the order listed here:

1. *Deciding on the problem.* At first this may be simply a subject or topic of interest. Eventually, however, it must be worded in the form of a specific research question or questions to provide a focus for the rest of the work.
2. *Reviewing the literature.* Usually others have conducted research on the same topic. Find their reports and use them to determine what you can accomplish through your own research.

3. *Formulating research questions.* The work of others points to questions that have not been answered. Your own interest, time, resources, and available methods help determine what specific questions within your broad topic you can actually tackle.
4. *Selecting a method.* Different questions require different types of data, which in turn suggests different methods of data collection and analysis. Some of those methods are described in detail in this chapter.
5. *Analyzing the data.* Data are analyzed at each stage of the research, not just when preparing the final report.

In this chapter, we act as if researchers in the social sciences always follow this step-by-step procedure. But do not assume that this procedure can be applied to every research question. One does not always progress easily from one step to another, and often one must go back and repeat earlier steps. In addition, some research questions require that we devise new ways of conducting research or new combinations of existing research methods.

Formulating Research Questions

Hypotheses Émile Durkheim's study of suicide (1951/1897) provides a good example of the process by which a sociologist converts a broad question about social change into the specific questions to be addressed in an empirical study—one that gathers evidence to describe behavior and to prove or disprove explanations of why that behavior occurs. These explanations are often, but not always, stated in the form of a **hypothesis**, a statement that expresses an informed (or "educated") guess regarding the possible relationship between two or more phenomena.

In his study of suicide, Durkheim challenged the intellectuals and scientists of his day. Through the presentation of verifiable statistical evidence on suicide rates in different societies, he demonstrated that he could predict where and when suicides would be more numerous. Psychological reasons might account for why a particular individual committed suicide, but Durkheim showed that "social variables" such as religion or fluctuations in economic conditions could explain differences in the number of suicides from one society or region to another.

The question that eventually led Durkheim to his empirical study of suicide actually had nothing to do with suicide. Like many educated people of his time, he was troubled by the consequences of large-scale social change in Western nations. In particular, he believed that industrialization and the rapid growth of

hypothesis: A statement that specifies a relationship between two or more variables that can be tested through empirical observation.

Suicide in the Public Spotlight The stock market crash of 1929 led to the suicides of people who had lost fortunes and faced immediate bankruptcy. More recently, the popular rock star Kurt Cobain killed himself. Both examples illustrate the key findings of Durkheim's pioneering studies of suicide. The suicide of an individual stockbroker or celebrity can be explained by the individual's state of mind at the time. However, the tendency for people to kill themselves when their fortunes change suddenly, or for celebrities to live out-of-control lives that often end in suicide, can be explained by social influences that cause people to feel lost or adrift. Durkheim labeled this feeling *anomie* and traced it to a lack of integration of the individual into social groups and communities.

© UPI/Bettmann/Corbis

© AP/World Wide Photos

cities weakened people's attachments to their families and communities. As people became increasingly anonymous and isolated, they were more likely to engage in a variety of self-destructive acts, the most extreme being suicide. In Durkheim's view, the act of suicide could be explained as much by social variables like rates of marriage and divorce as by individual psychological variables like depression or despair. Thus, for Durkheim, the study of suicide was a way of exploring the larger concept of integration or lack of integration into society: He sought to discover whether people who were less well integrated into society (that is, more isolated from other people) were more likely to commit suicide.

If this view were correct, Durkheim reasoned, the rates of suicide among various populations should vary along with measures of social integration. He therefore formulated these hypotheses, among others:

- Suicide rates should be higher for unmarried people than for married people.
- Suicide rates should be higher for people without children than for people with children.
- Suicide rates should be higher for people with higher levels of education (education emphasizes individual achievement, which weakens group ties).

- Suicide rates should be higher in Protestant than in Catholic communities (Protestantism places more stress on individual achievement than Catholicism does, and this in turn weakens group ties).

Variables Each of these hypotheses specifies a relationship between two variables that can be tested—that is, proved true or false—through empirical observation. In sociology, a **variable** is a characteristic of individuals, groups, or entire societies that can vary from one case to another. In the hypotheses just presented, the suicide rate is a social variable. Religion, education, marital status, and number of children are other variables in these hypotheses. The techniques Durkheim used to establish a set of hypotheses became a model for modern social-scientific research.

Let us examine one of Durkheim's classic hypotheses about suicide to see what it can teach us about applying the scientific method to social issues. In the hypothesis stating that suicide rates should be higher for unmarried people than for married people, Durkheim was

variable: A characteristic of an individual, group, or society that can vary from one case to another.

proposing that there is a relationship between two variables: incidence of suicide and incidence of marriage. The suicide *rate* is a measure of suicide that takes into account the size of the population. The marriage rate is also based on population size.

In this hypothesis, the suicide rate is referred to as the **dependent variable**; it is the variable that we are trying to explain. Our hypothesis suggests that variations in the suicide rate from one nation to another depend on the influence of another variable: marriage rates in different nations. Marriage rates are referred to as the **independent variable**. An independent variable is a factor that a researcher believes causes changes in the dependent variable. It is independent of the variable we are trying to explain. (We cannot, for example, imagine explaining variations in marriage rates in terms of changes in suicide rates.)

Extreme care must be used in making statements about causality, however. All social phenomena are caused by interactions among several variables rather than by only one or two. Thus we may say that lower marriage rates are one cause of suicide in a society. We know that there are other causes as well. (We discuss the issue of causality more fully later in the chapter.)

But, clearly, the story is more complicated than that. Some people commit suicide even though they are married and loved. Today, some people think about suicide as an alternative to a lingering hospital death. As Durkheim first showed a century ago, suicide can take many forms and have many causes in different societies. Since then, sociologists have argued that before anyone can develop a set of testable hypotheses about the causes of suicide (or any other complex social situation or behavior), it is necessary to know a great deal about the actual experiences of the people involved. Thus, even a sociologist who has never contemplated suicide or been part of a suicidal cult can attempt to see the world from the viewpoint of people in suicidal situations to find out what their daily lives are like. If they exist, such studies are usually the first ones consulted when a sociologist begins to conduct research on a particular problem.

Reviewing the Literature

Perhaps the insights one would need to understand the issues surrounding suicide are already available in "the literature"—in existing books or journal articles, published statistics, photos, and other materials. There is no need to conduct new research if the answers sought are already available. Most sociological research therefore begins in the library with a "review of the literature."

To stimulate our sociological imagination about suicide or any other social phenomenon, we need to look at a variety of studies that deal with this issue. But it takes some imagination just to think of the kinds of studies to look for. The various sociological perspectives described in Chapter 1 can help organize the search, especially when the perspectives are framed as questions.

Who? How Many? Where? Who are we talking about and where are they? This is a way of phrasing the ecological perspective, which suggests why that perspective is helpful in beginning research on an issue. In researching almost any subject involving human behavior, it is helpful to ask who is involved and in what numbers, and where the behavior in question occurs. In his pioneering research on suicide, Durkheim asked some very basic questions, including an important ecological one: Do suicide rates vary from one county to another within a nation? This question remains an active area of contemporary research, as we will see shortly.

The ecological perspective leads to two types of studies: community studies and demographic studies. *Community studies* are among the richest research traditions in sociology. They portray the typical day-to-day life of a particular population. An example of a community study of a suicidal group is John R. Hall's *Gone from the Promised Land: Jonestown in American Cultural History* (1987). Hall was not a member of the infamous People's Temple, but he had lived in several communes and religious communities, where he collected firsthand accounts of what it is like to be part of an intense community like the People's Temple. If a student wanted to explore what led members of the Heaven's Gate community to commit mass suicide in March 1997, Hall's study would be an excellent resource for ideas and comparisons. A search of the literature on the Heaven's Gate suicide would also turn up many firsthand accounts about what it is like to live among people who are obsessed with extraterrestrials and UFOs.

Demographic studies are also useful in studying the magnitude of a major social phenomenon such as suicide. Such studies provide counts of people in various relevant population categories—in this case, people who are isolated in one way or another, or who have actually attempted or succeeded in committing suicide, or who may be considering assisted suicide

---

dependent variable: The variable that a hypothesis seeks to explain.

independent variable: A variable that the researcher believes causes a change in another variable (that is, the dependent variable).

because of advanced illness. Demographic studies of suicide offer a wealth of information about this drastic behavior from nations throughout the world. In a recent study, for example, the French National Institute for Demographic Studies reported that "France has the sad privilege to be classed among the nations of the world with the highest rates of suicide. If one eliminates the states of the old communist bloc and China, France, with a suicide rate in the neighborhood of 20 per 100,000, is in the fourth position behind Finland, Denmark, and Austria. All the other nations have a lower suicide rate: Japan (15 per 100,000), Sweden (15), Germany (14), Norway (13), Canada (13), the United States (12), the United Kingdom (7), Italy (7), Greece (3)" (Constantine & Huberman, 2000; Kremer, 1998).

How Is the Situation Defined? Interactionist approaches to extreme phenomena such as suicide tend to look at how people who decide to take their own lives actually perceive their situation and how their interactions with others influence their behavior. In the case of suicide, such studies are not common, but sociological research can often draw on firsthand accounts. This is especially true of sensational suicides like the Heaven's Gate case: Since the 1997 mass suicide, numerous testimonials by former cult members have described how followers surrender their personal autonomy to the will of the leader and accept the leader's extreme definition of the situation (Bearak, 1997; Brooke, 1997). In the case of the Heaven's Gate cult, the definition of the situation was that all who killed themselves would be immediately transported to a better existence among the superior beings accompanying the Hale-Bopp comet. In fact, it was possible to follow the development of the deadly cult's beliefs on the Internet, which was also used to recruit new cult members (Levy, 1997).

What Groups or Organizations Are Involved? This question stems from the functionalist perspective. It asks how society is organized to deal with a social issue or problem. Functionalist sociologists are concerned with how social policies actually function, as opposed to how they are supposed to function. Studies of social interventions to prevent suicide, for example, would describe how community crisis hotlines are established and who operates them, and ideally would include some measures of how effective they are. These measures might include how many calls are received and the nature of the calls (for example, actual suicide threats, people in need of counseling, parents or guardians of possible teenage suicides, and so on). They might also include some description of how many times the potential suicide victim was saved through the actual intervention of crisis center personnel.

What Difference Does Power Make? This question arises from the conflict perspective. Studies that take a conflict approach to suicide would be likely to compare suicide rates for men and women around the world. The differences would be striking. In many nations, women are far more likely than men to kill themselves. Why this is so has a lot to do with men's power and women's lack of power in different societies. In remote parts of India, for example, a widowed woman is expected either to kill herself upon her husband's death or to submit herself entirely to the will of her husband's family—in which case she often serves essentially as a house slave, a condition that itself can drive a person to suicide. In some African and Latin American societies where marriages are often arranged or women are routinely sexually exploited, we also find higher suicide rates for women than for men. In consequence, many sociologists argue that changes in the relative power of women, resulting from greater literacy, access to occupations, and basic protections under the law, would improve the balance of power between the sexes and lead to more equal rates of suicide, in addition to many other benefits.

Is Gender Addressed Adequately? A problem with some social-scientific research literature is that the perspectives and experiences of men are assumed to apply to women as well. In fact, the same criticism applies to issues of race. The subjects of earlier research were often primarily white males, and their experience was often generalized to all others in the society, regardless of gender or race. The most glaring problems, however, occurred because male researchers tended to overlook the likelihood that women's experiences and perspectives would differ from those of their male subjects. As Sandra Harding, who has conducted extensive research on this problem, notes,

> Three decades ago women's movements around the globe were rigorously questioning the exclusion of women, their interests, and visions of the good life, from policy debates, including those of science. . . . What assumptions and processes had led male scientists and policy-makers to equate their own concerns and opinions with human perspectives in general? (1998, p. 1)

One such assumption, prevalent in mid-twentieth-century research, was that because women were subordinate to men, their membership in different social classes could be determined from the social class of their husbands. Men who were classified as members of the working class, for example, were assumed to have wives who were in the same class. Research on women who were not married or were working outside the home was extremely limited. However, as women enter the labor force in ever-greater numbers and as far more women conduct sociological research, this situation has changed; one can more easily find research and develop research projects in which women are fully represented in the research design and the analysis of results (Maynard, 2002).

Today, the literature on suicide includes an array of good studies on suicide among women, as well as studies that compare suicide rates by gender. And these studies often relate directly to the classic work of Durkheim. For example, a study of suicide among women in North Carolina found that suicide rates for white women were three times higher than those for African American women. The researchers explain their findings, in part, with a Durkheimian argument: They note that their data show that African American women are more involved in extensive family kinship groups than are many white women. The latter often arrive in the state from elsewhere and are not surrounded by family and friends of long standing, and this sense of relative isolation may help explain why their despair is not diffused by participation in social groups. Because about 6,000 women commit suicide annually in the United States, making suicide the second leading cause of death for women under 50, research directed specifically at suicide among women is clearly important in developing more effective intervention strategies (*Women's Health Weekly,* 2003).

The foregoing are only a few of the studies one would find in a review of the sociological literature on suicide. They are not cataloged according to the basic sociological perspectives. Rather, one needs to know what questions each of the sociological perspectives raises about an issue. For any research subject, if you ask questions about who and where and how people interact, what organizations and policies guide their actions, and who has the power, you will be well on your way toward a thorough review of what is known about the issue you plan to study.

THE BASIC METHODS

Once the researcher has specified a question, developed hypotheses, and reviewed the literature, the next step is to decide on the method or methods to be used in conducting the actual research. Sociological research methods are the techniques investigators use to systematically gather information, or *data,* to help answer a question about some aspect of society. The variety of methods is vast, with the choice of a method depending largely on the type of question being asked. The most frequently used methods are observation, experiments, and surveys. These are summarized in the For Review chart on page 29.

Qualitative and Quantitative Methods

Social researchers often distinguish between qualitative and quantitative fact-finding methods (Babbie, 2003). Qualitative methods, especially observation, focus group interviews, or case study interviews generate many facts about a relatively small number of cases. Quantitative research methods, in contrast, are based on systematic counting and statistical analysis of as many cases as possible. Quantitative researchers often rely on the categories or ideas developed through qualitative research and attempt to count the frequency of observations in different categories.

For example, we know that there are relatively high rates of suicide among veterans returning from the wars in Iraq and Afghanistan. A qualitative study of suicide among returning veterans might compare accounts of how different returning vets eventually took their own lives. What were their difficulties in adjusting to life back home? Did they seek treatment for mental conditions like posttraumatic shock? Did they have a family support group, or were they isolated and lonely? Questions like these generate a series of case studies of specific servicemen and women who took their lives. Such studies invariably yield more questions, such as: Are rates of suicide higher among returning veterans? Are they higher among vets who suffered from posttraumatic shock? These questions require quantitative research—that is, careful counting of the number of veterans who have committed suicide and the number who have reported recurring symptoms of posttraumatic shock.

Sociology & Social Justice

SUICIDE RATES AMONG VETERANS The case of returning veterans not only illustrates the value of both qualitative and quantitative research but also serves as an example of how sociological research addresses issues of social justice. Consider the following: More servicemen and -women took their own lives after serving in the Vietnam War than died on the battlefield (Hasws, 1998). A suicidal

The Basic Methods

Method	Description	Example
Participant Observation	A form of field observation in which the observer participates to some degree in the lives of the people being observed.	Martin Sanchez-Jankowski spends many months with youth gangs, gaining their trust and learning gang norms regarding violence.
Visual Sociology	Observational techniques that rely on the use of photography, video, or film to supplement or further document social behavior that can be observed.	Douglas Harper uses photos of hoboes riding the rails of American freight lines to document and record a vanishing way of life.
Controlled Experiment	An experimental situation in which the researcher manipulates an independent variable in order to observe and measure changes in a dependent variable.	Judith Gerone of the Manpower Research and Development Corporation conducts controlled experiments to find out what types of training and job programs are most effective in reducing state welfare rolls.
Field Experiment	An experimental situation in which the researcher observes and studies subjects in their natural setting.	Milton Rokeach meets three delusional mental patients, each of whom believes he is Jesus Christ. He introduces them to each other to learn how they incorporate conflicting information into their self-presentations.
Survey	Research in which a sample of respondents drawn from a specific population respond to questions either in an interview or on a questionnaire.	A team of National Opinion Research Center sociologists survey the public about their sexual behavior and attitudes.

rampage at Fort Hood, Texas, in 2009 revealed a history of neglect of recruits' and veterans' mental health problems, including suicide. Social-scientific data show that suicide rates among returning veterans are three times higher than those for the general population. These instances point to a significant social justice issue: Veterans are not receiving adequate mental health care. Revealing such facts is only the beginning of an effort to use that information to correct situations of grave social injustice, such as clearing troops with mental illnesses for service in Iraq (Olinger & Emery, 2008). Further research is clearly necessary, but so is a decisive effort by government and the military to provide appropriate mental health care for all members of the armed forces.

Observation

Participant Observation Much qualitative sociological research requires direct observation of the people being studied. Community studies like Hall's work on the People's Temple are based on lengthy periods spent observing a particular group. This research method is rarely successful unless the sociologists also participate in the daily life of the people they are observing—that is, become their friends and members of their social groups. Therefore, the central method of community studies is known as **participant observation**. The sociologists attempt to be both objective observers of events and actual participants in the social milieu under

study—not an easy task for even the most experienced researchers. In such situations, the observers faithfully record their observations and interactions in *field notes,* which supply the descriptive data that will be used in the analysis and writing phases of the study.

An excellent example of a study based on participant observation is Donna Gaines's *Teenage Wasteland* (1998). Gaines spent many months "hanging out" and speaking with teenagers in a relatively affluent suburban community adjacent to a large central city. Many of the teenagers lived lives devoted to school, family, and part-time jobs. Others became what Gaines called "suburbia's dead-end kids." They experimented with drugs, were attracted to satanism and heavy metal rock music, and flirted with violence and suicide, sometimes with tragic consequences. These young people, who often thought of themselves as "burnouts," described their lives to Gaines because she listened to them and showed that she could be trusted. In the beginning, they tested her by giving her information and telling her not to share it with other kids. Once they saw that she did not divulge secrets, she was able to learn much more about their behavior. By listening carefully, she also learned how some kids who seem to be at a dead end are actually struggling to create a better world for themselves, often against great odds.

---

participant observation: A form of field observation in which the researcher participates to some degree in the lives of the people being observed.

Another excellent example of a study based on participant observation is Douglas Harper's classic, *Good Company* (2006). Harper spent many months riding on freight trains and living in hobo "jungles." His goal was to describe how hoboes, or tramps or bums as they are often called, actually live and how they learn to trust or distrust one another in a world considered deviant by members of "respectable" society. Harper had to learn how to live the nomadic and often dangerous life of the hobo, which meant that he needed to establish a relationship with someone in that life who could serve as his mentor. The successful participant observer usually finds at least one such person. In Harper's case, a hobo named Carl made it possible for him to participate in, and therefore to understand and write about, the hobo's world.

Qualitative research carried out "in the field" where behavior is actually occurring is the best method for analyzing the processes of human interaction. If Donna Gaines had not spent a good deal of time with suburban teenagers, she would not have obtained nearly as many insights into their behavior. A shortcoming of this approach, however, is that it is usually based on a single community, group, or social system, which makes it difficult to generalize the findings to other social settings. Thus, community studies and other types of qualitative studies are most often used for exploratory research, and the findings serve as a basis for generating hypotheses for further research.

Visual Sociology An increasingly popular set of qualitative observational techniques involves the use of photography and videotape, or **visual sociology**. These techniques can be just as obtrusive as interviewing, if not more so, but they can also be used in unobtrusive ways, depending on the questions being asked.

One of the pioneers in the use of visual data in his research is Douglas Harper, whose study of wandering hoboes and homeless street people uses photography not as illustration but as data with which the sociologist can transport readers into the lives and cultures of people on society's rough margins. The cover of Harper's book, for example, presents a characteristic photo of Carl, the man who became one of Harper's most important mentors and informants about the vanishing world of the wandering hobo who rode the freights from one vagrant camp to another, following the crops or other forms of temporary labor (Harper, 2006).

Courtesy of Douglas Harper/Paradigm Publishers

Another example of visual sociology is taken from my own research with urban Roma gypsies living on the outskirts of Paris. For a summer during the 1960s I lived with a large family of Roma gypsy animal trainers who bought and sold animals for travelling circuses. This photo showing the family patriarch putting the bear through his paces is now a rare example of bear training, a practice that dates back to the middle ages but is now largely a lost form of urban entertainment in European cities.

Applying your sociological imagination to visual material can be both interesting and informative. Douglas Harper's work and the photos of Battaglia and Zecchin are only two

William Kornblum

This photo by the author is an example of visual sociology because it captures a form of urban entertainment that no longer exists in European cities. Here Alexander Ivanovich, the leader of a large family of Roma gypsies of the Boyash tribe, puts the family's trained bear through his paces as other members of the family look on.

visual sociology: Qualitative observation techniques that rely on the use of photography, video, or film to supplement or further document social behavior that can be observed.

examples of the rich vein of visual sociology that is captured in photographs, as well as in works of fine art. Similar examples are presented in the Visual Sociology boxes that appear throughout this book.

Experiments

Although, for both moral and practical reasons, sociologists do not have many opportunities to perform experiments, a large body of literature in the social sciences is based on experiments. Social scientists use two experimental models. The first and most rigorous is the controlled experiment conducted in a laboratory. The second is the field experiment, which is often used to test public policies that are applied to some groups but not to others.

Controlled Experiments In a **controlled experiment**, researchers manipulate an independent variable in order to observe and measure changes in a dependent variable. The experimenters form an **experimental group** that will experience a change in the independent variable (the "treatment") and a **control group** that will not experience the treatment but whose behavior will be compared with that of the experimental group. (The control group is similar to the experimental group in every other way.) This type of experiment is especially characteristic of studies at the micro level of sociological research.

Consider an example: Which line in Figure 2.1(b) appears to match the line in Figure 2.1(a) most closely? Could anything persuade you that a line other than the one you have selected is the correct choice? This simple diagram formed the basis of a famous series of experiments conducted by Solomon Asch in the early 1950s. They showed that the opinion of the majority can have an extremely powerful influence on that of an individual.

Asch's control group consisted of subjects[1] who were seated together in a room but were allowed to make their judgments independently. When they looked at sets of lines like those in Figure 2.1, the subjects in this group invariably matched the correct lines, just as you no doubt have. In the experimental group, however, a different result was produced by the introduction of an independent variable: group pressure.

Asch's experimental group consisted of subjects who were asked to announce their decisions aloud in the group setting. Each subject was brought into a room with eight people who posed as other subjects but were actually

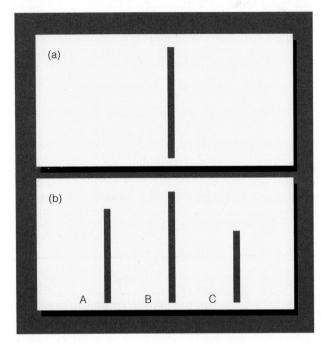

Figure 2.1 Lines Used in the Asch Experiment on Conformity. Cards like these were used in the Asch experiment. Subjects were asked to judge the lengths of various lines by comparing them with the three lines on the bottom card. The line on the top card quite obviously matches line B on the bottom card; all of the judgments were this simple.

confederates of the experimenter. When the lines were flashed on a screen, those "subjects" all chose a line that was not the matching one. When it was the real subject's turn to choose, he or she was faced with the unanimous opinion of a majority of "subjects" who had picked the wrong line. Thirty-two percent of the real subjects went along with the majority and chose the wrong line as well. Even among the subjects Asch called "independent" (the 68 percent of the real subjects who gave the correct response despite the pressure of the majority), there was a great deal of variation. Some gave the correct response at all times, whereas others gave it only part of the time. This conforming response was less likely when at least one other person in the group also went against the majority. By varying the number of people who said that the shorter line was longer, Asch was varying the degree of group pressure the subject experienced. Higher levels of the independent variable (group pressure) thus produced more "errors," or choices of the wrong line (Asch, 1966).

Field Experiments **Field experiments** are used extensively in evaluating public programs that address specific social problems. In these experiments, there is usually a "treatment group" of people who participate in the program and a control group of people who do not. In one example of this type of experiment, Angelo Atondo, Mauro Chavez, and Richard Regua of California's Evergreen Valley College attempted to test

---

controlled experiment: An experimental situation in which the researcher manipulates an independent variable in order to observe and measure changes in a dependent variable.

experimental group: In an experiment, the subjects who are exposed to a change in the independent variable.

control group: In an experiment, the subjects who do not experience a change in the independent variable.

field experiment: An experimental situation in which the researcher observes and studies subjects in their natural setting.

---

[1] The term *subject* refers to a person who participates in a controlled experiment.

the theory that if students at risk of failure are linked with adult mentors from their own communities, their chances of success will improve (Bashi, 1991). This popular theory stems in part from an influential article in the *Harvard Business Review* titled "Everyone Who Makes It Has a Mentor" (Collins & Scott, 1978). Many successful people from all walks of life, especially those who have escaped from impoverished backgrounds, say that they benefited greatly from the guidance of a mentor who steered them toward constructive goals. Most mentors do not actually tutor the individuals they spend time with; instead, they teach through conversation and example.

Researchers have found little evidence to prove that mentors actually make a difference. Unless one can compare students who had mentors with similar students who did not, it is impossible to say with any scientific certainty—that is, beyond individual stories of success or failure—whether the mentors made a difference. The Evergreen Valley College researchers therefore assigned 115 entering Latino students with low English proficiency to Latino mentors with excellent English skills. Their control group was composed of a comparable number of Latino students with similar characteristics; these students were not assigned to mentors. Both groups attended the same classes and took the same exams. At the end of the semester, 89 percent of the students with mentors passed the freshman English course, but only 46 percent of the students in the control group passed the same course. Later, more of the students with mentors went on to four-year colleges. Although much more controlled research of this nature needs to be done before we can fully understand how mentors help students overcome their educational problems, this example shows how a field experiment can advance our understanding of a social issue like success and failure in school (Bashi, 1991).

Sociologists can conduct "natural experiments" when two similar groups receive differing treatments and the results can be measured. An example of such an experiment deals with problems of teenagers, including early pregnancy, school failure, and suicide. Gregory J. Duncan and Jeanne Brooks-Gunn (1997) studied the lives of teenagers growing up in low-income neighborhoods. By comparing children from families with very similar social characteristics (for example, income, race, ethnic background, education), they could identify the effects of certain characteristics of their surroundings (for example, rates of poverty, crime, suicide, and drug abuse) on their behavior.

The Hawthorne Effect A common problem of experimental studies is that, just by paying attention to people in an experimental group, the researcher may be introducing additional variables. This problem was recognized for the first time in the late 1930s, when a team of researchers led by Elton Mayo conducted a famous series of experiments at Western Electric's Hawthorne plant. The purpose of the study was to determine the effects of various environmental and social conditions on the workers' productivity. One hypothesis was that improvements in the physical features of the workplace (such as better lighting) and improved social conditions (coffee breaks, different methods of payment, and so on) would result in greater productivity (the dependent variable). However, it appeared that *no matter what the experimenters did,* the workers' productivity increased. When the experimenters improved the lighting, the workers' productivity increased over that of the control group. When they dimmed the lights, productivity went up again. This was true even when the workers were subjected to somewhat worse conditions or were returned to their original conditions.

At first this series of experiments was considered a failure, and the researchers concluded that the variables they were introducing had little to do with the changes in productivity that resulted. On further reflection, however, they realized that the real independent variable was not better working conditions but, simply, *attention.* The workers liked the attention they were getting; it made them feel special, so they worked harder. This realization led the researchers to design experiments dealing with the effects of attention from supervisors and better communication between workers and their managers; those experiments led to a new philosophy of worker–management relations, described in Chapter 18. Today, the term **Hawthorne effect** is used to refer to any unintended effect that results from the attention given to subjects in an experiment.

The Hawthorne effect occurs in a variety of experimental situations. For example, in the Evergreen Valley College experiment described earlier, it is possible that the students with mentors were more successful than the students in the control group because they were getting extra attention from adults. Over time, the effects of that attention may disappear, and the experimenters may conclude that once the

Hawthorne effect: The unintended effect that results from the attention given to subjects in an experimental situation.

novelty of the extra attention wears off, students with mentors do not do appreciably better than others. Sophisticated evaluations of educational change try to anticipate the Hawthorne effect by making sure that the experimental group does not get more attention than the control group.

Survey Research

A sociological *survey* asks people to give precise information about their behavior, their attitudes, and at times the behavior and attitudes of others (for example, other members of their households). There is a world of difference between the modern sociological survey and the "social surveys" conducted around the turn of the twentieth century (see Chapter 1). Those surveys attempted to present an unbiased, factual account of the social conditions of a specific community; their findings could not be applied to other groups. In contrast, today's survey techniques make it possible to *generalize* from a small sample of respondents to an entire population. An example is election polls, in which a small sample of 1,000 or 2,000 respondents can be used to predict how the entire electorate will vote in a presidential election. Done properly, the modern survey is one of the most powerful tools available to social scientists.

National Censuses Surveys are the central method used in election polling, market research, opinion polling, television ratings, and a host of other applications. The most ambitious and most heavily used sociological surveys are national censuses. A *national census* is a full enumeration of every member of the society—a regular system for counting its people and determining where and under what conditions they live, how they work and gain income, their patterns of family composition, their age distribution, their levels of educational attainment, and related data. Without these measures, a nation cannot plan intelligently for the needs of its people. The United States conducts a national census every ten years. The 2010 census, the twenty-second in the nation's history, was the largest and most complex ever undertaken; it enumerated about 307 million people in about 130 million housing units.

Sample Surveys Once a national census has gathered the basic demographic and ecological facts about a nation's people, it is possible to add more information by using a smaller and far less costly **sample survey**. In the United States, this is done regularly through the Census Bureau's Current Population Survey (CPS). From the CPS, we get monthly estimates of employment and unemployment, poverty, births, deaths, marriages, divorces, social insurance and welfare, and many other indicators of the social well-being and problems of the American people.

Opinion Polls Another type of survey research, pioneered by sociologists earlier in this century, is the *opinion poll*. Today, opinion polling is a $5 billion industry. Marketing firms use opinion polls to help corporations make decisions about their products; political candidates and their staffs use them to measure the progress of their campaigns; and elected officials use polls to monitor public opinion on key political and economic issues.

A good example of an opinion poll is the Gallup Poll, conducted regularly by the Gallup Organization, one of the nation's oldest and most respected public opinion survey companies. In addition to this private polling corporation, research groups at a number of major universities conduct large-scale national opinion polls; an example is the National Opinion Research Center (NORC) at the University of Chicago.

The Gallup Poll illustrates how opinion surveys allow political leaders and the public to follow trends in the way people feel about social issues of all kinds. Table 2.1 shows how Americans rank different social issues and how these ratings changed between 2007 and 2008. Clearly, the greatest change has occurred in how Americans worry about the economy and unemployment. The threat of terrorism and fears about the availability or affordability of health care continue to worry the survey's respondents, but given the collapse of the domestic and world economies in 2008, it is not surprising that the greatest increase occurred in their worries about the state of the economy. Political commentators and professional political consultants watch these results extremely carefully because they often indicate how voters will react to these issues in political campaigns or in their feelings about what government officials are doing, or not doing, about these issues.

Polls like these generalize from a sample of 1,500 respondents, on average, to the entire U.S. population. How is it possible that a relatively small number of respondents can accurately represent the distributions of attitudes in a population of 307 million people? The answer is found in the techniques used to choose a random **sample** of the population in question. The term *sample* refers to a set of respondents selected from

---

sample survey: A survey administered to a selection of respondents drawn from a specific population.

sample: A set of respondents selected from a specific population.

TABLE 2.1

Recent Trends in Worry about Major National Issues
(percent who worry "a great deal")

	2007	2008	Change*
The economy	39%	60%	21%
Unemployment	25	36	11
Energy	43	47	4
Crime and violence	48	49	1
Possible terrorism against the U.S.	41	40	−1
Race relations	19	18	−1
Drug use	45	43	−2
Social Security	49	46	−3
The environment	43	40	−3
Health care	63	58	−5
Illegal immigration	45	40	−5
Hunger/homelessness	43	38	−5

*percentage points

Source: The Gallup Poll (2008).

a specific population. The first step in selecting a sample is to define the population to be sampled; in our example, the population consists of adult American citizens. The next step is to establish rules for the *random* selection of respondents. The goal of this procedure is to ensure that everyone within the specified population has an equal chance, or *probability,* of being selected to answer the survey questions. A sample in which potential respondents do not all have the same probability of being included is called a *biased sample.* To avoid bias, respondents must be selected by some process of random sampling.[2] In other words, any form of "volunteering" to be interviewed must be eliminated. (For example, researchers cannot intentionally select their friends to be part of the sample.) Only a random sample can be considered truly representative of a target population.

Surveys that are not based on random samples can be misleading. In a famous example from the early days of surveys, researchers predicted that candidate Alf Landon would defeat the incumbent president, Franklin D. Roosevelt, in the 1936 election. They based their prediction on data from a telephone survey, but at the time, many low-income households did not have telephones. The survey therefore overrepresented the opinions of those with higher incomes, who were less likely to vote for Roosevelt. Roosevelt won by a landslide, and since then political pollsters have been much more alert to the problem of sample bias. Many contemporary surveys also produce biased results, however. Most mail surveys and many telephone surveys that claim to be conducting research are actually marketing surveys designed to tap the opinions of particular subgroups rather than a representative sample of the entire population.

Sampling Error Before leaving the subject of opinion polling, we should mention one additional point. Polls are subject to *sampling error.* A sampling error of plus or minus 3 percentage points, for example, would be normal in a national sample of 1,500 to 2,000 respondents. This is important information for anyone who reads and interprets poll data. It means that a difference of 3 percent or less between the percentages of American adults expressing certain opinions could be due to chance rather than to a real difference in the distribution of opinions. In other words, it is possible that a higher percentage of people with a certain opinion were included in the sample just by chance than is actually the case in the total population. The possibility of sampling error is more critical when the difference between two sets of responses is small—for example, when 51 percent of the sample favors one presidential candidate and 49 percent favors another. In this example, the survey analysts would have to conclude that the election is "too close to call."

Questionnaire Design Just as important as careful selection of the sample to be interviewed is the design of the *instrument,* or questionnaire, to be used in the survey. Questionnaire design is both a science and an art. Questions must be worded precisely yet be easily understood. Above all, the questions must be worded in such a way as to avoid biasing the answers. For example, the General Social Survey asked whether the United States was spending too much, too little, or about the right amount on "assistance to the poor." About 66 percent of the respondents answered "too little." However, when "welfare" was substituted for "assistance to the poor," the findings were almost reversed, with slightly less than 50 percent responding that too much was being spent on welfare (Kagay & Elder, 1992). (See the Research Methods box on page 35.)

One of the first decisions to make in designing survey questions is whether they should be open or closed. A **closed question** requires respondents to select from a set of answers, whereas an **open question** allows them to say whatever comes to mind. The items in the Research Methods box are examples of closed questions. An example of an open question would be "Please tell me how you feel about your neighborhood." In answering this type of question, respondents can say that they do not like the neighborhood at all or give one or more reasons for liking it. The interviewer attempts to write down the answers in the respondents' own words.

[2] Random sampling is accomplished by a variety of statistical techniques that are discussed in more advanced courses.

closed question: A question that requires the respondent to choose among a predetermined set of answers.

open question: A question that does not require the respondent to choose from a predetermined set of answers; instead, the respondent may answer in his or her own words.

Avoiding Bias in Survey Questions

In *Asking Questions: A Practical Guide to Questionnaire Design* (1982), survey researchers Seymour Sudman and Norman Bradburn of the National Opinion Research Center describe a common abuse of survey methods: the use of biased questions. In a questionnaire mailed to them by a lobbying group, the authors noticed questions like these:

1. Do you feel there is too much power concentrated in the hands of labor union officials?
   yes ___ no ___

2. Are you in favor of forcing state, county, and municipal employees to pay union dues to hold their government jobs?
   yes ___ no ___

3. Are you in favor of allowing construction union czars the power to shut down an entire construction site because of a dispute with a single contractor, thus forcing even more workers to knuckle under to union agencies?
   yes ___ no ___

These questions violate even the simplest rules of objectivity in questionnaire design. As the authors point out, they are "loaded with nonneutral words: 'forcing,' 'union czars,' 'knuckle under.'" Such questions are "deceptive and unethical, but they are not illegal" (p. 3).

To eliminate bias, one would have to begin by rephrasing the questions. For example, the first question could read something like this:

1. Which answer best sums up your feeling about the amount of power held by labor union officials today?
   a. too much power
   b. about the right amount of power
   c. too little power
   d. don't know

Fair questions avoid nonneutral words and give respondents an opportunity to answer on either side of the issue or to say that they do not know. Whenever you read the results of a survey, ask yourself whether the questions are free of obvious bias, if they are specific or general, and if they are threatening or non-threatening to respondents. Such an evaluation can make a great difference in how you view the answers.

When researchers wish to include both forced-choice categories and freely given opinions in a single survey, the questionnaire usually begins with open questions and then shifts to closed questions on the same subject. Survey instruments that rely on open questions are often called *interview guides.* They guide a researcher's questions in certain directions and emphasize certain categories of information, but they also allow an interviewer to follow up respondents' comments with further questions.

Research Ethics and the Rights of Respondents

In asking any type of question in a survey or other type of sociological study, researchers must be aware of the rights of respondents. Much sociological research deals with the personal lives and inner thoughts of real human beings. Although most of that research seems relatively innocent, in some cases, the questions asked or the behaviors witnessed are embarrassing or even more damaging. In one famous example, Laud Humphreys (1970, 1975) studied interactions between men seeking casual sexual encounters in public restrooms. Many of his colleagues attacked him for invading the men's privacy. Others defended him for daring to investigate what had until then been a taboo subject (homosexuality and bisexuality). After all, they reasoned, Humphreys was careful to keep the men's identities secret. But he also followed some of the men home and conducted interviews with them there. They were not aware that he knew about their homosexual activities, and on reflection, Humphreys admitted that he had deceived his subjects.

Sociologists continue to debate the ethical dilemmas raised by Humphreys's research and by studies like the Asch experiment (in which the subject is duped and may feel embarrassed). As a result of such controversies, the federal government now requires that research involving human subjects be monitored by "human subjects review panels" at all research institutions that receive federal funding. A research review must be undertaken before a study can be funded or approved for degree credit. To pass the review, researchers must provide proof that they have taken precautions to protect the fundamental rights of human subjects. Those rights include privacy, informed consent, and confidentiality.

The right of **privacy** can be defined as "the right of the individual to define for himself, with only extraordinary exceptions in the interest of society, when and on what terms his acts should be revealed to the general public" (Westin, 1967, p. 373). **Confidentiality** is closely related to privacy. When a respondent is told that information will remain confidential, a researcher may not pass it on to anyone else in a form that can be traced to that respondent. The information can be pooled with information that other respondents provide, but extreme care must be taken to ensure that none of the responses can be traced to a particular individual.

---

privacy: The right of a respondent to define when and on what terms his or her actions may be revealed to the general public.

confidentiality: The promise that the information provided to a researcher by a respondent will not appear in any way that can be traced to that respondent.

Informed consent refers to statements made to respondents (usually before any questions are asked) about what they are being asked and how the information they supply will be used. It includes the assurance that their participation is voluntary—that is, that there is no compulsion to answer questions or give information. The respondent should be allowed to judge the degree of personal risk involved in answering questions even when an assurance of confidentiality has been given.

The ethics of social research require careful attention. Before undertaking research in which you plan to ask people questions of any kind, be sure you have thought out the ethical issues and checked with your instructor about the ethics of your methods of collecting data and presenting them to the public.

ANALYZING THE DATA

Survey research normally is designed to generate numerical data regarding how certain variables are distributed in the population under study. To understand how such data are presented and analyzed, refer to Tables 2.2, 2.3, and 2.4. We will use these tables as a framework for a discussion of some of the basic techniques of quantitative data presentation and analysis. Other types of sociological data and analytical techniques are explained at appropriate points in later chapters.

TABLE 2.2

Types of Households in the United States, 1980 and 2000 (in thousands)

Type of Household	1980	2000
All households	80,776	105,480
Nonfamily households	21,226	33,693
Families	59,550	71,787
With own children	31,022	34,558
Without own children	28,528	37,109
Married couple	49,112	54,493
With own children	24,961	24,836
Without own children	24,151	29,657
Male householder	1,733	4,394
With own children	616	2,215
Without own children	1,117	2,179
Female householder	8,705	12,900
With own children	5,445	7,562
Without own children	3,261	5,338

Source: Adapted from *Statistical Abstract*, 2010

informed consent: The right of respondents to be informed of the purpose for which the information they supply will be used and to judge the degree of personal risk involved in answering questions, even when an assurance of confidentiality has been given.

frequency distribution: A classification of data that describes how many observations fall within each category of a variable.

In approaching a statistical table, the first step is to read the title carefully. The title should state exactly what information is presented in the table, including the *units of analysis*—that is, the entity (for example, individual, family, group) to which a measure applies. In Tables 2.2 and 2.3, the units of analysis are households; in Table 2.4, they are married-couple families.

Also check the source of the information presented in the table. This is usually given in a source note at the end of the table. The source indicates the quality of the data (census data are considered to be of very high quality) and tells readers where those data can be verified or where further information can be obtained.

Frequency Distributions

As you begin to study the table, make sure you understand what kind of information is being presented. The numbers in Table 2.2 represent a **frequency distribution.** For each year (1980 and 2000), they show how various types of households were distributed in the U.S. population. Frequency distributions indicate how many observations fall within each category of a variable. Thus, within the category of "Married-couple family" for the variable "Type of Household," Table 2.2 indicates that there were 49,112,000 households in the United States in 1980 and that the total increased to 54,493,000 by 2000.

Table 2.2 can tell us a great deal about social change in the United States. For example, social scientists use the term *household* to designate all the people residing at a given address, provided that the address is not a hospital, school, jail, army barracks, or other "residential institution." A household is not necessarily a married-couple family, as is evident from the categories listed in Table 2.2. There are also many households in which a male, with or without his own children, or a female, with or without her own children, is the primary reporting adult. There are also millions of nonfamily households, in which the adults reporting to the census takers are not related. These include people who are roommates, people who are cohabiting, people who are sharing a house or apartment but do not consider themselves roommates, and many other possibilities.

In earlier times, it was considered unfortunate and a bit odd for a woman who was not a widow to head a household with children. It was assumed that a true family was composed of a male head, his wife, their children, and anyone related to them by blood who lived on their premises. The numbers in Table 2.2 show, however, that the number of female-headed households grew significantly between 1980 and 2000. The data also show an increase in the number of family households (from 59.6 million in 1980 to 71.8 million in 2000) and a similar increase in the number of nonfamily households (from 21.2 million in 1980 to 33.7 million in 2000).

Percent Analysis

Comparing the numbers for 1980 and 2000 in Table 2.2 can be misleading. Remember that these are absolute numbers. They show that the number of nonfamily households has increased and that the number of female-headed and male-headed households has increased. However, the number of married-couple families has also grown, as has the total number of households. How, then, can we evaluate the importance of these various changes?

To compare categories from one period to another, we need a way of considering changes in the overall size of the population. This can be done through **percent analysis.** Here is an example:

Table 2.2 shows that between 1980 and 2000, the number of male-headed households with one or more children rose from 616,000 to 2,215,000—a sixfold increase. But how important is this increase in view of the increase in the total population between 1980 and 2000? By using the total number of households in each year as the base and calculating the percentage of each household type for that year, we can "hold constant" the effect of the increase in the total number of households. In the case of male-headed households with one or more children, we find that the increase as a percentage of all households (from 0.84 percent in 1980 to 2.1 percent in 2000) is not as important as the increase in absolute numbers would suggest. The calculation is as follows:

$$616,000/87,776,000 \times 100 = 0.76$$
$$2,215,000/105,480,000 \times 100 = 2.10$$

Table 2.3 presents each household type as a percentage of total households in that year. This way of presenting the data eliminates the effect of the increase in overall population size between 1980 and 2000. The percentages in Table 2.3 reveal some significant changes that would not be evident from a comparison of absolute numbers. For example, although Table 2.2 seems to indicate that there has been an increase in married-couple families, Table 2.3 shows that when we control for the overall increase in households, married couples actually declined as a proportion of the total, from 60.8 percent in 1980 to 51.7 percent in 2000.

TABLE 2.3

Types of Households in the United States, 1980 and 2000 (as percentage of total households)

Type of Household	1980	2000
All households	100.0	100.0
Nonfamily households	26.3	31.9
Families	73.7	68.1
With own children	38.4	32.8
Without own children	35.3	35.3
Married couple	60.8	51.7
With own children	30.9	23.5
Without own children	29.9	28.1
Male householder	2.1	4.2
With own children	0.8	2.1
Without own children	1.4	2.1
Female householder	10.8	12.2
With own children	6.7	7.2
Without own children	4.0	7.2

The proportion of female-headed families, especially those with children, increased during the period. So did the proportion of nonfamily households.

We cannot yet view these findings as definite. We must examine them carefully to be sure we are making the right kinds of comparisons. Look again at Table 2.3. In the category of "Married-couple family," those with no children of their own decreased as a proportion of total households—from 29.9 percent in 1980 to 28.1 percent in 2000. However, this apparent decrease is due to the overall increase in other types of households. It is still not clear whether, among married-couple families, those without children decreased in relation to those with children. Indeed, if we take only the category of "Married-couple family" and compare subcategories within it, as shown in Table 2.4, we see that *as a percentage of married-couple households,* those without children actually increased from 49.2 percent to 54.4 percent.

This example should convince you to pay close attention to the comparisons made in numerical tables. In this example, we see that married

TABLE 2.4

Married-Couple Families, by Number of Own Children, 1980 and 2000 (as percentage of total households)

	1980	2000
Married-couple families	100.0	100.0
With own children	50.8	45.6
Without own children	49.2	54.4

percent analysis: A mathematical operation that transforms an absolute number into a proportion as a part of 100.

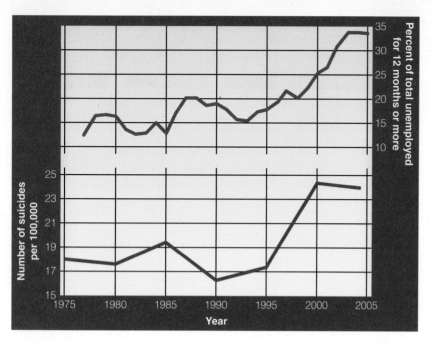

Figure 2.2 Japan Suicide and Long Term Unemployment Rates. Source: From "Suicide Rates in Japan" by Jon Peltier, in *Peltier Tech Blog*, Wednesday, July 30th, 2008 Peltier Technical Services, Inc., Copyright © 2010. Reprinted by permission.

couples with children accounted for a smaller percentage of all married-couple families in 2000 than in 1980, but we would not have seen this without making the additional comparison presented in Table 2.4.

Correlations

The term **correlation** refers to a specific relationship between two variables: As one varies in some way, so does the other. Figure 2.2 shows a correlation that leaders in Japan are particularly concerned about: between the long-term unemployment rate and suicide in that nation. We can observe from the graph that the two variables (rate of suicide and rate of unemployment) vary together: As one increases, so does the other. This correlation is another confirmation of Durkheim's original hypothesis that the rate of suicide in a society is related (correlated) to major changes in how people in the society behave (that is, whether they marry, find jobs, have children, and so on). No wonder, then, that according to the Japanese sociologists who gathered these data, the correlation between unemployment and suicide shows that suicide "is an expression of social distress, not only a matter of personal relationships" (quoted in Kremer, 1998, p. 8).

The search for correlations is a common strategy in many kinds of research. In market research, for example, the investigator seeks correlations between the use of certain products and other social variables. (Thus, the consumption of beer correlates with the proportion of male consumers in a target population such as baseball fans; the consumption of light beer correlates with the proportion of older, more weight-conscious male consumers; and so on.)

If you take other courses in sociology and statistics, you will learn techniques for calculating the strength of correlations among variables and for sorting out the effects of different variables on the dependent variable you are studying. Although the calculation of correlation is covered in statistics courses, for our purposes, it is helpful to know that the measure of correlation between two variables, termed the *correlation coefficient,* can vary between +1.0 and −1.0, with 0.0 representing no measured correlation at all. A correlation coefficient of +1 would indicate that the variables are positively and perfectly related—a change in one variable produces an equivalent change in the same direction (increase or decrease) in the other variable. A correlation coefficient of −1 would mean that the variables are perfectly inversely related—a change in one produces an equivalent change *in the opposite direction* in the other. In reality, most variables are not perfectly correlated, and correlation coefficients usually fall somewhere between these two extremes.

Correlation and Causation Although correlation can be useful in the analysis of relationships among variables, it must not be confused with causation. For example, the Japanese

---

correlation: A specific relationship between two variables.

sociologists who studied the correlations presented in Figure 2.2 do not claim that unemployment causes suicide; they only point out that the two variables are directly related in that as unemployment rises or falls, the suicide rate also rises or falls. Social scientists are cautious when making claims about causality because often a strong correlation that seems to indicate causality is in fact a spurious or misleading relationship. Take the example of storks and babies. The fable that storks bring babies seemed to be based on a statistical reality: Until recent decades, there was a correlation between storks nesting in chimneys in Dutch households and the presence of babies in those households. The more storks, the more babies. In fact, however, the presence of babies in the home meant more fires in the fireplace and more heat going up the chimney to attract storks to nest there. Storks do not bring babies; babies, in effect, bring storks. The real causal variable is heat, something that was not suggested in the original commonsense correlation. Correlations may suggest the possibility of causality, but one must be extremely careful in making the leap from correlation (or association) to causation.

> **Thinking Critically** The more young people tend to drink alcohol, the higher the rate of highway deaths among young people. Can we say that those highway deaths are the result of drinking and driving?

Mapping Social Data

Data about individuals or households can be mapped in precise ways, providing a powerful tool for analyzing sociological data (Monnier, 1993). In a sense, the mapping of data is a form of correlation. The variable under study—it could be poverty, crime, educational achievement, an illness such as acquired immunodeficiency syndrome (AIDS), or any other variable—is mapped over a given geographic area, thereby correlating the variable with location.

When Émile Durkheim conducted his early research to demonstrate the social (rather than the individual psychological) causes of suicide, he used some basic mapping techniques. With some simple maps of France, he showed that if one identified counties with high suicide rates and high rates of single, unmarried individuals, the results revealed clusters of counties with high and low rates on both variables. He believed that this clustering offered support for his theory that lack of integration of the individual into society—through marriage, employment, and participation in social groups

of all kinds—helps cause suicide (Durkheim, 1951/1897). But one of Durkheim's most important critics, Gabriel Tarde, proposed a more psychological explanation for the clustering. Tarde argued that the clusters that appeared on the map could be the result of imitation. He believed that where rates of suicide are high, people suffering from personal distress are likely to be imitating the suicidal behavior of others, which then explains why suicide rates appear in spatial clusters when one maps the incidence of suicide (Tarde, 1903).

Social scientists have been debating and seeking to test these competing theories ever since they were first stated. In their contemporary work on the problem, sociologists Robert D. Baller and Kelly K. Richardson have used some creative mapping techniques to tease out the relative contributions of integration versus imitation (Baller & Richardson, 2002). As shown in Figure 2.3, they rated each county in France on whether its suicide rate between 1872 and 1876 was high, low, or average. They also rated each county on whether the suicide rates of its neighboring counties were high, low, or average. This creates a ranking system for counties: They may be high-high (meaning that their suicide rate is high and so are those of their neighboring counties); low-low, meaning just the opposite; or high-low, meaning that while the county's suicide rate

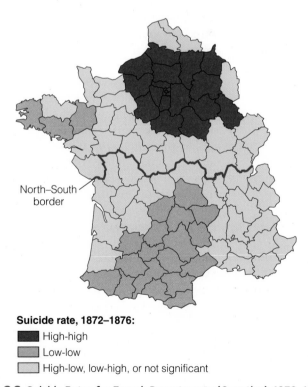

Suicide rate, 1872–1876:

■ High-high
■ Low-low
□ High-low, low-high, or not significant

Figure 2.3 Suicide Rates for French Departments (Counties), 1872–1876.

Source: From Robert D. Baller, Kelly K. Richardson, "Social Integration, imitation, and the graphic patterning of suicide." *American Sociological Review*, Vol. 67, No. 6 (Dec. 2002), Map 1, p. 881. Reprinted by permission of the authors and the American Sociological Association.

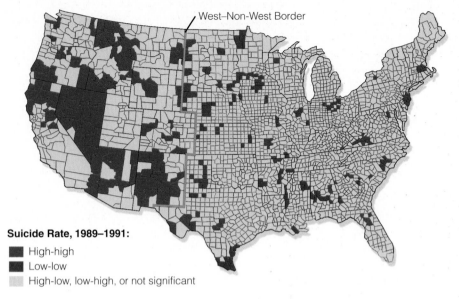

Suicide Rate, 1989–1991:

■ High-high
■ Low-low
▨ High-low, low-high, or not significant

Figure 2.4 Suicide Rates for U.S. Counties, 1989–1991. Source: Baller & Richardson, 2002.

is high, its neighboring counties' rates are low. (Of course, the opposite pattern, low-high, can exist, and there can be counties that rate neither high nor low, which means that they are "not significant.") They carried out the same procedure for all the counties in the United States between 1989 and 1991 (see Figure 2.4). In both cases, they were seeking to test Durkheim's most basic hypothesis: that suicide rates will tend to cluster in regions with differing levels of social integration. But they were also seeking to test Tarde's hypothesis that the clustering is due to imitation and that therefore neighboring counties will have similar suicide rates.

The two maps show some striking results, and these results offer stronger evidence for Durkheim's hypothesis than for Tarde's. In both maps, clustering of high and low rates of suicide by counties is evident. Yet both maps show many counties (those with no shading) that have high or low suicide rates but differ from their neighbors on these variables. If imitation is operating, as Tarde suggested in his hypothesis, it is certainly not uniform. There are clusters of counties with high suicide rates, which supports Tarde's imitation theory, but their locations offer far more support for Durkheim's integration theory.

As expected, suicide rates for France clustered in the region around Paris. In the late nineteenth century, the north of France was undergoing far more rapid urbanization than the south, with a great deal of new migration to Paris and surrounding towns. An exception that supports Durkheim's theory is Bretagne, in the northeastern corner, where traditional villages and high levels of social integration

help explain the clustering of low suicide rates that one sees clearly on the map. Similarly, the major clustering of low-low counties in southern France is also largely explained by the persistence of traditional villages and higher levels of social integration there.

When they turn to analysis of the U.S. map, the sociologists note that "High rates of suicide are clustered in the American West, with the notable exception of the state of Utah. Other significant clusters of high suicide rates are located in the western tip of Virginia and in Northern Florida" (Baller & Richardson, 2002, p. 881). The authors go on to point out that the western and eastern sections of the United States differ greatly, with the West (again with the exception of Mormon Utah) displaying higher rates of single, unmarried residents, more recent settlement, and many other indicators of weaker social integration than either the South or the East. Note that this analysis does not deny that imitation may be associated with some cases of suicides or argue that lack of social integration is the main cause of most suicides. The analysis shows that the association among the variables offers more support for one theory than for the other, but, clearly, a phenomenon as complex as suicide has many explanations, and individual cases are always explained by psychological as well as social factors. Remember, Durkheim and Tarde were arguing about why suicide rates might be high or low in a region, not about the causes of individual suicides. As we will see in the Mapping Social Change boxes in various chapters of this text, social scientists since Durkheim's time have found many new and creative ways to map social data.

Suicide as a Form of Protest

Suicide as a form of protest is one of the most profound and extreme acts imaginable. It is also a clear example of Durkheim's point that suicide can be a social as well as a psychological phenomenon. Any form of public death—be it an execution, a disaster, a brutal murder, or a suicide—sends shock waves through society and stimulates a great deal of emotion, especially among the witnesses. This has probably been true throughout history, but in our age of widespread mass media, sensational events are instantly visible to far greater numbers of people than ever before. A consequence of this change is that suicide deaths may have an immense impact on public feelings and opinions.

The suicide of a Buddhist monk by self-immolation during the Vietnam War was instantly broadcast around the world, not only as a photograph but also as a televised event. This act instantly conveyed the message that an important element of the South Vietnamese public opposed the actions of the U.S.-supported government then in power. For antiwar protesters in Europe and the United States, the event sent a clear moral message: Others who opposed the war must redouble their efforts to swing public opinion against the war.

Suicide bombings in public places by Palestinian militants are a central feature of the conflict in the Middle East. Without the power of a modern military with tanks and jet fighters, the Palestinians have relatively few means of striking out against the Israeli government. Fired by religious and political ideals, young Palestinians are sacrificing their lives to carry out bombing missions in much the same way that Japanese *kamikaze* pilots did during World War II. These bombings do nothing to ease the conflict between Muslim Palestinians and Jewish Israelis; in fact, they further escalate the bloodshed. So far, however, they have been an effective means of riveting world attention on the Palestinian struggle and the imbalance of power in the conflict.

From a sociological viewpoint, these examples represent a special type of suicidal behavior. These individuals, like all people who choose to end their lives, are no doubt motivated by many different feelings,

AP Photos

AP Photos

notably suffering and despair. Their decision to link their death to a social cause raises many other sociological issues, however. Is the protest justified? Should the event be widely publicized? Is the person a hero for having made the supreme sacrifice for a cause? These are all examples of the moral and sociological issues raised by protest suicides.

THEORIES AND PERSPECTIVES

The material presented in Tables 2.2, 2.3, and 2.4 is valuable because it shows how the nation's most important survey, the census, reveals fundamental changes in the structure of the population. But how do these facts about household composition relate to larger trends and issues? The census reveals trends, but it does not explain those trends unless an investigator armed with a set of hypotheses goes to work on the facts to make them prove or disprove a theoretical point.

A **theory** is a set of interrelated concepts that seeks to explain the causes of an observable phenomenon. Some theories attempt to explain an extremely wide range of phenomena, whereas others limit their explanations to a narrower range. In physics, for example, Newton's theory of gravitation related the force of gravity to the mass and distance between objects, but it did not try to explain why the force of gravity existed in the first place or how it was related to other forces, such as electromagnetism, that could be observed in nature. Einstein's theory of relativity, on the other hand, attempted to explain the relationships among all

theory: A set of interrelated concepts that seeks to explain the causes of an observable phenomenon.

natural forces, and it predicted forces that had not yet been observed, such as those released by nuclear fission.

It is often said that sociology lacks highly predictive theories like those of the physical sciences, but this is a debatable point. If theories are judged by their ability to explain observable phenomena and predict future events, then sociology has some powerful theories. At the beginning of the twentieth century, Émile Durkheim used his theory of social integration to predict the social conditions that would increase rates of suicide. The same theory of social integration predicted the conditions that would lead to the rise of totalitarian regimes such as that of the Nazis in Germany. Max Weber's theory of bureaucracy is valuable in explaining the experiences we are likely to have in organizations of all kinds. And Karl Marx's theory of class conflict is still the dominant explanation of the revolutions that occurred in feudal and early capitalist societies. No single sociological theory can explain all the complexities of human social life and social change, but in this respect, sociology is not very different from other social and physical sciences. A variety of economic theories compete for the attention of policy makers, and the theory of relativity has not fully explained the physical forces that produced the universe.

To cope with the many levels of social explanation, sociologists come to their work armed with a **theoretical perspective**: a set of interrelated theories that offer explanations for important aspects of social behavior. Like the methods of observation and analysis discussed in this chapter, theoretical perspectives are tools of sociological research. They provide a framework of ideas and explanations that help us make sense of the data we gather. The basic theoretical perspectives are the ones discussed in Chapter 1: interactionism, functionalism, and conflict theory. We also rely on the ecological perspective for descriptive data about communities and populations. At times, these perspectives offer competing explanations of social life; at other times, they seek to explain different aspects of society; and at still other times, they are combined in various ways.

theoretical perspective: A set of interrelated theories that offer explanations for important aspects of social behavior.

SUMMARY

How are the questions sociologists ask about social life the same as, and yet different from, the questions people from all walks of life normally ask?

- In all scientific research, certain basic steps must be completed: deciding on the problem, reviewing the literature, formulating research questions, selecting a method, and analyzing the data. Not all researchers follow these steps in the order given, and often one or more steps must be repeated.

- The first step in designing sociological research is formulating the question—that is, asking a question about a social situation that can be answered through the systematic collection and analysis of data.

- Often the research question is expressed in the form of a hypothesis. A *hypothesis* states a relationship between two *variables* that can be tested through empirical observation. The variable that is to be explained is the *dependent variable*. The other variable, a factor that the researcher believes causes changes in the dependent variable, is the *independent variable*.

- Before collecting new data, professional researchers review as much existing research and other data sources as possible. This "review of the literature" sometimes supplies all the data necessary for a particular study.

What basic scientific methods do sociologists employ in their research?

- The most frequently used research methods in sociology are observation, experiments, and surveys.

- Observation may take the form of *participant observation*, in which researchers participate to some degree in the life of the people being observed. It may also take the form of *unobtrusive measures*, or observational techniques that measure behavior but intrude as little as possible into actual social settings. Visual sociology involves the use of photography and videotape to observe people in a variety of settings.

- Sociological experiments can take one of two basic forms: the controlled experiment and the field experiment. In a *controlled experiment,* researchers establish an *experimental group,* which will experience the "treatment" (a change in the independent variable), and a *control group,* which will not experience the treatment. The effect of the treatment on the dependent variable can be measured by comparing the two groups. *Field experiments* take place outside the laboratory and are often used in evaluating public programs designed to remedy specific social problems. The treatment group consists of people who experience a particular social program, and the control group consists of comparable people who do not experience the program.

- A common problem of experimental studies is the *Hawthorne effect,* which refers to any unintended effect resulting from the attention given to subjects in an experiment.

- The third basic method of sociological research, the survey, asks people to give precise information about their behavior and attitudes. The most ambitious surveys are national censuses; the data obtained in a census can be supplemented by smaller, less costly *sample surveys. A sample* is a selection of respondents drawn from a specific population. To ensure that everyone in the specified population has an equal chance of being selected, the respondents must be selected by some process of *random sampling.*

- Questionnaire design is an important aspect of survey research. Questions must be precisely worded, easy to understand, and free of bias. *Closed questions* require respondents to select from a set of answers, whereas *open questions* allow respondents to say whatever comes to mind.

- Sociological researchers must always consider the rights of human subjects. *Privacy* is the right to decide the terms on which one's acts may be revealed to the public. *Confidentiality* means that researchers cannot use responses in such a way that they can be traced to a particular respondent. For respondents to give *informed consent,* they must be told how the information they supply will be used and must be allowed to judge the degree of personal risk involved in answering questions.

How do sociologists analyze the data they collect using the various research methods?

- The data gathered in a survey are usually presented in the form of statistical tables. In reading a table, it is important to know what the units of analysis are and what kinds of data are being presented.

- *Frequency distributions,* given in absolute numbers, reveal the actual size of each category of a variable, but to compare the numbers for different years, it is necessary to use *percent analysis.*

- Analyzing data often leads to the discovery of *correlations,* or specific relationships between two variables. Correlation should not be confused with causation.

- Mapping of data about individuals or households also provides a powerful tool for analyzing sociological data.

What is the difference between theories and perspectives?

- Once data have been presented and analyzed, they can be used to generate new hypotheses. The types of hypotheses that might be developed depend on the researcher's *theoretical perspective*—a set of interrelated theories that offer explanations for important aspects of social behavior. The functionalist, interactionist, and conflict perspectives result in quite different hypotheses.

The Kornblum Companion Website

www.cengagebrain.com

Supplement your review of this chapter by going to the companion website to take one of the tutorial quizzes, use the flash cards to master key terms, and check out the many other study aids you'll find there. You'll also find special features, such as GSS data and Web links that will put data and resources at your fingertips to help you with that special project or to do some research on your own.

Culture

FOCUS QUESTIONS

Could human beings exist without their cultures? What are some of the links between culture and social order?

How have humans evolved in ways that culture made possible?

Would we be truly human without our capacity to use language?

What are the main social processes that may occur when people come to live in a new culture as a result of immigration or other forms of social change?

How does the global influence of civilizations help bring about social change in other cultures?

For all Native Americans, the coming of Europeans to the New World marked the beginning of a long, drawn-out disaster. This is how sociologist Russell Thornton, himself a Native American, describes it:

> In the centuries after Columbus these "Indians" suffered a demographic collapse. Numbers declined sharply; entire tribes, often quickly, were "wiped from the face of the earth." This is certainly true of the American Indians on the land that was to become the United States of America. . . . The fires that consumed North American Indians were the fevers brought on by newly encountered diseases, the flashes of settlers' and soldiers' guns, the ravages of "firewater," the flames of villages and fields burned by the scorched-earth policy of vengeful Euro-Americans. (Thornton, 1987, pp. xv–xvi)

Others have shown that even when Indians and settlers tried to live as neighbors, it was difficult for them to do so (Cronon, 1983; Warhus, 1997). Seemingly simple differences could result in major conflicts. One major source of conflict could be found in the different ways in which Indians and settlers related to the land. The settlers tended to think of the bounty offered by their new environment as inexhaustible. They were not nearly as aware of the rhythms of nature as the Indians were, nor did they know how to live in harmony with nature without destroying its resources. So when they saw the Indians moving from one hunting or fishing ground to another as the seasons changed, they concluded that the Indians were not "improving the land" as proper farmers did. Over and over again, Europeans justified taking the Indians' forests and meadows by claiming that the Indians had no sense of ownership of the land because they did not clear it and farm it.

It would be difficult to overestimate the shock and confusion the Indians must have felt as the encroaching Europeans continually drove them from their traditional hunting and fishing grounds. The resulting conflicts were often bloody, with injuries and deaths on both sides, but the effects of the Europeans' technological superiority were especially devastating. The Europeans' cannons and rifles gave them the ultimate power to inflict their will on the Indians. No wonder that wherever they encountered the Europeans, the Indians began to covet the powerful new killing tools made of iron and steel. No wonder that the Europeans never doubted, even as they learned from the Indians how to survive in their new environment, that their own way of life was the only "true" civilization. Indeed, so powerful did the notion of European superiority become that even today we celebrate the "discovery" of the New World by European explorers. Too often, we forget that what happened in 1492 was not the discovery of a new world but the establishment of contact between two worlds, both already old (Jennings, 1975; Marks, 1998).

THE MEANING OF CULTURE

Was the European, or "Western," way of life really superior? This question remains a subject of stormy controversy throughout the world. Much of the resentment against Europeans and North Americans expressed by people in the Muslim world, for example, is based on the history of invasion, conquest, and domination by Western powers, a subject to which we return later in the chapter. For Native Americans in the Western Hemisphere, European invasion and settlement spelled the doom of some indigenous societies, but many others managed to survive and are struggling to maintain their way of life within the larger societies of Canada, the United States, Mexico, Bolivia, Brazil, and Peru. An example is provided by the Mayans of the Mexican highlands, as illustrated in the Visual Sociology box on page 51. And even when cultural contact

In this drawing, an ancient Mayan Indian is painting a mural depicting the daily life of his civilization. In 1492, Columbus and his crewmates narrowly missed encountering Mayan culture. If they had experienced it, their biased view of the Indians' culture as inferior would have been strongly challenged.

Peabody Museum of Archaeology and Ethnology

occurs on a more equal basis, as it does when people from different modern cultures meet, it is difficult to achieve understanding and arrive at mutually beneficial forms of cooperation. Why is this so? Many of the answers may be found in the study of human culture.

In everyday speech, the word *culture* refers to pursuits like literature and music. But these forms of expression are only part of the definition of culture. To the social scientist, "a humble cooking pot is as much a cultural product as is a Beethoven sonata" (Kluckhohn, 1949, quoted in Ross, 1963, p. 96). Culture is a basic concept in sociology because it is what makes humans unique in the animal kingdom. All the familiar forms of social organization, from the simplest family to the most complex corporation, depend on culture for their existence. Societies thus cannot exist without cultures. However, this does not mean that societies and cultures are the same thing. *Societies* are populations that are organized to carry out the major functions of life. A society's *culture* consists of all the ways in which its members think about

their society and communicate about it among themselves.

We can define **culture** as all the modes of thought, behavior, and production that are handed down from one generation to the next by means of communicative interaction—language, gestures, writing, building, and all other communication among humans—rather than by genetic transmission, or heredity. This definition encompasses a vast array of behaviors, technologies, religions, and so on—in other words, just about everything thought or made by humans. Among all the elements of culture that could be studied, social scientists are interested primarily in aspects that explain social organization and behavior. Thus, although they may analyze trends in movies and popular music (which significantly affect behavior in the modern world), they are even more likely to study aspects of culture that account for phenomena such as conflict between people with different languages and religions, or the conduct of people in different corporations that have their own ways of doing things ("corporate cultures").

Dimensions of Culture

The culture of any people on earth, no matter how simple it may seem to us, is a complex set of behaviors and artifacts. Robert Bierstedt (1963)

culture: All the modes of thought, behavior, and production that are handed down from one generation to the next by means of communicative interaction rather than by genetic transmission.

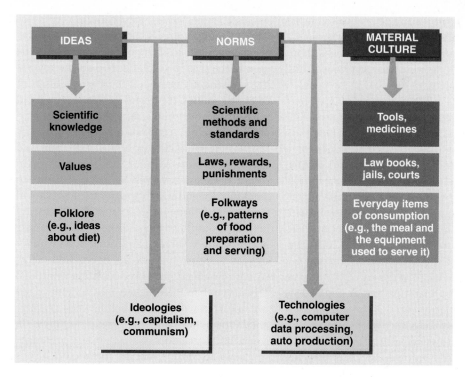

Figure 3.1 Dimensions of Culture.

suggested a useful framework for thinking about culture in which culture has three major dimensions:

1. **Ideas**, or ways of thinking that organize human consciousness;
2. **Norms**, or accepted ways of doing or carrying out ideas; and
3. **Material culture**, or patterns of possessing and using the products of culture.

Figure 3.1 presents some examples of these three dimensions, along with two aspects of culture—*ideologies* and *technologies*—that combine elements from more than one dimension.

Ideas As indicated in Figure 3.1, theories about how the physical world operates (scientific knowledge), strongly held notions about what is right and wrong (values), and traditional beliefs, legends, and customs (folklore) are among the most important types of ideas found in any given culture. Of these, values are especially important because people feel so strongly about them and because they often undergo changes that result in social conflict (as we can see in contemporary debates about "a woman's right to control her reproductive life" versus "a fetus's right to life").

Values are socially shared ideas about what is important. Thus, for most people in North America, education is a value; that is, they conceive of it as a proper and good way to achieve social standing. Loyalty to friends and loved ones, patriotism, the importance of religion, the significance of material possessions—these and other values are commonly found in our culture and many others, but of course there are wide differences in how people interpret these values and in the extent to which they adhere to them. Consensus on values tends to be far greater in small tribal and peasant societies than in large, complex industrial societies. In a large and diverse society such as ours, there is bound to be a good deal of conflict over values. Some people are satisfied with the way wealth and power are distributed, for example; others are less satisfied with the status quo. Some people feel that it is desirable to attempt to improve one's own well-being and that society as a whole gains when everyone strives to be well-off. Others assert that the value of individual gain conflicts with the values of community and social cohesion; too great a gap between the well-off and the not-well-off creates suffering, envy, crime, and other problems.

In societies undergoing rapid and far-reaching social change, such as the United States, there is bound to be conflict over values. Controversies over "family values," the sanctity of marriage, obligations to children, and similar issues tend to be particularly intense because people's individual experiences differ greatly. The more diversity there is among the people

ideas: Ways of thinking that organize human consciousness.

norms: Specific rules of behavior.

material culture: Patterns of possessing and using the products of culture.

values: The ideas that support or justify norms.

who share a basic culture, the more conflict there is likely to be over certain aspects of that culture. Such disputes often turn on ideas about how to behave—that is, on the norms governing behavior in a society.

Norms From people's beliefs in what is right and good—that is, from their values—are derived the norms, or rules of behavior, of a society. Values are more abstract than norms; they are the ideas that support or justify norms. Norms are more specific. They are "the adjustments which human beings make to the surrounding environment. We may think of them as solutions to recurring problems or situations" (Nisbet, 1970, p. 225). But norms involve more than behavior. Any given norm is supported by the idea that a particular behavior is correct and proper or incorrect and improper. Norms known as *taboos,* such as the prohibition against incest, are widely shared throughout a culture and strictly enforced, even if they are sometimes violated. Other norms, such as dress codes, are enforced less strictly and are often violated.

If we think about a complex aspect of everyday life such as driving a car, it is evident that, without norms, life would be far more chaotic and dangerous than it already is. When we drive, we keep to the right, obey traffic lights and speed limits, and avoid reckless behavior that could cause accidents. These are among the many norms that allow the automobile to be such an essential article of North American culture.

Examples of norms are easy to find. Take the college or high school classroom. The classroom is organized according to norms of educational practice: There should be a textbook or books; there should be class discussion; there should be assignments, exams, and grades. Note that these "shoulds" correspond to actual behavior. Norms usually refer to behavior that we either approve or disapprove of. However, members of a culture often disagree about how a particular norm operates. Thus, in the case of classroom organization, there is considerable debate about what educational practices are most effective. In Israel, for example, there is debate among Orthodox and liberal Jews over whether boys and girls should be separated in classrooms, and similar debates divide fundamentalist and more secular Muslims in other Middle Eastern societies.

In societies where values are in conflict or where changes are occurring rapidly, norms are also subject to change. This can readily be seen in the case of smoking. The norms governing smoking are changing rapidly as a result of changing values regarding health. Throughout most of the twentieth century, the norms governing smoking in the United States were very liberal. One could light up almost anywhere, and nonsmokers were expected not to complain. In recent decades, however, as the link between smoking (including exposure to secondhand smoke) and lung cancer and heart disease became scientifically established, the prevailing norms shifted. Tobacco smokers are now under intense pressure to "kick the habit," and places where smoking is permitted are under continual pressure to ban the practice. But the change has not occurred without conflict between smokers and nonsmokers and their representatives in courts and legislatures.

Laws are norms that are included in a society's official written codes of behavior. Laws are often developed by a specialized occupational group, such as priests in the ancient world and legislators, judges, and lawyers in the modern world. Some of the oldest examples of laws are the Code of Hammurabi, the ancient Babylonian code that specified punishments equal to the gravity of the crime (for example, an eye for an eye, a tooth for a tooth); the Ten Commandments, written by God, according to the Old Testament, on stone tablets at least 1,600 years before the birth of Christ; and the codes of Confucius, written and interpreted by the royal scribes of the ancient Chinese empire. In our time, laws are the special province of professional lawmakers, but all citizens participate in a network of laws and other written regulations that govern daily life. Basic behaviors like driving, going to school, marrying, and investing are greatly influenced by laws governing these activities.

At times, we may grumble or protest against laws that we find overly restrictive. And there is always some group planning to protest a law that is under consideration in a legislative body such as Congress. When we look beyond our own society, however, we see that the rule of law could save millions of lives if only it could be achieved in places like Iraq, Iran, Sudan, or many of the nations of the former Soviet Union. On the other hand, even in nations where the rule of law is generally accepted, social scientists often find that new laws have unintended consequences. For example, when extremely strict laws against adult drug dealing were passed, the dealers began recruiting juveniles to sell drugs. The lawmakers never intended the new laws

laws: Norms that are written by specialists, collected in codes or manuals of behavior, and interpreted and applied by other specialists.

The Once Glamorous Habit How quickly norms can change, and how much confusion those changes can bring. Smoking is a fine example. In less than four decades, this behavior changed from a glamorous activity to an act that many people consider deviant.

For most of the twentieth century, smoking was extremely popular. Most Americans smoked, and smoking was even more popular in England and on the European continent. Humphrey Bogart and Lauren Bacall smoked, as did most other movie stars, as well as many athletes and famous journalists. Smoking reached its peak of popularity during and just after World War II. Children often began smoking steadily by age fourteen, encouraged to do so by ubiquitous advertising that associated smoking with success, sensuality, and in the case of the famous Marlboro Man, masculinity.

Today, much has changed. Smoking has been shown to be extremely dangerous to health—not only the health of the smoker but also the health of people exposed to secondhand tobacco smoke. In consequence, the norms and values associated with smoking are being challenged throughout most Western nations. New norms restricting smoking in public places and in most offices are creating a deviant group of smokers who must linger outside their offices to satisfy their craving and enjoy a cigarette.

© Photofest

Michael Newman/PhotoEdit

to encourage teenagers to become dealers, but that is what happened in many states (Bayer, 1991; Benson, 2009). Laws often have unintended consequences because they were created in response to vigorous efforts by people who espouse particular ideologies.

Ideologies As indicated in Figure 3.1, **ideologies** comprise two dimensions of culture; they are sets or systems of ideas and norms. Ideologies combine the values and norms that all the members of a society are expected to believe in and act upon without question. A classic study of the emergence of an ideology was Max Weber's analysis of the link between Protestantism and capitalism, *The Protestant Ethic and the Spirit of Capitalism* (1974/1904). Weber noticed that the rise of Protestantism in

Europe coincided with the rise of private enterprise, banking, and other aspects of capitalism. He also noticed that a majority of the most successful early capitalists were Protestants. Weber hypothesized that their religious values taught them that salvation depended not on good deeds or piety but on how they lived their entire lives and particularly on how well they adhered to the norms of their "callings" (occupations). As a result, Protestants placed a high value on frugality and abstinence. To prove that they were worthy of salvation, they devoted themselves tirelessly to commerce and plowed their profits

---

ideologies: Systems of values and norms that the members of a society are expected to believe in and act on without question.

back into their firms. But Catholics, who did not share these values, were less single-mindedly dedicated to their business ventures. They often spent their profits on good deeds rather than investing them in their businesses. Weber attempted to show how a set of religious values and norms combined with economic norms to create the ideology of capitalism.

A contemporary example of an ideology may be found among religious fundamentalists. Christian fundamentalists in the United States generally share a set of ideas and norms of behavior that include the value of prayer, the value of family and children, the negative value of abortion and secular humanism (an ethical system based on scientific knowledge rather than on religious teachings), the belief in salvation and redemption for one's sins, and other values. In the Islamic world, conflict between religious fundamentalists and more secular people is especially widespread and bitter. And the fundamentalists often see the United States as a threat to their most cherished values. In part because of their prominence on global television, examples of life in the United States—especially those that feature sexual display, the consumption of alcohol, or gender equality (which is not permitted among Orthodox Muslims)—become symbols of moral corruption.

Values and Ideologies in Conflict Most societies and regions of the world experience conflict among people with differing values and ideologies. At times, this conflict can have serious outcomes, such as widespread violence among antagonistic factions or parties, as occurs in Northern Ireland between Protestants and Catholics, in Israel between ultraorthodox Jews and more secular populations, and in Islamic nations like Egypt or Algeria, between extremely orthodox Muslims and those who seek to forge compromises between secular values and religious ones. We return continually to the subject of conflicting values and resultant conflict among social groups because the subject is so important in understanding social life and social change. In discussing values and ideologies, however, it is important to point out that, in almost every known society, there are inherent conflicts in the values the society claims to support. In most democratic societies, for example, the values of liberty (freedom) and equality are cherished. But there is inherent conflict in these two central values. As Isaiah Berlin, one of the most influential social thinkers of the twentieth century, explained:

> Perfect liberty is not compatible with perfect equality. If man is free to do anything he chooses, then the strong will crush the weak, the wolves will eat the sheep, and this puts an end to equality. If perfect equality is to be attained, then men must be prevented from outdistancing each other, whether in material or in intellectual or in spiritual achievement, otherwise inequalities will result. (1998, p. 60)

We will see in later chapters that the desire of some groups to achieve total equality or freedom continues to produce ideologies in which the goals of action—that is, equality or freedom—lead people to assert that "the ends justify the means" used to achieve those ends, even if they include violating norms such as "Thou shalt not steal" or "Thou shalt not kill." Such ideologies led to some of the twentieth century's most atrocious episodes of mass violence and destruction. The inherent conflict among values can also be a creative aspect of society, however. As we seek to decrease the gap between the haves and the have-nots in the United States and elsewhere, while attempting to preserve as much individual freedom as possible, we realize that a just society is one in which people are free to seek the best peaceful means of balancing their conflicting interests and values (Cerulo, 2008).

Material Culture The third dimension of culture Bierstedt identified is material culture. Material culture consists of all the things a society produces. Mundane things, such as pots and pans or the wooden eating bowls of nonindustrial societies; immensely complex systems of things, such as the space shuttle; cherished items of religious worship, such as rosary beads or Indian fetish necklaces—all take their shape and purpose from the ideas of the culture that produces them. Members of societies that place a high value on science and efficiency are used to seeing these values expressed in material objects. For example, we design our houses to conserve energy and create desirable combinations of view and privacy. We may take older forms such as the ranch house or the Cape Cod bungalow and modify them to suit modern purposes; in this way, our houses combine tradition with usefulness. The particular form that appeals to us is usually a result of many different ideas, including what we know from our own upbringing, what we can afford, how much space we require, and the environmental conditions we anticipate.

The same is true throughout the world. If one travels in Africa, for example, one will find large areas in which people build square houses with thatched roofs or roofs of corrugated metal. Suddenly one will pass into a region where the

visual sociology

Living Maya The people in these photos are Mayans. They live in the mountains and jungles of Chiapas in southern Mexico and in part of Guatemala. They are descendants of the ancient Mayans, whose civilization flourished in Central America before 900 C.E.[1] The Mayan nobility was literate and, like the Egyptians, intensely interested in astronomy and the construction of calendars.

Contemporary Mayans live in villages, each of which is clustered around a market and a small but much-loved Catholic church. The men in one of the photos are holding a session of the village court. In the foreground, a man occupies the time by plaiting palm fronds to make a hat. The standing man is one of the parties to the dispute, as is the one arguing in the background. The seated men are civil officials.

Men's and women's lives are separated much of the time. In each village, the women engage in specialized crafts in addition to their other chores. In this highland region, most women are potters or weavers. Their crafts are an example of material culture passed from one generation to the next for centuries: The woven patterns and symbols on the cloth they are wearing can be seen in the clothing of ancient Mayans depicted on pottery and wall paintings.

Contemporary Mayans are extremely poor. Their subsistence agricultural and crafts economy is adequate to allow them to preserve their culture, but they crave access to better land. There are huge gaps in power and privilege between the Mayans and the wealthy, often absentee, landowners.

[1] C.E. stands for Common Era, which began at the birth of Christ. Likewise, B.C.E. stands for before the Common Era.

David Sutherland/Alamy

guatebrian/Alamy

John Cancalosi/Alamy

people build round houses with steeply pitched roofs. In another region, people may live in sprawling apartment-like complexes built of sand and mortar. When asked why they prefer a particular kind of dwelling, people usually say, "Because we have always built our houses this way." In other words, their answers are based on ideas of what is right for them. After further discussion, however, a traveler with a sociological imagination will discover that the different building forms also express functional ideas. The mud-walled "apartments" of the desert village seem cramped and dark to Western eyes, but the people of the village will explain that they often sleep outdoors and the dark, cool rooms are most useful on very hot days. On the other hand, the tribespeople in a society that builds round houses may give a spiritual explanation for the shape of their dwellings; they may believe that corners are places where evil spirits can lurk. In sum, different cultures may give different reasons for the form and function of a material element such as a house, but in all cultures, one will find a material culture that stems from the society's most important ideas (Rybczynski, 1987).

© Chenault Photography Studio, Decatur, Alabama

Quilt making is a highly social form of folk art. This quilt, created by University of Mississippi graduate student Patti B. Melton, is actually a sociology paper about quilt making reproduced on an original quilt. It blends ideas with material culture.

Technologies Technologies are another aspect of culture that spans two of Bierstedt's major dimensions. (Refer again to Figure 3.1.) **Technologies** are the things (material culture) and the norms for using them that are found in a given culture (Bierstedt, 1963; Ellul, 1964). Without the norms that govern their use, things are, at best, confusing and at worst useless or dangerous. In the United States, for example, new telecommunications technologies such as the Internet are powerful means of communicating information across long distances. But norms for using these new technologies—such as norms about privacy, freedom of speech, and personal accountability—are only now developing. There is serious conflict, for example, about whether pornographic messages directed at juveniles or email messages among colleagues in an office are protected under the norms of free speech or whether they are subject to public scrutiny and perhaps social sanction. Because the Internet is a relatively new form of communication, the norms (that is, laws) that apply to print and TV communication have not yet been fully worked out and applied to this new communications medium (Warner, 2008).

Guns present another case in which norms governing use of the things themselves are very much in dispute. Most Americans agree that people should be free to own and use hunting weapons—but what about handguns, assault rifles, machine guns, and rocket launchers? What, if any, restrictions should there be on the distribution and use of such weapons? As we will see in later chapters, the debate about appropriate norms for controlling the distribution and possession of deadly firearms that are clearly not designed for hunting is an indication of a deeper cultural conflict over the right to bear arms versus the state's monopoly over the use of force.

Norms and Social Control

Norms are what permit life in society to proceed in an orderly fashion without violence and chaos. Shared norms contribute enormously to a society's ability to regulate itself without constantly resorting to the coercive force of armies and police. The term **social control** refers to the set of rules and understandings that control the behavior of individuals and groups in a culture. Park and Burgess (1921) noted that throughout the world there are certain basic norms that contribute to social control: "All groups have such 'commandments' as 'Honor thy father and mother,' 'Thou shalt not kill,' 'Thou shalt not steal'" (p. 787).

The Normative Order The wide array of norms that permit a society to achieve relatively peaceful social control is called its **normative order**. These norms create what Morris Janowitz has termed "the capacity of a social group, including a whole society, to regulate itself" (1978, p. 3). This self-regulation requires a set of "higher moral principles" and the norms that express those principles. The norms define what makes a person a "good" member of the culture and society.

The most important norms in a culture are often taught as absolutes. The Ten Commandments, for example, are absolutes: "Thou shalt not kill," "Thou shalt not steal," and so on. Unfortunately, people do not always extend those norms to members of another culture. For example, the same explorers who swore to bring the values of Western civilization (including the Ten Commandments) to the "savage Indians" of the New World thought nothing of taking the Indians' land by force. Queen Elizabeth I of England could authorize agents like Sir Walter Raleigh to seize remote "heathen and barbarous" lands without viewing this act as a violation of the strongest norms of her own society (Jennings, 1975; Snipp, 1991). Protests by the Indians often resulted in violent death, but the murder of Indians and the theft of their land were rationalized by the notion that the Indians were an inferior people who would ultimately benefit from European influence. In the ideology of conquest and colonial rule, the Ten Commandments did not apply.

Sanctions In the social sciences, punishments and rewards for adhering to or violating norms are known as **sanctions**. Cultural norms vary according to the degree of sanction associated with them. Rewards can range from a smile to the Nobel Prize; punishments may vary in strength from a raised eyebrow to the electric chair. During the period of European colonial expansion, the murder of a native was far less strongly sanctioned than was the murder of another European.

Mores and Folkways The most strongly sanctioned norms are called **mores**.[1] They are norms that people consider vital to the

technologies: The products and the norms for using them that are found in a given culture.

social control: The set of rules and understandings that control the behavior of individuals and groups in a particular culture.

normative order: The array of norms that permit a society to achieve relatively peaceful social control.

sanctions: Rewards and punishments for abiding by or violating norms.

mores: Strongly sanctioned norms.

[1] This term, pronounced "morays," is the plural of the Latin word *mos*, meaning "custom."

These Iraqi women are carrying rifles to protect themselves and their children. When a society's civilians, especially women, arm themselves, it is a strong sign that the society's norms of social control are breaking down.

or informal. Laws and other norms, such as company regulations or the rules of games and sports, are known as *formal norms.* They differ from *informal norms,* which grow out of everyday behavior and do not usually take the form of written rules, even though they too regulate our behavior. For example, when waiting to enter a movie theater, it is usually permissible to have one member of a small group save a place in line for the others, who may arrive later. And in a "pickup" basketball game, a player can call a foul and the opposing player usually cannot contest the call. Of course, such norms are sometimes disputed, depending on how the people involved define the situation. In the case of the basketball game, when the player on whom the foul is called disagrees with the call—and the score is extremely close—different definitions of the situation can lead to conflict.

Thinking Critically

We all violate norms from time to time. But what are the consequences when many people begin violating norms—for example, running red lights or passing on the right-hand shoulder of the road?

Culture and Social Change

Cultures in all societies change, and much of social science is devoted to trying to understand and predict those changes. Norms and values that once were thought to be odd or criminal may come to be shared by the majority of the society's members, and the opposite can occur as well. In recent Olympic Games, for example, American women have excelled in many events in which they previously were weak performers. In the past, American women were often reluctant to lift weights and increase their strength for fear of being considered unfeminine. As the level of world competition increased, Americans found ways to overcome the older values that said women had to be soft to be feminine. Star female athletes continue to demonstrate to the American public that a woman can be muscular and athletic and still project a "feminine" identity.

This young woman has adopted a somewhat extreme version of a folkway that is fashionable among her crowd. Her supervisors at the bookstore where she worked insisted that she remove some of the piercings. She refused and was fired, the result of a clear difference in outlook about the significance of the norms involved.

continuation of human groups and societies, and therefore they figure prominently in a culture's sense of morality. **Folkways**, on the other hand, are far less strongly sanctioned. People often cannot explain why these norms exist, nor do they feel that they are essential to the continuation of the group or the society. Both terms were first used by William Graham Sumner in his study *Folkways* (1940/1907). Sumner also pointed out that laws are norms that have been enacted through the formal procedures of government. Laws often formalize the mores of a society by putting them into written form and interpreting them. But laws can also formalize folkways, as can be seen, for example, in laws governing the clothing worn in public places.

Sumner also pointed out that a person who violates mores is subject to severe moral indignation, whereas one who violates folkways is not. People who violate folkways such as table manners or dress codes may be thought of as idiosyncratic or "flaky," but those who violate mores are branded as morally reprehensible. Thus, among prisoners there are norms for the treatment of fellow prisoners, based on the types of crimes they committed "outside." Rapists and child molesters are moral outcasts; their offenses are the most reprehensible. Such offenders are often beaten and tormented by other prisoners.

The Research Methods box on pages 54–55 presents a typology of norms according to their degree of sanction and their mode of development—that is, whether they are formal

folkways: Weakly sanctioned norms.

Constructing a Typology of Norms

Typologies are ways of grouping observable phenomena into categories in order to identify regularities in what may appear to be a great variety of observations. For example, there are so many social norms that the average person has no hope of ever sorting them out without some kind of system for organizing them. In the chart presented here, the subject is norms of various types. The sociologist constructs the types by comparing various dimensions along which norms may differ. The norms listed in the Ten Commandments, for example, differ in many ways. The norm "Thou shalt not kill" not only is generally believed and passed along from one generation to the next but also is codified in laws. So is the commandment "Thou shalt not steal," which is an important part of the written laws of our society. But the commandment "Remember the Sabbath day, to keep it holy" is not a written law in the United States, at least not in the federal statutes.

Some states and communities have laws specifying that businesses must close on Sunday, but not all religious groups recognize Sunday as the Sabbath. In addition, ideas about what behaviors are appropriate on the Sabbath are changing, and laws governing those behaviors are being challenged. These differences indicate that norms may differ according to whether they are informally taught to new generations or whether they are formal, written "laws of the land."

A Typology of Norms

		Degree of Sanction	
		Relatively Weak	Relatively Strong
Mode of Development	Informal	Folkways, fashions	Taboos, mores
	Formal	Misdemeanor laws, some rules, guidelines, civil rights laws	Capital-offense laws, felony laws

When Gertrude Ederle (shown posing here in 1925) became the first woman to swim the English Channel, she was admired by other athletes but also became the butt of jokes about her supposed lack of femininity. Today, female sports stars like Venus Williams are far more widely admired both on and off the sports field.

Emphasis on normative behavior can lead one to neglect all the ways in which norms are constantly being debated and how they are changing (Wrong, 1999). Figure 3.1 and the Research Methods box present some key aspects of human culture, but they leave out all the social processes whereby elements of culture are produced and changed and diffused from one society to another. An unsuspecting reader might even conclude that all members of a society share its norms and values equally, which is hardly an accurate view of culture. Different people within a culture are likely to obey any given norm to varying degrees.

Consider the norms about marriage, fertility, and parenthood in the United States. All may agree that young couples should not have children out of wedlock. They may also agree that abortion is not a desirable means of birth control. Unmarried couples who are expecting a child resolve the conflict between these norms in different ways: Some obtain an abortion; others get married; still others decide to have the baby out of wedlock, perhaps giving it up for adoption. Some unmarried mothers decide to raise their

Another dimension along which norms may differ is the degree to which they are sanctioned—that is, the degree to which adherence is rewarded and violation is punished. The norm that men do not wear hats indoors is relatively weak. On the other hand, the norm that men and women do not casually display (or "flash") their genitals is strongly sanctioned.

Using these two comparative dimensions—mode of development (formal versus informal) and degree of sanction (weak versus strong)—we can create four categories: (1) norms that are informal and are weakly sanctioned (e.g., table manners, dress fashions), (2) norms that are informal but are strongly sanctioned (e.g., adultery), (3) norms that are part of the formal legal code but are weakly sanctioned (parking regulations, antismoking laws, and so on), and (4) norms that are formal laws and are strongly sanctioned (e.g., capital offenses like murder of a police officer or treason in wartime).

By juxtaposing the two dimensions, each with its two categories, we arrive at a fourfold classification of norms (see chart). But as Max Weber (1947) observed, such a classification is an "ideal-typical" arrangement of observations in a form that accentuates some aspects and neglects others. These ideal types (folkways, mores and taboos, misdemeanor laws, and felonies or capital offenses) are useful because they establish standards against which to compare the norms of other cultures. For example, not all cultures treat the norm about the Sabbath the way Americans do. In parts of Israel or the Islamic world, one could be arrested for violating the Sabbath laws. Typologies like this one also help identify areas of social change, as when one compares the way the Sabbath was treated in American laws early in the twentieth century, when stores were obliged to close on Sunday, to the way it is treated today, when Sunday is often viewed as another shopping day.

The aspects of society described by ideal types are rarely so uncomplicated in real life. Even the norms that seem most formal and unambiguous, such as the prohibition against murder, become murky under some conditions, as in cases of self-defense or in war, or in arguments about the death penalty and abortion. Studies of how actual behavior departs from the ideal-typical version invariably offer insights into how cultures and social structures are changing in the course of daily life.

children alone. One way or another, such events in people's intimate family lives require difficult choices among often contradictory norms.

Changing Norms of Intimacy Many critics of contemporary American society blame the rising divorce rate and the increase in single-parent families on changes in the norms of intimacy that occurred during the social movements and counterculture of the 1960s. They seek a return to "traditional" norms of intimacy, including premarital celibacy, monogamy, and heterosexuality. These and other critics of diversity in sexual norms view the 1950s as a "golden age" of traditional norms regarding the conduct of intimate relations.

Sociologists who study trends in American culture point out that the 1950s were by no means a golden age of traditional behavior. Changes in American culture brought on by mass advertising, television, the growing equality of men and women, and the increasingly widespread desire for individual happiness had been developing throughout the century. These longer-term changes, they reason, should be reflected in statistics on divorce rates. On the other hand, if changes in sexual norms were a result of the social movements of the 1960s, there should be evidence of such a connection in statistics on divorce rates over time.

Figure 3.2 shows that there is some justification for each of these positions, but it offers far more empirical support for the view that rising divorce rates are not a product of the tumultuous 1960s. Clearly, divorces have been increasing in frequency over a period going well back into the nineteenth century. These data call into question any "golden age" theory of stability in American couples.

But new norms never completely replace old ones in a large and diverse population like that of North America. In consequence, we will always see differences in how norms are understood and in how they guide actual behavior. And these differences, especially when they concern major areas of moral conduct, will continue to produce conflict over norms. (See Chapter 7 for further analysis of norms of intimacy and sexuality.)

Another important issue in understanding culture and social change is the question of the extent to which human culture is determined by biological factors, if at all. Are there any features that all human cultures share—any norms, for example, that are found in all known societies? And what happens when two cultures exist within the same society or nation? The remainder of this chapter explores these questions, beginning with one that has long been a subject of lively debate: the connection between culture and biology.

CULTURE, EVOLUTION, AND HUMAN BEHAVIOR

Of all the species of living creatures on this planet, human beings are the most widely distributed. The early European explorers—Columbus, Magellan, Cook, da Gama, and many others—marveled at the discovery of human life thriving, more or less, in some of the earth's most inhospitable environments. Nevertheless, they were convinced that the "savages" they encountered were inferior to the more powerful Europeans; in fact, they had difficulty accepting the idea that the native peoples were fully

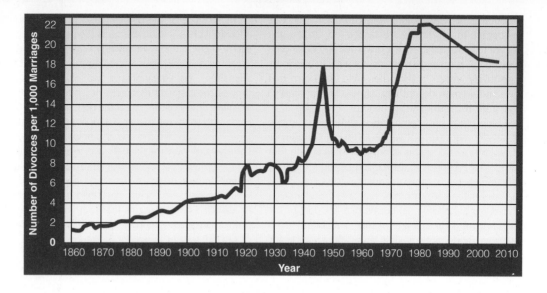

Figure 3.2 Divorce Rate, United States, 1860–2000. Figures for 1860–1920 represent the number of divorces per 1,000 existing marriages; figures for 1921–2000 represent the number of divorces per 1,000 married women age 15 and over. Source: Cherlin, 1981; Statistical Abstract, 2002.

human. In other words, they considered them biologically as well as culturally inferior.

Groups that are in conflict often accuse their enemies of being biologically inferior. The Nazis claimed that the Jews were a biologically inferior people. Arabs in Sudan often believe that Africans living in that nation's Darfur region are innately inferior, just as some highly prejudiced people in the United States believe that African Americans and other groups, such as Puerto Ricans, are inferior. These and other examples of racial and ethnic prejudice will be discussed more fully in later chapters. Here we need to emphasize that differences in culture often are wrongly viewed as innate biological differences between distinct peoples.

Students of sociology must see beyond this tendency to confuse cultural and biological differences. To do so, it is first necessary to understand how the famous naturalist Charles Darwin revolutionized modern thinking on these issues. Darwin's theory of natural selection is a starting point for scientific explanations of the differences among the animal species, including the human species (Eisley, 1961; Gould, 1996).

Darwin's Theory of Natural Selection

Darwin's theory of **natural selection** is the central explanation of how living species evolve to adapt to their changing environments. It is based on the observation that unexpected physical changes, or *mutations,* in organisms occur more or less randomly from one generation to

the next. When those mutations improve an individual organism's ability to survive in its environment, the new traits are "selected for"; that is, an individual that possesses those traits is more likely to survive, and hence more likely to reproduce and pass on its traits to the next generation, than individuals that lack the new traits. Over a few generations, such mutations can become so extensive that two species are created where formerly there was only one. This process of natural selection accounts for the great diversity of animal and plant life on the earth and for the ability of animals and plants to adapt to new environments.

Darwin's theory was based on his empirical observations of the natural world, especially those he had made as a scientist aboard the HMS *Beagle* during an extensive voyage of exploration in the early 1830s. On that voyage, Darwin had observed, in both fossils and living specimens, that some species of birds, turtles, and other animals had modified their physical form in ways that seemed to "fit" or "adapt" them to their environment. It took him more than two decades to decide to publish his findings. First, as a deeply religious person himself, he had to be sure in his own mind that the facts justified the revolutionary conclusions of his theory. Second, much of what he had observed directly challenged the fundamental beliefs of most of the religious and scientific leaders of his day. For when Darwin considered all the information, and especially his observations of similar species on different islands, it became clear to him that God had not created all the living things on earth at once—instead, they had been evolving through natural selection over many millions of years. And this theory could be applied to all other species, including humans.

natural selection: The relative success of organisms with specific genetic mutations in reproducing new generations with the new trait.

The Social Darwinists

The theory of natural selection had a dramatic impact on biological science but an even greater effect on the prevailing views of human society. Among the social thinkers who were most profoundly influenced by Darwin was Herbert Spencer of England, whose writings in sociology and philosophy were to dominate intellectual life in much of the Western world from 1870 to 1890.

According to Spencer, the fact that humans, unlike other species, have remained similar even on different continents must be explained by the fact that we adapt to changes in our environment through the use of culture rather than through biological adaptation. This process, which can be termed **cultural evolution**, parallels biological evolution in that the most successful adaptations are handed down to the next generation (Geertz, 1973). Spencer had this process in mind when he coined the phrase "survival of the fittest." He meant that the people who are most successful at adapting to the environment in which they find themselves—that is, the better-educated, wealthier, more powerful people—are most likely to survive and to have children who will also be successful.

Spencer concluded that it is impossible, by means of intentional action, to improve on the course of cultural evolution. The task of sociology, as he saw it, is to discover that course through empirical observation and analysis. Sociologists should not engage in efforts to reform society—to aid the poor, for example. To do so would be futile, and it could have the damaging effect of violating the principle of survival of the fittest in favor of the "artificial preservation of those least able to take care of themselves" (Spencer, 1874, p. 343).

Spencer's view, which came to be known as **social Darwinism**, claimed to explain why some people prospered during the Industrial Revolution while others barely scraped by. The people who were being pushed off the land and into the factories and slums of the cities were less well equipped culturally to succeed in an urban environment than people who could innovate and invent. There was not much for government to do, according to Spencer's theory, besides keep the peace and let the most competitive groups in society flourish. In so doing, those groups would give less competitive groups a chance to survive, if not to thrive.

By the end of the nineteenth century, efforts to apply the theory of natural selection to human societies had reached their logical extreme. Largely as a result of the work of Spencer and his followers, early sociologists in the United States

Many wealthy industrialists in the United States and elsewhere embraced the idea of social Darwinism in the nineteenth century. Their erroneous application of this notion to human inequality actively supported their belief that their success in life was a product of their superiority over the inferior masses. In their view, governmental or social intervention programs on behalf of the poor, such as laws limiting child labor in textiles and other industries, were misguided, since any improvement in the plight of the poor and their children would only cause the numbers of such "unfit" people in the population to increase.

and other Western societies favored the view that Western culture, with its emphasis on competition within the capitalist system, was clearly superior to all others and that the people who were most successful at competing within that system were to be considered superior human beings.

Sociology & Social Justice

SOCIOLOGY REJECTS SOCIAL DARWINISM The next generation of social scientists, who came to prominence around the turn of the twentieth century, rejected social Darwinism and its proponents. In Chicago and other U.S. cities where the Industrial Revolution was still occurring, they saw mounting evidence that wealth itself brings privilege to the children of the wealthy, regardless of whatever innate or learned traits they may possess. Cultural evolution, they argued, is a result of the development of more effective institutions and is not related to the innate qualities of individuals. Successful business firms, for example, are able to prosper in a highly competitive environment because of their superior organization. That is, their success can be explained without recourse to arguments about the genetic fitness of their leaders (Swedberg, 1994).

Sociobiology

The tendency to explain social phenomena in terms of biological causes such as physiology or genes is known as *biological reductionism*. For example, theories of crime proposing that there are genes that produce criminal behavior reduce the explanation of crime to biological causes. Some form of biological reductionism has emerged in every generation since Darwin's time.

---

cultural evolution: The process by which successful cultural adaptations are passed down from one generation to the next.

social Darwinism: The notion that people who are more successful at adapting to the environment in which they find themselves are more likely to survive and to have children who will also be successful.

The most recent version of biological reductionism is **sociobiology**. This term, coined by the Harvard biologist Edward O. Wilson (1975, 1998), refers to efforts to link genetic factors with the social behavior of animals. When applied to human societies, sociobiology has drawn severe criticism from both social scientists and biologists. Nevertheless, some sociologists support the sociobiological hypothesis that genes can explain certain aspects of human society and behavior (Barash, 2008; Radick, 2007). Because this hypothesis is so controversial, it deserves a closer look.

Let us take as an example the incest taboo, one of the strongest and most widespread norms in human life. The social scientist tends to explain the incest taboo as a cultural norm that is necessary for the existence of the family as a social institution. The family is an organized group with a need for well-defined statuses and roles. If the different statuses within the family became confused, as would undoubtedly happen if sexual intimacy were permitted between children and their parents or between brothers and sisters, it would be difficult to maintain the family as a stable institution (Davis, 1939; Malinowski, 1927).

Sociobiology takes a different view. For the sociobiologist, the incest taboo evolved in response to biological conditions—to

> . . . a deeper, more urgent cause, the heavy
> physiological penalty imposed by inbreeding.
> Several studies by human geneticists have
> demonstrated that even a moderate amount of
> inbreeding results in children who are diminished
> in overall body size, muscular coordination, and
> academic performance. More than 100 recessive
> genes have been discovered that cause hereditary
> disease . . . a condition vastly enhanced by
> inbreeding. (Wilson, 1979, p. 38)

Throughout most of the history of human evolution, sociobiologists point out, humans did not have any knowledge of genetics. Thus "the 'gut feeling' that promotes . . . sanctions against incest is largely unconscious" (Wilson, 1979, p. 40). Individuals with a strong predisposition to avoid incest passed on more of their genes to the next generation because their children were less likely to suffer from the illnesses that result from inbreeding. And over many centuries of natural selection of individuals who did not inbreed, humans developed "an instinct [to avoid inbreeding] which is based on genes" (Wilson, 1979, p. 40).

This leap, from the observation of a strong and persistent norm like the incest taboo to the belief that certain human behaviors are genetically programmed, is an example of the sociobiological hypothesis regarding human nature. Sociobiologists have proposed a hypothesis in which not only the incest taboo but also aggression, homosexuality, and religious feelings are genetically programmed, and they believe that future discoveries by geneticists will prove their hypothesis correct (Cecco & Parker, 1995).

Although it is true that genes set limits on human abilities and can be shown to influence many aspects of brain functioning, the hypothesis that genetic programming establishes complex forms of normative behavior is not supported by direct evidence: There is as yet no proof that such genes or sets of genes actually exist (Lewontin, 1995, 2003). Nevertheless, the rules of science require that we not reject the sociobiological hypothesis and that it remain an open area of investigation.

Culture Enhances Biological Evolution

The counterargument to sociobiology is that the human brain and other physical attributes are products of cultural and biological evolution and that, in the past 100,000 years of human evolution, there has been relatively little organic change in our species. Instead, the important developments in human life have occurred as a result of social and cultural change. This widely accepted view of culture denies that humans have innate instincts such as an instinct to avoid incest. It argues instead that the great advantage of culture in human evolution is its creation of a basis for natural selection that would not depend on genes but would allow humans to adapt relatively quickly to any physical or social environment (Geertz, 1973; O'Hear, 2005).

Archaeological evidence shows that humans were using tools for primitive agriculture and making jewelry and personal adornments more than 30,000 years ago (Stevens, 1988). No doubt they began using fire even earlier. Once humans could use fire, hand tools, and simple weapons, the ability to make and use these items increased the survival chances of those who possessed these skills. This would create conditions for natural selection in which the traits being "selected for" were those that involved the manipulation of cultural and social aspects of life. These abilities (dexterity, linguistic ability, leadership, social skills, and so on) would in turn influence the further development of the human brain, again through the process of natural selection.

---

sociobiology: The hypothesis that all human behavior is determined by genetic factors.

Research on primate behavior supports this view, showing that the higher primates use simple tools and that they teach this cultural technique to their young (Schaller, 1964). Jane Goodall (1968, 1994), for example, described how chimpanzees use sticks to probe for termites and chew leaves to produce a pulp to be used as a sponge to draw water out of tree stumps. These and many other instances of rudimentary culture among animal species demonstrate that culture is not unique to humans; recent research on chimpanzees has demonstrated that their societies have cultures that are passed down from one generation to the next and can adapt to new conditions in the chimps' environments (Friend, 2000). What happened in the case of humans that did not happen (or at least has not happened yet) in any other species is that, at a certain stage in human prehistory, our ability to alter our cultures in response to changing conditions developed so quickly that the human species entered a new realm of social life. In other words, culture became self-generating. And once it had been freed from genetic constraints, culture had no limits.

LANGUAGE AND CULTURE

Perhaps the most significant of the inventions made possible by culture is language. In fact, the learning of culture takes place through language. From our enormous capacity to use language is derived our collective memory (myths, fables, proverbs, ballads, and the like), as well as writing, art, and all the other media that shape human consciousness and store and transmit knowledge. In 2001, scientists identified a gene that appears to have a direct influence on human linguistic ability (Lai et al., 2001). They believe that this gene triggers the action of many other genes that contribute to the development of language ability; however, it may also be influenced by external variables such as the particular language the person is exposed to. Thus, although the capacity to learn language appears to be innate (Chomsky, 1965; MacWhinney, 1998), language does not occur outside a cultural setting and indeed is the most universal dimension of human cultures.

To return to the example at the beginning of the chapter, as savage as they may have appeared to Europeans, the Native Americans had language. They could have learned Spanish, just as the explorers could have learned their language if they had made an effort to do so. But without a common language that they could use to communicate, the Indians and the explorers misunderstood one another far more than they would have if they had possessed this powerful communication tool. Without language, neither side could explain to the other its strange behaviors and different ways of dealing with the physical world.

Research with Other Primates

What is unique about human language? Primatologists have shown that our closest evolutionary kin, the great apes (especially chimpanzees and pygmy chimpanzees), can learn language to some extent. Although their throats are not capable of producing the sounds that humans mold into language, apes do have the capacity to use language; that is, they can grasp the meanings of words as symbols for things and relationships. Through the use of sign language, or special languages using typewriters and other devices, apes can be taught a limited vocabulary (Radick, 2007).

Several examples of this capacity for learning have made headlines. Francine Patterson taught Koko, a female gorilla, American Sign Language. Koko learned more than 300 words. She also disproved the theory that apes can learn words but cannot invent new concepts. Koko invented sign words for *ring* ("finger bracelet") and *mask* ("eye hat") and was even able to talk in sign language about her feelings of fear, happiness, and shame (Hill, 1978). Subsequent research with a pygmy chimpanzee named Kanzi revealed that apes may actually be able to learn language through observation and imitation, the way a child does, rather than through long and difficult training (Eckholm, 1985). Sue Savage-Rumbaugh demonstrated that Kanzi could recognize English syntax, the patterns in the ordering of signs or words that give sentences different meanings. For example, Kanzi could distinguish between "Throw the potato at the ball" and "Throw the ball at the potato." Savage-Rumbaugh also showed that Kanzi could acquire new words simply by listening and watching as other chimps learn (Savage-Rumbaugh & Shanker, 1998). This kind of imitative language learning was formerly thought to be beyond the apes' mental capacities.

Fascinating as these experiments are, they only confirm the immense difference in communicative ability between humans and the apes. After months of training, an adult ape can use language with no more skill than an average 2½-year-old human infant (Bower, 2002; Dreifus, 1998). On the basis of the research conducted so far, primate researchers conclude that the apes' innate ability to learn language is severely restricted. No

Koko and her mentor, Penny Patterson.

amount of training can produce in apes the more advanced uses of language, including complex sentences containing abstract concepts, that are found in all normal humans regardless of their culture (Eibl-Eibesfeldt, 1989). For example, every human language allows its speakers to express an infinite number of thoughts and ideas that can persist even after their originators are gone. This property of human language, which is not shared by any other known species (Eisley, 1970; Greenspan, 2004), allows human groups to transmit elements of their culture from one generation to the next.

Does Language Determine Thought?

So complete is the human reliance on language that it often seems as if language actually determines the possibilities for thought and action in any given culture. Perhaps we are unable to perceive phenomena for which we have no nouns or to engage in actions for which we have no verbs. This idea is expressed in the **linguistic-relativity hypothesis**. As developed by the American linguists Edward Sapir and Benjamin Whorf in the 1930s, this hypothesis asserts that "a person's thoughts are controlled by inexorable laws or patterns of which he is unconscious. . . . His thinking itself is in a language—in English, in Sanskrit, in Chinese. And every language is a vast pattern-system, different from others" (Whorf, 1961, p. 135).

linguistic-relativity hypothesis: The belief that language determines the possibilities for thought and action in any given culture.

This observation was based on evidence from the social sciences, especially anthropology. For example, Margaret Mead's field research among the Arapesh of New Guinea had revealed that the Arapesh had no developed system of numbers. Theirs was a technologically simple society, and therefore complex numbering systems were not much use to them. They counted only "one, two, and one and two, and dog" (dog being the equivalent of four and probably based on the dog's four legs). To count seven objects, the Arapesh would say "dog and one and two." Eight would be "two dog," and twenty-four would come out as "two dog, two dog, two dog." It is easy to see that in this small society one would quickly become tired of attempting to count much beyond twenty-four and would simply say, "many" (Mead, 1971).

Other cultures have been found to have only a limited number of words for colors, and as a result, they do not make some of the fine distinctions between colors that we do. And in a famous example, Whorf argued that many languages have ways of referring to time that are very different from those found in English and other Indo-European languages. In English, we have verb tenses, which lead us to make sharp distinctions among past, present, and future time. In Vietnamese, in contrast, there are no separate forms of verbs to indicate different times; the phrase *tôi đi về*, for example, could mean "I'm going home" or "I went home" or "I will go home." The language of the Hopi also lacks clear tenses, and it seemed to Whorf that this made it unlikely that the Hopi culture could develop the systems of timekeeping that are essential to modern science and technology.

Thus, in its most radical form, the linguistic relativity hypothesis asserts that language actually determines the possibilities for a culture's norms, beliefs, and values. But there is little justification for this extreme version of the hypothesis. The Arapesh did not have a developed number system, but they easily learned to count using the Western base-10 system. Once they were exposed to the money economy of the modern world, most isolated cultures formed words for the base-10 number system or incorporated foreign words into their own vocabularies. They had no difficulty understanding the use of money, and this too became incorporated into so-called primitive cultures.

A more acceptable version of the linguistic relativity hypothesis recognizes the mutual influences of thought and language. For example, someone living in Canada or the northern parts of the United States is likely to have a much larger vocabulary for talking about snow (*loose powder, packed powder, corn snow, slush,* and so on) than a person from an area where snow is rare. A person who loves to watch birds will have a much larger vocabulary about bird habitats and bird names than one who cares little about bird life. To share the world of birders or of winter sports fans, we will need to learn new ways of seeing and of talking about what we see. So although the extreme version of the linguistic relativity hypothesis is incorrect, it has been a valuable stimulus toward the development of a less biased view of other cultures. We now understand that a culture's language expresses how the people of that culture perceive and understand the world and, at the same time, influences their perceptions and understandings.

 Thinking Critically Only cultures that used their language ability to develop forms of writing and reading could record their own history. Does this mean that nonliterate societies have no history, even though archaeologists can find and describe their settlements and ways of life?

CROSSING CULTURAL LINES

Global transactions of all kinds are a feature of the contemporary world. Success in such transactions often hinges on a person's ability to move comfortably among different cultures and subcultures. Unless we can see ourselves as others see us, however, we take for granted that our own cultural traits are natural and proper and that traits that differ from ours are unnatural and somehow wrong. Our ways of behaving in public, our food and dress and sports—all our cultural traits have become "internalized," so they seem almost instinctive. But once we understand another culture and how its members think and feel, we can look at our own traits from the perspective of that culture. This *cross-cultural perspective* has become an integral part of sociological analysis.

Ethnocentrism and Cultural Relativity

The ability to think in cross-cultural terms allows people to avoid the common tendency to disparage other cultures simply because they are different. However, most people live out their lives in a single culture. Indeed, they may go so far as to consider that culture superior to any other, anywhere in the world. Social scientists term this attitude **ethnocentrism**.

International study helps address problems of ethnocentrism by fostering cross-cultural contacts. In recent years, however, such exchanges have been hampered by the fear of terrorism.

Ethnocentrism refers to the tendency to judge other cultures as inferior in terms of one's own norms and values. The other culture is weighed against standards derived from the culture with which one is most familiar. The European explorers' assumption that the Native Americans could benefit from the adoption of European cultural traits is an example of ethnocentric behavior. But such obvious ethnocentrism is not limited to historical examples. We encounter it every day—for example, in our use of the term *American* to refer to citizens of the United States and not to those of Canada and the South American nations. Another example is our tendency to judge other cultures by how well they supply their people with the consumer goods we prize rather than by how well they adhere to their own values.

To get along well in other parts of the world, a businessperson or politician or scientist must be able to suspend judgment about other cultures, an approach termed **cultural relativity**. Cultural relativity entails the recognition that all cultures develop their own ways of dealing with the specific demands of their environments. This kind of understanding does not come automatically through the experience of living among members of other cultures; it is an acquired skill.

Limits to Cultural Relativity There are limits, however, to the value of cultural relativity. It is an essential attitude to adopt in understanding another culture, but it does not require that

ethnocentrism: The tendency to judge other cultures as inferior to one's own.

cultural relativity: The recognition that all cultures develop their own ways of dealing with the specific demands of their environments.

one avoid moral judgment entirely. We can, for example, attempt to understand the values and ideologies of the citizens who supported the Nazi regime, even though we abhor what that regime stood for. And we can suspend our outrage at racism in our own society long enough to understand the culture that produced racial hatred and fear. As social scientists, however, it is also our task to evaluate the moral implications of a culture's norms and values and to condemn them when we see that they produce cruelty and suffering (Mack, 2001). Consider the example of patriarchy.

A set of cultural values that appears in many places throughout the world emphasizes that women should be subservient to men. Women may be influential in the home, but in the larger world of the village, the city, and the nation, men should dominate and rule. The set of values and norms that support male dominance and the subordination of women is known as *patriarchy,* and societies that put the values of patriarchy into practice are known as patriarchal societies (Feldman, 2001).

Despite generations of struggle by women to break down the values and norms of patriarchy, boys in much of the world continue to be valued more highly than girls, a cultural value that leads to widespread suffering and neglect. Some patriarchal cultures forbid women to appear openly in public so that the honor of their fathers and husbands will not be put at risk. Other cultures set standards for women's appearance—small feet, for example, or large breasts and a small waist. Some cultures practice clitoridectomy, in which a woman's external genitalia are removed in order to reduce her desire for and enjoyment of sex. These and many similar practices are typical of patriarchal cultures or of cultures where patriarchal norms and values remain powerful despite women's assertion of their human rights.

A major feature of cultural change throughout the world is a movement by women to challenge the norms and values underlying patriarchy (Enloe, 2000). As women demand the right to determine how many children they bear; to read, drive, work in professions, and attend universities; and to enjoy other benefits often reserved for men, men perceive a threat to their power. And in many poor regions of the world, where men do not have much power or wealth themselves, the idea that women would challenge the status quo is seen by some as a direct threat to male honor and virtue and as a justification for many forms of violence against women. In such societies, women as well as men are often fearful of change and insist that traditional values must be upheld. Even in culturally diverse societies like those of the West, many women find themselves siding with men who want to return to the older values of patriarchy. But the dominant trend in the world, despite conflict and resistance, is for women to act as cultural innovators, challenge patriarchy, and assert their human rights.

Sociologists can be cultural relativists when they seek to understand the origins and strength of patriarchal norms in their own and other cultures. They can also recognize the limits of cultural relativity for understanding why many women reject values that insist on female subjugation. However, to argue that there are human values that transcend cultures, and that, for example, patriarchal cultures must change, is to risk imposing Western values on other cultures.

Cultural Hegemony: Myth or Reality?

When one culture's values, norms, and products become dominant and diminish the strength of another culture, the stronger culture is said to exert hegemony over the weaker one. The term **hegemony** refers to dominance or undue power or influence. A hegemonic culture is one that dominates other cultures, just as a hegemonic society is one that exerts undue power over other societies (Gramsci, 1992/1965, 1995). In Europe and parts of Asia, there is a lively debate about whether the culture of the United States endangers the continued existence of local cultures. English is the international language of business and science. American movies and popular music, fast food, and fashions are coveted around the world. Nations with less widespread languages and less powerful cultural institutions (for example, media industries, universities, fashion industries) fear that their young people will lose interest in their own cultures and embrace the values, language, and norms of Americans. Discussions of the global influence of McDonald's or the Disney Company appear often in social-scientific essays and cultural critiques in nations where American culture is popular. These issues have only gained in prominence since the tragic events of September 11, 2001.

Within the United States, too, sociologists and others worry that the powerful commercial cultural institutions that make certain types of

hegemony: Undue power or influence.

food, music, movies, fashions, and magazines popular throughout the nation endanger the existence of unique local subcultures. These critics fear that regional differences are decreasing and being replaced by a homogeneous culture that lacks diversity (Hardt & Negri, 2001; Oldenburg, 1997).

How realistic are these fears? Like all generalizations about social life, they contain some truth and much myth. Further research and more factual information are needed to understand whether American culture is indeed hegemonic, as its critics claim. But it is clear that there are societies in many regions of the world with extremely strong and distinct cultures that are able to produce their own adaptations of American commercial culture. India, for example, is the second leading producer of commercial movies in the world. There is a huge audience for Indian films, which combine the technology of Western films with the music, language, norms, and values of India's many subcultures (Palmer, 1995). And as Europe becomes more economically unified, it is likely that the strength of European cultural institutions will improve, despite the linguistic divisions that make some Europeans feel vulnerable to the influence of American culture.

At the same time, as the war in Iraq in 2003 demonstrated, there can be no doubt that the United States is more than just another culture on the world stage. It represents a powerful civilization whose cultural features have diffused across national boundaries to exert a lasting influence over the cultures of other societies. Does this mean that the United States exerts hegemony over other cultures? This question is a subject of continuing research and debate. Meanwhile, it is worth noting that much of what we think of as distinctive about American civilization originated in European, African, Asian, and Hispanic cultures. In consequence, we devote the rest of this chapter to a discussion of civilizations and cultural change.

CIVILIZATIONS AND CULTURAL CHANGE

Civilizations are advanced cultures. They usually have forms of expression in writing and the arts, powerful economic and political institutions, and innovative technologies, all of which strongly influence other cultures with which they come into contact. Some civilizations, like those of ancient Egypt and Rome, declined thousands of years ago and exist today mainly in museums and in the consciousness

These movie billboards in Madras, India, capture some of the excitement that can always be felt outside theaters where "Bollywood" films are featured.

of scholars, artists, and scientists. Others are living civilizations with long histories, like those of China and India, which were conquered by other civilizations and are rising again in forms that combine the old with the new. As we will see shortly, Islamic civilization is an example of such a civilization; so are the dominant civilizations of North America, Europe, the former Soviet Union, and Japan. They are dominant because they compete on a world scale to "export" their ideas and their technology—in fact, their entire culture or "blueprint for living."

The Concept of Civilization

Like most of the principal concepts in the social sciences, the concept of civilization can be elusive. It is used in many different contexts, in popular language as well as in social-scientific usage. In popular speech, the word *civilization* is often used to make negative comparisons between people who adhere to the norms of polite conduct, and are therefore said to be "civilized," and those who are "uncouth" and act like "barbarians" or "savages." This is how the word was understood by the explorers of Columbus's time, and colonial conquest was justified in part as an effort to civilize the barbarians.

In his effort to trace the origins of Western notions of what is civilized and what is barbaric behavior, the German sociologist Norbert Elias (1978/1939) showed that much of what we call "civilized" behavior is derived from the norms of the courts of medieval Europe. Elias based his study on accounts by medieval writers of the spread of "courtesy" (the manners of the court) to other levels of society. The following are some examples from a thirteenth-century poem on courtly table manners:

A man of refinement should not slurp his spoon when in company; that is the way people at court behave who often indulge in unrefined conduct.

It is not polite to drink from the dish, although some who approve of this rude habit insolently pick up the dish and pour it down as if they were mad.

A number of people gnaw a bone and then put it back in the dish—this is a serious offense. (p. 85)

Through his analysis of writings on manners, Elias demonstrated that Western norms concerning the control of bodily functions, from eating and sleeping to blowing one's nose, arose in the Middle Ages as the courts consolidated their power over feudal societies and exported the standards of courtly behavior to the countryside. Such behavior became a sign that a person was a member of the upper classes and not a serf or a savage.

In the social sciences, the most common use of the term *civilization* stems from the study of changes in human society at the macro level, which often requires comparisons among major cultures. In this context, a **civilization** is "a cultural complex formed by the identical major cultural features of a number of particular societies. We might, for example, describe Western capitalism as a civilization, in which specific forms of science, technology, religion, art, and so on, are to be found in a number of distinct societies" (Bottomore, 1973, p. 130). Thus Italy, France, Germany, the United States, Sweden, and many other nations that have made great contributions to Western civilization all have private corporations, and their normative orders, laws, and judicial systems are quite similar (Gramsci, 1995). Even though each may have a different language and each differs in the way it organizes some aspects of social life (European and North American universities define academic degrees differently, for example), they share similar norms and values and can all be said to represent Western civilization.

Islam is an outstanding example of a civilization that encompasses many distinct cultures and is still combining old and new cultural ways. The religion of Islam was founded by Mohammed, a prophet who was born in what is now Saudi Arabia in about 570 C.E. Sometime between 610 and 612 C.E., outside the holy city of Mecca, Mohammed is said to have received a visit from the Archangel Gabriel, who brought him the word of God. Mohammed became a charismatic religious leader whose followers developed his sayings and writings into the Koran, the Islamic sacred text.

Mohammed and his followers were a small band who sought to spread the new religion to the people among whom they lived. Most of the early Muslims were nomadic Arabs, primarily members of the Bedouin tribes who tended flocks in the arid and semiarid regions of Arabia. During his lifetime, Mohammed and his followers met with limited success in converting the various Arab peoples to the new faith. However, his successor, Omar, a man of great piety and military prowess, summoned believers to join in efforts to convert the tribes of pagan unbelievers. In a series of outstanding victories, Islamic armies conquered Syria (in 636), Iraq (in 637), Iran (in 637), Mesopotamia (in 641), and Egypt (in 642). These were all distinct societies and cultures, with those of the West part of the Greek and Byzantine civilization, with its written languages and highly developed cultures, while those of the East were influenced primarily by Persian civilization. For at least the next eight centuries, in fits and starts, Islamic civilization spread eastward and westward into the empires of India and China, over the old empires of Africa, and into Spain.

Compare the map of world cultures and civilizations in 1500 in Chapter 4 (Figure 4.4 on pages 84–85) with the map of the Islamic world as it existed around the same time (see Figure 3.3), after the wars between Christians and Muslims in what is now Israel and Jordan, and after the expulsion of the Islamic Moors from Spain. Both maps present some of the same facts, but Figure 3.3 suggests how important sea routes were in linking the far-flung regions of the Islamic world. Figure 4.4 emphasizes the many diverse societies, each with its own distinct culture, encompassed by Islamic civilization. Tribal peoples in Africa and Indonesia, nomadic Bedouin and African societies, residents of older empires in India and China—all shared, with many variations, the basic tenets of Islam while contributing their own cultural forms to a great and diverse civilization.

Effects of Cultural Contact

Although sociologists do not distinguish among cultures in terms of how "civilized" they are, this standard is often applied by members of

---

civilization: A cultural complex formed by the identical major cultural features of several societies.

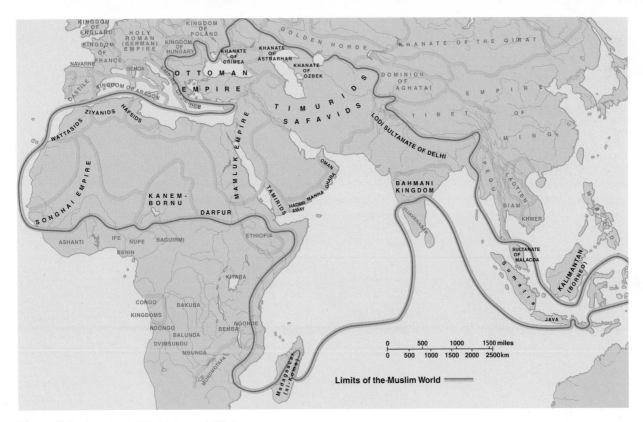

Figure 3.3 The Islamic World Around 1500.

different cultures when they come into contact with one another. The contact between Christopher Columbus and his crew and the natives of the West Indies is typical of such episodes. The explorers represented the relatively advanced civilization of Europe, while the natives whom the Europeans first encountered represented a much simpler culture, one that seemed totally "uncivilized" to the explorers. Throughout the period of world exploration, from the late fifteenth to the early nineteenth centuries, explorers and traders brought back reports of "savages" living in every part of the globe. These episodes of cultural contact (and often conflict as well) shaped the history of the next two centuries and continue to influence human existence.

Through such processes as exploration and conquest, civilizations invariably spread beyond their original boundaries. Figure 3.4 shows how, through much of the nineteenth century, England spread its version of Western civilization throughout the world as it conquered tribal peoples and established colonies in Africa and Asia. Colonial rule brought *cultural imperialism,* the imposition of a new culture on the conquered peoples. This meant that colonial peoples had to learn the languages of their conquerors, especially English, Spanish, French, Portuguese, and Dutch (or Afrikaans). Along with language came the imposition of ideologies like Christianity in place of older beliefs and religions.

The Crusades, or battles between Christian and Muslim forces, of the fourteenth and fifteenth centuries were a clash of civilizations that left enduring cultural legacies and memories of atrocities that continue to influence relations between Muslims and Christians to this day.

According to the historian Fernand Braudel, "The mark of a living civilization is that it is capable of exporting itself, of spreading its culture to distant places. It is impossible to imagine a true civilization which does not export its people, its ways of thinking and living" (1976/1949, p. 763). In his research on the contacts and clashes between the great civilizations surrounding the Mediterranean Sea during the 1500s, Braudel uses three important sociological concepts to explain the spread of civilizations

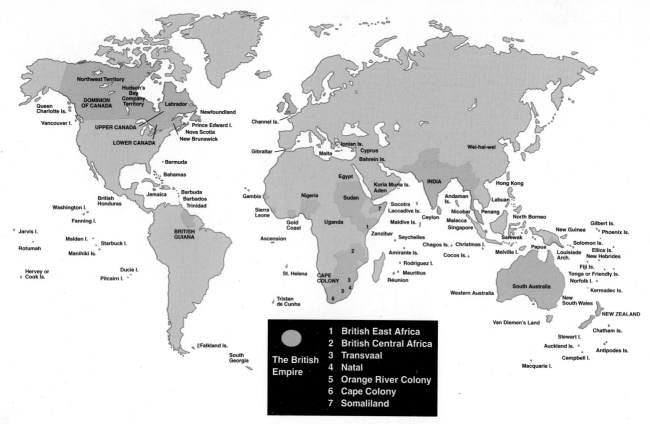

Figure 3.4 The British Empire in 1901. This map shows the extent of the British Empire in 1901. At the peak of its power, the empire had spread English civilization throughout the world. But is influence was strongest in peripheral, less industrialized lands. In a mere half-century, most of the empire would crumble, but the effects of English civilization on language and social institutions throughout the world would endure. Source: Eldridge, 1978.

Effects of Cultural Contact

Form of Contact	Description	Example
Acculturation	The process by which the members of a civilization incorporate norms and values from other cultures into their own.	Americans develop a taste for Italian, Chinese, and Mexican food, as well as other cuisines representing the cultures of the people who have joined their civilization.
Assimilation	The process by which culturally distinct groups within a civilization adopt the norms, values, and language of that civilization and gain equal status in its groups and institutions.	Spanish-speaking immigrants to the United States learn English and begin to move up the status hierarchy in education and jobs.
Accommodation	The process by which a smaller, less powerful society is able to preserve the major features of its culture even after prolonged contact with a larger, stronger culture.	Though conquered and forced onto reservations, Native Americans adapt by taking on many of the norms and values of the larger society while preserving aspects of their own culture.

around the world: acculturation, assimilation, and accommodation (see the For Review chart above).

Acculturation People from one civilization incorporate norms and values from other cultures into their own through a process called **acculturation**. Most acculturation occurs through intercultural contact and the borrowing or imitation of cultural norms. But there have been many instances of acculturation through cultural imperialism, in which one culture has been forced to adopt the language or other traits of a more dominant one. Thus, people in societies that were colonized in the nineteenth century were forced to learn the language of the conquering nation.

acculturation: The process by which the members of a civilization incorporate norms and values from other cultures into their own.

Aspects of our culture that we take for granted usually can be shown to have traveled a complicated route through other cultures to become part of our way of life. For example, Braudel's study of the Mediterranean civilizations shows that many of the plants and foods that are commonly found around the Mediterranean Sea, and later were imported to the New World, were borrowed from other cultures and incorporated into those of the Mediterranean societies. The process of acculturation can be traced in many other aspects of everyday life. As Ralph Linton has written:

> Our solid American citizen awakens in a bed built on a pattern which originated in the Near East. . . . He throws back covers made from cotton, domesticated in India. . . . He slips into his moccasins, invented by the Indians of the Eastern woodlands. . . . He takes off his pajamas, a garment invented in India, and washes with soap invented by the ancient Gauls. . . . He puts on garments whose form originally derived from the skin clothing of the nomads of the Asiatic steppes . . . and ties around his neck a strip of bright-colored cloth which is a vestigial survival of the shoulder shawls worn by the seventeenth-century Croatians. Before going out for breakfast he glances through the window, made of glass invented in Egypt, and if it is raining puts on overshoes made of rubber discovered by the Central American Indians and takes an umbrella, invented in southeastern Asia. (1936, p. 326)

The concept of acculturation can also be applied to how a newcomer people adopts the cultural ways of the host society. But acculturation is rarely a one-way process: At the same time that they are becoming more like their hosts in values and behavior, newcomers teach members of the host society to use and appreciate aspects of their own culture. Because of the cultural diversity of its population, examples of acculturation are especially prevalent in the United States. For example, even small towns in the interior of the country have at least one Chinese restaurant, but often the food served in those restaurants is a highly acculturated form of the cuisine of China. Indeed, most of the things we think of as part of the American way of life—from hamburgers, pizza, and baseball to democracy and free enterprise—originally were aspects of other cultures. That they have become incorporated into American culture through acculturation, and in the process have become changed from their original forms, does not deny the fact of their "foreign" origin (see the accompanying Global Social Change box).

global social change

North America's African Heritage All the important streams of migration to North America have combined to produce the culture of the United States and Canada. Examples of the influence of immigrant groups on American material culture abound: pizza (Italy), hamburgers (Germany), and French fries are among the most familiar. Of all these cultural influences, perhaps the least well understood is the legacy of the people who were brought to North America from Africa as slaves, especially during the last three decades of the eighteenth century and the first half of the nineteenth century.

One of the largest groups of slaves, captured in the region that is now Congo and Angola, was made up of people from many tribes who shared a common language system, Bantu. Many of these Bantu-speaking slaves were brought to the Carolinas. By the end of the Revolutionary War, the Carolinas had the largest concentration of people of African descent of any of the new states. In consequence, many of the names of towns and villages, rivers, streams, and other geographic features in those states are derived from Bantu words, often mixed with local Native American terms and transformed through long usage. Thus, Pinder Town in South Carolina is derived from the Bantu word for peanut, *mpinda*. George R. Stewart, an expert on place names, concludes that hundreds of small streams, swamps, and villages in the Carolinas, Mississippi, and Georgia received their names from the African slaves (Stewart, 1968; Vass, 1979).

The strongest evidence of African influence on contemporary American culture can be found in the Sea Islands along the Atlantic coast between Charlestown, South Carolina, and Savannah, Georgia. On the islands of Port Royal, Parris (famous for its Marine training base), Ladies, Hilton Head, and St. Helena reside the Gullah people. Also known as Geechi, they are descendants of slaves whose original tribes are believed to have lived in Angola (from which the word *Gullah* may be derived) and in the areas that are now Liberia and Sierra Leone. On the plantations, the Gullah slaves grew indigo and the long-fiber cotton that made the region famous in the period before the Civil War. Some of them escaped to become leaders of slave rebellions. Others were arrested during the Civil War for singing spirituals like "Roll, Jordan, Roll," to which they had added verses about their dreams of freedom. To this day, the religious practices, language, and place names of the Sea Island people bear witness to their African heritage (Creel, 1988).

A Gullah woman of the Sea Islands in a photograph taken early in the twentieth century.

Margaret Washington/Cornell University

Assimilation and Subcultures When culturally distinct groups within a larger civilization adopt the language, values, and norms of the host civilization and their acculturation enables them to assume equal statuses in the social groups and institutions of that civilization, we refer to that process as **assimilation**. When groups become assimilated into American society, for example, people often say that they have been "Americanized." Assimilation has been a major issue for immigrant groups in North America, as it still is for immigrants all over the world. It is no surprise, then, that we continually see articles in the press that ask such questions as: Will the various Hispanic peoples in America give up their language over time? Will American Jews marry members of other groups and lose their distinct identity? Will Italian Americans gradually forget their cultural heritage and come to think of themselves as "100 percent Americans" (Kasinitz, Mollenkopf, and Waters, 2008)? These kinds of questions form the subject matter of racial and ethnic relations in pluralistic societies like the United States.

When a culturally distinct people within a larger culture fails to assimilate fully or has not yet become fully assimilated, we say that it is a **subculture** within the larger culture. (The term is also applied to groups that have had significantly different experiences from those of most members of the society.) People who maintain their own subculture generally share many of the values and norms of the larger culture, but they retain certain rituals, values, traditions, and norms—and, in some cases, their own language—that set them apart. Thus we speak of African American, Latino, Native American, and a host of other subcultures in the United States. As explained in Chapter 12, these are also known as *ethnic groups* because their members have a sense of shared descent, a feeling of being "a people" with a history and a way of life that exists within a larger and more culturally diverse society.

Ethnic subcultures are created out of the experience of migration or invasion and subsequent adaptation to a host culture. But subcultures are also created out of the experience of people in complex societies who actively seek to create and maintain a way of life distinct from that of other members of their society (Fischer,

1987/1976; Gans, 1976/1967). For example, some subcultures in large cities are composed of artists and other people whose livelihood depends on the arts: theater people, rock musicians and record producers, visual artists, gallery owners, art critics, and curators. In many U.S. cities, and in some in Western Europe, there are quite distinct gay and lesbian subcultures. Because people who form homosexual relationships often experience hostility from heterosexuals, they tend to develop their own particular norms of communication and social behavior. When a subculture that challenges the accepted norms and values of the larger society establishes an alternative lifestyle, we call it a **counterculture**. The hippies of the 1960s, along with members of New Left political groups, activists in the women's movement, and environmental activists, formed a counterculture that had a significant influence on American politics and foreign policy during the Vietnam War years (Roszak, 1969).

Subcultures often are under intense pressure to conform to the society's most widely accepted norms and values or to adapt to new ways, new technologies, or the like. Social scientists often investigate the changing norms and values of such subcultures—the gay subcultures, the subculture of rock musicians and other celebrities, or the subcultures of occupational groups such as Appalachian miners, Wall Street lawyers, doctors, and construction workers, among others. This research is valuable in predicting trends in such areas as drug use, popular music, or patterns of labor–management conflict and cooperation (Boggs & Meyersohn, 1988; Flores, 2009; Volti, 2008).

Accommodation and Resistance Throughout history, many societies have withstood tremendous pressure to become assimilated into larger civilizations. But even greater numbers have been either wiped out or fully assimilated. Only a century ago, for example, one could still find many hundreds of hunter-gatherer societies throughout the world. Today, there are probably fewer than 100, and these live in the most isolated regions of the earth.

Larger and smaller societies do not usually develop ways of living together without the smaller one becoming extinct or totally assimilated into the larger one. But when the smaller, less powerful society is able to preserve the major features of its culture even after prolonged contact, **accommodation** is said to have occurred. For example, in the Islamic civilization of the Middle East before the creation of Israel in 1948, Jews and other non-Muslims usually found it rather easy to maintain their cultures

assimilation: The process by which culturally distinct groups in a larger civilization adopt the norms, values, and language of the host civilization and are able to gain equal statuses in its groups and institutions.

subculture: A group of people who hold many of the values and norms of the larger culture but also hold certain beliefs, values, or norms that set them apart from that culture.

counterculture: A subculture that challenges the accepted norms and values of the larger society and establishes an alternative lifestyle.

accommodation: The process by which a smaller, less powerful society is able to preserve the major features of its culture even after prolonged contact with a larger, stronger culture.

within the larger Arab societies. Compare this pattern of accommodation with the experience of the Jews in Spain, who were forced to leave in 1492 in one of the largest mass expulsions in history.

Accommodation requires that each side tolerate the existence of the other and even share territory and social institutions. The history of relations between Native Americans and European settlers in the Western Hemisphere is a complex story of resistance and accommodation. Throughout the period of conquest, expansion, and settlement by the Europeans, there was continual resistance by the native peoples. This resistance took many forms, including refusal to adopt Christianity, to speak English or Spanish, to sell goods and services to the settlers, and to fight in the Europeans' wars. Resistance did not save Native Americans from death by disease, military conquest, or famine, but it did allow them to maintain their cultures and to borrow from the settlers the cultural customs that were most advantageous to them. For example, the Plains Indians adopted horses from the Spanish explorers, which completely changed their culture, and much later they borrowed trucks from American culture, which helped them adapt to modern ranching.

Most cultures in the world today have had to confront the influence of at least one expanding civilization. In some instances, the result has been prolonged conflict and the eventual annihilation of the smaller society. However, throughout the world, it is as easy to find examples of failures in accommodation as it is to find examples of sharing and cooperation among cultures. Many of the conflicts that lead to severe social unrest, such as the conflict between Islamic and Eastern Orthodox Bosnians, or among warring tribal peoples in Rwanda, are based on cultural differences. Such conflicts may also occur

"Reading" Art for Social Meaning What does a careful look at this limestone carving of Rahotep and his wife Nofret, a royal Egyptian couple (carved by an official court artist around 2601 B.C.E) suggest to you about the culture of the ancient Egyptians? Think about norms of dress and hairstyle for men and women. Are they very different today? What about skin color? The sculpture suggests that the princely rulers of one of the world's first civilizations could have had either dark or light skin.

© The Art Archive/Egyptian Museum Cairo/Dagli Orti (A)/Picture Desk

Prince Rahotep and His Wife Nofret, artist unknown, about 2610 B.C.E. Painted limestone, height 47 inches.

within a culture. In the United States, for example, cultural differences are becoming more visible, and so are cultural conflicts. In no area of contemporary life is this more true than in the realm of intimacy and sexual behavior.

SUMMARY

Could human beings exist without their cultures? What are some of the links between culture and social order?

- In the social sciences, *culture* refers to all the modes of thought, behavior, and production that are handed down from one generation to the next by means of communicative interaction. Sociologists are concerned primarily with aspects of culture that help explain social organization and behavior.

- Culture can be viewed as consisting of three major dimensions: ideas, norms, and material culture.

- *Ideas* are the ways of thinking that organize human consciousness. Among the most important ideas are *values*, socially shared ideas about what is right.

- *Norms* are specific rules of behavior that are supported or justified by values; *laws* are norms that are included in a society's official written codes of behavior.

- *Ideologies* combine ideas and norms; they are systems of values and norms that the members of a society are expected to believe in and act on without question.

- A society's *material culture* consists of all the things it produces.

- *Technologies* combine norms and material culture; they are the things and the norms for using them that are found in a given culture.

- *Social control* refers to the set of rules and understandings that control the behavior of individuals and groups in a culture. The wide array of norms that permit a society to achieve relatively peaceful social control is called its *normative order*.

- *Sanctions* are rewards and punishments for adhering to or violating norms. Strongly sanctioned norms are called *mores;* more weakly sanctioned norms are known as *folkways*.

How have humans evolved in ways that culture made possible?

- One of the most hotly debated questions in the social sciences is how much, if at all, human culture is determined by biological factors.

- According to Darwin's theory of *natural selection,* mutations in organisms occur more or less randomly. Mutations that enable an individual organism to survive and reproduce are passed on to the next generation. This process permits animals and plants to adapt to new environments.

- Herbert Spencer and other social thinkers, who came to be known as *social Darwinists,* attempted to apply Darwin's theory to humans' ability to adapt to social environments. Spencer used the phrase "survival of the fittest" to describe this ability. People who were able to survive in the urban environment created by the Industrial Revolution were viewed as superior human beings.

- A more recent attempt to attribute social phenomena to biological processes is *sociobiology,* which refers to efforts to link genetic factors with the social behavior of animals. According to sociobiologists, such behaviors as incest avoidance, aggression, and homosexuality may be genetically programmed in human beings. As yet, there is no evidence that such genes or sets of genes actually exist.

- A more widely accepted view of culture denies that humans have innate instincts and states that human culture became self-generating at a certain stage in prehistoric times. Thus, human evolution is not dependent on genes; instead, cultural techniques allow humans to adapt to any physical or social environment.

Would we be truly human without our capacity to use language?

- The learning of culture is made possible by language.

- Although apes have been taught to use language to some extent, human language is unique in that it allows its speakers to express an infinite number of thoughts and ideas that can persist even after their originators are gone.

- According to the *linguistic-relativity hypothesis,* language also determines the possibilities for a culture's norms, beliefs, and values. A less extreme form of that hypothesis recognizes the mutual influences of culture and language.

What are the main social processes that may occur when people come to live in a new culture as a result of immigration or other forms of social change?

- The notion that one's own culture is superior to any other is called *ethnocentrism.*

- To understand other cultures, it is necessary to suspend judgment about those cultures, an approach termed *cultural relativity.*

- Throughout the world, women are at the forefront of cultural change, challenging norms of patriarchy that maintain the dominance of men over women.

- Recently, there has been much debate over whether the culture of the United States is exercising undue dominance or *hegemony* over other cultures throughout the world. Although more research is needed to answer this question, it is clear that, in many regions of the world, there remain extremely strong and diverse cultures that are able to maintain their own values and traditions.

How does the global influence of civilizations help bring about social change in other cultures?

- Similarities among cultures have resulted from the processes by which cultures spread across national boundaries and become part of a larger, more advanced culture called a civilization. A *civilization* may be defined as a cultural complex formed by the identical major cultural features of a number of particular societies.

- A key feature of civilizations is that they invariably expand beyond their original boundaries. The spread of civilizations can be explained by three processes: acculturation, assimilation, and accommodation.

- When people from one civilization incorporate norms and values from other cultures into their own, *acculturation* is said to occur.

- The process by which culturally distinct groups within a larger civilization adopt the language, values, and norms of the host civilization and gain equal statuses in its institutions is termed *assimilation*. (If a distinct people fails to assimilate fully, it is referred to as a *subculture,* but if it challenges the accepted norms and values of the larger society, it may become a *counterculture.*)

- When a smaller, less powerful society is able to preserve its culture even after prolonged contact with a major civilization, *accommodation* has taken place.

The Kornblum Companion Website

www.cengagebrain.com

Supplement your review of this chapter by going to the companion website to take one of the tutorial quizzes, use the flash cards to master key terms, and check out the many other study aids you'll find there. You'll also find special features, such as GSS data and Web links that will put data and resources at your fingertips to help you with that special project or to do some research on your own.

Martin Rose/FIFA/Getty Images

Societies and Nations

FOCUS QUESTIONS

How are societies organized in ways that help establish social order?

How do populations differ from societies, and what factors explain the rapid growth of the human population since the 1700s?

How does social organization influence your life and that of other individuals?

How does the existence of multiple societies in many nation-states threaten the stability of those nations?

It's almost 3 A.M. on a steamy summer Saturday. In the emergency room of Denver General Hospital, people of all ages and social types are slumped over chairs in the waiting room. Some are bent over in pain. Others are clutching injuries. Worried parents with a feverish child wait impatiently for their turn to finally pass through the doors where they will be seen by the medical staff. Outside the emergency room, incessant flashing lights signal the arrival of ambulances and police cars. They are bringing in the city's bloody harvest—its victims of speed, violence, and neglect. Those in danger of death or paralysis are rushed into treatment rooms for immediate, often heroic attention. Those with lesser injuries may sit in the corridors or even be shunted into the waiting room with the "walk-ins."

Clinical specialist Lori Jones often serves as the ER's triage nurse. Her job is to assess the gravity of patients' symptoms. "You cannot refuse care," she explains. "When you set a little sign outside with a light on it that says, 'Twenty-four-hour emergency room' you cannot refuse to

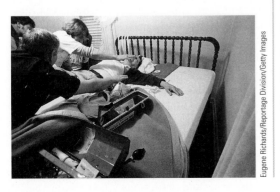

Eugene Richards/Reportage Division/Getty Images

see anybody. Then we turn around and teach everybody that in this day of limited resources and overcrowding, you can and must refuse care to those you don't believe need it" (Richards, 1989, p. 72). Jones has the unenviable job of deciding who gets care immediately, who gets it later, and who waits untreated through the long night. "The psych patient and the homeless person wants to stay in the ER to get seen immediately. He doesn't want to get put out in the waiting room or the walk-in clinic, where the wait could be two, four, five hours. And he knows from experience that the likelihood of getting a real bed upstairs is much greater if he comes in through the emergency room than the walk-in clinic" (p. 72). Most of the homeless street people who use the emergency room, Jones continues, "aren't bag people, don't have backpacks. Many of them don't push carts around and they don't carry big plastic garbage sacks of possessions around. They have only the clothes on their bodies. So if they can get into Denver General, it may be loud, but at least it's warm. At least we can't rob them. They will get a meal. Nobody will kick them or smash their head" (p. 73).

Inside the ER, the staff faces one crisis after another. Death lurks around the hospital gurneys, as many of us know from watching *ER,* one of the most popular programs in the history of American television. But Dr. Jim Rappaport has just witnessed the death of an elderly man with cardiac failure whom he thought he might be able to save. "Death is a very nebulous thing sometimes. . . . At what point does he die? I don't know. The heart can stop working, yet each cell is still alive." Dr. Rappaport and his team fought to revive this patient, but at some point they knew it was hopeless. Yet they continued their efforts and finally said, "Okay, he's dead. . . . But at some time, at some point, maybe even hours ago when he was being carried over here in the ambulance, his soul left him" (p. 94).

For Dr. Paul Thurman, director of emergency medical services at Denver General, the threat to the quality of emergency care, and to the ultimate survival of the

ER, is "without a doubt, selfishness. The reality is that about 20 percent of our society cannot or will not care for themselves. So the rest of society has got to make a decision about whether or not they're going to pick up the tab for that 20 percent." But as Dr. Thurman sees it, the society he lives in is increasingly indifferent to the needs of the poor and those who are otherwise incapable of caring for themselves. "So we have to decide what kind of society we want to live in. Do we want to live in a Calcutta where people die in the streets because there's nobody to help them and care for them? Or do we want to live someplace like Denver that has one of the finest indigent-care and trauma emergency systems in the country?" (p. 163).

The director and his staff of doctors, nurses, paramedics, ambulance drivers, and administrative personnel are struggling to provide care in the face of dwindling resources and budget cuts. The frustrations of trying to maintain an effective health care delivery organization in this negative environment weigh heavily on them. As Dr. Thurman says, "We're fighting desperately to keep alive, but I'm tired of working at a resource-poor hospital where we get punished for doing a good job."

Dr. Thurman and thousands of other health care professionals who serve less affluent segments of the U.S. population are now witnessing a relentless attack on public budgets. Federal tax cuts favoring more affluent citizens inevitably reduce the resources available for public health care and further erode the ability of hospitals to meet the health care needs of those who depend on their services (Higgins, 2003).

THE SOCIAL ORDER

Amid the chaos of incoming emergencies and frightened people, the emergency room staff maintain a definite social order. Despite the confusion and stress, all the professionals go about doing their jobs. As we see in the preceding vignette and in the Visual Sociology feature at the end of the chapter, these jobs are not always routine or easy. The triage nurse, for example, must watch as children and families wait anxiously for treatment while less deserving but more serious medical cases are placed ahead of theirs. The chief of medical services worries that even an effective social organization like that of the ER at Denver General is threatened by forces in the larger social environment within which the ER exists. Is the ER unique in facing these challenges? By no means. The members of every human social structure (organization, group, family, congregation, and so on) experience different kinds of stress as they seek to carry out their functions. The larger social environment is always changing and exerting new forces that shape and alter the social structure, whatever its nature. One of the main goals of this chapter is to indicate the common features of the many different types of human social structures and to show how they influence individual behavior.

Of course, humans are not the only animals capable of social organization. We can learn a lot about life in human societies by comparing ourselves with other social animals. If you look closely at an ant colony or a beehive, for example, you can see a remarkable amount of organization. The nest or hive is a complete social world with workers, warriors, queen, drones, and so on. Each individual ant or bee has something to do and seems to do it quite well. But the differences between human societies and those of social animals like bees are even more important than the similarities. An individual social animal, such as a worker bee, is able to perform only a certain number of innate (inborn) tasks. A human, by contrast, can learn an infinite number of tasks. Thus Lori Jones, the clinical specialist in the ER, can be a mother, a voter, a taxpayer, a congregation member, a driver, a singer, and pretty much whatever else she sets her heart on becoming.

We often assume that we know how society works and how to steer our way through it. But there will be times—especially when we are learning to adapt to changing social environments—when we will be unsure of what is expected of us and how we should perform. Worse still, there may be times when society itself is threatened, that its continued existence as we know it is endangered. We catch glimpses of the breakdown of society during riots or wars or severe economic recessions.

At earlier times in human history, plagues and famines were frequent reminders that people had little control over their own destinies. Today, plagues and famines still occur, but we are more often faced with real or potential crises of our own making: war, genocide, environmental disasters, drug addiction, criminal violence. Thus, if we are to continue to exist and thrive as a species, it is vital that we study societies and social structures—how they hold together, how they change, and why they sometimes seem to fall apart.

As in any science, we begin with some basic terms and definitions. The next few pages

Elements of Social Structure

Structural Element	Description	Example
Group	Any collection of people who interact on the basis of shared expectations regarding one another's behavior.	A discussion group; a Bible study class; a local union.
Status	A socially defined position in a group.	Orderly, practical nurse, registered nurse, resident, chief resident (all statuses in a hospital ward).
Role	The way a society defines how an individual is to behave in a particular status.	The doctor diagnoses and treats illnesses; the nurse provides care to patients under the doctor's supervision.
Role Expectations	A society's expectations about how a role should be performed, together with the individual's perceptions of what is required in performing that role.	A major league center fielder is expected to have a batting average over .300, drive in more than 75 runs, and cover the field with a minimum of errors.
Institution	A more or less stable structure of statuses and roles devoted to meeting the basic needs of people in a society.	The military is the primary institution devoted to providing national defense.

introduce the principal elements of social structure (see the For Review chart above). Later sections of the chapter apply these and related concepts to an analysis of how societies have developed since the beginning of human history and how differences among societies affect the lives of individuals. We also discuss the important distinction between societies and nation-states.

Society and Social Structure

The term **society** refers to a population of people (or other social animals) that is organized in a cooperative manner to carry out the major functions of life, including reproduction, sustenance, shelter, and defense. This definition distinguishes between societies and populations. The notion of a population implies nothing about social organization, but the idea of a society stresses the *interrelationships* among the members of the population. In other words, a population can be any set of individuals that we decide to count or otherwise consider, such as the total number of people living between the Rio Grande and the Arctic Circle, whereas a society is a population that is organized in some way, such as the population of the United States or Canada, or the Amish people of Pennsylvania. In the modern world, most societies are also (but not always) nation-states.

Social structure refers to the recurring patterns of behavior that people create through their interactions, their exchange of information, and their relationships. We say, for example, that the family has a structure in which parents and children and other relatives interact in specific ways on a regular basis. The larger society usually requires that family members assume certain obligations toward one another. Parents are required to educate their children or send them to schools; children are required to obey their parents until they have reached an age at which they are no longer considered dependent. These requirements contribute to the structure of relationships that is characteristic of the family. (The social scientist's method of diagramming family relationships, shown in the Research Methods box on page 76, creates a graphic depiction of family social structure.)

Participation in Multiple Social Structures

Throughout life, individuals maintain relationships in an enormous range of social structures, of which families are just one. There are many others. People may be members of relatively small groups like the friendship or peer group and the work group. They may also be members of larger structures like churches, business organizations, or public agencies. And they may participate in even more broadly based structures, such as political groups and party organizations, or interest groups like the National Rifle Association or Planned Parenthood. All these social structures are composed of groups with different degrees of complexity and quite different patterns of interaction. A military brigade, for example, is far more complicated than a barbershop quartet, and people behave quite differently in each, but both are social structures.

society: A population that is organized in a cooperative manner to carry out the major functions of life.

social structure: The recurring patterns of behavior that people create through their interactions, their exchange of information, and their relationships.

Using Kinship Diagrams to Portray Social Structure

Kinship diagrams like the one shown here provide a visual model of one type of social structure, that associated with family statuses extending over more than one generation. Such diagrams are used in both anthropology and sociology to denote lines of descent among people who are related by blood (children and their parents and siblings) and by marriage. To understand the chart, one must know the meanings of the symbols used; these are explained in the key to the chart. The chart applies these symbols to show how cross-cousin marriages between members of different bands link the bands together into a larger structure of kinship networks.

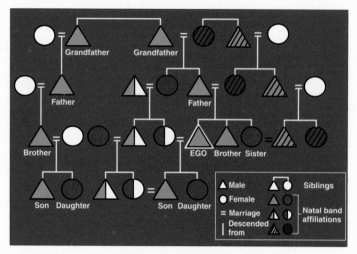

Source: Adapted from "The Origin of Society," by Marshall D. Sahlins. Copyright © September 1960 by Scientific America, Inc. All rights reserved. Reprinted with permission from Donald Garber.

The chart illustrates the social structure of a hunter-gatherer group. One individual, Ego, is used as a point of reference, and kinship links are traced from Ego's offspring, parents, grandparents, and more distant relatives.

In most hunter-gatherer societies, it is not possible for the entire society to travel and hunt within the same territory because of scarcities of game and edible vegetation. Instead, the society is divided into natal bands—in this example, those formed by the male line of descent (grandfather, father, Ego, Ego's son).

The striped coloring within the kinship symbols refers to another characteristic of hunter-gatherer societies: Marriages within the band cannot occur because of the incest taboo, and specific rules govern marriage outside the band. Note that Ego has married his father's sister's daughter, or his first cousin on his father's side. Ego's wife's brother has married a woman from outside their band. This couple has come to live in the husband's natal band. Their daughter has married Ego's son, another cross-cousin marriage. Ego's daughter, on the other hand, will eventually marry someone from another band and live outside her natal band. By means of this custom, the balance of women coming into and going out of the band is preserved.

Cross-cousin marriages, which are taboo in most Western societies, permit the hunter-gatherer band to develop a strong network of interfamily and interband kinship ties. This network widens over the generations and extends the ties of kinship throughout these small, mobile societies.

In such structures, our time, our activities, and even our thoughts may be "structured" according to the needs and activities of the group.

Elements of Social Structure

Groups The "building blocks" of societies are social groups. A **group** is any collection of people who interact on the basis of shared expectations regarding one another's behavior. One's immediate family is a group; so are a softball team, a seminar, a caucus, and so on. But a collection of people on a busy street—a crowd—is not a group unless for some reason its members begin to interact in a regular fashion. Usually a crowd is composed of many different kinds of groups—couples, families, groups of friends, and so on. They may be molded into a single group in response to an event that affects them all, such as a fire that creates the need for orderly evacuation of a building.

Statuses In every group, there are socially defined positions known as **statuses**. Father, mother, son, daughter, teacher, student, and principal are examples of familiar statuses in the family and the school. There are an infinite number of statuses in human societies. In a corporation, for example, the statuses range from president and chief executive officer to elevator operator and janitor. Between these two extremes, there could be thousands of other statuses. Moreover, the corporation can always create new statuses if the need arises. Thus, in the 1970s, when American society agreed to combat racism and sexism in business and government, many corporations invented the status of affirmative action director to provide equal opportunities for workers of both sexes and all racial and ethnic groups. In the 1990s, as the larger society became more deeply concerned about environmental pollution and drug abuse, corporations began to create new statuses like pollution control manager, drug counselor, or ethics officer (Kelley, 1998). Each of these statuses then became part of the "corporate structure."

Note that, like many other sociological terms, *status* has more than one meaning. It can refer to a person's rank in a social system and also to a person's prestige—that is, the esteem with which others regard him or her. The fundamental meaning, however, is the one we use here, in which *status* refers to a position in a social structure (family, team, company, and so on).

Adaptability of Social Structures Human societies rely heavily on the creation of statuses to adapt to new conditions like environmental

---

group: Any collection of people who interact on the basis of shared expectations regarding one another's behavior.

status: A socially defined position in a group.

pollution and drug use, and one can often observe this adaptation occurring in daily life. In the family, for example, it is increasingly common (though not universally condoned) for young adults to cohabit before marriage or after divorce. As a large-scale phenomenon, this is a relatively recent trend in American society, so new that we have not defined very well the new status of the person who participates in such a relationship. Do we say *boyfriend* or *girlfriend, lover, mate, significant other,* or what? These terms seem awkward to us because this is an emerging status that our society has not yet fully accepted or defined.

This example highlights an essential point about human social structure: It is never fixed or perfectly formed but instead is always changing and adapting to new conditions. Often the process of change involves much conflict and uncertainty, and often there is little consensus about how one should perform in a given status. Should the president of a corporation be an aloof, aggressive leader who directs subordinates with little regard for their feelings? Or should the president show concern for employees' feelings and personal needs and perhaps thereby gain greater loyalty and motivation? This is just one of thousands of dilemmas arising from questions about how we should act in a given social status. To clarify our thinking about statuses in groups and the behaviors associated with those statuses, sociologists make a distinction between a *status* and a *role.*

Roles and Role Expectations The way a society defines how an individual is to behave in a particular status is referred to as a **role**. In the ER, for example, Lori Jones has several distinct roles to play. She specializes in giving care to homeless and psychiatric patients, but she also often serves as triage nurse, and then her role involves deciding which patients must wait while others receive care for more serious conditions. It bothers her that the ER offers care to all but has to turn some away, yet she manages to perform her role to the satisfaction of her superiors in the organization. Clearly, the ways in which people actually perform a role may vary widely. They are the product of **role expectations**—that is, the society's expectations about how a role should be performed, together with the individual's perceptions of what is required in performing that role.

To appreciate the importance of role expectations, you need only think of the mothers and fathers of your close friends: All hold the same statuses, but how different their behavior is! Part of that difference is caused by personality, but another part is determined by how those mothers and fathers perceive what is expected of them in the statuses they hold. One mother and father may have been raised to believe that children should work to support the family. They will insist that their children get early job experience. Another couple may have been taught that childhood is too short and should be prolonged if possible. Other things being equal (for example, adequate family income), that couple will not encourage their children to find jobs before they are more or less the same age the parents were when they went to work.

Social change makes for even more debate and anxiety about role expectations, a subject we discuss in more detail later in the chapter. For now, consider the example of a mother who is also an attorney (something that was rather rare before women gained greater access to professional training). Because of the demands of her profession, she may be unable to take on a leadership position in the school PTA. However, being active in the PTA may have been one of her role expectations for motherhood before the opportunity for a professional career (and the income it provides) became attractive. Now she may feel harried by the pressure of her conflicting statuses of mother and attorney—to say nothing of her other possible statuses, such as wife, daughter, citizen, consumer, committee member, and so on. She may demand that her spouse perform tasks not traditionally associated with the status of husband, and he may or may not modify his original role expectations about that status; in either case, there is likely to be some conflict in the family as it adjusts to these changes.

Another family, in which the mother is also in the labor force but no father is in the home, has even more adjusting to do. The older children may take on parental roles far sooner than they might have in a two-parent family, but they may also resent this added responsibility and take out their anger on themselves and their siblings. In still another family, one that conforms to the tradition in which the mother is a homemaker and the father works outside the home, the pressures and conflict created by multiple role expectations may not occur in the early years. But what happens when the father retires and gives up his lifelong status of breadwinner? This is a time when traditional families often experience strain.

role: The way a society defines how an individual is to behave in a particular status.

role expectations: A society's expectations about how a role should be performed, together with the individual's perceptions of what is required in performing that role.

In sum, sociologists are interested in the way roles and statuses affect individual behavior. They also study how lack of roles—for example, lack of a job—can affect people's lives. They do not deny the importance of personality, but they place greater emphasis on the influence of social structure as an explanation of individual behavior.

Organization in Groups Groups vary greatly in the extent to which the statuses of their members are well or poorly defined. The family is an example of a group in which statuses are well defined. Although parents carry out their roles in different ways, the laws of their society define many of the obligations of parenthood, placing certain limits on what they can and cannot do. Other groups are much more informal. We may form groups for brief periods in buses, hallways, or doctors' offices. These groups also have roles and statuses, but they are variable and ill defined. All of the people riding on a bus are passengers, but they do not interact the way the members of a family do. At most, we expect civility and "small talk" from other passengers on a bus, but we demand affection and support from other members of our family.

Social change is constantly making the structure of groups more varied and complex. To continue with our earlier example, the family may seem to have well-defined statuses, but high rates of divorce and remarriage have increased the proportion of families in which one parent is a stepparent (see Chapter 15). Consider the difference in role expectations for "mother" or "father" in a situation in which each spouse must interact with a new spouse, a new set of children, and a new set of in-laws, as well as an ex-spouse and children in the first family, plus the old in-laws, plus his or her own parents. Balancing role expectations among the often competing demands of these groups can be a daunting task, one that was far less common when the norms of society made it difficult to divorce.

Groups Within Organizations Groups also vary greatly in how they are connected with other groups into a larger structure known as an *organization*. An army platoon, for example, includes the well-defined statuses of private, corporal, sergeant, and lieutenant, each with specific roles to play in training and combat. Platoons are grouped together under the leadership of higher officers to form companies; this pattern is repeated at higher levels to create the battalion and the brigade.

Figure 4.1 shows the formal structure of a typical army combat brigade. It does not show the brigade's informal organization, which consists

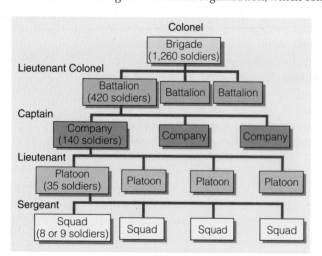

Figure 4.1 The Structure of a Typical Combat Brigade.

institution: A more or less stable structure of statuses and roles devoted to meeting the basic needs of people in a society.

of the ways its members actually relate to one another on the basis of friendships or animosities or mutual obligations of various kinds. (The importance of informal organization within larger groups, such as bureaucracies, is discussed more fully in Chapter 6.)

Social Institutions

The structure of most of the important groups in society is determined by shared definitions of the statuses and roles of their members. When such statuses and roles are designed to perform major social functions, they are termed **institutions.** In popular usage, the word *institution* generally refers to a large bureaucratic organization such as a university or hospital or prison (usually with cafeterias that serve "institutional food"). But although the word has this meaning in everyday language, the sociological use of the term should not be confused with this meaning.

 People often talk about how someone they know *performs* a role versus what their *expectations* of that role are. Listen to such conversations and see how often they deal with the distinction between role and role expectations.

Institutions and Basic Social Needs In sociology, an institution is a more or less stable structure of statuses and roles devoted to meeting the basic needs of people in a society. The family is an institution that controls reproduction and the training of new generations. The market is an institution that regulates the production and exchange of goods and services. The military is an institution that defends a society or expands its territory through conquest. Any particular family, corporation, or military unit is a group or organization within one of these institutions (Skolnick & Currie, 2004).

Within any given institution, norms specify how people in various statuses are to perform their roles. Thus, to be a general or a new recruit or a supplier of military hardware is to have a definite status in a military institution. But to carry out one's role in that status is to behave in accordance with a normative system—a set of mores and folkways that distinguish a particular institution from others. For example, in a classic research study about the process of becoming a military recruit, Sanford Dornbusch (1955) described how all the signs of a person's status in civilian life, from fashions in dress to the use of free time, are erased in boot camp. The new status of recruit must be earned by adhering to all the norms of military life. So it is with every social institution: Each has a specific

Social Structures and the Inner-City Emergency Room After several years as a clinical specialist and triage nurse, Lori Jones knows that to be happy in her roles and balance all the conflicting demands they make on her, she must do some serious soul-searching. As she explains it, "Actually, I've never done anything really great here, never saved anyone's life. The paramedics have. Yet patients have been lucky to have me as a nurse. No, I've never done things that are dramatic. I mean, I was happy because patients took their medicine as directed for 10 days. That's a feat in itself."

Documentary photographer Eugene Richards (1989) spent many months conducting participant observation research in the emergency room of Denver General Hospital, known by local residents as the "knife and gun club." His photos capture the physically and emotionally draining work of the ER staff and the stark drama of the emergencies they deal with. In the text accompanying the photos, he often refers to the social structure of the ER. The roles of the ER staff are well defined, so that even if the work stretches them to the limits of endurance, they know what they must do and how they must assist each other. Indeed, underneath its drama and emotion, the ER is a marvel of human social organization, without which it could not be effective.

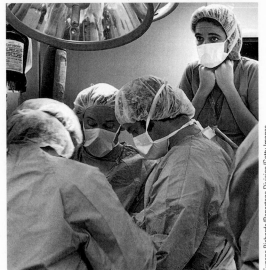

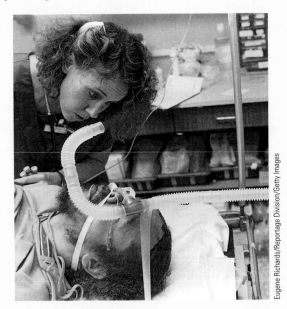

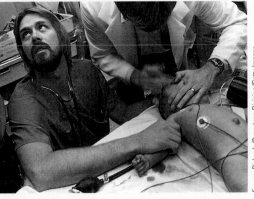

Eugene Richards/Reportage Division/Getty Images

set of norms to govern the behavior of people within that institution.

Institutions of religion, education, politics, and economics tend to become more complex over time as the society changes. It is helpful, therefore, to think of large-scale societies like the United States as having several institutional *sectors* (sets of closely related institutions). For example, the economy is an institutional sector that includes markets, corporations, and other economic institutions. Politics is an institutional sector containing legal, executive, legislative, and other political institutions. Each of these institutions, in turn, is composed of numerous groups and organizations. The Chicago Mercantile Exchange is a market organization; the U.S. House of Representatives is a legislative organization. In this book, as in much social-scientific writing, the term *institution* is used to refer to institutional sectors (for example, religion) as well as to the institutions included within a sector (for example, Buddhism or Christianity).

How Institutions Change The history of human societies is marked by the emergence of new institutions like the university or the laboratory. In fact, a dominant feature of human societies is the continual creation of new social institutions, a feature that social theorist Talcott Parsons (1951, 1966) labeled **differentiation**. By this term, Parsons meant the processes whereby sets of social activities performed by one social institution are divided among different institutions. In small-scale agrarian societies, for example, the family not only performs reproductive and training functions but is the primary economic institution as well. Over time, the family gradually loses some of its functions to new institutions. The processes of differentiation result in the emergence of new institutions designed to manage economic production (corporations), train new generations (schools), develop new technologies (science), or perform other important social activities. (Many of those institutions are discussed in detail in Part 4 of this book.)

differentiation: The processes whereby sets of social activities performed by one social institution are divided among different institutions.

Why does this differentiation take place? One answer is provided by critical sociologists like Michel Foucault, who argue that new institutions in society, such as the prison, the insane asylum, and the public medical clinic, have evolved as a means by which the powerful members of society can control unruly or rebellious members whose behavior threatens their comfort (Baber, 2001; Foucault, 1977). This critical perspective on institutions appears often in the work of contemporary sociologists, and we return to it in more detail in later chapters. A more functionalist explanation for the emergence of new institutions highlights the importance of population growth, which creates a demand for new ways to coordinate the activities of groups and individuals and to carry out new social functions (for example, education of large numbers of people or delivery of health care to people regardless of income). The need to perform these new functions for an ever-growing number of people places demands on existing institutions that they cannot fulfill, which leads to the development of new institutions—that is, new ways of performing important social functions (Halbwachs, 1960; Loomis & Rodriguez, 2009). We will see in the next section that new institutions are particularly likely to emerge when populations grow, the need for coordination of their activities increases, and new demands are placed on older institutions that they cannot entirely fulfill.

Thinking Critically: How is the institution of higher education changing? Remember that change in your own college or university represents change in an *organization*, not change in the institution of higher education as a whole.

POPULATIONS AND SOCIETIES

The remarkable growth of the world's human population is the great biological success story of the last million years (at least from the perspective of the human species). Throughout most of the evolution of the human species, the world's population remained relatively small and constant. At the end of the Neolithic period, about 8000 B.C.E., there were an estimated 5 to 10 million humans, concentrated mainly in the Middle East, East Africa, southern Europe, and a few fertile river basins in India, China, Latin America, and Central America. The overall range of human existence was much wider—populations were moving north as the last ice age receded about 12,000 years ago—but the human population was nowhere near as widely distributed as it is today. By the time of Jesus there were an estimated 200 million people on the earth, and by 1650, there were an estimated 500 million. By 1945, however, the population had reached about 2.3 billion, and it is now more than 6 billion. This trend is illustrated in Figure 4.2.

What explains the shape of this population curve—the gradual rise in the world's population during the late Stone Age, the increasing rate of growth in the early millennia of recorded history, the explosive growth after 1650? The answers have to do with the changing production technologies of societies and their ever-increasing ability to sustain their populations, not only those directly involved in acquiring food but the ever-larger numbers of people in other occupations as well. In other words, human population growth is related to the shift from hunting and gathering to agriculture and, later, to industrial production as the chief means of supplying people with the necessities of life.

The First Million Years: Hunting and Gathering

A human lifetime is no more than a twinkling in time. What are seventy or eighty years compared with the billions of years of the earth's existence? What is one generation compared with the millions of years of human social evolution? Yet many of us hope to leave some mark on society, perhaps to change it for the better, to ease some suffering, to increase productivity, to fight racism and ignorance . . . what vaulting ambitions! And what a radical change from the worldview of our ancestors! The idea that people can shape their society—or even enjoy adequate shelter and ample meals—is widely accepted today. For most of human history, however, mere survival was the primary motivator of human action, and thus a fatalistic acceptance of human frailty in the face of overwhelming natural forces was the dominant worldview.

For most of the first million years of human evolution, human societies were developing from those of primates. Populations were small because humans, like other primates, lived on wild animals and plants. These sources of food are easily used up and their supply fluctuates greatly; as a result, periods of starvation or gnawing hunger might alternate with bouts of gorging on sudden windfalls of game or berries. Thus, the hunting-and-gathering life that characterized the earliest human populations could support only extremely small societies; most human societies, therefore, had no more than about sixty members.

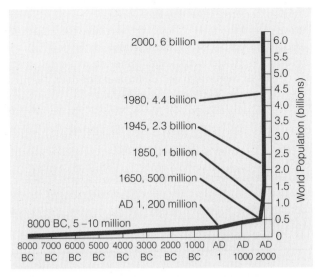

Figure 4.2 World Population Growth from 8000 B.C.E. to 2000 C.E. This chart shows world population growth since 8000 B.C.E. Note that the first, more gradual increase in world population is associated with the shift to agriculture but that its effects are dwarfed by the impact of the Industrial Revolution, which began in the eighteenth century. Source: Data from Office of Technology Assessment.

Archaeological evidence indicates that some hunting-and-gathering societies began to develop permanent settlements long before the advent of agriculture. The emergence of farming was one of the changes that accelerated human social evolution, but in parts of what is now Europe and the Middle East, there were stable settlements of hunter-gatherers as early as 30,000 to 20,000 years ago. These rather large and complex societies were most firmly established in the area that is now Israel at the end of the last ice age, some 13,000 to 12,000 years ago (Albright, 2006; Henry, 1989).

Despite the slow pace of human evolution until about 35,000 years ago, some astonishing physical and social changes occurred during that long period, changes that enabled human life to take the forms it does today. Among them were the following:

- The development of an upright posture (which freed the hands for eventual use of tools) and an enlarged cerebral brain, making possible vastly increased cognitive abilities and the development of language.
- Social control of sexuality through the development of the family and other kinship structures and enforcement of the incest taboo (see Chapter 3).
- The establishment of the band of hunter-gatherers as the basic territorial unit of human society, coupled with the development of kinship structures that linked bands together into tribes. Within the

band, the family became the primary economic unit, organizing the production and distribution of food and other necessities.

The chart in the Research Methods box illustrates how kinship ties link hunting-and-gathering bands together into a larger society. Anthropologists and sociologists often represent social structures with diagrams of this type. They show how people are related to one another and what their status is in relation to any other person in the social structure.

By the end of the last ice age, many aspects of this evolutionary process were more or less complete. Human societies had fully developed languages and a social structure based on the family and the band. To be cast out of the band for some wrongdoing—that is, to be considered a deviant person (see Chapter 8)—usually meant total banishment from the society and eventual death, either by starvation or as a result of aggression by members of another society (Salisbury, 1962). But warfare and violence were not typical of early human societies, as social anthropologist Marshall Sahlins has pointed out:

> Warfare is limited among hunters and gatherers. Indeed, many are reported to find the idea of war incomprehensible. A massive military effort would be difficult to sustain for technical and logistic reasons. But war is even further inhibited by the spread of a social relation—kinship—which in primitive society is often a synonym for "peace." Thomas Hobbes's famous fantasy of a war of "all against all" in the natural state could not be further from the truth. War increases in intensity, bloodiness, duration, and significance for social survival through the evolution of culture, reaching its culmination in modern civilization. (1960, p. 82)

On the other hand, one must not romanticize the hunter-gatherers. Their lives were far more subject to the pressures of adaptation to the natural environment than has been true in any subsequent form of society. Individual survival was usually subordinated to that of the group. If there were too many children to feed, some were killed or left to die; when the old became infirm or weak, they often chose death so as not to diminish the chances of the others. Thus, the frail Eskimo grandfather or grandmother wandered off into the snowy night to "meet the polar bear and the Great Spirit." Additional thousands of years of social evolution would pass before the idea that every person could and should survive, prosper, and die with dignity would even occur to our ancestors.

The Transition to Agriculture

For some time before the advent of plow-and-harvest agriculture in the Middle East and the Far East, hunting-and-gathering societies were supplementing their diets with foods acquired through the domestication of plants and animals. In this way, they were able to avoid the alternation of periods of feast and famine caused by reliance on animal prey. Karl Marx was the first social theorist to observe that social revolutions like the shift to agriculture or industrial production are never merely the result of technological innovations such as the plow or the steam engine. The origins of new forms of society are to be found within the old ones. New social orders do not simply burst upon the scene but are created out of the problems faced by the old order. Thus, as they experimented with domesticating animals and planting crops, some hunting-and-gathering societies were evolving into nomadic shepherding or **pastoral societies** in which bands followed flocks of animals. Others were developing into **horticultural**

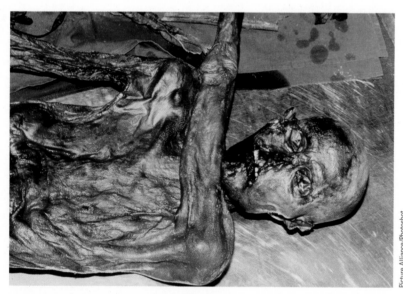

The "Ice Man," discovered in 1991 by hikers in the Italian Alps, is shown here. These remains, among the oldest samples of human flesh and organs ever discovered, are providing a wealth of knowledge about daily life in hunting-and-gathering societies 5,000 years ago.

pastoral society: A society whose primary means of subsistence is herding animals and moving with them over a wide expanse of grazing land.

horticultural society: A society whose primary means of subsistence is raising crops, which it plants and cultivates, often developing an extensive system for watering the crops.

societies in which the women raised seed crops and the men combed the territory for game and fish (Hicks & Beaudry, 2006).

Regarding this momentous change in the material basis of human survival, the historian William McNeill made the following observation:

> The seed-bearing grasses ancestral to modern cultivated grains probably grew wild eight or nine thousand years ago in the hill country between Anatolia and the Zagros Mountains, as varieties of wheat and barley continue to do today. If so, we can imagine that from time immemorial the women of those regions searched out patches of wheat and barley grasses when the seeds were ripe and gathered the wild harvest by hand or with the help of simple cutting tools. Such women may gradually have discovered methods for assisting the growth of grain, e.g., by pulling out competing plants; and it is likely that primitive sickles were invented to speed the harvest long before agriculture in the stricter sense came to be practiced. (1963, p. 12)

As a result of these and other innovations, agriculture became the productive basis of human societies. Pastoral societies spread quickly throughout the uplands and grasslands of Africa, northern Asia, Europe, and the Western Hemisphere, and grain-producing societies arose in the fertile river valleys of Mesopotamia, India, China, and—somewhat later—Central and South America (see Figure 4.3). Mixed societies of shepherds and marginal farmers wandered over the lands between the upland pastures and the lowland farms.

The First Large-Scale Societies We have reached the beginning of recorded history (around 4000 B.C.E.), which was marked by the rise of the ancient civilizations of Sumer, Babylonia, China and Japan, Benin, and the Maya, Incas, and Aztecs. The detailed study of these societies is the province of archaeology, history, and classics. But sociologists need to know as much as possible about the earliest large-scale societies because many contemporary social institutions (e.g., government and religion) and most areas of severe social conflict (e.g., class and ethnic conflict) developed sometime in the agrarian epoch—between 3000 B.C.E. and 1600 C.E.

From the standpoint of social evolution, the following dimensions of agrarian societies are the most salient (Braidwood, 1967):

- Agriculture allows humans to escape from dependence on food sources over which they have no control. Agrarian societies

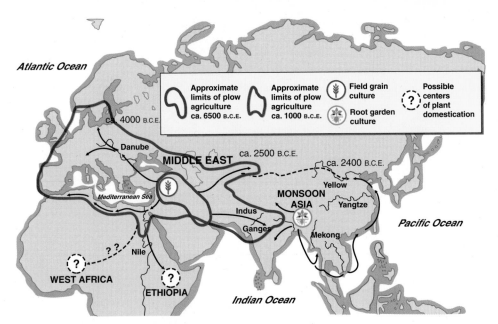

Atlantic Ocean

Approximate limits of plow agriculture ca. 6500 B.C.E.

Approximate limits of plow agriculture ca. 1000 B.C.E.

Field grain culture

Root garden culture

Possible centers of plant domestication

ca. 4000 B.C.E.

Danube

MIDDLE EAST

ca. 2500 B.C.E.

ca. 2400 B.C.E.

Yellow

MONSOON ASIA

Yangtze

Mediterranean Sea

Indus

Ganges

Mekong

Pacific Ocean

Nile

WEST AFRICA

ETHIOPIA

Indian Ocean

Figure 4.3 Agricultural Origins.

Source: From *The rise of the West: A history of the human community* by William McNeill. © 1963 University of Chicago Press. Used with permission.

produce surpluses that do not merely *permit* but *require* new classes of people who are not food producers to exist. An example is the class of warriors, who defend the surplus or add to it through plunder.

- Agriculture requires an ever-larger supply of land, resulting in conflicts over territory and in wars with other agricultural or pastoral societies.
- The need to store and defend food surpluses and to house the nonagrarian classes results in new territorial units: villages and small cities.

New Social Structures Freed from direct dependence on undomesticated species of plants and animals, agrarian societies developed far more complex social structures than were possible in simpler societies. Hunting-and-gathering societies divided labor primarily according to age and sex, but labor in agricultural societies was divided in new ways to perform more specialized tasks. Peoples who had been conquered in war might be enslaved and assigned the most difficult or least desirable work. Priests controlled the society's religious life, and from the priestly class emerged a class of hereditary rulers who, as the society became larger and more complex, assumed the status of pharaoh or emperor. Artisans with special skills—in the making of armaments or buildings, for example—usually formed another class. And far more numerous than any of these classes were the tillers of the soil, the "common" agricultural workers and their dependents.

The process whereby the members of a society are sorted into different statuses and classes based on differences in wealth, power, and prestige is called **social stratification** and is described in detail in Chapter 10. In general, societies may be **open**, so that a person who was not born into a particular status may gain entry to that status, or **closed**, with each status accessible only by birth. A key characteristic of agrarian societies is that their stratification systems became extremely closed and rigid. Because most people were needed in the fields, people had few opportunities to move from one level of society to another. These, therefore, were closed societies.

The emergence of agrarian societies was based largely on the development of new, more efficient production technologies. Of these, the plow and irrigation were among the most important. The ancient agrarian empires of Egypt, Rome, and China are examples of societies in which irrigation made possible the production of large food surpluses, which in turn permitted the emergence of central governments led by pharaohs and emperors, priests and soldiers. Indeed, some sociologists have argued that because large-scale irrigation systems required a great deal of coordination, their development led to the evolution of imperial courts and such institutions as slavery, which coerced large numbers of agrarian workers into forced labor (Wittfogel, 1957).

By the fifteenth century, the world's population was organized into a wide variety of societies with many different types of cultures and

social stratification: The process whereby the members of a society are sorted into different statuses.

open society: A society in which social mobility is possible for everyone.

closed society: A society in which social mobility does not exist.

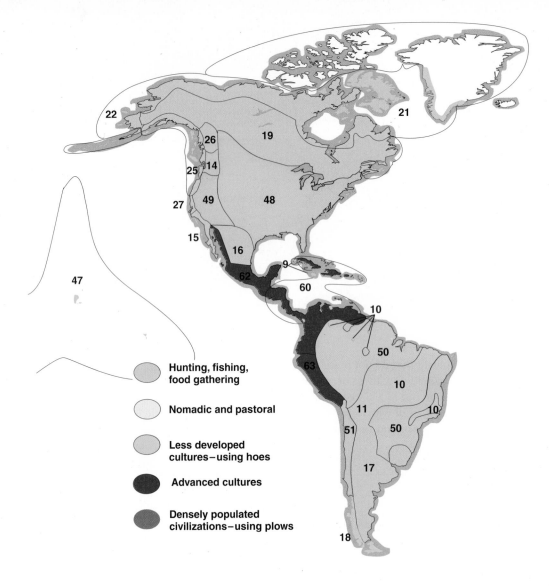

Figure 4.4 Civilization, "Cultures," and Primitive Peoples, ca. 1500. Source: Adapted from Hewes, 1954; Braudel, 1981.

Legend:

- Hunting, fishing, food gathering
- Nomadic and pastoral
- Less developed cultures—using hoes
- Advanced cultures
- Densely populated civilizations—using plows

1. Tasmanians
2. Congo Pygmies
3. The Vedda (Ceylon)
4. Andamanese
5. Sakai and Semang
6. Kubu
7. Punan (Borneo)
8. Negritos of the Philippines
9. Ciboneys (Antilles)
10. Gê-Botocudos
11. Gran Chaco Indians
12. Bushmen
13. Australians
14. Great Basin

15. Lower California
16. Texas and northeastern Mexico
17. Patagonia
18. Indians of the southern coast of Chile
19. Athabascans and Algonkin (northern Canada)
20. Yukaghir
21. Eastern and central Eskimos
22. Western Eskimos
23. Kamchadal, Koryak, Chukchi
24. Ainu, Gilyak, Gol'dy
25. Northwest coast Indians (United States and Canada)
26. Columbia Plateau

27. Central California
28. Reindeer-herding peoples
29. Canary Islands
30. Sahara nomads
31. Arabian nomads
32. Pastoral mountain peoples in the Near East
33. Pastoral peoples of the Pamir region and Hindu Kush
34. Kazakh-Kirghiz
35. Mongols
36. Pastoral Tibetans
37. Settled Tibetans
38. Western Sudanese
39. Eastern Sudanese

civilizations. Figure 4.4 depicts the spread of human life over almost 58 million square miles (more than 150 million square kilometers) of land in 1500, before the age of exploration and the Industrial Revolution. The seventy-six cultures and civilizations shown include hundreds of different societies, ranging from simple hunting-and-gathering societies to more advanced agrarian ones to civilizations that were already expanding and exporting their culture through conquest and trade. The areas numbered 1 to 27 on the map were inhabited by simple societies of hunters, gatherers, and fishers; those numbered 28 to 44 were inhabited by nomads and stockbreeders; areas 45 to 60 were inhabited by agricultural peoples, primarily peasants using hoes rather than plows; and the remaining areas were more advanced civilizations. With

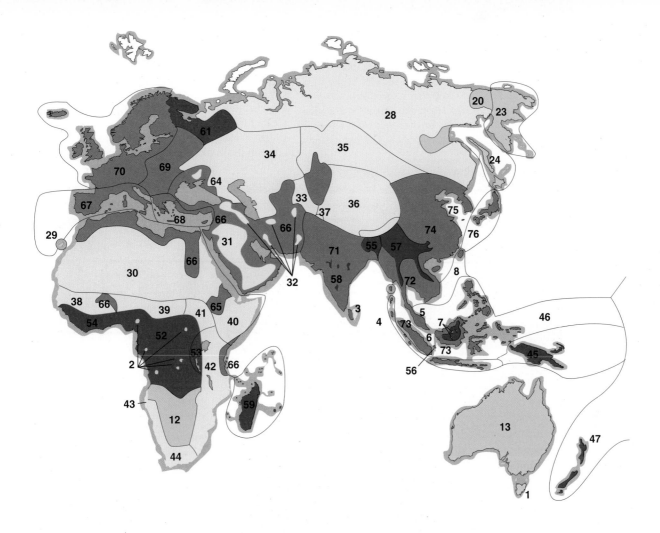

40. Somali and Galla of northeastern Africa
41. Nilotic tribes
42. East African stock-rearing peoples
43. Western Bantu
44. Hottentots
45. Melanesian Papuans
46. Micronesians
47. Polynesians
48. American Indians (eastern United States)
49. American Indians (western United States)
50. Brazilian Indians
51. Chilean Indians
52. Congolese peoples

53. Lake-dwellers of East Africa
54. Guinea coasts
55. Tribes of the Assam and the Burmese highlands
56. Tribes of the Indonesian highlands
57. Highland peoples of Indochina
58. Mountain and forest tribes of central India
59. Malagasy
60. Caribbean peoples
61. Finns
62. Mexicans, Maya
63. Peru and the Andes
64. Caucasians

65. Ethiopia
66. Settled Muslims
67. Southwestern Europe
68. Eastern Mediterranean
69. Eastern Europe
70. Northwestern Europe
71. India (this map does not differentiate between Muslims and Hindus)
72. Lowlands of Southeast Asia
73. Indonesian lowlands
74. Chinese
75. Koreans
76. Japanese

the exception of the Mayan and Incan civilizations, all of the latter were found in Asia, the Middle East, and parts of Europe.

The areas where the earliest civilizations emerged are to this day the most densely inhabited parts of the world. In 1500, these civilizations were marked by their ability to produce food surpluses through the use of the plow, domesticated animals, wheels, and carts. Above all, they are marked by the importance of towns and cities as centers of administration and religious practice. Although the Mayan and Incan civilizations did not have the wheel or the plow, their advanced systems of astronomy, art, and writing make a strong argument for their inclusion as world civilizations.

The explorations led by Portugal, Spain, and England in the sixteenth and seventeenth

The Valley of Mexico This detail from a famous mural in Mexico's Palacio Nacional shows what the Valley of Mexico looked like in 1519, when conquistador Hernán Cortés and a small band of Spanish soldiers began their conquest of the strife-weakened Aztec Empire. The empire's social structure was that of an agrarian society in which great effort was devoted to organizing irrigation projects and building temples and tombs. Power resided in the hands of the emperor Montezuma and a priestly caste that ruled over extended families of farmers and warriors. The Aztec capital, Tenochtitlan, was carefully planned to allow its inhabitants to enjoy the temperate climate of the high valley and was served by a sophisticated irrigation system (Iglesia, 1990).

Today, less than 500 years after the conquest, the Valley of Mexico is blanketed by Mexico City and the many urban communities surrounding it. The social structure of Mexico City is extremely complex because the city is both Mexico's largest commercial center and its national capital. The city is surrounded by ancient volcanoes, and the valley floor is prone to earthquakes. Despite the problems created by rapid growth in an environmentally sensitive area, Mexico City continues to grow and is now one of the largest metropolitan centers in the world.

© Nicholas Sapieha/Art Resource, NY

© John Neubauer/PhotoEdit

centuries resulted in conquest and European settlement in many parts of the Western Hemisphere and Africa. Conquest caused radical disruptions in the internal development of the world's cultures and civilizations. But even more drastic would be the changes caused by the Industrial Revolution of the eighteenth and nineteenth centuries.

The Industrial Revolution

In 1650, when Holland, Spain, and England were the world's principal trading nations (Block, 1990), the population of England was approximately 10 million, of which about 90 percent earned a livelihood through farming of one kind or another. Just 200 years later, in 1850, the English population had soared to more than 30 million, with less than 20 percent at work in fields, barns, and granaries. England had become the world's first industrial society and the center of an empire that spanned the world.

A similar transformation occurred in the United States. In 1860, on the eve of the Civil War, there were about 30 million people in the United States. Ninety percent of that population, a people considered quite backward by the rapidly industrializing English, were farmers or people who worked in occupations directly related to farming. A mere 100 years later, in 1960, only about 8 percent of Americans were farmers or agricultural workers, yet they were able to produce enough to feed a population exceeding 200 million. Today, less than 3 percent of Americans work on farms and ranches. These dramatic changes in England and America—and in other nations as well—occurred as a result of the Industrial Revolution.

The Industrial Revolution is often associated with innovations in energy production, especially the steam engine. But the shift from an agrarian to an industrial society did not happen simply as a result of technological advances. Rather, the Industrial Revolution was made possible by the rise of a new social order known as **capitalism**. This new way of organizing production originated in nation-states, which engaged in international trade, exploration, and warfare. Above all, the Industrial Revolution depended on the development of *markets*, social structures that would function to regulate the supply of and demand for goods and services throughout the world (Polanyi, 1944).

The transition from an agrarian to an industrial society affects every aspect of social life. It changes the structure of society in several ways, of which the following are among the most significant:

- The industrialization of agriculture allows many more people to be supported by each agrarian worker than ever before. Only a relatively small number of people live on the land; increasing numbers live in towns, cities, and suburbs.

capitalism: A system for organizing the production of goods and services that is based on markets, private property, and the business firm or company.

- Industrial societies are generally far more receptive to social change than agrarian societies. One result is the emergence of new classes like industrial workers and scientific professionals (engineers, technicians) and new social movements like the women's movement and the movement for racial equality.
- Scientific, technical, and productive institutions produce both unparalleled wealth and unparalleled destructive capacities.
- The world "shrinks" as a result of innovations in transportation and communication. A "global society" develops, but at the same time, the unevenness of industrialization leads to conflicts that threaten world peace.

These are only some of the important features of industrial societies. There are many others, which we analyze throughout this book. We will present research that shows how the types of jobs people find and the ways they spend their leisure time change as technology advances and their expectations change. We will also look into how other major institutions—especially the family—adapt to the changes created by the Industrial Revolution. In fact, we will continually return to the Industrial Revolution as a major force in our times.

The Theory of Postindustrial Society The United States and most other societies that experienced industrialization in earlier periods are now considered by sociologists to be "post-industrial societies" (Bell, 1999/1973). Even though these societies continue to be important producers of industrial goods like steel, glass, aluminum, and many other products, their industrial plants are highly automated and employment in manufacturing is no longer the largest sector of the economy. As sociologist Fred Block (1990) observes:

> [Ours is] a strange period in the history of the United States because people lack a shared understanding of the kind of society in which they live. For generations, the United States was considered as an industrial society, but that definition is no longer compelling. Yet no convincing alternative has emerged in its absence. (p. 1)

In this book, we continue to group the United States and the European nations among the world's industrial societies because they are major producers of manufactured goods. However, it should be noted that the businesses that produce those goods are dwindling in number because of the effects of the Industrial Revolution in other areas of the world. Also, for the first time in U.S. history, the number of jobs in services exceeds that of jobs in goods-producing industries. Far more jobs are created in restaurants and hotels each year than in steel mills and other kinds of factories (Sweet & Meiksins, 2008). In addition, there is an ever-increasing demand for highly educated and trained workers in new industries that deal with high-speed transmission of information. Computers and telecommunications technologies are surely produced by "industries," but they are not the type of mass-employment "smokestack" industries that were the hallmark of earlier stages of the Industrial Revolution.

The theory of postindustrial society explores why these fundamental changes have come about and their impact, not only on postindustrial societies themselves but also on the cultures of other societies. There are also many questions about what will happen as the world's societies become ever more interdependent and decisions made in metropolitan centers like Tokyo, Los Angeles, and London affect the lives of villagers in the most remote parts of the world. We will return to the issues raised by theories of industrialization and the emergence of postindustrial society throughout this book. For it is clear that while the most advanced and powerful nations are undergoing an economic transformation, the Industrial Revolution is in full swing in many other parts of the world, and in still other, more remote areas, older forms of agrarianism and tribalism continue to thrive.

The evolution of human societies continues at an ever more rapid pace. As we will see throughout this text, scientific and technological discoveries, wars, migration and immigration, disease, environmental change, and major social movements continually produce new social structures and ways of addressing social needs. Some of the major aspects of social evolution over the past 12,000 years are summarized in the For Review chart on page 88, which highlights important features of each type of society, from those of the hunter-gatherers to our own postindustrial society.

SOCIETY AND THE INDIVIDUAL

From Gemeinschaft to Gesellschaft

Imagine that you grew up in an agrarian village or a town in a slowly industrializing society and then came to live in the United States, as

Types of Societies

Type of Society	Historical Period	Energy/ Technology	Populations Sustained	Settlement Pattern	Social Organization	Examples
Hunting and gathering	Only type until about 12,000 years ago; a few examples remain but are threatened with absorption or extinction	Fire; crude weapons	Bands of 25 to 40 people	Nomadic	Family and clan centered; division of labor based on age and sex; little social equality	Eskimo; Pygmies of Central Africa; Aborigines of Australia
Horticultural and pastoral	From about 12,000 years ago, with rapidly decreasing numbers after about 4000 B.C.	Fire; hand tools for planting	Settlements of a few hundred, linked through clans and tribes to form societies with several thousand members	Horticulturists form small, relatively permanent settlements; pastoralists are nomadic, following herds of migrating animals (e.g., reindeer, caribou, buffalo) and engaging in some animal domestication	Family centered; religious systems encourage development of some specialized roles and statuses; emergence of social inequality based on prestige and wealth	Societies of the Fertile Crescent (now Iraq); Laplanders; Masai
Agrarian	From about 7,000 years ago; large but decreasing numbers today	Fire; animal power for plowing; extensive irrigation systems	Millions of people	Cities become common, but the vast majority of people live on the agricultural lands outside them	Family remains strong but loses some power as extensive religious, political, and economic institutions emerge; extensive division of labor and increased social inequality based on power, wealth, and religious prestige	Egypt under the pharaohs; Medieval Europe; ancient China and India
Industrial (manufacturing)	From about 1750 to present	Steam, electricity, gasoline power; railroads and auto/truck transport	Millions of people	Cities; agriculture is mechanized, and manufacturing absorbs displaced workers	Distinct and specialized religious, political, economic, educational, and family institutions; extensive division of labor and interdependence; social inequality persists but is subject to social reform movements and revolutions	China; Brazil; Eastern European nations; Argentina; Philippines; South Korea
Postindustrial	From about 1960 to present	Electricity, gasoline power, nuclear energy; information technologies (computers, satellites)	Millions of people; rates of high immigration	Cities and suburbs, forming sprawling metropolitan regions	Distinct religious, political, economic, educational, and family institutions; service industries replace manufacturing although industrial production continues	United States; most nations of Western Europe; Japan

Source: Adapted from Lenski & Lenski, 1982; Bell, 1999/1973.

millions of people do each decade. A difficult part of that experience would be getting used to the impersonality of modern American society when compared with the close relationships you had with people you had known all your life in your native society. American society would seem to be composed of masses of strangers organized into highly impersonal categories. You would have to get used to being a shopper, an applicant, a depositor, a sports fan, a commuter, and so on, and it would be necessary to shift from one to another of these roles several times in a day or even within a half hour.

Sociologists often describe this experience as a change from **gemeinschaft** (the close, personal relationships of small groups and communities) to **gesellschaft** (the well-organized but impersonal relationships of modern societies). These are German terms taken from the writings of the social theorist Ferdinand Tönnies (1957/1887). Complex industrial societies, Tönnies argued, have developed gesellschaft social structures like factories and office bureaucracies to such a degree that they tend to dominate day-to-day life in the modern world.

Primary and Secondary Groups The American sociologist Charles H. Cooley applied this distinction to the micro level of society in his discussion of primary versus secondary groups. "By **primary groups**," Cooley wrote,

> I mean those characterized by intimate face-to-face association and cooperation. They are primary in several senses, but chiefly in that they are fundamental in forming the social nature and ideals of the individual. [Such a group] involves the sort of sympathy and mutual identification for which "we" is the natural expression. (1909, p. 23)

Secondary groups, in contrast, are groups in which we participate for instrumental reasons—that is, in order to accomplish some task or set of tasks. Examples are school classrooms, town committees, and political party organizations. We do not have intimate relationships with other members of secondary groups; at least, we do not normally expect to do so and are not concerned about the fact that our relationships in such groups involve only a limited range of emotions.

Of course, when we speak of intimacy in primary groups, we do not always refer to love and warm feelings. The intimacy of the members of a family refers to a highly charged set of emotions that can range from love to hate but rarely includes indifference. But in a secondary group, indifference, or lack of personal involvement, is the norm. We therefore do not usually develop strong feelings toward all the other members of the organizations for which we work or the associations to which we belong, even though we may have one or more close friends among our workmates. The idea that people do not have to form primary attachments in work organizations also protects people with less power from those with more. Women in subordinate positions, for example, often need protection from men in more powerful positions who make sexual advances. Modern laws dealing with sexual conduct in the workplace classify such behavior as illegal harassment. But the fact that we need such laws points out the degree of conflict we often experience in performing our roles.

Role and Status in Modern Societies

Roles in Conflict Gesellschaft forms of social organization are more complex than gemeinschaft forms in terms of the number of statuses people hold and, thus, the number of roles they must perform. One result of this greater complexity is that roles in secondary groups or associations often conflict with roles in primary groups like the family. Much of the stress of life in modern societies is caused by the anxiety we experience as we attempt to balance the demands of our various roles. This anxiety is captured in the terms *role conflict* and *role strain.*

Role conflict occurs when, in order to perform one role well, a person must violate another important role. Parents who are also employees may experience this kind of conflict when their supervisors ask them to put in extra time, which cuts into the time they are able to spend with their children. The late Felicia Schwartz, a leader in research on women, work, and social policy, argued that many corporations informally classify their female employees into two categories: those who seem to make their career their primary life concern, and those who seek to balance career and family obligations. Often they reward the former and subtly discriminate against the latter by placing them

gemeinschaft: A term used to refer to the close, personal relationships of small groups and communities.

gesellschaft: A term used to refer to the well-organized but impersonal relationships among the members of modern societies.

primary group: A small group characterized by intimate, face-to-face associations.

secondary group: A social group whose members have a shared goal or purpose but are not bound together by strong emotional ties.

role conflict: Conflict that occurs when a person must violate the expectations associated with one role in order to perform another role well.

global social change

Women in the Sciences: The Harvard Controversy

Some of the world's most important scientific discoveries have been made by women. One need only think of Marie Curie's discovery of radium and radioactivity, for which she won two Nobel Prizes. Jane Goodall, whose work we discussed in Chapter 3, is widely recognized as one of the world's foremost specialists in primate behavior, to name only two well-known examples.

Lawrence H. Summers, former president of Harvard University, now director of the National Economic Council.

No wonder it was so infuriating for many women in science, when then-Harvard University President Lawrence Summers said in a public speech that genetic differences between men and women might be one reason that fewer women succeed in science and math careers. He also questioned the importance of discrimination in the relative lack of female professors in science and engineering at elite universities. Because Harvard is such an important institution of research and higher learning, one that attracts students, scientists, and scholars from throughout the world, President Summers's remarks were widely quoted. They also drew immediate opposition from prominent women on the Harvard faculty, and especially from two sociologists, Mary Waters and Theda Skocpol, both senior professors and faculty leaders. Professor Waters became the leader of an opposition movement that eventually influenced the university's board of directors to accept President Summers's resignation after more than a year of bitter campus and national controversy.

Why were these and other Harvard sociologists so directly involved in the controversy? Waters, chair of the sociology department, said that students who were upset about Summers's remarks had been coming to talk to her. She said his comments left her speechless. "Has anyone asked if he thinks this about African Americans, because they are underrepresented at this university? Are Hispanics inferior?

Mary C. Waters is the M. E. Zukerman Professor of Sociology at Harvard University. She specializes in the study of immigration, intergroup relations, the formation of racial and ethnic identity among the children of immigrants, and the challenges of measuring race and ethnicity.

Are Asians superior?" she said. "That's the road he's going down and I don't want to see any university go down that road." Waters remained outspoken during the controversy, especially because she felt that Summers had not adequately discussed any of the social obstacles that women in science face, especially the role conflict they experience if they choose to have children while pursuing high-level careers in the sciences. The demands of their roles as laboratory directors, for example, which often require that they work eighty or more hours a week in their labs, conflict with the demands placed on them as wives and mothers. A great university like Harvard has the resources to ease these conflicts by providing day care services, allowing women more maternity leave, and in many ways making their work environments more favorable for women who are attempting to balance domestic and career expectations. But this is not happening, and Waters and other women scientists were infuriated with President Summers for not emphasizing these issues but instead suggesting that they were genetically inferior to men (Bombardieri, 2005). They were also embarrassed for their female international students, who often face even greater obstacles to pursuing careers in science than do their peers in North America.

After the Harvard Faculty of Arts and Sciences issued a vote of no confidence, the embattled president stepped down. He was soon replaced with a new president, Drew Faust, a prominent scholar who has dedicated her administration to addressing, among other issues, the situation of women at Harvard and other research universities throughout the world (Bombardieri, 2005).

Three thousand foreign students (approximately 18 percent of the total student population) come to Harvard each year. An additional 2,000 foreign scholars also share the Harvard campus with their American peers.

on what some have called the "mommy track," in which the woman is allowed more flexibility to perform family roles at the cost of promotion to more demanding positions within the company. The 1993 Family and Medical Leave Act attempted to resolve this conflict between work and family roles by requiring corporations to provide options for family leave without penalties; nevertheless, role conflict stemming from the competing demands of family and work remains widespread, especially for working women but for men as well (Cramer & Boyd, 1995; Siegel, 1992). (See the Global Social Change box above.)

Research on how people behave in disasters also illustrates some of the effects of role conflict. For example, a police officer or firefighter who is on duty during a disaster that affects his or her own family may leave an assigned post to see to the safety of loved ones, thereby violating one role in order to fulfill another (Killian, 1952). Throughout the world, efforts are being made to train women in disaster-related skills so that some of the conflicts that male disaster workers experience may be eased through greater sharing of roles (Gramling & Krogman, 1997; Sapir, 1997).

Role strain occurs when people experience conflicting demands in an existing role or cannot meet the expectations of a new one. For example, when she performs the duties of a triage nurse, Lori Jones experiences severe role strain. She has been taught that care must be provided to those in need, but she must balance that role expectation against the fact that care is a scarce resource in a busy hospital and must be rationed (Chambliss, 1996; Goodman-Draper, 1995).

Role strain in the form of anxiety over poor performance is at least as common as role strain caused by conflicting expectations. For example, the unemployed head of a family feels severe stress as a result of inability to provide for the family's needs (Bakke, 1933; Jahoda, 1982; Jahoda, Lazarsfeld, & Zeisel, 1971). The mother of a newborn baby often feels intense anxiety over how well she can care for a helpless infant, a feeling that is heightened if she herself is young and dependent on others (Mayfield, 1984).

Ascribed Versus Achieved Statuses Role conflicts may occur in simpler societies, but they are far more common in societies undergoing rapid change. One reason for the relative lack of role conflict in simpler, more stable societies is that a person's statuses in such societies are likely to be determined by birth or tradition rather than by anything the person achieves through his or her own efforts. These **ascribed statuses** (peasant, aristocrat, slave, and so forth) usually cannot be changed and hence are less likely to be subject to different role expectations. Such statuses are found in industrial societies too (statuses based on race or sex are examples), but they tend to become less important in modern institutions than an **achieved status** like editor, professor, or Nobel Prize winner.

We expect to be able to achieve our occupational status, our marital and family statuses, and other statuses in the community and the larger society. Nevertheless, there is some tension in modern societies between the persistence of ascribed statuses and the ideal of achieved statuses. The empirical study of that tension and of efforts to replace ascribed with achieved statuses (for example, through equality of educational opportunity) is another major area of sociological research. We will show in later chapters that many of the social movements sociologists study arose as different groups in a society (for example, African Americans, wage workers, immigrants, and the poor) organized to press for greater access to economic and social institutions like corporations and universities.

Sociology & Social Justice

THE PROBLEM OF MASTER STATUSES One reason so many groups have had to organize to obtain social justice is related to the way statuses operate in many societies. Although any person may fill a variety of statuses, many people find that one of their statuses is more important than all the others. Such a status, which is termed a **master status**, can have damaging effects (Hughes, 1945). A black man, for example, may be a doctor, a father, and a leader in his church, but he may find that his status as a black American takes precedence over all of those other statuses. The same is often true for women. A woman may be a brilliant scientist and a leader in her community, and fill other statuses as well, but she may find that when she deals with men, her status as a woman is more important than any of the others.

The effects of a master status are also felt by people who have been in prison, by members of various racial and ethnic groups, by Americans overseas (Brownfield, Sorenson, & Thompson, 2001; Smith, 1996), and by members of many other types of groups. For example, recent studies have shown that people who are noticeably overweight or obese often find that their status as a "fat person" is a master status that denies them opportunities to be appreciated for their performance in other statuses, such as student or worker.

Patterns of discrimination and prejudice stemming from the problem of master statuses are not easily corrected. We will see at many points in this book that a major source of social

role strain: Conflict that occurs when the expectations associated with a single role are contradictory.

ascribed status: A position or rank that is assigned to an individual at birth and cannot be changed.

achieved status: A position or rank that is earned through the efforts of the individual.

master status: A status that takes precedence over all of an individual's other statuses.

change is the efforts of some groups to eliminate those patterns. When changes actually occur, however, they often take the form of new national laws, such as laws against racial segregation or gender discrimination. This raises the question of the distinction between societies and nations.

SOCIETIES AND NATION-STATES

For most people in the world today, the social entity that represents society itself is the nation-state. The assumed correspondence between society and nation can be seen in the fact that expressions like "the United States" and "American society" are often used interchangeably. Moreover, most people think of their society in terms of national boundaries; thus, if you were asked to name the society of which you are a member, you would be more likely to say "the United States" than "California" or "Chicago" or "the University of Texas." Yet as we will see shortly, societies and nations are by no means the same thing.

The State

To understand the distinction between society and nation, we need to begin with the concept of the state. In a lecture at Munich University, Max Weber described a state as

> a human community that (successfully) claims the monopoly of the legitimate use of physical force within a given territory.... The right to use physical force is ascribed to other institutions or to individuals only to the extent to which the state permits it. The state is considered the sole source of the "right" to use violence. (quoted in Gerth & Mills, 1958, p. 78)

The **state** thus may be defined as a society's set of political institutions—that is, the groups and organizations that deal with questions of "who gets what, when, and how" (Lasswell, 1936). The **nation-state** is the largest territory within which those institutions can operate without having to face challenges to their sovereignty (their right to govern). Weber was careful to note that the state has a monopoly over the use of force, which it grants to other agencies (for example, state and municipal governments) under certain circumstances. But the state gains

this right—the source of its power to influence the behavior of citizens—from the people themselves, from their belief that it is *legitimate* for the state to have this power. As we will see in later chapters, the concepts of power and legitimacy are essential to understanding the workings of the modern state; indeed, power is a basic concept at all levels of human social behavior. Here, however, we must examine the idea of a state as it operates in the concept of nation-state or, simply, nation.

The Nation-State

"One nation, under God, indivisible, with liberty and justice for all." We can say the words in our sleep, and often we do not bother to think about their meanings. After the tragic events of September 11, 2001, however, people of all ages began paying much greater attention to the significance of the Pledge of Allegiance. Repeating the pledge in schools and elsewhere is a highly significant action: It enhances the legitimacy of the state and thus helps create the nation—in this case, the nation known as the United States of America.

But do all the inhabitants of the United States of America think of themselves as members of one nation? To a large extent they do, and this is one of the greatest strengths of that nation. Yet at various times in American history, certain groups—African Americans, Native Americans, and the Amish, for example—have thought of themselves as separate peoples. And the idea of an "indivisible" nation was fought out in one of the bloodiest wars the world has ever known, the American Civil War.

In the century since the Civil War, the United States has not experienced any real tests of its national solidarity. Countries like Lebanon, Zimbabwe, Afghanistan, Northern Ireland, and the nations that were formerly part of Yugoslavia or the Soviet Union—to name only some—have been far less fortunate. For them, the issue of creating a national identity that can unite peoples who think of themselves as members of different societies remains a burning question.

Between 1945 and 1970, as peoples all over the world adjusted to the breakup of the European colonial empires, more than 100 new states were created. In these new nations, as in many of the nation-states that existed before the 1950s, the correspondence between national identity and society is often problematic, a situation that frequently produces social upheavals. In Nigeria, for example, the stability—indeed, the very existence—of the nation-state is endangered by animosities among the various societies included within its boundaries. Although the

state: A society's set of political structures.

nation-state: The largest territory within which a society's political institutions can operate without having to face challenges to their sovereignty.

mapping social change

Tribal Societies within the Modern State of Nigeria

Nigeria is the largest nation of Africa south of the Sahara desert. It is also a nation where social change is extremely rapid and unpredictable, for tribal and religious differences make it difficult to create a Nigerian identity among citizens, and the influence of Western oil companies seeking to exploit the huge oil reserves in the impoverished coastal region creates opportunities for both development and corruption.

As the detailed map shows, it is composed of many societies with their own cultures and languages. Some, like the Hausa, are spread out over vast territories in the northern savannas; others, like the Tiv and Ibo, are clustered in the more densely populated coastal plains and on the delta of the mighty Niger River. Nigeria was once considered a leader in the development of democratic political institutions. It had a lively cultural and intellectual life and produced some of Africa's best novelists and playwrights. Unfortunately, conflict between its many societies, often over the disposal of oil revenues from the Niger delta region, and rampant political and economic corruption continually threaten the viability of its democratic institutions.

Ken Saro-Wiwa was one of the leaders of the effort to create a modern Nigerian nation even as he fought for the well-being of his own region. He is now considered to be a martyr of Nigeria's democratic hopes. Despite protests throughout the world, this celebrated writer, environmental activist, and political dissident was executed by the Nigerian dictatorship in 1995. Saro-Wiwa was the founder and leader of the movement for the survival of the Ogoni people in Nigeria's oil-rich Niger delta region. Although the government and the Dutch Shell Corporation have taken billions of dollars' worth of oil from their lands, the Ogoni, a small tribal society in the Cross River region, still live in mud huts and dig for yams with bamboo sticks. Acid rain and oil pollution have destroyed many productive farms. Saro-Wiwa was attempting to organize a peaceful social movement to protest these conditions. His arrest and execution are an example of how environmental issues increasingly become confounded with intergroup politics in this land of tribal conflict.

In a speech about the heroism of Saro-Wiwa, Nigeria's Nobel laureate, the playwright Wole Soyinka, said that since his death "the state of ethnic consciousness has assumed a violent dimension." Indeed, the situation in the oil-rich Niger delta has deteriorated. Global oil companies are worried about continued oil production in the region, and guerilla actions against the oil companies continue. With the major Western powers preoccupied with problems in Iraq and the Middle East, there are limited resources to devote to solving ethnic tensions and problems of poverty in this strategic region of the world.

Source: Adapted from Kwamena-Poh, Tosh, Waller, & Tidy, 1982; Crowder, 1966.

nation has rich oil resources and is the largest of the sub-Saharan nations, its per capita income fell to $250 in 1993 from $1,000 in 1980. This decline is largely caused by conflicts among the major tribal groups and the failure of the nation-state's leaders to overcome those conflicts. Although there are scores of tribal and ethnic groups in the nation's population, the largest are the Hausa (21 percent), the Yoruba (20 percent), the Ibo (17 percent), and the Fulani (9 percent), each of which is afraid that the others might gain power at their expense (Olojede, 1995). The political instability of the nation is worsened by the ever-present threat of military dictatorship, which is to some degree a consequence of these ethnic divisions. (See the Mapping Social Change box on page 93.)

South Africa provides another example of differing levels of national and social identity. In 1994, the minority white government ended its policy of apartheid (racial separation) and yielded political power to the black majority under the leadership of Nelson Mandela and the formerly banned African National Congress (ANC). But black South Africans themselves often experience conflicts derived from differing tribal identities. The larger tribal populations, such as the Zulus, have their own kings and political leaders. Some are afraid that the incorporation of their people into a modern nation-state will cause them to lose their identity and rights as Zulus. As a result, there have been numerous episodes of conflict and violence between Zulu separatists and the members of the ANC.

The lack of a clear match between society and nation can be seen in the case of entire groups who think of themselves as "a people" (for example, the Ibo, the Jews, Native Americans, the Kosovars) as well as in smaller groups. You and your friends are part of "American society," by which we mean the populations and social structures found within the territory claimed by the nation-state known as the United States of America. But you are also part of a local community with smaller structures and more gemeinschaft, or primary, relationships. Indeed, this community level of society may have greater meaning for you in your daily life than society at the national level.

SUMMARY

How are societies organized in ways that help establish social order?

- A *society* is a population of people or other social animals that is organized in a cooperative manner to carry out the major functions of life.

- *Social structure* refers to the recurring patterns of behavior that create relationships among individuals and groups within a society.

- The building blocks of human societies are *groups*—collections of people who interact on the basis of shared expectations regarding one another's behavior. In every group, there are socially defined positions known as *statuses*. The way a society defines how an individual is to behave in a particular status is called a *role*.

- A social *institution* is a more or less stable structure of statuses and roles devoted to meeting the basic needs of people in a society. Within any given institution are norms that specify how people in various statuses are to perform their roles. New

institutions continually emerge through the process of *differentiation*.

How do populations differ from societies, and what factors explain the rapid growth of the human population since the 1700s?

- The growth of the world's population is directly related to the evolution of human social structures, which in turn is related to changes in production technologies.

- The first million years of human social evolution were characterized by a hunting-and-gathering way of life. During that time, the family and other kinship structures evolved, and the band became the basic territorial unit of human society.

- The shift to agriculture is commonly linked with the invention of the plow, but for centuries, human societies had been acquiring food through the domestication of plants and animals. Some became *pastoral societies* based on the herding of animals; others evolved into *horticultural societies* based on the raising of seed crops. However, the first large-scale agrarian societies evolved after the development of plow-and-harvest agriculture.

- Agrarian societies allow people to escape from dependence on food sources over which they have no control. In such societies, people produce surpluses that can be used to feed new classes of people who are not food producers, such as warriors. At the same time, these societies require increasing amounts of land, which may lead to conflicts over territory. The need to store and defend food supplies and to house larger numbers of people results in the growth of villages and small cities.

- The next major change in human production technologies was the Industrial Revolution, the shift from agriculture to trade and industry. This began in England around 1650, and spread to the United States and other nations during the next two centuries. Its impetus came not only from technological advances but also from the rise of a new social order: *capitalism.*

- The shift to industrial production affects social structure in several major ways. As a consequence of the industrialization of agriculture, relatively few people live on the land, whereas increasing numbers live in cities and suburbs. Greater openness to change results in the emergence of new classes and social movements. Scientific and technical advances produce tremendous wealth, and the world "shrinks" as a result of innovations in transportation and communication.

How does social organization influence your life and that of other individuals?

- For the individual member of a human society, adaptation to a more modern society involves a shift from *gemeinschaft* (close, personal relationships) to *gesellschaft* (well-organized but impersonal relationships).

- In modern societies, *primary groups* like the family are supplemented, if not replaced, by *secondary groups* (organizations or associations) whose members do not have strong feelings for one another.

- Roles in secondary groups often conflict with roles in primary groups, a situation known as *role conflict. Role strain* occurs when a person experiences conflicting demands within a single role.

- Another difference between simpler and more advanced societies is that in the former, almost all statuses are *ascribed* (determined by birth or tradition), whereas in the latter there is a tendency to replace ascribed statuses with ones that are *achieved* (determined by a person's own efforts).

- Sometimes a particular status takes precedence over all of an individual's other statuses; such a status is referred to as a *master status.*

How does the existence of multiple societies in many nation-states threaten the stability of those nations?

- When they think of a society, most people in the world today think in terms of the nation-state, or nation, of which they are members. (A *state* is a society's set of political structures, and a *nation-state* is the territory within which those structures operate.) But although the members of a society often think of themselves as members of a particular nation, this is not always so, and in extreme cases, the lack of a clear match between society and nation can result in a civil war.

The Kornblum Companion Website

www.cengagebrain.com

Supplement your review of this chapter by going to the companion website to take one of the tutorial quizzes, use the flash cards to master key terms, and check out the many other study aids you'll find there. You'll also find special features, such as GSS data and Web links that will put data and resources at your fingertips to help you with that special project or to do some research on your own.

Photosindia/Getty Images

Socialization

FOCUS QUESTIONS

What are the major experiences that account for the way an individual becomes a "social being"?

How much of the way people behave is determined by social learning and how much by genetic factors?

When sociologists study how people "construct a social self," does this mean they believe that only socialization influences human development?

In what ways do social environments influence the people we become as we mature in our social worlds?

How does learning to take social roles occur throughout a person's lifetime?

What do sociologists mean by gender identity, and in what ways can it be said to be socialized?

When a massive earthquake struck the island nation of Haiti in 2010, hundreds of thousands of children lost their parents. Because there were already more than 250,000 orphans in this poorest nation in the Western Hemisphere before the quake, the resulting population of children without parents became proportionally the largest in the world. But Haiti is not alone in attempting to provide care for an enormous number of orphaned children. AIDS in many African nations, and wars in Iraq, Afghanistan, the Congo, and the Balkans, have left thousands of children without one or more parents. According to the United Nations:

In the last decade alone, some 2 million of the world's youngest have died as a direct result of armed conflict. Millions of children have been seriously injured or permanently disabled. Even greater numbers have died as a result of malnutrition and disease, and more than 300,000 children under 18 years of age have been ruthlessly exploited as soldiers in government armed forces or armed opposition groups in ongoing conflicts. Increasingly, children are specifically targeted by combatants or abducted for use in forced labor or as sexual slaves. These shocking facts cannot begin to describe the damage done to an individual child who has survived such brutality. (United Nations, 2006)

Yet it is also true that humans, even the very young, are extremely resilient. They can recover from trauma, although it will take time and the help of others. Over our lifetimes we can change our sense of who we are and what we are capable of. These aspects of what we call our identity depend, as we will see in this chapter, on what we learn from parents, teachers, and mentors of all kinds. Children who have experienced war and disasters will always be marked by the experience, but in time, and with the help of caring adults and peers, they can resume more normal lives and remake their identities (Singer, 2006).

Parents throughout the world seek safe environments for their children to grow up in. We want safe schools and safe communities. We want our children to be well nourished and taught by experienced teachers and coaches who have their well-being foremost in mind. We often take for granted that the larger societies in which we raise our children will remain safe and stable. But calamity brings with it the nightmare of social chaos and the savagery of "might makes right." Rebuilding a society after a disaster or a war usually involves ministering to the needs of families and educators, especially by helping them care for children whose lives have been disrupted by violence and severe loss.

AP Photo/Ramon Espinosa

BECOMING A SOCIAL BEING

Earthquakes, wars, and other extreme social situations bring out the best and the worst in people. This leads observers to wonder why some people are endowed with qualities that we admire, such as courage or altruism, whereas others display qualities that we abhor, such as

By placing their efforts to help others above concerns about their own safety, these "first responders" to the catastrophic events of 9/11/2001 were demonstrating altruistic behavior at its best.

selfishness or cruelty. What are the influences of family, schools, peers, and other social groups on people's character and the way they behave toward others?

Socialization is the term sociologists use to describe the ways in which people learn to conform to their society's norms, values, and roles. People develop their own unique personalities as a result of the learning they gain from parents, siblings, relatives, peers, teachers, mentors, and all the other people who influence them throughout their lives (Corsaro, 2005; Handel, 2001). From the viewpoint of society as a whole, however, what is important about the process of socialization is that people learn to behave according to the norms of their culture. How people learn to behave according to cultural norms—that is, the way they learn their culture—makes possible the transmission of culture from one generation to the next. In this way, the culture is "reproduced" in the next generation (Gonzalez-Mena, 1998; Roer-Strier & Rosenthal, 2001).

Socialization occurs throughout life as the individual learns new norms in new groups and

situations. For purposes of analysis, however, socialization can be divided into three major phases:

1. *Primary socialization* refers to all the ways in which the newborn individual is molded into a social being—that is, into a growing person who can interact with others according to the expectations of society. Primary socialization occurs within the family and other intimate groups in the child's social environment.
2. *Secondary socialization* occurs in later childhood and adolescence, when the child enters school and comes under the influence of adults and peers outside the household and immediate family.
3. *Adult socialization* is a third stage, when the person learns the norms associated with new statuses such as wife, husband, journalist, programmer, grandparent, or nursing home patient (Lutfey & Mortimer, 2006).

There are several unresolved and highly controversial issues in the study of socialization, and we will explore some of them in this chapter. Chief among them are the following:

- *Nature and nurture.* What is the relative strength of biological (that is, genetic) versus social influences on the individual? This issue, often referred to as the nature–nurture problem, is raised most strikingly by the altruistic behavior of the firefighters and police officers who risked their lives attempting to rescue victims of the September 11 attacks. Was it some innate, biological aspect of their nature that made them act so selflessly ("nature"), or was their behavior a result of their home environment and professional training ("nurture")? Social scientists tend to lean toward explanations based on nurture, but as we will see in this chapter, biology plays a significant role in behavior and in forming an individual's personality. Neither biological nor sociological factors alone explain the complex behaviors involved in successful socialization (Wade, 2001).
- *The social construction of the self.* A second controversy in the study of socialization is the question of how a person's sense of self becomes established. We all learn to play many different roles, but how do the influences of others in our social world affect our role playing, and how do those experiences help form our sense of ourselves? How do people learn to conform to society's norms and to take the roles that society makes available to them?

socialization: The processes through which we learn to behave according to the norms of our culture.

- *Influences on socialization.* How do different social environments, such as the affluent suburban school or the slum neighborhood or the military boot camp, influence socialization? In other words, how do different social environments produce different kinds of people? What are the influences throughout life of different agencies of socialization and different experiences with other people?
- *Gender socialization and sexual identity.* Gender socialization refers to the ways in which we become the girls and boys and, gradually, the men and women of our society and culture. All the controversies over whether behavior is innate or learned are intensified when we consider the differences and similarities in the socialization of males and females. This is especially true with reference to the ways in which we are socialized to acquire a sexual identity.

In this chapter, we explore each of these questions in detail. We will be concerned primarily with the socialization of social beings—people who are able to perform roles, to feel empathy for others, to express emotions and yet control feelings that are antisocial, to nurture others and raise children who will also be able to nurture, and to take on new roles as they grow older. But the failures of socialization can also tell us a great deal about what is involved in creating the social being.

Consider the rash of hate crimes against people from immigrant backgrounds that have occurred recently in the United States, especially in border areas or where there are concentrations of new immigrants. Hate crimes against Latino immigrants have increased by over 30 percent since 2003 (Mock, 2010). In one typical case, prosecutors in Suffolk County, New York, say that 19-year-old Jeffrey Conroy, acting in an episode of anti-immigrant gang violence, fatally stabbed Marcelo Lucero in 2008. They argue that the stabbing death of Lucero near a train station was the culmination of an ongoing campaign of violence against Hispanics in an avocation the teens called "beaner-hopping" or "Mexican hopping." Available evidence indicates that Conroy and his friends were rather ordinary middle-class teenagers, but that they had listened to so many negative descriptions of immigrant men in their community that they began thinking such attacks were a form of sport. This behavior represents a failure in socialization: Although the teenagers are themselves directly to blame for their actions, the adults in their lives who preach hate against immigrants and directly or indirectly condone violence also must examine their consciences.

Sociology & Social Justice

SOCIALIZATION AND MORAL MOTIVATION The United States and most other Western societies are becoming become more and more diversified through continued immigration, but also through political changes that create divisions of opinion in a population. In increasingly diverse societies, the question of why individuals are motivated to act morally becomes much more important. Moral motivation, so lacking in the killing of Marcelo Lucero and other hate crimes, can be defined as individuals' readiness to abide by a moral rule that they understand to be valid, even when this rule is in conflict with other desires, such as the desire to exercise power and beat up people who are perceived as different or foreign. Research on how people are socialized to act morally show that moral norms, such as justice and fairness, are not simply imposed by society and accepted by the society's members. As situations change—as populations diversify, for example—they are reinvented in the context of cooperative, supportive relationships with parents, other adults, and even peers.

The issue in cases like the killing of Marcelo Lucero is not whether we should excuse what such individuals have done due to their socialization but whether we can learn from those crimes so that we can prevent others. And scientists, biological and social, still have a good deal to learn. Socialization is an extremely complex process. Some people who were abused and neglected can nonetheless become good parents, despite the odds. Others who seem to have experienced all the right influences can end up doing evil things, again despite the odds (Corsaro, 2005; Wrong, 1961). Even if we could trace all the social influences on a person's development, there would remain many unanswered questions about the combined influences of the person's genetic potential and his or her social experiences—that is, the relative importance of nature and nurture.

NATURE AND NURTURE

Throughout recorded history, there have been intense debates over what aspects of behavior are "human nature" and what aspects can be intentionally shaped through nurture or socialization. During all the centuries of prescientific thought, the human body was thought to be influenced by the planets, the moon, and the sun, or by various forces originating within the body, especially in the brain, heart, and liver. In the ancient world and continuing into the Middle Ages, blood, bile, phlegm, and other bodily fluids were thought to control people's moods and affect their personalities. Human behavior and health could also be affected by evil spirits or witches. Although they are no longer dominant explanations of human behavior, none of these ideas has ever entirely disappeared.

In the eighteenth century, a new and radical idea of the "natural man" emerged. The idea that humans inherently possess qualities such as wisdom and rationality, which are damaged in the process of socialization, took hold in the imaginations of many educated people. Social philosophers like Jean-Jacques Rousseau believed that if only human society could be improved, people would emerge with fewer emotional scars and limitations of spirit. This belief was based on the enthusiasm created by scientific discoveries, which would free humans from ignorance and superstition, and new forms of social organization like democracy and capitalism, which would unleash new social forces that would produce wealth and destroy obsolete social forms like aristocracy.

In the United States, Thomas Jefferson applied these ideas in developing the educational and governmental institutions of the new nation. Much later, in the nineteenth century, Karl Marx and other social theorists applied the same basic idea of human perfectibility in their criticisms of capitalism. A revolutionary new society, they predicted, would overturn the worst effects of capitalism and finally realize Rousseau's promise that a superior society could produce superior people. As these examples indicate, many of the most renowned thinkers of the last two centuries have rejected the belief that nature places strict limits on what humans can achieve.

The Freudian Revolution

Sigmund Freud was the first social scientist to develop a theory that addressed both the "nature" and "nurture" aspects of human existence (Nagel, 1994; Robinson, 1994). For Freud, the social self develops primarily in the family, wherein the infant is gradually forced to control its biological functions and needs: sucking, eating, defecation, genital stimulation, warmth, sleep, and so on. Freud shocked the straitlaced intellectuals of his day by arguing that infants have sexual urges and by showing that these aspects of the self are the primary targets of early socialization—that the infant is taught in many ways to delay physical gratification and to channel its biological urges into socially accepted forms of behavior.

Id, Ego, and Superego Freud's model of the personality is derived from his view of the socialization process. Freud divided the personality into three functional areas, or interrelated parts, that permit the self to function well in society.[1] The part from which the infant's unsocialized drives arise is termed the **id**. The moral codes of adults, especially parents, become incorporated into the part of the personality that Freud called the **superego**. Freud thought of this part of the personality as consisting of all the internalized norms, values, and feelings that are taught in the socialization process. In addition to the id and the superego, the personality as Freud described it has a third vital element, the **ego**. The ego is our conception of ourselves in relation to others, in contrast with the id, which represents self-centeredness in its purest form. To have a "strong ego" is to be self-confident and able to accept criticism. To have a "weak ego" is to need continual approval from others. (The popular expression that someone "has a big ego" and demands constant attention actually signals a lack of ego strength in the Freudian sense.)

The Role of the Same-Sex Parent In the growth of the personality, according to Freud, the formation of the ego or social self is critical, but it does not occur without a great deal of conflict. The conflict between the infant's basic biological urges and society's need for a socialized person becomes evident very early. Freud believed that the individual's major personality traits (security or insecurity, fears and longings, ways of interacting with others) are formed in the conflict that occurs when parents insist that the infant control its biological urges. This conflict, Freud believed, is most severe between the child and the same-sex parent. The infant wishes to receive pleasure, especially sexual stimulation, from the opposite-sex parent and therefore is competing with the same-sex parent. To become more attractive to the opposite-sex parent, the infant attempts to imitate the same-sex parent. Thus for Freud, the same-sex parent is the most powerful socializing influence on the growing child, for reasons related to the biological differences and attractions between the child and the opposite-sex parent.

Contemporary sociologists who are influenced by Freud's biologically and socially based theories have used his concepts of same-sex attraction and imitation of the same-sex parent's behavior to explain differences between men and women. Alice Rossi (2001), for instance, argues that women's shared experience of menstruation and childbearing creates a strong bond between mothers and daughters. Nancy Chodorow (2002, 2004) claims that women's earliest experiences with their mothers tend to convince them that a woman is fulfilled by becoming a mother in her turn; thus women are socialized from a very early age to "reproduce motherhood." Research on socialization has shown that men also are strongly influenced by the same-sex parent. Fathers often serve as models of behavior whom boys will emulate throughout their lives.

Freud's theory includes the idea that the conflicts of childhood reappear throughout life in ways that the individual cannot predict. The demands of the superego ("conscience") and the "childish" desires of the id are always threatening to disrupt the functioning of the ego, especially in families in which normal levels of conflict are either exaggerated or suppressed. Note, however,

[1] Freud never expected that actual physical parts of the brain that correspond to the id, ego, and superego would be discovered. Instead, he was referring to aspects of the functioning personality that are observed in the individual.

id: According to Freud, the part of the human personality from which all innate drives arise.

superego: According to Freud, the part of the human personality that internalizes the moral codes of adults.

ego: According to Freud, the part of the human personality that is the individual's conception of himself or herself in relation to others.

that Freud focused on the traditional family, consisting of mother, father, and children. The more families depart from this conventional form, the more we need to question the adequacy of Freudian socialization theory.

Behaviorism

In the early decades of the twentieth century, Freud's theory was challenged by a different branch of social-scientific thought, known as **behaviorism**. In contrast to Freud and others who saw many human qualities as innate or biologically determined (nature), behaviorists saw the individual as a *tabula rasa,* or blank slate, that could be "written upon" through socialization. In other words, individual behavior could be determined entirely through social processes (nurture). Behaviorism asserts that individual behavior is not determined by instincts or any other "hardware" in the individual's brain or glands. Instead, all behavior is learned.

Pavlov Behaviorism traces its origins to the work of the Russian psychologist Ivan Pavlov (1927). Pavlov's experiments with dogs and humans revealed that behavior that had been thought to be entirely instinctual could in fact be the result of **conditioning** by learning situations. Pavlov's dog, one of the most famous subjects in the history of psychology, was conditioned to salivate at the sound of a bell. The dog would normally salivate whenever food was presented to it. In his experiment, Pavlov rang a bell whenever the dog was fed. Soon the dog would salivate at the sound of the bell alone, thereby showing that salivation, which had always seemed to be a purely biological reflex, could be a conditioned, or learned, response as well.

Watson American psychologist John B. Watson carried on Pavlov's work with an equally famous series of experiments on "Little Albert," an 11-month-old boy. Watson conditioned Albert to fear baby toys that were thought to be inherently cute and cuddly, such as stuffed white rabbits. By presenting these objects to Albert at the same time that he frightened him with a loud noise (that is, a negative stimulus), Watson showed that the baby could be conditioned to fear any fuzzy white object, including Santa Claus's beard. He also showed that through the systematic presentation of white objects accompanied by positive stimuli, he could extinguish Albert's fear and cause him to like white objects again.

On the basis of his findings, Watson wrote:

Give me a dozen healthy infants, well-formed, and my own specified world to bring them up in and I'll guarantee to take any one at random and train him to become any type of specialist I might select—doctor, lawyer, artist, merchant-chief and, yes, even beggar-man and thief, regardless of his talents, penchants, tendencies, abilities, vocations, and race of his ancestors. (1930, p. 104)

For the behaviorist, in other words, nature is irrelevant and nurture all-important.

Skinner Behaviorists who followed Watson—the most famous being B. F. Skinner—developed even more effective ways of shaping individual behavior. Skinner (1976) and his followers reasoned that, to avoid failures in socialization, it is necessary to completely control all the learning that goes on in a child's social environment. Sociologists are critical of the notion that it is possible to control the world of a developing person. They argue that although the behaviorists may show us how some types of social learning take place, psychological research often does not deal with real social environments. It has very little to say about what is actually learned in different social contexts, how it is learned (or not learned), and the influences of different social situations on an individual throughout life (Corsaro, 2005). One type of situation that has been of interest to sociologists studying socialization processes is that of children reared in extreme isolation.

Isolated Children

The idea that children might be raised apart from society, or that they could be reared by wolves or chimpanzees or some other social animal, has fascinated people since ancient times. Romulus and Remus, the legendary founders of Rome, were said to have been raised by a wolf. The story of Tarzan, a boy of noble birth who was abandoned in Africa and raised by apes, became a worldwide best seller early in the twentieth century and has intrigued readers and movie audiences ever since. However, modern studies of children who have experienced extreme isolation cast doubt on the possibility that a truly unsocialized person can exist.

Throughout the nineteenth century and much of the twentieth, the discovery of a **feral** ("untamed") **child** always seemed to promise new insights into the relationship between biological

---

behaviorism: A theory stating that all behavior is learned and that this learning occurs through the process known as conditioning.

conditioning: The shaping of behavior through reward and punishment.

feral child: A child reared outside human society.

capabilities and socialization, or nature and nurture (Davis, 1947). Social scientists looked on each case of a child raised in extreme isolation as a natural experiment that might reveal the effects of lack of socialization on child development. Once the child had been brought under proper care, studies were undertaken to determine how well he or she functioned. These studies showed that victims of severe isolation are able to learn, but that they do so far more slowly than children who have not been isolated (Malson, 1972).

The case of Genie is representative of this line of research. Genie was born to a psychotic father and a blind and highly dependent mother. For the first eleven years of her life, she was strapped to a potty chair in an isolated room of the couple's suburban Los Angeles home. From birth, she had almost no contact with other people. She was not toilet trained, and food was pushed toward her through a slit in the door of her room. When child welfare authorities discovered her after the mother told neighbors about the child's existence, the father committed suicide. Genie was placed in the custody of a team of medical personnel and child development researchers.

In the first few weeks after she was discovered, everyone who observed Genie was shocked. At first glance, she looked like a normal child, with dark hair, pink cheeks, and a placid demeanor. Very quickly, however, it became clear that she was severely impaired. She walked awkwardly and was unable to dress herself. She had virtually no language ability—at most, she knew a few words, which she pronounced in an incomprehensible babble. She spit continuously and masturbated with no sense of social propriety. In short, she was a clear case of a child who had been deprived of social learning and in consequence was severely retarded in her individual and social development. She was alive, but she was not a social being in any real sense of the term. She had not developed a sense of self, nor had she the basic ability to communicate that comes with language learning.

For years, researcher Susan Curtiss (1977) studied Genie's slow progress toward language learning. Curtiss showed that Genie could learn many more words than a person with mental retardation would be expected to learn, but that she had great difficulty with the more complex rules of grammar that come naturally to a child who learns language in a social world. Genie's language remained in the shortened form characteristic of people who learn a language late in life. Most significant, Genie never mastered the language of social interaction. She had great difficulty with words such as *hello* and *thank you*,

S. Curtiss, Genie: A Psycholinguistic Study of a Modern Day "Wild Child" (Academic Press, 1977). Used by permission.

Genie drew this picture in 1977. At first she drew the picture of her mother and labeled it "I miss Mama." Then she drew more. The moment she finished, she took the researcher's hand and placed it next to what she had just drawn. She motioned for the researcher to write and said, "Baby Genie." Then she pointed under her drawing and said, "Mama hand." The researcher wrote the corresponding letters. Satisfied, Genie sat back and stared at the picture. There she was, a baby in her mother's arms.

although she could make her wants and feelings known with nonverbal cues.

Although tests showed that Genie was highly intelligent in many ways, her language abilities never advanced beyond those of a third grader. Genie gradually learned to adhere to social norms (for example, she stopped spitting), but she never became a truly social being. Eventually, the scientists who worked with her concluded that the most severe deprivation—the one that was the primary cause of her inability to become fully social and to master language—was her lack of emotional learning and especially her feelings of loss and lack of love. Genie expressed her inner longings in the picture here, which she drew at the age of about 16. Never fully capable of independent living, Genie has spent her life in a home for developmentally disabled adults (Newton, 2003; Rymer, 1994).

The Need for Love

All studies of isolated children point to the undeniable need for nurturance in early childhood. They all show that extreme isolation is

Harlow's primate research demonstrated that monkeys raised without maternal touching and affection often exhibited abnormal, self-stimulating behaviors, such as thumb sucking and rocking.

young. In some cases, they even crushed their babies' heads with their teeth before handlers could intervene. Although it is risky to generalize from primate behavior to that of humans, these studies of the effects of lack of nurturance bear a striking resemblance to studies of child abuse in humans. This research generally shows that one of the best predictors of abuse is whether the parent was also abused as a child (Ornish, 1998; Talbot, 1998).

Harlow's research was revolutionary because it overturned the Freudian and behaviorist assumptions that satisfying infants' biological needs is central to their development and that too much parental affection could actually impede development (Blum, 2002). Harlow's findings confirmed what many believed but could not prove: that parental nurturance and love play a profound, though still incompletely understood, role in the development of the individual as a social being. These and related findings offer support for social policies that seek to enrich the socialization process with nurturance from other caring adults—for example, in early-education programs. Yet many researchers and policy makers remain convinced that biological traits place limits on what individuals can achieve, regardless of the kind of nurturance they receive. In recent years, proponents of this view have focused on the controversial question of whether such complex traits as intelligence and sexual orientation are the result of genetic predisposition or childhood socialization.

The Debate over Genetic Influences

The role of genes in shaping traits such as intelligence and sexual orientation is a subject of continual research and much controversy. In neither case is there any definitive evidence that specific genes determine these highly significant aspects of human behavior. Research on the human genome promises to increase our understanding of how genes contribute to many aspects of human development and behavior, including intelligence and sexual orientation. At this writing, however, it is perhaps more important to understand the nature of the controversy.

Genes and Intelligence In an influential but scientifically flawed study titled *The Bell Curve*, biologist Richard Herrnstein and social psychologist Charles Murray (1994) attempted to show that IQ, one measure of intelligence, is an inherited trait that underlies inequality among different groups in the United States. Herrnstein and Murray do not believe that efforts to address

associated with profound retardation in the acquisition of language and social skills. However, they cannot establish causality, because it is always possible that the child may have been retarded at birth. Despite their lack of firm conclusions, studies of children reared in extreme isolation have pointed researchers in an important direction: They suggest that lack of parental attention can result in retardation and early death. This conclusion receives further support from studies of children reared in orphanages and other residential care facilities, which have shown that such children are more likely to develop emotional problems and to be retarded in their language development than comparable children reared by their parents (Blum, 2002).

In a series of studies that have become classics in the field of socialization and child development, the late primate psychologist Harry Harlow showed that infant monkeys reared apart from other monkeys never learned how to interact with other monkeys (Harlow, 1986, 2008); they could not refrain from aggressive behavior when they were brought into group situations. When females who had been reared apart from their mothers became mothers themselves, they tended to act in what Harlow described as a "ghastly fashion" toward their

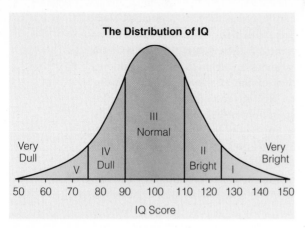

The Distribution of IQ

Very Dull

V

IV Dull

III Normal

II Bright

I

Very Bright

50   60   70   80   90   100   110   120   130   140   150

IQ Score

Figure 5.1 Defining the Cognitive Classes. Caution: The labels imposed on this IQ curve and the score used as boundaries between "cognitive classes" are those of Herrnstein and Murray and do not represent the thinking of many other social scientists. Source: Reprinted with the permission of The Free Press, a Division of Simon & Schuster, Inc., from *The Bell Curve: Intelligence and Class Structure in American Life* by Richard J. Herrnstein and Charles Murray. Copyright © 1994 by Richard J. Herrnstein and Charles Murray. All rights reserved.

inequalities through preschool programs, improvements in public education, and the like will ever successfully address the growing inequalities among individuals and groups in the United States or elsewhere. The real obstacle, they argue, is intelligence, or rather the lack of it in vast numbers of people. Because, by their reasoning, intelligence is an innate trait that is distributed unequally among various subgroups of the population, social intervention cannot do much to equalize the economic effects of differences in intelligence. In his most recent research, Murray argues that innate differences in intelligence make it useless to seek greater educational equality for African Americans and other poor minority groups (Feuer, 2009; Murray, 2008).

Like many traits that vary from one person to another, such as height and weight, scores on intelligence tests are distributed in the shape of a "normal" or bell curve. Most people's IQ scores fall near the center of the distribution, while some are at either extreme. About these facts there is no argument. Murray and Herrnstein further assert, however, that scores on the bell curve of intelligence are creating what they call "cognitive classes"—that is, categories of people whom they label "very bright," "bright," "normal," "dull," and "very dull" (see Figure 5.1). They note that 5 percent of the U.S. population falls within the left and right extremes of the curve, and another 20 percent are in Class II and Class IV. By this reasoning, approximately 50 million residents of the United States are classified in the lower cognitive classes.

Recent research on the human brain does indicate that the larger some regions of the brain

are, the more intelligent an individual is (Thompson et al., 2001). Despite these findings, however, most social scientists continue to oppose Herrnstein and Murray's conclusions, for several reasons. First, there has been much criticism of IQ as a single measure of intelligence. Many experts believe that intelligence is far too complex to be represented by a single measure like IQ (Gardner, 1983). Second, there is evidence of cultural and middle-class biases in the questions used to test IQ; examination of the test items reveals that they would be far more familiar to middle-class test takers than to those from a disadvantaged background. Third, the authors of *The Bell Curve* have been criticized for asserting that correlation is the same as causality. Just because IQ and poverty are correlated does not mean that IQ causes poverty. In Northern Ireland, for example, Catholic Irish individuals score lower on IQ tests than do Protestant Irish, but these differences in IQ are not found in the United States, where there are no differences in wealth or advantage between the two groups as there are in Northern Ireland. In other words, socialization in an impoverished environment could cause low IQ scores, rather than the other way around.

Most contemporary research on intelligence rejects narrow reliance on a single intelligence score like IQ in favor of a multidimensional conception of intelligence that corresponds far better to the great diversity of people we encounter in and outside of school. Howard Gardner (2002) has identified seven distinct types of intelligence, as follows:

- Visual/spatial intelligence
- Musical intelligence
- Verbal intelligence
- Logical/mathematical intelligence
- Interpersonal intelligence (ability to perceive other people's emotions and motivations)
- Intrapersonal intelligence (ability to understand one's own emotions and motivations)
- Bodily/kinesthetic intelligence (sometimes thought of in popular speech as "physical coordination" or natural athletic ability)

Once one begins to appreciate people for these multiple intelligences—for the intelligence that makes one a gifted artist or athlete, another a poet or someone with keen insights into people's feelings and motivations—it becomes impossible and even cruel to label people as average or dull on the basis of a far narrower definition of the highly complex set of natural and learned abilities known as intelligence. No doubt all these types of intelligence have a genetic basis, as well as the need for specific forms of socialization, but we are a long way from understanding how the complex interplay between nature and nurture produces people with the different intelligences that contribute so much to what we recognize as their personalities.

The Human Genome and Twin Studies In 2001, geneticists announced some major findings of the Human Genome Project, the first successful research effort to discover and describe all the genes in the human being (Baltimore, 2001). These results were awaited with great anticipation. Would they show that our development and behavior are influenced by our genes to a greater extent than we had previously thought?

One of the most stunning findings of the research was that the number of distinct genes in the human was about 30,000, far less than the number many scientists had expected. Some scientists immediately announced that this number was far too low to justify the conclusion that genes determine human personality and behavior. Others pointed out that even 30,000 genes would allow for a vast array of combinations and permutations of traits, which would result in enough variation to produce all the differences we see in humans. But all this leaves plenty of opportunity for the external environment created by parents and others to exert influences on development too. In short, so far the human genome project has primarily supported the theory that both genes and upbringing, both nature and nurture, play major roles in human development.

In recent decades, as we have learned more about the role of genes in shaping human behavior, the "nature" side of the argument has gained prominence. Human traits like aggression, extraversion, empathy, and many others appear to have genetic origins, although they cannot be directly traced to particular genes. The most powerful methods we have to study these issues come not from the Human Genome Project but from studies of identical twins.

Identical twins share the same exact genetic makeup, so if they experience different socialization, due for example to being reared in different homes, researchers can begin to trace the importance of environmental variables, or "nurture," while "nature" is held constant. These studies have shown that, to a limited degree, twins reared apart have similar personalities and are prone to develop similar mannerisms and likes and dislikes. But they are also capable of becoming quite different as they develop. The more complex the behavior studied, such as language ability, the greater the influence of home and other social environments. The less complicated the behavior, such as personal mannerisms or habits like smoking, the greater the influence of the twins' identical genes. In Chapter 7, we will see that the sexual preferences of twins reared apart appear to have a strong genetic origin, but that differences are observed often enough to allow for the importance of upbringing as well; this is especially true in females (Pinker, 2002; Van Gestel & Van Broeckhoven, 2003). In sum, although the results of the genome project and twin studies continue to help us tease out the facts in the age-old nature–nurture controversy, there are no simple answers and no end to the debates.

Thinking Critically Think about the conversations that take place in your family and among friends. Do you hear theories about human nature that emphasize genetic determinism, such as variations of the "bad seed" idea that a person can be "evil from birth"? Or do you hear "blank slate" theories that emphasize learning and experience, asserting that a person's behavior is entirely due to his or her upbringing?

THE SOCIAL CONSTRUCTION OF THE SELF

The self is the outcome of socialization; it may be defined as "the capacity to represent to oneself what one wishes to communicate to others" (Elkin & Handel, 1989, p. 47). Genie, an isolated, unsocialized child, did not develop a self in early childhood. She did not learn to formulate in her own mind the words that would express her feelings to others. Through socialization, most children learn to convert cries of discomfort, hunger, or fear into socially understandable symbols like "Want bottle" or "Go out now." These utterances show that the young child has learned to recognize his or her inner states and communicate them to others. "The child who can do this is on his or her way to becoming human, that is, to being simultaneously self-regulating and socially responsive" (Elkin & Handel, 1989, p. 47).

In sociology, the self is viewed as a social construct: It is produced or "constructed" through interaction with other people over a lifetime. Studies of how the self emerges therefore usually take an interactionist perspective.

Interactionist Models of the Self

The early American sociologist Charles Horton Cooley (1956/1902) was an astute observer of

Stages in the Development of Role Taking

Stage	Description	Example
Preparatory	The child attempts to mimic the behavior of people who are significant in his or her life.	A toddler tries to walk in her mother's shoes.
Play	Children play at being people who are significant in their lives, such as parents.	Three children play house: "You be the mommy, I'll be the daddy, and Joey can be the baby."
Game	Each child is able to become, in a symbolic sense, all the other participants in the game.	A group of neighborhood kids gather at a playground to play baseball.

human interactions. As he watched young children interact with adults and other children, he saw that they continually pay attention to how others respond to their behavior. He realized that they are seeking cues in other people's behavior that reflect their own—that tell them how they look, how well they express themselves, how well they are performing a task, and so on. Cooley reasoned that as we mature, the overall pattern of these reflections becomes a dominant aspect of our identity—that is, of how we conceive of ourselves. He proposed that, through these processes, we actually become the kind of person we believe others think we are. He called this identity the "looking-glass self." The looking-glass self is the reflection of our self that we think we see in the behaviors of others around us.

Language, Culture, and The Self Cooley's insight into the role of others in defining the self was the foundation for the view of the self proposed by George Herbert Mead. Like Cooley, Mead believed strongly that the self is a social product. We are not born with selves that are "brought out" by socialization. Instead, we acquire a self by observing and assimilating the identities of others (Grodin & Lindlof, 1996; Nisbet, 1970). The vehicle for this identification and assimilation is language. As Mead wrote, "There neither can be nor could have been any mind or thought without language; and the early stages of the development of language must have been prior to the development of mind or thought" (quoted in Truzzi, 1971, p. 272).

This view places culture at the center of the formation of the self. The kind of person we become is largely a result of the cultural influences that surround us during socialization. Through interaction with people who are Catholic, for example, one takes on the language, the jokes, and the style of a person of that religion. If the father is a firefighter and the mother a nurse, certain attitudes about service to society and about illness and danger will carry over to the child. If the same child plays on sports teams with children in the neighborhood, the norms and values of those children and their parents will become part of the child's experience and will be incorporated into his or her personality. Another child, growing up on a Sioux Indian reservation, would learn some of the same values—such as fair play, reward for achievement, and good citizenship—but would also learn the norms and values of the Sioux (for example, reverence for one's ancestors and for the natural environment).

As each person learns the norms of his or her culture and its various ways of communicating—whether through language, dress, or gestures—and as each experiences the influences of a particular family and peer group, a unique self is formed. The self, thus, is a product of many influences and experiences; every person emerges with a personality of his or her own, and each has incorporated the values of the larger society and of a particular subculture.

Role Taking: The Significant Other and The Generalized Other For Mead (1971/1934), two of the most important activities of childhood are play and games. In play, the child practices taking the roles of others. If you watch preadolescent children play, you will see them continually "trying on" roles: "You be the mommy and I'll be the teacher, and you'll come to school to find out why . . ." or "Pretend I'm Kobe Bryant and you're Lebron James and the score is tied in the fourth quarter." They are reenacting the dramas of winning and losing, or calling into question the behaviors of the schoolroom, or trying to understand sickness and death.

This idea of **role taking** is central to the interactionist view of socialization. It refers to the way we try to look at social situations from the standpoint of another person from whom we seek a response. Mead believed that children develop this ability in three stages, during which they gain their sense of self and learn to act as members of society. He labeled them the preparatory, play, and game stages (see the For Review chart above).

In the *preparatory stage,* children attempt to mimic the behavior of people who are significant in their life. **Significant others** are people who loom large in our lives, people who appear to be directly involved in winning and losing, achieving and failing. They tend to be people whose behavior we imitate—or whose behavior we seek to avoid. In the preparatory stage, the child's significant others are those who respond to calls for help and shape social behaviors such as language.

In Mead's second stage, the *play stage,* children play at being others who are significant in their lives:

> They play at being the others who are significant to them. They want to push the broom, carry the umbrella, put on the hat, and

role taking: Trying to look at social situations from the standpoint of another person from whom one seeks a response.

significant other: Any person who is important to an individual.

Graffiti and Belonging The teenagers who do it call it different things in different cities—writing, tagging, bombing, and even art. Most adults think of it as graffiti and consider it either a nuisance or, if it defaces their property, outright vandalism. Sociologists who study teenage behavior often explain that graffiti can send many kinds of signals. Gangs often use initials or symbols to mark their turf in an urban community. Aspiring young artists, influenced by hip-hop styles of dress and music, may become interested in graffiti as an art form (Giuffre, 2009; Valle & Weiss, 2010).

Photo by Rich Puchalsky

No matter how much we might understand and appreciate well-executed graffiti art, however, graffiti art is a defacement of public property. Its removal is costly; its effects can be demoralizing; and it can be read as a sign that a community's young people, or at least some of them, are alienated and angry. Realizing this, many youth workers and community leaders have found ways to reach out to graffiti artists and guide their energy and creativity into socially constructive or at least more acceptable channels. In New Orleans, for example, the Ya Ya Collective has made graffiti art into products that young people can reproduce and sell. This builds their self-esteem and provides funds for other constructive youth activities. In New York City, where graffiti art is extremely popular, a sympathetic industrialist has hired graffiti artists to paint enormous murals on the walls of an old factory building in an industrial section of the city. In Philadelphia, Chicago, Cleveland, and many other U.S. cities, graffiti artists and their friends are applying their skills to the creation of public murals commemorating the victims of gang violence or the drug trade, and in so doing, they are learning to use their art in positive ways.

do all the other things they see their parents do, including saying what their parents say. The story is told of the four-year-old playing "daddy" who put on his hat and his coat, said "good bye," and walked out the front door, only to return a few minutes later because he didn't know what to do next. He had taken as much of his father's work role as he could see and hear—the ritualized morning departure. (Elkin & Handel, 1989, p. 49)

As children grow in age and experience, they enter what Mead called the *game stage.* To take part in a game, a child must have already learned to become, in a symbolic sense, all the other participants in the game. Mead called this the ability to "take the role of the generalized other." Thus, in a baseball game:

> [The child] must know what everyone else is going to do in order to carry out his own play. He has to take all of these roles. They do not have to be present in consciousness at the same time, but at some moments he has to have three or four individuals present in his own attitude, such as the one who is going to throw the ball, the one who is going to catch it, and so on. (Mead, 1971/1934, p. 151)

When we are able to take the role of the generalized other, we know that rules apply to us no matter who we are. We know, for example, that rules about not smoking in the school building apply equally to students, parents, and teachers, and that those who violate the rules will not be excused because of their status (Mead, 1971/1934).

The **generalized other** is a composite of all the roles of all the participants in the game. A person who participates in a game like baseball, for example, has developed the capacity for role taking and now (again using Mead's phrase) "takes the role of the generalized other." When little children play team games, they often have a difficult time taking specific roles. Watch them play basketball or soccer, for example, and you will often find them clumping together in an effort to get the ball. As children mature, their growing sense of the generalized other makes them better able to understand all the roles on a team and to learn their own specific roles and positions in games and team sports.

The generalized other represents the voice of society, which is internalized as "conscience." For some people, the generalized other demands perfection and strict adherence to every rule. For others, the generalized other may be extremely demanding where sports and other games

generalized other: A person's internalized conception of the expectations and attitudes held by society.

are concerned but much more relaxed about achievement in school or adherence to the norms of property. For still others, the generalized other may insist on amassing large amounts of money as the primary indicator of success, or it may require community service and not value financial success at all. Within any given culture, such variations will be wide but will tend to follow certain easily recognized patterns.

Role Playing and "Face Work"

Why do Americans and Asians often have so much difficulty understanding each other even when they are playing the same role (for example, student) and speaking the same language? Why are adolescents so "hung up" about their dealings with other kids? And why is it that seemingly small insults are treated as major signs of disrespect in some communities? These are all examples of the kinds of problems that arise when people actually play the roles for which they have been socialized. They can be analyzed according to the rules of interaction known as *face work*.

As Erving Goffman (1965) defines it, *face* is the positive social value people claim for themselves by acting out a specific set of socially approved attributes (for example, politeness, humor, strength, cuteness, sensitivity). Through his close observations of seemingly routine greetings, formulas of politeness, and the give-and-take of small talk, Goffman identified the rules of interaction whereby people seek to present a positive image of themselves, their "face." Once they have established a specific image, they seek to defend it against any possible threat that might cause them to "lose face." The concept of saving or losing face is found in cultures throughout the world. Indeed, in most Asian cultures, the rules of face work are even more elaborate than in the West. (The importance of such unwritten rules of interaction is explored more fully in Chapter 6.)

Most of us take for granted that we want to maintain our self-respect and not lose face in social situations, but we do not give much thought to the actual interactions that serve to maintain face. Goffman explains:

> Just as the member of any group is expected to have self-respect, so also he is expected to sustain a standard of considerateness; he is expected to go to certain lengths to save the feelings and the face of others present, and he is expected to do this willingly and spontaneously because of emotional identification with the others and with their feelings. . . . The person who can witness another's humiliation and unfeelingly retain a

cool countenance himself is said in our society to be "heartless," just as he who can unfeelingly participate in his own defacement is thought to be "shameless." (1965, p. 10)

The way we apply these rules of face work in performing our roles differs according to the prestige of the people involved. When firing a junior clerk, for example, the boss does not go to nearly the same lengths to take into account the employee's feelings as he does when firing a vice president. Recent studies of role playing among adolescents in inner-city communities draw on these insights. Where there is very little prestige of any kind for people to share, "fronting"—or pretending to play roles that one cannot really perform (for example, great ballplayer, ladies' man)—is extremely common. So, therefore, are potential threats to one's face. Face work in such communities—among gang members, for example—can become a deadly business, especially in situations where one senses disrespect (that is, feels that one is being "dissed") (MacLeod, 1995/1987; Williams & Kornblum, 1994). Emotions play a strong part in these interactions. When people believe that they have seriously lost face, their feelings can run extremely high, especially if their peers do not hurry in to defend them.

Rapid social change tends to heighten the difficulties people have in playing the roles they have been socialized to perform. As Goffman points out:

> A person's performance of face-work, extended by his tacit agreement to help others perform theirs, represents his willingness to abide by the ground rules of social interaction. Here is the hallmark of his socialization as an interactant. If he and the others were not socialized in this way, interaction in most societies and most situations would be a much more hazardous thing for feelings and faces. (1965, p. 31)

As populations become more diverse through the processes of social change (for example, immigration and migration), the possibilities for cultural confusion over the rules of successful role playing in school or on the street can multiply dramatically.

This interactionist perspective on socialization emphasizes how people become social actors and how they intuitively adopt the rules and rituals of interaction, such as face work, that exist in their cultures. But it does not address how people acquire their notions of morality. We know that people generally learn their values and ideas of morality as young children in the family, but research on child development shows that the acquisition of morality is not a simple matter.

Theories of Moral Development

Suppose that a friend asks you for help on an exam. Suddenly rules about cheating come into conflict with the norm that friends help each other. We face such moral conflicts all the time, and they often involve higher stakes or greater risk. Throughout life, we face a wide variety of moral dilemmas, which have a significant effect on our personalities. In consequence, social scientists have devoted considerable study to the processes through which people develop concepts of morality. Among the best-known students of moral development are the Swiss child psychologist Jean Piaget and the American social psychologists Lawrence Kohlberg and Carol Gilligan.

Piaget Piaget stands with Freud as one of the most important and original researchers and writers on child development. In the 1920s, he became concerned with how children understand their environment, how they view their world, and how they develop their own personal philosophies. To discover the mental processes unique to children, he used what was then an equally unique method: He spent long hours with a small number of children, simply having conversations with them. These open-ended discussions were devoted to getting at how children actually think. In this way, Piaget discovered evidence for the existence of ideas that are foreign to the adult mind (Corsaro, 2005). For example, the child gives inanimate objects human motives and tends to see everything as existing for human purposes. In this phase of his research, Piaget also described the egocentric aspect of the child's mental world, which is illustrated by the tendency to invent words and expect others to understand them.

In the later phases of his research and writing, Piaget devoted his efforts to questions about children's moral reasoning—the way children interpret the rules of games and judge the consequences of their actions. He observed that children form absolute notions of right and wrong very early in life, but that they often cannot understand the ambiguities of adult roles until they approach adolescence. This line of investigation was continued by the American social psychologist Lawrence Kohlberg, whose theory incorporates Piaget's views on the development of children's notions of morality (Kohlberg, Levine, & Hewer, 1983).

Kohlberg Kohlberg's theory of moral development emphasizes the cognitive aspects of moral behavior. (By *cognitive,* we mean aspects of behavior that one thinks about and makes conscious choices about, rather than those that one engages in as a result of feelings or purely intuitive reactions.) In a study of fifty-seven Chicago children that began in 1957 and continued until the children were young adults, Kohlberg presented the children with moral dilemmas such as the following:

> A husband is told that his wife needs a special kind of medicine if she is to survive a severe form of cancer. The medication is extremely expensive, and the husband can raise only half the needed funds. When he begs the inventor of the drug for a reduced price, he is rebuffed because the inventor wants to make a lot of money on his invention. The husband then considers stealing the medicine, and the child is asked whether the man should steal in order to save his wife. (Kohlberg & Gilligan, 1971)

On the basis of children's answers to such dilemmas at different ages, Kohlberg proposed a theory of moral development consisting of three stages: (1) *preconventional,* in which the child acts out of desire for reward and fear of punishment; (2) *conventional,* in which the child's decisions are based on an understanding of right and wrong as embodied in social rules or laws; and (3) *postconventional,* in which the individual develops a sense of relativity and can distinguish between social laws and moral principles. Subjects in the preconventional and conventional stages often immediately assume that stealing is wrong in the dilemma Kohlberg has posed, but postconventional thinking in older children causes them to debate the fairness of rules against stealing in view of the larger moral dilemma involved.

Gender and Moral Reasoning Kohlberg's studies have been criticized for focusing too heavily on the behavior of boys and men from secure American families and not exploring possible alternative lines of moral reasoning that may prevail among females or people from differing cultural and racial backgrounds (Dawson, 2002; Garcia, 1996; Gilligan et al., 1998). Pioneering work by social psychologist Carol Gilligan, an early collaborator of Kohlberg's, has produced an impressive body of evidence that demonstrates the propensity of females to make moral choices on the basis of a somewhat different line of reasoning from that generally followed by males. Gilligan's research, and that of others who have followed her lead, shows that females are more likely than males to base moral judgments on considerations of caring as well as justice or law. More than their male counterparts, females tend to look for solutions to moral dilemmas that also serve to maintain relationships. Females, therefore, more often invoke caring solutions that consider the needs of both sides.

A good example of this difference appears in the work of D. Kay Johnston (1988), in which adolescent boys and girls were presented with dilemmas taken from Aesop's fables. The young people were read a fable that presents a moral dilemma and then asked what they understood the problem to be and how they would solve it. In the fable of the dog in the manger (see Figure 5.2), the problem is, clearly, that the dog has taken sleeping space from the deserving ox. Some adolescents judged the situation purely in terms of which animal had the right to the space, and made statements like "It's [the ox's] ownership and nobody else had the right to it." Others sought a caring solution that would take into consideration both animals' needs, and made statements like "If there's enough hay, well, this is one way, split it. Like, if they could cooperate."

The table in Figure 5.2 shows that boys were more likely than girls to give solutions based on rights, whereas girls were more likely than boys to choose solutions that emphasized caring. Some chose solutions that combined the two approaches. As Gilligan notes, "An innovative aspect of Johnston's design lay in the fact that after the children had stated and solved the fable problems, she asked, 'Is there another way to think about this problem?' About half of the children . . . spontaneously switched orientation and solved the problem in the other mode" (from Gilligan et al., 1998, p. xxi). On the basis of this and much subsequent research, Gilligan concludes that by age eleven most children can solve moral problems both in terms of rights (a justice approach) and in terms of response (a caring approach). The fact that a person adopts one approach in solving a problem does not mean that he or she does not know or appreciate others.

Gilligan and others who study moral development and gender point out that adolescence is a critical time in the development of morality and identity. However, in schools and elsewhere in society, the message that comes across is that norms, values, and the most highly esteemed roles require that there be a "right way" to feel and think. Most often this right way is associated with the justice focus—and the caring focus is silenced, along with the voices of girls and others to whom it appears to be a valuable alternative mode of moral reasoning (Wren,

The Dog in the Manger

A dog, looking for a comfortable place to nap, came upon the empty stall of an ox. There it was quiet and cool, and the hay was soft. The dog, who was very tired, curled up on the hay and was soon fast asleep.

A few hours later the ox lumbered in from the fields. He had worked hard and was looking forward to his dinner of hay. His heavy steps woke the dog, who jumped up in a great temper. As the ox came near the stall the dog snapped angrily, as if to bite him. Again and again the ox tried to reach his food, but each time he tried the dog stopped him.

Moral Orientation of Spontaneous Solution for the Dog in the Manger Fable, by Gender

Orientation	Female	Male
Rights (justice)	12	22
Response (caring)	15	5
Both	3	1

Source: Adapted from Johnston, 1988, p. 57.

Figure 5.2 The Dog in the Manger. Source: Adapted from Johnson, D.K. "Adolescents' solutions to dilemmas in fables: Two moral orientations—two problem solving strategies." In C. Gilligan, J.V. Ward, J.M. Taylor & B. Bardige (Eds.), *Mapping the Moral Domain*. Cambridge, MA: Harvard University Press.

1997). In adolescent girls and many minority students, this form of silencing can be detrimental to the development of the self in social situations, a subject of immense importance to which we will return at many points in later chapters (Taylor, Gilligan, & Sullivan, 1995).

Gender and Emotional Intelligence Women often complain that the men in their lives are "out of touch with their emotions." Men complain that women too often want to talk about feelings and are too quickly moved to tears. These differences in behavior are the subject of a good deal of research in psychology and sociology that attempts to help people understand why "men are from Mars [the warlike planet] and women are from Venus [the planet symbolizing love and the emotions]" (Gray, 1992; Tannen, 2001). Critics of such research point out that the evidence shows that both men and women can show great differences in their ability to understand and deal with their emotions and their interpersonal relations. Books that emphasize gender differences can make it seem as if nothing can be done about lack of emotional intelligence.

Jerome Kagan and others who study what Gardner identified as inter- and intrapersonal intelligence note that it can be found in both men and women, although women in our society are socialized to develop it more fully (Kagan, 2000; Kagan et al., 2001). Their research shows that—whether one is male or female—by better understanding what goes into emotional

TABLE 5.1

Skills of Emotional Competence

1. **Awareness of one's emotional state**, including the possibility that one is experiencing multiple emotions, and at even more mature levels, awareness that one might also not be consciously aware of one's feelings due to unconscious dynamics or selective inattention

2. **Skill in discerning others' emotions**, based on situational and expressive cues that have some degree of cultural consensus as to their emotional meaning

3. **Skill in using the vocabulary of emotion** and expression terms commonly available in one's subculture, and at more mature levels, skill in acquiring cultural scripts that link emotion with social roles

4. **Capacity for empathic and sympathetic involvement** in others' emotional experiences

5. **Skill in understanding** that inner emotional state need not correspond to outer expression, both in oneself and in others, and at more mature levels, understanding that one's emotional-expressive behavior may affect another and taking this into account in one's self-presentation strategies

6. **Skill in adaptive coping** with aversive [negative] or distressing emotions by using self-regulatory strategies that ameliorate the intensity or temporal duration of such emotional states (e.g., "stress hardiness")

7. **Awareness** that the structure or nature of relationships is in part defined by both the degree of emotional immediacy or genuineness of expressive display and the degree of reciprocity or symmetry within the relationship

8. **Capacity for emotional self-efficacy**, in which individuals view themselves as feeling, overall, the way they want to feel (that is, emotional self-efficacy means that one accepts one's emotional experience, whether unique and eccentric or culturally conventional)

Source: Adapted from Kagan et al., "The Need for New Constructs." *Psychological Inquiry.* Vol 12, No. 2, Jan 4, 2001. Copyright © 2001 Psychology Press. Reprinted by permission of the publisher (Taylor & Francis Group, http://www.informaworld.com).

Erik Isakson/Fancy/Photolibrary

intelligence, one can vastly improve one's personal relationships and one's understanding of why others react as they do in different social situations. The basic dimensions of emotional intelligence as researchers are studying them appear in Table 5.1.

SOCIAL ENVIRONMENTS AND EARLY SOCIALIZATION

Socialization occurs in many different settings, some of which are more desirable than others. Children are born into relatively affluent homes in American suburbs, the squalid slums of Calcutta, the icy wastes of Siberia, the extreme poverty of the Somalian desert, and myriad other environments—yet in every case, those who survive infancy are socialized to live and perhaps flourish in these diverse settings. Cross-cultural studies attempt to show how socialization processes vary in different societies and how they are affected by demographic trends (trends in population size and composition) within societies. This comparative view of socialization attempts to bring together the often competing viewpoints of biological, psychological, and sociological explanations of developmental patterns.

Within the United States, important differences exist from one region or state to another in the degree of stress children experience in their home environment and in the "success" of their socialization. A crude but extremely useful measure of stress in children's socialization environment is the percentage of children living in households with incomes below the official poverty line. Similarly, measures of the percentage of children living in single-parent households—which are more likely than other households to have below-average income and in which the parent may have less opportunity to interact with the child—are useful indicators of the outcomes of socialization. As shown in Figures 5.3 and 5.4, these indicators improved slightly during the boom years of the late 1990s, but have deteriorated dramatically since the economic crisis of 2008.

The circumstances reflected in measures such as these have many causes and significant consequences for social change. For example, because teenage mothers are more likely than nonteenagers to be single mothers and to be living in poverty—two negative stresses on a child's socialization environment—teenage pregnancy becomes an important issue. A major consequence of the high rate of teenage fertility in the United States is that proportionately more babies are born into homes in which they will not receive the attention and material benefits that will help them realize their full genetic potential (Furstenberg, 2000). In other

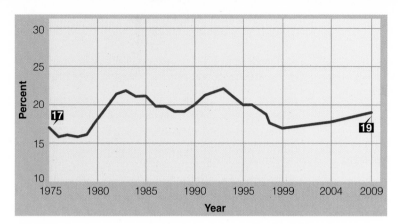

Figure 5.3 **Percentage of Children in Poverty, 1975–2004.** Source: Annie E. Casey Foundation, 2001; National Center for Children in Poverty, 2005.

words, these children start off at a disadvantage in that their mothers have not yet reached maturity themselves and are likely to be single parents who may be unable to provide their children with economic security, love, and discipline.

Urie Bronfenbrenner, one of the nation's most respected and original researchers in the field of childhood socialization, summarizes the conclusions of numerous studies with these two propositions:

> *Proposition 1.* In order to develop normally, a child needs the enduring . . . involvement of one or more adults in care of and joint activity with the child.
>
> *Proposition 2.* The involvement of one or more adults in care of and joint activity with the child requires public policies that provide opportunity, status, resources, encouragement, example, and, above all, time for parenthood, primarily by parents, but also by other adults in the child's environment, both within and outside the home. (1981, p. 39)

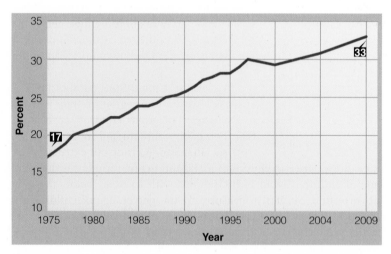

Figure 5.4 **Percentage of Families with Children Under 18 Headed by a Single Parent, 1975–2004.** Source: Anne E. Casey Foundation, 2001; National Center for Children in Poverty, 2005. These indicators are published annually by the Anne E. Casey Foundation of Baltimore, which is devoted to improving the quality of children's lives in the United States.

agencies of socialization: The groups of people, along with the interactions that occur within those groups, that influence a person's social development.

agents of socialization: Individuals who socialize others.

anticipatory socialization: Socialization that prepares an individual for a role that he or she is likely to assume later in life.

No society can pass legislation requiring that a growing child experience love in the family or household. It can, however, formulate public policies that support the efforts of families to raise healthy children and socialize them well. It can do this, in part, by helping provide children with the attention of caring adults outside the family, especially in agencies of socialization such as schools.

Agencies of Socialization

Agencies of socialization are the groups of people, along with the interactions that occur within those groups, that influence a person's social development throughout his or her lifetime (Elkin & Handel, 1989; Gonzalez-Mena, 1998). The most familiar agencies of socialization are the family, schools, socializing agencies in the community, religion, the peer group, and the mass media. **Agents of socialization** are individuals, such as parents and teachers, who socialize others. Later in life, adults may experience further socialization in the workplace, in universities, or in the military, to cite a few common examples. People who wish to change self-destructive behaviors such as drug addiction or alcoholism may be resocialized in "twelve-step" programs such as Alcoholics Anonymous. (Resocialization is discussed further in the next section.)

Within all agencies of socialization, numerous processes shape the individual's development. Among these are direct instruction, imitation or modeling of behavior, and reinforcement of particular behaviors (for example, through rewards or punishments). Many of these processes appear in what sociologists call **anticipatory socialization.** Whenever individuals play at a role that they are likely to assume later in life, anticipatory socialization is taking place. By dressing up in adult clothing, for example, or by playing house, children imitate the behavior of adults. Adolescents who attend a senior prom are being socialized in anticipation of a time when they will be expected to participate in formal social events. Workers in a corporate setting may rehearse before coworkers a presentation that they will later make to a potential client.

Another important aspect of agencies of socialization is that they must continually deal with social change. For example, as guns have become more readily available in large cities, peer groups have had to cope with the ever-present danger of violent death. As divorce and remarriage rates have risen, many families have had to deal with new structures that add much more complexity to family socialization patterns. As a result, the influence of social change

The New York Public Library, Photography Collection, Miriam and Ira D. Wallach Division of Art, Prints and Photographs

In the early decades of the twentieth century, when children routinely worked in textile mills and coal mines, the environment in which they were socialized forced them to take on adult roles at an early age. For some children growing up in North America, this is still true, for different reasons.

on socialization patterns has become an important area of sociological research.

In the following pages, we briefly consider several agencies of socialization: the family, the schools, the community, the peer group, and the mass media. We will encounter many of these agencies in later chapters as well.

The Family The family is the primary agency of socialization. It is the environment into which children are born and in which their earliest experiences with other people occur—experiences that have a lasting influence on the personality. Family environments vary greatly, not only in terms of such key variables as parents' income and education but also in terms of living arrangements, urban versus rural residence, number of children, relations with kin, and so on. Much contemporary research centers on the effects of different family environments on the child's development (De Visscher & Bouverne-De Bie, 2008; Eshleman, 2003).

Family Dinners and Child Development
Recent research confirms what many parents know intuitively: Enormous benefits are enjoyed by children raised in families whose members get together regularly around the dinner table. For example, a study by the National Center on Addiction and Substance Abuse (CASA; 2005) at Columbia University found that children who reported that they had family dinners were less

likely to develop drug abuse problems as teenagers and young adults and more likely to do well in school. The strength of this finding seems to diminish somewhat with family size because, with more children, there is less parental attention to go around, but the findings are striking nonetheless. In another study of 4,746 middle and high school students, social scientists at the University of Minnesota found that girls whose families rarely ate together were 75 percent more likely to use extreme dieting techniques and develop eating disorders than were girls from families who often had dinner together.

In a pioneering quantitative study of interactions around the dinner table, Michael Lewis and Candice Feiring studied mealtime in 117 American families. Their study, "Some American Families at Dinner" (1982), shows that a typical three-year-old child interacts regularly with a network of kin, friends, and other adults who may play a significant role in the child's early socialization (see the Research Methods box on page 114).

One of the most significant changes occurring in American society today is the increase in the proportion of children growing up in poor families (20 percent of all children under age eighteen) or in no family at all. In the early 1990s, nearly 10 percent of all children in the United States were growing up in households with no parent present. Most of these children (76 percent) were being socialized in the homes of grandparents and relatives; the remainder were living in group homes and with nonrelatives (Gross, 1992). There has also been a steady increase in the number of single-parent families, the vast majority of which are headed by women (Corsaro, 2005).

But changes in the way families are organized or how they cope with changing social conditions are not a new phenomenon. The values and ways of parents are never entirely valid for their children, although the degree to which this is true depends on how much social change is experienced from one generation to the next. Socialization creates the personalities and channels the behaviors of the members of a society, but that socialization is never entirely finished. Thus, in *Manchild in the Promised Land,* his masterpiece about growing up in Harlem during the period of rapid migration of blacks from the South, Claude Brown (1966) wrote that his rural-born parents did not "seem to be ready for urban life." Their values and norms of behavior made no sense to their son, who had to survive on Harlem's mean streets:

> When I was a little boy, Mama and Dad would beat me and tell me, "You better be good," but

Measures of Central Tendency When we say that the family is a primary agent of socialization, we mean not just a child's parents but all the people who make up the "family constellation"—that is, the people who are most often in the child's home and therefore capable of influencing his or her socialization. But how can we measure such a thing across many different families? In an innovative study of whom children interact with other than their parents, Lewis and Feiring observed ten very young children from different families and counted the number of people they saw each week. Here is an example of their data and the basic methodological tools they used to analyze it.

The following table presents information from Lewis and Feiring's study of three-year-old children's social networks:

© Spencer Grant/PhotoEdit

The Three-Year-Old Child's Social Network ($N = 117$)*

	X	Range
Number of relatives other than parents seen at least once a week	3.20	0–15
Number of child's friends seen at least once a week	4.43	0–13
Number of adults seen at least once a week	4.38	0–24

*N refers to the number of subjects in a study.

Source: Lewis & Feiring, 1982, p. 116.

To understand this table, you must know something about the common measures of central tendency. The measure used here is the *mean,* represented by X. To arrive at this number, each child's total number of friends, relatives, and other adults seen at least once a week is added to the totals for the other children in the sample. The grand total is then divided by the number of children in the sample. The resulting number can be used to represent the entire sample.

The other common measures of central tendency are the *median* and the *mode.* The median is the number that divides a sample into two equal halves when all the numbers in the sample are arranged from lowest to highest. The mode is simply the score that occurs most often in the sample. Sociologists make a point of specifying which of these measures they are using. They try to avoid the term *average* in statistical tables because it can refer to any one of these measures.

To illustrate all three measures of central tendency, let us use the following data from ten of the children in Lewis and Feiring's sample:

Child	Number of Friends Seen at Least Once a Week
1	2
2	3
3	3
4	7
5	12
6	2
7	3
8	1
9	0
10	5
	38

In this example, the mode would be 3 and the median would also be 3, but the mean would be 3.8.

I didn't know what being good was. To me it meant that they just wanted me to sit down and fold my hands or something crazy like that. Stay in front of the house, don't go anyplace, don't get into trouble. I didn't know what it meant, and I don't think they knew what it meant, because they couldn't ever tell me what they really wanted. The way I saw it, everything I was doing was good. If I stole something and didn't get caught, I was good. If I got into a fight with somebody, I tried to be good enough to beat him. If I broke into a place, I tried to be quiet and steal as much as I could. I was always trying to be good. They kept on beating me and talking about being good. And I just kept on doing what I was doing and kept on trying to do it good. (p. 279)

Brown's story is that of a young man whose parents' experience with social change created a severe disjunction in their lives. This left them ill equipped to socialize their son for the demands of a new environment, so he learned to survive on the streets. He became a thief and a gang fighter. But by his own account, later in his life he was greatly influenced by people who had studied the social sciences and created well-functioning institutions, special schools in particular, that could bring out his talents and socialize him for a more satisfying and constructive life than he had led as a child.

The Schools For most of us, regardless of what our home life is like, teachers are generally the first agents of socialization we encounter who are not kin. In some cases, children are also influenced by agents of socialization in the church (for example, priests or rabbis); for most children, however, the school is the most important agency of socialization after the family. Children experience many opportunities to perform new roles in school (for example, student, teammate, friend). No wonder, then, that in these first "public" appearances, they tend to be highly sensitive to taunts or teasing by other children. Even very young children can become distraught when they feel that they are not wearing the "right" clothes.

Schools are institutions where differences between the values of the family and those of the larger society come into sharp focus. In some families, for example, there may be great concern about any form of sex education in school. Much of the conflict over educational norms—what ought to be taught, whether there should be prayer in school, and the like—stems from differences between the values of some families and the values that many educators wish to teach. Such conflicts point to the exceedingly difficult situation of schools in American society. Research on school–family and school–community relations has shown that the schools are expected to conserve the society's values (by teaching ideals of citizenship, morality, family values, and the like) and, at the same time, to play a major role in dealing with innovation and change (by expanding the curriculum to include new knowledge, coping with children's perceptions of current events, addressing past patterns of discrimination, and the like) (Meier, 2008).

In Chapter 17, we return to questions of whether the schools help reproduce the status quo and maintain inequalities in society and whether they can actually affect the lives of disadvantaged students. Although many individuals credit schools (and even particular teachers) with helping them improve their lives in important ways, there is much debate on these questions in the social sciences.

The Community Schools may be the most important agency of socialization outside the family, but in most communities one can find many other significant agencies of socialization, including day care centers, scout troops, churches, recreation centers, and leagues of all kinds. These agencies engage in many and varied forms of socialization. Parades on Memorial Day and the Fourth of July, for example, reinforce values of citizenship and patriotism. Participation in team sports instills values of fair play, teamwork, and competitive spirit. The uniforms and equipment required for these activities stimulate shopping trips to suburban malls and downtown department stores, and those trips prepare children and adolescents for the time when they will be consumers (Bravo, Cass, & Tranter, 2008; Newman, 1988).

Of all these agencies of socialization, the day care center is perhaps the most controversial. Polls often show that Americans have doubts about the effects of day care on very young children, and scandals involving charges of child abuse by day care workers intensify those fears. But when both parents work outside the home, or when single parents must work to support their children, day care centers may play a

critical role in socialization. Many studies have shown that high-quality centers are not harmful to children and may be beneficial in some cases, but the norm requiring that mothers stay home to care for young children remains strong in many communities (Kammerman, 1995).

The central question that these studies address is the quality of the care received. A longitudinal study of 1,153 children at ten different child care centers revealed that, in the better centers (as measured by indicators such as the qualifications of the caregivers and the amount of individual attention the children receive), children actually made gains in language and thinking abilities that were superior to those made by children from similar backgrounds who were not enrolled in high-quality day care. This effect was especially strong for children from lower-income households (Garrett, 1997). On the negative side, however, the researchers found that the more time young children spent in centers with lower-quality care, the more likely they were to experience problems in forming close attachments with their mothers.

Conservative critics of day care often argue that mothers of young children should stay home and care for them. But about three of every five mothers with children under age three are in the labor force. They and their working spouses are dependent on various forms of child care. Whatever the reasons for this dependence, it is a reality of contemporary life. High-quality child care is becoming an increasingly pressing social need as a consequence of long-term changes in the economy that have resulted in large numbers of dual-earner families. At present, there are far too many low-quality day care facilities—as many as 40 percent of the total, according to one estimate—providing care to children in the United States (McLanahan, 2010).

Religion Religion may be involved in socialization in different ways throughout an individual's lifetime. For children and adults who attend church services, the shared expression of spirituality and the teaching of the religious leader can be an extremely powerful source of moral beliefs. These beliefs often shape one's behavior outside religious institutions. During World War II, for example, most of the people who risked their lives to rescue Jews, gypsies, gays, and political refugees had deep religious and moral convictions (Winik, 1996).

Sociological studies show, too, that among contemporary parents, those who attend religious services are half as likely to divorce as parents who never attend religious services. Again, the moral influence of religion, whether Christian, Muslim, Jewish, or other, is highly

then & now

The Indomitable Human Spirit Blind and deaf as a result of a serious illness when she was only nineteen months old, Helen Keller appeared incapable of learning. In the late nineteenth and early twentieth centuries, people like Helen were relegated to back wards of custodial care institutions or back bedrooms in affluent homes. However, through the patient instruction of a gifted teacher, Helen learned to understand the connection between objects and the words for them. Suddenly this profoundly handicapped girl had a future in the social world. Helen Keller became a world-famous symbol of the unconquerable human spirit, blazing a trail that other people with disabilities and special needs have followed ever since.

In a more recent example of the triumph of will over biological destiny, nineteen-year-old Craig Ludin is shown at his high school graduation. Craig was the first student with Down's syndrome to graduate from a regular high school in his school district.

AP Photos

Courtesy of the Ludin Family

associated with moral beliefs that may help married couples endure hard emotional times. To the degree that religious behavior reduces the likelihood of divorce, it strengthens the family, the most important agency of socialization (Clydesdale, 1997). Finally, among elderly people, religion often becomes a major source of solace as they face the continual deaths of loved ones and their own eventual demise.

Despite its contributions to lifelong socialization, religion can also be the source of innumerable conflicts. Disputes over child rearing, sexual conduct, the role of prayer in education, and many more rancorous conflicts are brought to public awareness by people with differing religious backgrounds. Should gay men and women serve as spiritual leaders? Should same-sex marriages be performed in churches and synagogues? Is prayer in the classroom a violation of other people's right to the constitutionally guaranteed separation of church and state? These questions hardly begin to exhaust the list of controversies that are based on differences in the moral teachings of various religious communities.

In 2010, President Obama issued an executive order that reversed the "don't ask, don't tell" policy of the U.S. armed forces concerning gay men and women in the military. That policy, in effect since the Clinton administration, tolerated gay enlisted personnel as long as they did not publicly acknowledge their homosexuality. With the approval of his close military advisors, the president reversed this policy, much to the scorn of many who oppose homosexuality and to the relief of gay leaders. We return to these issues in Chapter 16, but note here that sociologists throughout the world are among those in the forefront of efforts to find ways in which people with strong convictions on either side of these disputes can come together peacefully to resolve their differences or work out compromises that prevent moral conflicts from worsening (Vondra, 1996).

The Peer Group In the United States, the peer group tends to be the dominant agency of socialization in middle and late childhood. A **peer group** is an interacting group of people of about the same age. Among adolescents, peer groups exert a strong influence on their members' attitudes and values. Studies confirm the high degree of importance adolescents and adults alike attach to their friendship groups. Adolescents typically acquire much of their identity from their peers and consequently find it difficult to deviate from the norms of behavior that their peer group establishes (Homans, 1950; Lansford et al., 2009; Whyte, 1943). In fact, the peer group may become even more important than the family in the development of the individual's identity.

There is often a rather high level of conflict within families over the extent to which the peer group influences the adolescent to behave in ways that are not approved of by the family. Even where conflict is limited, the peer group usually provides the child's first experience with close friendships outside the family. The peer group becomes the child's age-specific subculture—that is, a circle of close friends of roughly the same age, often with shifting loyalties.

The peer group usually engages in a set of activities that are not related to adult society. The peer subculture may, for example, include games that adults no longer play and may have forgotten. British researchers Iona and Peter Opie (1969) conducted a classic study of the games children play. They identified hundreds of games that were known by children between the ages of six and ten all over England. They also identified the common elements in those games, such as chasing, pretending, or seeking, and found that the central problem in many of the games was to choose an "It," a seeker or chaser. The children's efforts to avoid being "It" seemed to express their desire to avoid roles that made them different from the others.

peer group: An interacting group of people of about the same age that has a significant influence on the norms and values of its members.

When children's peer groups are faced with conflict and social change—for example, in communities with high levels of poverty and demographic change, or with groups of culturally distinct people moving in and out—the group often organizes itself in gangs for self-defense and aggression. There are many such gangs in large American cities characterized by rapid immigration, widespread drug use, and the ready availability of cheap handguns. Under these conditions, both male and female peer groups may become extremely dangerous environments for socialization, yet they exert a strong attraction on adolescents who seek protection and companionship in what they perceive as a hostile world (Sanchez-Jankowski, 1991; Williams, 1989).

The Mass Media The most controversial agency of socialization in American society is the mass media. In debates about the effects of the media on socialization, television comes under the greatest scrutiny because of the number of hours children spend in front of the "electronic babysitter." Estimates of how much television children watch differ, depending on the methodology used to conduct the study (especially because simply having the set turned on does not mean that the children are actually watching). Many studies have shown that the amount of television viewed varies, depending on whether the child comes from a poor home with few alternatives or a more affluent one where other activities are available. Children from poor homes in urban communities often have the set turned on for seven or more hours a day, as opposed to children from more affluent homes, where the set is turned on for three or four hours daily (Corsaro, 2005).

The effects of all this television viewing on children and adolescents are a subject of intensive research. In particular, the effects of seeing violent acts on television or listening to violent music like gangsta rap are hotly debated. There is no doubt that the amount of violence shown on TV is immense. George Gerbner, one of the nation's foremost researchers on TV violence, has been monitoring the number of violent acts shown on TV. His data indicates that the average child will have seen about 100,000 acts of violence on TV before graduating from elementary school—and the number is far higher for children from poor neighborhoods (Gerbner, 1990).

Gerbner's research also suggests that children and adolescents are spending increasing amounts of time playing violent video games, which may also increase their own tolerance for violence (Gerbner, 2001). His concerns are confirmed in more recent research on video games and children's violence that shows a direct correlation between number of hours playing violent games and the propensity to act violently. These effects are particularly strong in newer games in which players are able to create their own, often violent, characters (Fischer, Kastenmüller, & Greitemeyer, 2010).

Studies by Dale Kunkel and associates have also looked at sexual content on television. Their analysis of the content of TV shows indicates that nearly two of every three shows (64 percent) contain some sexual content. Across all programs with any sexual content, an average of 4.4 scenes per hour involve sex. Their data reveal that "not only are sexual talk and behavior a common element in television programming, but . . . most shows including sexual messages devote substantial attention to the topic" (Kunkel et al., 2007, p. 605). Researchers are continuing to study the effects of watching many hours of television with high levels of sexual content on children's behavior, but this is a genuine concern for many parents.

It is important to recognize, however, that the effects of television on socialization are by no means all negative. Television, radio, and the

Researchers are finding that children who watch a great deal of violence or play many hours of violent video games are showing increases in violent thoughts as well as a tendency to view violence as a more normal response than is true of children who do not consume violent media.

movies provide windows into social worlds that most people cannot otherwise enter. Children, adolescents, and adults in modern American society learn far more about current events, social issues, and the arts than people did before the advent of television.

SOCIALIZATION THROUGH THE LIFE COURSE

Childhood socialization is the primary influence on the individual. Whatever happens to a person later in life, the early childhood experiences that shaped the social self will continue to influence that person's attitudes and behavior. But people are also affected by their *life course:* the set of roles they play over a lifetime and the ways in which those roles change as a consequence of social change. The life course of a person who came of age just before World War II, for example, would typically have involved service in the armed forces for men (perhaps including combat duty) and work in a factory (or noncombatant military service) for women. The life course of children born after the war is far less likely to include a long or intense experience of warfare. Natural disasters, major changes in the educational system, shifts in the political stability of their society, and other changes can influence the roles people play during their life course.

Changes in the culture of a society can also significantly alter the roles people play during their lives. In the first half of the twentieth century, for example, the role of wife was less likely to include periods of single parenthood than it is now. Fifty years ago, the role of an elderly person was far more likely to involve relative poverty and dependence on younger family members than it does now. The need to adjust to changes in society and culture, and the fact that so many people are able to make such adjustments (though often with great difficulty), are evidence that socialization is not finished in childhood but continues throughout the life course (Riley, Foner, & Waring, 1988; Settersten & Owens, 2002).

Adult Socialization

Two major sources of socialization after childhood are significant others (especially new friends) and occupational mobility (especially new jobs). When one moves into a new neighborhood, for example, one often makes new friends, and their influence may lead to new activities. A person who has always been uninvolved in politics may, through the influence of new friends, become committed to a social movement such as the women's movement. New friends can also introduce a person to dangerous or unlawful activities such as drug use. One's *core identity*—the part of the self formed in early childhood that does not change easily—may prevent one from being overly influenced by a given peer group later in life, or it may actually cause that influence to be stronger (Elkin & Handel, 1989).

Such factors as stress and changing physical health may also have an influence on socialization. In a study of how people change their patterns of behavior later in life, Marjorie Fiske and David A. Chiriboga (1990) found that everyday hassles and the boredom of a routine, predictable existence have a greater impact on the changes adults seek in their lives than earlier research indicated.

In any new activity, the newcomer or recruit must learn a new set of norms associated with the roles of the organization. Socialization associated with a new job, for example, often requires that a person learn new words and technical terms. The individual also needs to interact with a new peer group, usually composed of people whose work brings them together for long hours. And in most jobs, the person will eventually be faced with choices between loyalty to peers and loyalty to those higher in the organizational hierarchy who can confer benefits such as raises and promotions. In these and most other adult socialization experiences, the choices people make will be highly influenced by the individual's core identity, which continually shapes his or her responses to new situations and challenges (Jones-Correa & Ajinkya, 2007; Rosow, 1965).

Resocialization

Many adults and even adolescents experience the need to correct certain patterns of prior social learning that they and others find detrimental. This need usually cannot be met through the normal processes of lifelong socialization, especially because the individual's negative behaviors may be leading toward a personal crisis and causing immense pain for family members and friends. Such a condition often responds to specific efforts at resocialization.

Resocialization is a process whereby individuals undergo intense, deliberate socialization designed to change major beliefs and behaviors.

resocialization: Intense, deliberate socialization designed to change major beliefs and behaviors.

A Study in Altruism Whether it is the earthquake in Haiti, the tsunami in Indonesia, or the terrorist attack on the World Trade Center towers, as thousands of people flee for their lives, others rush in to rescue survivors. At the World Trade Center in 2001, for example, firefighters, emergency medical technicians, police officers, and civilians were the heroes of the tragedy. Theirs were acts of great altruism. They put their own survival at risk and sacrificed themselves to save others. We rightly honor anyone who displays such heroic altruism.

Surely some aspects of the ability to act in self-sacrificing ways on behalf of others are innate, but just as surely this behavior can be learned as well. In the case of firefighters, police officers, and others whose jobs require them to take risks to save others, much of what we understand as altruistic behavior is instilled during their rigorous training as recruits. A few weeks after the September 11 tragedy, the Fire Academy of New York City graduated 250 new firefighters. On the stage at the graduation ceremony were six empty chairs. Each was marked with an American flag or other symbol indicating that the seats were vacant because these probationary firefighters, or "probies" as they are called, were killed when the buildings collapsed. They were serving as trainees in firehouses that responded early to the alarms and thus had the greatest number of casualties. Like the more seasoned veterans who were killed in the collapse of the buildings, the probies were "just doing their jobs."

Another aspect of what we see as altruism comes from the intense bond of friendship that is formed by small groups who encounter highly dangerous situations. The self-sacrificing behavior that we understand as altruism also stems from each individual's belief that he or she must pitch in to help the group, not just to help strangers. The intensity of the peer bonds among firefighters, for example, helps explain the intense feelings they displayed when asked to withdraw from the Ground Zero site. The firefighters regard the area as sacred ground and did not want to be excluded from the search for the bodies of their missing comrades.

The altruistic behavior of so many on that day also provided a model of altruism for a generation of children throughout the world. The Halloween following the event saw a tremendous display of firefighter, nurse, and police costumes as children embraced and emulated the behavior of their new heroes.

It is often aimed at changing behaviors such as excessive drinking, drug use, and overeating, which are particularly common in affluent societies, where individuals are exposed to a great deal of choice and many pleasurable stimuli. Perhaps most widespread and pernicious among the addictions, alcoholism affects millions of people in North America and is involved in much of the marital strife that leads to family breakup. Alcoholics Anonymous (AA) is a voluntary group program that uses techniques of resocialization and peer example to help group members reject behaviors based on drinking and learn new, more positive ones that do not involve drinking. The AA approach to voluntary resocialization, which emphasizes personal change and positive growth as the individual passes through twelve steps of recovery, has been successfully applied to problems of obesity and overeating (Overeaters Anonymous) and narcotic use (Narcotics Anonymous). Many other self-help support groups assist people in dealing with almost any kind of problem that they previously struggled with in isolation (Cloud et al., 2007).

Much successful resocialization takes place in what is often called a **total institution** (Goffman, 1961). Total institutions are settings in which people undergoing resocialization are isolated from the larger society under the control of a specialized staff whose members may have experienced the same process of resocialization. All aspects of the patients' daily lives are controlled, and supervision is constant. Drug treatment centers, for example, are often set up as total institutions for the resocialization of people with addictions. Even the smallest rewards, like extra time alone or more freedom to walk around the grounds, are controlled by the staff and bestowed only on patients who have made progress toward the resocialization goals.

In the resocialization process, the staff first attempts to tear down the individual's former sense of self. This stage may include various

total institution: A setting in which people undergoing resocialization are isolated from the larger society under the control of a specialized staff.

TABLE 5.2

Erikson's View of Lifelong Socialization

Stage of Life	Conflict	Successful Resolution in Old Age*
Infancy	Trust versus mistrust	Appreciation of interdependence and relatedness
Early Childhood	Autonomy versus shame in the development of the will to be a social actor	Acceptance of the cycle of life, from integration to disintegration
Play Age	Initiative versus guilt in the development of a sense of purpose	Humor; empathy; resilience
School Age	Industry versus inferiority in the quest for competence	Humility; acceptance of the course of one's life and unfulfilled hopes
Adolescence	Identity versus confusion; struggles over fidelity to parents or friends	A sense of the complexity of life; merger of sensory, logical, and aesthetic perception
Early Adulthood	Intimacy versus isolation in the quest for love	A sense of the complexity of relationships; value of tenderness and loving freely
Adulthood	Generativity versus stagnation in interpersonal relationships	*Caritas* (caring for others) and *agape* (empathy and concern)
Old Age	Integrity versus despair	Wisdom and a sense of integrity strong enough to withstand physical disintegration

*Erikson did not believe that elderly people necessarily resolve these conflicts entirely, but when they do, the results are as shown here.

forms of degradation and abasement in which the individual is forced to reject the undesired thoughts and behaviors. In the next stage, the staff rewards the individual's attempts to build a new sense of self that conforms to the goals of the resocialization process (for example, a sober person who accepts responsibility for previous wrongs and achieves new interests and new awareness of personal strengths). However, although total institutions can be extremely powerful environments for resocialization, they sometimes run the risk of creating people who need to remain in their controlling environment—that is, people who have developed a need to be "institutionalized."

Erikson on Lifelong Socialization

The theories of social psychologist Erik Erikson are especially relevant to the study of adult socialization and resocialization. Erikson agreed with Freud that early-childhood experiences shape a person's sense of identity. But Erikson also focused on the many changes occurring throughout life that can shape a person's sense of self and ability to perform social roles successfully. He demonstrated, for example, that combat experiences can produce damaged identities because soldiers often feel guilty about not having done enough for their fallen comrades (this is sometimes called "survivor guilt"). For Freud, in contrast, war-produced mental illness was always related to problems that the soldier had experienced in early childhood.

In *Childhood and Society* (1963), Erikson's central work on the formation of the self, the concept of identification takes center stage. **Identification** is the social process whereby the individual chooses other people as models and attempts to imitate their behavior in particular roles. Erikson noted that identification with these role models occurs throughout the life course. He pointed out that even older people seek role models who can help them through difficult life transitions.

For Erikson, every phase of life requires additional socialization to resolve the new conflicts that inevitably present themselves. Table 5.2 presents the basic conflicts that each person must resolve throughout life and indicates how a positive role model can help in resolving those conflicts. (The major theories of socialization are summarized in the For Review chart on page 121.)

Resocialization at any age usually requires a strong relationship between the individual and the role model. In the military or in a law enforcement agency, for example, the new recruit is usually paired with a more experienced person who offers advice and can show the recruit the essential "tricks" of the new role—"how it's really done." Thus, in Alcoholics Anonymous, once the self-confessed "drunk" faces up to that negative identity, he or she is taught new ways of thinking, feeling, and acting that do not involve drinking. An important step in this process is the development of a relationship with a *sponsor,* a person who has been through the same experiences and has made a successful recovery. In fact, one of AA's aims is to enable the resocialized alcoholic to eventually perform as a sponsor—that is, as a role model—for someone else who is going through the same process.

identification: The social process whereby an individual chooses role models and attempts to imitate their behavior.

Theories of Socialization

Theorist	Description of Theory
Sigmund Freud	Through socialization, the infant is gradually forced to channel its biological urges into socially acceptable forms of behavior. Major personality traits are formed in the conflict between the child and its parents, especially the same-sex parent.
George Herbert Mead	Socialization is a process in which the self emerges out of interaction with others, not only in early infancy but throughout life. Role taking is central to socialization; role-taking ability develops through interaction during childhood in the preparatory, play, and game stages.
Jean Piaget and Lawrence Kohlberg	Children develop definite awareness of moral issues at an early age but cannot deal with moral ambiguities until they mature further. This insight is incorporated in Kohlberg's preconventional, conventional, and postconventional stages of moral development.
Erik Erikson	Throughout the life course, the individual must resolve a series of conflicts that shape the person's sense of self and ability to perform social roles successfully. Central to this process is identification, in which the individual chooses other people as role models.
Carol Gilligan	Children tend to develop different ways of resolving moral dilemmas. Some (more often male) tend to rely on strict rules of right and wrong; others (most often but not always female) tend to make judgments based on notions of fairness and cooperation. In societies where male voices are dominant, issues of cooperation and fairness are passed over, to the detriment of female socialization for leadership and achievement.

Thinking Critically

Once people have been socialized in families, schools, and peer groups, can they still change? Of course they can. They can learn new trades or skills, give up drinking and drugs, make new friends and join new groups, and much more. But can a shy, introverted person be resocialized to become an outgoing extravert, or a self-involved person to become a self-sacrificing leader?

GENDER SOCIALIZATION

No discussion of socialization would be complete without some mention of gender socialization. **Gender socialization** refers to how we learn our gender identity and develop according to cultural norms of "masculinity" or "femininity." Gender is not synonymous with biological sex; rather, "scholars use the word *sex* to refer to attributes of men and women created by their biological characteristics and gender to refer to the distinctive qualities of men and women (or masculinity and femininity) that are culturally created" (Epstein, 1988, p. 6). By *gender identity,* we mean "an individual's own feeling of whether she or he is a woman or a man, or a girl or a boy" (Kessler & McKenna, 1978, p. 10). Our ideas about what it means to be a man or a woman are different from our ideas about the anatomical definitions of *male* and *female.* These ideas are shaped by the values and socialization practices of our culture, but once they have become part of the self, they are usually extremely strong (Blakemore & Hill, 2008).

The strength and pervasiveness of gender socialization are illustrated by the case of Ron Kovic, the author of *Born on the Fourth of July* (played by Tom Cruise in the Oscar-winning movie directed by Oliver Stone). Kovic saw himself as a red-blooded American male, a typical product of socialization for boys raised in his community. He was taught to be patriotic, to want to defend his country, to want to prove his manhood in combat and sexual conquest. When he became paralyzed in Vietnam and lost his manhood (in the narrow, conventional sense of that term), he began to doubt everything—including himself, his country, and his friends. His story raises some important issues, not least of which is how we form our ideas of manliness and womanliness and whether those ideas are helpful in meeting the demands of a changing world.

The story of Ron Kovic is a striking example of how powerful gender socialization is and the effects it can have on a person's life. Elsewhere in this book (especially in Chapter 13), we present other examples of the importance of gender in explaining life chances. But here, let us look at a more mundane example of the results of gender socialization: the way men and women view their own bodies. Men tend to think of their own bodies as "just about perfect," but women tend to think of themselves as overweight, at least when they compare themselves

gender socialization: The ways in which we learn our gender identity and develop according to cultural norms of masculinity and femininity.

with a mental image of an attractive woman. These are the findings of a survey of 500 college-age men and women conducted by April Fallon and Paul Rozin (cited in Goleman, 1985).

A striking indication of the role of agencies of socialization in the social construction of gender identity is the influence of the mass media in promoting ideal body types for men and women. These influences have a lot to do with why men tend to be happy with their bodies whereas many girls and women are extremely unhappy with theirs. Men dominate the media—radio, television, newspapers, and magazines—that communicate American culture to mass audiences. Girls and women are continually exposed to images of women created by men, and those images more often than not portray women as sex objects, to be judged by the shape and size of their breasts, the length of their legs, and the tightness of their skin (Chernin, 1981; Fenton et al., 2010). It is little wonder, then, that women feel worse about themselves than do men—about whom it has been said that it is sexy to be a sagging warrior or a somewhat wrinkled philosopher.

Among the many momentous changes occurring in the world today, those affecting gender socialization are among the most profound. Research on socialization in Asia and Africa, for example, shows how norms that favor boys over girls act to the detriment of female development. Studies from Bangladesh, India, Nepal, and Pakistan reveal that boys receive more food than girls, even though girls are often required to expend more calories on field and household work. In many poor nations, it is common for boys to be fed before girls and for the higher-quality nutrients, especially proteins, to be consumed before the girls are allowed to eat. So strong are gender norms in those countries that most mothers are convinced that this practice is necessary when food is scarce (Jacobson, 1992). As a result of this and other socialization practices that place females at a disadvantage, the health and learning capacities of women are imperiled from an early age in many parts of the world.

SUMMARY

What are the major experiences that account for the way an individual becomes a "social being"?

- *Socialization* refers to the ways in which people learn to conform to their society's norms, values, and roles.

- Primary socialization consists of the ways in which a newborn individual is molded into a person who can interact with others according to the expectations of society. Secondary socialization occurs in childhood and adolescence, primarily through schooling, and adult socialization refers to the ways in which a person learns the norms associated with new statuses

How much of the way people behave is determined by social learning and how much by genetic factors?

- Among the most basic questions in the study of socialization is that of "nature" versus "nurture": To what extent does the development of a person depend on genetic factors, and to what extent does it depend on learning?

- The first social scientist to develop a theory that addressed this issue was Sigmund Freud. Freud believed that the personality develops out of the processes of socialization through which the infant is gradually forced to control its biological urges. He divided the personality into three functional areas: the *id*, from which unsocialized drives arise; the *superego*, which incorporates the moral codes of elders; and the *ego*, or one's conception of oneself in relation to others.

- In the growth of the personality, the formation of the ego or social self is critical. According to Freud, this takes place in a series of stages in which conflict between the demands of the superego and those of the id is always threatening to disrupt the functioning of the ego.

- *Behaviorism* asserts that all behavior is learned. It originated in the work of Ivan Pavlov, who showed that behavior that was thought to be instinctual could in fact be shaped or *conditioned* by learning situations. This line of research was continued by John B. Watson, whose experiments revealed the

- ability of conditioning to shape behavior in almost any direction.

- Studies of *feral children,* who have experienced extreme isolation or have been reared outside human society, show that such children are able to learn but that they do so far more slowly than children who have not been isolated in early childhood.

- Other studies have found that normal development requires not only the presence of other humans but also the attention and love of adults. Children raised in orphanages and other nonfamily settings are more likely to develop emotional problems and to be retarded in their language development than comparable children reared by their parents.

- The role of genes in shaping traits such as intelligence and sexual orientation is a subject of continual research and much controversy. In neither case is there any definitive evidence that specific genes determine these highly significant aspects of human behavior.

When sociologists study how people "construct a social self," does this mean they believe that only socialization influences human development?

- Interactionist models of socialization stress the development of the social self through interaction with others.

- One of the earliest interactionist theories was Charles Horton Cooley's concept of the "looking-glass self," the reflection of our self that we think we see in the behaviors of other people toward us.

- This concept was carried further by George Herbert Mead, who emphasized the importance of culture in the formation of the self.

- Mead believed that when children play, they practice *role taking,* or trying to look at social situations from the standpoint of another person. This ability develops through three stages. During the preparatory stage, children mimic the behavior of the *significant others* in their social environment. During the play stage, they play at being others who are significant in their lives. During the game stage, they

develop the ability to take the role of the *generalized other*—that is, to shape their participation according to the roles of other participants.

- In playing the roles for which they have been socialized, people adhere to rules of interaction known as "face work." They seek to present a positive image of themselves, their "face," and to avoid being embarrassed or "losing face."

- Lawrence Kohlberg proposed a three-stage sequence of moral development in which the child's moral reasoning evolves from emphasis on reward and punishment to the ability to distinguish between social laws and moral principles.

- Other sociologists, especially Carol Gilligan, have challenged Kohlberg's theory on the ground that it does not distinguish between moral reasoning based on rules and justice (most common in males) and moral reasoning based on fairness and cooperation (most common in females).

- Research on gender and emotion has shown that—whether one is male or female—by better understanding what goes into emotional intelligence, one can vastly improve one's personal relationships and one's understanding of why others react as they do in different social situations.

In what ways do social environments influence the people we become as we mature in our social worlds?

- Studies of the environments in which socialization occurs have found that normal development requires the involvement of one or more adults in the care of the child, as well as public policies that promote such involvement.

- *Agencies of socialization* are the groups of people, along with the interactions that occur within those groups, that influence a person's social development.

- Within all agencies of socialization, one finds a great deal of *anticipatory socialization,* in which individuals play at a role that they are likely to assume later in life.

- After the family, the most important agencies of socialization are schools. Other socializing agencies include day care centers, churches, leagues, and other associations.

- Religion may be involved in socialization in different ways throughout an individual's lifetime.

- The dominant agency of socialization outside the family is the *peer group,* an interacting group of people of about the same age. Peer groups exert a significant influence on individuals, from adolescence on.

- The mass media, especially television, are another significant agency of socialization in American society.

How does learning to take social roles occur throughout a person's lifetime?

- The roles a person plays over a lifetime are influenced by social and cultural change.

- Socialization after childhood often occurs as a result of occupational mobility and the influence of significant others.

- A person's core identity shapes that individual's responses to new situations and challenges.

- *Resocialization* may occur at any time during adulthood. Sometimes people undergo resocialization to correct patterns of social learning that they and others find detrimental.

- Erik Erikson focused on *identification,* the social process whereby the individual chooses adults as role models and attempts to imitate their behavior.

What do sociologists mean by gender identity, and in what ways can it be said to be socialized?

- An important aspect of socialization is *gender socialization,* or the ways in which we learn our gender identity and develop according to cultural norms of masculinity and femininity.

- Gender identity is an individual's own feeling of whether he or she is a male or a female.

The Kornblum Companion Website

www.cengagebrain.com

Supplement your review of this chapter by going to the companion website to take one of the tutorial quizzes, use the flash cards to master key terms, and check out the many other study aids you'll find there. You'll also find special features, such as GSS data and Web links that will put data and resources at your fingertips to help you with that special project or to do some research on your own.

Paul Chesley/Getty Images

6

The Importance of Groups

Characteristics of Social Groups

Interaction in Groups

Formal Organizations and Bureaucracy

Groups in Complex Societies

Interaction in Groups

What are some of the social groups that humans regard as most important in their lives?

What are the key characteristics of social groups?

A few basic principles tend to govern interaction in groups. What are these, and how do sociologists study these interactions?

Why are bureaucracies such a prominent feature of contemporary societies?

What roles do various kinds of social groups play in complex societies?

What did being a gang member mean to you?" criminologist Dana Nurge recently asked a young woman from Boston. This is how she responded:

Columbia University Press

> I was always left out in some way and then when I met them (this was before I graduated from high school) I was more than part of the family. . . . They had this need for me. Everyone wants to feel needed and I felt really at home with them. The other thing was I wasn't alone, anytime day or night, I could just pick up the phone and call someone. Financial-wise, also, I remember a couple times, like my first time moving out, I remember a couple times being worried about making rent and all the responsibilities. Guys would just come by with the groceries. . . . They helped me be able to go to college. . . . I couldn't have done it without them. (Nurge, 2003, p. 175)

After years of studying gangs in San Antonio, Texas, sociologist Avelardo Valdez found examples like the this one, in which gang members helped their affiliated girls achieve success in the larger society, but far more often, his observations showed the cost of gang membership, or just the association of women with men in gangs, as it increased the risk of violence and injury. The following excerpt from his study is typical. The speaker is a 14-year-old girl who has been hanging out with members of the Invaders, a Chicano gang in San Antonio:

> I was downtown coming from my grandpa's . . . I was by myself. This chick bumped into me so I was like, "Hey watch out, say excuse me or something next time." She started tripping [getting mad]. All of a sudden she goes, "Yeah I know you, you're with the . . ." I just see her coming with her homegirls. I saw a guy I knew from the gang. He goes, "Hey what's up?" I give him my purse to hold and we just strapped it on. I sort of brought it on to myself because I didn't have to say nothing but I still did. She pushed me, so I just hit her and I threw her to the floor. Then somebody kicked me in the mouth and I was like what, it hurt and I started bleeding. I got up and I went at her and I busted her nose. (Valdez, 2007, p. 103)

In a parallel study of women members of the Latin Queens, termed a gang by the police but known to its members as a "street organization," sociologists David Brotherton and Camila Salazar-Atias (2003) observe:

> Most Queens come from backgrounds where racism, sexism, poverty, and limited opportunities are prevalent. They joined the nation in search of empowerment, identity, and support to compensate for a life in which they have often been left to fend for themselves as children and then left with minimal support to take care of their own children. (p. 205)

In the organization, they encountered new forms of sexism and double standards, which they often had to resist and struggle against, and like the women who spoke to Dana Nurge, they also wished there had been other alternatives for them. But the power of some of the friendships they made is something they have always cherished. This and much previous research emphasizes that gang peer groups have a strong and lasting influence on the men and women who pass through them (Moore, 1991). But gangs are only one of many types of friendship groups. For insights into other types of friendship groups, and glimpses of the strong emotions friends often express within them, see the Visual Sociology box that appears later in the chapter.

THE IMPORTANCE OF GROUPS

The study of gangs or "street organizations" is rich in examples of how people interact in groups. It also shows how the strong norms or values of a peer group often lead gang members to behave violently in particular situations (Horowitz, 1985; Hunt & Joe-Laidler, 2001). Martin Sanchez-Jankowski, who has conducted extensive comparative research on gangs throughout the United States, concludes that, contrary to what people often believe, most gang members do not enjoy fighting. Indeed, those who are not among the gang's leaders fear violence. However, as he reports, "Fear of physical harm . . . is not sufficient to completely deter gang violence, because most gang members . . . believe that if you do not attack, you will be attacked" (Sanchez-Jankowski, 1991, p. 177).

The social fabric of modern societies is composed of millions of groups. Gangs are only one example. Some groups are as intimate as a pair of lovers. Others, like the modern corporation or university, are extremely large and are composed of many interrelated subgroups. We need to perform well in all of these groups, and to have this ability is to be considered successful by others. But the knowledge needed for success in a group with a formal structure of roles and statuses is very different from the "smarts" needed for success in intimate groups.

In this chapter, we analyze some important aspects of the way people behave in groups. We begin by describing the most common types of groups and showing how they can be linked together by overlapping memberships. This first portion of the chapter illustrates the functionalist view of group interaction. It looks at the structure of groups and the way they function in society. It also provides typologies of groups and insights into how they are organized. However, simply describing categories of groups and their structure does not explain how groups maintain the commitment of their members or why they change. In the middle portion of the chapter, we examine these questions, focusing on the give-and-take that occurs in groups of different types and on how these interactions produce group norms. This section illustrates interactionist research on behavior in groups. In the final section of the chapter, we show how formal organization and bureaucracy operate to control conflict and cope with change in groups. The concepts introduced here will be used over and over again in other chapters, for, as we saw in Chapter 4, groups of all kinds are the building blocks of society.

CHARACTERISTICS OF SOCIAL GROUPS

We sometimes group people together artificially in the process of analyzing society (for example, all the teenagers in a particular community). This results in a **social category**, a collection of individuals who are grouped together because they share a trait deemed by the observer to be socially relevant (for example, sex, race, or age). But there are also groups in which we actually participate and in which we have distinct roles and statuses. These are **social groups**. A social group is a set of two or more individuals who share a sense of common identity and belonging and who interact on a regular basis. Group members are recruited according to specific criteria of membership and are bound together by a set of membership rights and mutual obligations. Throughout this chapter, unless indicated otherwise, when we refer to groups we mean social groups.

Group identity, the sense of belonging to a group, means that members are aware of their participation in the group and know the identities of other group members. This also implies that group members have a sense of the boundaries of their group—that is, of who belongs and who does not belong. Groups themselves have a social structure that arises from repeated interaction among their members. In those interactions, the members form ideas about the status of each and the role each can play in the group (Hare et al., 1994; Homans, 1950; Kaplan & Martin, 1999). A skilled member of a work group, for example, may become the leader, whereas an unskilled member may become recognized as the person whom others instruct and send on errands.

Through their interactions, group members develop feelings of attachment to one another. Groups whose members have strong positive attachments to one another are said to be highly cohesive, whereas those whose members are not strongly attached to one another are said to lack cohesion. Groups also develop norms governing behavior in the group, and they generally have goals such as performing a task, playing a game, or making public policy. Finally, because groups are composed of interacting human beings, we must also recognize that all groups have the potential for conflict among their members; the resolution of such conflicts may be vital to continued group cohesion.

Membership criteria, awareness and boundaries, a clear social structure, cohesion, norms and goals, and the possibility of conflict are among the dimensions that define different groups. If we briefly compare two types of groups—say, members of a computer chat group and a married couple—the importance of these dimensions becomes apparent. Membership in the couple is exclusive, and cohesion is based on intimacy and trust. If he comes home and says, "Honey, I brought a new woman into

---

social category: A collection of individuals who are grouped together because they share a trait deemed by the observer to be socially relevant.

social group: A set of two or more individuals who share a sense of common identity and belonging and who interact on a regular basis.

our relationship—I think you'll like her," she is likely to throw him out—unless they happen to come from a culture that practices polygyny, and even then the sudden introduction of a new woman into the household will disrupt the cohesion of the original marital group. Among members of the chat group, in contrast, the only criterion for membership may be ownership of the proper equipment and knowledge that the chat room exists. Not all the members will know one another, and new members can be added without any sense of reduced cohesion as would occur in the case of the married couple.

Group Size and Structure

If one walks along a beach on a nice summer day, the variety of groups one sees is representative of the range of social groups in the larger society. Many of the groups on the beach are composed of family members. Others are friendship groups, usually consisting of young men and women of about the same age. Some of these are same-sex groups; others include both males and females. Then there are groups that form around activities such as volleyball. Still other groups may be based on occupational ties; the lifeguards and their supervisors or the food service workers are examples of such groups.

Primary and Secondary Groups The distinction made in Chapter 4 between primary and secondary groups helps clarify some of the differences among the groups one encounters on a sunny beach or anywhere else in society. Charles Horton Cooley defined **primary groups** as "those characterized by intimate face-to-face association and cooperation. They are primary in several senses, but chiefly in that they are fundamental in forming the social nature and ideals of the individual" (1909, p. 25). Out on the beach, the family groups are, of course, examples of primary groups. So are the groups of friends one observes lying near one another on the sand. Among the food service workers or the lifeguards, there are probably some smaller friendship groups or cliques within the larger occupational group.

The occupational groups serving the beachgoers are examples of **secondary groups** (a concept we also introduced in Chapter 4 but need to use again here). The concept of a secondary group follows from Cooley's definition of primary groups, although he did not actually use

Beaches are excellent places to observe the way people group themselves in public. Note the way people cluster in primary groups and tend to leave as much space as possible between themselves and other groups.

the term (Dewey, 1948). Secondary groups are characterized by relationships that involve few aspects of their members' personalities. In such groups, the members' reasons for participation are usually limited to a few specific goals. The bonds of association in secondary groups are usually based on some form of contract, a written or unwritten agreement that specifies the scope of interaction within the group. All organizations and associations, including companies with employers and employees, are secondary groups. (See the For Review chart on page 129.)

Within most secondary groups, one can usually find several primary groups based on regular interaction and friendship. Scottish clans and Native American tribes, for example, are secondary associations that link individuals together through a network of kin relations. But within the clan or tribe there are many primary groups, often composed of men or women of roughly the same age.

Group Size and Relationships As the number of people in a group increases, the number of possible relationships among the group's members increases at a greater rate. When there are two people in the group, there is just one

---

primary group: A social group characterized by intimate, face-to-face associations.

secondary group: A social group whose members have a shared goal or purpose but are not bound together by strong emotional ties.

Primary and Secondary Groups

Type of Group	Type of Interaction	Contributions to Society	Examples
Primary	Intimate, face-to-face; full range of emotions may be expressed	Help form individual's character and values	Family, peer group, gang
Secondary	Limited to specific roles (e.g., shopper, cashier); limited aspects of individuals' personalities and emotions are expressed	Carry out written and unwritten contractual relationships (e.g., buyer and seller)	PTA, teacher and students, hospital safety committee

relationship; but when a third person is added, there are three possible relationships (A-B, B-C, A-C). When a fourth person joins the group, there are six possible relationships; a group of only six people includes fifteen possible relationships. The Research Methods box on pages 130–131 describes this phenomenon in more detail.

Group Size and Conflict When a group has more than six members, the relationships among the members become so complex that the group is likely to break up into smaller groups, at least temporarily. One often sees this behavior at parties or other social gatherings. Five or six people can carry on a conversation in which all participate, but as more people arrive, the group tends to break up into several smaller groups of two, three, or four people. The difficulty of maintaining relationships as group size increases becomes greater when there are limited resources, such as seats in cars or at restaurants, and the group is forced to define the smaller groups that will interact more intimately. Intimacy and feelings of closeness to others decrease with increased group size.

The more possible relationships there are, the more likely it is that some of those relationships will be troubled by conflict and jealousy. In consequence, the larger the group, the more unstable it tends to be and the more likely it is to break up into smaller groups in which there are fewer possible relationships and a greater likelihood of positive bonds among members. If a larger group devoted to specific goals is to maintain its cohesion in the face of the pressures created by primary attachments among its members, it will need a formal, agreed-upon structure of leaders and rules. Thus, a group of friends who decide to meet regularly to play a favorite game (it could be golf, basketball, or any other activity) will find that as their numbers grow they need a formal structure—an elected or appointed leader, a calendar of events, a membership list, and written rules of conduct. In short, it will be transformed from a primary group into a secondary group or, more specifically, a formal organization.

Dyads and Triads The basic principles of group structure and size were first described by the pioneering German sociologist Georg Simmel (1858–1918). Simmel perceived the need to study behavior in small and larger groups because groups are the basic units of life in all societies. Among his other contributions, he pointed out the significance of what he termed a **dyad**, a group composed of only two people, and a **triad**, a group composed of three people. He recognized that the strongest social bonds are formed between two people, be they best friends, lovers, or married couples. But he saw that dyads are also vulnerable to breakup, because if the single relationship on which it is based ends, the dyad ends as well.

When the dyadic bond is strong (and the two who share it often jealously guard their intimacy), the introduction of a new person into the group often causes problems. Conventional wisdom has much to say about this matter ("Two's a couple, three's a crowd"), and if one observes children's play groups, it soon becomes clear that much of the conflict they experience has to do with the desire to have an exclusive relationship with a best friend. In families, too, the shift from a dyad to a triad

These Italian men have not seen each other for a long time. They clearly have a close dyadic relationship, and it appears that something quite joyous is leading them to express their affection at this moment. But what else can we glean from this photo? Do you think these men are gay? Is it acceptable for men to embrace this way in public?

dyad: A group consisting of two people.

triad: A group consisting of three people.

Diagramming Group Structure A variety of techniques, collectively known as *sociometry*, are used to study the structure of groups. The three most frequently used sociometric methods are basic group diagrams, diagrams that indicate the valence of group bonds, and sociograms, which chart individuals' preferences in groups.

At their simplest, group diagrams show the number of members in a group. Each member is represented by the same symbol, and the presence of a relationship

between two members (a dyad) is shown by a line. The group diagrams in the figure labeled A show how the number of possible relationships increases geometrically as the number of members in the group increases arithmetically. (The formula for calculating the number of possible dyadic relationships in a group, where R is the number of relationships and N is the total number in the group, is $R = N(N - 1)/2$.)

In a slightly more sophisticated type of diagram, a bond between group members is shown only if it actually exists in the group's interactions. Where two members have no relationship, no line is drawn between them. In a peer group, it is usual for all the members to have a relationship, even if they do not all share the same strong feelings of friendship. In work groups, however, it is not uncommon for people to have relationships that are based on cooperation in carrying out specific tasks. If their work does not bring them together, they may not have any relationship at all.

Diagram B represents a group in a school where the third-grade (3G) and fourth-grade (4G) teachers

(A)

2 people
1 relationship

3 people
3 relationships

4 people
6 relationships

5 people • 10 relationships

6 people • 15 relationships

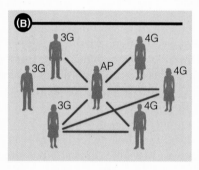

(B)

often creates problems. A couple who are experiencing a high level of conflict may believe that a baby will offer them a challenge they can meet together, thereby renewing their love for each other. At first, this may seem to be the case, but as the infant makes more demands on the couple's energy and time, the father may feel that the baby is depriving him of attention from his wife, and he may become jealous of the newcomer. For some couples, the addition of a child can actually increase the chance of divorce, resulting in a stable mother–child dyad and a single male.

The instability of triads can also be seen among siblings. The third child often has a rough time of it. Not only is the third child "the baby" who imagines it impossible ever to perform as well as the older siblings, but he or she has

committed the unintentional sin of breaking the bond between the two older children. Whenever one of the latter becomes intimate and loving toward "the baby," the other will demand attention and remind the third child that he or she is the newcomer. This point should not be exaggerated; sibling triads can be stable and happy. In general, however, the addition of a third child increases the possibility of conflict and coalitions within the family (Deaux & Wrightsman, 1988; Fivaz-DePeursinge et al., 2009).

Communities

At a level of social organization between the primary group and the larger institutions of the nation-state are communities of all descriptions. A **community** is a set of primary and secondary groups in which individuals carry out important life functions such as raising a family, earning a living, finding shelter, and the like. Communities may be either territorial or nonterritorial. Both include primary and secondary groups. In general, a **territorial community** is contained

community: A set of primary and secondary groups in which the individual carries out important life functions.

territorial community: A population that functions within a particular geographic area.

who teach language arts are members of a committee convened by an assistant principal (AP). The diagram shows that three of the teachers share relationships (either through cooperation in projects at work or through friendship), whereas the other teachers are linked to one another only through their work with the assistant principal.

Valence is the feeling that exists between any two people in a group. It refers to the value they place on their relationship. The most basic valence in a relationship is positive or negative—that is, like or dislike. In most friendship groups, there is a balance of valences; otherwise the group will likely break up. In diagram C, for example, Joe likes Dave and Marty and they reciprocate that liking, but Dave and Marty do not like each other. Joe must either convince the other two to like each other or choose between them.

A difference in valence between two people can be indicated with a double line and two different signs. In Diagram D, we see that Dave now likes Marty but Marty does not reciprocate. Perhaps he would like to have an exclusive relationship with Joe and views Dave as

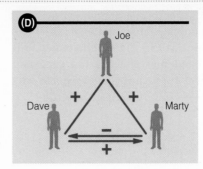

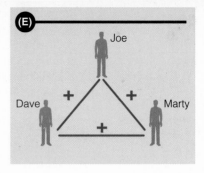

a threat. But now the pressure is all on Marty. If Joe can show him that Dave likes him and if Dave continues to try to get along with him, the group is likely to form a balanced triad, as shown in diagram E.

In a larger group, such as a school classroom or a club, the members express their preferences for one another in many ways. They call one another to talk about the events of the day or week; they gossip about one another; they give one another advice or criticism. These preferences result in a structure of cliques that exists within the larger group. When social scientists wish to chart such a structure, they simply ask all the members of the group to list the three or four individuals in the group with whom they would most like to spend time. By indicating each choice with a line, the researchers create a diagram known as a sociogram. Diagram F is a simple sociogram showing preferences among nine boys and girls. It reveals, among other things, that there is a clique centering on Dave and another centering on Latasha.

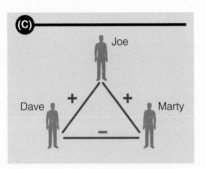

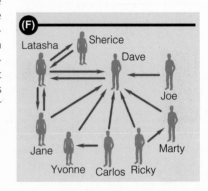

within geographic boundaries, whereas a **nonterritorial community** is a network of associations that form around shared goals. When people speak of a professional community, such as the medical or legal community, they are referring to a nonterritorial community.

Territorial communities are populations that function within a particular geographic area, and this is by far the more common meaning of the term as it is used both in everyday speech and in social-scientific writing (Lowry, 1993; Suttles, 1972). Territorial communities are usually composed of one or more *neighborhoods*. The neighborhood level of group contact includes primary groups (particularly families and peer groups) that form attachments on the basis of proximity—that is, as a result of living near one another.

In studies of how people form friendship groups in suburban neighborhoods, both Herbert Gans (1976/1967) and Bennett Berger (1961) found that proximity tended to explain patterns of primary-group formation better than any other variable except social class. Thus,

families that move into new suburban communities tend to find friends among others of the same social class who live near enough to allow ease of interaction and casual visiting; the same is true of students living in dorms on college campuses. On the international level, observers of the ethnic violence in Rwanda have reported that where members of the Hutu and Tutsi tribes live close to one another and cooperate in bringing in the harvest, they are far less likely to bear grudges and resort to violence than members of the same tribes who do not interact in their daily lives (Bonner, 1994).

That physical propinquity and social-class similarities explain how friendships form may seem self-evident to us. Moving into new neighborhoods or into dorms is a constant feature of life in a large urban society. But the idea that mere proximity, rather than kinship or tribal membership, can explain patterns of association

nonterritorial community: A network of relationships formed around shared goals.

would shock many people in simpler agrarian societies, where mobility is not a fact of everyday life. As Robert Park (1967/1925) noted, it is only under conditions of rapid and persistent social change that proximity helps explain why people form groups.

Networks

In modern societies, people form their deepest friendships in face-to-face groups, but these groups are, in turn, integrated into larger and more impersonal secondary-group structures that may extend well beyond the bounds of territorial communities. William F. Whyte opened *Street Corner Society* (1943), his classic study of street-corner peer groups, with this observation:

> The Nortons were Doc's gang. The group was brought together primarily by Doc, and it was built around Doc. When Doc was growing up, there was a kid's gang on Norton Street for every significant difference in age. There was a gang that averaged about three years older than Doc; there was Doc's gang, which included Nutsy, Danny, and a number of others; there was a group about three years younger, which included Joe Dodge and Frank Bonilli; and there was still a younger group, to which Carl and Tommy belonged. (p. 3)

Like the gangs Joan Moore studies in contemporary Los Angeles, the Nortons were active in a network of neighborhood-level peer groups. They would compete against other groups of boys of roughly the same age in a yearly round of baseball, bowling, and occasional interneighborhood fights (the latter occurring mainly in their younger teenage years). But the Nortons were also integrated into the community through some of its secondary associations. At election time, for example, they were recruited by the local political party organization, and some were recruited by the racketeers who controlled their low-income neighborhood. Party organizations and organized crime groups are examples of secondary associations that extend outside territorial communities.

In-Groups and Out-Groups The "corner boys" whom Whyte studied tended to be hostile toward similar groups from other street corners and even toward certain boys from their own territory. These distinctions between groups are common and are referred to as *in-group–out-group distinctions.* The **in-group** consists of one's own peers, whereas **out-groups** are those whom one considers to be outside the bounds of intimacy. Simmel observed that in-group–out-group distinctions can form around almost any quality, even one that many people would not consider meaningful at all. Thus, in a study of juvenile groups in a Chicago housing project, Gerald Suttles (1972) found that distinctions were made between boys who lived in lighter or darker brick buildings. Similarly, American children who have lived in foreign countries for several years may find themselves relegated to out-groups when they return to school in the United States (Smith, 1994, 1996).

In-group–out-group distinctions are usually based on such qualities as income, race, and religion. In "Cornerville," the community Whyte studied, group boundaries were based on educational background, particularly college versus noncollege education. One of Whyte's informants explained this distinction:

> In Cornerville the noncollege man has an inferiority complex. He hasn't had much education, and he has that feeling of inferiority. . . . Now the college man felt that way before he went to college, but when he is in college, he tries to throw it off. . . . Naturally, the noncollege man resents that. (Whyte, 1943, p. 79)

In-group–out-group distinctions often make it difficult for secondary associations to attract members from both groups. For example, when two ethnic or racial groups in a community make in-group–out-group distinctions, they often find themselves drifting toward different political parties or forming distinct factions within the same party rather than uniting to find solutions to common problems.

Reference Groups An important area of sociological research is the study of how the decisions we make are influenced by significant or relevant others—that is, by people we admire and whose behavior we try to imitate. The ways in which other people influence our attitudes and behavior are often discussed in terms of the **reference group**, a group that the individual uses as a frame of reference for self-evaluation and attitude formation (Deaux & Wrightsman, 1988; Merton, 1968). Some reference groups set and enforce certain values and norms, while others serve as a standard for comparison (Hare, 1992).

An example of the influence of reference groups can be seen in a study of students at Bennington College conducted by Theodore

in-group: A social group to which an individual has a feeling of allegiance; usually, but not always, a primary group.

out-group: Any social group to which an individual does not have a feeling of allegiance; may be in competition or conflict with the in-group.

reference group: A group that an individual uses as a frame of reference for self-evaluation and attitude formation.

Newcomb (1958). Newcomb found that conservative students often joined groups of like-minded peers who would affirm the values they had brought with them from home. They also tended to go home for visits more often than liberal students did. By their senior year, however, they were likely to have shifted their attitudes to match those of a more liberal reference group because most of the students and faculty at Bennington were liberal or radical.

This pioneering research has been replicated in many later studies. For example, studies of how students' religious beliefs change in college show that the more a student's religious beliefs differ from those of other students interviewed, the more likely it is that his or her beliefs will change. Conversely, the more a student's parents' religious views coincide with the dominant views of other students at the college, the less likely it is that his or her beliefs will change (Roberts, Koch, & Johnson, 2001).

The concept of reference groups has many practical applications. In market research, for example, it is common practice to determine where potential customers tend to seek help in forming their attitudes—which magazines, writers, or television commentators they turn to for guidance. These *opinion leaders,* as they are sometimes called, are the customer's reference group in a particular area of consumption (Katz, 1957).

Note that a reference group is not the same thing as a primary group. The friends with whom you spend much of your time are a primary group and are likely to act as a reference group as well. But we can have numerous reference groups, each of which influences us in different areas. As you choose a profession or career, you will be influenced by the values and norms of reference groups in that field. As you gain new statuses, such as that of parent, voter, or supervisor, you will look to other reference groups: child care experts, political commentators, and the like. A reference group thus can be quite limited in its relationship to you as an individual—its influence may act on what you are to become rather than on what you are now.

Social Network Analysis The study of whom people associate with, how those choices are made, and the effects of those choices on social structure and individual personality is known as *social network analysis.* The British social scientist Elizabeth Bott, one of the founders of this branch of sociology, describes a social network as follows:

> Each person is, as it were, in touch with a number of people, some of whom are directly in touch with each other and some of whom are

not. . . . I find it convenient to talk of a social field of this kind as a network. The image I have is of a set of points, some of which are joined by lines. The points of the image are people, or sometimes groups, and the lines indicate which people interact with each other. (1977, p. 256)

The diagrams in the Research Methods box illustrate social networks.

Social network analysis grew out of studies of the choices people make in becoming members of various social groups. Today, network analysis extends beyond this rather narrow subject to studies illustrating how the interconnectedness of certain members of a society can produce interaction patterns that may have a lasting influence on the lives of people both within and outside the network (Fleisher, 2005). Sociologist Aubrey Bonnett (1981), for example, studied *susus,* the rotating credit associations formed by West Indian ethnic groups in New York and other American cities. These associations are actually networks composed of numerous small groups of neighbors and kin who join with other similar groups, pay an entrance fee, and take turns borrowing funds from the association at far lower interest rates than a bank would charge. Bonnett showed that these associations bring together members of immigrant groups with different histories and values. As one *susu* organizer told him, "In my *susu* I have had members from all the islands. . . . You would be amazed to know how it has helped to get ourselves together, yes, this is really a West Indian thing" (p. 67).

Sociologists who study the rich and powerful have described the networks that provide the social contacts essential for success in entering and climbing upward in the corporate hierarchy (Cookson, 1997). In their book *Preparing for Power,* for example, Peter Cookson and Caroline Persell (1985) describe how young men from elite social backgrounds develop strong attachments with others like them at prestigious prep schools, primarily (but not entirely) located in New England or Virginia, such as Phillips Academy in Massachusetts (founded in 1778) and Woodberry Forest School in Virginia (founded in 1889). Through attendance at these schools, the children of the wealthy acquire a great deal of "cultural capital." All students gain some forms of cultural capital by mastering a range of academic and social skills, but prep school students acquire a special form of cultural capital that prepares them for life among the elite. They learn how to speak; how to behave at formal social gatherings; how to behave as a guest at a country home; what colleges, clubs, companies, and communities one should seek to enter; and what kind of spouse one should marry. In

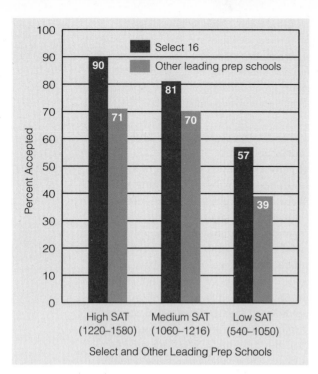

Figure 6.1 **Acceptance at Highly Selective Colleges, by Prep School Status and SAT Scores.** Source: Adapted from Cookson & Persell, 1985.

short, prep schools "integrate new brains with old wealth to revitalize the upper classes" (Cookson & Persell, 1985, p. 187).

Prep school students' social networks come into play when they apply to elite colleges, particularly Ivy League institutions such as Harvard, Yale, and Princeton. Many of the administrators and admissions officers at the most prestigious colleges were former prep school students themselves and tend to show preference for applicants from those schools. Figure 6.1 shows that when graduates of elite prep schools apply to the most competitive colleges, even those with low SAT scores are highly likely to gain admission. Later in life, they will continue to benefit from the social bonds formed at prep school. As can be seen in Figure 6.2, business managers

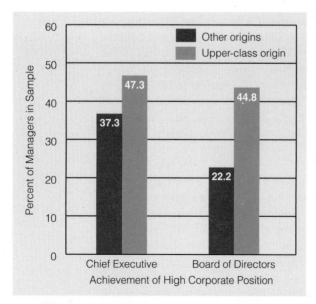

Figure 6.2 **Achievement of High Corporate Position, by Manager's Class Background*.** *Includes all college backgrounds. Source: Adapted from Cookson & Persell, 1985.

from upper-class backgrounds (those who are most likely to have attended elite prep schools) are more likely to become chief executive officers or to hold seats on the boards of major corporations than those from other social-class backgrounds.

Thinking Critically In some high schools, students don't seem to be able to have a social life unless they join a clique. Does it always have to be so? If you were a teacher or administrator in a school where there were serious conflicts between cliques, what would you do about it?

In other research on social networks, sociologists have demonstrated how the networks of individuals and groups within which people interact provide social support in times of stress or illness. Interestingly, this research often finds that people's social-support networks shrink as they become elderly and infirm and lose the ability to offer support or other resources in exchange (Fischer, 1982; Peek, Zsembik, & Cloward, 1997). Size limitations create a need for smaller groups, making multiple memberships more difficult, but by no means impossible. In fact, popularity in school often goes to people who are accepted in different cliques (Cacioppo, Fowler, & Christakis, 2009).

The Perspectives chart on page 135 summarizes the functionalist, interactionist, and conflict perspectives on groups.

INTERACTION IN GROUPS

In the preceding section, we discussed the basic building blocks of social structure: the primary group and the secondary group. We saw that these groups are embedded in more complex structures of groups, such as communities and social networks. We also saw how reference groups act to confirm an individual's values and behavior, and how the distinctions we make between in-groups and out-groups help us maintain the boundaries of the groups in which we participate. But these descriptions do not tell us what principles of interaction operate to produce cohesion or instability in groups. In this section, therefore, we take a closer look at interaction in groups.

Simply to observe that a society is a fabric of groups is to neglect two important questions: How can groups stay together in the first place, and how can a fabric of groups produce a cohesive society? All around us we see evidence that groups constantly fall apart. Romantic ardor cools; marriages fail; friends argue and seek new friends; successful rock bands split up;

Perspectives on Groups

Perspective	Central Issues Raised	Key Research Topics
Functionalist	What is the structure of groups of different sizes and functions? How do groups' functions and goals influence group structure?	Research on group size and number and strength of relationships; research on emergence of different roles and statuses as groups change their functions.
Interactionist	How do people interact in groups, especially in ways that enhance group cohesion and avoid conflict?	Studies of principles people apply to their interactions with others in groups; emergence of group leadership in relation to frequency of interaction; studies of "remedial work" in groups.
Conflict	Why do groups split up? How do power and influence operate in different types of groups and in different cultures?	Studies of how formal structure and bureaucracy emerge in groups in response to conflict.

businesses go bankrupt; entire nations are fragmented by social and economic upheavals. Certainly new groups are forming all the time, but others are continually breaking up.

When a group disintegrates, we often hear explanations like "I felt that I was only giving and not getting"; "We started to grow in separate directions"; "I no longer shared their values"; or "He [she, they] wanted too much of the [credit, money, power]." In short, when one listens to explanations of why groups fail to stay together, it sounds as if people are continually seeking more rewards for themselves or blaming others for taking too much of whatever rewards are available—too much attention, affection, approval, prestige, money, power, or whatever. If this is the case, it does not seem likely that groups could have much cohesiveness or stability.

One can imagine even more serious effects of human greed and self-serving behavior. The seventeenth-century English philosopher Thomas Hobbes was deeply concerned about the relationship between the individual and society. How is society, or any group within it, possible if every individual puts his or her own interests above those of everyone else? Hobbes warned that the sum total of everyone's selfish acts would be a "war of all against all." In *Leviathan,* his monumental treatise on this problem, he concluded that humanity needs an all-powerful but benevolent dictator to prevent social chaos.

Today, people throughout the world reject this notion. We strive to create a social order based on individual freedom, democracy, and the rule of law. Nevertheless, wherever we look, we see major threats to such a social order. Often such threats emerge when people feel that the groups they count on, such as political parties and legislatures, are not acting fairly. This brings us back to the question of how, given individual selfishness, humans can achieve order, cooperation, and trust in their associations rather than coercion, terror, and chaos.

Principles of Interaction

Social scientists have identified certain principles of interaction that help explain both stability and change in human groups. Among them are the pleasure principle, the rationality principle, the reciprocity principle, and the fairness principle.

The Pleasure Principle People seek pleasure and avoid pain. This basic principle implies nothing about what any particular person considers pleasurable or painful—as we will see shortly, this is a personal matter that depends on the norms that have been internalized by the individual. In fact, many people seem to find pain or discomfort desirable; examples include athletes and dieters, who seek pain or deprive themselves of pleasure in order to experience the thrill of athletic victory or the satisfaction of wearing a smaller size.

The pleasure principle does not address the question of what people value; all it states is that once one knows what kind of pleasure one wants, one will seek more of it and less of other values that yield less personal pleasure (Bentham, 1789; Homans, 1961). When applied to behavior in groups, the pleasure principle simply means that, over time, people will continue to interact in a group in which they experience the pleasure of reward but will withdraw from groups in which the pain they experience outweighs their pleasure.

The Rationality Principle In social life, a gain in pleasure may require some pain as well. For example, when two members of a group of friends become very close and find pleasure in their intimacy, the others may feel left out, and this causes some pain both to the two intimate friends and to those who feel left out. In larger groups, when a rule such as a ban on smoking is made, some members will derive more pleasure than others from the cleaner air. Those whose smoking pleasure is denied will experience an increase in pain. Consciously or not, people in groups calculate whether they think they will benefit personally from continued interactions with others in the group. Thus, the smokers may decide that their loss of pleasure outweighs the gains they get by being in the group, and they may decide to withdraw. In a small friendship group, the other friends may feel that the pain of their jealousy outweighs the pleasure of their friendship with the intimate pair, and they may exclude them from the group.

The rationality principle means that people change their behavior according to whether they think they will be worse or better off as a consequence. In other words, in making decisions about their interactions with others, people tend to make rough calculations of costs and benefits.

Does the idea that people seek a net gain from their interactions mean that they are greedy or self-centered or lacking in motives other than materialism? Not at all. Of course, there are extremes: Hedonists are people who can enjoy only the pleasures of the body. Egotists are people who especially enjoy praise and have little capacity for giving to others. Altruists are people who enjoy giving to others, often at the expense of other pleasures. But most people strive to behave in ways that give them physical pleasure and ego gratification ("boosts" or "strokes") and also give them the satisfaction of helping others. How far they go in either direction depends on what other members of the group are doing (for example, how giving the others are, how selfish, how self-sacrificing), which brings us to the principle of reciprocity in interaction.

The Reciprocity Principle In group interactions, people usually adhere to a norm stating that what others do for you, you should try to do for others. This is one of the world's most commonly followed rules of interaction; it is known as the norm of reciprocity (Homans, 1950). The Roman statesman Cicero recognized this principle when he wrote, "There is no duty more indispensable than that of returning a kindness. All men distrust one forgetful of a benefit." But it is not necessary to go back to ancient Rome for evidence that reciprocity is vital to continuing group interaction. Suppose that you and two classmates meet at a local pub. You pick up the check, and the other two agree to do so on future occasions. The next time, one of them pays the check. The third time, the third friend pleads a lack of cash and you each end up paying your own share. You and your second friend will feel that this is an embarrassing lapse on the part of your third friend. If such behavior continues, you will feel that the third friend takes more than he or she gives or thinks his or her company is worth the extra money you paid. It is unlikely that this

threesome will stay together very long if such imbalances continue.

Under what conditions could such a group continue despite the third friend's violation of the norm of reciprocity? One condition involves a loss of prestige for the third friend. You might ask him or her to make up the debt by doing something else for you. In that case, the third friend takes a subordinate position in the group, doing your bidding to make up for the inability to reciprocate on an equal basis (Homans, 1961; Mauss, 1966/1925). However, suppose that the nonreciprocator is a movie star or a powerful politician. Then the other two members of the threesome might continue to pay in the hope that the third member would continue to spend his or her valuable time with them and that there would be an even greater reward for them in the future. This kind of relationship is quite common in social life. Social climbers seek ways to flatter and please someone who can confer prestige on them. Power seekers do the same in attempting

Rembrandt depicts the traveler, a Jew who had been beaten and robbed and left on the road between Jerusalem and Jericho, as he is welcomed into the home of a Samaritan. Jesus used the parable of the Good Samaritan (Luke 10:25-29) to teach us to treat our neighbors as we would treat ourselves. This "golden rule," though usually associated with Jesus's teachings, appears in other world religions as well. *The Good Samaritan*, 1630. Oil on canvas: 68.5 × 57.3 cm.

to gain positions in which they will have power over others. A suitor may forgo reciprocity for quite a while in the hope of eventual conquest. In these situations and others like them, the participants in the interaction have an intuitive sense of what they are giving and what they are receiving or hope to receive in the future.

In the Bible, the norm of reciprocity is stated in the form of the Golden Rule: "Do unto others as you would have others do unto you" in effect urges individuals to go beyond expectations of immediate reciprocity and set examples of higher moral behavior in the hope that others will follow them. In real life, however, people tend to prefer that their gains be as immediate as possible.

The Fairness Principle We see ample evidence that people tend to expect certain kinds of treatment from others and that they tend to become angry when they do not receive it, especially when they feel that they have done what is expected of them in the situation. We want the rules to apply equally to everyone in the game, be it a friendly game of pool or the more complex game of corporate strategy. When we are not rewarded in the same ways as others, we say that we are being treated unfairly.

The unfair condition in which a person or group has come to expect certain rewards for certain efforts, yet does not get them while others do, is called *relative deprivation*. In times of economic depression, for example, all groups in society must make do with less, and all feel deprived in comparison to their previous condition. If the economy begins to improve and some groups begin to get higher wages or more profits while others experience only limited improvement, members of the less successful group will feel deprived relative to others and will likely become angry even though their own plight has actually improved somewhat (Merton & Kitt, 1950). The concept of relative deprivation is useful in explaining some aspects of group behavior and appears again in Chapter 9 when we discuss how major social changes like revolutions occur.

People's ideas about what is fair in their interactions with others often conflict with simple calculations of gain and loss. The fact that they might come out ahead in an interaction does not guarantee that they will feel good about it and continue the interaction. This was demonstrated in a series of studies by Daniel Kahneman and his associates (1986). A large number of respondents were asked to judge the fairness of this situation:

A landlord rents out a small house. When the lease is due for renewal, the landlord learns that the tenant has taken a job very close to the house and is therefore unlikely to move. The landlord raises the rent $40 more than he was planning to.

An economist might argue that the tenant should think of the situation in terms of whether the rent increase is offset by the savings in the cost of travel to work (in terms of money, time, convenience, and so on). If the tenant still comes out ahead, it will make sense to sign the new lease (the rationality principle). But more than 90 percent of the respondents in Kahneman's survey said that the landlord was being unfair. In real life, when people feel that a transaction like this is unfair, they often end the interaction, even at some cost to themselves. In this case, people intuitively felt that the landlord was unfairly taking advantage of a gain made by the tenant (the new job) without adding anything to the property to earn the right to raise the rent.

There are many examples of situations in which notions of fairness outweigh the principle of rationality. Barbers and beauticians do not charge more for haircuts on Saturday, nor do ski resorts usually charge more for lift tickets on holidays. Although they might like to profit from the higher demand at those times, they are afraid that people would regard the higher prices as unfair (Frank, 1988). However, services are often offered at higher rates during peak seasons, or bargain offers such as early-bird specials or off-peak fares are used to encourage people to accept services at times other than those they might prefer.

Applications of the fairness principle can be seen in a great many group situations. In the classic study conducted by Elton Mayo and his associates at Western Electric's Hawthorne plant (see Chapter 2), the observers often noticed that workers used various forms of joking and sarcastic teasing to enforce the group's norms. They had a strong sense of how much work they should turn out, both individually and as a group, in order to merit their pay: a fair day's work for a fair day's pay. In the plant's Bank Wiring Room, where workers hand-wired electrical circuits, it was common to see the members of a work group gang up and hit a fellow worker on the shoulder in a joking way when he produced more than the other members of the group. This practice was called "binging." A worker was binged by his coworkers if he produced too much in a day or if he did not produce enough, according to their definition of a fair day's work.

The Hawthorne researchers quickly realized that binging was part of the work group's culture. That culture—the group's norms of conduct and the ways in which its members interpreted those norms—was unique to that group and applied only when the group was at work. The culture of this work group had evolved over many years, its norms functioning to control the workers' responses to the demands of the company's managers. Thus, in a later analysis of the Bank Wiring Room observations, George Homans (1951) made the following comments:

> [The workers] shared a common body of sentiments. A person should not turn out too much work. If he did, he was a "rate-buster." The theory was that if an excessive amount of work was turned out, the management would lower the piecework rate so that the employees would be in the position of doing more work for approximately the same pay. On the other hand, a person should not turn out too little work. If he did he was a "chiseler"; that is, he was getting paid for work he did not do. (p. 235)

The Economic Person versus the Social Person

The four basic principles of interaction in groups—pleasure, rationality, reciprocity, and fairness—contain some obvious (and fortunate) contradictions. People calculate their individual gain and loss in interactions; they act according to their own interests in many situations. Does this mean that people will always act selfishly? Not always, for we have seen that people also have deeply ingrained notions of reciprocity and fairness, which often prompt them to act in favor of the needs of the group or of society as a whole. Indeed, many situations that arise in the social world demonstrate the "war of all against all" that would be created in a society whose members always acted in what they perceive as their self-interest.

Consider this example. It's the Friday evening rush hour. The streets downtown are crowded with cars, their drivers eager to get home. Although the light is about to turn red, some drivers move their cars into the intersection. They expect the traffic ahead of them to move forward so that they will be clear of oncoming traffic from the cross street. But alas, the traffic ahead does not move and the intersection is blocked. Now the cars on the cross street have a green light, but they cannot move. Fearing that they will miss their first opportunity to move, and amid much horn blowing and gnashing of teeth, some of these drivers also squeeze into the intersection. This chaotic and frustrating situation is known as "gridlock." It is the result of a number of drivers acting in what they perceive as their own interests and not thinking about the needs of all the drivers as a group. The gridlock example is one among many that could be cited in which it is actually advantageous for individuals to behave in a prosocial manner rather than attempt to maximize their immediate advantage (Hechter & Kanazawa, 1997; Sen, 1997; Shotland, 1985).

Situations like gridlock are made worse by problems of communication among individuals in groups. When people can freely share information about what they are experiencing, and when they trust each other's communications, positive outcomes are more likely than when information is not shared or communication is thwarted. It is not surprising, therefore, that communication and behavior in groups is an active area of sociological research.

Communication and Behavior in Groups

"Human interaction," Herbert Blumer observed, "is mediated by the use of symbols, by interpretations, or by ascertaining the meaning of one another's actions" (1969b, pp. 179–180). By "mediated," Blumer meant that we do not normally respond directly to the actions of another person. Instead, we react to our own interpretations of those actions. When we see other drivers moving into an intersection against the light, for example, we begin to feel that everyone is out for themselves and we would be fools to hold back. These interpretations are made in the interval between the stimulus (the other person's action) and the response (our own action). Thus, in the account at the beginning of the chapter, the gang member must frequently decide whether her honor is being attacked or, perhaps more important, whether her gang sisters and family members will demand that she defend her honor. Her decision to fight or not to fight will be guided by her interpretation and by her need to gain standing in the eyes of her peers.

What factors explain how people decide how to act and whether they should act for their own benefit or according to some notion of what is good for society? This question has been the subject of much research in the social sciences. Let us therefore take a brief look at some research on how people define their needs in social situations.

Definitions of the Situation "Situations we define as real, are real in their consequences," stated W. I. Thomas (1971/1921), a pioneer in the study of social interaction. By this he meant that our understandings or definitions of what is occurring around us—whether they are correct or

This aerial photo of a traffic gridlock in China could have been taken in almost any major metropolitan city in the world.

Patrick Durand/Sygma/CORBIS

not—guide our subsequent actions. Thus, in a study of how dying patients are treated by medical groups in hospital emergency rooms, David Sudnow (1967) found that when a patient was brought in with no heartbeat, the person's age had a great deal to do with what happened next. The arrival of a younger patient who seemed to be dying would produce a frenzy of attempts to restart the heart. At times, the entire group of medical personnel would become involved. But an old person with no heartbeat was far more likely to be pronounced dead on arrival, with little or no mobilization of the medical group. The patient's aged appearance caused the members of the group to define the situation as one that did not require urgent and heroic efforts.

In another study of medical groups, Charles Bosk (1979) found that surgeons developed subtle techniques for covering up the mistakes of residents in training, depending on how a particular situation was defined. They would invoke those techniques when they believed that a mistake was caused by lack of knowledge. But mistakes they believed to be caused by carelessness would often be exposed. The guilty resident would be held up to public scorn, often at a high cost to the new surgeon's status among other doctors.

Both of the foregoing studies began by questioning how definitions of the situation account for how people interact. The emergency room teams that Sudnow observed were performing an unofficial cost–benefit analysis in deciding when to apply heroic measures to heart attack victims. The surgeons in Bosk's study were resolving a conflict between their teaching and medical functions: Young surgeons need to learn and practice, but patients need to receive the best possible care. Surgeons in training need to maintain a good reputation among other doctors and nurses, even if they make mistakes. Such conflicts are resolved by norms that define the legitimacy of mistakes in different situations.

As these examples show, the subtle cues that define each situation are communicated through phrases, gestures, and other symbolic behavior, as well as through explicit evaluations of what was good and bad about a particular operation. Several subfields of sociology and psychology study the intricacies of this micro world of human interaction. Among the most powerful of these are ethnomethodology and the dramaturgical approach.

Ethnomethodology Studies based on the interactionist perspective use a variety of techniques. One of these is **ethnomethodology**, the study of the underlying rules of behavior that

---

ethnomethodology: The study of the underlying rules of behavior that guide group interaction.

guide group interaction. Harold Garfinkel (1967) used this approach in a classic series of studies of how we use verbal formulas to create a flow of communication that we feel is normal and that we can understand. Garfinkel asked his students to engage in conversations with friends and family members in which they violated some of the simple norms that most people follow in carrying on a conversation. Here is an example:

> **Victim:** [*waving cheerily*] How are you?
>
> **Student:** How am I in regard to what? My health, my finances, my schoolwork, my peace of mind, my . . . ?
>
> **Victim:** [*red in the face and suddenly out of control*] Look! I was just trying to be polite. Frankly, I don't give a damn how you are. (Garfinkel, 1967, p. 44)

As this brief example shows, when the norms of spoken interaction that "entitle" us to continue or maintain a conversation are violated, the conversation often cannot continue because there is no longer a mutual area of interaction within which both parties tacitly agree to make sense.

Such "experiments" reveal the taken-for-granted rules of interaction. For Garfinkel (1991), they show that social order is in fact the result of interactions in which people strive to create coherence, consistency, order, and meaning in their ordinary lives. Garfinkel's analysis of the case of Agnes is another important example of the contributions of ethnomethodology to our understanding of social behavior at the micro level.

Agnes participated in Garfinkel's research because she wanted to understand herself better and share her experience in life with professional researchers. Outwardly she was a woman; however, she had a penis, which she thought of as a mistake of biology. As a child, she had suffered because her parents insisted on socializing her as a boy and required her to play boys' sports. In her deepest feelings, however, she knew that she was a girl, and when she developed breasts during puberty, this new biological fact confirmed her beliefs about herself. You may be wondering what this has to do with how we behave in groups. Garfinkel noted that Agnes's gender, as opposed to her sex, was based on her feelings and behaviors, especially

her choice of female friends, all of which Agnes viewed as "natural facts" but, according to Garfinkel, were also *social performances* that established her gender in the groups she joined (Garfinkel, 2000).

By recognizing that most roles, not just those of boy or girl, are performances, you will expand your sociological imagination. People often feel that when someone they know acts out a new role, they are being untrue to their original identity, or "phony." But even the most modest star athlete learns to perform as a star off the field because that is what many groups in the school and the community expect; and it is not easy to remain modest in this new role. Similarly, in interactions with others, the homecoming queen adopts a socially defined way of behaving that she may adapt to her own personality but that she also borrows from others. We don't always realize the extent to which our behaviors in society are performances. Ethnomethodology helps us go beyond the limited notion of "phoniness" to see how we borrow many of our routines from the behaviors of others in groups of various kinds.

The Dramaturgical Approach Another technique used in research on interaction in groups is the **dramaturgical approach**. This approach is based on the recognition that much social interaction depends on how we wish to impress those who may be watching us. For example, Erving Goffman (1965) observed that people change their facial expressions just before entering a room in which they expect to find others who will greet them. Couples who are fighting when they are "backstage" often present themselves as models of friendship when they are "onstage"—that is, in the presence of other people. And many social environments, such as hotels, restaurants, and funeral parlors, are explicitly set up with a front and a back "stage" so that the public is spared the noisy and sometimes conflicted interaction occurring "behind the scenes." Using these and other strategies to "set a stage" for our own purposes is known as **impression management**.

In a well-known study that applied this dramaturgical view of group interaction, James Henslin and Mae Briggs (1971) drew upon Briggs's extensive experience as a gynecological nurse at a time when most gynecologists were men. Their research showed how doctors and nurses play roles that define the situation of a pelvic examination as unembarrassing and attempt to save each participant's "face," or self-image. According to Henslin and Briggs, the pelvic examination is carried out in a series of scenes. First the doctor and the patient discuss

---

dramaturgical approach: An approach to research on interaction in groups that is based on the recognition that much social interaction depends on the desire to impress those who may be watching.

impression management: The strategies one uses to "set a stage" for one's own purposes.

the patient's condition. If the doctor decides that the patient needs a pelvic examination, the doctor will leave the room; this is the end of scene 1. In scene 2, the nurse enters. Her role is to work with the patient to create a situation in which the body of the patient is hidden behind sheets and her depersonalized pelvic area is exposed for clinical inspection. Through these preparations, the patient becomes symbolically distanced from the doctor, allowing the next scene, the pelvic examination itself, to be desexualized. The props, the stage setting (the examining room), and the language used help define the situation as a nonsexual encounter, thereby saving everyone involved from embarrassment.

Altruism and the Bystander Effect Intentional efforts to set a stage for purposes of impression management are common in hospitals, restaurants, and many other social settings where patterns of behavior are predictable and regular. But in other public settings, especially streets, parks, and public transportation, people often try to hide behind their anonymity. They shy away from behavior that will be noticed, as illustrated in the following example of the *bystander effect.*

Late one night in 1964, Kitty Genovese was returning home from a social gathering when she was attacked by a man who had been following her. He caught up with her in front of a bookstore in the shadow of her apartment building. As she screamed for help, he stabbed her in the chest. Lights went on in some of the apartments, but no one intervened or called the police. One neighbor shouted from a window: "Let that girl alone." The stalker retreated and waited, and when the neighbor no longer seemed to be watching, he resumed his attack. He raped and stabbed the young woman to death in a stairwell of her apartment house, but the police did not receive a call for help from any of her neighbors until thirty minutes after she had first cried out. Interviewed later, the neighbors said things like, "I was tired" or "I assumed others were helping" or "Frankly, we were afraid."

This incident is often cited to illustrate the callousness and indifference of urban dwellers (Frank, 1988). It might seem from this and similar incidents that when faced with danger and other costs of helping (for example, time in court, dealing with the police, loss of time from work), people make fairly selfish calculations of gain and loss. But what about all the unselfish acts of heroism one reads about? Victims are dragged from fires and submerged cars and broken ice by self-sacrificing civilians, many of whom die in their efforts to help others. Did Kitty Genovese simply have a bad group of neighbors? Empirical evidence suggests that the situation is more complicated than that (Batson, Batson, & Todd, 1995).

Studies by Bibb Latané and John Darley (1970) have shown that bystanders are likely to offer help to victims in many experimental situations. But their research also shows that the presence of other people who do not help increases the chances that no one will help. In one simple experiment, a graduate student begins interviewing the subject. The interviewer excuses himself or herself for a moment and enters an adjacent room. Suddenly the waiting subject hears cries for help: "Oh my God, my foot . . . I . . . I can't get this thing off!" In about 70 percent of the trials, the subject rushes into the room and offers help. But when there is another person in the first room who just sits there and seems to be associated with the experimenter in some way, only 7 percent of the subjects intervene to help the "victim."

In the experiment just described, the helper can see that no one else has offered help. In the Kitty Genovese case, however, people could conveniently assume that others must be helping. The point is that when responsibility seems to be diffused rather than falling on a particular individual, people are more likely to avoid helping others. But when people cannot avoid defining the situation as involving them, they are likely to throw the rationality principle to the winds and intervene altruistically despite the potential costs to themselves (Deaux & Wrightsman, 1988; McIntyre, 1995).

Is she begging, ill, or praying? Will anyone stop to find out? The bystander effect helps explain why no one is stopping in this photo. The presence of others diffuses responsibility. Also, we are socialized to behave in correct and socially acceptable ways. When others fail to react, we often take this as a signal that a response is not needed or not appropriate.

Interaction and Group Structure

The examples just discussed involve decisions and strategies that individuals use in deciding whether to behave more selfishly or more altruistically in a given situation. As members of groups, however, we often explain our actions on the basis of our status in the group rather than on the basis of our individual expectations. "I must do this because I'm expected to lead," we might say, or "I can't decide that without talking to the others and especially to [the group's leader]." These explanations suggest that the structure of groups is a major factor in explaining behavior.

In William F. Whyte's landmark study of street corner groups in an Italian American neighborhood of Boston, the author systematically observed interactions among the boys in different gangs, especially one known as the Nortons because they "hung out" on a corner near the Norton Street Settlement House. When the Nortons went bowling together at the local lanes, they were quite competitive among themselves. A boy's status in the group was closely related to his bowling skill; Doc, their leader, was the best bowler. One day, when the group was involved in a hot bowling match among themselves, a boy known as Long John seemed to be "bowling over his head," according to the other boys. Long John was not very popular in the group, and his presence among them depended mainly on his friendship with Doc. When he seemed likely to beat even Doc, the others began to heckle him. Soon his score fell to its usual lower level. Sometime later, Doc left the group. Long John remained in it, but whenever the boys went bowling, his scores were even lower than they had been before. When Doc rejoined the group, Long John's performance improved again, but never became higher than Doc's. Whyte interpreted this, along with much related evidence from other competitive games the boys played, as demonstrating that status in the group determined performance, rather than the other way around.

The implications of this finding are far-reaching. Status in a group can have a strong influence on performance. This means that when other group members expect an individual to perform in a certain way, or at a given level of skill, that person's actual performance often conforms to those expectations, regardless of his or her actual skill. Doc was a leader in the group because of his skills at bowling and other competitive sports, but his leadership was also due to the fact that he was the person most of the other boys went to for information or advice. Doc was at the center of most of their

One night sociologist William Whyte was out gambling with the Nortons. He asked one of the big-time gamblers running the game a question that the man disliked and seemed to avoid. Later, Doc had to remind Whyte that, as Doc explained, you can just hang around and you'll learn the answer in the long run without even having to ask the question. This was an important lesson in group interaction and, for Whyte, in the art of participant observation research.

interactions. When someone of lesser status, like Long John, seemed to be challenging him, the group acted in ways that "put him in his place."

The Emergence of Group Leaders Doc was the undisputed leader of the Nortons. He initiated and received by far the most communications from other group members. But what makes the person who initiates and receives the most communications the most respected member of the group? The reasons vary with the specific tasks the group is performing, but if we assume competence at those tasks, usually the person at the center of the communication is spending a lot of time helping others as well as helping the group accomplish its goals. That person not only is concerned with his or her own performance (the rationality principle) but also takes pleasure in helping the others—giving approval, voicing criticism, making suggestions, and so on (the pleasure principle). In exchange for this help, the other members of the group give approval, respect, and allegiance to the central individual (the reciprocity principle).

Is the person who initiates the most interactions respected simply because he or she talks a lot? Common sense tells us that this is far from true. People who talk a lot but have little to contribute soon find that no one is paying attention to them. Their rate of interaction then drops sharply. Other patterns have also been identified. In their research on small groups, for example, Robert F. Bales and Philip Slater (1955) observed that the person who initiated the most interactions—both in getting tasks done and in supporting the suggestions of others—often came to be thought of as a leader. The group began to orient itself toward that person and to expect leadership from him or her. A second member of the group, usually the one who initiated the second-highest number of interactions, was often the best-liked person in the group.

Bales and his colleagues concluded that groups often develop both a "task leader" (or instrumental leader) and another leader, whom they called a "socioemotional leader" (or expressive leader). The former tends to adhere to group norms and to take the lead in urging everyone to get the work done, but this often leaves other members with ruffled feelings. The socioemotional leader is the person who eases the group over rough spots with jokes, encouragement, and attention to the group's emotional climate. Thus, most classes have a class clown whose antics help release the tension of test situations, and most teams have a respected member who can also joke around and get people to relax under pressure. At times, the same person performs both of these roles, but generally there is an informal division of labor between the two kinds of leaders (Beebe, 1997).

Recall the Bank Wiring Room observations mentioned earlier. The workers at the equipment wiring tables are an example of small-group structure. Their leader was the person who could turn out the most work in the shortest time and could also help the others do their work. The workers with higher status in the group were those who showed greater competence at their jobs and stronger allegiance to the norms of the group. Newer workers with lesser skills and less shared experience with the others were spoken to less and were less central to the group's interactions. Over time, as these workers were "binged" for producing too little or too much, the other workers would see how they handled the criticism and would form opinions of them accordingly.

Remedial Work in Groups

Erving Goffman, a master at observing interaction in groups, pointed out that some individuals are constantly working to maintain the group's smooth functioning. These people often take on the informal role of "socioemotional leader," and they are especially attuned to injuries or slights to the self-esteem of one or more group members. They engage in "remedial work," striving to smooth ruffled feathers and keep the group working toward a consensus or a high level of productivity. In a recent study of a conservative political discussion group, for example, Jason Clark-Miller and Jennifer Murdock (2005) observed that disagreements about ideas frequently threatened the discussion. A group member with particularly extreme or conspiratorial views, such as someone who constantly issued warnings about the black helicopters of the so-called New World Order, would consistently be overlooked when hands were raised. Or, in the case of prolonged quarrels, the moderator would appeal to group members' sense of civility in smoothing over arguments that threatened to break up the group.

In her research on children's playgroups, Barrie Thorne found that remedial work in playground settings is more common among girls than among boys. In fact, she found that when quarrels interfere with girls' play, the participants often stop the activity and go off in smaller groups to repair the emotional damage. When boys argue during a game, the shouting may become intense, but at some point, they come to a decision regardless of hurt feelings and continue with the game. Remember, however, that Thorne was observing playground groups in which boys tended to play versions of ball games that are highly competitive, whereas girls were more likely to be playing less competitive games in which winning or losing was perhaps less important than maintaining smooth interactions and having fun (Thorne, 1993). When girls play team sports, their groups tend to be just as competitive and quarrelsome as those of boys, and the remedial work they do does not appear to differ significantly from that of boys (Grasmuck, 2005).

Effects of Group Size

One of the most striking aspects of the Hawthorne studies is that they revealed how norms about fairness in the work group are enforced through a particular kind of joking ritual (binging). But why are workers binged when they produce too much? That behavior points to the conflict between the norms of the small work group (which are based on fairness) and those of the larger corporation in which the group is embedded (which are based on rationality). The company would like to obtain greater productivity from the workers, but the workers have their own ideas about what is fair and their own ways of enforcing the group norm.

This example brings us to the principle that as group size increases, or as the number of groups in an organization increases, the leaders of the larger organization must invest in ways of controlling the behavior of groups within it (Hechter, 1987). This investment often takes the form of agents of control such as foremen, accountants, inspectors, and the like. These roles become part of the formal structure known as bureaucracy.

FORMAL ORGANIZATIONS AND BUREAUCRACY

An **informal organization** is a group whose norms and statuses are generally agreed upon but are not set down in writing. The Bank Wiring Room observations are a classic example of the dynamics of informal organizations. The members of the work groups in the wiring room followed group norms that limited their output. Usually such groups have leaders who help create and enforce the group's norms but have no formal leadership position in the company.

A **formal organization** has explicit (often written) sets of norms, statuses, and roles that specify each member's relationships to the others and the conditions under which those relationships hold. Organization charts and job descriptions are typical of such organizations. Formal organizations take a wide variety of forms. For example, the New England town meeting is composed of residents of a town who gather to debate and discuss any issues the members wish to raise. The tenants' association of an urban apartment building is also composed of people who reside in a specific place, but its scope of action is usually limited to housing issues. Both of these are formal organizations because clearly stated rules define who may participate and the scope and manner of that participation. As is true in many formal organizations, the members of town meetings and tenants' associations try to arrive at decisions through some form of democratic process—that is, by adhering to norms that allow the majority to run the organization but not to infringe on the rights of the minority.

Voluntary Associations

A familiar type of formal organization is the **voluntary association**. People join groups like the PTA or the Rotary Club to pursue interests they share with other members of the group. Voluntary associations are usually democratically run, at least in principle, and have rules and regulations and an administrative staff. Churches, fraternal organizations, political clubs, and neighborhood improvement groups are examples of voluntary associations often found in American communities. Sociologists study these associations in order to understand how well or poorly people are integrated into their society.

Bureaucracies Bureaucracies are another common type of formal organization. A **bureaucracy** is a specific structure of statuses and roles in which the power to influence the actions of others increases as one nears the top of the organization; this is in marked contrast to the democratic procedures used in other kinds of organizations. Voluntary associations, for example, usually have some of the elements of bureaucracies, but they are run as democratic structures in which power is based on majority rule rather than on executive orders as is the case in pure bureaucracies like General Motors or the U.S. Army.

We owe much of our understanding of bureaucracies to the work of Max Weber, who identified the following typical aspects of most bureaucratic organizations:

1. *Positions with clearly defined responsibilities:* "The regular activities required for the purposes of the organization are distributed in a fixed way as official duties."

2. *Positions ordered in a hierarchy:* The organization of offices "follows the principle of hierarchy; that is, each lower office is under the control and supervision of a higher one."

3. *Rules and precedents:* The functioning of the bureaucracy is governed "by a consistent system of abstract rules" and the "application of these rules to specific cases."

4. *Impersonality and impartiality:* "The ideal official conducts his office . . . in a spirit of formalistic impersonality . . . without hatred or passion, and hence without affection or enthusiasm."

5. *A career ladder:* Work in a bureaucracy "constitutes a career. There is a system of 'promotions' according to seniority, or to achievement, or both."

6. *The norm of efficiency:* "The purely bureaucratic type of administrative organization . . . is from a purely technical point of view, capable of attaining the highest degree of efficiency" (Weber, 1958/1922). (See the For Review chart on page 145.)

Weber believed that bureaucracy made human social life more "rational" than it had ever been in the past. Rules, impersonality, and the norm of efficiency are some of the ways in which

---

informal organization: A group whose norms and statuses are generally agreed upon but are not set down in writing.

formal organization: A group that has an explicit, often written, set of norms, statuses, and roles that specify each member's relationships to the others and the conditions under which those relationships hold.

voluntary association: A formal organization whose members pursue shared interests and arrive at decisions through some sort of democratic process.

bureaucracy: A formal organization characterized by a clearly defined hierarchy with a commitment to rules, efficiency, and impersonality.

bureaucracies "rationalize" human societies. By this, Weber meant that society becomes dominated by groups organized so that the interactions of their members will maximize the group's efficiency. Once the group's goals have been set, the officials in a bureaucracy can seek the most efficient means of reaching those goals. All the less rational behaviors of human groups, such as magic and ritual, are avoided by groups organized as bureaucracies. But the people in a bureaucracy may not take full responsibility for their actions. For this and other reasons, therefore, Weber had some misgivings about the consequences of the increasing dominance of bureaucratic groups in modern societies.

Bureaucracy and Obedience to Authority

In Chapter 2, we discussed the experiments on conformity conducted by Solomon Asch. Asch's demonstration of the power of group pressure raised serious questions about most people's ability to resist such pressure. It also led to many other studies of conformity. None of those studies is more powerful in its implications and disturbing in its methods than the series of experiments on obedience to authority conducted by Stanley Milgram.

Milgram's study was designed to "take a close look at the act of obeying." As Milgram described it:

> Two people come to a psychology laboratory to take part in a study of memory and learning. One of them is designated as a "teacher" and the other as a "learner."

The experimenter explains that the study is concerned with the effects of punishment on learning. The learner is conducted into a room, seated in a chair, his arms strapped to prevent excessive movement, and an electrode is attached to his wrist. He is told that he is to learn a list of word pairs; whenever he makes an error, he will receive electric shocks of increasing intensity.

The real focus of the experiment is the teacher. After watching the learner being strapped into place, he is taken into the main experimental room and seated before an impressive shock generator. Its main feature is a horizontal line of 30 switches, ranging from 15 volts to 450 volts, in 15-volt increments. There are also verbal designations ranging from SLIGHT SHOCK to DANGER—SEVERE SHOCK. The teacher is told to administer the learning test to the man in the other room. When the learner responds correctly, the teacher moves on to the next item; when the learner gives an incorrect answer, the teacher is to give him an electric

Characteristics of Bureaucracy

Characteristic	Example
Positions Clearly Defined	Job or position descriptions detail the responsibilities of each job, avoiding confusion about the duties of each jobholder.
Positions Ordered in a Hierarchy	Positions in the company or agency are ranked from top to bottom so that each position reports to another and supervisory responsibilities are clear.
Rules and Precedents	"The way we do things" in an organization is often written down, becoming "the book" of rules to be followed by its members.
Impersonality and Impartiality	At least in principle, actual performance on the job, rather than personal likes or favors, determines each individual's performance rating.
Career Ladder	New jobs are posted so that employees have a chance to move up in the organization on the basis of achievement, seniority, or both.
Norm of Efficiency	The company continually seeks to increase its efficiency by increasing productivity per employee; this can lead to layoffs or downsizing.

shock. He is to start at the lowest shock level (15 volts) and to increase the level each time the learner makes an error, going through 30 volts, 45 volts, and so on (1974, pp. 3–4).

The "learner" is an actor who pretends to suffer pain but receives no actual shock. The subject ("teacher") is a businessperson or an industrial worker or a student, someone who has been recruited by a classified ad offering payment for spare-time work in a university laboratory.

Milgram was dismayed to discover that many of his subjects were willing to obey any order the experimenter gave. In the basic version of the experiment, in which the "learner" is in one room and the "teacher" in another from which the "learner" is visible but cannot be heard, 65 percent of the subjects administered the highest levels of shock, while the other 35 percent were obedient well into the "intense shock" levels.

Milgram used a functionalist argument to explain the high levels of obedience revealed in his experiment: In bureaucratic organizations, people seek approval by adhering to the rules, which often absolve them of moral responsibility for their actions. But he also explored the conditions under which conflict will take place—that is, the conditions under which the subject will rebel against the experimenter. In situations in which subjects are forced to confront the consequences of their behavior, their ability to rely on "duty" to justify that behavior seems to be diminished. This effect is presented in graphic form in Figure 6.3.

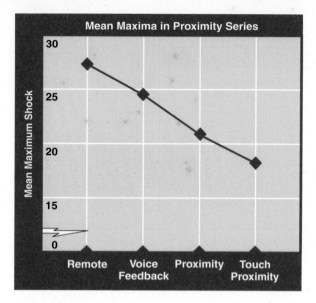

Figure 6.3 Mean Maximum Shocks in Four Variants of Milgram's Experiment. This graph illustrates the relationship between the maximum shocks administered and the four variants of proximity between the "teacher" and the "learner" in Milgram's experiment. The variants were: (1) remote (different rooms, in which the learner could be seen but not heard); (2) voice feedback (different rooms, in which the learner could be seen and heard); (3) proximity (both in the same room); and (4) touch proximity (both in the same room, in which the teacher was told to force the learner's hand onto the electrode). Source: Adapted from *Obedience to Authority* by Stanley Milgram, 1974.

When Milgram published the results of these experiments, he was criticized for deceiving his subjects and, in some cases, causing them undue stress. The controversy this study created was one of the factors leading to the establishment of rules to protect human subjects in social-scientific research (see Chapter 2).

Commitment to Bureaucratic Groups

Another question related to the impact of bureaucracy on individuals is how bureaucratic groups—which are based on unemotional and "rational" systems of recruitment, decision making, and reward—can sometimes attract strong commitment from their members. One possible answer is ideology: People believe in the goals and methods of the bureaucracy. Another explanation is that people within a bureaucracy form primary groups that maintain their commitment to the larger organization. There is considerable research evidence showing that both explanations are valid—both ideology and primary-group ties may operate to reinforce people's commitment to the organization. Several studies have addressed this question.

Primary Groups in Bureaucracies In one of those studies, Morris Janowitz and Edward Shils were assigned by the U.S. Army to study the attitudes of German soldiers captured during World War II. They were attempting to discover what had made the German army so effective and committed that isolated units continued to fight even in the face of certain defeat. As Shils explained it, he and Janowitz "discovered the influence of small, closely knit groups on the conduct of their members in the performance of tasks set them from the outside" (Janowitz & Shils, 1948, p. 48).

The soldiers of the German army behaved in this way not because of their ideological commitment to the Nazi cause but because of their loyalty to small combat units whose members had become so devoted to each other that to continue to fight even in the face of defeat, so as not to be dishonored as a group, seemed the only possible course of action. The close primary-group structure of the German army was one of the secrets of its success. The other was an extremely efficient organization. Ideology figured very little in its effectiveness. This finding has been replicated many times in studies of armies in combat situations (Salo & Siebold, 2008).

Ideological Primary Groups Primary groups would seem to have no place in a "pure" bureaucracy, yet the studies just cited, together with

the Hawthorne experiments and other empirical studies, show that no bureaucracy completely eliminates primary groups. In the case of the German army, it appears that the development of primary groups among combat troops increased the effectiveness of the organization as a whole. Another important study of this subject was Philip Selznick's (1952) analysis of what made the Bolsheviks so effective that they won out in the fierce competition for political dominance in Russia. Selznick's data show that the ideologically based primary group was the key element in the Bolsheviks' organization. Small, secret "cells" of devoted communists were organized in neighborhoods, factories, army units, farms, and universities. Through years of tense and often dangerous political activity, the members became extremely devoted to each other and to their revolutionary cause. According to Selznick, this doubling of ideological and primary-group cohesion made the Bolsheviks themselves an "organizational weapon."

In the contemporary world, we see many examples of ideological groups with extremely high levels of cohesion. These groups are often devoted to the leadership and teachings of a forceful and highly attractive leader whose ideas exert a powerful influence over the behavior of the group's members. We often focus on the extremes of such behavior, such as the suicidal terrorism of the al Qaeda network or the mass suicide by the members of Heaven's Gate, but many ideological groups can have beneficial effects on their members. This is especially true of groups devoted to personal or spiritual growth and positive behavior change. Indeed, the ability of people to form groups freely and to leave them if they so desire is one of the distinguishing features of societies in which the rule of law protects the individual's right of association. A problem of some ideological groups, however, is that they may not grant their members this freedom (Elshtain, 1997).

GROUPS IN COMPLEX SOCIETIES

As we saw in Chapter 4, in simpler tribal or peasant societies, all social structures are primary groups. The tribal village is usually so small that everyone knows everyone else. The family, the extended family, the age- and sex-segregated peer groups, and even the village itself all allow for everyone to know everyone else and to interact on an almost daily basis. Typically these societies lack a secondary-group level of life. In traditional peasant societies, for example, no associations existed to bring soldiers or farmers or artisans together in groups that extended beyond the limited bounds of the village. Yet that level of social life is what makes contemporary societies so efficient.

The differences between simpler and more complex societies can also make interactions among members of those societies confusing and frustrating (see the Global Social Change box on this page). For example, when they were first assigned to peasant villages in Africa and Asia, American Peace Corps volunteers were unprepared to cope with situations in which villagers came to them with requests for money or help with marital and family problems. For the young Americans, such requests were appropriate only among members of primary groups—and the villagers did not know how to engage in interactions that are typical of secondary groups (Luce, 1964).

Effects of Bureaucratization

In his most famous theoretical work, *The Division of Labor in Society* (1964/1893), Émile Durkheim reasoned that as a society becomes larger and more complex, there is a vast increase in interdependence among its members as the labor needed to feed, house, educate, communicate with, transport, care for, and defend them becomes more complex. Faced with more choices regarding what work to prepare for, whom to interact with, and how to live life, the individual gains much more freedom than is available to a member of a simpler society, who (for better or worse) is confined to a narrow set of village primary groups. But Durkheim also saw that modern people could be overwhelmed by the choices open to them and might find it difficult to decide what groups to join and what norms to conform to.

global social change

Communication and Interaction Global social change continually produces waves of refugees and immigrants. For example, the new freedom to travel in the former Soviet satellite nations has produced thousands of ambitious immigrants. Civil strife in Central and Latin America, repression in some Asian societies (for example, China and North and South Korea), growing poverty in the Philippines and Albania, and continuing turmoil in Cambodia and the Middle East have pushed families and individuals out of their homes and into the growing tide of worldwide migration. All of these changes are having profound effects on the linguistic diversity of North America and on how people in the Americas communicate and interact.

The United States continues to hold a global reputation as a place where immigrants can work to achieve success—and because it remains the nation with the most liberal immigration policies, it is no surprise that more than 500,000 legal immigrants and refugees (and an untold number of illegal entrants) seek to establish themselves in the "golden land" each year. In California alone during the 1990s, about 3.4 million immigrants brought their languages and cultural norms to the nation's most diverse state. But no part of the nation missed the increases in cultural and linguistic diversity that result from immigration. Between 1990 and 2000, for example, the number of foreign-language speakers in the state of Georgia (not known as a primary destination of immigrants) increased from 285,000 to 696,000. All told, by 2010, more than 46 million residents of the United States did not speak English in their homes (U.S. Census Bureau, 2010).

The need for English as a second language (ESL) courses in schools and community centers throughout the nation is only one indication of the kinds of demands that increased linguistic diversity is placing on communities where immigration rates are high. In time, most Americans come to appreciate the diversity of languages spoken in the United States as a great cultural resource. But the welter of different languages spoken in some communities can also be a source of conflict. The extreme rapidity of the change has created numerous controversies over how, and more specifically in what language, people should be expected to interact. For example, in 1989, a federal judge ruled that the Los Angeles suburb of Pomona had violated the civil rights of Korean and Chinese businesspeople by requiring them to place English signs on their shops. In the same community, however, Filipino nurses lost a suit against a hospital that would not allow them to speak their native language, Tagalog, among themselves at work.

Elsewhere in the nation, there is an active movement to make English the "official" language of the United States. This is accompanied by a crusade against the offering of bilingual education for children from other language groups. The problems of interaction across linguistic divides are greatest in states with large proportions of foreign-language speakers, such as New Mexico (37 percent), California (40 percent), Texas (32 percent), Hawaii (28 percent), and New York (29 percent). Conversely, these states benefit the most from the wealth of new phrases and styles of communication stemming from their growing linguistic diversity.

Max Weber shared Durkheim's concern about the prospects for human happiness in the modern world. According to Weber (1958/1922), the fundamental paradox of modern society was the rise of bureaucracies, which greatly expanded the efficiency and rationality of society but "disenchanted the world." He meant that the modern formal organization, which is so effective in science, the military, the state (government), economic production, and almost every other aspect of modern life, at the same time deprives human life of a sense of magic

then & now

From Haymarket Square to Waco

Although they occurred almost a century apart, the bombing in Chicago's Haymarket Square in 1886 and the tragic assault and fire at Waco, Texas, in 1993 reveal many similarities in the way radical groups and agencies of the state become enmeshed in situations that lead to tragedy. The Haymarket bombing took place during a demonstration calling for an eight-hour workday held by a small, intensely cohesive group of committed anarchists, most of whom were immigrants from Europe, especially Germany. The anarchists were devoted to the destruction of the capitalist system of production and private property in the United States and elsewhere. Their leaders, including Albert Parsons, were speaking at the rally when Chicago police began forcing them to disperse. A bomb was thrown into the crowd, killing seven police officers and three bystanders and wounding at least 100 others. Authorities immediately blamed the anarchists and their followers for the explosion, and their leaders were jailed. Four were hanged; one committed suicide. But their guilt was never well established, and after intense lobbying by labor groups and civil rights activists, Illinois Governor John Peter Altgeld pardoned the remaining anarchists in 1893.

The followers of David Koresh, known as the Branch Davidians, were a dedicated religious group, with little concern for destroying capitalism but firm ideas about the need to live a highly spiritual life apart from what they perceived as a corrupt society. They believed that they needed to arm themselves against a hostile world. Accusations of child abuse and felonies by their leader only confirmed the group's fears and desire to protect themselves. A long and sensational armed standoff between federal law enforcement agencies and the Branch Davidians ended in gunfire and the deadly fire pictured here. The Waco tragedy, like the Haymarket bombing, became symbolic both of the excesses of ideologically motivated and isolated extremist groups and of the blundering authoritarianism of law enforcement agencies (Niebuhr, 1995).

Bettmann/Corbis

AP Photos

and spirituality. The rise of fundamentalist movements in the Islamic world, for example, is in part a reaction against the influence of bureaucratic rationality in the institutions of secular states such as Egypt and Algeria.

Even in Western societies, where the process of bureaucratization is extremely advanced, people may feel overwhelmed by bureaucracy. Individuals often feel like tiny cogs in a huge set of interlocking organizations. They fill specific statuses and enact clearly defined roles yet feel more and more oppressed. Weber deplored the type of person that he felt the organizations of modern society were producing: "a petty routine creature, lacking in heroism, human spontaneity, and inventiveness" (Weber, 1922, in Gerth & Mills, 1958, p. 50). In the United States, however, democracy was always linked with citizens' freedom of association and their tendency to join groups that could protect them against the excesses of large bureaucracies, on the one hand, and offer them the intimacy of small-group sociability on the other.

Sociology & Social Justice

INDIVIDUALS VERSUS BUREAUCRACIES What rights do individuals have in a bureaucracy? If you believe that you have been unfairly disciplined, or fired, or discriminated against because of some aspect of your social status (for example, race, gender, ethnicity, religion), do the rules of the organization you are dealing with grant you any rights of appeal? As a student, if you are accused of some infraction of the rules, such as plagiarism, must you accept the punishment of the academic department, or can you at least appeal to some higher authority in the administration? Universities and colleges almost always have such an appeals process. Companies in the private sector often do not. The vast majority of workers in U.S. companies believe that they cannot be fired without just cause, but this is simply not true (Freeman & Rogers, 2002). If their company does not have an appeals process whereby their firing can be reconsidered, they must accept the initial decision and leave. On the other hand, when employees are represented by a union on their job site, they usually have recourse to a grievance process whereby they can have their case heard on appeal. If the parties do not agree on a reinstatement or some other remedy, an impartial judge or arbitrator can be called in to render a final decision under the terms of the labor–management contract.

The traditional rights of private property assign ultimate power to the employer, whose property the company is. The employer can decide to hire and fire workers according to whatever criteria seem appropriate to that owner. But when an employer has signed a labor–management contract with a union, the individual worker is offered additional protection from arbitrary firing or other forms of discipline by the appeals process. This is why so many employers make such strong efforts to prevent unions from being formed in their companies, and why workers who realize how powerless they are as individuals often band together to invite a union to represent them in their dealings with the company bureaucracy.

A Nation of Joiners

Early in the nineteenth century, the French aristocrat and social theorist Alexis de Tocqueville visited the United States. Like other European intellectuals, he considered the new nation a bold and controversial social experiment. He had doubts about the capacity of common people to govern themselves without the guiding hand of a strong aristocracy. As he traveled through the former colonies, however, he was pleasantly surprised by what he observed. He was particularly impressed by the number of associations and small groups to which people belonged. Americans, he wrote, are "a nation of joiners" (1956/1840).

This phrase is often quoted in descriptions of democratic societies. Democratic theory states that when people are given the freedom to associate with whomever they choose, they will form a complex civil society—that is, a society characterized by a never-ending and always changing array of freely formed nongovernmental groups (see Chapter 9). In fact, as we will see in later chapters, dictators who wish to curtail individual freedom invariably attempt to control the groups people can belong to.

The number of voluntary associations in the United States and other democratic nations has increased steadily since de Tocqueville's day. But what kinds of groups are people forming? What needs are they seeking to satisfy through their group interactions? A study by sociologist Robert Wuthnow directly addresses these questions. Wuthnow's findings, described in his book *Sharing the Journey* (1994), are based on a large survey carried out by the Gallup Poll in which respondents were asked about their group affiliations and the benefits they derived from them. Table 6.1 presents Wuthnow's estimates of the major types of groups to which people in the United States belong. Religious groups engaging in Bible study and Sunday school classes are by far the largest category; self-help groups in which people attempt to resocialize themselves, to replace old and often destructive patterns of behavior with new and more desirable ones, are the second largest category.

Wuthnow's study revealed that people who join small groups are owning up to a deeply felt need to be part of a face-to-face primary group in which they attain a sense of community with others. There are often wide gaps between people's needs and the ability of group interactions to actually satisfy those needs. But these gaps do not seem to produce widespread disillusionment. In their search for a feeling of belonging and community, people continue to join groups of all kinds.

 Thinking Critically

We often think of bureaucracy in negative terms and complain about "the bureaucrats in Washington." But bureaucracy can be a rational way of organizing groups. If bureaucratic rules are necessary in some kinds of groups, does that mean that we have to accept "red tape" and other abuses of bureaucratic organization as inevitable?

TABLE 6.1

Groups in the United States

	Estimated Numbers	
	Members	Groups
Sunday school classes	18–22 million	800,000
Bible study groups	15–20 million	900,000
Self-help groups	8–10 million	500,000
Special interest groups		
Political/current events	5–10 million	250,000
Book/discussion groups	5–10 million	250,000
Sports/hobby groups	5–10 million	250,000

Friendship Groups These photos capture some aspects of the universal experience known as friendship. They show friendship groups in societies throughout the world—from the young men and women in an urban street gang, to a group of wealthy English boarding school students, to scenes from African villages. They convey the emotional quality of interaction among friends, but to the sociological observer, they also reveal some characteristic features of friendship groups.

First, note the size of the groups. None numbers more than six or seven members. As Georg Simmel first pointed out in his writing on friendship and intimacy, the larger a group is, the more opportunities there will be for conflict and schism. Increasing size brings with it a greater likelihood of jealousy, competition for attention, and conflicting values, and these tensions tend to break up larger friendship groups into smaller, more intimate ones like those shown here.

A second observation that we can draw from these photos has to do with gender differences in friendship groups. Children's friendship groups are involved in anticipatory socialization, in taking roles that will later be incorporated into adult statuses. Thus, the girls are seen holding dolls and practicing the interactions that occupy so much of their mothers' time. And unlike the typical male friendship group, female groups often have the responsibility of caring for younger siblings, as we see in at least one of these photos. Males more typically engage in activities that involve physical prowess and changes in the environment. We also see in these photos that gender segregation is normally the rule in friendship groups, although there may be many exceptions, such as when games are played in school.

Finally, these photos remind us of the emotional complexity of friendship. Friends can fight and make up, but people who are less than friends often cannot fight without ending the relationship.

SUMMARY

What are some of the social groups that humans regard as most important in their lives?

- The social fabric of modern societies is composed of millions of groups of many types and sizes.

- Unlike a *social category*, a collection of individuals who are grouped together because they share a particular trait, a *social group* is a set of two or more individuals who share a sense of common identity and belonging and who interact on a regular basis. Those interactions create a social structure composed of specific statuses and roles.

What are the key characteristics of social groups?

- A *primary group* is characterized by intimate, often face-to-face, association and cooperation. *Secondary groups* are characterized by relationships that involve few aspects of the personality; the members' reasons for participation are usually limited to a small number of goals.

- As the number of people in a group increases, the number of possible relationships among the group's members increases at a faster rate. A group composed of only two people is known as a *dyad*. The addition of a third person to form a *triad* reduces the stability of the group.

- At the level of social organization between the primary group and the institutions of the nation-state are *communities.* *Territorial communities* are contained within geographic boundaries; *nonterritorial communities* are networks of associations formed around shared goals. Territorial communities are usually composed of one or more neighborhoods in which people form attachments based on proximity.

- Groups formed at the neighborhood level are integrated into networks that may extend beyond geographic boundaries.

- Key factors in the formation of networks are *in-group–out-group* distinctions. Such distinctions can form around almost any quality but are usually based on such qualities as income, race, or religion.

- Another type of group is the *reference group,* a group that an individual uses as a frame of reference for self-evaluation and attitude formation.

- The study of whom people associate with, how those choices are made, and the effects of those choices is known as *social network analysis.*

A few basic principles tend to govern interaction in groups. What are these, and how do sociologists study these interactions?

- Social scientists have identified certain principles of interaction that help explain both stability and change in human groups. Among them are the pleasure principle, the rationality principle, the reciprocity principle, and the fairness principle.

- The balance among these principles varies from one situation to another, with economic motives dominating in some instances and social needs winning out in others.

- From an interactionist perspective, an important factor determining how people behave in a given instance is their definition of the situation.

- Studies of interaction in groups use a variety of techniques. *Ethnomethodology* is the study of the underlying rules of behavior that guide group interaction. The *dramaturgical approach* regards interaction as though it were taking place on a stage and unfolding in scenes.

- The strategies that people use to set a stage for their own purposes are known as *impression management.*

- Research on the bystander effect has shown that when responsibility seems to be diffused, people are more likely to avoid helping others. But when people define the situation as involving them, they will intervene despite the potential costs to themselves.

- Research on small groups has shown that they tend to develop two kinds of leaders: a "task leader" who keeps the group focused on its goals and a "socioemotional leader" who creates a positive emotional climate.

Why are bureaucracies such a prominent feature of contemporary societies?

- *Informal organizations* are groups with generally agreed-upon but unwritten norms and statuses, whereas *formal organizations* have explicit, often written, sets of norms, statuses, and roles that specify each member's relationships to the others and the conditions under which those relationships hold.

- A *voluntary association* is a formal organization whose members pursue shared interests and arrive at decisions through some sort of democratic process.

- A *bureaucracy* is a formal organization characterized by positions with clearly defined responsibilities, the ordering of positions in a hierarchy, governance by rules and precedents, impersonality and impartiality, a career ladder, and efficiency as a basic norm.

- One effect of the increasing dominance of bureaucracies in modern societies is the possibility that individuals will not take full responsibility for their actions. A study of obedience to authority conducted by Stanley Milgram raised serious questions about people's ability to resist pressure to carry out orders for which they are not personally responsible. Milgram also found, however, that rebellion against authority is more likely when individuals who rebel have the support of others.

- Other studies have found that commitment to bureaucratic organizations is greatest when it is supported by ideology or by strong primary-group attachments.

What roles do various kinds of social groups play in complex societies?

- As a society becomes larger and more complex, it tends increasingly to be characterized by secondary groups and organizations. These make the society more efficient, but they can also cause confusion and unhappiness.

- Durkheim pointed out that members of complex societies have greater freedom to choose what groups to join and what norms to conform to, but can be overwhelmed by the choices open to them.

- Weber examined the effects of the rise of bureaucracies in modern societies, finding that individuals often come to feel like tiny cogs in a huge set of interlocking organizations.

- Democratic theory states that when people are given freedom of association, they will form a complex civil society. This theory is supported by research on group membership in the United States. Religious and self-help groups are two types of groups that Americans join in large numbers.

The Kornblum Companion Website

www.cengagebrain.com

Supplement your review of this chapter by going to the companion website to take one of the tutorial quizzes, use the flash cards to master key terms, and check out the many other study aids you'll find there. You'll also find special features, such as GSS data and Web links that will put data and resources at your fingertips to help you with that special project or to do some research on your own.

Jason Stitt, 2010/Used under license from Shutterstock.com

7

Perspectives on Sexuality: An Overview

Biological and Social Views of Sexuality

Sex versus Gender

Sociological Perspectives on Sexuality and Sexual Institutions

Sexuality

FOCUS QUESTIONS

What are the major sociological perspectives on sexuality?

How do biological and social factors interact to shape human sexuality?

What are the key distinctions between sex and gender? Between sexual orientation and sexual behavior?

What are some current trends and issues in the study of sexuality and sexual institutions?

One of the oldest love poems in Western culture is the fifth book of the Old Testament, the famous *Song of Songs* of Solomon. Wise beyond his years, and enormously wealthy, Solomon ruled Israel from 971 to 931 B.C.E. His poem is both erotic and spiritual. On the surface, it is about the love, erotic and deeply heartfelt, that Solomon and a young Shulamite girl felt for each other. More symbolically, it is often interpreted as a testimony to the love of God for Israel, or as an anticipation of the love of Christ for the church. In our own time, however, passages from the *Song of Songs,* like the one quoted here, are often recited in marriage ceremonies among both religious and nonreligious couples and families. Here is the third chapter of this timeless love poem:

Behold, thou art fair, my love; behold, thou art fair; thou hast doves' eyes within thy locks: thy hair is as a flock of goats, that appear from mount Gilead.

Thy teeth are like a flock of sheep that are even shorn, which came up from the washing; whereof every one bear twins, and none is barren among them.

Thy lips are like a thread of scarlet, and thy speech is comely: thy temples are like a piece of a pomegranate within thy locks.

Thy neck is like the tower of David builded for an armoury, whereon there hang a thousand bucklers, all shields of mighty men.

Thy two breasts are like two young roes that are twins, which feed among the lilies.

Until the day break, and the shadows flee away, I will get me to the mountain of myrrh, and to the hill of frankincense.

Thou art all fair, my love; there is no spot in thee.

Come with me from Lebanon, my spouse, with me from Lebanon: look from the top of Amana, from the top of Shenir and Hermon, from the lions' dens, from the mountains of the leopards.

Thou hast ravished my heart, my sister, my spouse; thou hast ravished my heart with one of thine eyes, with one chain of thy neck.

How fair is thy love, my sister, my spouse! how much better is thy love than wine! and the smell of thine ointments than all spices!

Thy lips, O my spouse, drop as the honeycomb: honey and milk are under thy tongue; and the smell of thy garments is like the smell of Lebanon.

The poem is also an expression of the many connections between carnal and spiritual love. It celebrates the lover's sensual feelings for the loved one and passionately inventories the loved one's body, comparing its forms and parts to beloved places, or animals, or to the delicious perfumes of nature or the human world. No wonder the *Song of Songs* has often been used to justify more open expression of sexuality against those who would censor or suppress the erotic imagery of love and sex.

The poem has also been cited against those who would deny the possibilities of love and sexual union between people of different backgrounds and social statuses. Elsewhere in the poem, one of the lovers talks about being black and comely, desirable to his or her lover. In nations where interracial marriage was forbidden, such as South Africa under apartheid or some states in the United States before the civil rights movement, these passages were read as biblical precedents and cited in opposition to such laws.

But the *Song of Songs* cannot be read entirely as a celebration of mutual fidelity in marriage. King Solomon and his love lived in a society in which the norms governing sexuality and marriage were quite different from those of most societies today. Solomon was said to have had 700 wives and 300 concubines. Perhaps he wrote the *Song of Songs* before he became polygamous, as is often assumed, but certainly he and his contemporaries conducted their intimate lives according to norms that differed greatly from those we are familiar with. We can embrace the poem as an ancient expression of love, intimacy, and sexuality, but as with so much scriptural writing, we cannot take it as a universal blueprint for how everyone should behave. As we will see in this chapter, sexuality and the norms of intimacy are among the most deeply held beliefs of people in many cultures, and sexuality is at the heart of much that occurs within and among societies, even if it is not always acknowledged.

PERSPECTIVES ON SEXUALITY: AN OVERVIEW

If sexuality, intimacy, and love are so deeply interconnected, why is it that sexual behavior is often the source of severe conflict in societies? Why is so much interpersonal violence related to sexual behavior? Why are different factions of our own society so divided over issues of homosexual marriage, legalization of prostitution, or the exposure of a female breast on television? Why are issues of sexuality and the rights of men and women so hotly contested in the United States and many other societies? These are among the central questions asked by sociologists who study sexuality, and we will explore them from a number of perspectives in this chapter.

The Biological Perspective

From a biological perspective, sexuality is part of the larger process known as **sexual selection**, in which individuals seek, consciously and unconsciously, to transmit their genes to the next generation through mating and procreation. Sociologists agree, of course, that sexuality is how individuals and societies reproduce

themselves, but their research on sexual behavior often focuses on the ways in which sexuality and sexual behavior create dilemmas that all societies seek to resolve, in different ways, through their cultures and social institutions.

Sociologists who study sexuality from a biological perspective ask questions like these: What are the unique aspects of human sexuality, and how does our biological nature act to shape our social behavior and institutions? As we will see later in the chapter, pioneering work by Krafft-Ebing, Freud, Kinsey, and many contemporary researchers has sought to answer empirical questions about the varieties of human sexual behavior. In so doing, these researchers have helped to change our ideas about sexuality throughout the life course.

The Functionalist View

Functionalist social scientists ask how societies seek to influence or control the sexuality of their members through specific cultural norms, such as the incest taboo or the requirement of premarital celibacy, or through institutions like marriage and prostitution. From a functionalist perspective, for example, the fact that prostitution has appeared in so many historical periods and in so many different societies inevitably raises the question of whether societies may actually benefit in some ways from a commerce in sex that they generally condemn as deviant.

Conflict Theory

Sociologists who take a conflict perspective ask why sexuality is such a frequent source of conflict both within and between societies, and how power in social relations affects sexual behavior—especially in the case of rape or patriarchic control over women's bodies. Those who stress the importance of conflict as a cause of social change, for example, are likely to ask why sexuality—especially rape—is so often a feature of warfare, or why so much conflict within American society can be traced to differing norms about appropriate sexual behavior. Before the pathbreaking work of feminist scholars like Susan Brownmiller, for example, the dominant view of rape was that it was the consequence of unchecked male passions and that, in the absence of social controls during warfare, these passions resulted in rape. But in her classic book *Against Our Will* (1975), Brownmiller showed that rape, "the use of the penis as a weapon," whether in wartime or not, is primarily an act of hostility against women and the brutal exercise of power to cause fear and to dominate.

The Interactionist Perspective

How do people conduct their sexual lives through interactions with others? In what ways does their sexuality shape their behavior and body image? Interactionist research tends to reveal the ways in which sexuality is socially constructed—that is, not passed along without change from one generation to the next, but rather, subject to all kinds of changing norms and ideas about beauty, sexiness, personal identity, and many other social ideas that are relevant to intimate behavior. For example, interactionist research on body image in different cultures shows that in Western nations before the twentieth century, the ideal feminine body image was of a voluptuous woman. It seeks to explain how that ideal was

---

sexual selection: Conscious or unconscious effort of individuals to transmit their genes to the next generation.

replaced by an emphasis on slenderness that is unattainable for most women (Engeln-Maddox, 2006).

Global Feminist Perspectives

Can universal norms of human rights ever supersede regional cultural norms that oppress women, even when they are based on religious values? How can efforts to eliminate sex tourism and trafficking in women and boys be supported? How does a feminist human rights perspective apply to the global fight against sexually transmitted diseases like AIDS? These research questions are responses to the concerns of women and men who are active in the global movement to address problems of gender inequality and political repression of women.

Sociology & Social Justice

SEXUALITY AND PERSECUTION Throughout the world people are persecuted because their sexual behavior or preferences differ from the norms of their culture. They may be homosexuals attacked by peers at school, or women who have been raped or sexually exploited in ways that cause them stigma, dishonor, and death. Or they may be men who have been raped in prisons; the varieties of sexually based injustice are endless. Examples of ways in which people are deprived of control over their own bodies, a basic violation of their human rights, vary throughout the world. After reviewing many of these wrongs that people suffer because of their gender and sexuality, feminist sociologist Martha Nussbaum asks, "What does it mean to respect the dignity of a human being? What sort of support do human capacities demand from the world, and how should we think about this support when we encounter differences of gender or sexuality" (Nussbaum, 1999)?

Nussbaum and other feminist social thinkers are addressing extremely complex and emotional issues. Sexual behavior is highly norm-governed, but the norms can be vastly different from one culture to another. The sociological stance in the face of so much variation and so many moral dilemmas is to first try and understand the cultural norms in a given situation. What explains why so many Muslim women accept the veil (Lazreg, 2009)? What explains why women on European beaches go topless? Why do brothers kill their sisters to salvage the honor of the family name? What explains why some cultures practice female circumcision, which may deprive women of sexual pleasure?

But to understand a culture and its norms does not mean that those norms should not be subjected to more universal ideas of human rights. Feminist sociologists like Nussbaum point out that many of the problems women experience, such as rape, honor killings, poverty, and illiteracy, stem from the fact that in many cultures they are still considered to be the possessions of fathers and husbands, who can treat them as if they had no rights as human individuals. So although feminist social scientists can plead for sensitivity to the norms of other cultures, and understand why women may prefer to be veiled in some cultures, they can also insist that women require the same opportunities as men to choose their behaviors without fear of violence or retribution. And they must have *agency*; that is, they must feel that the choices they make can matter in the world, so that their sisters and daughters can also have the same choices. We will have reason to return to these issues throughout this text, and especially in Chapter 13.

The issues raised by women (and men) who seek to enhance the right of women to choose their own fate—rather than being treated more or less as the possessions of fathers and husbands who make the decisions that affect their lives—are central problems of world politics. They are especially relevant to the conflicts that arise between orthodox religious leaders and more secular political activists. It is somewhat paradoxical, therefore, that one of the major findings of global feminist research is that most women in traditional patriarchal societies do not resent male power. A recent survey of women in Islamic societies, for example, showed that the majority do not reject male authority and do not feel oppressed by strict Islamic dress codes and norms that discourage them from being seen in public. Women and men who have been socialized in a particular culture tend to support its norms, especially when they are strongly backed by religious leaders. But the central issue for those who seek greater gender equality is how the minority of women who are desperate for education and freedom from the control of men can be empowered to express their views.

In this section, we have merely touched on some of the important issues in contemporary sociological research on sexuality. Many of these issues are discussed further in later sections. The Perspectives chart on page 157 summarizes the different assumptions and theoretical issues that are the focus of each research perspective and offers examples of empirical questions that arise at the micro, middle, and macro levels of social analysis. Note, however, that sociologists almost always study sexuality, like many other social phenomena, at more than one level of analysis and from more than one perspective.

Perspectives on Sexuality

Perspective	Key Focus	Questions at Different Levels of Analysis
Biological	The unique aspects of human sexuality, and how our biological nature shapes our social behavior and social institutions	*Micro:* Is there a biological basis for sexual attraction? *Middle:* How does sexual selection work in different families? *Macro:* What are the demographic and biological implications of birth control?
Functionalist	How societies seek to influence or control the sexuality of their members through norms, such as the incest taboo or premarital celibacy, or through institutions, such as marriage and prostitution	*Micro:* What sex roles are found in families? *Middle:* What social functions does celibacy serve in organizations such as the Catholic Church? Do colleges and universities perform "mate selection" functions? *Macro:* How does globalization foster the rise of sex tourism?
Conflict	Why sexuality can be a source of conflict with and among societies, and how power in social relations bears on sexual behavior, especially in reference to rape or the patriarchal control of women's bodies	*Micro:* Can conflict actually help couples move toward more companionate sexual relationships? *Middle:* What processes help communities that are divided over issues of sexual morality avoid rancor and schism? *Macro:* To what degree are issues of sexuality and gender equality involved in tensions between Western and Islamic societies?
Interactionist	How people conduct their sexual lives through interactions with others; how their sexuality shapes their behavior and body image	*Micro:* Do changing "sexual scripts" influence the intimate behavior of couples? *Middle:* What interactions over issues of sexuality shape school policies on sex education? *Macro:* How do activities in conservative and liberal social movements frame issues of gender equality and homosexual rights?
Global feminist	Whether universal norms of human rights can ever supersede cultural norms that oppress women; how sex tourism and trafficking in women and boys can be eliminated; how a feminist perspective emphasizing gender equality can be applied to the global fight against sexually transmitted diseases like AIDS	*Micro:* What satisfactions do women experience in highly gender-stratified cultures and societies that help legitimate continuing gender inequality? *Middle:* How does women's poverty in different regions of the world contribute to the spread of sexually transmitted diseases like AIDS? *Macro:* How can women gain more power and influence over global debates about women's reproductive rights?

BIOLOGICAL AND SOCIAL VIEWS OF SEXUALITY

Sex is the source of many of our most intense pleasures. It can also be the source of some of our most intense feelings of self-doubt or shame and, in some cases, the worst conflicts we ever experience. The basic biological facts of human sexuality create problems that human societies and cultures have been addressing for thousands of years. And much of this is related to the fact that humans have some of the most unusual sexual characteristics in the animal world.

Evolution and Sexuality

Human beings, unlike all other mammals, can enjoy sex at all times of the female's estrus (menstrual) cycle. In addition, only human females show no outward physical sign that they are receptive to mating during ovulation (which in most cases is *not* the time when menstrual bleeding occurs, contrary to what many people believe). The females of other primate species boldly display some physical sign, such as swollen and brightly colored genitalia, that they are biologically ready for mating.

Humans are also unique for having sex in private rather than in front of others of their species, as is the case with other animals, and humans are the only species we know of that has sex while the female is pregnant and even after the woman has reached menopause (also a biological trait unique to humans) and can no longer bear children. In other species, the male seeks to inseminate as many females as he can, a strategy designed to place as many of his gene sets into the next generation as possible. Among humans, the male might be inclined to do likewise but typically is constrained from doing so by strong cultural norms and sanctions.

This relationship between human sexuality and culture has been shaped by a long process of evolution. Some of the basic aspects of human sexuality, such as the potential readiness of the female throughout her lifetime or the propensity for humans to engage in sex even after female menopause, are biological traits, but they exist within a well-defined set of cultural norms that have also evolved over thousands of generations of human existence. Among these norms are the incest taboo, sanctions for sexual infidelity, the consensual nature of sexual behavior (as opposed to rape or other forms of sexual coercion), and norms dealing with procreation versus birth control. Most of these norms, as we will see in this chapter, function to strengthen the bonds between sexually active partners (Lippa, 2002).

From a biological perspective, the evolution of human sexuality to be independent of the female estrus (or menstrual) cycle makes sense in that it permits sexual intimacy to become a part of the mating pair's bond rather than a strictly procreative activity. There are many dimensions to love, but without an active sexual relationship, couples find it more difficult to sustain their loving relationship throughout the many stresses they encounter as their families grow.

Another key benefit of the evolutionary connection between human biology and culture is the tendency of human parents to care for their infants for a much longer period than is typical of most other animals. The physical and cognitive development of human infants requires far more nurturing, over longer periods, than is true for most other species (pandas, kangaroos, elephants, whales, and dolphins are notable exceptions). A baby calf walks hours after birth and can follow its mother and feed from her teat as she grazes. But even after it walks and begins talking, a human infant is entirely dependent on adult care; its ever more curious mind is likely to get it into serious trouble without the watchful eyes of caring adults (Hyde, 2006).

Sexual Dimorphism and Sexual Selection

Why do males and females of most animal species, including humans, exhibit *sexual dimorphism*—that is, why do they have different secondary sex characteristics? *Primary sex characteristics* are the male and female genitals. *Secondary sex characteristics* are physical and behavioral features, other than the genitalia, that differ in males and females. In humans, secondary sex characteristics develop at puberty and include the female's breasts and the male's greater average size and larger amount of body hair, as well as differences between females and males in average body size and strength and in aggressiveness. In other animals, differences in secondary sex characteristics are often far more spectacular, especially in the male. Examples include the immense and colorful tail of the male peacock, or the brilliant coloring of male mallard ducks. These differences between males and females of many species, including humans, are termed **sexual dimorphism**.

Charles Darwin was the first biologist to develop an explanation for the differences between males and females that create sexual dimorphism. In many animal species, especially among mammals, males produce far more reproductive gametes (sperm cells) than do females (ovarian egg cells). From an evolutionary perspective, Darwin argued, each individual, male or female, has an equal need to prove its fitness—that is, to successfully reproduce his or her genes in the next generation. But because females produce far fewer eggs, and, therefore, fewer sets of genes in the reproductive process, they have much more reason to be prudent about whom they choose to mate and produce offspring with. Males, on the other hand, have more reason, from a strictly biological viewpoint, to scatter their sperm far and wide and to mate with more than one female, whereas females have more reason to choose males who appear to be as fit as possible. To attract as many females as possible, males of many species display outward signs of their fitness, such as the peacock's tail or the huge antlers of the moose.

Cultural Change in Gender Display

Darwin's theory of sexual selection, a key part of his more comprehensive theory of natural selection, explains the evolution of sexual dimorphism. Sexual dimorphism in humans, however, is evolving most rapidly in ways that are far more related to changing social constructions of gender—that is, to human cultures' changing ideas of maleness and femaleness—than to secondary sex differences or other biological traits, such as aggression. Over hundreds of thousands of years of evolution, humans have developed cultural norms that emphasize forms of sexual dimorphism that are independent of physical differences. We supplement the basic biological dimorphism of men and women with gender-specific clothing and facial adornments, as can be seen in photos of male and female tribal members, or in the typical dress of males and females during the Victorian era.

In the United States and other urban-industrial societies, cultural norms of dress and makeup that emphasize sexual dimorphism have tended to weaken over the past century. Women in business settings wear clothing that, though different from that of men, is far more similar to male attire than ever before. And as more women engage in physical conditioning and competitive sports, gender differences in physical strength and endurance are becoming somewhat less marked (Wang et al., 2010). Such changes point to a remarkable feature of our species: Although male and female humans differ in primary and secondary sexual characteristics, we have the ability through our cultures to de-emphasize the physical differences

sexual dimorphism: Differences in secondary sex characteristics between males and females of a species.

This painting, *Windsor Castle in Modern Times*, by Edwin Henry Landseer, depicts the youthful Queen Victoria (who reigned from 1837 to 1902) and Albert, the Prince Consort, "at home" in the 1840s. Their clothing emphasizes sexual dimorphism, with her gown cut to show the curve of her shoulders and chest, and his clothes emphasizing the strength of his shoulders and legs, and the prominence of his riding boots. Victorian fashion also introduced the crinoline and bustle, fashion accessories that exaggerate the contrast between a woman's waist and her posterior. Do you think current fashions for men and women emphasize or de-emphasize sexual dimorphism?

between the sexes, or, if we wish, to make them more prominent. Moreover, what we consider "normal" gender or sexual behavior may be normal in our culture but not in a different culture or historical period. In reality, there are endless variations in dress and sexual behavior in different cultures. These variations demonstrate the immense flexibility and adaptability of human culture. This uniquely human capacity to shape the way our bodies look leads us to the important distinction between gender and sex.

SEX VERSUS GENDER

Masculine and *feminine* are among the most confusing words used both in everyday speech and in the social sciences. The confusion stems largely from the fact that our ideas of what is male and female in human individuals are based on overlapping influences from biology and culture. As in all such situations, it is difficult to sort out the relative contributions of each source of influence. But the underlying concepts are not hard to grasp: One's sex is primarily a biological quality, whereas one's gender

is formed largely (but not entirely) by the cultural forces we experience through socialization from infancy on (Connell, 2005).

Sex refers to the biological differences between males and females, including the primary sex characteristics that are present at birth (that is, the presence of specific male or female genitalia) and the secondary sex characteristics that develop later (facial and body hair, voice quality, and so on). Note that these biological qualities differ considerably among individuals, so the differences between the sexes are not always as marked as the male–female distinction suggests. Moreover, some people are born as **hermaphrodites**; their primary sexual organs have features of both male and female organs, making it difficult to categorize the person as male or female. Hermaphrodism is uncommon

---

sex: The biological differences between males and females, including the primary sex characteristics that are present at birth.

hermaphrodites: Individuals whose primary sexual organs have features of both male and female organs, making it difficult to categorize the person as male or female.

What is the ideal body for the contemporary woman? Americans spend billions of dollars a year on diets, health club fees, and other goods and services designed to give them the physical appearance they crave. And although women were once expected to be soft and curvaceous, men today are just as likely to be attracted to women who have developed their physiques through exercise and bodybuilding.

(estimates vary from one in 20,000 to one in 40,000 births), but it occurs in societies throughout the world.

Another ambiguous sexual category consists of the many thousands of **transsexuals**, people who feel strongly that the sexual organs they were born with do not conform to their deep-seated sense of what their sex should be. Transsexuals, or *intersexuals*, a term many prefer, sometimes undergo a course of endocrine hormonal treatments to change their secondary sex characteristics and may eventually have irreversible sex-change operations. Most such operations are performed to change the individual from a male to a female.

One of the most famous transsexuals in the contemporary world is the writer Jan Morris. As a male, James Morris reported on an expedition

to Mount Everest and engaged in many activities that are associated with masculinity. But as Morris explains it: "I was born with the wrong body, being feminine by gender but male by sex, and I could achieve completeness only when the one was adjusted to the other" (quoted in Money & Tucker, 1975, p. 31). In making this distinction, Morris is calling attention to the difference between her biologically determined sex and her socially and emotionally influenced gender.

Gender refers to the culturally defined ways of acting as a male or a female that become part of an individual's personal sense of self. Most people develop a "gut-level" sense of themselves as male or female, boy or girl, early in life; however, as we will see shortly, some people's gender identities are more ambiguous. It is not entirely clear exactly how a person's gender identity is formed, but much of the research done to date suggests that the assignment of a gender at birth has a strong influence during the early years of life. In other words, children's feelings of being a boy or a girl are defined more by how they are treated by their parents than by their actual biological sex characteristics (Sanlo, 2005).

Gender and Sexuality

Sexuality refers to the manner in which a person engages in the intimate behaviors connected with genital stimulation, orgasm, and procreation. Like most areas of human behavior, sexuality is profoundly influenced by cultural norms and social institutions such as the family and the school, as well as by social structures like the class system of a society. Sociologists and historians believe that major changes in sexuality have occurred as a result of changes in the economy, politics, and the family. Over the past 300 years, for example, the meaning and place of sexuality in American life have changed: from a family-centered system for ensuring reproduction and social stability during the colonial period; to a romantic and intimate sexuality in nineteenth-century marriage, with many underlying conflicts; to a commercialized sexuality in the modern period, when sexual relations are expected to provide personal identity and individual happiness apart from reproduction (Bhattacharyya, 2002; Rossi, 1994).

These changes are important and controversial, but they hardly begin to exhaust the range of sexual norms in cultures throughout the world. Polygyny (the practice of having more than one wife), which is illegal in North America and Europe, is condoned throughout much of the Islamic world. Marriage between adult men and preadolescent girls is practiced in some Asian cultures but would be considered a form of child abuse in the West. Women bare

transsexuals: People who feel strongly that the sexual organs they were born with do not conform to their deep-seated sense of what their sex should be.

gender: The culturally defined ways of acting as a male or a female that become part of an individual's personal sense of self.

sexuality: The manner in which a person engages in the intimate behaviors connected with genital stimulation, orgasm, and procreation.

their breasts on the beaches of Europe but may be issued summonses for the same behavior on the beaches of North America. On American television screens, women and men expose their bodies in a fashion that is horrifying to people in many parts of Asia. Catholic priests and nuns are expected to remain celibate, whereas Protestant ministers are encouraged to have families, as are Jewish rabbis. With all of these contrasting cultural norms governing sexuality, can it be said that there are any universal sexual norms?

Universal cultural norms exerting social control over sexuality include the *incest taboo, marriage,* and *heterosexuality,* but even these norms include variations and differing degrees of sanction. As noted in Chapter 3, the incest taboo is known in every existing and historical society and serves to protect the integrity of the family. Incest norms prohibit sexual intimacy between brothers and sisters and between parents and children, and usually specify what other family relatives are excluded. However, the specific relatives involved vary in different cultures; for example, sex between first cousins is permissible in some societies and strictly taboo in others.

Every society also has marriage norms that specify the relationships within which sexual intimacy is condoned. Marriage norms protect the institution of the family, confer legitimacy on children, and specify parental rights and obligations. However, there are vast cultural differences in how these norms operate and how strenuously they are enforced. Figure 7.1 summarizes the basic human mating systems. Although monogamy is by far the most common mating (and marriage) norm, polygyny is widespread in some regions of the world, especially parts of Africa and the Middle East. Polyandry—in which one woman shares intimacy and family bonds with more than one man—exists in some smaller societies but is rare in the contemporary world. Where it has existed in recent memory, as in Tibet and Nepal, it appears as "fraternal polyandry," in which brothers live with one wife, but this form of polyandry is now forbidden in those cultures (Borgerhoff Mulder, 2009).

Promiscuity is found among small minorities of men and women, but it is an extremely risky behavior in an era of virulent sexually transmitted diseases. Far more common is a variant of monogamy known as "serial monogamy," in which partners mate and either marry or maintain a relationship, but then move on to another monogamous relationship. In addition to these variations, there are some important variations with regard to the strength of sanctions on adultery (sexual intimacy outside marriage).

Most cultures prohibit adultery, but the sanctions vary widely. Former French president François Mitterrand, who died in 1996, specified in his funeral instructions that both his wife and his mistress were to be present at his funeral. Although adultery is not openly condoned in France, it is far less strongly sanctioned than in the United States, where evidence that a presidential candidate may have had an extramarital affair often becomes a major issue in the campaign (Herzog, 2008).

Sexual Behavior

Since the groundbreaking research by Alfred Kinsey and his associates in the 1940s (see the Research Methods box on pages 162–163), Americans have been fascinated by studies of sexual behavior among people like themselves. Although our own sexuality is a private matter, we are often curious to know how our sexual behavior corresponds to that of others. Reliable and nonjudgmental sexual knowledge was extremely difficult to come by in the mid-twentieth century, which helps explain why Kinsey's rather dry and scientific volumes became best sellers. Contemporary sociologists who study sexuality using social-scientific methods find their work in great demand, even if their subject still raises eyebrows among those who believe that sexuality should remain a strictly private matter. Pepper Schwartz, one of the nation's leading sociological researchers on sexuality and relationships, observes that when told what her work is about, "At first, people get a bit embarrassed, but then comes the torrent of questions! When I was younger and single it was a bit of a double-edged sword: Men were fascinated, but sometimes intimidated" (Schwartz, 2006).

Before the "sexual revolution" of the mid-twentieth century, however, research and public discussion like that of Schwartz and others in her field was even more controversial.

The Sexual Revolution The idea of separating sexual intercourse from reproduction has an extremely long history. There are prehistoric cave paintings in Europe drawn by preliterate humans over 12,000 years ago that depict the use of primitive forms of condoms. In England during the Industrial Revolution, there were marked declines in births that were largely the result of sexual abstinence and delay of marriage (Lewis, 2005). With widespread use of improved techniques of birth control in the second half of the twentieth century, sexual behavior among couples was more effectively separated from reproduction. Through the use of improved condoms and then, in the 1960s and early 1970s, the development of the birth control pill, millions of men and women in the United States and other nations were able to engage in sexual behavior without worrying about unwanted

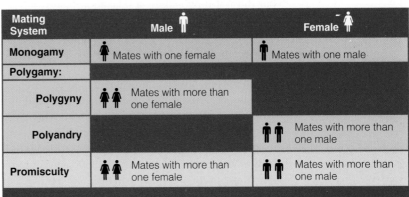

Figure 7.1 Terminology of Mating Systems. This figure applies to hetero-normative mating systems (in which homosexual mating is not sanctioned), a situation that is changing in many societies, including the United States, but can still be found in many others, especially among the developing nations. Source: Kenyon, 2006.

Asking Questions about Sexual Behavior

The pioneers of sex research were courageous in their determination to use scientific methods to learn about human sexual behavior. Every major study of human sexuality has been greeted with scorn by those who believe that sex should not be a subject of public discussion, or that it should be subject to religious norms that specify "proper" sexual conduct—implying that there is no reason to know more about actual sexual behavior (Gagnon, 2004).

Early social-scientific research on human sexuality typically addressed questions about sexual deviance and crimes. The Viennese psychiatrist Richard Freiherr von Krafft-Ebing published his landmark study, *Psychopathia Sexualis,* in 1886. It was based on an analysis of 286 cases of people who engaged in compulsive sexual behaviors; many of his subjects had been committed to mental hospitals for sex crimes. Krafft-Ebing conducted extensive examinations and interviews of these patients in an effort to identify specific patterns of sexual perversity. He coined the terms *sadism* and *masochism*, and identified many now-familiar types of sexual compulsions and crimes. His book was also the first to carefully describe male and female orgasm and to emphasize the importance of clitoral stimulation and female sexual pleasure. It also concluded that homosexuality is not an illness, a theme that would reappear in the sex research of Sigmund Freud.

Like Krafft-Ebing, Freud based his research on extensive case studies and lengthy sessions with patients who consulted him for assistance in alleviating or curing mental problems, some of which were derived from their sexual behavior. As he developed the first comprehensive theory of the development of the human personality, he drew on the extensive data his patients had provided about their earliest childhood experiences. Freud had also observed in painstaking detail the development of his own children, and based on these observations and data, he began to argue that infantile sexuality must be seen as an integral part of a broader developmental theory of human personality. He argued that early-childhood sexual experiences—from the pleasure babies obtain from genital stimulation to the potentially traumatic experience of seeing their parents engage in sex—are crucial factors in the development of the adult personality.

Freud published his observations and theories in *Three Essays on the Theory of Sexuality* (1905). This volume was also among the very first to present medical arguments that homosexuality is not a mental illness and that homosexuals should not be obliged to deny their feelings and attempt to live as heterosexuals.

The publication of his essays on sexuality brought Freud immense international acclaim and criticism. Many religious conservatives viewed writings on sexuality as serving to legitimate sexual liberties and perversions. Among social scientists and sex researchers, these essays are considered on a level with the revolutionary work of Albert Einstein in physics or Karl Marx in sociology. All changed the way educated people thought and felt about human existence. They shaped the thought and politics of the twentieth century as intellectual turning points in the development of the modern world.

Anna Freud and Sigmund Freud.

pregnancies. To some of the activists and writers of the tumultuous 1960s, it seemed that a sexual revolution was occurring in which people could have multiple sexual partners with limited responsibility for the consequences of sexual intimacy.

During the middle decades of the century, these changes in reproductive technologies, sexual attitudes, and behaviors contributed to—but did not fully cause—a relaxation of restrictive norms governing sexuality, leading to more explicit discussion of sex and to a tendency to engage in sexual behavior at younger ages. Sexuality became a political issue and helped mobilize new social movements, including the gay and lesbian movement and the second great wave of the women's movement. There was also increased commercialization of sexuality through pornography and the mass media. In the heady atmosphere of the 1960s, ideas of free love, promiscuity, love-ins, public nudity, and public eroticism thrilled many and shocked many others.

This aspect of the sexual revolution ended rather abruptly in the 1980s with the spread of AIDS and other sexually transmitted diseases. Although it was often erroneously perceived as a "gay disease," the epidemic caused increased public concern about responsibility and security in sexual relations. Public education about the need to practice safe sex and the dangers of promiscuous sex led to renewed emphasis on the connections between sexual intimacy and lasting partnerships (Connell & Hunt, 2006). In the United States today, a conservative reaction against the sexual permissiveness of earlier decades has resulted in the withdrawal of funds for family planning, opposition to gay marriage, and attacks on sex education in the public schools, to cite only a few of many possible examples. But the lasting effects of the changes in sexuality that took place in the mid-twentieth century continue. As we see in Figure 7.2 on page 164—and despite campaigns for abstinence in many communities—sexual behavior among teenage girls and young adult women increases rapidly between the ages of fifteen and nineteen (Mosher, Chandra, & Jones, 2005). Figure 7.3 on page 164 indicates that women are using contraceptive methods, and in the case

As these early studies of sexuality became better known, researchers turned their attention to the sexual behavior of "normal" people, partly in response to a growing public demand for more scientific knowledge about sexuality. The research on male and female sexual response by a fearless University of Indiana biologist, Alfred Kinsey, and his associates was the most ambitious and influential study of this kind.

Kinsey and his assistants traveled throughout the United States to interview men and women about the most intimate aspects of their sexual lives. They conducted over 18,000 interviews and published their findings in *Sexual Behavior in the Human Male* in 1948 and *Sexual Behavior in the Human Female* in 1953. Both books became national best sellers. Kinsey believed that face-to-face interviews were the best way to get honest answers from respondents. He taught himself and his colleagues to conduct interviews and analyze them statistically in a way that maintained the respondents' trust. They never displayed their own feelings and always conveyed a nonjudgmental attitude that enabled their subjects to describe any sexual behavior, however stigmatized it might be. Kinsey was also extremely careful to convince respondents that their records would remain completely confidential. Over the years, that promise has been kept: No individual records were ever revealed.

Kinsey's methods borrowed heavily from interviewing techniques developed by sociologists, and in time the field of research on sexuality became more of a subfield of sociology and psychology than a subfield of biology. However, controversies over Kinsey's research and contemporary studies of sexual behavior still influence public policy. Opponents of sex surveys often make claims like the following:

> Kinsey's work has been instrumental in advancing acceptance of pornography, homosexuality, abortion, and condom-based sex education, and his disciples even today are promoting a view of children as "sexual beings." Their ultimate goal: to normalize pedophilia, or "adult-child sex." (Concerned Women for Justice, 2006)

The most recent survey research on sexual behavior in the United States was conducted in the mid-1990s by Edward Laumann, John H. Gagnon, and their associates at the University of Chicago. It was attacked vigorously by former Senator Jesse Helms and other members of Congress. Senator Helms said that his "opposition to funding surveys of human sexual behavior was based on his conviction that they are intended solely to 'legitimize homosexual lifestyles.' He argued for research that 'presents a clear choice . . . between support for sexual restraint among our young people or, on the other hand, support for homosexuality and sexual decadence'" (Laumann et al., 1994, p. 4).

Sex researchers are undeterred by such criticism and are far more modest about the results of their findings. Sociologist John H. Gagnon, himself a leading sex researcher, writes, "Sex research and the sex researcher have played an important role in providing benchmarks for sexual practices, illuminating general understanding, and providing the content for ideological debates about the right and wrong of sexuality in the society" (Gagnon, 2004, p. 26).

Alfred Kinsey interviewing a woman for his survey on female sexuality.

of condom use, so are their male lovers, but one of the results of recent campaigns against contraception and early sexual activity is an increase in withdrawal during intercourse, a risky contraceptive behavior, especially in inexperienced teenagers (Mosher, Chandra, & Jones, 2005).

Today there is a great deal of sexual information and discussion in the popular media, and far more open, frank talk about sexual behavior than was true a few decades ago, especially among women. Think of the magazines at supermarket counters that feature articles like "Seven Ways to Make Him Go Wild in Bed (after he comes home from fishing)." But the spread of sexually transmitted diseases like AIDS, as well as the links between sexuality and health, contribute to the need for scientific knowledge about sexual behavior. To gather data about intimate issues that are independent of the moral crusades that surround issues of sexuality, we must rely on large-scale sample surveys like the ones most recently conducted by an international team of researchers on changing sexual behavior in the United States and other parts of the world (Gagnon, 2004).

In 2006, a team of sociologists led by Edward Laumann and John H. Gagnon released the preliminary results of an international survey of sexual behavior. This is the first major international study to include large numbers of respondents from diverse religious traditions, including Christian, Jewish, Muslim, Buddhist, and other Asian religions, as well as atheists. The survey was conducted in twenty-nine nations, in every major region of the world, and gathered information from 27,500 adults between the ages of forty and eighty. Designed to look at the relationships between health and sexuality in older people, it has already produced a number of extremely important findings. The international survey supplements a major survey of sexual behavior among Americans conducted by the same research team in the mid-1990s (Gagnon, 2004; Laumann et al., 1994; Laumann et al., 2006) and the National Survey of Family Growth, an ongoing study of sexual and reproductive behavior conducted under the auspices of the U.S. Centers for Disease Control and Prevention (Mosher, Chandra, & Jones, 2005).

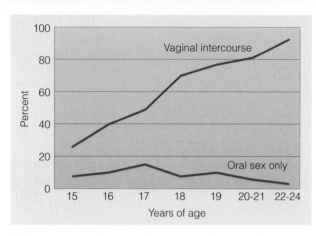

Figure 7.2 Percentage of females 15–24 years of age who have had vaginal intercourse, and percentage who have had oral sex with a male but not vaginal intercourse, by age: United States, 2002. Source: Mosher, Chandra, & Jones, 2005.

Sexual Satisfaction and Gender Equality The international study was intended to draw out people's subjective evaluation of the role of sex in their relationships with partners. Although it includes data on people who were not sexually active in the previous twelve months, the most important findings pertain to those who said that they had engaged in sexual activity. The survey included questions about how physically or emotionally satisfying their relationships are and how important sex is to them. Interviewers from each nation asked respondents about their overall happiness; physical and mental health circumstances, including sexual dysfunction; their attitudes toward sex; and their attitudes toward various social and demographic factors, including age, education, income, and religious affiliation.

The most significant finding of the survey is that couples who live in Western nations, where there is greater equality between men and women, report the most satisfaction with their sex life. In contrast, couples living in nations where men are dominant over women—especially in East Asia and, to a somewhat lesser extent, the Middle East—report less satisfaction with the physical and emotional quality of their sex life. Men and women reported the greatest sexual satisfaction

in five countries, led by Austria and followed by Spain, Canada, Belgium, and the United States. At the low end of the satisfaction scale were couples in Japan and Taiwan. Countries such as Turkey, Egypt, and Algeria were in the middle. The results from Japan are of particular interest to sex researchers because Japan is the most affluent nation in East Asia. The findings imply that the lower levels of sexual satisfaction in this region of the world are not due merely to relative poverty (Wills, 2006).

Sociologist Edward Laumann, who directed the research, observed that in relationships based on equality, couples tend to develop sexual habits that are more in keeping with both partners' interests. "Male-centered cultures where sexual behavior is more oriented toward procreation," his team found, "tend to discount the importance of sexual pleasure for women" (quoted in University of Chicago, 2006).

Sexual Satisfaction and Health Status Figure 7.4 shows another important finding of the study: People's subjective feelings of sexual well-being are closely correlated with their self-reported health status. Of the respondents who said that their health was fair, for example, 32 percent reported a sexual relationship with their partner that was "very/extremely pleasurable in the past year." But 68 percent of those who said that their health was excellent enjoyed very or extremely pleasurable sex with their partner. Note here that we cannot be sure of the direction of causality in these relationships: Are people healthy because they have pleasurable sexual relationships, or do they have such relationships because they are healthy? No doubt both directions of causality may be at work, although it is more likely that people who are ill or perceive themselves as ill find scant pleasure in any aspect of their bodily functions. For the researchers, the important finding is that subjective feelings of physical and sexual well-being are closely related.

There were large gender differences—across all the societies studied—in sexual well-being. On average, the levels of sexual health and well-being reported by men were at least 10 percent higher than those reported by women. In Western nations, two-thirds of men and women reported that their sexual relationships were satisfying, and 80 percent said that they were satisfied with their ability to have sex. About half of the men, but only one-third of the women, said that sex was extremely or very important in their lives.

In Middle Eastern nations, 50 percent of men and 38 percent of women found their sex lives satisfying. About 70 percent said that they

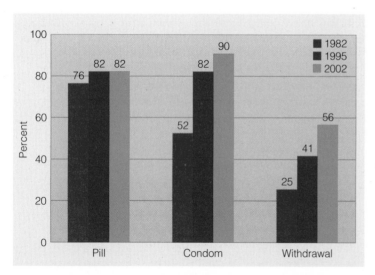

Figure 7.3 Percentage of sexually experienced women 15–44 years of age who have ever used the specified contraceptive method: United States, 1982, 1995, and 2002. Source: Mosher et al., 2004.

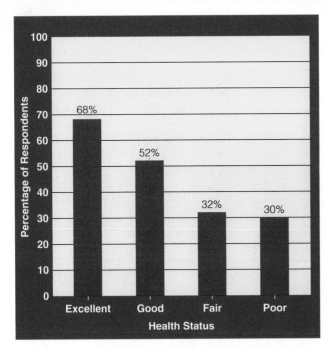

Figure 7.4 Percentage of respondents who say their physical relationship with their partner has been very/extremely pleasurable in past year, by health status of respondent. Notes: Overall health based on a 5-point scale, where 5 is "extremely satisfied," and 1 is "not at all satisfied." Satisfaction based on a 4-point scale, where 4 is "excellent" and 1 is "poor." Source: Global Study of Sexual Attitudes and Behaviors funded by Pfizer, Inc. Copyright 2002 Pfizer, Inc. Used with permission from Pfizer Inc.

were satisfied with their ability to have sex. Sixty percent of men and 37 percent of women said that sex is an important part of their overall lives. In East Asian countries, including China, only about one-quarter of men and women reported physical and emotional pleasure with sex, whereas two-thirds of the men and half of the women reported satisfaction with their ability to have sex. Twenty-eight percent of the men said that sex was important to them, compared to only 12 percent of the women (Laumann et al., quoted in University of Chicago, 2006).

Note that with very few exceptions, the international study reports on sexual behavior among male–female couples, while a great deal of the controversy that swirls around research on sexuality involves distinctions between heterosexual, homosexual, and bisexual behavior, to which we now turn.

Heterosexuality, Homosexuality, and Bisexuality Heterosexuality refers to sexual orientation toward members of the opposite sex, in contrast to **homosexuality**, which refers to sexual orientation toward members of the same sex, and **bisexuality**, which refers to sexual orientation toward members of either sex. (**Sexual orientation** refers to an enduring emotional, romantic, sexual, or affectional attraction that a person feels toward another person.) Heterosexuality is

a norm in every society, but it too is subject to wide variations. Norms of heterosexuality function to ensure that genital sexual intercourse occurs between men and women in the interest of population replacement and growth. The Shakers of colonial North America were an example of a society that attempted to curtail or eliminate heterosexuality in favor of celibacy in the interest of religious piety. Because Shaker society could continue only through recruitment of new members, it eventually declined and essentially disappeared.

Although heterosexuality is practiced in some form in all societies, about one-third of them actively ban homosexual practices, while many other societies tolerate them in some form (Murdock, 1983). Historically, the best-known examples of heterosexual societies that condoned some forms of homosexuality and bisexuality were ancient Greece, Confucian China, and precolonial Hawaii. In most Western cultures, the norm of heterosexuality is strong, but older taboos against homosexuality and bisexuality are changing as well.

The intensity of debate over homosexuality in the United States is an indication, on one hand, of how controversial the behavior remains and, on the other, of the growing, if grudging, acceptance of alternatives to traditional norms of heterosexuality. In the mid-twentieth century, homosexuality was a taboo subject. Homosexual men and women remained secretive, or "in the closet," about their true sexual feelings. Today, there is at least one homosexual member of Congress and many other openly homosexual elected officials and public figures. At the same time, fear and loathing of homosexuals—gay men and lesbian women—is a major social problem that continues to cause immense human suffering.

Estimates of the incidence of homosexuality—that is, the proportion or number of women and men in a population who are sexually attracted to people of the same sex—raise several social-scientific and political problems. In his studies of sexual behavior, Alfred Kinsey interviewed thousands of Americans about their sexual preferences and behavior. Kinsey described sexuality as a continuum extending from exclusive heterosexuality to equal attraction for the same and the opposite sex (bisexuality) to exclusive homosexuality. By his estimates, about 4 percent of men and 2 percent of women were exclusively homosexual, whereas far more, perhaps as many as one-third of men and about 12 percent of women, reported having had a homosexual experience leading to orgasm at least once in their lives (Kinsey, Pomeroy, & Martin, 1948, 1953). Subsequent reinterpretations of Kinsey's data led social scientists to estimate that 10 percent of the population was homosexual or had strong homosexual tendencies. This estimate became a commonly cited statistic in the politics of the gay rights movement.

The most recent major study of sexuality in the United States, conducted by the federal Centers for Disease Control and Prevention (CDC), found this estimate to be too high, at least for males. As can be seen in Table 7.1 on page 167, 6.0 percent of males and 11.2 percent of females reported having had same-sex contact at some time in their lives. Researchers used many techniques to ensure confidentiality and anonymity for

sexual orientation: An enduring emotional, romantic, sexual, or affectional attraction that a person feels toward another person.

heterosexuality: Sexual orientation toward members of the opposite sex.

homosexuality: Sexual orientation toward members of the same sex.

bisexuality: Sexual orientation toward members of either sex.

Changing Gender Expectations in Japan

Like most modern nations, Japan has experienced major changes in the conduct of sexual behavior between men and women. In preindustrial Japanese society, women were subordinate to their husbands and, with few exceptions, were not expected or even allowed to pursue their own independent careers. One of the few careers open to women outside the marital relationship was that of the geisha. The geisha was a highly trained professional who learned all the traditional Japanese performing arts, as well as ways of pleasing others, especially the man or men who could afford to pay for her expensive clothing and lifestyle. In return, she provided social and sexual pleasure, but always in a very discreet fashion. Today the tradition of the geisha still exists, as we see in the photo on the left, but the institution's importance is much diminished as modern Japanese women have become educated and capable of living independently, as does the professional city administrator in the photo on the right, who is also a single woman.

© Bob Krist/CORBIS

John Nordell/Index Stock Imagery/Photolibrary

respondents, but all agreed that questions about sensitive sexual subjects often yield data that underestimate the frequency of behaviors that are disapproved of by a majority of the population. Also, in this 2002 survey, the questionnaire failed to ask respondents about mutual masturbation, a common sexual practice among young males engaging in same-sex behavior as well as among heterosexual couples. The 1994 National Opinion Research Center (NORC) study conducted by Laumann, Gagnon, and others produced results similar to those of the CDC study, but it asked more explicit questions about sexual identification. In Figure 7.5, note that the proportion of respondents who identified themselves as homosexual or bisexual was far lower than the proportion reporting same-sex desires or experiences (Laumann et al., 1994; Mosher, Chandra, & Jones, 2005).

Given the anxiety that many people feel about discussing their sexual orientation and behavior, however, some people believe that the NORC study may underestimate the incidence of homosexuality in the population (Gould, 1995). It should also be noted that the ambitious study would have included a far larger sample and somewhat greater scientific accuracy if Congress had not withdrawn the federal funds that were scheduled for its budget.

Only about 0.8 percent of men and 0.9 percent of women in the NORC sample said that they were bisexual. But when they were asked more indirectly whether they were sexually attracted to both sexes, the proportions who said yes increased to 4.1 percent of women and 3.9 percent of men. Martin S. Weinberg observes that many people "would say that the person who feels the sexual attraction to both sexes but never acts on it is not bisexual, but in my definition they are" (quoted in Gabriel, 1995, p. A12). The findings of Weinberg's recent study of bisexuality refute the common assumption that bisexuality is a stage leading to eventual homosexual orientation. His sample included women and men who had come to their bisexuality from both heterosexual and homosexual orientations and for whom bisexuality appeared to be a stable lifestyle.

Many homosexuals, along with an increasing number of biologists and social scientists, believe that homosexuality will be shown to have genetic origins (Rosario, 2001; Tobach & Rosoff, 1994). In accounts of their earliest sexual feelings, homosexual men, and lesbian women to a lesser degree, recount their experiences of sexual attraction to members of the same sex. These experiences convince many gay people that their sexual orientation is not merely a "lifestyle choice," as critics often claim. To date, however, there is insufficient evidence from genetic studies to resolve this issue (Kemper, 1990; Kluger, 2001).

Critics of research on sexuality have consistently opposed federal funding of studies like those discussed in this chapter. Congress and major national foundations have continued to fund the research, however (Laumann et al., 1994). Surveys of sexual behavior, whether international or limited to particular societies, are of

TABLE 7.1

Number of Males and Females 15 to 44 Years of Age and Percentage Who Have Ever Had Specified Type of Sexual Contact, by Selected Characteristics

Males	Opposite-sex sexual contact				Any same-sex contact
	Any	Vaginal intercourse	Oral sex	Anal sex	
All males 15–44 years of age	90.8	87.6	83.0	34.0	6.0
15–19 years	63.9	49.1	55.1	11.2	4.5
20–24 years	91.4	87.6	82.2	32.6	5.5
25–44 years	97.3	97.1	90.1	40.0	6.5

Females	Opposite-sex sexual contact				Any same-sex contact
	Any	Vaginal intercourse	Oral sex	Anal sex	
All females 15–44 years of age	91.7	89.2	82.0	30.0	11.2
15–19 years	63.3	53.0	54.3	10.9	10.6
20–24 years	91.3	87.3	83.0	29.6	14.2
25–44 years	98.4	98.2	88.3	34.7	10.7

Source: From Global Study of Sexual Attitudes and Behaviors, 2002. Used with permission from Pfizer Inc.

vital importance for a number of reasons. Above all, it is increasingly important to know about how widespread risky sexual behaviors are in large populations. Risky sexual behaviors—such as unprotected sex, anal intercourse (which can rupture small blood vessels), and casual sex with multiple partners—may lead to greater exposure to sexually transmitted diseases, including the HIV virus, which causes AIDS. The more knowledge we have about sexual behaviors, the more effective public education and other policies we can develop to address issues of sexual health, reproduction, and the development of sexual orientation.

Teenage Sexual Behavior and Social Control Recent trends in sexual activity among teenagers, as shown in Figure 7.6, indicate that young people in the United States are heeding messages about being more responsible and safe in their sexual behavior. The figure shows that, since the 1980s, a decreasing proportion of teenage males and females say that they have engaged in sexual intercourse. Note that the decline is far greater among young men than among young women, and the issue is sexual intercourse rather than sexual activity, which includes many sexual behaviors short of actual intercourse. The results demonstrate that teenagers, and their parents, are more concerned about the control of early sexuality, although it is also true that the data show that almost a third of contemporary unmarried teenagers are experiencing sexual intercourse (National Center for Health Statistics, 2002). The implications and consequences of this behavior are considered further in later chapters.

Sociologist Peter Bearman and his associates have presented important findings showing that parents and their teenage children are approaching issues of sexuality with more concern for safety and responsibility. One of Bearman's studies, for example, looks at the impact of virginity pledges. In response to an initiative of the Southern Baptist Church in 1993, about 2.5 million teenagers have taken public "virginity pledges," in which they promise to abstain from sexual intercourse until they marry. In a thorough analysis of data from teenagers who took this pledge, as well as from the National Longitudinal Study of Adolescent Health, the researchers found that the pledges reduce teenage sexual intercourse well below the 30 percent found in the general population. But such pledges are less effective among older adolescents than among younger ones, as one might suspect. Second, pledging delays intercourse most effectively in situations in which teens perceive the pledge to make them special

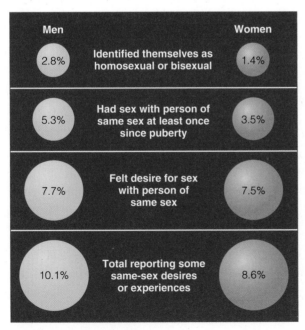

Figure 7.5 Defining "Gay". Source: From Laumann et al., *The Social Organization of Sexuality*. Copyright 1994 University of Chicago Press. Reprinted with permission.

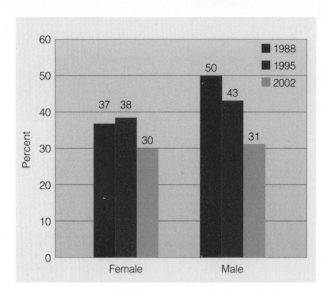

Figure 7.6 Percent of never-married females and males 15–17 years of age who have ever had sexual intercourse: United States, 1988–2002. Source: National Center for Health Statistics, 2002.

and different. In cases in which most of the teenagers whom they know are taking the pledge, the effects are far less robust. Bearman explains this finding as a consequence of the fact that the pledge is part of an identity movement, a movement that is designed to make teenagers feel especially moral and good about their identity as sexual abstainers. When the pledge is the norm in the teenager's social network, however, this feeling of positive identity is not as strong and the pledge is more likely to be broken.

Another finding of Bearman's research on virginity pledges is more troubling. Pledge breakers are significantly more likely to engage in unprotected sex and thus put themselves and their partners at risk of sexually transmitted diseases and unwanted pregnancies. It appears that teenagers who break their pledges do so in moments of passion, but because they typically have not planned to have sex or learned safe sexual practices, their sexual behavior is more often unprotected by contraception than is sexual behavior among teenagers who did not take the pledge (Bearman & Brückner, 2001).

Sexual Orientation and Individual Development As we saw earlier, the term *sexual orientation* refers to an enduring emotional, romantic, sexual, or affectional attraction that a person feels toward another person. We can describe sexual orientation as falling along a continuum. This means that an individual does not have to be exclusively homosexual or heterosexual but can feel varying degrees of attraction for people of both sexes over his or her lifetime. Different people may realize at different points in their lives that they are heterosexual, gay, lesbian, or bisexual.

Sexual behavior is not the same as sexual orientation. Adolescents— as well as many adults—may identify themselves as homosexual or bisexual without having had any sexual experience. Other young people may have had sexual experiences with a person of the same gender yet do not identify themselves as gay, lesbian, or bisexual. This is particularly important for adolescents because adolescence is a time when young people engage in a good deal of sexual experimentation.

All teenagers face developmental challenges, including the need to develop social skills, make career choices, and become members of peer groups. Gay, lesbian, and bisexual youth also discover that they must cope with prejudice, discrimination, and even violent behavior in their families, schools, and communities. These experiences may have negative effects on their mental health and education. These students are more likely than heterosexual students to report missing school due to fear, being threatened by other students, and having their property damaged at school (Haugen & Box, 2006).

Homophobia Fear of homosexuals and same-sex attraction is known as **homophobia**. Although it is common in the United States, its causes are not entirely clear. Most social-scientific (as opposed to religious or ideological) explanations of homophobia hinge on an analysis of the problems of masculinity and the male role in Western societies.

Identification with the male gender and the ability to take male roles, as these are defined for young boys and later for men, are two different aspects of what it means to "be a man." Every society has its notions of what distinguishes men from women and seeks to teach boys and girls how to perform the roles assigned to their gender. In Western societies, the male role is often depicted in movies, on television, and in advertising as distinct from that of the female and imbued with more strength, power, and rationality—a subject to which we return later in the chapter. Because as children they lack the power and strength they admire in images of masculinity, young boys typically become concerned about being seen as lacking in masculinity and the ability to take on male roles. Thus, their vocabulary is rich with terms of abuse for anyone they believe to be lacking in masculinity: *wimp, nerd, turkey, sissy, lily liver, yellowbelly, candy ass, ladyfinger, cream puff, mama's boy, dweeb, geek,* and so on (Connell, 2005). As they grow older, a small but socially significant proportion of boys may become violently homophobic and go out of their way to abuse males whom they consider effeminate or "queer."

Girls may develop homophobic attitudes early in life, often through identification with their male siblings, but homophobia among females also occurs during the teenage years. Female socialization stresses the ability to attract men and appear feminine; anxiety over these aspects of the female role may lead young women to reject other women whom they do not regard as feminine enough.

These early patterns can have lasting consequences. Throughout their lives, the early habit of avoiding emotions for fear of seeming effeminate tends to make men in the United States and other Western societies wary of expressing their

homophobia: Fear of homosexuals and same-sex attraction.

feelings and of admitting vulnerability: "Men are thus denied an important part of their . . . well-being when they cannot touch and cannot express their tender feelings for other men. Needless to say, gay men are damaged by the negative implications of homophobia" (Blumenfeld, 1992, p. 37). And, of course, homophobia causes untold suffering for gay men and women (Airton, 2009).

Many people are troubled by homosexuality, and by sexuality as it is represented in the media, which they consider pornographic. Among those who are most offended are people who hold fundamentalist religious beliefs. For example, 77 percent of fundamentalist Baptists disagree or strongly disagree with the statement "Even if homosexuality is wrong, the civil rights of gays should be protected." They believe that homosexuality is immoral and inimical to the propagation of the species. In the United States, it is common for prominent fundamentalist Christian leaders to state that AIDS is God's punishment for the homosexual lifestyle (Ammerman, 1990). (Fundamentalist Muslims and deeply orthodox Jews hold similar views.) Despite the vehemence of such religiously motivated attacks, gay men and women in smaller towns and communities with fundamentalist congregations are increasingly asserting their right to live as they choose as long as they do not violate the rights of others (Lampman, 2001; Miller, 1989).

Controversies over male and female roles, homosexuality, norms of sexual conduct, and the origins of sexual orientation all point to the extent to which these central areas of social life are subject to change and reactions to change. In the study of gender inequalities, the origins and patterns of change in the condition of women and the continuing inequalities between the sexes are a central focus, which we discuss more fully in Chapter 13.

Of course, sociological research on sexuality is not limited to the sexual behavior of individuals or to surveys of sexual behavior in large populations. Sociological research on sexuality often addresses issues of changing sexual behavior and sexuality in social institutions like religion, the family, or economic markets, and it is to some of these subjects that we now turn.

SOCIOLOGICAL PERSPECTIVES ON SEXUALITY AND SEXUAL INSTITUTIONS

As noted at the beginning of the chapter, the major sociological perspectives tend to raise different questions about sexuality and social institutions, and to offer differing explanations for sexual phenomena. Consider, for example, the issue of celibacy in religious institutions, especially the Catholic Church. This is a subject of enormous interest in the contemporary world because of issues such as pedophilia among a small minority of Catholic priests and continuing debates about whether the norm of priestly celibacy still serves the church well. This subject has been addressed especially well by functionalist sociologists.

Celibacy and the Catholic Church

In 2005, after some lively debate among the attending bishops, the World Synod of the Catholic Church voted by a wide majority to uphold the norm of priestly celibacy, which has been a feature of the church since the early Middle Ages. The church explains the norm of priestly celibacy and celibacy for nuns in religious terms. Those who consecrate their lives to serving God should not be distracted by the demands of the flesh. Catholics who hope to change the long-standing norm of priestly celibacy argue that it is increasingly difficult to recruit young men for the priesthood when they know that they will be required to remain celibate for the remainder of their lives. And because the bishops are not willing to accept women as priests, the problem is especially serious in Catholic parishes throughout the world. This problem has become a subject of intense controversy as accusations of sexual abuse by priests have made headlines in recent years. In 2010, a major scandal rocked the church hierarchy as victims of sexual abuse came forward to accuse priests in Germany, Ireland, and Italy. To make matters worse, the pope himself was implicated in the possible cover-up of offending priests while he was a cardinal.

Manifest versus Latent Functions of a Norm Sociologists have addressed priestly celibacy by first insisting on the distinction between what they term manifest versus latent functions (Elster, 2000; Merton, 1968). The *manifest function* of a norm is the public explanation for why it should be respected—in this example, the church's argument that priests and nuns should not be distracted, either spiritually or physically, from their devotion to God. The *latent functions* of a norm are the advantages and, in some cases, the disadvantages it brings that are not part of the official explanation for the norm. A latent function of priestly celibacy is that unmarried priests do not have wives and children to divide their loyalties or distract them from their commitment to the work of the church. The unmarried priesthood is therefore less likely to challenge the hierarchy of the church and the ultimate authority of the pope. Another latent function, this one also a disadvantage to the church, is that the norm against marriage for Catholic clerics may make the clergy somewhat more attractive to men and women with a predisposition toward same-sex sexuality.

Survey data from American Catholics and American Catholic clerics indicate that about 66 percent of lay Catholics support changes in the social structure of the church, including marriage for priests and elevation of women to the priesthood. Among American priests, 72 percent of those who responded to the most recent poll supported priestly marriage, and 49 percent favored permitting women to become priests. Similar differences between official church teachings and the clergy exist with respect to other issues of sexuality. Although in the United States the church has had some success in recruiting young priests who support its strictures on sexuality, recent surveys show that only 50 percent of active priests think that homosexual sex is always wrong, and only 27 percent fully support the church's teaching that the use of birth control is wrong (Greeley, 2004). As the debate over these norms widens within the church itself, pressure mounts on the highest church authorities to reconsider its teachings on sexuality. But the Catholic Church is a global institution with huge populations in regions outside the United States where people have opinions that differ from those of the majority of American Catholics. For this and many other reasons, change in this venerable religious institution occurs very slowly, if at all (Schoenherr, 2002).

Prostitution: A Functionalist Explanation

Prostitution is described in some of the earliest written documents in human history. It appears in the Old and New Testaments of the Bible, in the Koran, and in ancient Indian and Chinese texts. No wonder it is often called "the world's oldest profession." Defined simply as the sale of one's sexual services, prostitution may be practiced by men as well as women, by children as well as adults. Kathleen Barry, who has studied prostitution throughout the world, finds that poverty is the most important condition leading women and men into prostitution, but drug addiction, slavery, and widowhood are also prominent causes in some cultures). Widespread as prostitution is, however, global statistics on the numbers of prostitutes or people who make their living in the expanding global sex industry are inadequate. In an attempt to obtain such data, the International Labour Organization of the United Nations studied the sex trades in a number of Southeast Asian nations where sex tourism has been a major economic activity. The study estimated that in Indonesia, Malaysia, the Philippines, and Thailand alone, about 800,000 women and men, both children and adults, trade sex for money and can be classified as prostitutes (Lim, 1998).

Prostitution is widely condemned as morally wrong. Why, then, is it so ubiquitous? Sociological and historical research on the subject has found that wherever it occurs, it is both reviled and tolerated, and although the exchange of sex for money has always defined its essence, the institution of prostitution has adapted itself to new societies and new social conditions, most recently the rise of global communications and travel.

In the ancient Roman city of Pompeii, prostitution was conducted in a special building, or brothel, known as the *lupanar* (McGinn, 2004), as well as in rooms hidden away on the upper floors of taverns and lodging houses scattered throughout the city. In contemporary American cities, the ancient institution takes a greater variety of forms (Bernstein, 2001, 2004). At the bottom of the status hierarchy of prostitution are female streetwalkers and male prostitutes (including some transvestites). Streetwalkers are the most visible sex hustlers and often the ones most exploited by pimps and drug dealers. Old-fashioned brothels, often run by an enterprising madam who typically relies on organized crime for protection, operate in many cities, but they are fewer in number than the escort services now accessible through the Internet. At the highest level of the sex trade are private call girls, many of whom have a roster of wealthy customers and command high fees for their services. All but the street prostitutes operate largely outside the public view and rather easily avoid the negative sanctions of polite society.

Bernstein's recent research shows that among younger clients—a large proportion of whom are single or divorced—there is an increasing demand not only for sexual services but also for an emotional connection, so that prostitutes increasingly see themselves as providing important services beyond mere physical sex (Bernstein, 2004). A recent study of English prostitutes (Sanders, 2006) shows that most believe that their profession serves important functions in contemporary society. They see themselves as providing sexual experience and knowledge to inexperienced and handicapped individuals, helping to prevent adulterous relationships, helping to spread knowledge about

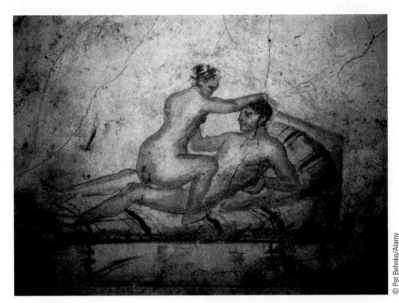

This is one of the wall paintings of prostitution that can still be seen in the *lupanar*, the public brothel in the ancient Roman city of Pompeii. The city was destroyed by a massive eruption of Mount Vesuvius in 79 c.e., but extensive excavations have uncovered the brothel and revealed the paintings on its walls.

according to Marx, "is only a specific expression of the general prostitution of the labourer," and the pimp or madam or entrepreneur who controls the prostitute's livelihood is also a "fallen" person, both a beneficiary and a victim of the sale of the prostitute's sexual services. Note that this critical perspective moves researchers to take the side of the prostitute or sex worker, whereas the functionalist perspective is less likely to offer explanations for why a man or woman might become a prostitute. Critical research on prostitution is far more likely to examine the influence of poverty, powerlessness, and lack of alternatives in explaining how, despite the risks, hardships, and moral condemnation they experience, so many women and young men are driven or attracted into the "oldest profession." Contemporary sociologists who apply the critical and feminist perspectives are most likely to document the global trends toward sex trafficking and sex tourism, a subject to which we will return shortly (Pryce, 2006).

sexually transmitted diseases, and promoting sexual health. For these reasons, most prostitutes who consider themselves professionals (rather than amateurs who sell sexual services for drugs, or people in need of cash who sell sex on occasion) would prefer to be known by the more neutral term *sex workers*.

The prostitutes in the English study offered functionalist explanations for why their ancient profession exists. Sociologists have long made similar arguments. The functionalist sociologist Kingsley Davis, who began writing about prostitution in the 1930s, was one of the first to argue that prostitution exists as an institution that is considered morally outside respectable society but is tolerated to varying degrees because it helps societies control the sexual behavior of their members, especially men. According to Davis, prostitution functions to provide alternatives for men who are not sexually fulfilled in their marriages, or who want sexual activity with multiple women. Because men typically have the power to enforce or not enforce the norms of their societies, they can choose to enforce sanctions against prostitution either verbally or with occasional crackdowns on prostitutes, while looking the other way most of the time (Davis, 1937, 1965).

Arguing from a far more critical perspective, Karl Marx viewed prostitution as merely another form of capitalism in which people sell their labor power (along with their creativity and, in essence, a part of themselves) to those who own the means of production. Prostitution,

Sexuality and Power—A Conflict Perspective

Think of how many conflicts over issues of sexuality and gender relations there are in the world today. Many religious fundamentalists are deeply offended by pornography, public displays of sexuality, and explicit discussions of sexuality and birth control. People who strongly believe in gender equality and feel that sex should be mutually satisfying often express dismay at patriarchal societies in which women must submit to the will of their fathers, brothers, and husbands, dress in ways that hide their physical features, and accept arranged marriages in which they are treated essentially as male possessions.

Social scientists have their own feelings about these conflicting values, but their contribution is to provide unbiased facts about gender inequality and sexuality. However, in cases of extreme forms of violence against women, such as female genital mutilation and rape, social scientists usually are not value-neutral. These forms of social control over women's bodies and lives may be condoned in a given culture, but they are condemned in the larger community of nations, as reflected in the Universal Declaration of Human Rights. Sociological research on these issues, in consequence, does not attempt to offer cultural justifications for these behaviors. Rather, it seeks to convey information about them as social phenomena in the hopes that the more we know and understand about why they exist, the better prepared we will be to decrease and perhaps even eliminate them (Kimmel, 2006; McPhillips, 2005).

Rape and Power Rape is a widespread sexual phenomenon that sociologists have long associated with issues of control and power rather than merely with sexuality. At present, more than half a million cases of rape and violent sexual harassment of women and men are reported each year in the United States, where the Federal Bureau of Investigation (FBI) estimates that a rape occurs every 5.6 minutes (Kornblum and Julian, 2007). Although rape takes many forms, including marital rape, gang rape, rape in prisons, and rape during warfare, the two most important forms of rape are (1) rape of a woman by a man seeking to satisfy his personal desires and exert his power, and (2) rape carried out in situations of warfare, where it

These teenage prostitutes found their way from small towns to a large city, where they were recruited into prostitution by some young pimps whom they met and partied with for some days. The police often pick them up for soliciting on the streets, and they find themselves sleeping in the station house or lockup while waiting for their pimps to bail them out.

© Leonard Freed/Magnum Photos

may be an explicit policy designed to intimidate the enemy population. The first form is illustrated by the situation in the United States before the Civil War. When a master "took" one of his female slaves for sexual pleasure, he did not necessarily think he was committing rape; in fact, however, because the slave had no power to resist, the behavior qualifies as sexual assault, or rape. The second form is illustrated by the current situation in the Darfur region of Sudan, where the Janjaweed militia systematically uses rape to terrorize civilian populations and, in many cases, force them to flee their lands and homes. Mass rape of this kind has also been used by the Soviets against the German population in World War II, by the Serbs against Muslim civilians in Bosnia, and by warring factions in Rwanda and Congo.

At the level of individual behavior, rape usually occurs in intimate situations, but it still involves issues of power and control. According to the federal Office on Violence Against Women (2006),

> Sexual assault is a crime of power and control and offenders often choose people whom they perceive as most vulnerable to attack or over whom they believe they can assert power. Sexual assault victims come from all walks of life. They can range in age from the very old to the very young. Many victims of sexual violence are under 12. Sixty-seven percent of all victims of sexual assault reported to law enforcement agencies were juveniles (under the age of 18); 34% of all victims were under age 12. One of every seven victims of sexual assault reported to law enforcement agencies were under age 6. Men and boys are sexually assaulted. Persons with disabilities are also sexually assaulted. Assumptions about the "typical" sexual assault victim may further isolate those victimized because they may feel they will not be believed if they do not share the characteristics of the stereotypical sexual assault victim.

There are many widespread myths about rape. One often hears the argument, for example, that women invite rape by dressing in provocative or revealing ways, or that a woman who enters a man's room and engages in kissing and petting is indicating willingness to engage in further sexual activity. Again, the legal definitions and rules are based on social-scientific evidence about the relationships between power and the desire to control another person through sexuality:

> Rape and sexual assault are crimes of violence and control that stem from a person's determination to exercise power over another. Neither provocative

dress nor promiscuous behavior are invitations for unwanted sexual activity. Forcing someone to engage in nonconsensual sexual activity is sexual assault, regardless of the way that person dresses or acts. Even if a person went voluntarily to someone's residence or room and consented to engage in some sexual activity, it does not serve as a blanket consent for all sexual activity. If a person is unsure about whether the other person is comfortable with an elevated level of sexual activity, the person should stop and ask. When someone says "No" or "Stop," that means STOP. Sexual activity forced upon another without consent is sexual assault.

Rape as a Tactic of War In every armed conflict of the new century, rape, torture, and other forms of violence against women have been reported by sociologists and neutral observers of international conflict (Renzetti & Bergen, 2005). When used as a tactic of war, rape is employed to produce intimidation, humiliation, and political terror, to extract information, to reward soldiers, and to carry out "ethnic cleansing." Amnesty International, which monitors human rights progress and abuses throughout the world, affirms the basic link between violence against women and the widespread traditional view of women as property and the use of women as sexual objects. The organization also emphasizes that violence against women, sexual and otherwise, is an effective terror tactic:

> Around the world, women have long been attributed the role of transmitters of culture and symbols of nation or community. Violence directed against women is often considered an

Thousands of women and girls have been raped and sexually abused in the conflict in the Darfur region of western Sudan. "Women will not tell you easily if they have been raped," says one. "In our culture, it is a source of shame. Women hide this in their hearts so that men don't hear about it."

© AP/Wide World Photos/Ron Haviv/VII

attack against the values or "honor" of a society and therefore a particularly potent tool of war. Women therefore experience armed conflicts as sexual objects, as presumed emblems of national and ethnic identity, and as female members of ethnic, racial, religious, or national groups.

Women who are victims of sexual violence, whether in war or in other situations, experience grave and, in many cases, lasting effects, including chronic medical problems, trauma and other psychological damage, life-threatening diseases like AIDS, forced pregnancy, stigmatization, and often rejection by family members and the community. In consequence,

> The Rome Statute of the International Criminal Court recognizes rape and other forms of sexual violence by combatants in the conduct of armed conflict as war crimes. Many acts of sexual violence—including rape, gang rape, abduction and sexual slavery, forced marriage, forced pregnancy, forced maternity, and sexual mutilation [excision of the clitoris, usually as part of a ritual of "female circumcision" that makes the woman incapable of experiencing sexual pleasure]—constitute torture under customary international law. These acts are considered war crimes and constitute grave breaches of the Geneva Convention. When rape and sexual violence are committed as part of a widespread or systematic attack directed against any civilian population, they are considered crimes against humanity, and in some cases may constitute an element of genocide.

The Social Construction of Sexualities—An Interactionist Perspective

In wartime, a man who ordinarily respects women and behaves honorably finds himself in a platoon that has invaded a village. In the heat of the moment, he participates in a gang rape of a helpless teenage girl. An elderly widow who thought her sexual life had ended discovers herself in a new relationship that includes sexual activities she never imagined before. A gay man who was used to having different sexual partners in his youth has remained faithful to the same partner for over a decade, and now they wish to enjoy the benefits of a socially sanctioned marital relationship. A teenage girl and her boyfriend are enjoying their growing sexual intimacy, but they are experiencing conflict about "going all the way." These are only a few examples of the infinite variety of sexual situations that people throughout the world might experience. Sociologists have had great success in applying the idea of cultural scripts to explain the ways in which people conduct their sexual lives and how they negotiate the sometimes sensitive and embarrassing aspects of intimate sexual relations.

Sexual Scripts Sociologists who apply the interactionist perspective have furthered our understanding of both conventional and unconventional sexual behaviors by emphasizing how cultural norms and ideas about ideal sexual experiences shape actual sexual behavior. They show that a person's sexuality is "socially constructed" according to his or her society's ideas about sexual behavior, which are communicated through the popular media, religious teachings, family teachings, and much more. The resulting "constructions"—which naturally differ from one society to another—take the form of **sexual scripts**, or ideas and fantasies about what our sexual experiences should or could be like. Developed by William Simon and John H. Gagnon (1969, 2003), the concept of a sexual script is a metaphor that helps explain differences between sexual expectations and actual sexual conduct. The researchers know that most people have not memorized actual "scripts" to guide their sexual activity, but they do have definite ideas about sexual conduct that influence the way they behave sexually. The concept of a sexual script emphasizes the social influences on sexual behavior rather than relying only on the concepts of biological or "natural" sexual behavior found in the work of earlier sex researchers like Freud and, to a lesser degree, Kinsey.

Interactionist research distinguishes among three kinds of sexual scripts: cultural scenarios, intrapsychic sexual scripts, and interpersonal sexual scripts. *Cultural scenarios* are held by many, if not all, members of a given culture. They are collective ideas about sexual goals, proper behaviors, and ideal outcomes—cultural guidelines for how, where, when, why, and with whom to be sexual (Irvine, 2003). They are historically specific and change in response to changes in the culture in which they are found. For example, in the Massachusetts Bay Colony's laws on sexual offenses, which were in effect between 1641 and 1660, we read this article: "If any man lyeth with mankinde as he lyeth with a woeman, both of them have committed abhomination, they both shall surely be put to death" (Peiss, 2002, p. 71). That Massachusetts became the first state to allow same-sex marriage is a powerful commentary on the likely direction of continuing change in our cultural scenarios about sexuality, despite reactions against more liberal sexual scripts.

Intrapsychic sexual scripts exist in our minds, in the form of fantasies and notions about how we would like to have sex and with whom. They are influenced by our culture, but they are also influenced by the particulars of our upbringing and our actual experiences. *Interpersonal scripts*, on the other hand, are developed between or among specific groups of people as ways of actually being sexual with each other. They are based on each other's expectations and ideas about what constitutes appropriate and pleasurable sexuality.

Differences between intrapsychic and interpersonal sexual scripts tell us a great deal about the conflicted aspects of our sexuality and how cultural change influences sexual conduct. Oral sex, for example, was considered "kinky" or somewhat deviant in the mid-twentieth century, and there were laws against it in some states. Today, as we see in Table 7.1, it is a common aspect of couples' sexual conduct, which means that it has become a widespread aspect of cultural scenarios and intrapsychic and interpersonal sexual scripts. Earlier in the last century, it was quite common for men's sexual scripts to include ideas about seduction, in which a woman's refusal to have sex could be ignored because it was assumed that when she said no, she really meant yes. Although such conduct still occurs, it can lead to well-founded charges of rape.

Double standards of sexual conduct, in which men's desires are assumed to take precedence over those of women, remain strong in our culture, but their strength is diminishing. As women and their lovers arrive at more mutually satisfying interpersonal sexual scripts, sexuality can help foster greater gender equality and tolerance of diversity in other aspects of society as well (Dworkin & O'Sullivan, 2005).

Thinking Critically

Do double standards about male and female sexual behavior still exist in your social world? In your high school, were girls who experimented sexually thought of as sluts? What about the boys? Were boys who talked to peers about their sexual conquests thought of as jerks or studs? How about the types of explanations that people apply to sexual behavior? Do they think of men as "needing" to play around and of women as "needing" to find a guy to marry?

Sexuality in a "Shrinking" World—A Global Feminist Perspective

At the historic Fourth World Congress on Women, held in Beijing in 1995, delegates—the majority of whom were women activists from almost all the world's nations—agreed that protection of girls and women from sexual, economic, and political repression was essential. As the conference's *Platform for Action* states:

> Available indicators show that the girl child is discriminated against from the earliest stages of life, through her childhood and into adulthood.

sexual scripts: Ideas and fantasies about what our sexual experiences should or could be like.

In some areas of the world, men outnumber women by 5 in every 100. The reasons for the discrepancy include, among other things, harmful attitudes and practices, such as female genital mutilation, son preference—which results in female infanticide and prenatal sex selection—early marriage, including child marriage, violence against women, sexual exploitation, sexual abuse, discrimination against girls in food allocation and other practices related to health and well-being. As a result, fewer girls than boys survive into adulthood. (United Nations, 1995)

This statement is an example of current thinking, particularly among feminist intellectuals and social scientists, about how closely related sexual violence and exploitation are to global gender inequality, especially the powerlessness of women in many developing nations. In the years since the 1995 gathering, there have been many setbacks for the cause of greater gender equality, especially the rise of global terrorism, the continuing spread of AIDS, and the gathering strength of regimes that oppress women and girls rather than treating them as citizens with rights equal to those of men. But there have also been some positive developments. For example, women in Kuwait can vote in national parliamentary elections and even run for office, having won the right to vote in 2005. Rola Dashti and other political activists in Kuwait now use their new political influence to address the issue of violence against women (Fattah, 2006).

The Globalization of Sexual Commerce

Travel for purposes of sexual pleasure, often known as *sex tourism*, is associated with globalization and the rapid growth of cities in the Third World. In Bangkok, Manila, Jakarta, and many other South Asian cities, the widening gap between the rich and the poor, along with the modernization of rural areas, account for a huge migration of families with young children into the squalid slums of these booming cities. Often desperate for a means of survival, families may sell their children into the sex trades, or children who are orphaned by the death of their parents due to AIDS and other diseases may prostitute themselves, again as a means of survival. In some cultures, the norm that specifies that widows are not eligible to remarry may also essentially force impoverished women into lives of prostitution and the production of pornography. Similar migration and recruitment into the sex trades occur along the U.S. border with Mexico and throughout much of southern Africa, where the large-scale displacement of

men from their villages and towns in search of work increases contacts with prostitutes and accounts for much of the spread of the HIV virus from one African nation to another (Altman, 2001).

In her analysis of globalization and the rise of sex industries (the "sex sector") in Southeast Asia, sociologist Lin Lean Lim writes, "Sex tourism represents a high-profile segment of the sector, with males from Western countries, Japan, and Taiwan coming to the Philippines and Thailand on specially organized sex tours" (1998, p. 6). Until recently, such tours were organized as package deals involving airlines, hotels, and tour operators, many of them international corporations. In recent years, however, international campaigns spearheaded by women's groups have discouraged sex tours.

Although the United Nations estimates that there are hundreds of thousands of adult and child prostitutes, and others engaged in supporting their work, as well as people involved in the production and sale of pornographic materials, Lim and others who study the globalization of commercial sex emphasize the importance of collective action and women's activism, both in their own nations and on a global scale, in addressing these problems (Lim, 1998).

Similar findings apply to efforts to stem the spread of AIDS throughout the world. The activities of women's groups and their allies have maintained political pressure on governments to deliver on their promises to give women greater freedom to manage their own sexual and reproductive lives. Here again we see the links between empowerment of women and more successful approaches to social problems stemming from sexual behaviors. We return to this subject in Chapter 9, which discusses collective behavior and social movements.

Thinking Critically

Can you think of some positive as well as negative influences of globalization on sexuality in different parts of the world? Do you think television, films, and other communications media influence women to seek empowerment as full citizens in their societies?

Kuwaiti women's rights activist Rola Dashti, left, shares a light moment with Kuwait's Prime Minister Sheikh Sabah Al Sabah, right, at the entrance hall of the National Assembly building, Kuwait City, after the Parliament session of April 19, 2005. Kuwait's lawmakers approved in this session a Municipal Council law allowing women to vote and run in the Council's elections.

AP Photo/Gustavo Ferrari

Sex and the American City

Why was *Sex and the City* such a popular show? Since its first season in 1998, the stories of Sarah Jessica Parker's character and her friends' love affairs and efforts to connect with men in loving relationships in their urban world was among the most watched shows on television, especially among teenage girls and young adults (male and female).

One sophisticated sociological answer can be found in recent studies of sexuality in the city by a research team at the University of Chicago. As the study's director, Edward Laumann, explains, "As more people remain single for longer periods of time, or become single because of divorce, elaborate 'markets' to facilitate people in their search for companionship and sex have developed in major cities."

In previous generations, people tended to marry shortly after entering the workforce, and they tended to remain married to the same partner. Today's marriages occur later in life and are often briefer, resulting in changes in the ways in which people meet and form relationships.

The sexual "markets" are important means for people to organize their lives, the researchers found. "Sex is a powerful force in relationships. It provides fulfillment and contributes to a high quality of life," Laumann observed.

The researchers found that people seek partnering in two kinds of sexual markets: one transactional and the other relational. The transactional market consists of encounters primarily for the purpose of forming a short-term relationship. Transactional marketplaces are located in bars, health clubs, and other venues where people who don't know each other can meet and form relatively uncommitted relationships. The relational market functions in a more complicated way and is often guided by friends and family, who may introduce people with common interests and common backgrounds. "People in this market determine who gets in. People know more about the two people being introduced and the arrangement that develops is meant to be a long-term one," Laumann said. Churches and synagogues sometimes provide settings where the relational marketplace works as well, he added.

Richard Levine / Alamy

© Kayte M. Deloma/PhotoEdit

SUMMARY

What are the major sociological perspectives on sexuality?

- From a biological perspective, sexuality is part of the larger process known as *sexual selection*, in which individuals seek, consciously and unconsciously, to transmit their genes to the next generation through mating and procreation.

- Functionalist social scientists ask how societies seek to influence or control the sexuality of their members through specific cultural norms, such as the incest taboo or the requirement of premarital celibacy, or through institutions like marriage and prostitution.

- Sociologists who take a conflict perspective ask why sexuality is such a frequent source of conflict both within and between societies, and how power in social relations affects sexual behavior—especially in the case of rape or patriarchal control over women's bodies.

- Interactionist research tends to reveal the ways in which sexuality is socially constructed—that is, not passed along without change from one generation to the next, but rather subject to all kinds of changing norms and ideas about beauty, sexiness, personal identity, and many other social ideas that are relevant to intimate behavior.

- From a global feminist perspective, efforts to enhance the right of women to choose their own fate—rather than being treated more or less as the possessions of fathers and husbands who make the decisions that affect their lives—are central to world politics.

How do biological and social factors interact to shape human sexuality?

- The relationship between human sexuality and culture has been shaped by a long process of evolution. Some of the basic aspects of human sexuality are biological traits, but they exist within a well-defined set of cultural norms that have also evolved over thousands of generations of human existence.

- Males and females of most animal species, including humans, exhibit *sexual dimorphism*—that is, they have different secondary sex characteristics.

- From a biological perspective, sexuality is part of the larger process known as *sexual selection*, in which individuals seek, consciously and unconsciously, to transmit their genes to the next generation through mating and procreation.

- In the United States and other urban-industrial societies, cultural norms of dress and makeup that emphasize sexual dimorphism have tended to weaken over the past century.

What are the key distinctions between sex and gender? Between sexual orientation and sexual behavior?

- *Gender* refers to the culturally defined ways of acting as a male or a female that become part of an individual's personal sense of self. Most people develop a "gut-level" sense of themselves as male or female, boy or girl, early in life. *Sexuality* refers to the manner in which a person engages in the intimate behaviors connected with genital stimulation, orgasm, and procreation.

- Global research on sexuality has found that couples who live in Western nations, where there is greater equality between men and women, report the most satisfaction with their sex life. In contrast, couples living in nations where men are dominant over women—especially in East Asia and, to a somewhat lesser extent, the Middle East—report less satisfaction with the physical and emotional quality of their sex life.

- *Heterosexuality* refers to sexual orientation toward the opposite sex, in contrast to *homosexuality,* which refers to sexual orientation toward the same sex, and *bisexuality,* which refers to sexual orientation toward either sex. *Sexual orientation* refers to an enduring emotional, romantic, sexual, or affectional attraction that a person feels toward another person. Fear of homosexuals and same-sex attraction is known as homophobia.

What are some current trends and issues in the study of sexuality and sexual institutions?

- The issue of celibacy in religious institutions, especially the Catholic Church, is a subject of enormous interest in the contemporary world because of issues such as pedophilia among a small minority of Catholic priests and continuing debates about whether the norm of priestly celibacy still serves the church well. This subject has been addressed especially well by functionalist sociologists.

- Some functionalist sociologists argue that prostitution exists as an institution that is considered morally outside respectable society but is tolerated to varying degrees because it helps societies control the sexual behavior of their members, especially men. Critical research on prostitution is more likely to examine the role of poverty, powerlessness, and lack of alternatives in explaining how women and young men are driven or attracted into the "oldest profession."

- Conflict theorists point out that rape is a widespread sexual phenomenon associated with issues of control and power, rather than merely with sexuality.

- Sociologists who apply the interactionist perspective have furthered our understanding of both conventional and unconventional sexual behaviors by emphasizing how cultural norms and ideas about ideal sexual experiences shape actual sexual behavior. They show that a person's sexuality is "socially constructed" according to his or her society's ideas about sexual behavior, which are communicated through the popular media, religious teachings, family teachings, and much more.

- Feminist intellectuals and social scientists emphasize how closely related sexual violence and exploitation are to global gender inequality, especially the powerlessness of women in many developing nations.

The Kornblum Companion Website

www.cengagebrain.com

Supplement your review of this chapter by going to the companion website to take one of the tutorial quizzes, use the flash cards to master key terms, and check out the many other study aids you'll find there. You'll also find special features, such as GSS data and Web links that will put data and resources at your fingertips to help you with that special project or to do some research on your own.

Deviance and Social Control

FOCUS QUESTIONS

How do societies define deviant behavior?

What are some important dimensions of deviance?

What are the major biological and sociological explanations of social deviance?

How is deviance prevented from threatening social stability?

see them lining up for breakfast well before seven in the morning. They look exhausted and threadbare. Some have slept all night in chairs in the "day room." Others are drifting in from the street, where they have a secret hiding place to sleep—perhaps over a grate that will blow warm air on them and keep them from freezing in winter. Since the terrorist attack on New York City and the resulting loss of thousands of jobs, there is also a growing number of displaced workers, many of whom had recently struggled to get off the welfare rolls and into the labor market. But most of the homeless seeking help have been living on the streets for other reasons.

This morning, for example, I see Willie W. and his partner Samantha. They are crack addicts who roam the streets at night in search of soda and beer cans, which they will sell to other street folk for half of the 5-cent deposit price because they need immediate cash for a "hit" of cocaine. This morning, I also spot the loan shark who waits outside the Drop-in Center to see if anyone who owes him money is around so he can collect on debts or threaten those who are delinquent in their loan payments.

As I climb the stairs to the Center, I am greeted by several staff members who were formerly homeless but are now on their way to work in the Center's kitchen or its various recycling and building maintenance programs. They greet me as "The Professor." The reason I appear so frequently in their midst is that I am the chair of the Center's board of directors. Its founder and executive director is a former doctoral student at my university. Two of my closest friends are also deeply involved in its programs. Although I have no shortage of responsibilities to occupy my time, I could hardly refuse when I was asked to serve on the board. It made sense for me to take my turn leading the board and participating in the Center's activities. Nor am I at all unusual in this regard. Most professional sociologists find that their research and writing lead to invitations to take active roles in organizations that seek to put into practice what they have recommended in their research. I chose to make a commitment of time and energy to the Grand Central Neighborhood Social Services Agency because I have been involved in research with the city's street population for almost three decades.

I believe that the problems of homeless adults in urban centers raise fascinating sociological issues about how a good society might deal with people who deviate from its norms and are "down and out." I find it sad that, in one of the world's richest cities, people are living on the streets. Often they are mentally ill or addicted, or both. Equally often, they are younger people who have been pushed out of crowded apartments after the breakup of a family. On the streets, they find many opportunities to make some money through various forms of hustling.

Race, ethnicity, and gender also play a part in the story. Two generations ago, the city's homeless were primarily white males. Today, most street people are black and Hispanic men from extremely poor backgrounds. An increasing number of street people are women; about 20 percent of those who come to the Center are female. All the people who spend time at the Center deviate from the norms of success that Americans value so highly. Many have served time in prison—but in fact, they frequently come to the Center to escape from predatory felons on the streets. At the Center, we offer a helping hand, but we insist that everyone take responsibility for his or her behavior. Those who can do so often begin the long and difficult process of rehabilitation.

Like many other volunteers, I spend time at the Center because it gives me a sense of gratification to share my skills with those in need. My sociological "beat" for the past thirty-five years has been the canyons of Midtown Manhattan, where the homeless center is located, only a few blocks from my office at the university. I accumulate knowledge about the city through my experiences with the people who own the skyscrapers, those who work there, and those who wander the city streets in search of whatever scraps of opportunity come their way. We can learn a great deal about society by understanding how the center of our largest city can be a caring community as well as a place where people seek to profit from their interactions with one another, and how those who live on the streets, the homeless and the deviant, are treated. As a professional sociologist, I also serve because it is a way to gain knowledge about society's outcasts and the approaches the city is developing to address their needs.

WHAT IS DEVIANCE?

In this chapter, we look at some current trends in deviance and crime and examine issues of social control—that is, what societies do to attempt to curb deviance. Research on these and related issues is a major field within sociology. No one can offer definitive answers, but sociologists are frequently looked to for research on trends in deviance and criminal behavior and for explanations of these trends.

Deviance, broadly defined, is behavior that violates the norms of a particular society. All of us violate norms to some degree at some time or other, however, so for purposes of this chapter we need to distinguish between *deviance* and *deviants*. Deviance can be something as simple as dyeing one's hair purple, wearing outrageous clothing, or becoming tipsy at a stuffy party. Or it may be behavior over which the individual has little control, such as being homeless and living on the street, or it may consist of more strongly sanctioned departures from the society's norms—such acts as rape, mugging, and murder. Not all deviance is considered socially wrong, yet it can have negative effects for the individual. For example, "whistle-blowers," who publicize illegal or harmful actions by their employers, deviate from the norms of bureaucratic organizations and are often threatened with the loss of their jobs. Yet they benefit the public by calling attention to dangerous or illegal activities (Stevenson, 1992).

A deviant person, by contrast, is someone who violates or opposes a society's most valued norms, especially those that elite groups value. Through such behavior, deviant individuals become disvalued people, and their disvalued behavior provokes hostile reactions (Davis, 1975; Goffman, 1963; Sagarin, 1975; Schur, 1984). *Deviant* may be a label attached to a person or group. Or the word may refer to behavior that brings punishment to a person under certain conditions.

The Social Construction of Deviance

Violations of social norms by people in powerful positions, such as influence peddling or extramarital sexual liaisons by government officials, raise an important question: What are the conditions under which violations of norms are punished? Here is an area of conduct in which there is often some uncertainty about what is legal and what is merely sleazy. Such questions reveal that deviance is not absolute. As sociologist Kai Erikson explains it, deviance "is not a property inherent in certain forms of behavior; it is a property conferred upon these forms by the audience which directly or indirectly witnesses them" (1962, p. 307). Some of us may believe that influence peddling is deviant, whereas others may believe that it is acceptable. Which of our views become the norm—and which are to be enforced through rewards and punishments—is just as important as the behavior itself. This point is illustrated in Erikson's classic study of deviance among New England's Puritans during the infamous Salem witchcraft trials:

> No one knows how the witchcraft hysteria began, but it originated in the home of the Reverend Samuel Parris, minister of the local church. In early 1692, several girls from the neighborhood began to spend their afternoons in the Parris kitchen with a slave named Tituba, and it was not long before a mysterious sorority of girls, aged between nine and twenty, became regular visitors to the parsonage. (Erikson, 1966, p. 141)

deviance: Behavior that violates the norms of a particular society.

The girls quickly drew the concerned attention of Salem's ministers and its doctor. Unable to understand much about their hysterical state or to deny their claim that they were possessed by the devil, the doctor pronounced the girls bewitched. This gave them the freedom to make accusations regarding the cause of their unfortunate condition.

Tituba was the first to be accused and jailed. She was followed by scores of others as the fear of witches swept through the community. Soon women with too many warts or annoying tics were accused, tried, and jailed for their sins. Then the executions began. In the first and worst of the waves of executions, in August and September of 1692, at least twenty people were killed, including one man who was pressed to death under piled rocks for "standing mute at his trial."

For the sociologist, the Salem witch hunt of 300 years ago has meaning for today's world. Erikson showed that the punishment of suspected witches served as a defense against the weakening of Puritan society. By casting out the "witches," the Puritans were reaffirming their community values: strict adherence to religious devotion, fear of God, abstinence from the pleasures of secular society (for example, drink, sex, music, dance), and the like.

For Erikson, the trial and punishment of the so-called witches illustrates Émile Durkheim's earlier discovery that every society creates its own forms of deviance and in fact needs those deviant acts. The punishment of deviant acts reaffirms the commitment of a society's members to its norms and values and thereby reinforces social solidarity. On the surface, Durkheim argued, deviant acts may seem to be harmful to group life, but in fact, the punishment of those who commit such acts makes it clear to all exactly what deviations are most intolerable.

By Durkheim's reasoning, the stark images of punishment—the guillotine, the electric chair, the syringe, the wretched life behind bars—become opportunities to let the population know that those who threaten the social order will be severely judged.

The historic 2003 Supreme Court decision that, by a 6–3 vote, overturned a Texas law forbidding homosexual consensual sexual behavior is another outstanding example of the way ideas of what is deviant change as the norms of a society change. In his majority opinion, Justice Anthony Kennedy wrote that most states had long abandoned criminal penalties for consensual sexual relations conducted in private because (1) such prohibitions encourage disrespect for the law by penalizing conduct that many people engage in, (2) such statutes regulate private conduct that is not harmful to others, and (3) such laws tend to be arbitrarily enforced and thus invite the danger of blackmail. Texas was one of thirteen states that still had antisodomy laws on their books (Greenhouse, 2003).

The Court's historic decision came as a result of years of social change in which gay Americans had fought in the courts and in the streets, through social movement activities like those described in Chapter 9, to convince the public that such laws were an invasion of their privacy and their civil rights. "The state cannot demean their existence or control their destiny by making their private sexual conduct a crime," the Court's opinion stated (Greenhouse, 2003, p. 1; Mixner & Bailey, 2002). In a dissenting opinion, however, Justice Antonin Scalia warned that the Court had taken sides in a divisive "culture war" within American society and that the decision would open the way to legalization of same-sex marriages. Scalia's opinion reflects the fact that an important minority of Americans still believe that homosexual sex is immoral and will continue to argue that government should intervene in people's private lives for that reason.

In 2010, Secretary of Defense Robert Gates announced the formation of a committee to examine how the military could change enforcement rules for the military's "don't ask, don't tell" policy. This rule essentially insists that gay men and women remain closeted while serving in the military; if their sexual orientation is revealed, they can be, and often are, discharged. Gates's committee will seek ways to make it more difficult for such discharges to occur, and Gates has made it clear that he would like to see Congress withdraw the rule. That may not occur, but the fact that the nation's highest military officer is opposed to the policy is an indication of the force behind this change in norms about sexuality (Ackerman, 2010).

The Trial of George Jacobs, August 5, 1692. Oil on canvas by T. H. Matteson, 1855. Jacobs was accused of witchcraft.

Essex Institute, Salem, Massachusetts/Peabody Essex Museum

Tyrone Garner, left, and John G. Lawrence, right, are greeted by a supporter at a rally. In 2003, a Supreme Court decision overturned a Texas law forbidding homosexual consensual sexual behavior, under which Garner and Lawrence had been taken to court.

As this example and that of the Puritans suggest, the study of deviance reveals a great deal about directions of social change in a society. Indeed, the study of deviance is central to the science of sociology, not only because deviance results in major social problems like crime but also because it can bring about social change. The Puritans were a deviant group in England at the time of their emigration to the New World. They challenged the authority of the king and many of the central norms that upheld the stratification system of feudal England. In the Massachusetts Bay Colony, they created their own society, one in which they were no longer deviant. However, they also created their own forms of deviance, which reflected their unique problems as a society on a rapidly changing social frontier.

In fact, if you think about it, every aspect of daily life—driving, walking, going to a show, shopping, church attendance, political behavior, or sports—is controlled by the norms of a given culture. And thousands of deviant acts, such as cutting in front of someone in a line, are frowned on by people asserting their normative expectations ("Excuse me, I was ahead of you here"). All these norms and assertions constitute a system of informal social control without which society would dissolve into chaos, confusion, and hostility (Gibbs, 1994).

Forms of Social Control

The ways in which a society prevents deviance and punishes deviants are known as **social control**. As we saw in Chapter 3, the norms of a culture, the means by which they are instilled in us through socialization, and the ways in

which they are enforced in social institutions—the family, the schools, government agencies—establish a society's system of social control. In fact, social control can be thought of as all the ways in which a society establishes and enforces its cultural norms. It is "the capacity of a social group, including a whole society, to regulate itself" (Janowitz, 1978, p. 3).

The means used to prevent deviance and punish deviants are one dimension of social control. They include the police, prisons, mental hospitals, and other institutions responsible for applying social control, keeping order, and enforcing major norms. But if we had to rely entirely on official institutions to enforce norms, social order would probably be impossible to achieve (Chwast, 1965). In fact, the official institutions of social control deal mainly with the deviant individuals and groups that a society fears most. Less threatening forms of deviance are controlled through the everyday interactions of individuals, as when parents attempt to prevent their children from piercing their tongues or wearing their hair in dreadlocks.

In this chapter, we look first at the dimensions of deviance and at how its meanings emerge and change as a society's values change. Then we explore the major sociological perspectives on deviance and social control. The final section deals with the ecology of deviance—its distribution among people in different subcultures and income categories—and ends with an introduction to the organization and functioning of institutions of social control such as prisons.

DIMENSIONS OF DEVIANCE

Deviance is an especially controversial topic: There is usually much disagreement not only about which behaviors are deviant and which are not but also about which behaviors should be strongly punished and which should be condoned or punished only mildly. The debate over whether abortion should be legal is a good example of such disagreement. As Erikson noted, "Behavior which qualifies one man for prison may qualify another for sainthood since the quality of the act itself depends so much on the circumstances under which it was performed and the temper of the audience which witnessed it" (1966, pp. 5–6).

social control: The ways in which a society encourages conformity to its norms and prevents deviance.

Consider former South African president Nelson Mandela. Mandela was released from a maximum-security prison in 1990, after serving almost thirty years of a life sentence for his leadership of the movement to end apartheid, South Africa's racial caste system. The dominant white minority in South Africa regarded Mandela and other opponents of apartheid as criminals. But the black majority viewed him as a hero and a martyr; indeed, black South Africans revere Mandela the way people in the United States revere George Washington. Until the black majority began to mobilize world opinion and other nations began to enforce negative sanctions against South Africa, the white minority was able to define Mandela's opposition to apartheid as a criminal activity.

Power: Who Decides What Is Deviant?

The power to define which acts are legal and which are illegal is an important dimension of deviance. In the United States, the power of some groups to define certain acts as deviant helps explain why, for example, cultivating marijuana is a crime but influence peddling is not. In this connection it is interesting to note that during the 1960s, when the children of powerful members of society began smoking marijuana, legislators found themselves under pressure to relax the enforcement of marijuana laws.

The fact that people in power can define what behavior is deviant and determine who is punished fails to explain differences in definitions of deviance in societies with different cultures. Behavior regarded as normal in one society may be considered highly deviant in another. For example, in the United States, drinking alcoholic beverages is considered normal; in orthodox Islamic culture, it is forbidden. Even within a society, members of certain social groups may behave in ways that others consider deviant. Thus, the official norms of the Catholic Church do not permit the use of artificial methods of birth control, yet large numbers of Catholics use condoms, birth control pills, and other contraceptives. Similarly, the norms of Judaism require that the Sabbath be set aside for religious observance, yet many Jews work on the Sabbath and do not attend synagogue; in parts of Jerusalem, orthodox Jews have occasionally attacked Jews for violating that norm. Clearly, differences in values are another important source of definitions of deviance and of disagreements about those definitions.

Deviance: Ascribed versus Achieved

Another dimension of deviance concerns attributes that are ascribed unavoidably at birth (for example, race or physical appearance) or that a person cannot control (such as having a convict as a parent), in contrast to statuses that are achieved through actual behavior, which is usually voluntary (see the discussion of ascribed versus achieved statuses in Chapter 4). A criminal is deviant in ways that a person with a mental illness is not, and criminal behavior is more costly to society. In many situations, however, people with mental illnesses are labeled as deviant, and this label may actually drive them toward criminality. A related issue is how people who deviate from generally accepted norms manage to survive in societies where they are considered outsiders. In fact, deviant people often form their own communities with their own norms and values, and these deviant subcultures sustain them in their conflicted relations with "normal" members of society.

An important point is that deviant subcultures, which engage in prostitution, gambling, drug use, and other deviant behaviors, could not exist if they did not perform services and supply products that people in the larger society secretly demand. It would be wise, therefore, not to draw the distinction between deviant and normal people too sharply. Many people deviate from the norm, and their deviations create opportunities for others whose identities and occupations are deviant.

In sum, three dimensions of social life—power, culture, and voluntary versus involuntary behavior—create the major forces operating in any society to produce the forms of deviance that are typical of that society.

 Thinking Critically Many people believe that the most serious deviant acts are those identified in religious scriptures or moral codes. Sociologists, however, observe that different societies may define deviance in different ways. From this perspective, can the moral codes of one culture be applied to other cultures?

Deviance and Stigma

To narrow the range of phenomena we must deal with in discussing deviance, let us keep in mind Erving Goffman's (1963) distinction between stigma and deviance. "The term **stigma,**" Goffman stated, "refers to an attribute that is deeply discrediting" and that reduces the person "from a whole and usual person to a tainted and discounted one" (p. 3). People may be

---

stigma: An attribute or quality of an individual that is deeply discrediting.

stigmatized because of ascribed statuses such as mental illness, eccentricity, or membership in a disvalued racial or nationality group. In some instances, the stigma is visible and obvious to all, as in the case of a disfigured person like Joseph Merrick, commonly known as the Elephant Man. Suffering from a disease that grossly distorted his face, Merrick was rejected by society even though he was a highly intelligent person. In other cases, stigma is revealed only with growing acquaintance, as in the stigma attached to the children of convicts. For the stigmatized person, the disqualifying trait defines the person's *master status* (Becker, 1963; Scull, 1988) (see Chapter 4). A blind person, for example, may be an excellent musician and a caring parent, but the fact that he or she is blind will outweigh these achieved statuses except in unusual cases like that of Stevie Wonder.

Stigmatized people deviate from some norm of "respectable" society, but they are not necessarily social deviants. The term *deviant*, Goffman argued, should be reserved for people "who are seen as declining voluntarily and openly to accept the social place accorded them, and who act irregularly and somewhat rebelliously in connection with our basic institutions" (1963, p. 143). Among the people Goffman classified as social deviants are "prostitutes, drug addicts, delinquents, criminals, jazz musicians, bohemians, gypsies, carnival workers, hoboes, winos, show people, full time gamblers, beach dwellers, homosexuals, and the urban unrepentant poor." Goffman's list is somewhat whimsical and leaves out more recent groups that are considered socially deviant, such as Internet pornographers; his point, however, is that these are examples of social groups that "are considered to be engaged in some kind of collective denial of the social order. They are perceived as failing to use available opportunity for advancement in the various approved runways of society" (1963, p. 144).

According to the definition of *stigma*, the population of social deviants is smaller than that of stigmatized individuals; only some stigmatized behaviors are socially deviant. Deviant behaviors are characterized by denial of the social order through violation of the norms of permissible conduct. This point should be kept in mind as we continue our discussion of criminals and other people who are considered social deviants.

Deviance and Crime

Much of the study of social deviance focuses on crime. **Crime** is usually defined as an act, or the omission of an act, for which the state can apply

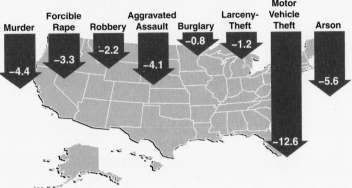

According to the FBI's latest Preliminary Semiannual Uniform Crime Report, violent crime in the nation dropped 3.5 percent and property crime declined 2.5 percent during the first six months of 2008, compared to the same period in 2007.

Figure 8.1 Declining Crime Rates. Source: The Federal Bureau of Investigation, "Good News: Crime is Declining." January 2009. http://www.fbi.gov/page2/jan09/ucr_statistics011209.html.

sanctions. Those sanctions are part of the criminal law, a set of written rules that prohibit certain acts and prescribe punishments to be meted out to violators (Kornblum & Julian, 2007). But the questions of which specific behaviors constitute crime and how the state should deal with them are often controversial.

In every society, there are some behaviors that almost everyone will agree are criminal and should be punished and other behaviors that some consider criminal but others do not. All societies punish murder and theft, for example, but there is far more variation in the treatment of adultery, prostitution, and pornography. Indeed, the largest number of "crimes" committed in the United States each year are so-called public order crimes such as public drunkenness, vagrancy, disorderly conduct, prostitution, gambling, drug addiction, and certain homosexual interactions. Many sociologists claim that these are "victimless crimes" because they generally cause no physical harm to anyone but the offenders themselves (Silberman, 1980; Thio, 1998). Not all social scientists agree with this view, however. Some point out that crimes like prostitution actually inflict damage on society because they are usually linked with an underworld that engages in far more serious and costly criminal activities (Wilson, 1977).

In the United States, crime is considered one of the nation's most serious social problems. According to the Federal Bureau of Investigation (FBI), one violent crime occurs every twenty-two seconds and one property crime every three seconds. The most serious, most common, and most likely to be reported crimes are called *index crimes* by the FBI because they are included in its Crime Index, a commonly used measure of crime rates. These categories of crime, along with crime rates, are shown in Figure 8.1. The graph shows a downward trend in crime, but at this writing, it is too soon to tell what effect the recent severe recession will have on crime rates.

For many decades, criminologists and other social scientists have criticized the FBI's crime statistics on the ground that they do not reflect differences in effectiveness among crime-reporting agencies. For example,

crime: An act, or omission of an act, that is prohibited by law.

STRENGTH OF SANCTION		
	Weak	**Strong**
DEGREE OF CONSENSUS — Weak	• Recreational drugs • Homosexuality • Abortion	• Sales of whiskey during Prohibition • Prostitution • Abortion before 1973 Supreme Court ruling
DEGREE OF CONSENSUS — Strong	• Schizophrenia • Driving while intoxicated • Public drunkenness • Corporate crime • Wife or child beating	• Major crimes (felonies) • Treason • Terrorism

Figure 8.2 A Typology of Deviance in the United States.

a more professional law enforcement agency in one community could show higher crime rates than an agency in another community that has done a haphazard job of collecting the statistics. Another serious problem is that there is no way of knowing the crime rates for the total population (rather than just for those who actually reported crimes). Since 1973, therefore, the Census Bureau and the Law Enforcement Assistance Administration have conducted semiannual surveys in which respondents are asked whether they or their businesses have been victims of robbery, rape, assault, burglary, or other forms of theft. In general, these *victimization surveys* show that the overall rate of serious crime is actually between two and three times higher than the reported Crime Index.

Deviance and Changing Values

One factor that makes it especially difficult to gather and interpret statistics on crime is that definitions of crime and deviance are constantly changing as the society's values change. Almost every week, for example, there are reports in the media of men who have battered and even killed their wives, often after serving brief prison sentences for previous attacks on their spouses. Why, many ask, are law enforcement agencies unable to prevent and punish spouse abuse? Advocates of new laws and greater investment in prevention point out that, until fairly recently, a married woman was viewed as the husband's household possession. Although both husband and wife had sworn to "love, honor, and cherish" their partner, these norms were applied far more forcefully to women than to men. This double standard is changing as men and women become more equal before the law, but in the minds of many judges and juries this change is not yet complete, and men who abuse their spouses often receive relatively light sentences (Heitzeg, 1996).

In any society, agreement on particular aspects of crime and deviance can range from weak (in cases in which there is much controversy) to strong (in cases in which there is little disagreement). Negative sanctions, or punishments, can also range from very weak to very strong. Capital punishment is the strongest sanction in the United States, followed by life imprisonment. Minor fines or the suggestion that a person undergo treatment for a behavior viewed as deviant are relatively weak sanctions. Nor are all sanctions formal punishments meted out according to law. Some deviance can be punished by means of shunning or the "silent treatment," and milder infractions can be controlled by simply poking fun at the person.

Figure 8.2 uses these distinctions to construct a typology of deviance as it is generally viewed in the United States today. Figure 8.3

STRENGTH OF SANCTION		
	Weak	**Strong**
DEGREE OF CONSENSUS — Weak	• Killing Native Americans • Lynching of African Americans • Prostitution • Wife or child beating	• Abortion • Indebtedness • Illegitimacy • Divorce • Adultery
DEGREE OF CONSENSUS — Strong	• Political corruption • Corporate crime	• Major felonies against whites • Treason • Homosexuality

Figure 8.3 Deviance in the United States Before the Civil War.

Disgraced financier Bernard Madoff enters a federal court in New York City on March 12, 2009. When he was riding high as a major Wall Street financier, customers begged him to invest their fortunes for them. For years, however, he was using money from new investors to make payments to earlier ones, a classic financial scheme known as a Ponzi scheme after the Italian swindler Charles Ponzi (1882–1949). When the financial bubble burst in 2008, Madoff could not cover his losses or pay his creditors. Investors lost approximately $50 billion in the largest swindle in recorded history.

suggests what such a typology would have looked like before the Civil War. Together they highlight some of the continuities and changes in patterns of deviance in the United States over the last 150 years. You will probably find points in them to argue with—which is further evidence of the difficulty of classifying deviant behaviors in a rapidly changing society.

Driving while intoxicated (DWI) is classified as a deviant behavior, but only in the last two decades have DWI laws been enforced systematically. Why? We all fear drunk drivers and agree that driving while intoxicated is dangerous, but this norm is frequently violated because so many people are social drinkers. Consequently, we have placed relatively mild sanctions on DWI except when the outcome is a fatal accident. In recent decades, however, organized groups opposed to drunk driving have helped change public attitudes, and the sanctions against DWI have become stronger. In a study of this issue, Joseph Gusfield (1981) found that the highway death rate had steadily decreased after 1945, but that the public's perception of the relationship between highway deaths and drunk driving had sharpened as a result of media coverage and lobbying by citizens' groups.

As discussed in Chapter 7, our perception and treatment of homosexuality is another example of significant change in Americans' attitudes toward a deviant behavior and the strength of the sanctions invoked. A century ago, homosexuality was widely regarded as a serious form of deviance.

Homosexuals were persecuted whenever they were exposed. Today there is far less agreement on this subject (Stolberg, 1997). A recent survey by the Kaiser Family Foundation (2001) found that 76 percent of lesbians, gays, and bisexuals believe there is greater acceptance of their sexual orientation, but almost equal numbers (74 percent) report that they have experienced prejudice and discrimination. About one-third (32 percent) report that they have been the target of physical violence because someone believed they were gay. The decision by the Wal-Mart Corporation, the largest private employer in the United States, to accord gay and lesbian employees and their companions the same rights and benefits as other Wal-Mart employees followed closely on the 2003 Supreme Court decision in *Lawrence and Garner* v. *Texas*. It, along with the earlier decision by the Disney Corporation to extend health benefits to same-sex partners of its employees, is another indication of the increasing acceptance of gay rights.

Elsewhere in the world, the growing emphasis on religious orthodoxy in Islamic nations has led to attacks on writers and intellectuals who question Muslim norms. The Egyptian Nobel Prize winner Naguib Mahfouz, the Iranian novelist Salman Rushdie, and Bangladeshi physician, poet, and essayist Taslima Nasrin are among those who have experienced such attacks (Weaver, 1994). In the case of Nasrin, because she speaks and writes against the traditional norms that require women to remain in the home, to be bought as brides, and otherwise to be treated as male possessions, fundamentalist religious leaders have offered a reward (about $2,500) to anyone who succeeds in killing her. Nasrin is now in hiding in Europe. "I know if I ever go back I'll have to keep silent, stay inside my house," she says. "I'll never lead a normal life in my country, until my death" (quoted in Weaver, 1994, p. 60).

Deviant Subcultures

Even when the majority of the population can be said to support a particular set of values, it will contain many subcultures whose lifestyles are labeled deviant by the larger population (Adler & Adler, 2006). A deviant subculture includes a system of values, attitudes, behaviors, and lifestyles that are opposed to the dominant culture of the society in which it is found (Bucholtz, 2002). The members of the subculture are also members of the larger society; they have families and friends outside the subculture with whom they share many values and norms. Within the subculture, however, they pursue values that are opposed to those of the larger or

Taslima Nasrin, a world-renowned feminist author, doctor, and poet from Bangladesh, is in hiding to avoid arrest or mob violence because of her criticism of Islamic gender norms. She is considered a courageous intellectual by many in the West but a deviant criminal by key political and religious authorities in her own country.

"mainstream" culture. Subcultures evolve their own rather insulated social worlds or communities, with local myths ("X is a slick dealer who once beat the tables in Vegas and was asked to leave the casino"), ways of measuring people's reputations ("he's a punk," "he's cool"), rituals and social routines, particular language or slang, formal and informal uniforms (for example, the attire worn by bikers), and symbols of belonging such as tattoos (Simmons, 1985).

Subcultures of drug users and dealers provide many illustrations of the ways in which people who engage in deviant activities often develop their own vocabulary and norms. In his fascinating accounts of the subculture of crack dealers and crack cocaine addicts, urban sociologist Terry Williams (1989, 1992) shows how everyone in the crack subculture—addicts, dealers, and hangers-on—used the language of the famous TV series *Star Trek* in referring to the drug and its use. The drug itself was often called "Scotty," and smoking the drug was referred to as "beaming up." As addicts and dealers became more deeply enmeshed in the frantic world of crack cocaine use, they spent increasing amounts of time in the company of people who spoke their language and saw the world as they did.

School Shootings and Deviant Peer Groups

Powerful examples of the strength of deviant subcultures come out of recent sociological research commissioned by the U.S. Congress in order to better understand the tragic phenomenon of school shootings.

Katherine Newman (2004) and her team of researchers studied school shootings in more than thirty U.S. communities. They reviewed the records and conducted more than 150 interviews with family members of victims and perpetrators alike, especially in two of the communities where juveniles went on rampages at their schools, leaving a wake of death and shattered lives among teachers and peers. The researchers reported their findings to Congress and in Newman's book, *Rampage*. Their two guiding research questions were: How could these low-crime, family-centered, white communities have spawned such murderous violence? How did these particular families, known and respected by neighbors, teachers, and preachers, produce rampage killers? The researchers discovered three important conditions that help explain the episodes. In these rural or suburban communities, young people of high school age are just as involved as urban kids in forming cliques and in-groups (jocks and cheerleaders, nerds, Goths, and so on), but these small communities provide few alternatives for students who are excluded from the status groups in their high school. School shooters—all of whom are boys—are often excluded from the school's peer groups and strive to demonstrate their value or daring to peers in groups that they admire. Most often, the shooters become increasingly desperate to win status in the school's alienated, outsider group. They begin bragging and indirectly suggesting that they intend to do something violent in the school. Typically, one or more of the shooters has a history of cruelty, often exhibited by torturing or killing animals. Among adolescents in the schools, the norm of silence and sanctions against "snitchers" are strong, so no one reports these situations. In such small and close-knit towns, even parents who suspect that something is amiss do not want to disturb the peace of their neighbors by calling attention to suspect juvenile behavior. When it is too late, everyone begins to see where they failed to notice the warning signs.

In the years since the shootings at Columbine High School and elsewhere, a number of planned shootings have been uncovered and prevented. The key lessons for teachers and parents are to better understand the dynamics of youth subcultures, to be ready and willing to intervene when signs of deeply troubled youth appear, and to help their children understand why it is necessary, on occasion, to deviate from the norms of their peer groups and warn adults about dangerous peers.

Tattoos in Ritual and Fashion Throughout history, as people from different cultures have come into contact with one another, norms that are part of the core culture of one society have been adopted by deviant subcultures in other societies. In time, these deviant symbols tend to become less deviant, mildly acceptable, and even fashionable. A fascinating example of this process is the spread of tattooing into the cultures of Europe and the United States, where tattoos are a major fashion statement.

European and American sailors who traveled through the Pacific Islands and Oceania in the eighteenth century were impressed by the *tattaw* worn by people of high status like the Maori chief shown here. The sailors adopted tattoos as a subcultural symbol of their "exotic" experience.

Later, criminals and prostitutes in European cities took to tattoos as a way of signifying their allegiance to their trades. With almost endless time on their hands, prisoners made amateur tattoos (requiring hours of painful jabs with an inked pen) into symbols of their opposition to the "screws" and "pigs"—their jailors and tormentors. Soldiers of the colonial powers, such as the French legionnaire shown here, also adopted tattoos as a way of boasting of their various campaigns.

In the United States in recent decades, the tattoo has become a sign of membership in certain subcultures, such as those of skinheads and bikers. But in the world of fashion, what was once thought of as deviant becomes daring and may eventually be fashionable, as can be seen in the contemporary use of the tattoo as a fashion statement among young people at almost all levels of society.

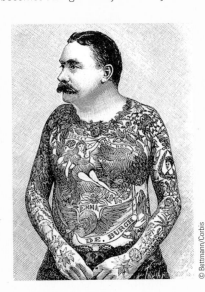

Organized Crime and Criminal Subcultures

Professional criminals—including major drug dealers, bookmakers and gamblers, hired killers, and loan sharks—who think of their illegal activities as occupations, often become part of secret crime organizations. The best known of these (but hardly the only one) is the Mafia. Originally a Sicilian crime organization, the Mafia spread to the United States via Sicilian immigration in the late nineteenth century. Today, families associated with the Mafia are active in most large cities in North America and elsewhere and often extract millions of dollars from legitimate businesses in exchange for "protection." In recent years, some successful crusades have been waged against the Mafia, but competing crime organizations—some with global reach—are always ready to engage in lucrative criminal activities as the Mafia is weakened (Ianni, 1998; Reichel, 2005).

The subculture of organized crime is not unique to North America. In Japan, rates of homicide and mugging are small fractions of those in U.S. cities, yet organized crime thrives there (Kaplan, 1998). The Yakuza, as the crime gangs are called, control many illegal businesses, just as they do in other countries, and they often resort to violence to discipline their members. As in the United States, organized crime in Japan is most directly associated with behaviors that are considered deviant, such as prostitution, gambling, and drug use, and a distinct subculture has emerged among those who engage in such behaviors.

In sum, many deviant subcultures flourish because they provide opportunities to engage in behavior that is pleasurable to some people but is considered deviant in "respectable" (conventional) society. Clearly, therefore, the line between what is normal and what is deviant is not nearly as distinct as one might believe from official descriptions of the norms of good conduct. However, one must recognize that membership in a deviant subculture, especially for those without money and power, often leads to exploitation and early death. In view of these problematic aspects of deviant subcultures, sociological and other explanations of deviant behavior take on special importance.

THEORETICAL PERSPECTIVES ON SOCIAL DEVIANCE

Biological Explanations of Crime

Early in the twentieth century, Italian criminologist Cesare Lombroso (1911) claimed to have proved that criminals were throwbacks to primitive, aggressive human types who could be recognized by physical features like prominent foreheads, shifty eyes, and red hair. Although modern researchers have thoroughly refuted Lombroso's theory, efforts to link body type with crime seem destined to reappear from time to time. In the 1940s, for example, the psychologist and physician William Sheldon announced that body type was correlated with crime. He believed that human beings could be classified into three types: *ectomorphs,* or thin people; *endomorphs,* or soft, fat people; and *mesomorphs,* or people with firm, well-defined muscles. The latter were most prone to crime, according to Sheldon, but he neglected to account for the possibility that mesomorphs simply might have led harder lives than less muscular individuals and therefore were better equipped to commit crimes that required strength (Glueck & Glueck, 1950).

The more modern biological view is represented in the influential work of social scientists Richard J. Herrnstein and James Q. Wilson. Their research has led them to conclude that both biology and social environment play important roles in producing criminals. In commenting on a recent study by researchers at the U.S. Department of Justice who reported that more than one-third of adult criminals in U.S. jails have at least one parent who is or was a criminal, Herrnstein argues that whatever determines criminality "is transmitted both genetically and environmentally. So kids brought up in criminal families get a double exposure" (quoted in Butterfield, 1992, p. A16).

A study by Deborah W. Denno offers support for this view. In her research on 1,000 African American boys from poor neighborhoods of Philadelphia, Denno (1990) found that a disproportionate number of individuals who had become criminals had histories of childhood hyperactivity. This genetically transmitted condition makes it difficult for children to concentrate and succeed in school. It also can make it hard for them to hold jobs. Thus, Denno sees the causal chain as one of biological hyperactivity leading to social failure and then to crime.

To many social scientists, however, such studies are not convincing. Marvin E. Wolfgang, one of the nation's leading experts on criminality, points out that all the studies deal with people from poor neighborhoods, where a high proportion of people will be sent to jail whether or not they are related and whether or not they were hyperactive as children. Hyperactivity is also found in children from wealthier families but is not so often associated with criminal careers because the families can afford special schools and programs to deal with the condition (Reichel, 2005).

Social-Scientific Explanations of Deviance and Crime

Early sociological explanations of social deviance were also influenced by the biological concept of disease (Thio, 1998). The earliest of these explanations viewed crime as a form of social pathology, or social disease, that could be attributed to the evils of city life. Corruption, criminality, and depravity were thought to be bred in slums and to infect innocent residents of those grim communities, just as typhoid spread in unclean surroundings and was passed by contact from one victim to another (Bulmer, 1986). Thus, many early juvenile training schools were built in rural areas where they would be isolated from the "corrupting" influence of cities (Platt, 1977).

The early sociological view of deviance and crime as resulting from "social pathologies" tended to rely on subjective terms and circular reasoning. To refer to crime or other forms of deviance as *pathologies,* for example, does not provide a theory of causality. It merely attaches a scientific-sounding term to behavior that the larger society finds wrong or offensive. It does not offer any insight into why the behavior occurs, what benefits it might provide for those who engage in it, and how it evolves from other behaviors or social conditions. In consequence, the view of deviance as a kind of disease was replaced by more objective and verifiable theories derived from the functionalist, interactionist, and conflict perspectives of modern sociology.

The Functionalist Perspective

The functionalist theorist Robert K. Merton (1938) developed a useful typology of deviance based on how people adapt to the demands of their society. Merton's aim was to discover "how some social structures *exert a definite pressure upon certain persons in society to engage in nonconformist rather than conformist conduct*" (p. 672). "Among the elements of social and cultural structure," Merton continued, "two are important for our purposes. . . . The first consists of culturally defined goals, purposes, and interests. . . . The second . . . defines, regulates, and controls the acceptable modes of achieving these goals" (pp. 672–673).

Merton's explanation of deviance is based on the concept of **anomie**, or normlessness. In his view, anomie results from the frustration and confusion people feel when what they have been taught to desire cannot be achieved by the legitimate means available to them. Merton believes that people in North America and other modern societies exhibit high levels of anomie because they are socialized to desire success in the form of material well-being and social prestige. For many, however, the means of attaining these culturally defined ends, such as hard work and saving, seem to be out of reach. This is especially true for people who experience rapid social changes, such as the closing of factories, and find themselves deprived of opportunities to attain what they have come to expect from life.

Consider some examples. Possession of money is a culturally defined goal. Work is a socially acceptable mode of achieving that goal; theft is not. Mating is also a culturally defined goal. Courtship and seduction are acceptable means of achieving it; kidnap and rape are not. But if theft and rape are unacceptable means of achieving culturally defined goals, why do they exist? According to Merton, the gap between culturally defined goals and acceptable means of achieving them causes feelings of anomie, which in turn make some people more likely to choose deviant strategies of various kinds.

Through socialization, we learn the goals and acceptable means of our society. Most of us would love to be rich, powerful, or famous. We accept these goals of our culture. We also accept the legitimate means of achieving them: education, work, the electoral process, plastic surgery, acting school, and so on. We are conformists. However, not everyone accepts either the cultural goals or the accepted means of achieving them. Some become "innovators" in that they explore (and often step over) the frontiers of acceptable goal-seeking behavior; others (for example, hoboes) retreat into a life that rejects both the goals and the accepted means; some rebel and seek to change the goals and the institutions that support them; and still others reject the quest for these precious values while carrying out the rituals of social institutions. (In many bureaucracies, for example, one can find ritualists who have given up the quest for promotion yet insist on receiving deference from people below them in the hierarchy.) Merton's typology thus is based on whether people accept either the cultural goals of their society, the acceptable means of achieving them, or neither. Figure 8.4 presents this typology, and Figure 8.5 gives examples of all the deviant types in Merton's framework.

You may be wondering how Merton's theory explains why some poor people resort to crime while so many others do not. The answer lies in the way people who experience anomie gravitate toward criminal subcultures. For example, adolescents who choose to steal to obtain things that their parents cannot afford must learn new norms—they must learn how to steal successfully and must receive some approval from peers for their conduct (Cloward & Ohlin, 1960; McNamara, 1994). As a result, they drift toward deviant peer groups. But this explanation presents a problem for the functionalist perspective: If there are deviant subcultures, the idea that there is a single culture whose goals and means are shared by all members of society is called into question. The presence of different subcultures within a society suggests another possible cause of deviant behaviors: conflict between groups.

Conflict Perspectives

Cultural Conflict In an influential essay, "Crime as an American Way of Life," Daniel Bell (1962) observed that at the turn of the twentieth century, cultural conflict developed between the "Big City and the small-town conscience. Crime as a growing business was fed by the revenues from prostitution, liquor, and gambling that a wide-open urban society encouraged and that a middle-class Protestant ethos tried to suppress with a ferocity unmatched in any other civilized society" (p. 128). This example of conflict between the official morality of the dominant

anomie: A state of normlessness.

MODE OF ADAPTATION		CULTURAL GOALS	INSTITUTIONAL MEANS
	Conformity	+	+
	Innovation	+	−
	Ritualism	−	+
	Retreatism	−	−
	Rebellion	+/−	+/−

Figure 8.4 The Merton Typology—Modes of Adaptation by Individuals Within the Society or Group. Merton explained that "+" signifies acceptance, "−" signifies elimination, and "+/−" signifies rejection and a substitution of new goals and standards. The line separating "Rebellion" from the other roles signifies that the individual no longer accepts the society's culture and structure; other individuals, though they may deviate, continue to accept the society's culture and structure.

American culture and the norms of subcultures that do not include strictures against gambling, drinking, prostitution, and the like encouraged the growth of criminal organizations. In the United States, as in many other societies, such organizations thrive by supplying the needs of millions of people who appear to be law-abiding citizens yet engage in certain deviant behaviors.

The prohibition of alcoholic beverages in the United States from 1919 to 1933 is an example of how cultural conflict can lead to situations that encourage criminal activity. Prohibition has been interpreted as an effort by the nation's largely Protestant lawmakers to impose their version of morality on immigrant groups for whom consumption of alcohol was an important part of social life. Once they were passed, however, laws against the production and sale of alcoholic beverages created opportunities for illegal production, bootlegging (smuggling), and illegal drinking establishments (speakeasies). These in turn supported the rise of organized crime syndicates. The current laws against drugs like marijuana and cocaine have similar effects in that they lead to the clandestine production and supply of these illegal but widely used substances. Similar opportunities to make illegal profits arise when states pass laws raising the drinking age (Gusfield, 1966, 1996; Kornblum & Julian, 2007).

To recognize these effects of cultural conflict is not to condone the sale and use of illegal drugs. Rather, it is to be aware that whenever laws promoted by the powerful impose a set of moral standards on a minority, illegal markets are created that tend to be supplied by criminal organizations. Such organizations often act ruthlessly to control their markets, as can be seen in the murder of Mexican border guards who surprise drug smugglers, the killing of cocaine traffickers by rival dealers in Miami or Los Angeles, or the periodic "wars" that break out among organized crime families in many large cities.

Marxian Conflict Theory For Marxian students of social deviance, the cultural conflict explanation is inadequate because it does not consider the effects of power and class conflict. Marxian sociologists believe that situations like Prohibition do not occur just because of cultural conflict (Quinney, 2001). They happen because the powerful classes in society (that is, those who own and control the means of production) wish to control the working class and the poor so that they will produce more. From the Marxian perspective, as criminologist Richard Quinney (1980) points out, "crime is to be understood in terms of the development of capitalism" (p. 41), as it was in Marx's original analysis. From this perspective, most crime is essentially a form of class conflict—either the have-nots taking what they can from the ruling class, or the rich and their agents somehow taking what they can from the poor (Young, 2002).

The economic "robber barons"—John Jacob Astor, John D. Rockefeller, J. P. Morgan, Leland Stanford, Andrew Carnegie, and many others—amassed huge fortunes in the period of booming industrial growth following the Civil War. But they often resorted to illegal means in pursuing the culturally approved goal of great wealth. Among other tactics, they used violence to drive settlers off land they had purchased or to break strikes by workers, and they were not

MODE OF ADAPTATION		THE POOR	THE MIDDLE CLASS	THE RICH
	Conformity	The working poor	The suburban family	The wealthy civic leader
	Innovation	The mugger	The embezzler	The stock manipulator
	Ritualism	The chronic welfare recipient	The resigned bureaucrat	The hedonist
	Retreatism	The wino or junkie	The skidding alcoholic	The bohemian
	Rebellion	The bandit	The anarchist	The fascist

Figure 8.5 Examples of Social Roles Based on the Merton Typology. These examples are meant to illustrate possible outcomes, not stereotypes. Perceptions of approved goals and means may vary. In some cases, for example, the very rich, who feel that they could be even richer were it not for legal obstacles in their path, may choose to bend the rules. Also, a number of cultural values besides wealth shape the likelihood that a member of a given class will behave as he or she does.

above manipulating prices in order to drive out competitors and monopolize the markets for steel, oil, coal, precious metals, food products, and numerous other goods (Chernow, 1998). In Merton's typology of deviance, their actions would classify them as "innovators" (see Figure 8.5), but from a Marxian viewpoint, they were merely carrying out "the logic of capitalism," which was based on the exploitation of the poor by the rich and powerful.

Marxian students of deviance point out that legal definitions of deviant behavior usually depend on the ability of the more powerful members of society to impose their will on the government and to protect their actions from legal sanctions. Thus the crimes of the robber barons almost always went unpunished. Definitions of what is criminal and who should be punished are generally applied more forcefully to the poor and the working class than to the middle and upper classes (Quinney, 2001; Reiman, 2002).

Marx and his collaborator Friedrich Engels recognized that the working class (or *proletariat,* as they called it) would resort to individual crimes like robbery when driven to do so by unemployment and poverty, but they believed that the workers would be more likely to form associations aimed at destroying capitalism. The chronic poor, however, would form a class that Marx called the *lumpenproletariat,* people who were unable to get jobs in the capitalist system or were cast off for not working hard enough or for being injured or sick. Marx did not believe that members of this class would join forces with the proletariat. Instead, they would act as spies, informers, and thugs whose services could be purchased by the rich to be used against the workers. Marx agreed with other thinkers of his time who called the lumpenproletariat the "dangerous class" created by capitalism; from its ranks came thieves, prostitutes, gamblers, pickpockets, con artists, and contract murderers.

It can be helpful to think of crime and deviance as symptoms of the class struggles that occur in any society and to show that laws that define and punish criminal behaviors are often imposed by the powerful on those with less power. But to attribute crime as we know it to the workings of capitalism is to suggest that if capitalism were abolished, crime would vanish. This clearly is not the case. In China, for example, there are thousands of prisoners in a vast chain of prison camps. Many of those prisoners were convicted of the deviant act of opposing the dominant regime (Ning, 1995; Wu & Wakeman, 1995). And in Cuba, homosexuality is considered a serious crime. It is severely punished because the society's leaders think of it as

Do they look like criminals? The United States has less than 5 percent of the world's population, but it has almost a quarter of the world's prisoners, and more than half are young black or Hispanic males. A critical perspective on crime and deviance in the United States sheds light on why this nation leads the world in producing prisoners. Americans are locked up for crimes—from writing bad checks to using drugs—that would rarely produce prison sentences in other countries. They are kept incarcerated far longer than prisoners in other nations. Released prisoners, especially African Americans, typically suffer major disadvantages on the job market for the rest of their lives.

an offense against masculinity and a symbol of a "decadent" subculture. As these examples indicate, societies with Marxian ideologies are hardly free from their own forms of deviance and crime.

The Interactionist Perspective

Functionalist theories explain deviance as a reaction to social dysfunctions; conflict theories explain it as a product of deviant subcultures or of the type of class struggle that occurs in a society in a particular historical period. Neither of these approaches accounts very well for the issues of recruitment and production (Rubington & Weinberg, 1999). *Recruitment,* in this context, refers to the question of why some people become deviants while others in the same social situation do not. *Production* refers to the social construction of new categories of deviance in a society—that is, to the ways in which people and groups are defined as deviant according to the changing perceptions of normal and abnormal behavior by members of a society, especially those with significant amounts of power (Adler & Adler, 2006).

Recruitment Through Differential Association In 1940, sociologist and criminologist Edwin H. Sutherland published a paper, "White-Collar Criminality," in which he argued that official crime statistics do not measure the many forms of crime that are not correlated with poverty. Outstanding among these are *white-collar crimes*—that is, the criminal behavior of people in business and professional positions. In Sutherland's words:

> White-collar criminality in business is expressed most frequently in the form of misrepresentation in financial statements of corporations, manipulation in stock exchange, commercial bribery, bribery of public officials directly or indirectly in order to secure favorable contracts and legislation, misrepresentation in advertising and salesmanship,

embezzlement and misapplication of funds, short weights and measures and misgrading of commodities, tax frauds, misapplication of funds in receiverships and bankruptcies. These are what Al Capone called "the legitimate rackets." These and many others are found in abundance in the business world. (1940, pp. 2–3)

Sutherland was pointing out that an accurate statistical comparison of the crimes committed by the rich and the poor was not available. But his paper on white-collar crime also set forth an interactionist theory of crime and deviance:

White-collar criminality, just as other systematic criminality . . . is learned in direct or indirect association with those who already practice the behavior; and . . . those who learn this criminal behavior are segregated from frequent and intimate contacts with law-abiding behavior. Whether a person becomes a criminal or not is determined largely by the comparative frequency and intimacy of his contacts with the two types of behavior. This may be called the process of differential association. (pp. 10–11)

The concept of **differential association** offered an answer to some of the weaknesses of functionalist and conflict theories. Not only did it account for the prevalence of deviance in all social classes, but it also provided clues to how crime is learned in groups that are culturally distinct from the dominant society. For example, in the 1920s, sociologist Clifford Shaw (1929) observed that some Chicago neighborhoods had consistently higher rates of juvenile delinquency than others. These were immigrant neighborhoods, but their high rates of delinquency persisted regardless of which immigrant groups lived there at any given time. Sutherland's theory explained this pattern by calling attention to the culture of deviance that had become part of the way of life of teenagers in those neighborhoods. According to Sutherland, the teenagers became delinquent because they interacted in groups whose culture legitimated crime. It was not a matter of teenage delinquents deviating from conventional norms because the approved means of achieving approved goals were closed to them. Rather, they acted as they did because the culture of their peer group made crime an acceptable means of achieving desired goals.

In an empirical study that tested Sutherland's theory, Walter Miller (1958) found that delinquency in areas with high rates of juvenile crime was in fact supported by the norms of lower-class teenage peer groups. In three years of careful observation, Miller found that delinquent groups had a set of well-defined values: *trouble, toughness, smartness, excitement, fate,* and *autonomy.* Whereas other groups felt that it was important to stay out of trouble, the delinquent groups viewed trouble—meaning fighting, drinking, and sexual adventures—as something to brag about, as long as they didn't get caught. Toughness as shown by physical prowess or fearlessness; smartness as evidenced by the ability to con or outsmart gullible "marks"; the excitement to be found in risking danger successfully; one's fate as demonstrated by luck or good fortune in avoiding capture; and the autonomy that crime seemed to provide in the form of independence from authorities—all were values of delinquent groups that distinguished them from nondelinquent groups in the same neighborhoods.

A study undertaken by criminologist Donald R. Cressey (1971/1953) supported the theory of differential association with the finding that embezzlers generally had to learn how to commit their crime by associating with people who could teach them how to commit it with the greatest likelihood of avoiding suspicion. Moreover, Cressey also found that people who became embezzlers often had serious personal problems (marital difficulties, gambling, and the like) that directed them toward people who would influence them to commit this form of white-collar crime. When bank officials had problems of this sort, they were vulnerable to suggestions of deviant means of gaining money and solving personal problems while maintaining the outward signs of respectability.

Are women more or less susceptible than men to influences that can lead toward crime and imprisonment? Criminologist Leanne Alarid and her term of investigators spoke to young men and women ages seventeen to twenty-five who had been convicted of a felony and were doing prison time. In fact, they were inmates of shock treatment boot camps for first offenders. Along all the independent variables—age, education, social class—they were similar. But the women were more likely to have committed drug offenses, whereas the young men were more likely to have committed crimes against property, such as automobile theft. The same proportion, about 20 percent, of both groups were locked up for violence against another person. When asked about the criminal backgrounds of their peers, family members, or other associates, male and female felons did not differ; each had been influenced by others with criminal records, although

differential association: A theory that explains deviance as a learned behavior determined by the extent of a person's association with individuals who engage in such behavior.

the actual criminal histories differed somewhat by gender (more involvement in drug deals for women, somewhat more involvement in street crimes for men) (Alarid, Burton, & Cullen, 2000).

Not all deviants are people whose means of achieving success have been blocked or who are acting out some form of class struggle or have associated with a deviant group. For example, many alcoholics and drug users are not thought of as deviant, either because their behavior is not considered serious or because other people do not witness it. From an interactionist perspective, the key question about such people is how others understand their behavior (Rubington & Weinberg, 1999). The central concepts that attempt to answer this question are *labeling* and the *deviant career*.

Labeling According to symbolic-interactionist theory, deviance is produced by a process known as **labeling**, meaning a societal reaction to certain behaviors that labels the offender as a deviant. Most often, labeling is done by official agents of social control like the police, the courts, mental institutions, and schools, which distribute labels like "troublemaker," "hustler," "kook," or "blockhead" that stick to a person, often for life (Becker, 1963; Lerman, 1996; McCorkel, 2003). Here is what occurs in the labeling process, according to Howard Becker:

> Social groups create deviance by making the rules whose infraction constitutes deviance and by applying those rules to particular people and labeling them as outsiders. From this point of view, deviance is not a quality of the act the person commits, but rather a consequence of the application by others of rules and sanctions to an "offender." The deviant is one to whom that label has been successfully applied; deviant behavior is behavior that people so label. (1963, p. 9)

In a famous experiment that tested the effects of labeling, David Rosenhan (1973) and eight other researchers were admitted to a mental hospital after pretending to "hear voices." Each of these pseudopatients was diagnosed as schizophrenic. Before long, many of the patients with whom the pseudopatients associated recognized them as normal, but the doctors who had made the diagnosis continued to think of them as schizophrenic. As the pseudopatients waited in the lunch line, for example, they were said to be exhibiting "oral-acquisitive behavior." Gradually, the pseudopatients were released with the diagnosis of "schizophrenia in remission," but none was ever thought to be cured.

Rosenhan and his researchers also observed that not only did the diagnosis of schizophrenia label the patient for life, but the label itself became a justification for other forms of mistreatment. The doctors and hospital staff disregarded the patients' opinions, treated them as incompetent, and often punished them for infractions of minor rules. The hospital's social atmosphere was based on the powerlessness of the people who were labeled mentally ill.

Rosenhan's study accelerated the movement to reform mental institutions and to deinstitutionalize as many mental patients as possible. But studies of deinstitutionalized mental patients (many of whom are homeless), together with research on the problems of released convicts, have shown that the labels attached to people who have deviated become incorporated into their definitions of themselves as deviant. In this way, labeling at some stage of a person's development tends to steer that person into a community of other deviants, where he or she may become trapped in a *deviant career* (Bassuk, 1984).

Deviant Careers "There is no reason to assume," Becker (1963) points out, "that only those who finally commit a deviant act actually have the impulse to do so. It is much more likely that most people experience deviant impulses frequently. At least in fantasy, people are much more deviant than they appear" (p. 26). Becker suggests that the proper sociological question is not why some people do things that are disapproved of but rather, why "conventional people do not follow through on the deviant impulses they have" (p. 26). According to Becker and other interactionists, the answers are to be found in the individual's commitment to conventional institutions and behaviors.

Commitment means adherence to and dependence on the norms of a given social institution. The middle-class youth's commitment to school, which has developed over many years of socialization in the family and the community, often prevents him or her from giving in to the impulse to play hooky. "In fact," Becker asserts, "the normal development of people in our society (and probably in any society) can be seen as a series of progressively increasing commitments to conventional norms and institutions" (p. 27).

Travis Hirschi studied how people become committed to conventional norms. Such commitment, he found, emerges out of the interactions that create our social bonds to others. When we are closely tied to people who adhere

---

labeling: A theory that explains deviance as a societal reaction that brands or labels people who engage in certain behaviors as deviant.

to conventional norms, we have little chance to deviate. And as we grow older, our investment in upholding conventional norms increases because we feel that we have more to protect. This process of lifetime socialization in groups produces a system of social control that is internalized in most members of the society. (See Chapter 5 for more on socialization over the life course.)

By the same token, people who once give in to the impulse to commit a deviant act and are caught, or who become members of a deviant group because it has recruited them, gradually develop a commitment to that group and its deviant culture. The reasons for taking the first step toward deviance may be many and varied, but from the interactionist perspective, "one of the most crucial steps in the process of building a stable pattern of deviant behavior is likely to be the experience of being caught and publicly labeled as deviant" (Becker, 1963, p. 31).

In a well-known empirical study of youth gangs, William J. Chambliss (1973) applied both the conflict and interactionist perspectives. For two years, Chambliss observed the Saints, a gang of boys from rather wealthy families, and the Roughnecks, a gang of poor boys from the same community. Both gangs engaged in car theft and joyriding, vandalism, dangerous practical jokes, and fighting—and in fact, the Saints were involved in a larger overall number of incidents. The Roughnecks, however, were more frequently caught by authorities, described as "tough young criminals headed for trouble," and sent to reform school. The wealthier youths were rarely caught and were never labeled delinquent.

Chambliss observed that the Saints had access to cars and could commit their misdeeds in other communities, where they were not known. The Roughnecks hung out in their own community and performed many of their antisocial acts there. A far more important explanation, according to Chambliss, pertained to the relative influence of the boys' parents. The parents of the Saints argued that the boys' activities were normal youthful behavior; the boys were just "sowing their wild oats." Their social position enabled them to influence the way their children's behavior was perceived, an influence the Roughnecks' parents did not have. Thus, the Roughnecks were caught and labeled and became increasingly

committed to deviant careers, whereas the Saints escaped without serious sanction.

In commenting on Chambliss's study, sociologists often note that members of both gangs engaged in deviant acts, referred to as **primary deviance**, but that only the Roughnecks were labeled delinquent by the police and the juvenile courts. As a result of that labeling, many of the Roughnecks went on to commit acts that sociologists call **secondary deviance**—that is, behaviors appropriate to someone who has already been labeled delinquent. (For example, in juvenile detention centers, teenage offenders often learn deviant skills such as drug dealing.) The distinction between primary and secondary deviance is useful because it emphasizes that most of us deviate from cultural norms in many ways; but once we are *labeled* deviant, we tend to commit additional deviant acts in order to fulfill the negative definitions that society has attached to us (Goode, 1994; Rubington & Weinberg, 1999).

Reasonable as the labeling perspective appears, it has not always been borne out by empirical research. Some studies have found that people who have been labeled as delinquent after being caught and convicted of serious offenses go on to commit other deviant acts, but other studies have found that labeling can lead to decreased deviance and a lower probability of further offenses (Rubington & Weinberg, 1999). Some recent research explores the limits of labels and shows how people attempt to resist and change the negative labels assigned to them in prison. Research by Jill McCorkel (2003), for example, has shown that women in prison form peer groups that help their members forge new identities that reject the labels assigned to them by police and prison guards. (The various theoretical approaches to crime that we have discussed are summarized in the Perspectives chart on page 197.)

CRIME AND SOCIAL CONTROL

We have seen that crime as defined by a society's laws is only one aspect of the range of behaviors included in the study of deviance. However, both the general public and policymakers in government are most concerned with crime and its control. In this section, therefore, we touch on some of the most hotly debated issues in the control of crime and the treatment of criminals.

At the beginning of the chapter, we defined social control in broad terms as all the ways in which a society establishes and enforces its

primary deviance: An act that results in the labeling of the offender as deviant.

secondary deviance: Behavior that is engaged in as a reaction to the experience of being labeled as deviant.

Theoretical Perspectives on Deviance

Perspective	Description	Critique
BIOLOGICAL THEORIES	Crime and other forms of deviance are genetically determined.	There is evidence of effects of socialization.
SOCIOLOGICAL THEORIES Social Pathology	The deviant or criminal person is a product of a "social sickness" or social disintegration.	A popular notion, but discredited among professionals because it suggests no real theory of causality.
Functionalism	Deviance and crime result from the failure of social structures to function properly. Every society produces its own forms of deviance.	Does not explain why some people drift into deviant subcultures.
Conflict Theories *Cultural conflict*	Cultural conflict creates opportunities for deviance and criminal gain in deviant sub-cultures (e.g., prohibitions create opportunities for organized crime).	Explains a narrow range of phenomena.
Marxian theory	Capitalism produces poor and powerless masses who may resort to crime to survive. The rich employ their own agents to break laws and enhance their power and wealth.	Crime exists in societies that have sought to eliminate capitalism.
Interactionist Theories *Differential association*	Criminal careers result from recruitment into crime groups based on association and interaction with criminals.	Excellent explanation of recruitment but not as effective in explaining deviant careers.
Labeling	Deviance is created by groups that have the power to attach labels to others, marking particular people as outsiders. It is extremely difficult to shed a label once it has been acquired, and the labeled person tends to behave in the expected manner.	Not always supported by empirical evidence. People can also use labels (e.g., drunk) to change their behavior.

cultural norms. Certainly it is true that without socialization and the controlling actions of social groups like the family, schools, the military, and corporations, there would be much more anomie, crime, and violence. But in considering a society's means of controlling crime, sociologists most often study what might be called "government social control"—that is, the society's legal codes; the operation of its judicial, police, penal, and rehabilitative institutions; and the ways in which its most powerful members promote their views of crime, deviance, and social control (Marwell, 2004; Scull, 1988). The rates of crime and the ways in which crime affects communities in different parts of the society establish the need for these institutions of social control, as we will see in the following analysis.

Ecological Dimensions of Crime

Figure 8.6 shows the rates of violent and property crimes by major regions of the United States. Rates of property crimes, including burglary and automobile theft, are significantly higher in the South and West than in the Northeast or Midwest. The same differences can be seen in rates of violent crimes; the South ranks far higher in violent crime rates than any other region of the nation. Ecological data like these are an essential starting point in research on the incidence of crime and provide some insights into why such regional differences exist. In the South, for example, high rates of gun ownership contribute to higher homicide rates. The higher average age of the population in the Northeast contributes to lower rates of rape, assault, and other violent crimes in that region, while greater disparities between rich and poor lead to a higher rate of robbery.

Another contribution of ecological analysis can be seen in data on the relationship between crime and the age composition of a population. Some political leaders and social scientists favor larger police forces and tougher law enforcement, and they argue that these policies are responsible for the decrease in the U.S. crime rate over the past several years. However, it must not be forgotten that the age composition of the U.S. population is changing rather

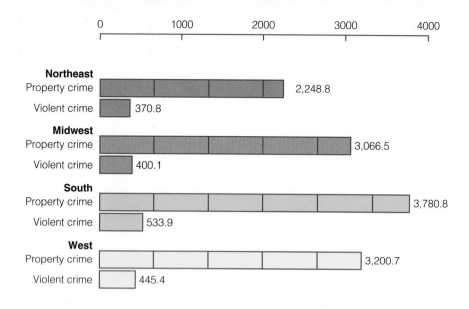

| | 0 | 1000 | 2000 | 3000 | 4000 |

Northeast
Property crime — 2,248.8
Violent crime — 370.8

Midwest
Property crime — 3,066.5
Violent crime — 400.1

South
Property crime — 3,780.8
Violent crime — 533.9

West
Property crime — 3,200.7
Violent crime — 445.4

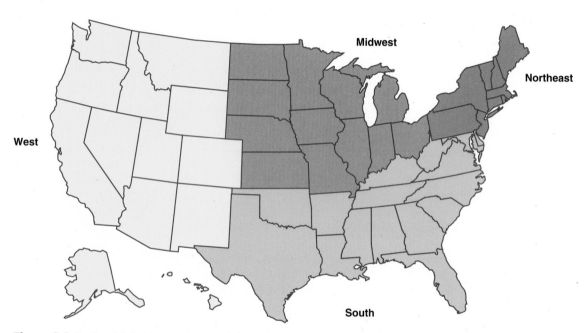

Midwest

Northeast

West

South

Figure 8.6 Regional Crime Rates, 2004—Violent and Property Crimes per 100,000 Inhabitants.

quickly. Among other things, the baby boom generation has moved out of the young adult years, those in which people are most likely to commit crimes. (About half of all reported violent crimes are committed by males between the ages of fifteen and twenty-four.) Thus, before taking credit for reducing crime rates, law enforcement agencies and advocates of tough anticrime measures must determine whether lower crime rates are not simply the result of a reduction in the most crime-prone portion of the population.

Still another variable that lends itself to ecological analysis is the effect of possession of firearms on homicide rates. The United States far outpaces other industrialized nations in number of homicides. No other nation even comes close. This fact is often thought to reflect a greater overall level of violence in the United States than in other comparable nations. However, in a study comparing rates of violent crime in Denmark and northeastern Ohio, investigators found that rates of assault did not differ in the two areas.

What did differ were rates of homicide using firearms; such crimes were much more frequent in Ohio. In Denmark, the possession of handguns is banned, whereas in the United States, "50 percent of all households have guns and one in five has a handgun" (Mawson, 1989, p. 239). (See the Mapping Social Change box on page 199.)

As mentioned earlier in the chapter, studies of criminal victimization, which ask people about their actual experiences with crime, have provided a major breakthrough in our understanding of crime in complex societies. Of every 1,000 crimes, slightly more than half are reported to the police. Of those, about 15 percent result in arrests. Twenty-five years ago, less than

Worldwide Homicide Rates

Although homicide rates in the United States have declined in the past few years, there is little likelihood of a decline to levels characteristic of the European democracies. Currently, two major nations, Brazil and Russia, have far higher homicide rates than do the United States, and so do the Bahamas and Ecuador. All are undergoing rapid social change with drastic increases in inequality. None enjoy the degree of affluence found in the United States. In Ecuador, high homicide rates are associated with rapid urbanization in cities like Guayaquil, where murders occur at extremely high rates in sprawling slums. In the Bahamas, the international drug markets, which often include stops in this island nation, are an important influence. In Russia, the abrupt changes accompanying the transition from communism to capitalism have produced an increase in homicide rates. The United States shares certain characteristics with each of these nations. Like Ecuador and other Latin American nations, the United States has a frontier tradition that fosters and still maintains norms of rugged individualism, including the use of guns. It also has a widening gap between rich and poor, with the poor often concentrated in ghetto slums. Drug activity contributes significantly to homicide rates, as in the Bahamas. And as in Russia, much of the deadly violence occurring in the United States stems from the activities of criminals in organized gangs.

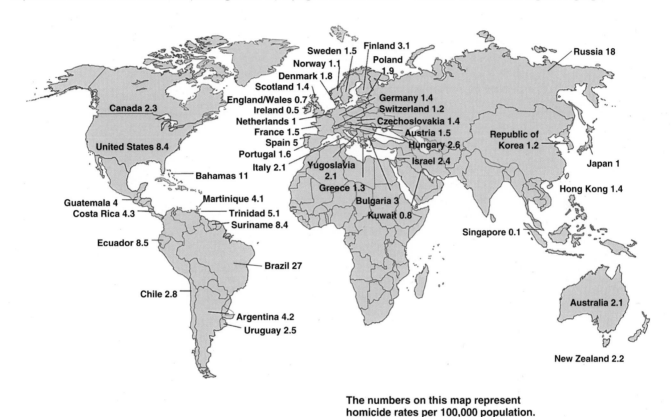

The numbers on this map represent homicide rates per 100,000 population.

half of all arrests led to convictions, and less than half of those convicted were sentenced to custody. Today, with more forceful law enforcement strategies, including tougher sentencing guidelines and mandatory jailing of third-time offenders in California and other states, these figures have improved. Nevertheless, the probability that a criminal who is arrested will be sentenced to prison remains low. On the other hand, although the probability of arrest for a single property crime, such as theft of an automobile, is low, criminals who commit many crimes do tend to be arrested eventually; this is especially true for those who commit violent crimes (Butterfield, 1998).

Recidivism Most sociological studies of **recidivism** (the probability that a person who has served a jail term will commit additional crimes and be jailed again) indicate that there is an overall probability of 50 percent that recidivism will occur. Recidivism is most frequent among young minority males from poor backgrounds who are addicted to drugs and/or alcohol and have been imprisoned on numerous previous occasions for crimes against property (Land, 1989).

recidivism: The probability that a person who has served a jail term will commit additional crimes and be jailed again.

In recent years, there have been forceful efforts to reduce the types of recidivism that pose the greatest threat to society. People who have been convicted of violent offenses and people who have served prison sentences for sex crimes are especially likely to be refused parole. In the case of sex offenders, community notification policies, in which neighbors are informed that a former sex offender has moved into their community, are being implemented in many states (Zonana, 1997). The high crime rates of recent decades, together with public support for longer sentences and less lenient parole policies, have resulted in record high rates of imprisonment, as we will see shortly. Despite the decline in crime rates in the last few years, states are finding it necessary to build new prisons even as corrections professionals search for alternatives to imprisonment, especially for nonviolent offenders.

Institutions of Coercive Control: Courts and Prisons

When the United States consisted mainly of small agrarian communities, there were jails and courts, but a great deal of social control of deviance and crime was carried out by the local institutions of the family or the church. Parents, for example, were expected to control their children; if they did not, they would lose the respect of other members of the community. As the size, complexity, and diversity of societies increase, however, the ability of local institutions to control all of the society's members is diminished (Wirth, 1968/1938). Societies therefore develop specialized, more or less coercive institutions to deal with deviants. Courts, prisons, police forces, and social welfare agencies grow as the influence of the community on the behavior of its members declines.

Capital Punishment: The Global Debate

Capital punishment provides an illustration of the issues raised by the use of coercive forms of social control. Only recently (during the twentieth century in some nations) has there been ongoing debate about the morality of capital punishment (Phillips, 1998). In simpler societies and earlier civilizations, execution was more than a penalty carried out on an individual; it was also an occasion for a public ceremony. People attended beheadings and hangings, and hawkers sold food and other goods as the crowd waited for the bloody pageant. The villain's death reaffirmed their common values, their solidarity as a people who could purge evil elements from their midst.

Today, by contrast, the value of capital punishment is a matter of heated debate, as we have seen recently in the case of Timothy McVeigh, who was executed on June 11, 2001, for the Oklahoma City bombing in which 168 people were killed. Although many people believe that the death sentence is necessary as a means of discouraging some individuals from committing terrible crimes, others are convinced that there is no justification for putting another person to death, the more so because it is always possible that the person did not commit the crime for which he or she was executed. One study of this grim possibility provides evidence of as many as eighteen wrongful executions in the United States since 1900 (Radelet & Bedau, 1992). In 2000, faced with DNA evidence that helped overturn almost half the pending death sentences in his state, former Governor George Ryan of Illinois suspended the death penalty in that state.

Sociology & Social Justice

RACE AND CAPITAL PUNISHMENT In one of a series of groundbreaking studies of the equity of capital punishment, criminologists Marvin Wolfgang and Marc Riedel (1973) found that blacks were more likely than whites to be executed for the same crimes and that people who could afford good lawyers were far more likely to escape execution than those who lacked the means to hire the best legal defenders. As we see in Figure 8.7 on page 202, although the absolute number of white inmates on death row is higher than that of African Americans by about 500, in proportional terms—that is, relative to their proportion of the overall population—African Americans are extremely overrepresented on death rows in the United States. The graphs also show that since capital punishment was reinstated in many states, the overall number of inmates on death row has increased dramatically.

In 1972, swayed by evidence like that gathered by Wolfgang and Riedel and by arguments that capital punishment could be considered "cruel and unusual punishment" according to the Constitution (most other Western nations had already banned the death penalty), the Supreme Court ruled in a 5-to-4 decision that, in the absence of clear specifications for when it might be used, the death sentence violated the Eighth Amendment to the Constitution. Congress has since passed new legislation that legalized the death sentence, and the Court has not overruled it. As a result, capital punishment has been reinstated in many states; there are now between fifteen and thirty executions a year in the United States, with Texas, Florida, Mississippi, and Arizona leading in numbers of executions carried out. Indeed, in many states, the debate about capital punishment has shifted away from its morality to questions about

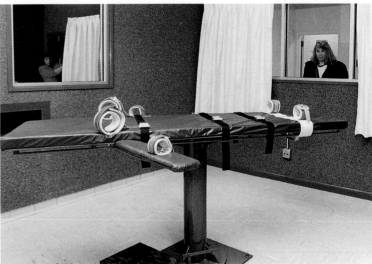

In earlier periods of history, executions were public spectacles and often considered a form of entertainment. In modern societies that still practice execution, there is much less emphasis on public spectacle and much more on the technology employed to perform the execution.

technology, such as the use of lethal injections versus the electric chair.

Does capital punishment have the effect of deterring people from committing murder, as its advocates claim? Much of the evidence on this subject is negative. Little encouragement is found in comparisons among states that have the death penalty, states that have had it with interruptions, and states that have never had it. For example, despite its highly visible and frequent use of capital punishment, the murder rate in Texas is almost 6 per 100,000 people, whereas the rate in Massachusetts, which has not had capital punishment, is 3 per 100,000 people (*Statistical Abstract,* 2010). Yet in interviews with criminals charged with robbery, James Q. Wilson (1977) found evidence that fear of the death penalty discouraged them from carrying guns. There is still considerable opposition to the death penalty among people who feel that it represents cruel and unusual punishment, but surveys indicate that the tide of public opinion has turned toward support for it. Surveys also show, for better or worse, that the public's increased concern about crime has led to greater emphasis on use of the death sentence as retribution for murders that have already been committed and less emphasis on the possible deterrent effect of capital punishment (Koretz, 2003).

Today, the value of capital punishment is a matter of heated debate throughout the world. In 2009, at least 714 prisoners were executed in eighteen countries, four of which—Iran, Iraq, Saudi Arabia, and the United States—accounted for 88 percent of the total. (China,

which refuses to report its numbers, is believed to have executed thousands.) All the European nations and most of the former states of Russia have abolished capital punishment. Many countries decided to suspend or abolish capital punishment as a way of turning over a new leaf after long periods of political repression. This is true of South Africa and several Latin American nations (Argentina, Brazil, Peru, Nicaragua, and El Salvador).

Plea Bargaining: A Revolving Door? Capital punishment addresses a small but sensational part of the crime problem. Many more criminals will commit other types of crimes, and for them, too, the issue of punishment and rehabilitation is controversial. At present, judges and police officers feel hampered by the relatively limited array of treatments for criminal offenders, as well as the backlogs in the courts and the overcrowding of prisons. One means of reducing the pressure on the judicial system is **plea bargaining**, in which a person who is charged with a crime agrees to plead guilty to a lesser charge and thereby free the courts from having to conduct a lengthy and costly jury trial. The shorter sentences that result from this system somewhat diminish the size of prison populations. Plea bargaining has been criticized as "revolving-door justice"; however, criminologists estimate

plea bargaining: A process in which a person charged with a crime agrees to plead guilty to a lesser charge.

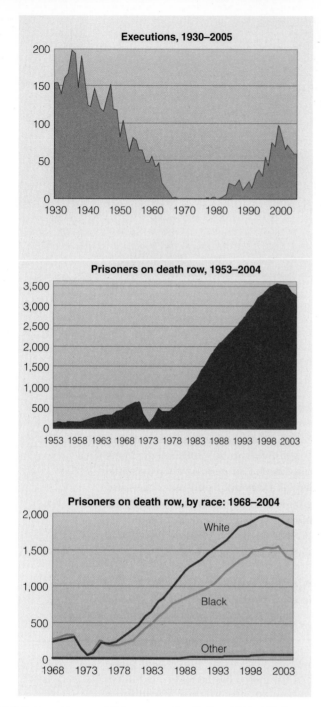

Figure 8.7 Executions and Prisoners on Death Row, United States.
Source: Bureau of Justice Statistics.

the United States has the highest rate of incarceration in the world, with Russia trailing by a substantial margin.

The largest proportion of new prisoners were arrested for the sale or possession of felonious quantities of narcotics and were imprisoned under mandatory sentencing laws that were passed as part of the national war on drugs. The arrests were by no means equally distributed among drug users and dealers, however. According to researchers with the Sentencing Project, for young Americans from ghetto communities and many other impoverished areas, "going to prison has become as inevitable . . . as going to college is for middle-class kids" (quoted in Butterfield, 1992, p. E4). Although there is widespread agreement on the need to punish drug dealers and users, there is growing debate over the desirability of sending them to prison, because it is highly likely that a young person who is imprisoned will learn from fellow inmates more diverse and devious ways of breaking the law. Indeed, critics often refer to prisons as "schools for crime." As early as 1789, the English philosopher and economist Jeremy Bentham wrote that "prisons, with the exception of a small number, include every imaginable means of infecting both body and mind. . . . An ordinary prison is a school in which wickedness is taught by surer means than can ever be employed for the inculcation of virtue" (pp. 351–352). In the intervening years, many studies have attempted to show under what conditions this is true and what can be done to increase the deterrent effects of prison and prevent it from becoming a community that socializes criminals.

Felony Disenfranchisement Another serious concern about the way inequalities in imprisonment affect specific groups in the society is the denial of voting rights to some ex-felons, a situation known as *felony disenfranchisement*. Most states do not allow imprisoned felons to vote, but fourteen states bar ex-offenders who have fully served their sentences from voting throughout their lifetime.

The scale of felony disenfranchisement in the United States is greater than anywhere else in the world. Approximately 5.3 million U.S. citizens are disenfranchised, including more than 1 million who have fully completed their sentences (Sentencing Project, 2010). In its research report on the issue, the Sentencing Project of the Human Rights Watch reported that

> [t]he racial impact of disenfranchisement laws is particularly egregious. Thirteen percent of African American men—1.4 million—are disenfranchised, representing just over one-third (36 percent) of the total

that an increase of only 20 percent in the number of offenders who are tried and imprisoned would place an intolerable burden on the correctional system (Reid, 1993; Vogel, 2008).

The Punishing Society Despite the small proportion of crimes resulting in arrest and imprisonment, prisons remain the social institution with the primary responsibility for dealing with criminals. From 1980 to 2003, the total number of Americans in prisons of all kinds almost quadrupled (see Figure 8.8). In addition to the prison population, more than 4.5 million people are on probation or parole. It is estimated that within a decade, more than 7.5 million people will be under some form of law enforcement surveillance in the United States. Figure 8.9 shows that

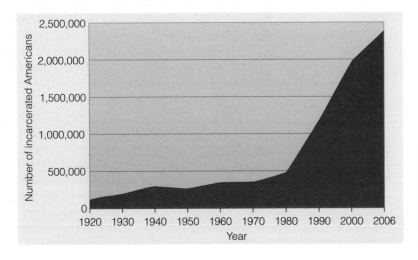

Figure 8.8 Incarcerated Americans 1920–2006. Sources: Justice Policy Institute Report: The Punishing Decade, & U.S. Bureau of Justice Statistics Bulletin NCJ 219416 – Prisoners in 2006.

disenfranchised population. In two states, our data show that almost one in three black men is disenfranchised. In eight states, one in four black men is disenfranchised. If current trends continue, the rate of disenfranchisement for black men could reach 40 percent in the states that disenfranchise ex-offenders.

Disenfranchisement laws in the United States are a vestige of medieval times when offenders were banished from the community and suffered "civil death." Brought from Europe to the colonies, they gained new political salience at the end of the nineteenth century when disgruntled whites in a number of southern states adopted them and other ostensibly race-neutral voting restrictions in an effort to exclude blacks from the vote (Losing the Vote: The Impact of Felony Disenfranchisement Laws in the United States, October 1998).

Although these laws are partly holdovers from the era of enforced racial segregation, particularly

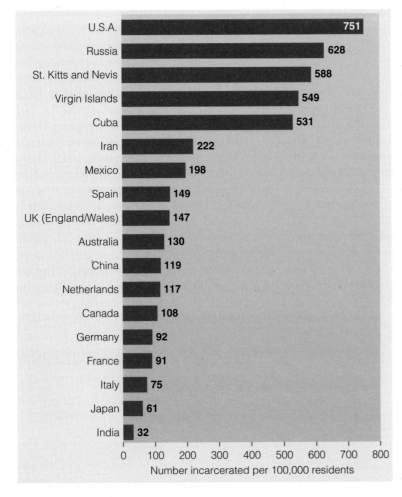

Figure 8.9 Incarceration Rates, Selected Countries (number incarcerated per 100,000 residents). Source: From *FACTS ON POLICY: Incarceration Rate* March 4, 2008. Copyright 2008 Hoover Institution Press. Reprinted by permission.

in southern states like Florida, it is unlikely that the laws will be changed soon, because they benefit many incumbent political leaders.

 Conservatives accuse liberals of being "soft on crime" and failing to consider the feelings of victims. Liberals often accuse conservatives of being mean-spirited and using stiff penalties, such as felony disenfranchisement, for political purposes. How can sociological study moderate these debates?

Functions of Prisons

The arguments used to support reliance on prison as the central institution of punishment and reform are mainly functionalist in nature. The functions of prisons are said to be *deterrence, rehabilitation,* and *retribution* (that is, punishment) (Goode, 1994; Hawkins, 1976). As Bruce Jackson, a well-known student of prison life, explains it, a prison is supposed to deter criminals from committing crimes ("its presence is supposed to keep those among us of weak moral strength from actions we might otherwise commit"); it is supposed to rehabilitate those who do commit crimes ("within its walls those who have, for whatever reason, transgressed society's norms are presumably shown the error of their ways and retooled so they can live outside in a more acceptable and satisfactory fashion"); and it is supposed to punish criminals—the only function that it clearly fulfills (1972, p. 248). (See the For Review chart below.)

Sociologists who study prisons note their similarity to other kinds of total institutions—social environments in which every aspect of the individual's life is subject to the control of authorities. In these institutions, inmates are deprived of their former statuses through haircuts, uniforms, and the like (Goffman, 1961). (See Chapter 5.) The goal of this process is to socialize the inmates to behave in ways that suit the organization's needs. In most such organizations, however, there is also a strong inmate subculture consisting of norms that specify ways of resisting the officials in favor of such values as mutual aid and loyalty among the inmates (Rubington & Weinberg, 1999). Thus, in Bruce Jackson's (1972) collection of case histories of criminals and prison inmates, a seasoned convict summed up the world of the total institution as follows: "A penitentiary is like a prisoner of war camp. The officials are the enemy and the inmates are captured. They're on one side and we're on the other" (p. 253). The same prisoner also described another aspect of life in total institutions. Often, as the following passage reveals, the inmates develop "institutional personalities." As they try to conform to the norms of the organization, they become dependent on the routines and constraints of minimal freedom and, hence, minimal responsibility:

> I like things to be orderly. You might say I'm a conservative, I like the status quo. If something's going smooth I like that. . . . Some people . . . kind of like it. Get institutionalized. You know, there's a lot of security in a place like this for a person. . . . They'll tell you when to get up, they'll tell you when to go to bed, when to eat, when to work. They'll do your laundry for you. You don't have to worry about anything else. (p. 253)

Only relatively recently, beginning mainly in the second half of the twentieth century, has the goal of rehabilitation—the effort to return criminals to society as law-abiding citizens—been taken seriously as a function of prisons. Critics of the prison system, many of whom argue from a conflict perspective, often claim that far from rehabilitating their inmates, prisons in fact function as "schools for crime" (Califano, 1998).

Yet much sociological research defends prisons as a means of deterrence and a necessary form of retribution (Allen, 1997). James Q. Wilson (1977) holds to the functionalist position: Societies need the firm moral authority they gain from stigmatizing and punishing crime. He believes that prisoners should receive better forms of rehabilitation in prison and be guaranteed their rights as citizens once they are outside again. But, he states, "to destigmatize

Justifications for Punishment

Justification	Function	Critical Issues
Deterrence	To prevent crime and protect society from criminal predation	Does prison, as a dominant form of punishment, deter crime or socialize criminals?
Rehabilitation	To resocialize criminals so that they can reenter society	What forms of rehabilitation actually work to prevent recidivism?
Retribution	To take revenge on the criminal and make punishment show society's anger about the crime	Do extreme punishments reduce all members of society to the level of the criminal?

visual sociology

The Tunnel People

During the first few decades of the twentieth century, homeless people, especially homeless men, were shunted into the "skid row" areas of cities in the United States. Skid row was, and in some cities still is, a downtown area where there were cheap hotels and cafeterias and religious missions, all of which served the needs of homeless men, day laborers, migrant workers, chronic alcoholics, and other destitute people who sought to survive in the city while maintaining socially disapproved lifestyles. But skid row areas have been no more immune to major social change than other parts of cities. Changes in the need for cheap labor in seaports and regional shipment centers have decreased the demand for day-rate workers. The population of hoboes has decreased as well, and so has the propensity for some men to go on drunken binges for weeks at a time, drifting into skid row to enjoy its cheap bars and anonymous hotels. Most recently, the rising value of downtown real estate has driven out the older skid row institutions. So where can men and women who are homeless, alcoholic, drug-addicted, mentally ill, or destitute live in the contemporary city?

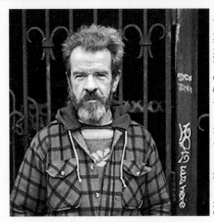

© Margaret Morton, author/photographer The Tunnel (Yale University Press, 1995)

"Coke and amphetamines started coming back to me because I had an unlimited supply [as a cook in a hospital]. . . . And I gave up everything, gave up everything that I had. And I moved into a place called the Amtrak Tunnel."

© Margaret Morton, author/photographer The Tunnel (Yale University Press, 1995)

"I got my little family [18 cats], and that's enough. I don't need any more. They are all different, and if you're feeling bad they make you feel better. They're not like people, they're not two-faced. They don't have one side and then have another the next time. So that's why I love my animals. And they know they can depend on me."

Very often, these people still live in the downtown area, under boxes, or on the periphery of the city, in vacant lots or alleys. In the largest cities, however, these people are often driven from such visible spots. Some find shelter underground, in railroad tunnels and abandoned water tunnels. Photographer Margaret Morton and sociologist Terry Williams have conducted extensive studies of the lives of homeless people underground. These photos are from Morton's photo essay about the tunnel people of Midtown Manhattan. They show how desperately people strive to create the semblance of a "normal" life under the most discouraging conditions.

© Margaret Morton, author/photographer The Tunnel (Yale University Press, 1995)

"I have always done most of the cooking. In most cases, people will heat up stuff down there. But cooking stews, soups—I'm an excellent cook. I love to cook. People say, "That was amazing!" I've never cooked professionally, I've never had a job as a cook. They thought I was in the military or something, which I wasn't."

crime would be to lift from it the weight of moral judgment and to make crime simply a particular occupation or avocation which society has chosen to reward less (or perhaps more) than other pursuits" (p. 230). Some states have taken the desire to get tough on crime to the extreme of reinstating chain gangs.

Whatever their sociological perspective, all students of the American prison system agree that by far the least successful aspect of prison life is rehabilitation. Research conducted in the 1970s consistently found that the only effective rehabilitation programs in prisons, measured by lower rates of recidivism, involve job training and education (Martinson, 1972). Indeed, during the 1980s, as prison populations more

than doubled, sociologists and criminologists found that men from very poor and minority backgrounds were beginning to regard prison as the only way they could obtain job training, health care, and other social services. Prisons, after all, provide more adequate housing than the large urban shelters for the homeless; inmates also receive three meals a day and, in some cases, a chance to improve their skills and self-confidence.

The conservative mood of the United States has had a profound impact on rehabilitation programs in federal and state prisons. Throughout the United States, legislative and administrative actions reduced educational opportunities for offenders, especially by limiting federal and state funding for prison postsecondary education programs (Koretz, 2003; Tewksbury & Taylor, 1996). Some legislators, showing far more interest in punishment than in rehabilitation, supported a proposal that would void all court-ordered efforts to correct inhumane conditions in prisons (Kunen, 1995). As one of the nation's leading researchers on crime and prisons has observed, in our society, "we have an exaggerated belief in the efficacy of imprisonment. The real problem is that we make life really terrible for some people and then blame them when they become dangerous" (Norval Morris, quoted in Butterfield, 1992).

SUMMARY

How do societies define deviant behavior?

- *Deviance*, broadly defined, is behavior that violates the norms of a particular society. The deviant label is attached to a person who violates or opposes a society's most valued norms.

- The ways a society encourages conformity to its norms and prevents deviance are known as *social control.*

- There is usually much disagreement not only about which behaviors are deviant but also about which ones should be condoned or punished only mildly.

What are some important dimensions of deviance?

- An important dimension of deviance is the power of some groups in society to define which acts are legal and which are illegal.

- Deviance should be distinguished from *stigma.* A stigmatized person has some attribute that is deeply discrediting, such as a disfiguring disease, but is not necessarily a social deviant.

- The study of deviance is concerned with social deviants—that is, people who voluntarily violate the norms of their society. In particular, the study of deviance focuses on criminal deviance: acts or omissions of acts for which the state can apply sanctions.

- As a culture's values and norms change, so do its notions of what kinds of behavior are deviant and how they should be sanctioned.

- The extent to which the members of a society agree on whether a particular behavior is deviant can range from weak (in cases in which there is much controversy) to strong (in cases in which there is little disagreement). Negative sanctions, or punishments, can also range from very weak to very strong.

- A deviant subculture includes a system of values, attitudes, behaviors, and lifestyles that are opposed to the dominant culture of the society in which it is found. Many deviant subcultures are harmful to society because they sustain criminal occupations. Others provide opportunities to engage in behavior that is pleasurable to many people but is considered deviant in "respectable" society. The boundaries between what is normal and what is deviant are not distinct.

What are the major biological and sociological explanations of social deviance?

- Biological explanations of deviance relate criminality to physical features or body type.

- These explanations influenced the earliest sociological theories of deviance, which viewed crime and other forms of social deviance as varieties of "social pathology" that could be attributed to the evils of city life. This view has been replaced by more objective and verifiable theories drawn from the basic perspectives of modern sociology.

- Functionalist theories of deviance include Robert Merton's typology based on how people adapt to the demands of their society. In this view, through socialization people learn what goals are approved of in their society and the approved means of achieving those goals. Individuals who do not accept the approved goals and/or the legitimate means of achieving them are likely to engage in deviant behaviors.

- Functionalist theories have been criticized for assuming that there is a single set of values that all the members of a society share. Conflict theorists stress the relationship between cultural diversity and deviance.

- The two main types of conflict theories are cultural conflict theories and Marxian theories. Cultural conflict theories concentrate on how conflicting sets of norms result in situations that encourage criminal activity. Marxian theories place more emphasis on class conflict, explaining various types of crime in terms of the social-class position of those who commit them.

- Interactionist theories of deviance focus on the issues of recruitment (the question of why some people become deviant whereas others do not) and production (the creation of new categories of deviance in a society).

- Edwin H. Sutherland's theory of *differential association* holds that whether people become deviant is determined by the extent of their association with criminal subcultures.

- Interactionists believe that deviance is produced by the process of *labeling,* in which the society's reaction to certain behaviors is to brand or label the offender as a deviant. Once acquired, such a label is likely to become incorporated into the person's self-image and to increase the likelihood that he or she will become committed to a "deviant career."

How is deviance prevented from threatening social stability?

- Studies of the incidence of crime often start with ecological data on crime rates.

- Studies of criminal victimization, which ask people about their actual experiences with crime, have made a major contribution to the understanding of crime in complex societies.

- The methods used to control crime change as societies grow and become more complex. In larger, more diverse societies, the ability of local institutions to control all of the society's members is diminished. Such societies tend to develop standardized, more or less coercive institutions to deal with deviants. Among the most prominent institutions of social control in modern societies are courts and prisons.

- The primary functions of prisons are said to be deterrence, rehabilitation, and punishment. However, prisons do not seem to deter crime, and only recently has the goal of rehabilitation been taken seriously. Numerous studies have found that prisons are not successful in rehabilitating their inmates and in fact often serve as "schools for crime." The only rehabilitation programs that appear to be effective are those that give inmates job training and work experience.

The Kornblum Companion Website

www.cengagebrain.com

Supplement your review of this chapter by going to the companion website to take one of the tutorial quizzes, use the flash cards to master key terms, and check out the many other study aids you'll find there. You'll also find special features, such as GSS data and Web links that will put data and resources at your fingertips to help you with that special project or to do some research on your own.

Collective Behavior, Social Movements, and Mass Publics

FOCUS QUESTIONS

How does collective behavior become channeled into social movements?

How do collective behavior and social movements bring about social change?

Why do some people get involved in social movements while others refuse to do so?

What are mass publics, and how do they set the stage for public opinion?

When the Arizona State legislature passed a law intended to crack down on illegal immigration, it unleashed a storm of protests on both sides of the issue. The law, which both proponents and critics said was the broadest and strictest immigration measure in generations, would make failure to carry immigration documents a crime and give the police broad power to detain anyone suspected of being in the country illegally. Opponents argued that the law would lead to blatant racial profiling and discrimination against people of Mexican origin, among others. The law's vocal supporters answered that it would send a clear message that illegal aliens will be arrested and deported. Even before Arizona Governor Jan Brewer signed the bill, President Obama strongly criticized it. Speaking in the White House Rose Garden at a naturalization ceremony for twenty-four active-duty service members who were becoming citizens, the president called for a federal overhaul of immigration laws to avoid "irresponsibility by others." The Arizona law, he added, threatened "to undermine basic notions of fairness that we cherish as Americans, as well as the trust between police and our communities that is so crucial to keeping us safe" (Archibold, 2010).

Rather than face a new wave of state legislation on the explosive issue of immigration law reform, congressional leaders promised to move the debate to the national level by considering immigration reform as a top priority.

All this renewed activity, from protests in the streets to the arguments of opposing parties in Congress, was a reflection of how passionate people can become over the issue of immigration and its control. Throughout the United States, and in other nations as well, the Arizona law stimulated a wide variety of protests, as well as public demonstrations of support. In a church social hall in New York, at an all-night vigil for immigration reform more than 2,000 miles from Arizona, Marco Peralta, a construction worker and former amateur boxer, said that he would like to meet the Arizona governor. Peralta, like others at the church gathering, did not have violence in mind.

"I would like to pray with her, to change her heart," said Peralta, a fifty-six-year-old undocumented worker who added that he moved to New York from Peru a decade ago to send his son and daughter to college.

Rather than demonstrate in a large public space, as many others did across the nation, Peralta and his group stayed in the church for a marathon of songs, dances, stories, and prayers. A sign posted outside the Roman Catholic church, behind a statue of Christ on the cross, said, "People of all faiths are invited to participate" (Schweber, 2010).

THE MEANINGS OF DISORDER

What do public protests and demonstrations accomplish? What motivates people to speak out and march for a cause? Are such events a sign of disorder and lack of social solidarity? In this chapter, we look closely at various types of collective behavior, many of which may be far more vocal and even violent than the prayer gathering Marco Peralta attended. Other gatherings may be playful or joyful, as when a local team wins an important championship. Momentous episodes, like the riots in Los Angeles after the trial of the police officers accused of beating Rodney King, are the most severe form of collective behavior. The race riots that occurred during the civil rights movement in the United States, the huge rallies and demonstrations that ushered in the Nazi regime in pre–World War II Germany, the strikes and picketing that spurred the labor movement, the demonstrations that enlivened the women's movement in the 1970s, the demonstrations and riots that occurred during the global trade talks in Seattle in 2000, and most recently, the outburst of patriotic feeling and anger among masses of Americans after the events of September 11, 2001, the beginning of the war against the Taliban in Afghanistan, the rise of the tea party movement against "big government" and rising federal deficits, and the continuing demonstrations for and against immigration reform—these are all examples of episodes of collective behavior that have shaped history in our time.

But there are less extensive and less dangerous incidents of collective behavior in modern societies that are nonetheless important. Think of the panic that occurs when a local bank or savings and loan association is rumored to be short of funds, or the joyful anticipation of the crowds that form outside ticket offices when the home team makes it to the World Series. These events also have economic and social significance: Banks fail when panicked depositors suddenly withdraw their savings; rock promoters wait in gleeful anticipation of full houses as the lines of ticket buyers form. The huge slump in the tourism industry throughout the world since the terrorist attacks of 2001 is another example of how collective behavior can have profound effects on societies.

Many sectors of the American economy profit from the behavior of large numbers of people. The fast-food industry, for example, depends on the behavior of the millions of Americans who stop for burgers and fries as they stream home from beaches or concerts or movies. Every year, the toy industry produces, or tries to produce, gimmicks that capture the attention and spending of millions of eager consumers—Tamagotchi virtual pets, Polly Pockets, Furbys, bobble head dolls, Beanie Babies, and so much more. Every year, fortunes are made and lost on the accuracy of predictions about whether the public will embrace such notions. The computer industry also has profited from fads and crazes, and it has also seen what happens when crazes wane rapidly. During the economic boom of the 1990s, the hottest stocks were those of Internet start-up companies like iVillage.com and major computer equipment suppliers like Cisco Systems. But many of these companies were benefiting from a classic stock "bubble" in which people with little knowledge of the stock market were rushing to buy in the hope of instant profits. As all stock bubbles do, this one burst as the world economy began to slide into recession

and war in 2001. Millions of people who were still holding highly overvalued stocks lost their investments. Similarly, in 2008, as the bubble in housing prices threatened to burst, thousands of people who had been counting on selling their homes at high prices suddenly placed them on the market, thus increasing the supply of available real estate and decreasing the prices of private homes—yet another example of how collective behavior exerts an immense influence on the economy and other institutions of society.

The study of mass or collective behavior encompasses a wide range of phenomena and presents many problems of classification and explanation. Yet this subject deserves our attention because of its influence on social change. The student occupation of Beijing's Tiananmen Square; the strikes and protests in Poland, Czechoslovakia, Hungary, and Romania that toppled communist dictatorships; the antitax movements occurring throughout the United States—all attest to the immense power of people who are moved to engage in collective action for social change.

In our attempt to understand the nature and importance of these behaviors, we turn first to the concepts that define their similarities and differences. Then we turn to theories regarding why these episodes occur, how social movements mobilize, and with what consequences. Finally, we discuss the concept of mass publics, especially the leisure and consumption behavior of large populations.

The Nature of Collective Behavior

The term **collective behavior** refers to a continuum of unusual or nonroutine behaviors that are engaged in by large numbers of people. At one extreme of this continuum is the spontaneous behavior of people who react to situations they perceive as uncertain, threatening, or extremely attractive. The violence that occurred in race riots in Los Angeles and Cincinnati is one example of spontaneous collective behavior. Another is the sudden action of coal miners who decide that conditions in the mines are unsafe and walk off their jobs in a sudden and unplanned wildcat strike. Such behaviors are not governed by the routine norms that control behavior at the beach or on the job (Smelser, 1962).

At the other extreme of the continuum of collective behaviors are rallies, demonstrations, marches, protest meetings, festivals, and similar events. These involve large numbers of people in nonroutine behaviors, but they are organized by leaders and have specific goals. When workers in a union plan a strike, for example, their

collective behavior: Nonroutine behavior that is engaged in by large numbers of people responding to a common stimulus.

picketing and rallies are forms of organized collective behavior whose purpose is to demonstrate their solidarity and their determination to obtain their demands. When blacks marched in commemoration of those who died in the 1919 race riot in Chicago, the event was organized to build solidarity and publicize a new determination to resist racism. In such cases, the organization that plans the event and uses collective behavior to make its feelings or demands known is a social movement.

Social Movements A **social movement** is an intentional effort by a group in a society to create new institutions or reform existing ones. Such movements often grow out of more spontaneous episodes of collective behavior; once they are organized, they continue to plan collective events to promote their cause (Blumer, 1978; Genevie, 1978). Some movements, like the antiabortion movement, resist change; others, like the civil rights movement and the labor movement, have brought about far-reaching changes in social institutions. Still others, like the gay and lesbian movement, seek to gain the rights other groups have won. In this sense, they are movements for full inclusion in society by people who believe they are discriminated against. We will see in later chapters that many social movements in a multiethnic and culturally diverse society challenge discrimination and deviant labels (Tarrow, 1994, 2008).

The American labor movement provides many examples of how spontaneous episodes of collective behavior and the social movements associated with them can change the course of a society's development. During the nation's stormy transition from an agrarian to an industrial society, workers fought against the traditional right of employers to establish individual wage rates and to hire and fire workers as they pleased. They demanded instead the right to organize unions that could negotiate a collective wage rate (collective bargaining) for each category of workers. They also demanded better working conditions and benefits. Mass picketing, sit-down strikes, and pitched battles between workers and company agents or the police were everyday events in the 1890s.

Often the workers joined in walkouts, rioting, or other spontaneous collective actions when faced with mine disasters or intolerable working conditions. From these episodes of bitter conflict emerged the modern labor movement. In the ensuing decades, culminating in the period of New Deal legislation in the 1930s, the workers' demand for collective bargaining was institutionalized. By this we mean that the right to join unions and bargain collectively was incorporated into the nation's laws. Labor unions thereby became recognized organizations, and collective bargaining became the recognized means of settling labor disputes. The unions, in turn, had to discipline their workers by specifying in labor agreements the conditions under which collective behavior such as strikes and other actions could be used.

By the end of World War II, labor unions and collective bargaining had become legitimate social institutions. In recent decades, however, drastic declines in industrial employment in North America and Europe, combined with renewed attacks on labor unions by employers, have led to a decrease in the ranks of organized labor. But major social movements that affect the lives of millions of people often go through cycles of growth and decline followed by a rebirth of activity. It remains to be seen whether this will be true of the labor movement (Geoghegan, 1991, 2007).

Spontaneous Collective Behavior These are unplanned outbreaks of collective behavior from which social movements often arise, and they can take many forms. Such behaviors range from the demonstrations and riots that mark major revolutions to the fads, fashions, crazes, and rumors that sweep through modern societies with such rapidity that what is new and shocking one day will be a subject of nostalgia the next. Some of these types of collective behavior become associated with social movements, but many do not. Each of them, however, can result in extremely important changes in behavior among large numbers of people, with consequences that can dramatically alter thousands or even millions of lives.

Fads and Fashions Fads and fashions are closely related forms of spontaneous collective behavior. They tend to be somewhat short-lived, but they may also result in more permanent changes in behavior.

Fads are new forms of behavior that are pursued with enthusiasm by large numbers of people once they are said to "catch on," but then fade as people begin to see them as "not the newest thing." Adults often disparage fads—which has the effect of making fads quite popular among the young, who are often casting about for new ways of expressing themselves and asserting their unique identities. There are language fads (with music exerting a strong

social movement: Organized collective behavior aimed at changing or reforming social institutions or the social order itself.

then & now

When Studying Social Movements Is Controversial

The hippies of the 1960s were an excellent example of an expressive social movement. But the hippies did not themselves invent what became known as the hippie lifestyle. Like participants in all social movements, the hippies adopted behaviors of earlier movements. Among these was the German naturist movement of the early twentieth century, which believed in returning to nature, avoiding meat, practicing nudity, and getting away from "the urban rat race." One member of the movement, Bill Pester, is shown here at his palm log cabin in Palm Canyon, California, in 1917. With his "lebensreform" philosophy, nudism, and raw foods diet, he was one of the many German immigrants who "invented" what became the hippie lifestyle more than a half century later. He left Germany in 1906, at age nineteen, to avoid military service.

American history texts often use examples like this one to discuss the hippie movement of the 1960s. They also discuss the civil rights movement, the labor movement, the women's movement, and the environmental movement, all of which drew on older ideas and earlier efforts to bring about what the movements' participants viewed as progress and social justice. Current texts frequently present analyses of expressive cultural movements, like rock and roll or hip-hop, that have attracted the support of millions of young people throughout the world over the past four generations.

But what one thinks about social movements can be highly political. For example, conservatives in Texas and elsewhere often question whether the history and consequences of social movements are represented in an unbiased way. Some resent what they see as a bias in the way these social movements are taught in the schools. In 2010, the Texas State Board of Education, one of the nation's most important because it influences how school texts are written throughout the United States, voted by an 8–7 margin to modify the high school history curriculum (McKinley, 2010). A slight majority on the board voted to see the issue of separation of church and state questioned in the high school curriculum, greater emphasis on the Christian origins of the nation's founders, and emphasis on the violence of the Black Panthers along with material on the nonviolent philosophy of Martin Luther King, Jr., and voted against placing special emphasis on the historical contributions of Latinos. A minority member of the board, Mary Berlanga, accused the majority of "trying to rewrite history," and staged a demonstration of her own in which she made it plain how she viewed the curriculum changes.

The board also rejected the addition of hip-hop to a list of important cultural and social movements. In the commercial, media world of hip-hop, the Texas vote offered a minor public relations boost at best. Rap may be fading somewhat as an expressive movement, but breakout albums and sound business investments will sustain hip-hop kings for some time. Rap star Jay-Z, shown here, earned $34 million in 2010; his nearest competitor, 50 Cent, earned $32 million, not far ahead of Sean "Diddy" Combs, with $28 million. What do these earnings tell us about what we value? Banning the subject from the curriculum cuts off such discussion, at least in the schools. But it makes the subject even more likely to be talked about among young people themselves, which is what the people in the hip-hop movement would most appreciate.

Mary Helen Berlanga accused fellow members of the Board of Education of "rewriting history."

Jay-Z, born Shawn Carter, released his eleventh studio album in 2009.

Bill Pester, shown here in front of his squatter cabin in Palm Canyon, California, in 1917, was a German immigrant and an early founder of what became the hippie movement in the 1960s.

influence on the emergence of new slang); dress and adornment fads; fads in toys, recreation (scooters, yo-yos, hula hoops), food, and much more. Some fads may have important lasting effects, such as the unprecedented popularity of the Harry Potter books, which are influencing millions of young people around the world to become more interested in reading for pleasure. Fads tend to rise in popularity and then decline in predictable ways once they become so widespread in the population that it appears that "everyone is doing it" (Meyersohn & Katz, 2004).

Fashions are currently accepted styles of appearance and behavior that are thought of as temporary in that they will soon be replaced by new fashions. In small-scale, simpler societies people wear traditional garb and otherwise tend to behave in the same ways most of the time, so fashion is unknown to them. Fashion is one of the defining characteristics of modern societies, where people are often more concerned about the future than the past, where dress and other behaviors mark one's status relative to others, and where powerful commercial interests want to stimulate the creation of new desires at all times. Marketing and advertising are major industries devoted to manipulating fashion, but they are not always successful, a fact that makes the study of fads and fashion particularly important in modern economies (Berger, 2000).

Rumor and Gossip Rumor and gossip are common and closely related forms of communication in modern societies. *Rumor* is information from an anonymous source that is transmitted informally—that is, from person to person—at least initially, although it may later be "picked up" by formal channels of communication such as newspapers and television.

Gossip is information spread informally that usually centers on other people whom one knows or knows about (in the case of gossip about celebrities). Although people often sneer at gossip, especially of the malicious variety, English social scientists who study interpersonal communication have found that about 65 percent of conversations—whether the speakers are men or women, neighbors, coworkers, homemakers, or executives—can be categorized as gossip about others. The research shows that gossip is a way for people to develop relationships, establish trust and social status, and create a community based on inclusion and exclusion (Karras, 2003). In modern societies like the United States, gossip can also become a commodity that is collected by specialists (gossip columnists) and disseminated in specialized publications such as *People* magazine, which are part of the contemporary "celebrity industry."

In the post-9/11 world, governments are trying with renewed vigor to understand and control the spread of rumors that might cause panic or lead to needless endangerment of large numbers of people. Through its Department of Homeland Security, for example, the U.S. federal government tries to monitor rumors concerning terrorism and related subjects so as to be in a position to provide confirmation or refutation of the information they contain. By posting the latest information on public-access websites, and through its color-coded system for warning the public about the danger of possible terrorist activities, the Department of Homeland Security seeks to remain in control of public opinion and public awareness rather then let unverified rumors develop into panic and mass hysteria.

Panic and Mass Hysteria *Panic* is a form of collective behavior in which people respond to a perceived threat in uncoordinated and unpredictable ways, usually involving fear, flight, rage, or other emotions and irrational behaviors as well. Cooperative social relationships break down under panic, and behavior is uncoordinated, as illustrated by the surge of desperate people who streamed into the New Orleans Superdome during Hurricane Katrina. Panic swept through the gathered crowd as rumors of imminent rescue or impending collapse of the roof alternately elated or terrified the helpless refugees. Panics also occur among people who are not in contact with one another, as we see in the case of stock market panics, in which people begin selling their stocks to protect their own interests, but their mass behavior results in further panic selling and a possible market crash (Bodenhorn, 2002).

Mass hysteria is a form of widespread collective anxiety, often caused by an unfounded belief, that results in panic behaviors of various kinds. The Salem witch hunt discussed in Chapter 8 is a famous example of mass hysteria, and there were isolated outbreaks of mass hysteria and panic directed against Arab and Indian Americans (mistaken for Arabs) after the 2001 attacks on the Pentagon and the World Trade Center. In fact, a goal of terrorism is to create mass fear, anxiety, and hysteria that result in panic. The better prepared communities are to quickly address public fears, through effective forms of communication and early intervention by skilled law enforcement agents and educated citizens, the more difficult it is for terrorism to succeed in fomenting mass hysteria and panic (Coates, 2002).

A Typology of Spontaneous Collective Behaviors

Sociologists who study collective behavior try to discover the conditions under which different types of spontaneous collective behavior occur and why these events do or do not become linked to social movements. Typologies that classify these phenomena are an important first step toward understanding why collective behavior assumes particular forms and develops through recognizable stages.

Crowds and Masses Students of spontaneous forms of collective behavior often begin by examining the structure of such behavior—that is, by discovering whether the people who engage in a particular kind of behavior are in close proximity to one another or whether they are connected in a more indirect way. In this regard, it is useful to distinguish between crowds and masses. A **crowd** is a large number of people who are gathered together in close proximity to one another (for example, at a demonstration or a football game). A **mass** is more diffuse; it does not occur in a physical setting. A mass is a large number of people who are all oriented toward a set of shared symbols or social objects (Lofland, 1981); an example is the audience for a particular television program. Collective behavior can occur in crowds, in masses, or in both at once.

Motivating Emotions The actual behavior a crowd or mass generates depends largely on the emotions the people involved feel are appropriate

> **crowd:** A large number of people who are gathered together in close proximity to one another.
>
> **mass:** A large number of people who are all oriented toward a set of shared symbols or social objects.

to express in a particular situation. They may be motivated by many desires—for excitement, for a change, for material gain, and so on—but outwardly, most will express an emotion that the norms of the event suggest is appropriate. In the witch hunt described in Chapter 8, the people involved expressed fear and hysteria (although, as is true of all such events, some were "just going along with the crowd"). Hostility aroused by anger, desire for revenge, or enraged hatred is another common motivating emotion in episodes of collective behavior. Lynch mobs are one kind of hostile crowd. The crowd of angry strikers that violently opposes "scabs" trying to cross a picket line is another. At the mass level, the outbreak of animosity toward Islamic people immediately after the suicide attacks on the World Trade Center and the Pentagon in 2001 is an example of collective behavior provoked by hostility.

Joy is a third important emotion that motivates crowds and masses. The Mardi Gras celebrations in Latin America and New Orleans involve large, joyful crowds that create a "moral holiday" in which behaviors that would not be acceptable at other times are tolerated. At the mass level, the immense interest and joy that spread through the U.S. population during the 1998 baseball season, when Mark McGwire and Sammy Sosa were locked in a battle to break the single-season home run record, raised the sport to new levels of popularity and stimulated many examples of joyful collective behavior, especially in St. Louis and Chicago.

By cross-classifying the most significant emotions that motivate collective behavior (fear, hostility, and joy) with the structural dimensions of crowd and mass, sociologist John Lofland created a typology that includes a wide range of spontaneous collective behaviors. This typology is presented in Figure 9.1. Can you think of examples of your own for the various types? Where, for example, would you place the Boston Tea Party, the riots that often occur during spring break in Florida, or the crowds who came out to pay last respects to firefighters and police officers killed in the line of duty on September 11, 2001?

The Lofland typology is useful as a means of classifying some types of collective behavior, but remember that these categories often overlap: Mass behavior may turn into crowd behavior; spontaneous collective behavior may generate a social movement. For example, when the stock market crashed in 1929, at the beginning of the Great Depression, crowds of panic-stricken investors spilled onto the streets of the financial districts of New York, Chicago, and San Francisco. There, within sight of one another, people found their fears magnified. And those fears quickly spread through the mass of Americans. The panic started among people who had made investments on credit, but it spread to people who merely had savings in local banks. Crowds of terrified savers descended on the banks, which were unable to handle the sudden demand for withdrawals because they also had been investing in stocks that were suddenly worthless. As a result, millions of Americans lost all their savings.

In the aftermath of the stock market crash and the resulting mass panic, social movements arose to seek reforms like investment insurance that would protect investors against similar episodes in the future. During the depression that followed the crash, citizens' fears about their lack of retirement or pension savings led to the development of new institutions like Social Security and deposit insurance. Today, AARP, an organization that represents the interests of older people who once protested as an unorganized mass, is the largest single lobbying group in American politics.

Dimensions of Social Movements

There are as many types of social movements as there are varieties of spontaneous collective behavior. Just think of how many "causes" can motivate people to take organized action. We have a social-welfare movement,

an antitax movement, a civil rights movement, a women's movement, a gay rights movement, an environmental protection movement, a consumer movement, an animal rights movement, and on and on. Within each of these movements, numerous organizations attempt to speak for everyone who supports the movement's goals. Actually, any large social movement is likely to include several different social movement organizations (sometimes referred to as SMOs) that may have different, and at times competing, ideas about how to achieve the movement's goals (Rodgers, 2009). As we will see shortly, the way such organizations operate and the goals they seek depend largely on the characteristics of the social movement itself.

Classifying Social Movements Social movements can be classified into four categories based on the goals they seek to achieve, as follows:

1. *Revolutionary movements* seek to overthrow existing stratification systems and institutions and replace them with new ones (McAdam, McCarthy, & Zald, 1996). The Russian Bolsheviks who founded the Communist Party were a revolutionary social movement that sought to eliminate the class structure of Russian society. They also believed that the institutions of capitalism—the market and private property—must be replaced with democratic worker groups, or soviets, whose efforts would be directed by a central committee. Meanwhile, certain other existing institutions, especially religion and the family, had to be either eliminated or drastically altered.

2. *Reformist movements* seek partial changes in some institutions and values, usually on behalf of some segments of society rather than all. The labor movement is basically reformist. It seeks to alter the institutions of private property by requiring the owners of businesses to bargain collectively with workers concerning wages and working conditions and to reach an agreement that will apply to all the workers in the firm. Collective bargaining replaces the old system of individual contracts between workers and employers, but it does not eliminate the institution of private property.

3. *Conservative movements* seek to uphold the values and institutions of society and generally resist attempts to change them, unless their goal is to undo undesired changes that have already occurred. The conservative movement in the United States

	SETTING	
BASIC EMOTION	**Crowd**	**Mass**
Fear	• Panic exodus from burning theater • Hostages taken by terroist groups aboard an airliner	• Natural calamities • Red scares • Crime waves • Three Mile Island • Salem witch-hunts
Hostility	• Political rallies, marches • Lynch mobs • Race riots	• Scapegoating of public figures • Waves of cross burnings
Joy	• Revival meetings • Rallies in Nazi Nuremberg • Mardi Gras carnival • Rock concerts • Sports events	• Gold rushes • Punk fashion • Jogging • Disco • *Star Wars, Star Trek* • Pokémon

Figure 9.1 Lofland's Typology of Spontaneous Collective Behaviors. Source: Adapted from *Social Psychology: Sociological Perspectives*, edited by Morris Rosenberg and Ralph H. Turner. Copyright © 1981 by The American Sociological Association.

seeks to reinforce the values and functions of capitalist institutions and to support the traditional values of such institutions as the family and the church.

4. *Reactionary movements* seek to return to the institutions and values of the past and, therefore, to do away with some or all existing social institutions and cultural values (Cameron, 1966; McAdam, McCarthy, & Zald, 1996). The Ku Klux Klan is part of a reactionary social movement that seeks a return to the racial caste system that was supported by American legal institutions (laws, courts, and the police) until the 1954 Supreme Court decision that declared the "separate but equal" doctrine unconstitutional.

Other Types of Movements Although they cover some of the movements that have had the greatest impact on modern societies, the preceding four categories do not include all the possible types of movements. Herbert Blumer (1969a), for example, identified a fifth category: *expressive social movements,* or movements devoted to the expression of personal beliefs and feelings. Those beliefs and feelings may be religious or ethical or may involve an entire lifestyle, as in the case of the punk movement of the 1980s. An important aspect of expressive social movements is that their members typically reject the idea that their efforts should be directed at changing society or the behavior of people who do not belong to their movement. Their quest is for personal expression, and if others choose to "see the light," that is welcome. Expressive social movements often peter out when aspects of self-expression are

adopted in the larger society (for example, when people copy styles of fashion or language) and the original adherents must use new means of identification. For this reason, among others, the slang expressions that groups use are always changing as the original slang terms become part of everyday speech.

Some of the various New Age groups and organizations now popular in the United States fit well into the expressive social movement category. Non–Native Americans who appropriate some aspects of Native American spirituality (including the use of sweat lodges or the practice of vision quests), but do not proselytize, are examples of people involved in an expressive social movement. So are men who follow the teachings of poet Robert Bly, practicing collective drumming to bond together and express suppressed emotions. But, like the followers of any social movement, there can be those in expressive social movements whose extreme beliefs lead to tragedy, as in the 1996 case of the seven-year-old pilot whose parents encouraged her to follow her vision at all costs. Such advice resulted in her death, as well as the deaths of her father and her flight instructor.

Other social scientists who study social movements identify still more categories or develop subcategories that combine elements of those just presented. (See the For Review chart on page 216.) For example, in his classic study *The Pursuit of the Millennium* (1961), Norman Cohen analyzed what are known as *messianic* or *millenarian* movements, which are both revolutionary and expressive. A millenarian movement promises its followers total social change by miraculous means. These movements envision a perfect society of the future, "a new Paradise on earth, a world purged of suffering and sin, a Kingdom of the Saints" (p. xiii). Most millenarian movements are secular, but the term is derived from the Christian concept of the millennium, the return of Christ as the messiah who would save the world after a thousand years. Often millenarian movements begin as small movements with a charismatic leader (someone who seems to possess special powers) and a few devoted followers.

An example of a millenarian movement is the People's Temple, whose 900 members committed suicide by drinking cyanide-laced Kool-Aid in Jonestown, Guyana, on November 18, 1978. In his insightful study of the events leading up to the tragedy, sociologist John Hall (1987) demonstrates that popular interpretations of this shocking story are inadequate. The conclusion reached in most previous accounts is that the movement's leader, Jim Jones, was a deranged personality who acted as the Antichrist

Types of Social Movements

Type	Description	Example
Revolutionary	Seeks to overthrow and replace existing stratification systems and institutions	Bolsheviks
Reformist	Seeks partial changes in some institutions and values, usually on behalf of some segments of society rather than all	Labor movement, civil rights movement
Conservative	Seeks to uphold the values and institutions of society and resists attempts to change them	Antitax, antigovernment movements
Reactionary	Seeks to return to the institutions and values of the past and do away with existing institutions and values	Ku Klux Klan
Expressive	Devoted to the expression of personal beliefs and feelings	Punk movement
Millenarian	Combines elements of revolutionary and expressive movements	People's Temple

and led his misguided followers to their deaths. Such accounts, according to Hall, may be comforting expressions of moral outrage, but by exaggerating the influence of an individual personality, they lead to wrong ideas about how such tragedies can occur.

In his analysis of the Jonestown tragedy, Hall shows that the members of the People's Temple were part of a cohesive social movement. Many were of modest means, some quite poor; in addition, most were black. They felt alienated from the larger society and saw the Temple as their community and often as their family. It is only by understanding the strength of their attachments to one another and to their belief in Jones's vision of a better world to come that one can make sense of what appears to be a senseless, insane episode of collective behavior. Faced with what they perceived as imminent attacks on the religious community they were building in the jungle, some Temple members worked themselves into a frenzy in which they became convinced that they were "stepping over" into a better world by drinking the poison. Others may or may not have felt these emotions; we cannot ever know. Probably some people thought the occasion was a practice suicide. In any case, Hall's study shows that Jones planned the act of "collective martyrdom" but that most of the Temple members followed to the end because of their unshaken belief in the goals of their social movement.

Collective Behavior and Social Change

Collective behaviors and social movements can be placed along a continuum according to how much they change the societies in which they occur. At one end of the continuum are revolutions; at the other are fads or crazes. In the twentieth century, three major revolutions occurred: the communist revolutions in Russia and China and the fascist revolution that brought Hitler to power in 1933. These can be placed at the high end of the continuum. At the low end are fads and crazes that may excite us for a while but usually do not bring about lasting change in the major structures of society (Turner, 1974). Some people become so completely involved in a fad or craze (as has occurred, for example, among collectors of Beanie Babies) that they devote all their energies to the new activity.

Between the two extremes of revolutions and fads are social movements that capture the attention of masses of people and have varying effects on the societies in which they occur. The women's movement, the civil rights movement, and the environmental protection movement are examples of movements with broad membership in many nations and the continuing power to bring about social change. The gay liberation movement, the antiabortion movement, the evangelical movement, and the hospice movement (see Chapter 14) are examples of more narrowly focused movements that are concerned with a single issue or set of issues but also exert powerful pressure for change in the United States and other societies.

Other movements attract large numbers of people for a brief time but then begin to falter and have a limited impact. An example is the men's spiritual and social renewal movement, which found expression in the Million Man March by African Americans and the mass rallies held by the Promise Keepers. The latter was a social movement organization led by a former University of Colorado football coach. It called on Christian men to become "promise keepers, not promise breakers" and promoted racial reconciliation, family values, and male leadership within the family. After much initial success in attracting large crowds of men to stadiums throughout the United States, the group faltered as its leaders tried to find other ways of raising money besides charging high admission prices for its rallies. Unable to cover expenses, the organization was forced to fire at least two-thirds of its staff as its leaders struggled to attract donations (Sahagun & Stammer, 1998; Wheeler, 1998).

Theories of collective behavior and social movements often seek to explain the origins and effects of revolutionary and reform movements (Crosley, 2002; McAdam, McCarthy, & Zald, 1988). In the remainder of this section, we discuss several such theories. We begin by examining the nature of revolutions and revolutionary movements. Then we consider theories that attempt to explain why major revolutions and revolutionary social movements have arisen and what factors have enabled them to succeed. Note that these are macro-level theories: They

Tannen Maury/EPA/Landov

The tea party movement, which gained force during the 2010 congressional election campaigns, is a conservative social movement calling for less government and lower taxes, among other things. Many of its followers view the extension of access to health care through mandatory insurance as a governmental intrusion on individual liberty. As a conservative movement it is unusual, however, for its conscious adoption of a form of protest associated with the protest rallies of the major social movements of the twentieth century.

explain revolutions and the associated social movements as symptoms of even larger-scale social change.

SOCIAL MOVEMENT THEORY

Early theories of collective behavior were derived from the notion that hysteria or contagious feelings such as hatred or fear could spread through masses of people. One of the first social thinkers to develop a "mass contagion" theory of collective behavior was Gustave Le Bon. An aristocratic critic of the emerging industrial democracies, Le Bon believed that those societies were producing "an era of crowds" in which agitators and despots are heroes, "the populace is sovereign, and the tide of barbarism mounts" (1947/1896, pp. 14, 207). In its usual definition,

he pointed out, "the word 'crowd' means a gathering of individuals of whatever nationality, profession, or sex, and whatever be the chances that have brought them together." From a psychological viewpoint, however, a crowd can create conditions in which "the sentiments and ideas of all the persons in the gathering take one and the same direction, and their conscious personality vanishes" (quoted in Genevie, 1978, p. 9).

Le Bon attributed the strikes and riots that are common in rapidly urbanizing societies to the fact that people are cut off from traditional village social institutions and jammed into cities, creating a mass of strangers. This view of early industrial societies was clearly exaggerated; research has shown that urban newcomers seek to form new social attachments, often drawing on networks of people who came to the city from the same rural area or small town. But more important, research on crowds and riots has shown that despite the appearance of a "mob mentality" and psychological "contagion," people who participate in riots, demonstrations, and other forms of collective behavior actually have many different motivating emotions and may only *appear* to be following along blindly as if they were part of a herd. Also, the contagion theories of Le Bon and others are not useful in understanding the full range of collective behaviors and social movements that arise from the strains and conflicts characteristic of rapidly changing industrial societies (Rochon, 1998).

Theories of Revolution

Sociologists often distinguish between "long revolutions," or large-scale changes in the ecological relationships of humans to the earth and to one another (for example, the rise of capitalism and the Industrial Revolution), and revolutions that are primarily social or political in nature (Braudel, 1984; Wolf, 1984a, 1984b). The course of world history is shaped by long revolutions, whereas individual societies are transformed by social revolutions like those that occurred in the United States, France, and Russia (Bobrick, 1997).

Political and Social Revolutions Sociologist Theda Skocpol (1979) makes a further distinction between political and social revolutions. A **political revolution** is a transformation in the political structures and leadership of a society that is not accompanied by a full-scale rearrangement of the society's productive capacities, culture, and stratification system. A **social revolution**, on the other hand, sweeps away the old order. Social revolutions not only change the institutions of government but also bring about basic changes in social stratification. Both political and social revolutions are brought about by revolutionary social movements as well as by external forces like colonialism.

The American Revolution, according to Skocpol's theory, was a political revolution. The minutemen who fought against British troops at the Battle of Concord were a political revolutionary group that did not at first seek to change American society in radical ways. On the other hand, the revolutions that destroyed the existing social order in France during the entire nineteenth century, and then in Russia and China in the twentieth century, were social revolutions. Their goal was to transform the class structure and institutions of their societies.

political revolution: A set of changes in the political structures and leadership of a society.

social revolution: A complete transformation of the social order, including the institutions of government and the system of stratification.

Why do revolutionary social movements arise, and what makes some of them content mainly with seizing power whereas others call for a complete reorganization of society? For most of the twentieth century, the answer was some version of Marxian conflict theory. This theory uses detailed knowledge of existing societies to predict the shape of future ones. According to Marx, the world would become capitalist; then capitalist markets would come under the control of monopolies; impoverished workers and colonial peoples would rebel in a generation of mass social movements and revolutions; and a new, classless society would be created in which the workers would own the means of production. (Marx's theory of class conflict leading to revolution is discussed in Chapters 7 and 10.)

Relative Deprivation The idea that the increasing misery of the working class would lead workers to join revolutionary social movements was not Marx's only explanation of the causes of revolution. He also believed that under some conditions, "although the enjoyments of the workers have risen," their level of dissatisfaction could rise even faster because of the much greater increase in the "enjoyments of the capitalists, which are inaccessible to the worker" (Marx & Engels, 1955, vol. 1, p. 94). This is a version of the theory known as relative deprivation (Stouffer et al., 1949). According to this theory, the presence of deprivation (that is, poverty or misery) alone does not explain why people join revolutionary social movements. Instead, it is the feeling of deprivation relative to others. We tend to measure our own well-being against that of others, and even if we are doing fairly well, if they are doing better, we are likely to feel a sense of injustice and, sometimes, extreme anger. This feeling of deprivation relative to others may result in revolutionary social movements.

Alexis de Tocqueville came to the same conclusion in his study of the causes and results of the French Revolution. He was struck by the fact that the revolution did not occur in the seventeenth century, when the economic conditions of the French people were in severe decline. Instead, it occurred in the eighteenth century, a period of rapid economic growth. Further, Tocqueville concluded that "revolutions are not always brought about by a gradual decline from bad to worse. Nations that have endured patiently and almost unconsciously the most overwhelming oppression often burst into rebellion against the yoke the moment it begins to grow lighter" (1955/1856, p. 214). The Zapatista rebellion in the Mexican state of Chiapas is an example of a revolutionary movement that began among very poor Indian peasants but became a broader movement during a time of relative economic prosperity (Preston, 1998; Vilas, 1993).

Revolutions are most likely to occur when a period of improvement in economic and social conditions is followed by a sudden reversal. Figure 9.2 expresses this relationship in terms of expectations that rise steadily over time. When expectations are rising, people do not generally rebel, either through outbreaks of mob action or through organized social movements. But when their situation begins to improve and their expectations continue to rise as well, the extent to which their expectations are actually satisfied takes a sharp downturn. Then the conditions for rebellion are enhanced. At some point, the affected population can no longer tolerate the gap between what it expects and what it is getting, and rebellion breaks out.

Think of Figure 9.2 as applying to any form of collective protest, not just to revolutions. For example, the unrest and revolutionary sentiments in many Islamic nations, which had been developing for decades but were brought home most clearly after the terrorist attacks on the United States in 2001, lend themselves to analysis using the "J-curve" model of Figure 9.2. Growing populations of educated males in nations like Pakistan, Saudi Arabia, and Algeria create rising expectations. The failure of these nations' economies to continue to grow and to provide careers for more educated young people creates a gap between needs and their satisfaction. Those who feel the greatest resentment and are most likely to join radical movements are not the impoverished people from the countryside but those in cities who feel that the gap between their expectations and their prospects is intolerable. They may blame forces outside their nation, such as Israel, the United States, or Western culture in general, but the J-curve model helps explain who joins the radical movements and why.

Thinking Critically | How can the concept of relative deprivation help us understand why poor people in rich nations feel deprived and often join protest movements?

Clearly, the theory of relative deprivation can help us identify the conditions that lead to collective behavior. However, it cannot tell us when the gap between expectations and reality will become intolerable or what kinds of

relative deprivation: Deprivation as determined by comparison with others rather than by some objective measure.

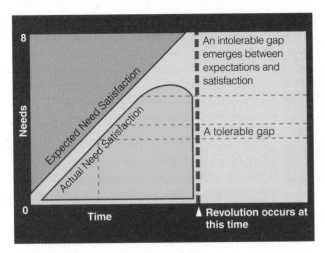

Figure 9.2 Need Satisfaction and Revolution. Source: Davies, 1962.

collective behavior will actually occur. To know those things, we must be far closer to the actual situation and be skilled at observing and interpreting it.

Valuable as they are as broad theories of revolution, neither Marxian theory nor Tocqueville's analysis can fully explain the rise of revolutionary movements throughout the contemporary world, nor do they help much in understanding the changes major social movements go through as their goals and organizations evolve. These and related questions are analyzed more directly in recent theories of protest cycles and changes within social movements, to which we now turn.

Protest Cycles, Collective Action Frames, and Charisma

In analyzing the revolutions and social movements that have occurred during the past 500 years, sociologist Charles Tilly concludes that there are no "neat formulations of standard, recurrent conditions for forcible transfers of state power" (1993, p. 237). However, he agrees with other social scientists who have pointed out that revolutionary social movements and other forms of protest often occur in cycles or waves of unrest. He cites the waves of protest that swept through Europe and North America in the late 1960s, and the current wave of fundamentalist movements in much of the contemporary world. During such waves, he observes:

> One set of demands seems to incite another, social movement organizations compete with each other for support, demands become more extreme for a while before subsiding. As this happens, activists often experiment with new ways of organizing, framing their demands, combating their enemies and holding on to what they have. (1993, p. 13)

Protest Cycles No theory adequately explains why particular waves of protest sweep through societies. But Sidney Tarrow (1994), another astute student of protest cycles, argues that waves of protest often result from major social shocks such as an international economic recession, the end of a period of warfare, or hasty actions by one or more governments. Such an episode occurred in Poland in the early 1980s, when the communist regime announced large increases in the price of bread. The movement that ensued, known as Solidarity, demonstrated the weakness of that regime (McAdam, Tarrow, & Tilly, 2001).

During cycles of protest, social conflicts of all kinds increase dramatically, often resulting in violent episodes of collective behavior—riots, demonstrations, and counterdemonstrations. The heightened sense that change is in the air, often brought on by the activities of a few "early risers" among protest leaders (for example, Lech Walesa and Mahatma Gandhi), triggers a variety of processes, including imitations of the protest elsewhere (for example, Martin Luther King, Jr., and Nelson Mandela, who followed the example set by Gandhi). The wave of protest may also build momentum through diffusion, often spreading from larger cities to smaller towns and rural areas. New ideologies arise to justify collective action and help the movement mobilize a following (Meyer & Tarrow, 1998).

Collective Action Frames The old saying that "the pen is mightier than the sword" refers to the power of ideas to shape our view of society and why it needs to be changed. In the twentieth century, we saw the revolutionary power of ideas embodied in *The Communist Manifesto* of Karl Marx and Friedrich Engels, Hitler's *Mein Kampf* ("My Struggle"), and the "Little Red Book" containing the sayings of Chairman Mao Zedong. In our own time, such books as Newt Gingrich's *To Renew America* and William Bennett's *Book of Virtues* have presented ideas that motivate various conservative movements in the United States. Millions of people latch on to these concepts, not because they are incapable of forming their own ideas, but because all of us are continually trying out ideas that seem to explain our times and help us make sense of our experiences in society. When faced with quandaries about what to do, what actions we should take or support, we often seek out concepts that are phrased as *collective action frames*, sets of beliefs and interpretations of events that inspire and justify social movements (Snow & Benford, 1992). There are many examples of collective action frames in the seminal work of William Gamson on this subject.

Gamson (1992) is one of the leading students of political activism in the United States. His research on social movements focuses on the mental frameworks people develop in their thinking about social issues. He notes that social movements strive to offer one or more collective action frames. These are ways of thinking that justify the need for and desirability of a particular course of action. After studying the writings and speeches of activists in many different social movements, Gamson concluded that collective action frames have three important components:

1. *An injustice component,* which includes a sense of moral indignation or outrage against a perceived injustice and the people who are said to be responsible for the condition.
2. *An agency component,* which embodies the idea that not only is there injustice, but together "we" can do something about the condition.
3. *An identity component,* which defines who "we" are, usually in opposition to a "they" with different tastes and values.

Antiabortion and pro-choice activists confront each other at a demonstration outside a clinic in Wichita, Kansas. At least in this example, the two are speaking rationally to each other, each from his or her own action frame based on his or her stance on the issue.

Gamson studies collective action frames by recording conversations in which people discuss social issues. In exploring the injustice component, for example, he listened carefully for statements like "That is just wrong," "That's unfair," "That really burns me up," or "That pisses me off." Such statements are marked by "explicit moral condemnation, unqualified by offsetting arguments and unchallenged by other group members," as in this example:

Characters:

Marjorie, a waitress, in her forties.
Judy, a data entry clerk, in her thirties.
Several others who don't speak in this scene.
(The group is near the end of a discussion of nuclear power.)

Marjorie: They shouldn't have taken all that money to nuclear power and everything—and you've got kids starving in America.

Judy: Yeah.

Marjorie: You've got homeless people. Where's your values? *[pause]* They suck. They really do suck.

Judy: You're on tape.

[laughter]

Marjorie: I don't care what I'm on. Still—it's obvious—when you pay millions and millions of dollars in nuclear plants when people in America are starving—

Judy: Right.

Marjorie: And you've got homeless people, no matter what they are—whether they're drunks or they're—whatever they are. Mentally ill people and you've got them living on a street.

And you've got a family of five people—I worked for Legal Services of Greater Boston and I had people in the Milner Hotel, mothers with five kids in one room. Living. And we don't have places for them, but we have places to build nuclear plants. That's garbage. That's garbage! (Gamson, 1992, pp. 48–49)

Sociologists listen carefully to the way people like Marjorie frame issues. In this case, the frame is injustice, and the point being made is that too much money is being spent on nuclear plants while other, more pressing human needs go unmet. In this dialogue, Marjorie is outspoken and not afraid of having her opinions recorded, whereas Judy is far more passive and concerned about voicing her opinions. The sociologist would probably conclude that, within her circle of friends and coworkers, Marjorie is a leader who often frames issues for others. As you listen to conversations like this one, observe how the leaders frame issues for the other participants. (Remember, however, that leaders do many other things in addition to creating or modifying collective action frames.)

Charisma Leadership plays an extremely important part in the success or failure of revolutionary social movements. The leaders of such movements are often said to have almost supernatural powers to inspire and motivate masses of followers. Max Weber (1968) called this ability **charisma**. Charismatic leaders appear to possess extraordinary "gifts of the body and spirit" that mark them as specially chosen to lead. The voices that Joan of Arc heard, which convinced her to take up arms to save France, gave her the inspired commitment to her cause that is a central feature of charisma. Mahatma Gandhi, probably the twentieth century's greatest example of a charismatic leader, swayed the Indian masses with his almost supernatural spirituality and courage. Similarly, the extraordinary gifts of oratory, faith, and energy displayed by Martin Luther King, Jr., allowed him to emerge as the most influential leader of the American civil rights movement.

A movement's followers also may attribute special powers to their leader. This process may produce additional myths about the leader's powers and add to whatever personal magnetism the leader originally possessed. For example, Hitler was no doubt a fiery orator, but his followers' frenzied reception of many of his ideas contributed to the perception that he possessed exceptional charisma.

Institutionalization of Charisma The charisma of the leader may eventually pose problems for the movement. Every social movement

charisma: A special quality or "gift" that motivates people to follow a particular leader.

Bettman/Corbis

Mahatma Gandhi, probably the most revered and charismatic political figure of the twentieth century, is shown here during a light moment with his granddaughters. Gandhi's techniques of nonviolence and passive resistance inspired Martin Luther King, Jr. and other leaders of political protest movements. But his followers often endured great personal sacrifice and physical danger, for his political strategies called for nonviolent civil disobedience in the name of higher moral values.

however, that revolutions are not made with slogans and visions alone (Schumpeter, 1950). The unification of the workers required an organization that not only would carry forward the revolution but also would take the lead in mobilizing the Russian masses to work for the ideals of socialism. The charismatic leader who built this organization was Vladimir Ilyich Lenin.

Under the leadership of Lenin (and far more ruthlessly under his successor Joseph Stalin), any Bolshevik who challenged party discipline was exiled or killed. Because the goals of the Bolsheviks were so radical and the need for reform so pressing, Lenin and Stalin argued that the party had to exert total control over its members in order to prevent spontaneous protests that might threaten the party's goals. By the same logic, other parties had to be eliminated and dissenters imprisoned. Such efforts by an elite to exert control over all forms of organizational life in a society are known as *totalitarianism*. All social movements that believe the ends they seek justify the means used to achieve them risk becoming totalitarian because they sacrifice the rule of law and the ideal of democratic process to gain power (Harrington, 1987; Howe, 1983).

REVOLUTIONARY SOCIAL MOVEMENTS Neither political nor social revolutions necessarily advance the level of a society's social justice. The political revolution in the United States required years of struggle, civil war, and continuing social movements to approach its ideals of social justice as expressed in the Declaration of Independence and the Constitution. Social revolutions, such as the French Revolution or the communist revolutions of Russia and China, produced immense changes in their societies but failed by themselves to address issues of inequality, especially gender and racial inequality. The history of most revolutionary social movements includes episodes in which opportunistic and power-hungry leaders gain control of the movement. The goals of the revolution, such as the French Revolution's ideals of liberty, equality, and fraternity, or the communist revolutions' ideal of a democratic socialist society, were continually being thwarted by political factions motivated primarily by the desire for power.

Revolutionary social movements appear to be especially prone to dominance by ruthless, power-driven leaders like Stalin, Robespierre, or Hitler who can quickly mobilize force and exert terror to quell protest. By suppressing the rights of individuals to freely express themselves, they also begin to dismantle the institutions necessary for the rule of law. At the same time, the fierce beliefs of the revolutionaries at the core of the movement frequently lead them to accept the leader's claims that the goals (ends) of the revolution justify whatever means are necessary to win. Time after time, this "the ends justify the means" reasoning has led down a trail of murder and further terror as those who are suspected of belonging to the wrong class or of not being loyal to the leader are eliminated. As mentioned earlier, this process occurred in Russia after 1917, making it almost impossible for democratic institutions of self-rule, including local courts, schools, legislative bodies, or trade unions, to exist independent of the communist state. In consequence, the ideal of social justice that gave rise to the revolution was subverted, and new inequalities and injustices replaced or coexisted with prerevolutionary ones until the collapse of the corrupt communist political system in 1989. The lesson from the previous century's revolutions must be that social movements that abandon their democratic principles and practices will never long succeed in advancing social justice goals.

must incorporate the goals and gifts of its leaders into the structure of the movement and eventually into the institutions of society. But how can this be done without losing track of the movement's original purpose and values? Scholars who follow Weber's lead in this area of research have termed this problem the *institutionalization of charisma* (Shils, 1970). Weber recognized that the more successful a movement becomes in taking power and assuming authority, the more difficult it is to retain the zeal and motivation of its charismatic founders. Once the leaders have power and can obtain special privileges for themselves, they tend to resist continued efforts to eliminate inequalities of wealth and power. Perhaps the most instructive example of this problem is the institutionalization of the ideals of Marx and Lenin in the organization of the Bolshevik Party in Russia after the revolution of 1917.

"Workers of the world, unite," Marx urged in *The Communist Manifesto;* "you have nothing to lose but your chains" (Marx & Engels, 1969/1848). The vision of a society in which poverty, social injustice, repression, war, and all the evils of capitalism would be eliminated formed the core of Marxian socialism. The founders of the Bolshevik Party realized,

Co-Optation Weber's theory predicts that social movements will institutionalize their ideals in a bureaucratic structure. Moreover, the more successful they are in gaining the power to change society, the more they

This famous photo from the Holocaust shows Jewish families being rounded up for transport to concentration camps. Under Hitler, the National Socialists (Nazis) often resorted to public acts of terror to establish their absolute rule over German society. Such totalitarian regimes often use terror to destroy mutual trust among citizens.

will be forced to control the activities of their members. In fact, once a movement is institutionalized, with an administration and bureaucratic rules, its leaders tend to influence new charismatic leaders to become part of their bureaucracy, a process known as *co-optation*. In the labor movement, for example, maverick leaders who demand reform and instigate protests in their plants are often given inducements such as jobs in the union administration to keep them under control (Geschwender, 1977; Kornhauser, 1952). In politics, local party leaders may identify and co-opt a local leader who they think might pose a threat in the future, offering inducements such as support in an election campaign in return for loyalty to the party. In this way, an outspoken environmentalist and history professor named Newt Gingrich was co-opted by the Republican leaders of Carrollton, Georgia, early in his political career.

SOCIAL MOVEMENTS AND CIVIL SOCIETY

Most of us will probably never belong to a revolutionary social movement, at least not one devoted to the violent overthrow of a government or society. Nor will many of us experience the deadening force of totalitarian rule. Most of us therefore will not find ourselves helping to rebuild the voluntary associations and communities, the leagues, the hobby groups, the unions and auxiliaries of our societies, as the people of many Eastern European nations are doing today. But many readers of this book will become members of organizations and groups of all kinds, which in turn may become affiliated with broader social movements.

In contemporary societies, an ever-growing and changing array of groups and organizations are formed in the public sphere of social activity. Known as **civil society**, this sphere of nongovernmental, nonbusiness social activity is composed of millions of church congregations, sports

civil society: The sphere of nongovernmental, nonbusiness social activity carried out by voluntary associations, congregations, and the like.

leagues, amateur arts groups, charitable associations, ethnic associations, and much more. Mexican sociologist Carlos M. Vilas explains it as follows:

> Civil society refers to a sphere of collective actions distinct from both the market and "political society"—parties, legislatures, courts, state agencies. Civil society is not independent of politics, but clearly, when people identify themselves as "civil society," they are seeking to carve out a relatively autonomous sphere for organization and action. (1993, p. 38)

In parts of the world where the gap between rich and poor is widening, people who participate in civil society are often recruited into movements calling for social and economic justice. Vilas observes, for example, that in Latin America, activity in social movements "creates a confluence of the poor and middle classes . . . to confront the traditional alliance of the rich and powerful with the state" (1993, p. 41).

Even people who never join social movements will surely find that their lives are changed by them. The social critic H. L. Mencken once said that if three Americans are in a room together for more than an instant, two of them will try to change the morals of the third. Campaigns against smoking, drinking, rock lyrics, violence on television, wearing of furs, abortion, and many other behaviors would seem to confirm this notion. So too would patterns in the growth of different types of voluntary associations. Since 1980, the number of nonprofit voluntary associations devoted to educational and cultural issues has increased by about 33 percent, and the number devoted to public affairs has risen by 74 percent (*Statistical Abstract*, 2005).

Resource Mobilization and Free Riders

What is the relationship between civil society and social movements? In sociology, this is known as the *resource mobilization* question, referring to the fact that social movements need to mobilize existing leaders and organizations rather than simply relying on the participation of people who happen to be moved to action (Klandermans, 1997; Zurcher & Snow, 1981). A second, related issue is known as the *free rider problem.* This refers to the tendency of many people not to lend their support and resources—time, money, and leadership—to social movements but to reap the benefits anyway (Marwell & Ames, 1985; Olson, 1965).

In any democracy, the involvement of young people in the political process is essential

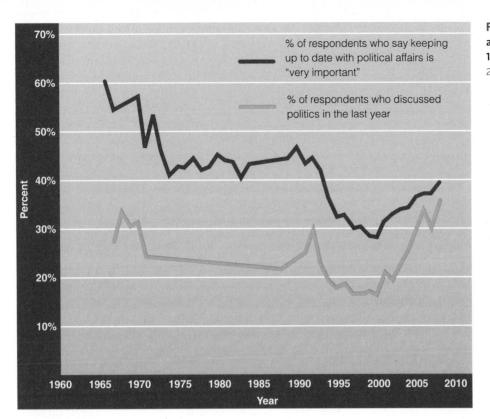

Figure 9.3 Interest in Politics among American College Freshmen, 1966–2008. Source: Sander & Putnam, 2010.

Legend:
% of respondents who say keeping up to date with political affairs is "very important"

% of respondents who discussed politics in the last year

yet always somewhat problematic because the young are not necessarily integrated into the political system as voters, and often do not see why they should be. This has been a serious issue for the United States, but as political sociologist Robert Putnam (Putnam, 2000; Sander & Putnam, 2010) has pointed out,

> Whether they were in college, high school, or even grade school when the twin towers and the Pentagon were hit, the members of the 9/11 generation were in their most impressionable years and as a result seem to grasp their civic and mutual responsibilities far more firmly than do their parents. While the upswing in volunteering [Putnam] observed in the mid-1990s may have been largely an effect of school-graduation requirements or the desire to gain an edge while seeking admission to selective colleges, the years since 9/11 have brought an unmistakable expansion of youth interest in politics and public affairs. For example, young collegians' interest in politics has rapidly increased in the last eight years, an increase all the more remarkable given its arrival on the heels of thirty years of steady decline. From 1967 to 2000, the share of college freshmen who said that they had "discussed politics" in the previous twelve months dropped from 27 to 16 percent; since 2001, it has more than doubled and is now at an all-time high of 36 percent. (See Figure 9-3.)

Putnam observes further that in his 2008 presidential election campaign,

> Barack Obama ably surfed this wave of post-9/11 youthful civic engagement. Though the initial ripple had been visible years before he became a national figure, he and his campaign mightily amplified it. Some credit Internet-based social networking for bolstering youthful interest in politics and community life, but the advent of the well-known social-networking sites Facebook (2004) and Twitter (2006) occurred years after the initial upturn in civic engagement by young people. Nonetheless, the Obama campaign adroitly deployed classic organizing techniques to expand the impact of such new technologies.

Are these issues important? One need only think of all the American workers who have benefited from the gains won by the labor movement. Or consider the rapid growth of the environmental protection movement in the late 1960s and early 1970s. What explains the speed with which that movement captured a central place in public policy debates? In studying the growth of the movement for environmental quality, sociologist Carol Kronus (1977) found that out of a sample of 209 existing organizations in a midwestern city, more than half gave material and moral support to the new movement. She also found that the groups that did link up with the movement were ones whose goals agreed with those of the movement and whose members believed in taking action to improve the quality of life in their community. Free rider groups, on the other hand, agreed that "something should be done about pollution" but felt that other groups should take action before they were able to do so.

The free rider hypothesis predicts that some proportion of the potential members of a social movement will not lend their energies to the movement but will nonetheless hope that the movement will provide them with the benefits they desire. But how extensive will the free rider problem be? How many people will act out of self-interest and try to reap the rewards of the movement's activities without making any personal sacrifices? A growing body of research findings addresses these empirical questions.

One study of the free rider problem is Edward J. Walsh and Rex H. Warland's (1983) study of the social movement that emerged after the near-meltdown of a nuclear reactor on Three Mile Island (TMI) near Middletown, Pennsylvania, in 1979. After the accident, antinuclear organizations grew in strength throughout the TMI area. Many of them opposed efforts to start up a second reactor at TMI, which had not been damaged. Walsh and Warland found that of those who opposed the start-up of the undamaged reactor, 87 percent were free riders in that they had not contributed in any way to the organizations opposing the start-up. Of those in favor of the start-up, 98 percent were free riders. Only a very small percentage— 6 percent of those in opposition and 1 percent of those in favor—actually participated. Although this proportion is small, it does show that there are people who are ready to be mobilized as activists on an important social issue. Not everyone is an apathetic or alienated free rider.

The TMI research suggests that most people will be free riders and will not join a social movement even when they strongly desire it to succeed and to win victories that will benefit them directly. Other research, however, indicates that the tendency to be a free rider is far less prevalent than the TMI study suggests (Garner & Tenuto, 1997). Data show that more than 85 percent of American households make charitable contributions of some kind. And public television and radio stations do succeed, after much persuasion, in raising the funds they need to continue operating, even though many people who enjoy listening or viewing would just as soon be free riders. Findings from thousands of laboratory experiments conducted since the early 1970s also contradict the free rider hypothesis, showing instead that many people make commitments and sacrifice individual gain in order to advance a cause. Still, the relatively low proportion of people who join a cause, vote in elections, or voluntarily support institutions like National Public Radio promises to make the free rider problem a subject of continuing social-scientific research (Frank, 1988; Hechter, 1987).

In predicting the success of any given social movement, resource mobilization theory points to the need to consider how well a movement enlists or mobilizes the resources available to it—for example, how well it deals with gender differences in recruiting members or how well it keeps recruits actively committed. Research by Doug McAdam (1992) on the experience of men and women in the civil rights movement during the early 1960s shows that organizations in the movement were constantly struggling with the problem of gender. White women who volunteered as civil rights workers often had more experience in organizations and civil rights activities than white male applicants but were less likely to be selected to participate. Those who were selected were often assigned jobs that were considered "women's work" unless they protested against such treatment. Indeed, McAdam found that many of the women who participated in the civil rights movement later became active in the women's movement, in part because of the discrimination they had encountered in the earlier movement. Studies of participation in other social movements during the 1960s have yielded similar findings: Women are available to be mobilized by social movements, but once they become involved they often find themselves faced with issues of gender inequality (McAdam, Tarrow, & Tilly, 2001).

Mobilization of the law is another aspect of resource mobilization theory that has been studied extensively in recent years (Hoffman, 1989). For example, in his research on how activists attempt to use laws and the courts to advance their cause, sociologist Robert Burstein (1991) shows that activists seeking to end discrimination against minorities in the labor force are increasingly petitioning the courts under the equal employment opportunity (EEO) laws. Burstein and other researchers have found that social movements are more likely to use the tactics of protest and demonstration when they are outside the framework of the law, but when laws have been passed in response to their demands, they become more likely to initiate lawsuits to achieve their goals. Burstein shows that the more a group has mobilized its resources as a protest movement, the more it will be able to mobilize the law in its favor as well. Thus, racial minorities and women's groups, two well-mobilized civil rights groups, do much better at winning enforcement of EEO laws than religious or ethnic minorities, which have less experience in organizing and mobilizing members and money for their causes.

MASS PUBLICS AND PUBLIC OPINION

Most of the examples of social movements discussed in this chapter have had a lasting impact on the societies in which they occurred. In this section, we apply the insights gained from earlier sections to some of the less world-shaking, yet still exciting, aspects of collective behavior in American life.

Over the past century, North America has been transformed from a continent of wide-open spaces and agrarian settlements to the earth's technologically most advanced and socially most mobile region. For the rest of the world, North America is synonymous with both the best and the worst aspects of what has come to be known as "modern" society.

According to Reinhard Bendix (1969), **modernization** consists of all the political and economic changes that accompanied industrialization. Among the indicators of modernity are urbanization, the shift from agricultural to industrial occupations, increasing literacy, and the demand for greater political participation by the masses. Bendix is quick to point out that his definition does not necessarily apply everywhere in the world. Modernization may not always mean what it has meant in Western societies. It is probable, however, that despite setbacks in some nations, modernization will always involve the extension of rights, values, and opportunities from the elites to the masses in a society.

None of these changes in the organization of societies came about without collective protests and social movements. For every radical movement, there was a reactionary one; for every revolution, there was a counterrevolution or at least an attempt at one. Social movements continue to exert pressure for change in our norms and institutions. The extension of voting rights to women and blacks had to be won through political struggle. Nor did the benefits of new technologies come automatically to American workers; they had to be won by the labor movement. Likewise, the rapidly expanding cities of North America became scenes of collective protest as new populations of immigrants and new occupational groups fought for "a piece of the pie" (Lieberson, 1980). It is hardly surprising, therefore, that so much sociological research is devoted to studying how modern society came about, discovering how much further some groups have to go to gain their share of the benefits of modernity, and analyzing the role of collective behavior in these major social changes.

Consider just one important consequence of the labor movement of the late nineteenth and early twentieth centuries: the change in the length of the workweek. "Our lives shall not be sweated, from day until night closes," went the famous refrain of the Industrial Workers of the World (known as the Wobblies). "Hearts starve as well as bodies; give us bread but give us roses." Figure 9.4 shows how the demand for enough leisure to enjoy the finer things of life produced dramatic changes in the hours of work. Today, we take the eight-hour day and the five-day week as the standard for full-time employment, but in 1919, American steelworkers shut down the entire industry in protest

against the twelve-hour day and the six-day week (Brody, 1960). The concessions they and other workers won were an essential part of the extension of the benefits of modernity to working people (McAdam & Snow, 1997). The increase in leisure created by these changes in hours and days of work, along with greater social mobility as a result of increased affluence, contributed to the development of mass publics.

Mass Publics

A **mass public** is a large population (regional or national) of potential spectators or participants who engage in collective behavior of all kinds. Typically this behavior consists of the formation of crowds, audiences, or streams of buyers and voters, but it can also include crazes, panics, and the spreading of rumors. Thus a massive traffic jam can turn into a dangerous panic; a joyful victory celebration can become a violent riot like the one that occurred after the Chicago Bulls won the National Basketball Association championship in 1992. The panic selling that occurred on Wall Street in 1987 and again in 1998, resulting in a massive loss of investors' capital, is a reminder that mass publics continue to exert important effects even when supposed safeguards are in place. For this reason alone, social scientists, urban planners, and governmental leaders increasingly recognize the need to cooperate in anticipating the behaviors of mass publics.

This recognition has come after some painful experiences. A famous case of failure to anticipate the possible reaction of mass publics occurred with the 1938 radio dramatization of H. G. Wells's novel *The War of the Worlds*. This broadcast, which vividly described an invasion of the earth by Martians, was presented to the radio audience in documentary fashion, beginning with a fictitious news flash and continuing with reports of a spacecraft landing in a New Jersey field and descriptions of the invading Martian army. Not realizing that the broadcast was only a dramatic presentation, hundreds of thousands of Americans began gathering in panicky crowds, while thousands of others jammed telephone lines in efforts to reach loved ones.

A few years later, when the dramatization was broadcast by Radio Quito using Ecuadorian place names, "the initial reaction was the same as in the United States. Multitudes poured out

---

modernization: A term used to describe the changes that societies and individuals experience as a result of industrialization, urbanization, and the development of nation-states.

mass public: A large population of potential spectators or participants who engage in collective behavior.

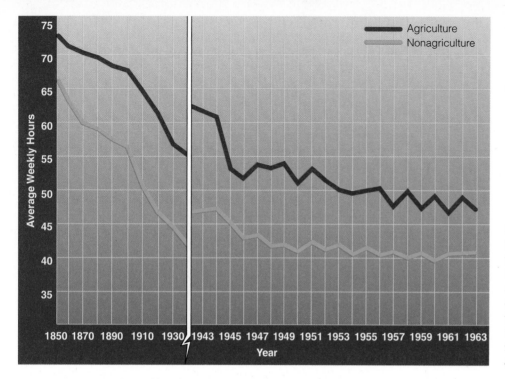

Figure 9.4 **Average Weekly Work Hours, 1850–1963.** There was a brief increase in average hours of work between 1940 and 1943 (not shown on the chart), due to the labor shortages that occurred during World War II. Source: Data from Bureau of Labor Statistics; adapted from Dankert, Mann, & Northrup, 1965.

of the city in all directions, running and driving as far and as fast as they could" (Klass, 1988, p. 48). When listeners found out that the program had been a dramatization, they were so angry at the radio station that they burned it to the ground. As a result of experiences like these, broadcasters became far more careful in presenting fictitious accounts on the radio (Cantril, 1982/1940).

The availability of more time for leisure pursuits, the development of a mass market for automobiles, and the technological revolution in communications and the mass media have all exerted an immense influence on the lifestyles of mass publics, who in turn shape the society in which they live. For example, through their demands for roads, leisure facilities, and services that cater to a highly mobile lifestyle, Americans have transformed the physical landscape. Rural scenes of farms and small towns still exist, but throughout the nation, they are being enveloped by networks of suburbs, shopping malls, and pleasure grounds (stadiums, amusement parks, and so on) linked together by a labyrinth of highways (Carter, 1975; Flink, 1975; Thomas, 1956).

Mass publics have also created the conditions that make whole new industries possible. The hot dog, for example, was an innovation that allowed large numbers of people to eat while strolling along the boardwalk at Coney Island. But fast food soon became an industry and even, as McDonald's founder Ray Kroc (1977) described it, an art form:

> Consider, for example, the hamburger bun. It requires a certain kind of mind to see beauty in a hamburger bun. Yet, is it any more unusual to find grace in the texture and softly curved silhouette of a bun than to reflect lovingly on the hackles of a favorite fishing fly? Or the arrangement of textures and colors in a butterfly wing? . . . Not if you regard the bun as an essential material in the art of serving a great many meals fast. (p. 99)

The thousands of teenagers and young adults who work in the fast-food industry may not sing rhapsodies to buns and burgers; nevertheless,

their industry owes its existence to the behavior of hungry, mobile multitudes—that is, mass publics.

 What physical changes in the United States and other modern nations during the twentieth century made mass publics and mass events like giant rallies or the Super Bowl more important than ever?

Public Opinion

The presence of mass publics sets the stage for the emergence of public opinion as a powerful force in modern societies. **Public opinion** refers to the values and attitudes of mass publics. The kinds of behavior that develop out of public opinion include fads, fashions, demands for particular goods and services, voting behavior, and much more.

Public opinion is shaped in part by collective behavior, especially social movements. An example is the rise of the conservative movement in the United States in the 1980s. The impact of this movement can be seen in the contrast between the results of the 1964 and 1984 presidential elections. In 1964, Barry Goldwater, who espoused many of the ideals of the current conservative movement, was overwhelmingly defeated by Lyndon Johnson, an outspoken proponent of liberal social-welfare policies. Twenty years later, the pendulum of public opinion had swung to the other extreme, and the staunchly conservative Ronald Reagan defeated a liberal candidate, Walter Mondale, by an even greater

public opinion: The values and attitudes held by mass publics.

The NAMES Project Quilt Many people have come to understand the personal meaning of the AIDS epidemic through an extraordinary social movement—the NAMES Project—and the famous quilt it has produced. This movement uses visual sociology in the most creative way one can imagine. As AIDS activist Elizabeth Taylor writes:

> The quilt is a moving depiction of stories of the loved ones of human beings who have died of AIDS. It reflects the true spirit of America. In their tragedy and grief over loss immeasurable, contributors to the quilt have used art and love to keep the spirit of their loved ones alive. (quoted in Ruskin, 1988, p. 7)

© 1987 Matt Herron/Take Stock

"There is nothing beautiful about AIDS. It is a hideous disease.... With the Quilt, we're able to touch people in a new way and open their hearts so that they no longer turn away from it but rather understand the value of all those lost lives."

The NAMES Project and the quilt are the inspiration of Cleve Jones (2002). When the actor Marvin Feldman, his dear friend of fourteen years, died of AIDS at age thirty-three, Jones went to his backyard with paint, stencils, and a sheet to create a memorial. He remembers spending the entire day thinking about Marvin: "I thought about why we were best friends and why I loved him so much. By the time I finished the piece, my grief had been replaced by a sense of resolution and completion." Jones realized that if people got together to make quilts as

© 1987 Matt Herron/Take Stock

memorials to lost loved ones, the activity could be a redeeming and mobilizing way of confronting the AIDS epidemic.

A Quaker with years of political activism in his background, Jones became a gay activist when he was eighteen. In 1986, he appeared on *60 Minutes* and discussed his reactions to the discovery that he was HIV positive. Shortly thereafter, in Sacramento, he was attacked by two men who stalked him, calling him "faggot," and stabbed him in the back. During his convalescence, he thought more about the idea of the NAMES Project and the quilt; after his recovery, he joined with AIDS activists throughout the world to advertise the project. Before long, hundreds of groups of friends and family members were producing quilt panels to commemorate the loss of their loved ones.

On October 11, 1987, the quilt was displayed on the Mall in Washington, D.C.; it covered an area larger than two football fields. It was subsequently exhibited throughout the United States, focusing attention on the need for increased funding for research and the allocation of additional medical resources to the AIDS epidemic. Today, the quilt is constantly on display around the world, especially on December 1 of each year, which has been designated World AIDS Day by the United Nations and the World Health Organization.

© 1987 Matt Herron/Take Stock

"The saddest thing, I believe, for Roger Lynn," says his devoted friend, Cindy McMullin, "was how unfair it was to have to die without knowing how it—AIDS—is going to end."

---

percentage of the electoral vote than Johnson had won in 1964.

Public opinion and interest in political issues is also shaped by more fleeting shifts in the national mood as citizens respond to experiences that are shared by all, especially through the mass media. For example, no one knows for sure whether the severe droughts that burned crops in the United States and Canada in 1987 and 1988 were a symptom of global warming, but scientists' warnings about this possibility, and their explanations of the "greenhouse effect" caused by high levels of carbon dioxide and other gases in the atmosphere, were made far more vivid by media coverage of the drought. Public support for environmental protection was stimulated still further by media coverage of the oil spill by the Exxon *Valdez* in Alaska in 1989.

Public Opinion, Collective Behavior, and Social Change In the contemporary world, major efforts are under way to sway public opinion on issues of immense social importance. At this writing, for example, the populations of European nations are angry about the demands being placed on their national budgets to help prevent the poorer nations of the European Union, Greece and Spain in particular, from defaulting on their outstanding loans. The European bailout is on a par with the one that occurred in the United States in 2008, at the end of the Bush administration, and like that one, it has generated a wave of negative public opinion about banks and government deficits. Moreover, slow recovery from a severe recession has meant that unemployment is relatively high and new job growth is slow in both Europe and the United States. All this translates into negative public opinion about Congress's ability to address these and other pressing issues. Because at this writing the Democrats are in power in both the Senate and the House of Representatives, this means that many incumbent Democrats fear that they will be turned out of office in the elections of November 2010. This is similar to the situation that the Republicans faced in 2006, when a wave of negative public opinion resulted in severe losses for them at the polls.

Immigration, its control, and the issue of naturalization for those who have entered the country without proper documents are also subject to wide fluctuations in public opinion. Immigration policy arouses passions that result in demonstrations on both sides of the issue, as we saw in the account at the beginning of the chapter. Volatile public issues like immigration are also prone to be exploited by individuals who seek to use the fears and passions of the electorate for their own purposes. Known as demagogues, they are people with access to the mass media, especially television and radio, and thus have the potential ability to sway public opinion. Demagogues tend to focus on feelings and fears rather than on fact and rationality in making arguments that attract people to demonstrations. This is not to say that public demonstrations and protest movements are always instigated by demagogues; far from it. But demagogues often use the opportunities that confusion, fears, and rapid fluctuations in public opinion bring about to enhance their own reputation and power. Commenting on the lack of civility in U.S public statements about political issues, President Obama spoke at numerous graduations and gatherings in 2010, calling for greater tolerance in a "poisonous political climate." He criticized both extremes of the political spectrum for using words such as *socialist, fascist,* and *Soviet-style takeover,* and said he regretted that such thinking has begun "to creep into the center of our discourse" (quoted in Slevin, 2010).

SUMMARY

How does collective behavior become channeled into social movements?

- The term *collective behavior* is used to refer to a continuum of unusual or nonroutine behaviors that are engaged in by large numbers of people. At one end of the continuum is the spontaneous behavior of people reacting to situations they perceive as uncertain, threatening, or extremely attractive; these behaviors include fads and fashions, rumor and gossip, and panic and mass hysteria. At the other end are events that involve large numbers of people in nonroutine behaviors but are organized by leaders and have specific goals.

- The set of organizations that plan such events is a *social movement.*

- The study of spontaneous forms of collective behavior often begins by distinguishing between crowds and masses. A *crowd* is a large number of people who are gathered together in close proximity to one another. A *mass* is a large number of people, not necessarily in close proximity, who are all oriented toward a set of shared symbols or social objects.

- Collective behavior can occur in crowds, in masses, or in both at once. The actual behavior that a crowd or mass generates depends largely on the emotions people involved in the events feel are appropriate to express in those situations.

- The most significant categories of emotions that motivate collective behavior are fear, hostility, and joy.

- Social movements have been classified into four types based on the goals they seek to achieve: (1) revolutionary movements aim to overthrow existing stratification systems and social institutions; (2) reformist movements seek partial changes in some institutions and values; (3) conservative movements attempt to uphold the existing values and institutions of society; and (4) reactionary movements seek to return to the institutions and values of the past.

- In addition, there are expressive social movements, or movements devoted to the expression of personal beliefs and feelings, and millenarian movements, which are both revolutionary and expressive.

- Within any large social movement, there are likely to be several different social movement organizations, or SMOs.

How do collective behavior and social movements bring about social change?

- Early theories of collective behavior were based on the notion that hysteria or contagious feelings like hatred or fear could spread through masses of people. Gustave Le Bon attributed the strikes and riots that are common in rapidly urbanizing societies to a mob mentality created by the presence of large numbers of strangers in crowded cities.

- Sociologists often distinguish between "long revolutions," or large-scale changes in the ecological relationships of humans to the earth and to one another, and revolutions that are primarily social or political.

- *Political revolutions* are transformations in the political structures and leadership of a society that are not accompanied

- by a full-scale rearrangement of the society's productive capacities, culture, and stratification system.

- *Social revolutions* not only change the institutions of government but also bring about basic changes in social stratification.

- According to Marx, revolutions would occur as a result of the spread of capitalism: Impoverished workers and colonial peoples would rebel against the capitalists and create a new classless society.

- Also, Marx and Tocqueville pointed to the role of *relative deprivation,* noting that the feeling of deprivation relative to others—not the presence of deprivation itself—may result in revolutionary social movements.

- More recent analyses have shown that revolutionary social movements often occur in cycles or waves. Waves of protest may result from major social shocks. During such periods, social conflicts increase dramatically, often motivated by collective action frames, or sets of beliefs and interpretations of events.

- Successful leaders of social movements are often said to have almost supernatural powers to inspire and motivate their followers. Max Weber called this ability *charisma.*

- Over time, however, those leaders' goals must be incorporated into the structure of the movement, a process that is referred to as the institutionalization of charisma.

- However, the more successful the movement, the more difficult it is to maintain the zeal of its founders. In extreme cases, the process can end in totalitarianism, or efforts by an elite to control all forms of organizational life in a society.

Why do some people get involved in social movements while others refuse to do so?

- The sphere of public nongovernmental, nonbusiness social activity is termed *civil society.*

- People who participate in civil society are often recruited into social movements. The resource mobilization question refers to the need to mobilize existing leaders and organizations to achieve a movement's goals.

- A related issue is the *free rider problem,* the tendency of many people not to lend their support and resources to social movements but to reap the benefits anyway.

- In recent years, activist groups have made increasing use of laws and the courts. Social movements are more likely to use the tactics of protest when they are outside the framework of the law; when laws have been passed in response to their demands, they become more likely to initiate lawsuits to achieve their goals.

What are mass publics, and how do they set the stage for public opinion?

- *Mass publics* are large populations of potential spectators or participants who engage in collective behavior.

- Such factors as increased leisure time, the almost universal use of automobiles, and the technological revolution in communications and the mass media have had an immense influence on the lifestyles of mass publics, which in turn shape the society in which they live.

- The presence of mass publics makes possible the emergence of *public opinion,* or the values and attitudes of mass publics.

- The behavior that develops out of public opinion can take a variety of forms, including fads, fashions, and demands for particular goods and services.

- Public opinion is shaped in part by collective behavior, especially social movements. It is also affected by experiences that are shared by all the members of society through the mass media.

The Kornblum Companion Website

www.cengagebrain.com

Supplement your review of this chapter by going to the companion website to take one of the tutorial quizzes, use the flash cards to master key terms, and check out the many other study aids you'll find there. You'll also find special features, such as GSS data and Web links that will put data and resources at your fingertips to help you with that special project or to do some research on your own.

10

The Meaning
of Stratification

Stratification
and the Means
of Existence

Stratification
and Culture

Power, Authority,
and Stratification

Stratification
in the Modern Era

Theories of
Stratification

Stratification
and Global Poverty

Mike Goldwater/Alamy

Stratification and Global Inequality

FOCUS QUESTIONS

How are a society's patterns of inequality structured and reproduced from one generation to the next?

How are stratification and inequality related to the way people in a society gain their livelihood?

How are a society's inequalities supported by the norms and values of its culture?

In what ways do power and authority work to maintain existing stratification and inequality?

How does stratification in the modern era account for global patterns of inequality?

How do theories of stratification explain global inequality and major aspects of inequality in individual societies?

What is the impact of globalization on stratification in rich and poor nations?

One out of every six people on earth lives in extreme poverty. That amounts to more than 1 billion people who live on less than $1 a day, the basic measure used to assess the extent of dire poverty, the condition in which parents and children routinely go to bed hungry. At the same time, the number of wealthy and middle-class people is rapidly increasing in the poor nations of the developing world, so that India, whose population is now reaching beyond 1.1 billion, now has a larger middle-class population than the entire population of the United States. In the developed nations, one-fifth of the world's population enjoys a far greater proportion of the world's total wealth. And yet, even in the United States and other affluent nations, poverty persists. The gap between the rich and the poor is continually widening, both in rich and in poorer nations. Severe global recessions help widen the gap and reveal the hardships faced by people at the middle and lower levels of the economic hierarchy. But it often requires a natural disaster, such as the earthquake in Haiti in 2010, or in China in 2009, or Hurricane Katrina in New Orleans in 2005, to reveal the depths of hardship faced by people at the bottom of the social and economic ladders in their societies.

The enormous earthquake that struck the capital of Haiti, Port-au-Prince, in 2010, brought the world's media spotlight on the poorest nation in the Western Hemisphere and once again raised more universal questions about possible solutions to poverty and the responsibility of the more affluent nations toward Haiti and nations like it. Disasters that focus world attention on issues of poverty and global inequality also lead to questions about what can be done. What are the most effective ways in which affluent nations can help relieve poverty and distress? Can aid be effective in societies with extremely corrupt and ineffective political leadership? What can we, as private citizens who may not be wealthy but who live in the richer nations, do to help? The United States currently spends $450 billion on its military, but only about $16 billion in official development assistance, yet wars are one of the primary causes of poverty, as we have seen in the cases of Iraq and Afghanistan. Meanwhile, in sub-Saharan Africa, more than 15 of every 100 children die before the age of 5. In western Kenya, fertilizer costs more than twice what it costs in France or the United States. Ethiopia is so deforested that rural households cannot use manure as fertilizer because they need it as cooking fuel. Because of HIV/AIDS, life expectancy in crisis countries like Botswana has dropped to below 40 years. For about $3 billion from the rich nations, 2 million malaria deaths could be averted; $25 billion a year would be enough to deliver life-saving health services to low-income countries. The United States has recently given over $700 billion to its major banks and financial institutions to prevent their collapse and possible world financial disaster, while rich financiers profited from the economic debacle of 2008.

Grim facts like these, and accompanying visual images of dire poverty and disaster throughout the world, can lead readers to despair and withdraw rather than to take action. A natural response to extreme poverty and suffering is denial, to try to shut it out from view, to become resigned and apathetic. Sociological understanding of global inequality offers a way out of such retreat and denial. First, an understanding of the causes and consequences of global poverty provides insight into how to fight hunger, disease, corruption, and apathy. Study of how people in their communities are finding ways to help themselves become more economically secure, for example, offers clues to how a government and its global partners can best support these efforts (Sen, 2009).

THE MEANING OF STRATIFICATION

Numerous social-scientific studies have demonstrated that all human societies produce some form of inequality (Bendix & Lipset, 1966; Harris, 1980; Murdock, 1949; Sen, 1992). In the simplest societies, this inequality may exist because one family's fields produce more than another's do, because one family has accumulated a greater herd than others have, or because one family has produced more brave warriors and thus has received more esteem from the other families in the tribe. In other societies, periods of scarcity caused by earthquakes, droughts, famines, or changes in the migration patterns of game animals may allow some clans or large families who were already relatively well off to become wealthier and more powerful. These advantages may then be passed on to the next generation and become part of the society's social structure (Sen, 2001, 2009). But as societies become increasingly complex, encompassing ever-larger populations and more elaborate divisions of labor, these simple forms of inequality are replaced by more clearly defined systems for distributing rewards among members of the society. And those systems result in the classification of families and other social groups into rather well-defined layers, or *strata*. In each society, the various strata are defined by how much wealth people have, the kinds of work they do, whom they marry, and many other aspects of life.

Colonialism, Capitalism, and Global Inequality

Patterns of inequality in most societies are complicated by histories of war and conquest. Civil wars, invasions by other societies or nations, ethnic and racial oppression of all kinds, enslavement of populations viewed as "inferior" or "backward," and rule by foreign or colonial powers are all ways in which some groups have been relegated to inferior positions in their societies while others have found their fortunes enhanced by historical events. American Indians, for example, especially those living on reservations, were conquered by people of European origin in the nineteenth century. The inequalities of income and wealth that they still experience are derived in important ways from that historical experience. The same is still true of African Americans whose families trace their origins to slavery. Similarly, there are families in the United States whose land and wealth originated in the plantation economy and its slave labor, while others, who lost their land in the same war, became mechanics and tradespeople. But this is not unusual. Throughout the world, the origins of rich and poor can often be traced to histories of conquest and exploitation. In contemporary China, it is common for people with power and

wealth to have had strong social ties to highly placed members of the Communist Party or the army, whose revolutionary cadres overturned the ancient forms of Chinese inequality and created new ones. In India, where rule by English colonial officials often replaced but sometimes worked through indigenous royal families, power and wealth often remains in the hands of Indian families who cooperated with their British rulers. In every colonial society, there were indigenous people and families that flourished and, through education in England, gained skills in the languages and business practices of their colonizers that contributed to their current wealth and power after their country won independence in the aftermath of World War II.

The Western colonial powers, holding vast regions of Africa, Asia, and Latin America, used their colonies to produce primary products—from plantations, mines, and forests in particular. Colonial administrators and their agents in businesses and the military disrupted and changed the smaller-scale tribal and village societies that they colonized. Through trade and commerce, they brought these societies into far greater contact with the modernizing world of the nineteenth and twentieth centuries. As we see in the contemporary world, however, the legacies of colonialism are enormous and continuing. They account in many ways for the unfavorable situation of the nations of Asia and Africa relative to the rest of the world (Hobsbawm, 1989). Those that still depend on the export of primary products are also those with high rates of rural poverty. The overall poverty of a nation cannot always be blamed on a history of conquest, colonial domination, and its place in the system of global capitalism, but these historic processes clearly continue to shape contemporary patterns of inequality, wealth, and power (Wallerstein, 2004).

In this chapter, we will see how inequality in societies results in systems of social stratification that are a major feature of every society's social structure. Stratification is nonrandom in that each infant is born into a family that already has a particular place in the society's system of inequality, although some societies provide more opportunities than others for that same infant to move upward in the hierarchies of wealth, power, and honor (prestige). We will also see that although issues of poverty are especially important in the study of stratification, many other values besides wealth and poverty are distributed unequally in societies.

In every society, regardless of its history, **social stratification** is a society's system for ranking people and distributing rewards according to such attributes as income, wealth, power,

---

social stratification: A society's system for ranking people hierarchically according to such attributes as wealth, power, and prestige.

prestige, age, sex, ethnicity, race, religion, and even celebrity. These ranking systems correspond to actual social structures, such as castes or classes, and they account for a great deal of what people experience in life—their pleasures and pains, their patterns of health and illness, their level of education, and much more. In this chapter, we will deal primarily with stratification by wealth, power, and prestige. The final section of the chapter is devoted to global stratification, with an emphasis on global poverty.

Caste, Class, and Social Mobility

In every society, people are grouped into different categories according to how they earn their living. This produces an imaginary set of horizontal social layers, or *strata,* that are more or less closed to entry by people from outside any given layer. A society that maintains rigid boundaries between social strata is said to have a **closed stratification system**; a society in which the boundaries are easily crossed is said to have an **open stratification system**.

Social Mobility In open societies, it is possible for some individuals and their families, and even entire communities, to move from one stratum to another; such movement is termed **social mobility**. A couple whose parents were unskilled workers may become educated, learn advanced job skills, and be able to afford a private house instead of renting a modest apartment as their parents did. Such a couple is said to experience **upward mobility**. If they have enough wealth to make their parents comfortable and to help other family members, the entire family may enjoy upward social mobility. If everyone with the same education and skills and the same occupation experiences greater prosperity and prestige, the entire occupational community is said to be upwardly mobile. But a family or community's fortunes can also decline. People with advanced skills—in engineering or higher education, for example—may find that there are too many of them around. They may not be able to afford the kind of housing, medical care, education for their children, and other benefits that they have come to expect. When this occurs, they are said to experience **downward mobility**.

Castes The best examples of closed societies are found in caste societies. **Castes** are social strata into which people are born and in which they remain for life. Membership in a caste is an **ascribed status** (a status acquired at birth) rather than an **achieved status** (one based on the efforts of the individual). Members of a particular caste cannot hope to leave that caste. Slaves

and plantation owners formed a caste society in the United States before the Civil War. The slaves were captives; runaway slaves were pursued and returned to their masters. Their children were born into slavery. Plantation owners, on the other hand, had amassed great wealth, especially in the form of land, and this wealth was passed on to their children. On occasion, a plantation family might lose its wealth, or another family might acquire a plantation and the wealth and prestige that went with it, but this form of social mobility was rather infrequent and did not alter the caste nature of the plantation system.

Today, much of modern India remains influenced by caste-based inequalities. South Africa, on the other hand, is an example of a society that is moving away from a rigid caste system. Under apartheid, blacks in South Africa were a racially defined caste that was kept at the bottommost rungs of the stratification order by violent repression and laws mandating racial segregation. Although this situation has been changing since the remarkable transition to majority African rule, the extreme poverty of the black population indicates that most aspects of the caste system are far from ended.

During the Persian Gulf War in 1991, many Americans discovered that Kuwait and other nations of the Middle East have strong ruling castes. In Kuwait, the ruling caste is composed mainly of members of a single large, immensely wealthy family—the al-Sabahs—many of whom live on the Italian and French rivieras for much of the year. The al-Sabahs rule Kuwait as a closed leadership caste, and after the Gulf War, they were careful to reimpose their rule and forestall any democratic movement that could challenge their authority.

In India, the caste system remains extremely strong, but it is changing. The lowest castes are defined by the types of work they are assigned, usually work that is thought of as particularly

---

closed stratification system: A stratification system in which there are rigid boundaries between social strata.

open stratification system: A stratification system in which the boundaries between social strata are easily crossed.

social mobility: Movement by an individual or group from one social stratum to another.

upward mobility: Movement by an individual or group to a higher social stratum.

downward mobility: Movement by an individual or group to a lower social stratum.

caste: A social stratum into which people are born and in which they remain for life.

ascribed status: A position or rank that is assigned to an individual at birth and cannot be changed.

achieved status: A position or rank earned through the efforts of the individual.

unclean or shameful, like preparing the dead for cremation or cleaning toilets. Formerly known as the untouchables, these castes are now termed "scheduled" castes to signify their rights to affirmative action under the Indian constitution. In the more egalitarian and open society India is trying to establish, the scheduled castes are to gain rights and opportunities previously denied to them, but in the old India that persists in thousands of rural villages and even in the cities, the old notions of untouchability remain strong (De Zwart, 2000; Deshpande, 2000).

Classes and Status Groups Classes, like castes, are social strata, but they are based primarily on economic criteria such as occupation, income, and wealth. England is famous for its social classes and for the extent to which social class defines how people are thought of and how they think of themselves. George Orwell, the author of *Animal Farm* and *1984,* also wrote about the English class system earlier in the twentieth century, when he lived among the homeless and the migrant workers at the very bottom of the class system, and among miners whose lives were sacrificed to the dangers and diseases of the coal mines. In the following passage, he described how class affected his own life:

> All my notions—notions of good or evil, of pleasant and unpleasant, of funny and serious, of ugly and beautiful—are essentially middle class notions; my taste in books and food and clothes, my sense of honour, my table manners, my turns of speech, my accent, even the characteristic movements of my body, are the products of a special kind of upbringing and a special niche halfway up the social hierarchy. (quoted in Campbell, 1984)

Classes are generally open to entry by newcomers, at least to some extent, and in modern societies, there tends to be a good deal of mobility between classes. Moreover, the classes of modern societies are not homogeneous—their members do not all share the same social rank. Within economically defined social classes, there are wide variations in social status—that is, in how much prestige individuals or groups are accorded in the wider society. Among very

rich people, for example, one can identify different **status groups** such as the Kennedys, organized crime families, and the families of film stars or professional athletes.

The concept of status groups is illustrated by "high society." The nobility of England is one of the world's most prestigious status groups, despite the well-publicized marital troubles of the royal family. In the United States, people with names like Rockefeller, Du Pont, Lowell, Roosevelt, Harriman, and many others, who are of western European Protestant descent, often have more prestige than people with just as much wealth who are of Italian, Jewish, or African American descent. In cities throughout the United States, these very rich and prestigious families, who form a status group defined by wealth and reputation, also tend to form groups that interact among themselves and play significant roles in their communities. The society pages of metropolitan newspapers devote most of their gossip to the philanthropic activities and private affairs of these families. Yet most of the old wealthy families in North America came from quite modest origins, as revealed by the underlying meanings of their names. Rockefeller, for example, means "a dweller in rye fields"; Du Pont, "one who lives near the bridge"; Harriman, "a manservant"; Roosevelt, "one living near a rose field." Moreover, people with prestigious family names typically enjoy fortunes gained from activities that were once considered too lowly to permit entry into high society. Thus, the Fords were looked down on by high society because their fortune was linked to "smelly gasoline"; the Whitneys, whose fortune a century ago spread over five states and 36,000 acres of palatial homes, were looked down on by older millionaire families because their funds were derived from 5-cent trolley fares rather than from railroad freight charges. In sum, both money and family prestige—gained by living expensively and engaging in public philanthropy—are required for entry into the highest levels of upper-class society (Domhoff, 1983; Hacker, 1997).

Life Chances

Rankings from high to low are only one aspect of social stratification. The way people live (often referred to as their *lifestyle*), the work they do, the quality of their food and housing, the education they can provide for their children, and the way they use their leisure time are all shaped by their place in the stratification system.

The way people are grouped with respect to access to scarce resources determines their **life chances**—that is, the opportunities they will

---

class: A social stratum defined primarily by economic criteria such as occupation, income, and wealth.

status group: A category of people within a social class, defined by how much honor or prestige they receive from society in general.

life chances: The opportunities an individual will have or be denied throughout life as a result of his or her social-class position.

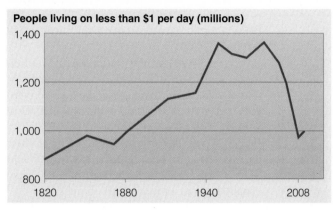

Figure 10.1 World Poverty, 1820–2008. Source: World Bank, 2003, 2008.

poverty as a condition in which people are forced to live on the equivalent of $1 per day or less. But there are billions more living in societies throughout the world who subsist at poverty levels just above extreme poverty. Figure 10.2 indicates the proportions of the world's population living at income levels from $1 a day to $10 a day. Each bar includes the entire global human population of 6.46 billion in 2008. Moving slightly above the level of extreme poverty, we see that almost half the total — over 3 billion people — live on less than $2.50 a day, and fully 80 percent live on $10 a day or less. At $2.00 or $2.50 per day, most members of impoverished families may find enough food to continue working, but not enough to afford to send their children to school beyond the primary grades, or to afford health care or adequate shelter.

have or be denied throughout life: the kind of education and health care they will receive, the occupations that will be open to them, how they will spend their retirement years—even where they will be buried. The place in a society's stratification system into which a person is born (be it a comfortable home, with access to good schools, doctors, and places to relax, or a home that suffers from the grinding stress of poverty) has an enormous impact on what he or she does and becomes throughout life. A poor child may overcome poverty and succeed, but the experience of struggling out of poverty will leave a permanent mark on his or her personality. And most people who are born poor will not attain affluence and leisure even in the most open society.

More than any other social condition, poverty has the greatest impact on people's life chances throughout the world. Figure 10.1 shows that, as measured by the number of people living on less than $1 a day, the level of poverty in the world rose rapidly during the twentieth century, reaching a peak after World War II. During the global economic boom of the 1990s, world poverty began to decline. With the recent global recession, however, there remain at least 1.2 billion people in the world who must subsist on less than $1 a day. Imagine trying to live on such a small amount! Even in the most impoverished regions of the world, where most people have relatively little, this is difficult. To subsist on so little means that one's life chances are greatly diminished, especially because of lack of medical care, schooling, and an adequate diet. Income is by no means the only indicator of life chances, but as people have more money at their disposal, their daily lives improve, as do their chances of mobility in their societies. People living in abject poverty, however, must spend most of their waking hours securing the bare necessities to remain alive.

As we have seen, the United Nations and other international agencies define extreme

STRATIFICATION AND THE MEANS OF EXISTENCE

The principal forces that produce stratification are related to the ways in which people earn their living. In the nonindustrial world, most people are small farmers or peasants. When they look up from their toil in the fields, they see members of higher social strata—the landlords, the moneylenders, the military chiefs, the religious leaders. These groups control the peasants' means of existence: the land and the resources needed to make it produce. Even when farmers or peasants are citizens of a modern nation-state with the right to vote and to receive education, health care, and other benefits, the landowners still take their full share when harvests are poor whereas the peasants must make do with less. The system of stratification that determines their fate depends on how much the land can yield. In contrast, in the stratification systems of industrial societies, whether capitalist or socialist, most people are urban wage workers whose fates are determined by the managers of public or private firms. If the firms are no longer productive or consumers no longer desire their products, urban workers may lose their jobs and suffer economic hardship.

The stratification systems of the United States and Canada, where less than one-twentieth of the population works the land, are most relevant to understanding people's life chances in modern industrial societies. To understand the conditions of life for most of the world's population, we must study inequalities in rural villages, where more than two-thirds of the earth's people till the soil and fish the rivers and oceans (see Figure 10.3).

Stratification in Rural Villages

Throughout the world, most impoverished people live in rural villages. Of course, not all rural agriculturists are impoverished, but in most of the developing world, their situation is extremely difficult. In India and China, more than 1.2 billion people spend their lives coaxing an existence from the soil. Millions of other rural villagers squeeze a modest livelihood from the land in Central and South America, Africa, and Southeast Asia. Social divisions in these villages are based largely on land ownership and agrarian labor. Yet even in rural villages, inequalities of wealth and power are increasingly affected by world markets for agricultural goods and services, as shown in the Global Social Change box on page 237.

In peasant societies, the farm family, which typically works a small plot of land, is the basic and most common productive group.

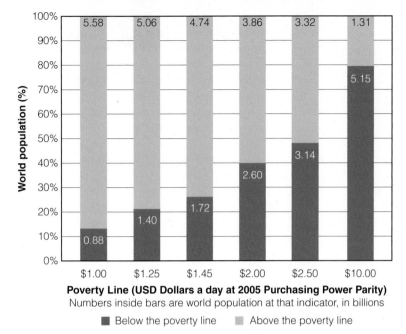

Figure 10.2 **Percent of people in the world at different poverty levels, 2005.** Source: World Bank Development Indicators 2008.

even as India is touted as a model for free-market economic development in the third world (Mishra, 2006).

In China before the communist revolution of the late 1940s and early 1950s, a feudal system of stratification organized the lives of peasants and gentry alike. Among the peasants, there were four broad strata (Tawney, 1966/1932). The rich peasants had enough land to meet their basic needs and to produce a surplus that could be converted into cash at local markets. They usually had one or more draft animals, and they often hired less fortunate villagers to help in their fields. The "middle peasants" were the second stratum of Chinese village life. They had a small plot of land, barely adequate shelter, and enough food and fuel to get through the winter. A small surplus in good years allowed them to own their own houses and even to have a few animals to provide meat on feast days. Only a very few of these "luxuries" were available to the third stratum, the poor peasants, and virtually none were to be had by the fourth and worst off, the tenants and hired laborers. Most Chinese peasants were in this impoverished stratum.

At the top of the stratification system of prerevolutionary China (that is, above the rich peasants) were the gentry, the class of landowners whose holdings were large enough to allow them to live in relative comfort and freedom from labor:

> What made the lives of the gentry so enviable to the working peasants was the security they enjoyed from hunger and cold. They at least had a roof over their heads. They had warm clothes to wear. They had some silk finery for feast days, wedding celebrations, and funerals. . . . The true landlords among them did no manual labor either in the field or in the home. Hired laborers or tenants tilled the fields. Servant girls and domestic slaves cooked the meals, sewed, washed, and swept up. (Hinton, 1966, p. 37)

Such families can be found in the villages of modern India. The Indian village reveals some important dimensions of stratification in third world societies. For example, women are assigned to hard work in the fields and are also expected to perform almost all of the household duties. This is true even in families that are well off. Men from the higher castes may be innovators, but women of all castes and male members of the "backward" castes do most of the productive work (Myrdal, 1965; Redfield, 1947).

India is often considered a "capitalist success story" because it has a rapidly growing and well-educated urban middle class of over 250 million people and its major entrepreneurs are investing heavily in the older industrialized nations. But as some grow richer and their lifestyles become highly visible on television and other media, the gap between the haves and the have-nots grows wider, creating the age-old problem of relative deprivation. India still has 380 million people who live on less than a dollar a day, and each year 2.5 million Indian children die, accounting for about one of every five infant deaths worldwide. For these reasons, the potential for political unrest, violence, and revolutionary activity is increasing

As our examples from contemporary India and prerevolutionary China show, the facts of daily life in rural villages are determined largely by one's place in the local system of agricultural production. The poor peasant family, with little or no land, hovers on the edge of economic disaster. Work is endless for adults and children alike. Meals are meager; shelter is skimpy; there is not much time for play. Among the gentry, who have large land holdings and hired help to ease the burdens of work, there are "the finer things of life": education, ample food and shelter, music and games to pass the time. Of course, wealth brings additional responsibilities. Participation in village or regional politics takes time away from pleasure, at least for some. So does

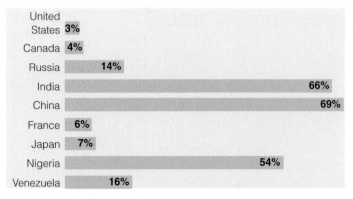

Figure 10.3 **Percentage of Workforce Engaged in Agriculture, Various Countries.** Source: Lye, 1995; World Bank, 2000.

Sweetness and Power Sugar, tobacco, tea, coffee, cocoa, pepper—these and other substances derived from plants that thrive in tropical climates were once unknown in the colder regions of the earth. Of them all, sugar has had perhaps the greatest influence. Sugar can be combined with cocoa to produce chocolate, fermented to produce rum, refined and sold in its pure form, or baked in candy and confections of all kinds. An exotic crop known mainly to European royalty in the fifteenth century, sugar had become an addiction of the masses by the eighteenth century. The craving for sugar and rum stimulated the growth of slavery in the New World and the transformation of entire Caribbean islands into sugar plantations (Mintz, 1985).

Most people do not realize that even today the production of sugar requires extremely dangerous and debilitating labor. This is especially true in Florida, where soft soils prohibit mechanized harvesting of the cane. Let a Jamaican sugar harvester who was brought to Florida as a contract worker continue the description:

> In America, we work the roughest way to make a living. Coming over, they ask if you're willing to work seven days a week, willing to wash your own clothes, willing to eat poor, willing to obey, work in all style of weather, eat rice seven days. . . .
>
> A lot of accident happen in the cane. When you working, your hand is sweaty, maybe the bill [cane knife] slip and fly up, maybe take your hand or your foot. Maybe take your partner. . . . Your hand become like part of a machine. Look at my palm, the bill rest here—it make a channel for its shape. When you go home, that one hand you have been cutting with you can't use. It's no good for anything else; you got to use the other one until it heal. . . .
>
> We got to watch dangerous things . . . snake, bobcat, ants climbing up your pants leg to sting you, the cane stalk strike you in the eye, pierce your eardrum. Sometime to make money we got to eat no lunch, we got no time for it, sick and still working. . . . You got to be wise and understand yourself and be quiet, otherwise they send you home. It is very disappointing, the way we are treated. We are slaves. They pay us money, but really they buy us. (quoted in Wilkinson, 1989, pp. 56–57)

Investigations of working conditions for contract labor in Florida and other cane-producing areas of the United States have spurred sporadic efforts to improve conditions on the plantations, but little has changed. In fact, the officials of the cane-producing companies make free use of arbitrary firings and deportations, and they prefer foreign workers who can be easily controlled and summarily dismissed and sent home. A section of a report published by the Florida Fruit and Vegetable Growers Association, "Why Foreign Workers Are More Productive," explains how raw power maintains the flow of sugar:

> The "unique and awesome form of management power" that sugar cane growers exert over their foreign workers provides a supermotivated workforce. As a Vice President of U.S. Sugar once said, "If I had a remedy comparable to breaching"—that is, firing and deporting—"an unsatisfactory worker which I could apply to the American worker, they'd work harder too." (quoted in Wilkinson, 1989, p. 64)

Catherine Karnow/CORBIS

charity work. However, the power that comes with such activities provides opportunities to amass still more wealth. Thus, with few exceptions, the poor remain poor while the rich and powerful usually become richer and more powerful. And between the rich and the poor, there are other strata—the middle peasants, the middle castes with skills to sell—whose members look longingly at the pleasures of the rich and console themselves with the fact that at least they are not as unfortunate as the humble poor in the strata below them.

How does this scheme of social stratification compare with that found in industrial societies? We will see that there is more mobility in industrial societies but that, just as in rural societies, to be born into the lower strata is to be disadvantaged compared with people who are born into higher strata.

Stratification in Industrial Societies

Structural Mobility The Industrial Revolution profoundly altered the stratification systems of rural societies. The mechanization of agriculture greatly decreased the number of people needed to work on the land, thereby largely eliminating the classes of peasants and farm laborers in some societies. This dimension of social change is often called **structural mobility**: An entire class is eliminated as a result of changes in the means of existence. As noted in Chapter 4, the Industrial Revolution transformed the United States from a nation in which almost 90 percent of the people worked in farming and related occupations into one in which less than 10 percent did so. Similar changes took place in England and most of the European nations and are now taking place in many other parts of the world.

But structural mobility did not end with the Industrial Revolution. Today automation, foreign competition, and technological advances are creating new patterns of structural mobility. Older smokestack industries like steel and rubber have been steadily losing factories and jobs; newer industries based on information and communication technologies have been creating plants and jobs, but the people who have become superfluous as a result of the closing of their plants are not always willing to move to new jobs and often are not trained to meet the demands of such jobs. Thus, structural mobility often leads to demands for social policies like job training programs and unemployment insurance (Bluestone, 1992).

structural mobility: Movement of an individual or group from one social stratum to another caused by the elimination of an entire class as a result of changes in the means of existence.

Spatial Mobility A second major change brought on by the Industrial Revolution was a tremendous increase in **spatial mobility** (or geographic mobility). This term refers to the movement of individuals, families, and larger groups from one location or community to another. The increase in spatial mobility resulted from the declining importance of the rural village and the increase in the importance of city-centered institutions such as markets, corporations, and governments. Increasingly, one's place of work became separate from one's place of residence; people's allegiance to local communities was weakened by their need to move, both within the city and to other parts of the nation; and social strata began to span entire nations as a result. Working-class people created similar communities everywhere, as did the middle classes and the rich (Janowitz, 1978; Seligson & Passe-Smith, 1993).

Despite these immense changes, our relationship to the means of existence is still the main factor determining our position in our society's stratification system. We continue to define ourselves to one another first and foremost in terms of how we make a living: "I am a professor; she is a doctor; he is a steelworker." Once we have dealt with the essentials of our existence—essentials that say a great deal about the nature and the quality of our daily lives—we go on to talk about the things we like to do with our lives after work or after educating ourselves for future work.

Global Stratification

Global stratification—the unequal distribution of wealth, well-being, consumable energy, and social justice—among the world's people is vividly illustrated in Figure 10.4. In this chart, virtually everyone in the United States falls within the richest 20 percent. (Refer back to Figure 10.2, which shows that over 80 percent of the world's population subsists on $10 a day ($3,650 a year) or less.)

In the urbanizing world of sprawling cities, the physical distances between wealth and poverty may be measured by a fence. Social stratification and inequality are made visual here in concrete and tar paper.

Anne-Marie Palmer/Alamy

spatial mobility: Movement of an individual or group from one location or community to another.

Compared to the rest of the world, the people of Europe and North America, Japan, and Australia are at the center not only of consumption but also of wealth and power. The stark differences in consumption of what we value, as shown in Figure 10.5, are mirrored in the statistics of every major world commodity, particularly energy and access to ample amounts of clean water.

Stratification systems, whether at the global level or at the level of particular societies and cultures, are always changing through revolutions, wars, natural disasters, and, with great effort and frequent failure, intentional actions such as the United Nations' Millennium Development Goals.

Sociology & Social Justice

MEETING THE WORLD MILLEN-NIUM GOALS In 2000, at the dawn of a new century and a new millennium, the United Nations approved a set of goals for reducing poverty and its related miseries. Known as the Millennium Development Goals, these ambitious goals have become accepted throughout the world as critical paths toward increasing global social justice. World leaders agreed that, without adequate living conditions, education, and health care, among other values, it is impossible for people to realize their human potential, or even to continue existing, a situation that is inherently unjust. In their most succinct form, the Millennium Development Goals are as follows:

Goal 1: Eradicate extreme poverty and hunger.
Goal 2: Achieve universal primary education.
Goal 3: Promote gender equality and empower women.
Goal 4: Reduce child mortality.
Goal 5: Improve maternal health.
Goal 6: Combat HIV/AIDS, malaria, and other diseases.
Goal 7: Ensure environmental sustainability.
Goal 8: Develop a global partnership for development.

Agencies of the United Nations, and relevant nongovernmental organizations (NGOs) that deal with global poverty and related issues, report that progress toward reaching the goals has been extremely uneven. There are nations that have achieved some or even many of the goals, while others are not on track to realize any. China, whose economy is now among the fastest growing in the world, has reduced its poor population from 452 million to 278 million. Similarly, India has almost halved its poverty rate. On the other hand, areas needing the most reduction in poverty rates, such as the sub-Saharan regions of Africa, have yet to make

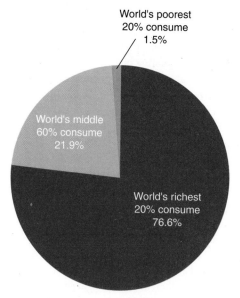

Figure 10.4 Share of world's private consumption, 2005. Source: World Bank Development Indicators 2008.

World's poorest 20% consume 1.5%

World's middle 60% consume 21.9%

World's richest 20% consume 76.6%

consciousness incorporated, for example, in every world religion. From the Koran: "All of us have been raised from dust and we shall be reduced to dust. There is no difference among the particles of dust; hence, why should there be any difference between somebody else and me?" From Christianity: "The joys and the hopes, the griefs and the anxieties of the men of this age, especially those who are poor or in any way afflicted, these are the joys and hopes, the griefs and anxieties of the followers of Christ."

The value of social justice underlying the Millennium Development Goals is also incorporated in more secular versions of the same ideal of social justice, such as the Declaration of the Rights of Man and Citizen of 1793, a primary document of the French Revolution, and the Declaration of Independence of 1776.

STRATIFICATION AND CULTURE

Why do people accept their "place" in a stratification system, especially when they are at or near the bottom? One answer is that they have no choice; they lack not only wealth and opportunities but also the power to change their situation. But lack of power does not prevent people from rebelling against inequality. Many people also believe that their inferior place in the system is justified by their own failures or by the accident of their birth. If people who have good cause to rebel do not do so and instead support the existing stratification system, those who do wish to rebel may feel that their efforts will be fruitless.

Another reason people accept their place in a stratification system is that the system itself is part of their culture. Through socialization, we learn the cultural norms that justify our society's system of stratification. The rich learn how to act like rich people; the poor learn how to survive. Women and men learn to accept the places assigned to them, and so do the young and the old. Yet despite the powerful influence of socialization, large numbers of people sometimes rebel against their cultural

any real progress toward reducing poverty or improving quality of life.

The ideals of social justice incorporated in the Millennium Development Goals are criticized by those who insist that meeting these goals is the responsibility of individual nations. Why should people unaffected by poverty have to sacrifice through higher taxes so that other nations might address their problems? These are legitimate arguments, but they are countered by a global moral sense or

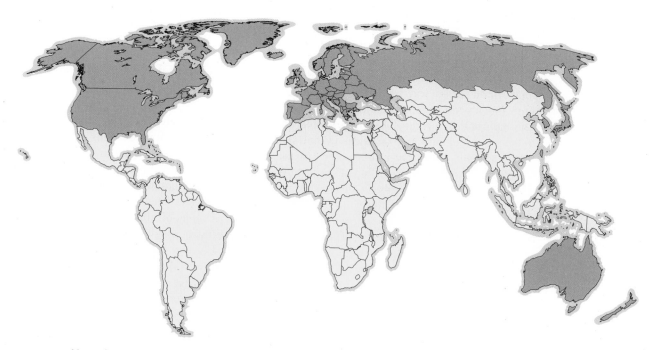

Note: Green includes G8 states and developed/first world states.

Figure 10.5 The north-south divide. Source: From http://en.wikipedia.org/wiki/North-South_divide

conditioning. To understand their reasons for doing so, we need to examine the cultural foundations of stratification systems.

The Role of Ideology

You will recall from Chapter 3 that an ideology is a system of ideas and norms that all the members of a society are expected to believe in and act on without question. Every society appears to have ideologies that justify stratification and are used to socialize new generations to believe that existing patterns of inequality are legitimate. In the United States, for example, most people embrace the ideology of the American dream, the idea that in America anyone who works hard can achieve success and wealth. At the same time, we know that the odds of achieving great wealth are very low. This is one reason people love to hear stories about poor or hard-working people of modest means whose lives are transformed by sudden lottery winnings. In the nineteenth and early twentieth centuries, the "rags-to-riches" stories of Horatio Alger were extremely popular. In Alger's first novel, *Ragged Dick,* the central character is a poor but honest boy who comes to the city looking for work. As he walks the streets, he sees a runaway carriage; he leaps onto the horses and stops them. In the carriage is a beautiful young woman who turns out to have a rich father. The father takes Dick into his business, where he proves his great motivation and becomes highly successful.

The people of prerevolutionary China were also guided by ideology; they believed in the teachings of Confucius (551–479 B.C.E.), which emphasized the need to accept one's place in a well-ordered, highly stratified society (McNeill, 1963). Similarly, the castes of Hindu India are supported by religious ideology. The *Rig-Veda* taught that Hindu society was, by divine will, divided into four castes, of which the Brahmans were the highest because they were responsible for religious ceremonies and sacrifices (Majundar, 1951; McNeill, 1963). Over time, other castes with other tasks were added to the system as the division of labor progressed and new occupations developed. Still another powerful ideology had its origins in Europe before the spread of Christianity. Tribal peoples in what is now France and Germany associated their kings with gods, and that association became stronger in the feudal era (Dodgson, 1987).

Religious teachings often serve as the ideologies of civilizations, explaining and justifying the stratification systems associated with them. But this relationship has not held in every historical period or for every religious movement. Originally, for example, the teachings of Jesus opposed the stratification systems of both the Roman Empire and the Jewish people. Jesus preached a gospel of love and claimed that "The last shall be first" and "It is easier for a camel to pass through the eye of a needle than for a rich man to enter heaven." He showed sympathy for prostitutes and outcasts such as moneylenders. No wonder his teachings appealed to the poor and downtrodden and enraged the wealthy and powerful. Over many centuries, however, Christ's teachings were incorporated into church doctrine and organization, and by the Middle Ages, Christianity was the ideology underlying the stratification system of kings, lords, merchants, and peasants. The vicars of the church upheld that system by affirming its legitimacy in coronations and royal weddings. They also presided over the execution of heretics who challenged the system, which was viewed as divinely ordained.

In our own era, the civil rights movement in the United States, the movement to end apartheid in South Africa, the struggle of the Northern Irish Catholics for independence from Britain, and other social movements often invoke the ideology of radical Christianity. "We Shall Overcome," the theme song of the civil rights movement, was borrowed from an African American Baptist spiritual, "I Shall Overcome," and transformed into a moving song of hope and protest with religious overtones.

Stratification at the Micro Level

These relationships between religious ideologies and the stratification systems of civilizations are macro-level examples of how culture maintains stratification systems from one generation to the next. But we can also see the connection between culture and stratification in the micro-level interactions of daily life. The way we dress—whether we wear expensive designer clothes, off-the-rack apparel, or second-hand clothing from the Salvation Army—says a great deal about our place in the stratification system. So does the way we speak, as anyone knows who has been told to get rid of a southern or Brooklyn accent in order to "get ahead." Our efforts to possess and display **status symbols**—material objects or behaviors that convey prestige—are encouraged by the billion-dollar advertising industry. Many other examples

status symbols: Material objects or behaviors that indicate social status or prestige.

could be given, but here we will concentrate on two sets of norms that reinforce stratification at the micro level: the norms of *deference* and *demeanor* (Goldhamer & Shils, 1939).

Deference By **deference**, we mean the "appreciation an individual shows of another to that other" (Goffman, 1958, pp. 488–489). In popular speech, the word *deference* is often used to indicate how one person should behave in the presence of another who is of higher status. Formulas for showing deference differ greatly within different subcultures. Among teenage gang members, for example, deference is shown through greeting rituals. Not performing those rituals properly may cause a gang member to feel "dissed" (disrespected), which can lead to violence. (Recall the discussion of "face work" in Chapter 5.) These formulas for showing deference illustrate how our society's stratification system is experienced in everyday life. In the United States, for example, we learn to address judges as "Your Honor" and feel embarrassed for the plaintiff who begins a sentence with "Excuse me, Judge." In most European countries, with their histories of more rigid stratification, people who want to show deference go further and address the judge as "Your Excellence."

Deference is not a one-way process, however. Erving Goffman has pointed out that the act of paying deference to someone in a higher status often obligates that other person to pay some form of deference in return: "High priests all over the world seem obliged to respond to offerings [of deference] with an equivalent of 'Bless you, my son'" (Goffman, 1958, p. 489). The point here is that deference is often symmetrical, in that both participants defer according to their place in the stratification system. Through deferent behavior and the appropriate response, both parties affirm their acceptance of the stratification system. Intuitively, we all know this. When we are stopped by a police officer, we may become deferential, using the most polite forms of address ("Yes, sir," "No, sir," and the like) in order to avoid embarrassment. The officer, in turn, may attempt to find out our place in the stratification system and use the appropriate forms of address in speaking to us.

Demeanor Our **demeanor** is the way we present ourselves—our body language, dress, speech, and manners. It conveys to others how much deference or respect we believe is due us. Here again, the interaction can be symmetrical.

In a photography show organized by the Chinese Embassy, deference to government power emerged as a dominant theme in images of Szechuan earthquake relief efforts. Shown here is a tent school opening ceremony.

The professor must make the first move toward informality in relations with students, for instance, but among professors of equal rank, there is far more symmetry. The move toward informal demeanor, such as the use of first names, can be initiated by whoever feels most comfortable in his or her status. On the other hand, asymmetry in the use of names can be used to reinforce stratification, as in an office where secretaries are addressed by their first names and supervisors by their last names plus a title such as Mr., Dr., or Professor.

Stratification and Social Interactions These largely taken-for-granted aspects of how we carry out or "construct" social stratification in our social interactions can have far-reaching effects. Maya Angelou, Richard Wright, and other African American writers vividly describe the extreme shuffling and obsequious deference demanded of their parents in the era of segregation. Even to lift one's eyes to admire a white woman could cause trouble for a black man in the Jim Crow regions of the United States. And as we saw in Chapter 8, failure to carry out the rules of demeanor—to twitch, to have a runny nose, to encroach on another person's space—can cause a person to be labeled deviant and to be cast out of the "acceptable" strata of society.

Deference, demeanor, and other ways in which we behave according to the micro-level norms of stratification further reinforce our sense of the correctness of those norms. Thus, in times of rebellion against a society's stratification system, those norms are explicitly rejected. During the 1960s, for example, students from wealthy families often wore jeans and tie-dyed shirts in social situations that would normally call for suits or dresses. Long hair and beards, refusal to wear a bra, and other symbolic acts in violation of generally accepted norms of demeanor were intended to communicate rejection of the society's stratification system.

deference: The respect and esteem shown to an individual.

demeanor: The ways individuals present themselves to others through body language, dress, speech, and manners.

Do Americans defer to people who are rich and powerful? We like to think of ourselves as egalitarian, democratic, individualistic people who do not bow and scrape for anyone. But what does the popularity of shows like *The Apprentice* or the older *Lifestyles of the Rich and Famous* tell us about our culture's values concerning equality and inequality?

POWER, AUTHORITY, AND STRATIFICATION

When the macro dimensions of social stratification change, the changes may be reflected in the behavior of people at the micro level, and those changes, in turn, accelerate change throughout the entire society. Thus, for example, the civil rights and women's movements have altered the norms of demeanor for blacks and women. Blacks insist on being referred to as blacks or African Americans if their ancestors were of African origin, rather than as Negroes. Women refuse to be called "girls" by men and especially by their supervisors, although some women continue to use the term among themselves. These may seem to be trivial matters, but when people demand respect in everyday interactions, they are demonstrating their determination to create social change at a more macro level—in other words, to bring about a realignment of social power.

Max Weber defined **power** as "the probability that one actor within a social relationship will be in a position to carry out his own will despite resistance" (1947, p. 152). This is a very general definition; it applies equally to a mugger with a gun and to an executive vice president ordering a secretary to get coffee for a visitor. But there is a big difference between the types of power used in these examples. In the first, illegitimate power is asserted through physical coercion. In the second, the secretary may not want to obey the vice president's orders yet recognizes that they are legitimate; that is, such orders are understood by everyone in the company to be within the vice president's power. This kind of power is called **authority**.

Even when power has been translated into authority, there remains the question of how authority originates and is maintained. This is a basic question in the study of stratification.

As we saw earlier, the fact that people in lower strata accept their place in society requires that we examine not only the processes of socialization but also how power and authority are used to maintain existing relations among castes or classes.

A brief review of the causes of the French Revolution of 1789 provides a case study in the relationship between power and authority on one hand and stratification on the other: Why did the French people accept the rule of absolute monarchs for so long before they finally ended that rule in a bloody revolution? In answering this question, we can apply all the aspects of stratification discussed so far. We will begin with the feudal stratification system.

The major strata of French society, called *estates,* were the nobility, the clergy, the peasantry, and the bourgeoisie (merchants, shopkeepers, and artisans). Each estate had its own institutions and culture, causing its members to feel that they were part of a unified community with its own norms and values. The estates were linked together by the time-honored norms of feudalism: vassalage and fealty. A *vassal* was someone who received a grant of land from a lord. In return, he swore *fealty,* an oath of service and loyalty, to that lord. The lord, in turn, swore fealty to a more powerful lord, until one reached the highest level of society, the king. (The feudal stratification system is described in detail in the Global Social Change box on page 244.) In such a well-ordered and legitimate system, where were the seeds of revolution? The answer lies in how power, authority, and changing modes of production combined to shake the foundations of feudal France.

In any system of stratification, there are likely to be conflicts. Those conflicts may be caused by the ambitions of a particular leader or group, but sometimes the characteristics of the system—and especially of the way its powerful members try to cope with change—create conflict. Both types of conflict existed in pre-revolutionary France. In order to compete with the other major European powers, the king had to raise armies and send fleets abroad, both of which cost huge sums of money and required thousands of men. But the vow of fealty extended only from the king to the highest level of the nobility. The nobles, in turn, were responsible for seeing that their vassals provided more money and men. This system placed severe constraints on the king, who could not easily make direct demands on all his subjects. The king was dependent on the nobles and, through them, on the lesser lords. In return, the lesser lords often demanded more power, thereby challenging the king's authority. In order to reduce the power

power: The ability to control the behavior of others, even against their will.

authority: Power that is considered legitimate both by those who exercise it and by those who are affected by it.

In this engraving from the era of the French Revolution, a group of nouvellistes are discussing the news in the Luxembourg Gardens. These were people who spread gossip and small pamphlets with news of the Bourbon court. Note the Parisian women and men listening to the discussions. Most were illiterates who depended on verbal news, much of which was gossip about the sexual conduct of the royalty, not very different in content from today's celebrity gossip.

of the lords, the king went to the clergy and the bourgeoisie for assistance.

As the cities grew and the trading and manufacturing capacities of the bourgeoisie expanded, the king was increasingly successful at drawing on the wealth of the bourgeoisie and undermining the power of the feudal lords. The court flourished; in fact, it seemed that the king's power had become absolute. Beneath the pomp and display of the court, however, the feudal stratification system was in ruins. The weakened nobility could no longer hold the fealty of the peasants. Throughout the countryside, the peasants groaned under the heavy debts imposed by their feudal masters, and they began to question the right of the nobility to tax them.

Meanwhile, in the cities and towns, the rapidly growing bourgeoisie, along with the new social stratum of urban workers, clearly understood that the king's power derived from their productive activities. They felt that they were not adequately rewarded for their support. At the same time, another new stratum, the intellectuals—people who had been educated outside the church and the court—was creating a new ideology and promoting it through new institutions like the press. This new ideology—"liberty, equality, fraternity"—helped spur revolutionary social movements, but the underlying cause of the revolution was a drastic change in the stratification system, in which the new middle class of wealthy business owners (the bourgeoisie) eventually challenged the ancient aristocracy.

In our own era, revolutions, social upheavals, and the wars often associated with them continue to occur in response to extremes of social stratification. In the early 1990s, for example, peasants in the impoverished province of Chiapas, Mexico, staged an uprising that shook the

These Afghan men, members of the Hazara society, are loyal followers of Muhammed Karim Khalili, who led them in the fight against the Taliban. Each has sworn a vow of fealty to his lord. They are listening in rapt attention to his homecoming speech after he participated in the formation of a new government in Afghanistan in December 2001.

Feudal Relationships Imagine a peasant in a feudal society whose daughter wishes to marry a young man from another village. (The society could be medieval England in the time of King Arthur or present-day Afghanistan, where daily life for many people in rural areas is still governed by feudal relationships.) She cannot marry without the permission of the lord of the manor, who controls the destiny of everyone dwelling on his land. The peasant must wait for the day when the lord holds court, settling disputes and hearing petitions. Finally, the day arrives. The peasant waits nervously until the reeve signals him to come forward. He approaches timidly, his wife following close behind. With the utmost humility he bows, pressing the knuckles of both hands against his forehead, while his wife makes the deepest possible curtsy. He makes his request as briefly as possible and silently awaits the lord's reply.

Feudalism is a form of social organization that has appeared throughout the world as societies have experienced the so-called agrarian revolution. It took somewhat different forms in China, Africa, Japan, and Europe, but it has always been characterized by a fundamental principle: the subordination of one person to another. As social historian Marc Bloch (1964) wrote, the essence of feudalism is "to be the 'man' of another man. . . . The Count was the 'man' of the king, as the serf was the 'man' of his manorial lord" (p. 145). In this system, known as *vassalage*, it was understood that women shared the loyalties of their male masters.

The swearing of *homage* signified and cemented the relationship of vassalage. In Bloch's words:

> Imagine two men face to face; one wishing to serve, the other willing or anxious to be served. The former puts his hands together and places them, thus joined, between the hands of the other man—a plain symbol of submission, the significance of which was sometimes further emphasized by a kneeling posture. At the same time, the person proffering his hands utters a few words—a very short declaration—by which he acknowledges himself to be the "man" of the person facing him. The chief and subordinate kiss each other on the mouth, symbolizing accord and friendship. Such were the gestures—very simple ones, eminently fitted to make an impression on minds so sensitive to visible things—which served to cement one of the strongest bonds known in the feudal era. (1964, pp. 145–146)

In the basic relationship of feudalism, signified in the act of homage, the vassal exchanges

independence for protection, but this relationship can also entail more material exchanges: The lord can give his vassal a manor with the associated land and villagers, and the vassal can provide wealth and soldiers when the lord requires them.

Sociologists study feudalism and its relationships not only because some institutions derived from feudalism still exist today, but also because the exchange of independence for protection continues to appear in modern social institutions. For example, organized crime "families" often resemble feudal systems, as do many political party organizations and some business corporations. As people in contemporary societies like Afghanistan fight for greater equality in economic and legal institutions, they often demand an end to what they describe as "paternalistic, feudal relations" in those institutions.

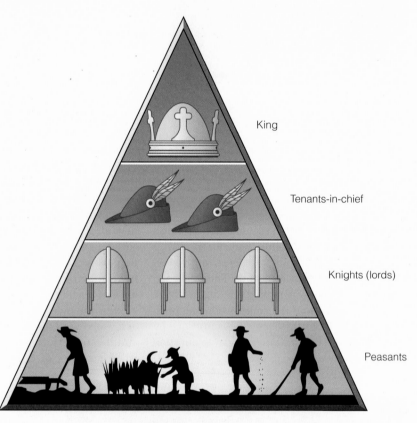

Feudal system

---

political and economic institutions of Mexican society. The Chiapas peasants and their rebel leaders continue to press for reforms in the regional economy and government.

STRATIFICATION IN THE MODERN ERA

In the preceding section, we saw that, under the leadership of the bourgeoisie, a new social order known as *capitalism* largely destroyed the feudal system. Capitalism is a form of economic organization based on private ownership and control of the means of production (land, machines, buildings, and so on). Whereas land was viewed as belonging to a family as a matter of birthright under feudalism, land can be bought and sold like any

other commodity under capitalism. And whereas feudal labor systems were based on institutions like serfdom, in which peasants were bound to a particular lord, in capitalist systems workers are free to sell their labor to the highest bidder and to endure unemployment when work is unavailable. Capitalism is also an ideology based on the value of individual rights; the ideology of feudalism, in contrast, stressed the reciprocal obligations of different strata of society.

In descriptions by Karl Marx and other observers of nineteenth-century society, capitalism seemed almost to explode onto the world scene. Capitalism made possible the dramatic change in production methods that we call the Industrial Revolution, which transformed England and Europe from a world of towns and villages, courts and cathedrals, to a world centered on markets, factories, and crowded cities. A few places where the physical structures of feudalism remain, such as Mont-Saint-Michel in France or Dubrovnik in Croatia, continue to illustrate the contrast between the old order and the new.

The Industrial Revolution began in England and other parts of Europe in the late seventeenth century. Today it is still occurring in other parts of the world, including China, India, and Africa. The former colonial outposts of England, France, and Germany are only now undergoing the transformation from rural agrarian societies to urban industrial states in which an ever-decreasing portion of the population is engaged in farming. Although not all of these industrializing societies have capitalist economies, they all are developing a class of managers, entrepreneurs, and political power brokers who resemble the bourgeoisie of early capitalism.

The Great Transformation

It is not enough, however, to describe the Industrial Revolution in terms of the spread of industrial technology and urban settlement. Technological innovations such as the steam engine, the railroad, the mechanization of the textile industry, and new processes for making steel and mining coal were accompanied by equally important innovations in social institutions.

In the words of sociologist and historian Karl Polanyi (1944), the Industrial Revolution was a "Great Transformation." For the first time in human history, the market became the dominant institution of society. By *market,* we do not refer only to the places where villagers sold produce or traded handicrafts. Rather, the market created by the Industrial Revolution was a social network that gradually extended over the entire world and linked buyers and sellers in a system that governed the distribution of goods of every imaginable type, services of all kinds, human labor power, and new forms of energy like coal and fuel oil.

Among other key elements of the Great Transformation were these:

- Goods, land, and labor were transformed into commodities whose value could be calculated and translated into a specific amount of gold or its equivalent—that is, money (Marx, 1962/1867; Schumpeter, 1950; Weber, 1958/1922).
- Relationships that had been based on ascribed statuses were replaced with relationships based on contracts. A producer hired laborers, for example, rather than relying on kinship obligations or village loyalties to supply workers (Polanyi, 1944; Smith, 1965/1776).
- The business firm or corporation replaced the family, the manor, and the guild as the dominant economic institution (Weber, 1958/1922).
- Rural people, displaced from the land, began selling their labor for wages in factories and commercial firms in the cities (Davis, 1955).
- In the new industrial order, demands for full political rights and equality of opportunity, which originated with the bourgeoisie, slowly spread to the new class of wage workers, to the poor, and to women. This was especially true in societies in which revolutions created more open stratification systems (Bendix, 1969; Tocqueville., 1980/1835; Mannheim, 1941).

The last point deserves further comment. The demand for full political rights originated with the bourgeoisie, which was to become the capitalist class in industrial societies—the owners of the factories, heavy equipment, and other means of producing wealth. To enhance their ability to do business and make profits, members of the bourgeoisie were interested in removing various feudal barriers to economic activity. They therefore called for an end to feudal restrictions on the sale of land, greater freedom for workers to move from one employer to another, the elimination of aristocratic privileges like the right to charge tolls for the use of roads and waterways that happened to cross a lord's property, and an end to voting rights based solely on aristocratic birth. These demands for economic and political rights then spread to the new class of wage workers and to other groups in society. Although the bourgeoisie originally sought rights for itself, it had to form alliances with workers and professionals in order to topple the feudal aristocracy. (Note that these demands originally applied only to men. As we will see in Chapter 13, similar movements for equal participation by women have taken many more generations to develop and gain influence.)

Class Consciousness and Class Conflict

The Great Transformation had a profound impact on the stratification systems of modern societies. It produced new and powerful social classes and thereby changed the way people thought about their life chances and the legitimacy of their society's institutions.

As Marx wrote in the last chapter of *Capital*, "The owners merely of labor-power, owners of capital, and landowners, whose respective sources of income are wages, profit, and ground rent . . . constitute the three big classes of modern society based upon the capitalist mode of production" (1962/1867, pp. 862–863). Here, as elsewhere, Marx defined classes in terms of the "modes of production" that are characteristic of a society in a given historical period. The workers, by far the largest class in modern societies, must sell their labor to a capitalist or landowner in return for wages. The capitalists and landowners are far less numerous than the workers, but because they own and control the means of production, they command more of everything of value than the workers do.

Of course, Marx recognized that, in societies that had not yet undergone the Great Transformation, there were classes that were very different from those of capitalist societies. The peasants of Russia, the slaves of America, and the exploited Indians of Latin America were "precapitalist" classes. Marx predicted that these strata would eventually be transformed into what he termed the *proletariat*, workers who did not own the means of production and had to survive by selling their labor.

Marx believed that the class of wage workers created by capitalism would inevitably rise up against the capitalist class and create a classless socialist society. As noted earlier, this prediction was proved wrong: Proletarian revolutions never occurred in the most industrialized nations. Yet Marx's description of how members of the working class become conscious of their situation as a class remains an important subject in the study of stratification.

Sources of Class Conflict "The history of all hitherto existing societies is the history of class struggles," wrote Marx and his collaborator Friedrich Engels in *The Communist Manifesto* (1969/1848, p. 11). In the modern era, "society as a whole is more and more splitting up into two great hostile camps, into two great classes directly facing each other, bourgeoisie and proletariat" (p. 11). Why did Marx and other observers of the capitalist system believe that conflict between the bourgeoisie and the proletariat was inevitable? The answer can be found by taking a closer look at Marx's analysis of the evolution of capitalism.

To begin with, Marx observed the misery of industrial workers (many of whom were children) in the smoky factories of industrial England. He saw that however wretched the workers were, there would always be what he called a "reserve army" of unemployed people who were willing to work for lower wages than those who already had jobs. He noticed that the capitalists (and the intellectuals they paid to argue in their defense) always blamed the workers themselves for the miserable conditions in which they were forced to exist. If they were hungry, the capitalists claimed, it was because they did not work hard enough or spent their pay on too much alcohol or could not curb their sexual passions and bore too many children. Thus the capitalists refused to accept blame for the misery of the working class under capitalism.

Marx also argued that business competition would eliminate less successful firms and result in monopolies, which would control prices and wages and thereby contribute still more to the impoverishment of the workers. Moreover, the capitalists had the power to determine who ran the government and who controlled the police and the army. If the workers were to rebel, the armed forces and police would intervene as agents of the capitalists. Through these means, the workers and the unemployed would be forced to remain a huge, helpless population that could be manipulated by the capitalists. Over time, according to Marx and his followers, these masses of people would become increasingly conscious of their plight and would unite in a revolution that would destroy the power of the capitalists and their allies.

But why would workers become more conscious of their situation as a class, rather than merely remaining miserable in their impoverished families and communities? This problem of class consciousness became a central issue in Marxian thought. The study of changing patterns of class consciousness remains a major subject in sociology to this day.

Objective and Subjective Classes In thinking about how a social class might be able to take collective action, revolutionary or otherwise, Marx distinguished between *objective* and *subjective* classes. All capitalist societies, according to Marx, have **objective classes**—social classes that an observer can identify by simply looking at people's visible, specific relationships to the means of production in their society. The workers, for example, are easily identified as an objective class that does not own the means of production and sells its labor power in return for wages. The capitalists are the objective class that owns the means of production—the machines and property and railroads—and buys the labor power of the working class.

objective class: In Marxian theory, a social class that has a visible, specific relationship to the means of production.

These facts would be visible to an observer equipped with an understanding of what makes a social class, but they might not always be evident to the members of those classes themselves. Thus, Marx identified **subjective class** as the extent to which the people in a given stratum of society actually perceive their situation as a class. For example, if the workers in the chicken processing industry in the southern United States are not aware that their low wages and miserable working conditions are similar to the conditions faced by millions of workers in other industries elsewhere in the world, and if they do not understand that they can improve their fortunes only by taking some of the power of another class (the capitalists), they are not yet a subjective class. Without this awareness of their situation, the workers are said to lack **class consciousness**. And without class consciousness, they cannot form the political associations that will allow them to fight effectively against the capitalists.

In an often-quoted passage, Marx described the peasants of France as an objective class because of their shared experience as agriculturists with small landholdings, but he was doubtful about their ability to form a subjective class:

> A small holding, a peasant and his family; alongside them another small holding, another peasant and another family. A few score of these make up a village, and a few score of villages make up a Department. In this way, the great mass of the French nation is formed by simple addition of homologous magnitudes, much as potatoes in a sack form a sack of potatoes. . . . In so far as there is merely a local interconnection among these small holding peasants, and the identity of their interests begets no community, no national bond and no political organization among them, they do not form a class. (1963/1869, p. 124)

The working class, Marx believed, would be different from the peasantry because its members would become conscious of their shared interests as a class.

The Classless Society Marx and other observers of early capitalism believed that the growing conflict between the working class, or proletariat, and the capitalist class, or bourgeoisie, would produce revolutions. In those revolutions, the proletariat and its allies would depose the bourgeoisie and establish a new social order known as socialism. Under socialism, the key institutions of capitalism—private ownership of the means of production, the market as the dominant economic institution, and the nation-state controlled by the bourgeoisie—would be

Bolsheviks and Capitalists Much of the social thought and ideology of the twentieth century was shaped by the Russian Revolution. The Bolsheviks, shown here storming the Winter Palace, put an end to the reign of the Russian czars and kindled hopes for a classless society not just among their own followers but among impoverished and politically repressed people throughout the world. However, the theories and ideals of socialism, when put into practice by dictatorial leaders like Lenin and Stalin, soon became corrupted. Although many impoverished Russians benefited from the revolution, the classless society never emerged. Instead, a new Soviet class system emerged, with the Communist Party elite at the top. Today, as Russia transforms itself into a capitalist nation, the old party elite and the newly rich class of entrepreneurs are combining to form a powerful new class at the top of Russian society. Other classes, especially blue-collar workers who were protected under Soviet socialism, are experiencing downward mobility, at least during the painful transition to a new social order.

abolished. The new society would be classless because the economic institutions that produced classes would have been eliminated and all the members of society would collectively own the means of production.

In fact, Soviet-style socialism as it developed under the Bolshevik dictatorship of Lenin and then Joseph Stalin did not produce a classless society. After the revolution of 1917, a new class of Communist Party officials, military generals, and government bureaucrats, known as the *nomenklatura,* replaced the older aristocratic ruling class. As the communist elite used its power to secure privileges that were not granted to those in the lower classes, the Marxian dream of a classless society gave way to cynicism. And in the years since the demise of the Soviet Union, the extremely rapid conversion of the economy from socialism to capitalism has created an entirely new class of self-made millionaires. Elsewhere in the world, however, in societies that did not experience the abuses and rigidity of Soviet communism, the Marxian view of class struggle remains one of the powerful forces generating social movements to reduce inequality.

Social Mobility in Modern Societies: The Weberian View

Well before the failure of Soviet communism, Max Weber took issue with Marx's view of social stratification. Marx had defined social class in

---

subjective class: In Marxian theory, the way members of a given social class perceive their situation as a class.

class consciousness: A group's shared subjective awareness of its objective situation as a class.

Marx and Weber on Stratification

	Origins of Stratification	How is Stratification Maintained?	What Changes Structures of Stratification?
Karl Marx	Economic: based on different populations' relationship to means of production in a given society	By the power of dominant castes or ruling classes over all other groups and institutions	Class conflict and revolution due to the rise of new classes (e.g., the bourgeoisie in feudal or aristocratic societies) or conflict between the proletariat and the bourgeoisie in capitalist societies
Max Weber	Economic and cultural: based on relationship to the dominant means of production and degree of honor and power given to one's occupation or family in the community or culture	By the power of ruling castes and classes, as well as by the honor and prestige conferred on the rich and powerful in religion and politics	Not just class conflict but also social mobility and new means of gaining power (e.g., through party politics in democracies)

economic terms; classes are based on people's relationship to the means of production. Weber challenged this definition of social class. People are stratified, Weber reasoned, not only by how they earn their living but also by how much honor or prestige they receive from others and how much power they command. A person could be a poor European aristocrat whose lands had been taken away during a revolution, yet his prestige could be such that he would be invited to the homes of wealthy families seeking to use his social status to raise their own. Another person could have little money compared with the wealthy capitalists, and little prestige compared with the European aristocracy, yet could command immense power. The late Mayor Richard J. Daley of Chicago (the father of the former mayor Richard M. Daley) was such a person. He was born into a working-class Irish American family, and although he was rather well-off, he was not rich. Yet his positions as mayor and chair of the Cook County Democratic Party organization made him a powerful man. Indeed, Mayor Daley's fame as a politician who led a powerful urban party organization but did not enrich himself in doing so helped his son continue the family tradition of political leadership: In 1989, his son, Richard M. Daley, was elected mayor of Chicago.

For Weber and many other sociologists, therefore, wealth or economic position is only one of at least three dimensions of stratification that need to be considered in defining social class. Prestige (or social status) and power are the others. We need to think of modern stratification systems as ranking people on all of these dimensions. A high ranking in terms of wealth does not always guarantee a high ranking in terms of prestige or power, although they often go together.

Other challenges to Marx's view of stratification focus on social mobility in industrial societies. Contrary to Marx's prediction, modern societies have not become polarized into two great classes, the rich and the poor. Instead, there is a large middle class of people who are neither industrial workers nor capitalists (Wright, 1989, 1997), and there is considerable social mobility, or movement from one class to another. (See the For Review chart above.)

Forms of Social Mobility Social mobility can be measured either within or between generations. These two kinds of mobility are termed *intragenerational* and *intergenerational* mobility. (See the For Review chart on page 249). **Intragenerational mobility** refers to one's chances of rising to or falling from one social class to another within one's own lifetime. **Intergenerational mobility** is usually measured by comparing the social-class position of children with that of their parents. If there is a great deal of stability from one generation to the next, one can conclude that the stratification system is relatively rigid.

Unfortunately, not all mobility is upward. Downward mobility involves loss of economic and social standing. It is a problem for families and individuals at all levels of the class structure in the United States and elsewhere. In its broadest sense, downward mobility can be defined as "losing one's place in society" (Newman, 1988, p. 7). In fact, the term encompasses many different kinds of experiences. The married woman who works part-time and then is divorced, loses the family house, and must move with her children to a small apartment and work full-time is experiencing downward mobility. The couple who have been living with the wife's mother in public housing and are forced out during a check of official rosters and must seek refuge in a homeless shelter also experience downward mobility. The affluent young couple living in a downtown condominium experience downward mobility when he loses his job at a brokerage firm and she must go on maternity leave while he is looking for a new job. Often entire groups of people experience downward mobility, as happened to U.S. citizens of Japanese descent when their property was confiscated during World War II, and to people of Indian descent when they were expelled from Kenya and Uganda in the 1970s.

The General Social Survey, conducted annually by the National Opinion Research Center (NORC), charts changes in people's opinions about their prospects for mobility in either direction, using questions like these:

intragenerational mobility: A change in the social class of an individual within his or her own lifetime.

intergenerational mobility: A change in the social class of family members from one generation to the next.

Types of Social Mobility

Type	Description	Example
Structural	Entire occupations emerge and grow rapidly while others decline because of technological or other major changes.	Workers in high-tech industries (e.g., computers); farmers.
Individual or Family *Intergenerational*	Children experience mobility in either direction in comparison with their parents.	Upward—father a blue-collar worker, son a lawyer; daughter a professor, mother a factory worker. Downward—parents professionals, children waiters and aspiring actors.
Intragenerational	A person experiences an upward or downward shift in economic and social status.	A steelworker becomes a janitor, earning less; his wife goes to work to help make ends meet.

Compared to your parents when they were the age you are now, do you think your own standard of living now is much better, somewhat better, about the same, somewhat worse, or much worse than theirs was?

When your children are at the age you are now, do you think their standard of living will be much better, somewhat better, about the same, somewhat worse, or much worse than yours is now? (NORC, 2000)

Fear of downward mobility weighs heavily on the minds of many Americans. A central feature of the American dream is that our standard of living, and that of our children, will continue to improve. But that vision is challenged by the realities of global competition and the "export" of jobs to lower-wage regions of the world (Block, 1990; Harrison, Tilly, & Bluestone, 1986). Indeed, the findings of the General Social Survey show that more respondents claim to have experienced an improvement in their standard of living compared to their parents than believe their children will also enjoy such mobility. These results are presented in Figure 10.6. We will see in Chapter 11 that a significant minority of Americans believe that they will soon be rich; this belief in the possibility of individual mobility favors those who already enjoy great wealth and advantage in American society.

THEORIES OF STRATIFICATION

"The first lesson of modern sociology," wrote C. Wright Mills in his study of white-collar workers, "is that the individual cannot understand his own experience or gauge his own fate without locating himself within the trends of his epoch and the life-chances of all the individuals of his social layer" (1951, p. xx). We have examined the concepts of stratification and life chances and have seen that they are related to the way people earn their living in different societies. But what about the origins of "social layering" or stratification? That subject will be discussed in the next chapter with specific reference to the social-class system of the United States and other nations. In this section, we examine some of the basic theories of social stratification. Each of the major sociological perspectives continues to generate important insights into how stratification and inequality emerge and operate in different societies.

Conflict Theories

As mentioned earlier, Marx's theory of stratification asserts that capitalist societies are divided into two opposing classes, wage workers and capitalists, and that conflict between these two classes will eventually lead to revolutions that will establish classless socialist societies. However, Marx's prediction was not borne out in any society that attempted to implement his ideas on a large scale. Deny it as they might, all of those societies developed well-defined systems of social stratification (Aronowitz, 2003; Djilas, 1982; Szelenyi, 1983). Each had an elite of party officials; an upper stratum of professionals, scientists, managers of economic enterprises, local party officials, and high police officials; a middle level of well-educated technical workers and lower professionals; a proletariat of industrial and clerical workers and military personnel; and a bottom layer of people who were disabled, criminals, or political outcasts.

No persuasive evidence shows that class conflict is heightening the division between workers and owners of capital in capitalist societies. Conflict does exist, but the industrial working class is shrinking and the new occupational groups do not always share the concerns of the industrial workers. Moreover, reforms of capitalist institutions have greatly improved the workers' situation, thereby reducing the likelihood that the revolution predicted by Marx will occur.

Modern conflict theorists agree with Marx's claim that class conflict is a primary cause of social change, but they frequently debate both the nature of the class structure and the forms taken by class conflict. Thus, Erik Olin Wright (1979, 1997) notes that Marxian theorists agree that workers who are directly engaged in the production of goods are part of the working class. However, he makes the following observation:

> There is no such agreement about any other category of wage-earners. Some Marxists have argued that only productive manual workers should be considered part of the proletariat. Others have argued that the working class includes low-level, routinized white-collar workers as well. Still others have argued that virtually all wage-laborers should be considered part of the working class. (1979, p. 31)

This disagreement stems, of course, from the fact that there is far greater diversity within all the classes of modern societies than Marx or his

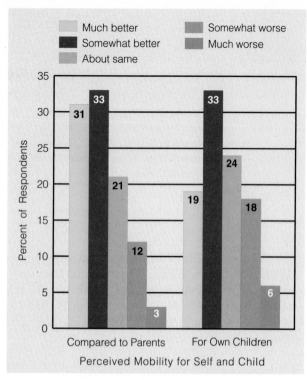

Figure 10.6 Perceptions of Mobility Opportunities (Respondent's Own and Children's). Source: NORC, 2000.

contemporaries imagined there would be. In addition to the bourgeoisie (the owners of large amounts of capital) and the petit bourgeoisie (the owners of small firms and stores), there is a constantly growing professional class, a class of top managers and engineers, another class of lower-level managers, and a class of employees with special skills (for example, computer specialists, nurses and other medical personnel, and operating engineers). Perhaps many of these people should think of themselves as part of the working class, but they normally do not because they are earning enough to enable them to live in middle-class communities.

Some conflict theorists focus on other aspects of social stratification besides class conflict. Melvin Tumin (1967/1984), for example, pointed out that stratification systems "limit the possibility of discovery of the full range of talent available in a society"; that they create unfavorable self-images that further limit the expression of people's creative potential; and that they "function to encourage hostility, suspicion, and distrust among the various segments of a society" (p. 58). Problems like wasted talent and poor self-image are among what Richard Sennett and Jonathan Cobb (1972) term "the hidden injuries of class," meaning the ways in which a childhood of poverty or economic insecurity can leave its mark on people even after they have risen out of the lower classes. In the next chapter, we look more specifically at social stratification in the United States and examine its consequences in human terms.

The Functionalist View

The functionalist view of stratification was originally stated by Talcott Parsons (1937, 1940) and Kingsley Davis and Wilbert Moore (1945). This theory holds that social classes emerge because an unequal distribution of rewards is essential in complex societies. Such societies need to reward talented people and channel them into roles that require advanced training, personal sacrifice, and extreme stress. Thus, the unequal distribution of rewards, which allows some people to accumulate wealth and deprives others of that chance,

is necessary if the society is to match the most talented individuals with the most challenging positions.

Research by sociologists in the former Soviet Union often provided support for the functionalist idea that extreme "leveling" deprives people of the motivation to achieve more than a minimum of skills. On the basis of extensive surveys of workers in Russia and other former Soviet republics, sociologist Tatyana Zaslavskaya (cited in Aganbegyan, 1989) argues that unless incentives are provided in the form of high wages and other advantages (such as better housing in areas where good housing is in short supply), engineers and scientists will resent their situation and will not work hard. In the United States, the same arguments are used to justify higher salaries for doctors, lawyers, and other professionals. Too much equality, it is said, reduces the incentive to master difficult skills, and the entire society may suffer from a lack of professional expertise as a result.

Critics of the functionalist view of inequality and stratification point to many situations in which people in positions of power or leadership receive what appear to be excessive benefits. For example, the late Roberto Goizueta, former chief executive officer of the Coca-Cola Company, was criticized for taking home more than $10 million in salary and bonuses. Many other executives were earning equally high salaries even when their companies were not doing as well as they might in international competition. With the annual pay of top Japanese executives often running 5 to 10 times less than that of their American counterparts, it became difficult for the American corporate elite to argue that these enormous rates of executive compensation are "functional." And in 2008, when the latest stock market bubble burst and small stockholders lost money while many company executives and insiders reaped huge profits by selling before the market declined, it was again difficult to show that this behavior was justified by the functional need for creative business entrepreneurs.

From the point of view of those who are critical of social inequality, the large sums paid to a few people seem wrong, especially when so many others are struggling just to survive. Indeed, the heads of large corporations in the United States often earn more than 50 times as much as the average employee of those corporations. Conflict theorists argue that such great disparity is not "functional" when it produces enormous gaps between the very rich and the working classes. Functionalist theory claims that the disparity is functional because top executives in a free enterprise system will seek the firms that are most willing to reward them for their talents. Those firms will benefit, and so will their workers.

The Interactionist Perspective

Conflict theory explains stratification primarily in economic terms. So does functionalist theory. Both trace the existence of certain classes to the central position of occupation, income, and wealth in modern life. But neither goes very far toward explaining the prestige stratification that occurs *within* social classes. Among the very rich in America, for example, people who have stables on their property tend to look down on people with somewhat smaller lots on which there is only a swimming pool. And rich families who own sailing yachts look down on equally rich people who own expensive but noisy powerboats. The point is that, within economic classes, people form status groups whose prestige or honor is measured not according to what they produce or how much wealth they own but according to what they buy and what they communicate about themselves through their purchases. Rolex watches and BMW cars are symbols of membership in the youthful upper class. Four-wheel-drive vehicles equipped with gun racks and fishing rods are symbols of the rugged and successful middle- or working-class male. Armani suits are symbols of urbane professionalism; tweed suits and silk blouses are signs that a woman is a member of the "country club set." All of these symbols of prestige and group membership change as groups with less prestige mimic them, spurring a search for new and less "common" signs of belonging (Dowd, 1985).

Our tendency to divide ourselves up into social categories and then assert claims of greater prestige for one group or another is of major significance in our lives. The interactionist perspective on stratification, therefore, may not be very useful in explaining the emergence of economic classes, but it is essential to understanding the behaviors of the status groups that form within a given class. Those behaviors, in turn, often define, reinforce, or challenge class divisions. The stratification system, in this view, is not a fixed system but is created over and over again through the everyday behaviors of millions of people.

Thinking Critically — From a functionalist viewpoint, the fact that a corporate executive makes over 100 times as much money as the average employee is justified by the important work (functions) that only he or she can perform. Do you agree? What would someone with a critical perspective on inequality and stratification argue about such a disparity in earnings?

STRATIFICATION AND GLOBAL POVERTY

Today it is no longer possible to think of a society's stratification system separately from those of other societies. More than ever before, the life chances of people in the United States and other highly developed nations are influenced by changes in the structure of industrial employment and the worldwide distribution of jobs. Early in the new millennium, for example, many nations of Southeast Asia were struck by economic crises. Indonesia experienced widespread social unrest in the wake of financial instability. Japan's economy, one of the strongest in the world, faltered badly. In 2001, the war against terrorism led by the United States began at the same time that another global economic turndown was occurring; this further impeded the recovery of many economies in poorer nations. These crises worsened the plight of millions of impoverished people in the developing nations and throughout the world.

Globalization opens new regions to increased trade and investment, to increased communications from other parts of the world, and to greater movement of people across national boundaries. But the benefits of these trends are not shared equally. The richest 20 percent of the world's nations command far more than their proportional share of wealth, information, and the energy needed to fuel development. The graphs in Figure 10.7 demonstrate that greater progress has come in developing new ways of doing business and enhancing the growth of world trade than in developing the human potential of poor populations (UNDP, 2003). These trends become even more evident when we consider the implications of world poverty for human health.

Inequality, Poverty, and Health

Differences in income and wealth can sometimes mean the difference between life and death. In some developing nations, health conditions are changing for the better as public health measures such as improved water supply and sewage systems take effect. Nevertheless, the correlation between poverty and death remains extremely high in many nations, and in recent years it has increased in the United States. These trends are so important that social scientists and international planning experts often use health statistics as a barometer of how well a nation is doing in improving the gap between the haves and the have-nots in its population.

Three indicators—mortality of children younger than age five, the adult illiteracy rate, and the incidence of income poverty—are often used as primary measures of social change when comparing different nations (UNDP, 2003). As we see in Figure 10.8, infant mortality rates are two or three times higher in families in which the mother has no education than in families in which the mother has secondary or higher education.

The single most important factor in determining survival is income. At the extreme of poverty, in Africa, with annual per capita income of US$800, life expectancy is fifty-two years. At the other end of the income spectrum, in the United States, with annual per capita income of US$20,000, average life expectancy is seventy-five years. Some low-income regions, however, have invested in the health and safety of their populations (for example, some states of India), and some affluent regions still rank relatively low on measures of health and survival (Sen, 1993; UNDP, 2005).

In impoverished nations like Ethiopia and Bangladesh, the primary causes of death before old age are diseases such as amoebic dysentery (severe diarrheal diseases), which result from contaminated water supplies and poor sewage systems. In more affluent nations, the rate of death before age sixty increases for adolescents and young adults, especially males,

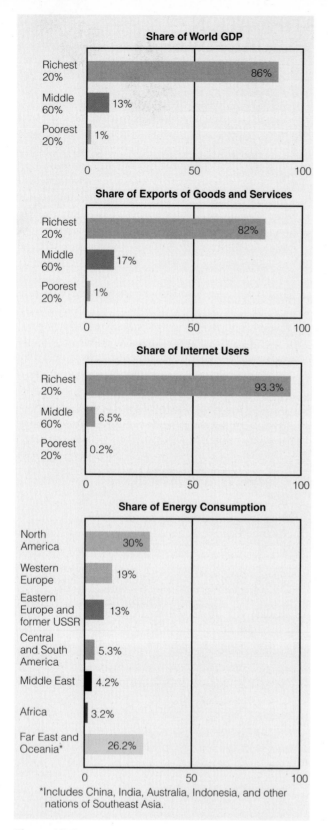

Share of World GDP

Richest 20%: 86%
Middle 60%: 13%
Poorest 20%: 1%

Share of Exports of Goods and Services

Richest 20%: 82%
Middle 60%: 17%
Poorest 20%: 1%

Share of Internet Users

Richest 20%: 93.3%
Middle 60%: 6.5%
Poorest 20%: 0.2%

Share of Energy Consumption

North America: 30%
Western Europe: 19%
Eastern Europe and former USSR: 13%
Central and South America: 5.3%
Middle East: 4.2%
Africa: 3.2%
Far East and Oceania*: 26.2%

*Includes China, India, Australia, Indonesia, and other nations of Southeast Asia.

Figure 10.7 Disparities Among Rich and Poor Nations.
Source: UNDP, 2003; *Statistical Abstract*, 2000.

Studies of health and inequality in the United States show that the health and mortality gap between the rich and the poor has widened considerably over the past four decades. From 1960 to the mid-1980s, the degree of inequality in mortality rates more than doubled. Racial inequality compounds this difference. Medicaid, the federal insurance program that serves more than 30 million people in the United States, has significantly improved the life expectancy of poor people, but poor diet, poor environmental conditions, high risk of contracting contagious diseases (especially new strains of tuberculosis), and high rates of interpersonal violence have helped widen the mortality gap. Because similar trends have been occurring in affluent nations of Europe, where universal health insurance is the norm, medical sociologists warn that better insurance systems for the United States will not automatically narrow the gap in the absence of improvements in eating and smoking habits and other lifestyle changes among the poor (Pear, 1993).

Gender Disparities

No contemporary society treats its women as well as its men. The United Nations Development Programme's annual *Human Development Report* summarizes the situation as follows:

> In developing countries there are still 60% more women than men among illiterate adults, female enrollment even at the primary level is 13% lower than male enrollment, and female wages are only three-fourths of male wages. In industrial countries unemployment is higher among women than men, and women constitute three-fourths of the unpaid family workers. (1997, p. 39)

In the world's poorest nations, women face double deprivation: Human development problems such as lack of medical care, lack of clean water, extreme poverty, and unemployment affect all but the comfortable elite, but women in these countries face even higher levels of risk and deprivation.

Figure 10.9 presents the findings of recent research on maternal mortality in childbirth. Deaths of women in childbirth are highly correlated with lack of access to health care services during pregnancy, to female illiteracy, and to unequal political rights for women. But in recent years, thanks in part to efforts to reach the UN Millennium Development Goals, the gains have been robust and positive, especially since 2000. The graph also shows that, were it not for the continuing grave impact of AIDS in the poorer nations, the results would have been even more encouraging.

because of the greater risks they take in their leisure activities (dangerous sports, drug use, and so on) and the higher rates of interpersonal violence in some of these nations (especially the United States).

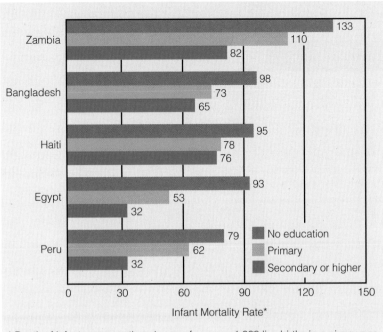

Figure 10.8 Mother's education and infant mortality in selected countries. Source: Gelbard, Haub, & Kent, 1999.

Infant Mortality Rate*

* Death of infants younger than 1 year of age per 1,000 live births in a given year.

We will see in Chapter 13 that political empowerment of women is a key to achieving greater equality between men and women throughout the world. In this regard, there are also some encouraging developments. Some developing nations now outperform much richer nations in achieving gender equality in political, economic, and professional activities: "Barbados is ahead of Belgium and Italy, Trinidad and Tobago outranks Portugal, and the Bahamas leads the United Kingdom. France lags behind Suriname, Colombia, and Botswana and Japan behind China, Guatemala, and Mexico" (UNDP, 1997, p. 41).

Achieving Greater Equality

The *Human Development Report* points out that some members of the upper classes have a vested interest in perpetuating poverty. The poor form "a mobile pool of low-paid and unorganized workers" who are useful for doing the dirty, dangerous, and difficult work that others often refuse to do. In industrial nations like the United States, many such jobs are taken by people from the poorest backgrounds and often by immigrants, legal and illegal. In many nations—including Russia and Nigeria, to mention two notable examples—patterns of corruption among the elite make it even more difficult to narrow the gap between rich and poor.

Reports about worldwide inequality and poverty often seem to suggest that the problems are so great that improvements are unlikely. But despite the magnitude of the challenge, skepticism and cynicism are unwarranted. After all, the report notes that since the 1990 World Summit for Children, the first-ever summit on human issues, rates of child mortality have declined in all regions of the world. Seven million more children's lives are being saved each year than in 1980. Immunization against major diseases like polio now covers over 80 percent of the world's children. On other fronts, such as reducing poverty among the aged or ensuring greater access to medical care, efforts to improve conditions are having demonstrable success. Although AIDS and other major diseases continue to wreak havoc in many regions, the record shows that when the world's nations make concerted efforts to address social ills related to poverty and disease, they can have major successes.

What is the price tag for poverty eradication? The *Human Development Report* estimates the cost of achieving basic social services for all people in the developing countries at about $40 billion per year over the ten-year period between 1995 and 2005 (see Table 10.1). However, as the

Figure 10.9 **A Long-Term Trend.** A broad survey of data from 181 countries has found that the number of women dying from childbirth or pregnancy worldwide has been declining steadily. Source: Reprinted from *The Lancet*, Volume 375, Issue 9726, Margaret C. Hogan, Kyle J. Foreman, Mohsen Naghavi, Stephanie Y Ahn, Mengru Wang, Susanna M. Makela, Alan D. Lopez, Rafael Lozano, Christopher JL Murray, "Maternal mortality for 181 countries, 1980—2008: a systematic analysis of progress towards Millennium Development Goal 5." Pages 1609–1623, 8 May 2010. Copyright © 2010, with permission from Elsevier.

— Estimated maternal deaths worldwide

----- Estimated deaths if H.I.V. epidemic had not occurred

TABLE 10.1

Cost of Achieving Universal Access to Basic Social Services

Need	Annual Cost (in U.S. dollars)
Basic education for all	$7.2 billion
Basic health and nutrition	$15.6 billion
Reproductive health and family planning	$15.4 billion
Low-cost water supply and sanitation	$0.8 billion
Total for basic social services	$48 billion

Source: Adapted from Box 6.4, p. 112 from "Human Development Report 1997" by UNDP. Updated with data from "Human Development Report 2000." Copyright © 1997 by Oxford University Press. Reprinted by permission of the publisher.

report concludes that lack of political commitment, not financial resources, is the real obstacle to poverty eradication. Eradicating absolute poverty is eminently affordable (UNDP, 2009).

Fighting Poverty, Fighting to Learn

What can a single person do to address the harsh realities of inequality? One answer that sociologists often give to this question is: "Try teaching, at least for a while, and teach somewhere where people are desperate for dedicated teachers."

In the face of extreme poverty and injustice, and against the systematic exploitation of those who have little by the few who seem to have everything, what can a teacher accomplish? We can never fully answer this question, but there are teaching situations around the world and in our own country where it will be obvious, once you are doing the work, that teaching the

basics of reading and writing is a primary route to social change. Consider, for example, the research of sociologist John L. Hammond among the peasants of El Salvador, where literacy instruction was a key element in the struggle for equality and democracy.

Hammond's focus was on the nation's teachers, particularly those known as "popular teachers." Unlike the teachers in the regular schools, who relied on traditional methods of rote memorization and drill and taught only students enrolled in the official schools, the popular teachers worked among the peasants and taught basic literacy. As Hammond (1998) observes: "Popular educators endorsed participatory, liberating teaching and criticized the authoritarian, rote-learning classes of traditional education" (p. 141). They threw themselves with great energy into their work among the children and adult peasants who came to them after their work in the fields. The popular teachers relied on people like Hammond, with extensive university educations, to help them train other teachers who could go to the fields and the refugee camps to teach basic literacy. By teaching illiterate children to read and write, they were giving them the tools needed to begin to understand the social conditions that made them poor and prevented them from taking a full part in building the democratic institutions of their societies.

As the accompanying illustration suggests, popular education in Central America was often critical of the status quo, and therefore at times it became the target of oppression by

Popular school, El Sitio del Lago, Cuscatlán.

Lois Kleffman, Sand Gap, KY

LEAMOS Y PLATIQUEMOS

El analfabetismo nos aprieta.

En nuestro país hay injusticia social

Pocos tienen mucho: grandes casas, haciendas, comida abundante, trabajan cuando quieren, sólo se la pasan ordenando.

En fin, viven como reyes.

La mayoría de nosotros tenemos poco, ni siquiera podemos estudiar.

HAY ANALFABETISMO por que HAY INJUSTICIA SOCIAL

Reprinted with permission of Funcionamiento CIAZO

This page from a literacy manual offers people in rural El Salvador a political analysis for why they are illiterate. "Illiteracy is squeezing us," it says. "A few have much: big houses, ample food, work when they wish—and live like kings. The majority of us have little, and cannot pursue learning. There is illiteracy because there is social injustice."

The Four Horsemen of the Apocalypse

Disasters and world poverty—pestilence, war, famine, and death—are visualized in the New Testament's Book of Revelations as the Four Horsemen of the Apocalypse and have often been the subject of artistic visions, such as this one painted in 1887 by the Russian artist Viktor Vasnetsov.

The Four Horsemen of the Apocalypse, 1887 (oil on canvas), Vasnetsov, Victor Mikhailovich (1848-1926)/Museum of Religion and Atheism, St. Petersburg, Russia/The Bridgeman Art Library

© Sebastiao Salgado/Contact Press Images

Contemporary visualizations of extreme poverty also document the way natural and man-made disasters contribute to global poverty. The noted Brazilian photographer Sebastiao Salgado, for example, has used his camera to document the consequences of drought and famine throughout the world, as in this photo of famine due to desertification in Africa.

Salgado spends long periods living with the people he photographs and using his camera to document their work and their efforts to overcome extreme poverty. "Salgado photographs from the inside," claims the Uruguayan writer Eduardo Galeano. "He remained in the Sahel desert for fifteen months when he went there to photograph hunger. He traveled in Latin America for seven years to garner a handful of photographs" (quoted in Salgado, 1990, p. 11).

Children's ward in the Korem refugee camp, Ethiopia, 1984.

© Sebastiao Salgado/Contact Press Images

Able-bodied men have left for town looking for work and food. They leave behind them their families and the "drought widows." Here people are walking on what used to be Lake Faguibine, Mali, 1985.

Photos and video reports from major disasters like the Haitian earthquake immediately galvanize world opinion and motivate millions of people to try to help, even if that means sending a contribution to an aid agency like the International Red Cross or a private charity. Some people with the means and moral motivation to make large sacrifices take on even greater responsibilities. One among many examples is that of Sean Penn, the Oscar-winning Hollywood actor, who not only rushed to Haiti immediately after the earthquake but also remained extremely active in working with Haitians as relief efforts continued. Penn has run a camp for displaced Haitians in the aftermath of the quake.

Penn also appeared before the U.S. Senate Committee on Foreign Relations to appeal for increased aid to earthquake-ravaged Haiti. "I come here today," he said, "in hope that we will address with bold clarity the razor's edge upon which Haiti lies, so that all in our own country, all that our country has given in sacrifice and service, will not be washed away with this rainy season and leave bright and dancing Haitian eyes to go still in death from disease and flood and, God forbid, the man-made disaster of violent unrest."

AP Photo/Ramon Espinosa

those in power. But teaching literacy, whether to rural children and adults in Central America, women in the Islamic world, or people in poor communities in our own society, always entails some controversy and risk. That is what makes it such a valuable endeavor and such a unique learning experience for students and teachers as well.

How are a society's patterns of inequality structured and reproduced from one generation to the next?

- *Social stratification* refers to a society's system for ranking people hierarchically according to various attributes such as wealth, power, and prestige.

- Societies in which there are rigid boundaries between social strata are said to have *closed stratification systems,* whereas those in which the boundaries are easily crossed are said to have *open stratification systems.*

- Movement from one stratum to another is known as *social mobility.*

- Most closed stratification systems are characterized by *castes,* or social strata into which people are born and in which they remain for life.

- Membership in a caste is an *ascribed status* (a status acquired at birth), as opposed to an *achieved status* (one based on the efforts of the individual).

- Open societies are characterized by *classes,* which are social strata based primarily on economic criteria.

- The classes of modern societies are not homogeneous; within any given class, there are different groups defined by how much honor or prestige they receive from society in general. Such groups are sometimes referred to as *status groups.*

- The way people are grouped with respect to their access to scarce resources determines their *life chances*—the opportunities they will have or be denied throughout life.

How are stratification and inequality related to the way people in a society gain their livelihood?

- The principal forces leading to social stratification are created by the means of existence in a given society.

- For small farmers or peasants (the majority of the world's population), social strata are based on land ownership and agrarian labor, with the members of the lowest strata doing the hardest work and those at the top of the stratification system living in relative comfort.

- Modern industrial societies are characterized by *structural mobility* (the elimination of entire classes as a result of changes in the means of existence) and *spatial mobility* (the movement of individuals and groups from one location to another).

How are a society's inequalities supported by the norms and values of its culture?

- People accept their place in a stratification system because the system itself is part of their society's culture.

- The facets of culture that justify the stratification system are learned through the processes of socialization. The system is often justified by an ideology.

- At the micro level, the norms of everyday interactions, especially *deference* and *demeanor,* reinforce the society's stratification system.

In what ways do power and authority work to maintain existing stratification and inequality?

- Changes in stratification systems may have as much to do with realignments of social power as with economic or cultural changes.

- *Power* has been defined as "the probability that one actor within a social relationship will be in a position to carry out his own will despite resistance."

- Legitimate power is called *authority* and is a major factor in maintaining existing relationships among castes or classes.

How does stratification in the modern era account for global patterns of inequality and global poverty?

- According to Karl Marx, capitalism divided societies into classes based on ownership of the means of production. The largest of these classes, the workers, must sell their labor to capitalists or landowners in return for wages.

- In time, the workers would become conscious of their shared interests as a class and would rebel against the capitalist class. The outcome of the revolution would be a classless society.

- Marx defined social class in economic terms. Max Weber took issue with this definition and pointed out that people are stratified not only by wealth but also by how much honor or prestige they receive from others and how much power they command.

- Marx's view of stratification is also challenged by studies of social mobility in industrial societies, which have shown that there is considerable movement between classes.

How do theories of stratification explain global inequality and major aspects of inequality in individual societies?

- Modern conflict theorists, like Marx, believe that class conflict is a primary cause of social change. They disagree, however, on the nature of the class structure of capitalist societies.

- Functionalist theorists believe that classes emerge because an unequal distribution of rewards is necessary in order to channel talented people into important roles in society.

- This view has been criticized because it fails to account for the fact that social rewards in one generation tend to improve the life chances of the next generation; nor does it explain why talented people from lower-class families often are unable to obtain highly rewarded positions.

- From the interactionist perspective, the stratification system is not a fixed system, but rather, one that is created out of everyday behaviors.

What is the impact of globalization on stratification in rich and poor nations?

- Globalization opens new regions to increased trade and investment, to increased communications from other parts of the world, and to greater movement of people across national boundaries, but the benefits of these trends are not shared equally.

- Social scientists and international planning experts often use health statistics as a barometer of how well a nation is doing in narrowing the gap between the haves and the have-nots in its population.

- In the world's poorest nations, women face double deprivation: Human development problems such as lack of medical care, lack of clean water, extreme poverty, and unemployment affect all but the comfortable elite, but women in these countries face even higher levels of risk and deprivation.

The Kornblum Companion Website

www.cengagebrain.com

Supplement your review of this chapter by going to the companion website to take one of the tutorial quizzes, use the flash cards to master key terms, and check out the many other study aids you'll find there. You'll also find special features, such as GSS data and Web links that will put data and resources at your fingertips to help you with that special project or to do some research on your own.

11

Dimensions of
Social Inequality in
America

Social Class and
Life Chances in the
United States

More Equality?

Still the Land of
Opportunity?

George Robinson/Alamy

Inequalities of Social Class

FOCUS QUESTIONS

How is inequality structured in the United States?

What kinds of social classes are found in the United States, and how do they compare to classes
in other societies?

How are social classes changing, and what effects do the changes have on members of these classes?

Do people still believe in the American dream of hard work and eventual upward mobility?

© Jose More

"What I always wonder," he says with a slight shrug, "is how the young people are going to find a way to get ahead now that the mills are gone." Ed Sadlowski is standing in the economically depressed streets of South Chicago, the neighborhood where he was born, where he raised his five children, and where most of them are now living and raising their own children.

"You know, Bill," he continues as we walk slowly past the deserted mill gates, "the United States makes as much steel as we did back when these plants had thousands of workers in them. It's just that we make it so much more efficiently. The productivity of the individual workers is so much greater because of the new technologies, computer control of machines, robotization, and all the rest. So the old system of mass employment in industries like steel is never going to return. The industrial proletariat that Marx wrote about is gone, at least in this country, but not elsewhere in the world where industry is newer."

Edward Sadlowski is among the nation's leading experts in industrial history and the sociology of labor–management relations. He is a renowned figure in Chicago politics and was recently featured in a book about people who sustain that city's gritty blue-collar culture (Kotlowitz, 2004). We met many years ago when I was working in a local steel mill while gathering material for a book about this famous industrial community. Sadlowski's grandfather came to Chicago as an immigrant from Poland, drawn like millions of others by the chance to use his strong back and hands in the search for a new and better life in America. His father worked in the mills too. He was one of the union's early organizers. From an early age, Ed was taught that the unions in steel, railroads, and mining were responsible for bringing workers a fairer share of the profits of their industries. To do so, they had to constantly struggle against the mill owners, who often seemed ready to kill them rather than relinquish their control or share the wealth.

When Sadlowski was eighteen, he too entered the mills, but he was able to join an apprenticeship program in which he learned a skilled trade as a machinist. He immediately became active in the union and was elected to represent his workmates as their grievance officer. This meant that he was a shop floor lawyer who assisted workers when they were disciplined or fired. Their union contract afforded the workers a common pay scale and scheduled increases in their hourly wages, as well as health and other benefits they had not had before the unions gained legal recognition in the 1930s. Overtime pay and the right to a semblance of due process in disciplinary matters were also among the union's achievements.

Eventually Sadlowski was elected president of the union at U.S. Steel's South Works, Chicago's largest steel mill until it was torn down in the late 1980s. Later in his career, he ran unsuccessfully for president of the U.S. and Canadian steelworkers' union, but he has remained a staunch union supporter throughout his working life.

As we walk through the "brownfields"—acre after acre of abandoned railroad tracks and ore docks on the shore of Lake Michigan—Sadlowski wonders about the future of manufacturing in the United States. There are so many groups whose mobility depended on well-paid jobs in plants with union contracts until very recently. In Chicago, there are thousands of African Americans and Chicanos whose progress toward greater economic security has been stymied by plant closings. And from his travels throughout industrial America, Sadlowski understands only too well the consequences of rapid deindustrialization. "Too many people are looking for too few good jobs," he comments. "You know, the kind that pay a decent wage you can raise a family on, and that pay for adequate health care."

DIMENSIONS OF SOCIAL INEQUALITY IN AMERICA

The Sadlowskis and other families in South Chicago are struggling to adapt to a set of social changes that are global in scope. People in different parts of the world are experiencing these changes in differing ways. South of the U.S. border, for example, parents in hundreds of communities miss their children, who have migrated to manufacturing towns along the border. In the industrial centers of North America, people mourn the loss of jobs to countries or regions where labor costs are lower. These changes are forcing many people to go back to school and gain new skills and credentials so that they can be more competitive in changing labor markets. Union members in manufacturing industries are being forced to end strikes and go back to work to avoid losing their jobs to "replacement workers." But these are only some of the realities of life in a stratified and rapidly changing industrial society. The U.S. economy also generates immense wealth and a growing number of jobs. The society boasts an enormous middle class, many of whose members are far more affluent than their parents and grandparents ever dreamed of becoming. Thus, social mobility and class conflict, the subjects of this chapter, remain important aspects of life in the United States and elsewhere in the world. No wonder they are central issues in contemporary sociology.

In the United States, inequalities of class and status have worsened despite rates of employment that are generally higher than those of other Western industrialized nations (Shapiro & Greenstein, 2001). In the nation's poorest communities—in urban ghettos, on Indian reservations, and in the hill country of the South—many people doubt whether the conditions of poverty and depression in which they live can ever be improved. Sociological research has shown that the rich are doing better than ever and that the gap between the haves and the have-nots is widening. In different regions of the nation, the details of social inequality differ, but the basic patterns of wealth, prestige, and power are the same (Aronowitz, 2003; Gans, 1995).

This chapter further develops many of the terms and concepts defined in the preceding chapter and applies them to the study of inequality and stratification in the United States. First we discuss how sociologists measure inequality and their changing views of how inequalities result in the formation of social classes. Then we examine social classes and life chances in the United States today and analyze the influence of social-class position on opportunities for upward mobility. We will see that there is nothing static about social inequality in the United States; changes in the class structure of North America are occurring even as this chapter is being written.

Measures of Social Inequality

The basic and most readily available measures of inequality in any society are income, wealth, occupational prestige, and educational attainment. Of these measures, income is most often used to give an initial view of social-class inequality.

Income Inequality A common method of comparing incomes in a large population like that of the United States is to divide the entire population into equal fifths, or *quintiles,* on the basis of personal or household income. The researcher starts at the lowest level of reported income and continues upward until one-fifth of the population has been accounted for. This establishes the upper and lower income boundaries for that quintile. The same procedure is used to identify each of the other four quintiles; the average income for each quintile can then be compared with the average incomes for the other four quintiles. Table 11.1 shows that income inequality increased between 1970 and 2008, and that the income gap between the highest income quintile and the lowest grew most rapidly in the last five years of that period. In 2008, the table shows, households in the top 20 percent of the income range enjoyed almost 50 percent of all income, whereas those in the lowest and next-to-lowest quintiles received income shares of 3.5 percent and 9.0 percent, respectively. The table also demonstrates that households whose earnings put them in the top 5 percent gained income most dramatically, from 16.6 percent of all income in 1970 to 27.3 percent in 2008.

The Gini Index, an International Measure of Inequality Social scientists often use other measures to study inequality between different nations or populations. Among these, the Gini index is the most commonly used. Developed by the Italian statistician Corrado Gini in 1912, the Gini index or coefficient measures income

TABLE 11.1

Share of Aggregate Income Received by Each Quintile and Top 5 Percent of Households

Year	Quintiles					Top 5%
	Lowest	Fourth	Third	Second	Highest	
1970	4.1	10.8	17.4	24.5	43.3	16.6
1975	4.4	10.5	17.1	24.8	43.2	15.9
1980	4.3	10.3	16.9	24.9	43.7	15.8
1985	4.0	9.7	16.3	24.6	45.3	17.0
1990	3.9	9.6	15.9	24.0	46.6	18.6
1995	3.7	9.1	15.2	23.3	48.7	21.0
2000	3.6	8.9	14.9	23.0	49.6	21.9
2004	3.4	8.7	14.7	23.2	50.1	23.6
2008	3.5	9.0	14.8	28.0	49.7	27.3

Source: U.S. Census Bureau, 2008.

inequality on a scale from 0 to 1, where 0 represents perfect equality, with everyone having the exact same income, and 1 represents perfect inequality, with one person having all income. (Scores are often multiplied by 100 to make them easier to understand.) According to United Nations calculations, Gini index ratings for countries range from 24.7 in Denmark to 74.3 in Namibia. Most urban industrial nations have a Gini coefficient in the high 20s to mid-30s. That of the United States fluctuates around 40, the highest among the nations to which it is usually compared.

The United States is not the only advanced industrial nation in the world where inequality as measured by the Gini index has risen very quickly. In Japan, it rose from 24 in the early 1990s to about 39 in 2007, a much faster increase

and a larger increase than that of the United States. The same has happened in China and much of East Asia, which has seen very fast economic growth (United Nations, 2006). Figure 11.1 presents a sample of recent changes in the Gini index for industrial nations and demonstrates the impact in the United States of the increasing concentration of income among a smaller number of individuals and households.

Wealth Inequality If one studies the distribution of wealth rather than income, the picture is even more skewed in favor of the rich. If we measure wealth in terms of the net financial worth of households—that is, their total assets— only about 6 percent of American households have a net worth of $250,000 or more, whereas over 25 percent have a net worth of less than

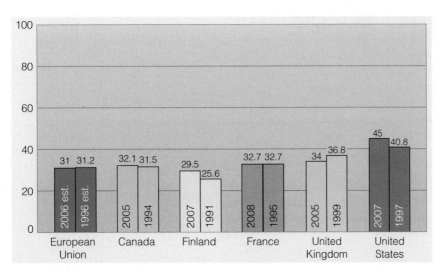

Figure 11.1 Change in Real Average Income, by Quintile, 2003–2004. Source: U.S. Census Bureau, 2005.

TABLE 11.2

Distribution of Net Worth and Financial Wealth in the United States, 1983–2007

Year	Total Net Worth		
	Top 1 percent	Next 19 percent	Bottom 80 percent
1983	33.8%	47.5%	18.7%
1989	37.4	46.2	16.5
1992	37.2	46.6	16.2
1995	38.5	45.4	16.1
1998	38.1	45.3	16.6
2001	33.4	51.0	15.6
2004	34.3	50.3	15.3
2007	34.6	50.5	15.0

Total assets are defined as the sum of: (1) the gross value of owner-occupied housing; (2) other real estate owned by the household; (3) cash and demand deposits; (4) time and savings deposits, certificates of deposit, and money market accounts; (5) government bonds, corporate bonds, foreign bonds, and other financial securities; (6) the cash surrender value of life insurance plans; (7) the cash surrender value of pension plans, including IRAs, Keogh, and 401(k) plans; (8) corporate stocks and mutual funds; (9) net equity in unincorporated businesses; and (10) equity in trust funds.

Source: From *Who Rules America? Power in America: Wealth, Income and Power,* by G. William Domhoff, September 2005 (updated August 2010). http://sociology.ucsc.edu/whorulesamerica/power/wealth.html. Reprinted by permission of the author.

$5,000 (U.S. Census Bureau, 2010). Table 11.2 shows that wealth is more unequally distributed than income. The top 20 percent of American households enjoys over 45 percent of all wealth (real estate, possessions, stocks, businesses, and so on). Within that top fifth, the richest 1 percent own almost 35 percent, a figure that the table shows has been creeping upward over the first decade of this century.

Recent studies show that the United States has replaced Great Britain and Ireland as the Western industrial nation with the largest gap between rich and poor (Hacker, 1997; Stewart, 2002). The percentage of total private wealth owned by the richest 1 percent of the U.S. population is rising quickly, but the opposite is true in Great Britain, which, unlike the United

For the upper classes, food becomes a symbol of status; its cost and the artistry of its presentation assume an importance well beyond mere nutrition.

educational attainment: The number of years of school an individual has completed.

States, has never been thought of as an egalitarian society (Hacker, 1997; Keller, 2005).

Remember, too, that inequality expressed in dollars of income or possession of property and stocks translates into large differences in what people can spend on health care, shelter, clothing, education, books, movies, trips to parks and museums, vacations, and other necessities and comforts.

Inequality in Educational Attainment The distribution of educational attainment and occupational prestige is more nearly equal than the distribution of wealth and income. **Educational attainment,** or number of years of school completed, has become more equal among major population groups in the United States, but important inequalities remain. The educational attainment of the nation's black population, for example, has risen dramatically in the past generation. In 1960, only 20 percent of African Americans (who were age twenty-five or older at the time) had completed four years of high school or more, whereas the comparable figure for whites was 43 percent. Today, 83.0 percent of African Americans have completed four years of high school or more, compared to 82.1 percent of whites; however, only 19.6 percent of African Americans have completed four years or more of college, compared to 29.8 percent of whites. For Latinos, the gap remains far wider. Approximately 62.3 percent of Latinos in the U.S. population (age twenty-five or older) have completed four years of high school or more, and 13.3 percent have completed four years of college or more (*Statistical Abstract,* 2010).

Women, on the other hand, have gained parity with men in educational attainment through high school and are rapidly pulling even in higher education, although they lag in scientific and technical fields like science and engineering. In 2003 (the latest year for which national data are available), 85 percent of women and 84 percent of men were high school graduates, but 29 percent of men had finished college, whereas the proportion was 26 percent for women. Throughout the nation, there has been a marked increase in women's college enrollments while the rate of male enrollment has remained more or less constant for the past decade. And women are outperforming men in many college and university majors. As we will see in Chapter 13, however, women's income at their first jobs after college averages only about 80 percent of what male graduates earn in equivalent jobs.

The past two decades have seen an important increase in the relationship between educational attainment and income. The advantage enjoyed by individuals with college educations

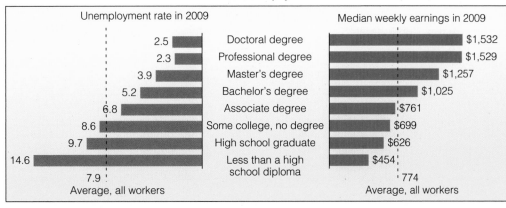

Education pays

Unemployment rate in 2009		Median weekly earnings in 2009
2.5	Doctoral degree	$1,532
2.3	Professional degree	$1,529
3.9	Master's degree	$1,257
5.2	Bachelor's degree	$1,025
6.8	Associate degree	$761
8.6	Some college, no degree	$699
9.7	High school graduate	$626
14.6	Less than a high school diploma	$454
7.9 Average, all workers		774 Average, all workers

Figure 11.2 Education pays. Source: Bureau of Labor Statistics, Current Population Survey.

over those with only high school diplomas has always been significant, but as shown in Figure 11.2, the less education one has, the greater the chance of being unemployed, and the lower one's earnings are likely to be. In a rapidly changing economy like that of the United States, educated people are in increasing demand, and education thus becomes a more common route to upward mobility. Improvements in **educational achievement**—in basic reading, writing, and computational skills—are increasingly vital to individual careers.

EDUCATION AND HUMAN CAPITAL For the entire society, improvements in educational achievement represent improvements in the nation's "human capital," the wealth-producing capacity of its people (Harbison, 1973). But educational attainment and educational achievement are not always correlated. When students are moved through the school system without meeting certain minimum achievement criteria, attainment figures mask significant gaps in achievement. As a result, large segments of the population are unable to achieve nearly as much as they might, either for themselves or for their society.

If citizens are unable to improve their human capital because their families cannot afford to keep them in school and out of the labor market, or because they are girls and their families do not believe they need as much education as boys do, or because they are disadvantaged by racial or ethnic discrimination, most people would agree that such conditions are unfair. But what does a concern for social justice suggest a society should be doing to address this aspect of inequality? The majority in most developed nations, including the United States, has supported taxation systems that ask the well off to pay higher proportions of their income than the less well off. This is known as

progressive taxation. Many conservatives do not believe this is a legitimate form of income redistribution or even that the goal of greater equality is legitimate. They wonder how a concern for social justice can result in the government's taking away income and wealth from some to give to others. An answer to this criticism is that there are times when investment in the needs of the entire population, by improving the human capital of disadvantaged children and their families, for example, justifies some sacrifice by the well off. In democracies, this kind of dispute over social justice is settled at the ballot box, and often the conservative view prevails. At the global level, however, how effectively issues of inequality are addressed has a lot to do with how a society is regarded in the community of world nations (Sen, 2000).

Inequality and Occupations Another aspect of social inequality, **occupational prestige**, is measured using surveys of how people throughout a society rate different jobs. Every society and culture ranks its occupations from low to high. Dirty jobs, work that others reject, and low-paying work of all kinds are ranked low in esteem or prestige. Work that is well paid, technically sophisticated, or involves caring for others tends to be thought of quite positively. Sociologists measure the prestige of occupations by asking samples of respondents to rate a wide variety of occupations. Such surveys find that the prestige people attach to an occupation is heavily influenced by the education required for the job or the authority it offers its holder, as well as by income (Blau & Duncan, 1967; Bose & Rossi, 1983; Davis, Smith, & Marsden, 2007). Thus, many occupations that do not pay extremely high salaries are nonetheless highly rated. The prestige of an occupation is ranked from 100 (highest) to 0 (lowest). Typically, prestige scores range from 95.8 for "physician" to 4.6 for "rag picker." Sociologists rank about 75; skilled construction work tends to rank in the 50s; while service workers such as stock clerk rank about 24 and waitress/waiter 22, at the lower end of the scale. Table 11.3 shows how selected occupations are ranked according to their prestige.

Sociologists use occupational prestige rankings to study many other aspects of work and stratification (Zhou, 2005). They find, for example, that occupational prestige is highly related to satisfaction with one's work, but the satisfaction is not always derived from either high occupational

educational achievement: Mastery of basic reading, writing, and computational skills.

occupational prestige: The honor or prestige attributed to specific occupations by adults in a society.

TABLE 11.3

Prestige Scores of Selected Occupations

Occupation	Prestige	Occupation	Prestige
Physician	95.8	Beautician	42.1
Mayor	92.2	Piano tuner	41.0
Lawyer	90.1	Landscape gardener	40.5
College professor	90.1	Truck driver	40.1
Architect	88.8	House painter	39.7
City superintendent of schools	87.8	Hairdresser	39.4
Owner of a factory employing 2,000 people	81.7	Pastry chef in a restaurant	39.4
Stockbroker	81.7	Butcher in a shop	38.8
Advertising executive	80.8	Washing machine repairman	38.8
Electrical engineer	79.5	Automobile refinisher	36.9
Building construction contractor	78.9	Someone who sells shoes in a store	35.9
Chiropractor	75.3	Cashier	35.6
Registered nurse	75.0	File clerk	34.0
Sociologist	74.7	Dress cutter	33.3
Accountant	71.2	Cattle driver, working for own family	33.0
High school teacher	70.2	Cotton farmer	32.4
Manager of a factory employing 2,000 people	69.2	Metal container maker	31.1
Office manager	68.3	Hospital aide	29.5
Administrative assistant	67.8	Fireman in a boiler room	29.2
Grade school teacher	65.4	Floor finisher	28.8
Power house engineer	64.5	Assembly-line worker	28.3
Hotel manager	64.1	Book binder	28.2
Circulation director of a newspaper	63.5	Textile machine operator	27.9
Social worker	63.2	Electric wire winder	27.6
Hospital lab technician	63.1	Vegetable grader	27.4
Artist	62.8	Delivery truck driver	26.9
Electrician	62.5	Shirt maker in a manufacturing plant	26.6
Insurance agent	62.5	Person who repairs shoes	26.0
Private secretary	60.9	Fruit harvester, working for own family	26.0
Floor supervisor in a hospital	60.3	Blacksmith	26.0
Supervisor of telephone operators	60.3	Housekeeper	25.3
Plumber	58.7	Flour miller	25.0
Police officer	58.3	Stock clerk	24.4
Manager of a supermarket	57.1	Coal miner	24.0
Car dealer	57.1	Boardinghouse keeper	23.7
Practical nurse	56.4	Warehouse clerk	22.4
Dental assistant	54.8	Waitress/waiter	22.1
Warehouse supervisor	54.2	Short-order cook	21.5
Assembly-line supervisor in a manufacturing plant	53.8	Babysitter	18.3
Carpenter	53.5	Rubber mixer	18.1
Locomotive engineer	52.9	Feed grinder	17.8
Stenographer	52.6	Garbage collector	16.3
Office secretary	51.3	Box packer	15.1
Inspector in a manufacturing plant	51.3	Laundry worker	14.7
Housewife	51.0	Househusband	14.5
Bookkeeper	50.0	Salad maker in a hotel	13.8
Florist	49.7	Janitor	12.5
Tool machinist	48.4	Yarn washer	11.8
Welder	46.8	Maid (F)/household day worker (M)	11.5
Wholesale salesperson	46.2	Bellhop	10.6
Telephone operator	46.2	Hotel chambermaid (F)/hotel bed maker (M)	10.3
Auto mechanic	44.9	Carhop	8.3
Typist	44.9	Person living on welfare	8.2
Typesetter	42.6	Parking lot attendant	8.0
Postal clerk	42.3	Rag picker	4.6

Source: Adapted from Bose & Rossi, 1983; Davis, Smith, & Marsden, 2007.

prestige or high income. Instead, it is often related to the sense that an occupation expresses one's creativity (painters, sculptors, authors) or that one is engaged in helping others (teachers, psychologists). (See Table 11.4.) Many other occupations, however, such as registered nurse or accountant, rank quite high on the prestige scale but rank only in the middle of the job satisfaction range because of other factors such as the difficulty or long hours involved. The same is true of police detectives and reporters or editors. Here are some examples:

Job Satisfaction	Mean Score % Very Satisfied
Police and Detectives	59.3
Registered Nurses	53.1
Editors and Reporters	52.9
Accountants	49.7

People and groups seek many other types of prestige or honor besides occupational prestige. One can attain prestige through a family name or reputation, achievement in a sport, or service to one's community, to name only a few other types of prestige. But occupational prestige is especially important because, along with income, wealth, and education, it figures prominently in the way people think about their position and that of others in their society's system of social-class rankings.

These measures of wealth, income, education, and occupational prestige indicate that not all Americans share equally in "the good life." But what are the larger consequences of these patterns of inequality, both for individuals and for society? Do they combine to form an identifiable system of social-class stratification in the United States? Do marked patterns of inequality contradict the American dream, the vision of the United States as a place where hard work and sacrifice will lead to success and material comfort? We explore these questions in the rest of this chapter, but before doing so, it is helpful to look at how views of inequality have changed over the past 150 years, as the United States was transformed from an agrarian society to an urban industrial society and, finally, to a postindustrial society.

Inequality, the American Paradox Why does the richest, most powerful nation on earth exhibit the highest measures of inequality and the highest rates of poverty of all the urban industrial nations with which it is often compared? And why do so many Americans tolerate and even defend these inequalities? The seeming contradiction between the nation's achievements—its productivity, its liberties,

TABLE 11.4

Top Occupations in Job Satisfaction

Rank	Occupations	Mean Score	% Very Satisfied
1	Clergy	3.79	87.2
2	Physical therapist	3.72	78.1
3	Firefighters	3.67	80.1
4	Education administrators	3.62	68.4
5	Painters, sculptors, related	3.62	67.3
6	Teachers	3.61	69.2
7	Authors	3.61	74.2
8	Psychologists	3.59	66.9
9	Special education teachers	3.59	70.1
10	Operating engineers	3.56	64.1
11	Office supervisors	3.55	60.8
12	Security and financial services salespeople	3.55	65.4

Bottom Occupations in Job Satisfaction

Rank	Occupations	Mean Score	% Very Satisfied
1	Roofers	2.84	25.3
2	Waiters/servers	2.85	27.0
3	Laborers, except construction	2.86	21.4
4	Bartenders	2.88	26.4
5	Hand packers and packagers	2.88	23.7
6	Freight, stock, and material handlers	2.91	25.8
7	Apparel clothing salespeople	2.93	23.9
8	Cashiers	2.94	25.0
9	Food preparers, misc.	2.95	23.6
10	Expediters	2.97	37.0
11	Butchers and meat cutters	2.97	31.8
12	Furniture/home furnishing salespeople	2.99	25.2

Job Satisfaction: On the whole, how satisfied are you with the work that you do? Would you say that you are very satisfied, a little dissatisfied, or very dissatisfied?

Mean score runs from 1 for someone who is very dissatisfied to 4 for someone who is very satisfied.

Source: Smith, 2007.

its entrepreneurial risk taking, its scientific leadership, to name a few (see Figure 11.3 for others)—and its growing inequality (and worsening health indicators like child mortality and homicide) is often called *the American paradox* (Halstead, 2003). In their attempts to understand the origins of this paradox, sociologists have conducted studies of how Americans' views of inequality have developed and changed in different

This list of "bests" and "worsts" is based on a variety of sources—including statistics from the United Nations, the Organization for Economic Cooperation and Development, and a number of other groups and experts—but the basic criteria are consistent. Among advanced democracies, the United States had to rank in the top three for a category to be listed under "bests," and in the bottom three for a category to be listed under "worsts." (Where applicable, all rankings were determined on a rate basis or as a percentage of population.)

Bests	Worsts
Gross domestic product	Poverty
Productivity	Economic inequality
Business start-ups	Carbon-dioxide emissions
Long-term unemployment	Life expectancy
Expenditure on education	Infant mortality
University graduates	Homicide
R&D expenditure	Health-care coverage
High-tech exports	HIV infection
Movies exported	Teen pregnancy
Breadth of stock ownership	Personal savings
Volunteerism	Voter participation
Charitable giving	Obesity

Figure 11.3 The American Paradox: Two Faces of the United States. Source: Halstead, 2003.

historical periods and different types of communities. They also conduct research on how individuals and groups perceive inequality and how it affects their life chances, subjects to which we turn in the next section.

Changing Views of Social Inequality

The dominant view of American society before the Industrial Revolution was the Jeffersonian view, which envisioned a society in which most families lived on their own farms or ran small commercial or manufacturing enterprises (Lipset, 1979; Pyle, 1996). This view emphasized the value of economic self-sufficiency through hard work—a value that has persisted to the present time. Yet even in Jefferson's time, the pattern of agrarian and small-town independence was not universal. In the early nineteenth century, American society was already developing a rather complex and diverse set of stratification systems. In New England and the Middle Atlantic states, the Jeffersonian ideal of rural farmers and town-dwelling tradesmen, all with similar degrees of power and prestige, was reflected in reality. In the larger cities, a landless wage-earning class was emerging, as was a class of factory owners, bankers, and entrepreneurs. Throughout most of the southern states, in contrast, the major classes were the plantation owners and their families and staffs, the town merchants and tradesmen, and the marginal "poor white" farmers; in addition to these classes, there was a caste: the slaves.

Native Americans were not included in any of these stratification systems. By the 1830s, they were already a *pariah group*—excluded from the stratification system and considered too different or inferior to be eligible for citizenship. As a consequence of their pariah status in many parts of the nation, together with the settlers' desire to inhabit their lands, most Native Americans were steadily pushed west of the Mississippi.

The Impact of the Depression A century later, in the mid-1930s, a different view of inequality had emerged. In the midst of the most severe economic and social depression in world history, President Franklin D. Roosevelt claimed that "one third of the nation [was] ill-housed, ill-clad, and ill-nourished." Faced with mass poverty and unemployment, with able-bodied workers selling apples on street corners, the grandparents of many of today's college students questioned the reality of the American dream. Many observers saw heightened conflict between workers and the owners of capital. To many intellectuals, it seemed that the dire predictions of Karl Marx were coming true and that only a communist revolution could save millions of people from starvation. Largely as a result of the depression and fear of social disorder, legislators in the United States passed reform measures like the Social Security Act to alleviate the worst effects of poverty.

Muncie, Indiana, made famous by Helen and Robert Lynd in their books *Middletown* (1929) and *Middletown in Transition* (1937), provided numerous examples of the gap in income and lifestyle between the working class and the owners or managers of capital. In the decades since 1880, industrialization had transformed Muncie from an agricultural trade center with 6,000 inhabitants into a bustling city of more than 36,000. According to the Lynds (1937), Muncie's businessmen "lived in a culture built around competition, the private acquisition of property, and the necessity for eternal vigilance in holding on to what one has" (p. 25). Below them was a middle class of managers, teachers, and clergymen who believed in the right of the business class to control wealth and power. In contrast, "across the railroad tracks from this world of businessmen [was] the other world of wage earners—constituting a majority of the city's population" (p. 25).

When the Lynds first studied Muncie, they became convinced that social-class divisions, especially those between the working class and the owners of local firms, were reaching a state of crisis. When they returned during the Great Depression, they continued to find evidence that the business class dominated the city's social institutions. In one famous passage, a local factory worker described the influence of the "X family": "If I'm out of work I go to the X plant; if I need money I go to the X bank, and if they don't like me I don't get it; my children go to the X college; when I get sick I go to the X hospital" (Lynd & Lynd, 1937, p. 74).

The X family (actually the Ball family) owned the large glassworks in Muncie. The Ball brothers had built a fortune making glass jars and then had expanded into real estate, railways, banking, and retail stores. They worked so hard that they had little time to enjoy the pleasures their wealth could obtain. But their children appeared to be creating a distinctive upper class with exclusive country clubs, farms where their horses could be maintained, and private planes.

As a result of the depression, there was more hostility between economic classes in Muncie in the late 1930s than there had been when the Lynds first studied Muncie's class system a few years before the depression. The workers and their families were far less likely to believe that the business class had a right to its wealth, and they were increasingly drawn to social movements, especially the labor movement, that challenged the power of the business class. The business class itself was far less unified because the managers of the large corporations moving into the city did not feel obligated to join the clubs and churches of the established elite.

Yankee City and the Black Metropolis

In the mid-twentieth century, several studies of inequality in American communities refined our knowledge of how income and prestige combine to produce stratification in modern communities. The research of William Lloyd Warner on social class in a New England town and Drake and Cayton's classic study of the caste system in an urban ghetto are important examples because they describe class relations more clearly than earlier studies did.

Warner's study was conducted in the seaside town of Newburyport, Massachusetts. Once famous for its whaling fleet and Yankee sea captains, Newburyport had become a small city with 17,000 inhabitants. It had several textile and shoe factories and had managed to hang on to many of its Yankee families, descendants of the hardy seafarers of the nineteenth century. For this reason, Warner dubbed the community "Yankee City."

Here is how Warner began his description of the class structure of Yankee City:

> Studies of communities in New England clearly demonstrate the presence of a well-defined social-class system. At the top is an aristocracy of birth and wealth. This is the so-called "old family" class. The people of Yankee City say the families who belong to it have been in the community for a long time—for at least three generations and preferably many generations more than three. "Old family" means not only old to the community but old to the class. (Warner, Meeker, & Calls, 1949, p. 12)

The new families of Yankee City's upper class, who possessed wealth but had not gained "old family" status, "came up through the new industries—shoes, textiles, silverware—and finance" (p. 12). Below this upper class, people in Yankee City identified an upper-middle class of highly respectable people who "may be property owners, such as storekeepers, or highly

TABLE 11.5

Prestige Classes in "Yankee City" and Proportion of Income Spent on Necessities of Life, 1941*

Prestige Class	Percent of Total Population	Percent Spent on Necessities
Upper-upper	1.4	33
Lower-upper	1.6	35
Upper-middle	10.2	51
Lower-middle	28.1	59
Upper-lower	32.6	66
Lower-lower	25.2	75

*"Necessities of life" include food, shelter, and clothing.

Source: *The Social Life of a Modern Community*, by W. Lloyd Warner and Paul S. Lunt, copyright © 1941. Reprinted by permission of Yale University Press.

educated professionals, but they do not have the wealth or family status to be included in the upper class" (p. 13).

Below these classes, Warner's respondents identified three "common man" levels. These made up the lower-middle class, which was composed primarily of clerks, skilled tradesmen, other white-collar workers, and some skilled manual workers. Directly below this class, and most difficult for the respondents to separate from those above and below it, was the "upper-lower class," or the people Yankee City residents identified as "poor but honest workers, who more often than not are only semi-skilled or unskilled" (p. 14).

In Yankee City, the lowest class—which Warner called the "lower-lower class"—was described as the "low-down Yankees [and immigrants] who live in the clam flats" and have a bad reputation among the classes above them. "They are thought to be improvident and unwilling or unable to save money . . . and therefore often dependent on the philanthropy of the private or public agency and on poor relief" (p. 14). The six "prestige classes" that Warner identified on the basis of the Yankee City interviews are listed in Table 11.5, along with the percentage of the city's population in each class and the proportion of income spent on necessities by members of each class. It is interesting that Warner's terms, such as *upper-middle class* and *lower-middle class,* have become part of our everyday vocabulary.

At about the same time that Warner was conducting his study, other social scientists were attempting to describe stratification in larger, more complex communities. Among those efforts was the first full-scale study of Chicago's African American community. It was carried out by St. Clair Drake and Horace Cayton in the 1940s and published under the title *Black Metropolis.*

Studies of inequality in southern communities had already documented the existence of a racially based caste system in the South (Dollard, 1937). But did this racial caste system also operate in northern cities like Chicago? At the time of Drake and Cayton's study, Chicago's black population was almost entirely concentrated in a single urban ghetto. A **ghetto** is a section of a city that is segregated either racially

ghetto: A section of a city that is segregated either racially or culturally.

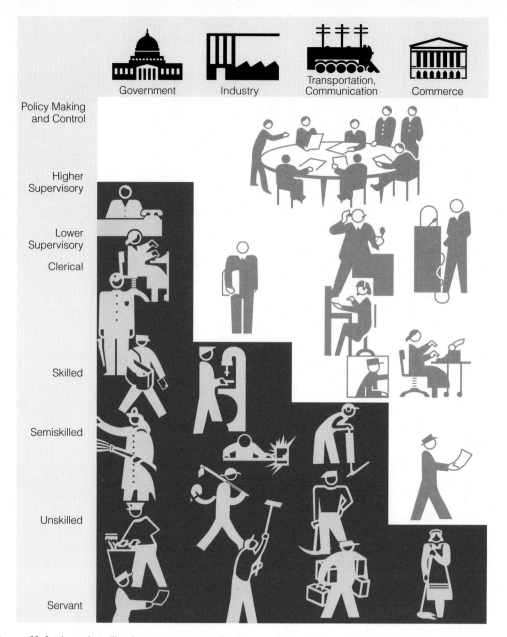

Figure 11.4 The Job Ceiling in Government and Private Industry Before the Civil Rights Movement. Source: Illustration from *Black Metropolis, a Study of Negro Life in a Northern City*. Copyright 1970 by St. Clair Drake and Horace Cayton, reproduced by permission of the author's estate.

or culturally (for example, by race, religion, or ethnicity). Large numbers of blacks worked outside the ghetto, but even those who could afford housing elsewhere were excluded from other neighborhoods by long-established patterns of residential segregation. In addition, despite the integration of workers in industrial occupations, especially steel and meatpacking, Drake and Cayton (1970/1945) showed that Chicago's blacks were barred from more comfortable, better-paying jobs in other fields. They called this pattern of occupational segregation, in which blacks could rise only so far, the *job ceiling*. In their view, the job ceiling, together with the segregation of blacks in the ghetto,

constituted a northern form of the racial caste system (see Figure 11.4). Many of the persistent racial inequalities in contemporary U.S. society have their origins in this urban job ceiling and the ways in which blacks and other people of color adapted to this form of stratification early in the twentieth century. We return to this subject in the next chapter.

Middletown Revisited In the late 1970s, Theodore Caplow, Howard Bahr, Bruce Chadwick, and their collaborators returned to Muncie to see how the community had changed over more than forty years. They found that the city's stratification system had changed a great deal since the depression. The population had increased from about 40,000 to more than 80,000. This size increase brought new patterns of mobility:

> The local dominance of a handful of rich families that looked so threatening in 1935 quietly faded away during the decades of prosperity that followed World War II. Hundreds of fortunes were made in the old ways and new—building subdivisions and shopping centers; trading in real estate; selling insurance, advertising, farm machinery, building materials, fuel oil, trucks and automobiles, furniture. . . . Middletown's new rich . . . lived much less ostentatiously than their industrial predecessors, and much of their money was spent away from Middletown (for yachts in Florida, condominiums in Colorado, boarding schools for their children, and luxury tours to everywhere for themselves). . . . The handful of families whose wealth antedated World War II adopted the same style. The imitation castles of the X, Y, and Z families were torn down or converted for institutional uses. (1983, p. 12)

The distinctive upper class that the Lynds saw emerging in Muncie in the 1930s had vanished by the 1970s:

> Meanwhile, at the lower end of the socioeconomic scale, life-styles were becoming more homogeneous. The residential building boom that began after World War II continued, year after year, to submerge the flat, rich farmlands at the edge of town under curved subdivision streets bordered by neat subdivision houses with various exteriors but nearly identical interiors. They all had central heating, indoor plumbing, telephones, automatic stoves, refrigerators, and washing machines. (pp. 12–13)

By the 1970s, the factory workers of Muncie were much better off than they had been in 1935. They enjoyed job security, health insurance, and paid vacations, and their average incomes were higher than those of many white-collar workers. These changes had come about largely as a result of the activities of labor unions, which had been excluded from Muncie's factories in 1935 but were accepted soon afterward (Caplow et al., 1983). But in the 1980s, the tide turned again, and Muncie, like many other industrial cities, began losing manufacturing jobs and gaining more low-paying service jobs.

Today, the community of Muncie is far less self-contained than it was in the early decades of the twentieth century (Caccamo, 2000). It is subject to the influence of outside forces such as the shift of manufacturing jobs to other regions and even to other countries. Members of all social classes are more dependent on impersonal institutions such as corporations, national labor unions, and international markets. As a result, the old upper class has less influence, and the class structure is less cohesive and much less clear-cut than it was in the 1930s.

Social Classes in Postindustrial Society In the decades since these studies were conducted, stratification in the United States has been strongly affected by technological and social changes. The study of these changes, which are thought to be producing a "postindustrial" society, is a lively area of sociological theory and research. The black workers whose struggles for dignity and economic security are described in *Black Metropolis,* the autoworkers whom the Lynds encountered in Muncie, the members of Ed Sadlowski's family who worked in the steel mills—all were part of an earlier form of industrial society in which masses of workers were employed in large-scale manufacturing enterprises. By 2000, however, over three-quarters of all workers were engaged in providing services rather than producing goods of any kind (*Statistical Abstract,* 2002). According to Daniel Bell (1973), the principal theorist of "postindustrial society," this change represents a "revolution taking place in the structure of occupations and, to the extent that occupation determines other modes of behavior . . . it is a revolution in the class structure of society as well" (pp. 125–26). Bell and other sociologists believe that we now live in a postindustrial society in which what counts is not "raw muscle power" but scientific, technical, and financial information. The dominant person in this new order of society is the professional, who is equipped by education and training to supply the skills and information that are in most demand.

Although Marx argued that social-class divisions in industrial societies would become sharper and that class conflict would become more bitter, many American sociologists find evidence to the contrary. Their research shows that changes in the structure of the American economy, and especially in the types of jobs it produces, have resulted in a blurring of class lines and an easing of class conflict between industrial workers and the owners and managers of the means of production. More people work in offices, fewer in factories. More people live in metropolitan suburbs, fewer in industrial communities surrounding large factories. The union electrician or plumber wears a work shirt but maintains a home in a suburb with neighbors who work in offices. The computer operator wears a white shirt and has graduate training in computer science but may receive an hourly wage. The social-class membership of these employees and many others appears to be more than a simple matter of worker versus owner. Some sociologists argue, therefore, that the various dimensions of stratification—educational attainment, occupational prestige, income, wealth, and family status—overlap in complicated ways that make it difficult for people to form well-defined ideas about social-class membership (Hodge & Treiman, 1968; Wolf, 1998). Moreover, such factors as race, ethnicity (national origin), religion, and lifestyle crisscross economic class divisions and further blur what may once have been clearly defined class boundaries (Jackman & Jackman, 1983; NORC, 2002).

Other sociologists, however, are convinced that inequalities in income, wealth, and family status still produce clear social-class divisions in the American population (Piven, 1997). Although industrial workers make up a declining proportion of the labor force, they are still a large segment of the population, as are the poor below them. And these populations continue to demand social-welfare policies like unemployment insurance and medical benefits, in keeping with their interests as less-advantaged classes. Members of the more affluent middle and upper classes also recognize the interests they share; they may demand such benefits as tax reductions or aid to college students (Fischer & Hout, 2007; NORC, 1998). Thus, some sociologists contend that the American public does recognize class divisions and that class inequalities do play an important role in determining life chances. Although this debate continues, the empirical evidence shows that people in the United States have a good idea of the social class to which they belong.

Thinking Critically: How do you perceive the social class structure of your own community? Do your neighbors have particular ways of expressing class differences (for example, the country club set, yuppies, poor folks, lunch pail guys, Joe Sixpack)? Listening to talk about inequality can challenge notions that ours is not a "class society."

Class Awareness in America Today

As mentioned in the preceding chapter, Marx reasoned that social class has both subjective and objective dimensions. In his view, objective social class depends on a person's relationship to the means of production (whether the person owns capital or land, sells his or her labor power, engages in a profession, and so on). Subjective social class is determined by how a person thinks about his or her membership in a social class. In Marx's view, many people who objectively are members of the working class because they sell their labor power and do not own capital in fact identify with the concerns of the owners of capital because they help run the businesses that employ them; in other words, they think of themselves as members of a higher class (subjective class), but their occupation places them in a lower class (objective class).

Subjective Versus Objective Measures

Research on class awareness distinguishes between these two views of social class. The *subjective method* of measuring social-class membership uses interviews in which respondents give their opinions about their class rankings. This method of assigning people to classes is based on personal opinions. The *objective method* uses indicators of rank such as occupational prestige (see Table 11.3), place of residence and type of home, level of education, and income, which are combined to form a composite index. These measures are not affected by the respondents' opinions. Researchers have found high correlations between the data generated by both methods (Hollingshead, 1949; Kahl, 1965). However, people who work in blue-collar occupations, and therefore are in the working class by objective measures, tend to identify themselves as members of the middle class (NORC, 1998; Vanneman & Cannon, 1987).

Sociologists often use opinion surveys when they study subjective perceptions of class membership, but they use the objective method when they are interested in the effects of class membership on other aspects of a person's life, such as political views or access to health care.

Over the last twenty-five years, the National Opinion Research Center's (NORC's) General Social Survey has found that an average of about 5 percent of Americans perceive themselves as being in the lower class (NORC, 1998). In contrast, sociologists Robert and Mary Jackman found that a higher proportion of respondents perceive themselves as poor. This difference probably occurs because the term *lower class* has a negative connotation for many people.

The NORC measures subjective social class by asking thousands of respondents this question: "People talk about social classes such as the lower class, the working class, the middle class, and the upper class. Which of these classes would you say you belong to?" Most people in the survey—at least 97 percent—assign themselves to a social class. The data in Table 11.6 show that almost 45 percent of men and 46 percent of women classify themselves as members of the middle class, whereas far fewer black

TABLE 11.6

Social Class by Sex and Race

	Social Class by Sex					
	Lower Class	Working Class	Middle Class	Upper Class	No Class	Total
Male						
Percent	4.5	47.4	45.0	3.1	0	100
N	607	6,423	6,087	425	0	13,542
Female						
Percent	5.7	45.0	46.2	3.1	0.01	100
N	977	7,681	7,899	524	1	17,082

	Social Class by Race					
	Lower Class	Working Class	Middle Class	Upper Class	No Class	Total
White						
Percent	4.1	44.2	48.7	3.0	0	100
N	1,049	11,380	12,537	783	1	25,750
Black						
Percent	11.6	58.6	28.2	3.6	0	100
N	482	2,350	1,170	150	0	4,152
Other						
Percent	7.3	51.8	38.6	2.2	0	100
N	53	374	279	16	0	722
Total number of cases	30,624					

Source: NORC, 2007.

Americans (28 percent) do so. And even though the category "lower class" has a negative connotation for many people, we see in the table that almost 6 percent of women rank themselves in this social class and almost 12 percent of African Americans do so. These and many similar survey studies led the Jackmans to conclude that, "at their basis, classes take shape in the public awareness as clusters of people with similar socioeconomic standing" (1983, p. 217).

Socioeconomic Standing The term *socioeconomic standing* (usually referred to as **socioeconomic status**, or **SES**) requires further definition. When people think of social-class divisions in American society, they first assign various occupations to broad class ranks. When there is confusion about how an occupation is ranked, people tend to think of other aspects of social class, such as family prestige, education, and earned income. And as they think about these factors, they tend to reach a consensus about what social class they belong to and what classes others should be assigned to.

But that consensus is by no means perfect. Some blue-collar occupations (for example, many skilled trades, such as plumber and electrician) are relatively well paid; in fact, they are better paid than many office jobs. This leads people to classify the holders of such jobs as members of the middle class, whereas they may assign holders of jobs in offices and stores to the working class. Remember also that the distinction between the lower-middle and upper-middle classes was the most difficult one for Warner's respondents in Yankee City and the other communities he studied.

Social Class in Rural Areas In addition to this ambiguity in how people distinguish among social classes, there is the problem of how classes are perceived in rural, as opposed to urban, areas. Most studies of social inequality in America exclude the farm population, which is relatively small—about 6 million people, or less than 2.2 percent of the total population (*Statistical Abstract,* 2010). But since the early years of the nation's history, farmers have occupied a special place in American culture and the U.S. economy. The problems farmers face in a society that is steadily eliminating family farms have caused them to develop a strong class identity despite differences in their actual wealth. Hence, in

socioeconomic status (SES): A broad social-class ranking based on occupational status, family prestige, educational attainment, and earned income.

the next section, which describes America's major social classes, we will include farmers and people living in rural areas dominated by agriculture, forestry, and mining.

SOCIAL CLASS AND LIFE CHANCES IN THE UNITED STATES

The rich have far more money than the poor, and they tend to have more education and a great deal more wealth, as measured by the value of homes, cars, investments, and much else. These facts are clear from the tables presented earlier in the chapter. But what differences do these inequalities make in people's lives? Social scientists often answer this question by analyzing the life chances of people born into different social classes. By *life chances*, they mean the relative likelihood that individuals will have access to the opportunities and benefits that the society values. Compare, for example, the life chances of a child born into a family in which the mother and father earn slightly more than the minimum wage by working in restaurants and supermarkets with the life chances of a child born into the home of a police officer and a teacher, and compare both of these with the life chances of a child born into the home of a successful banker. Will these differences in the circumstances of birth affect the quality of education each child is likely to receive? Will the children's access to quality health care differ? Will the differences in the social class of their families influence whom they are likely to marry? Will it make a difference in the likely length of their lives? The answer to all of these questions is, emphatically, yes. But being born into a given social class does not determine everything about an individual's life chances. The more a society attempts to equalize differences in life chances—by improving health care for the poor, for example, or by creating high-quality institutions of public education—the more it reduces the impact of social class on life chances.

Numerous social-scientific studies have shown that one's social class tells a great deal about how one will behave and the kind of life one is likely to have. Later in this section, we will discuss specific social classes, but first we present some typical examples of the relationship between social class and daily life. These examples apply primarily to American society, but many of the same conditions can be found in other societies as well.

Social Class in Daily Life

Class and Health A child born into a rich upper-class family or a comfortable middle-class family is far less likely to be premature or have a low birth weight than one born into a working-class or poor family. And a baby born into a family in which the parents are working at steady jobs is far less likely to be born with a drug addiction, AIDS, or fetal alcohol syndrome than a child born to parents who are unemployed and homeless. These and many other disparities contribute to what is often called "the socioeconomic status (SES) gradient in health" (Sapolsky, 1998).

The SES gradient in health exists in all of the world's nations and is based on a complex combination of social class and culture, but the gradient is particularly marked in the United States. The poorer you are in the United States, the more likely you are to suffer heart disease, respiratory disorders, ulcers, psychiatric diseases, and various cancers. This situation is reflected over and over again in surveys. In a national survey sponsored by *The New York Times*, for example, only 18 percent of poor respondents said that their health was excellent, whereas 37 people of people in households making over $50,000 and 56 percent of people in

households making over $150,000 a year rated their health as excellent. Because people in the lower quintiles of the income distribution also are much more likely to lack health insurance, it is no surprise that their lives are shorter and more likely to be marked by chronic illnesses (Scott, 2005).

Among adults, a salaried member of the upper class who directs the activities of other employees is less likely to be exposed to toxic chemicals or to experience occupational stress and peptic ulcers than wageworkers at their machines and computer terminals. Those workers, in turn, are more likely to have adequate health insurance and medical care than the working poor—dishwashers, migrant laborers, temporary help, low-paid workers, and others whose wages for full-time work do not elevate them above the poverty level (Keller, 2005). The working poor are the largest category of poor Americans, and like those who lack steady jobs, they often depend on local emergency rooms for medical care and report that they have neither family doctors nor health insurance. The same poor and working-class population is also more likely to smoke, consume alcohol, and be exposed to homicide and accidents—while receiving less police protection—than members of the classes above them.

Education Children of upper-class families are more likely to be educated in private schools than children from the middle or working classes. Sociologists have shown that education in elite private schools is a means of socializing the rich. As we saw in Chapter 5, Cookson and Persell's (1985) study of socialization in elite American prep schools found that "preppies" develop close ties to their classmates, ties that often last throughout life and become part of a network they can draw on as they rise to positions of power and wealth. The segregation of upper-class adolescents in prep schools also limits their dating and marriage opportunities to members of the same class.

Although middle-class parents are more likely than rich parents to send their children to public schools, they tend to select suburban communities where the schools are known to produce successful college applicants. The public schools that serve the middle classes spend more per pupil than the schools attended by working-class and poor children, and they offer a wider array of special services in such areas as music, sports, and extracurricular activities. Children in the middle and upper classes also tend to have parents who insist that they perform well in school and who can help them with their schoolwork. Moreover, children from working-class and poor families are more likely to drop out of school than children from upper-class families.

Politics The poor and members of the working class are more likely to vote for Democratic Party candidates, whereas upper- and middle-class voters are more likely to choose Republican candidates. Throughout the industrialized world, voters with less wealth, prestige, and power tend to vote for candidates who promise to reduce inequalities, whereas voters with higher SES tend to choose candidates who support the status quo. Thus, studies of social-class identification find that about 50 percent of poor respondents and 40 percent of respondents who assign themselves to the working class believe the federal government should be doing more to achieve full employment and job guarantees, as opposed to only about 25 percent of upper-middle-class respondents. They also found that about 48 percent of working-class respondents believe that "some difference" in levels of income (though less than currently exists) is desirable to sustain people's motives to achieve, whereas almost 50 percent of upper-middle-class respondents believe that a "great difference" is desirable. Members of all classes tend to agree that whatever differences there are ought to be based on individual achievement rather than on advantages inherited at birth. Members of the working class and the poor, however, are more likely than members of other classes to vote for candidates who propose measures that would increase equality of opportunity.

Many other examples could be presented to illustrate the influence of social class on individuals in American society. But social-class divisions also affect the society as a whole. Let us turn, therefore, to an examination of some key characteristics of the major social classes in the United States, and to some of the consequences of the different life chances these classes experience.

The Upper Classes

For the past fifteen years or more, the General Social Survey has found that approximately 4 percent of Americans classify themselves as members of the upper class (NORC, 2007). And as we saw at the beginning of the chapter, the richest members of the upper class, about 1 percent of the total population, control almost 40 percent of all personal wealth in the United States. However, sociologists regard the upper class as divided roughly into two subgroups: the richest and most prestigious families, who constitute the elite or "high society" and tend

then & now

From Food Baskets to Food Stamps

The late nineteenth-century drawing and the modern photograph shown here represent two quite different approaches to aiding the poor. The drawing depicts a well-off family bringing a basket of food to the poor. This form of individual charity, which the rich often thought of as their responsibility and obligation, frequently engendered hostility rather than gratitude. The modern welfare system replaces the individual gift with "entitlements" (that is, payments and other services to poor people who qualify for them). But the modern system is inadequate to provide a dignified quality of life. And as evidenced by the antiwelfare sentiment prevailing in the United States in recent years, the notion of entitlement to welfare is an easy political target because many middle- and working-class voters are persuaded that the welfare system unfairly takes money from them to sustain "undeserving" poor people.

Bettmann/Corbis

ROD LAMKEY JR/AFP/Getty Images

to be white Anglo-Saxon Protestants, and the "newly rich" families, who may be extremely wealthy but have not attained sufficient prestige to be included in the communities and associations of high society.

The Hyper-Rich Not since the "Roaring Twenties" have the very richest people in the United States, those in the top 0.1 percent of households, enjoyed such rapid increases in income and wealth (*Statistical Abstract,* 2010). Their average annual income is about $3 million a year, two and a half times the $1.2 million this small group of about 150,000 taxpayers earned in the early 1980s. In Figure 11.5, we see that the ranks of the hyper-rich increased by over 400 percent between 1983 and 2001, while the total number of households increased by only 27 percent. The tax policies of the Bush administration greatly favored this small class of hyper-rich and, to a lesser extent, those who are merely rich. The top 10 percent of households share a full 53 percent of the value of the tax cuts that will remain in effect at least until 2012. As we will see at the end of the chapter, Americans still cling to the belief that they will experience upward social mobility and even become rich, a belief that is reinforced by the rapid increase in the numbers of rich and hyper-rich individuals in U.S. society, even though the actual odds of getting rich are decreasing rapidly (NYT Correspondents, 2005).

At the beginning of the twentieth century, families whose wealth came from railroads and banking looked down on "upstarts" like Rockefeller and Ford, whose money came from oil and automobiles. More recently, the great manufacturing families—the Du Ponts (chemicals), the Rockefellers (oil), the Carnegies and Mellons (steel and coal)—have questioned the upper-class status of families like the Kennedys, whose wealth originally came from merchandising and whose Irish descent marked them off from the rich Protestant families. Throughout much of the twentieth century, the number of family-owned fortunes declined as large corporations like General Motors, Boeing, and IBM gained greater economic power. But major family fortunes still represent billions of dollars of capital controlled by members of immensely wealthy families, many of whom are not descended from earlier Yankee families.

The newest and largest family fortunes in the United States have developed from the new digital technologies and from high finance rather than from older manufacturing technologies. For example, Bill and Melinda Gates, founders of Microsoft, and financier Warren Buffett control the largest private fortunes in the Unites States. While retaining immense wealth for themselves and their children, they have created a foundation that will fund many worthy research and action projects throughout the world. Meanwhile, in emerging capitalist nations like India, new fortunes are still being made in steel and railroads, and some of the older family dynasties, such as the Tata family, originally derived their wealth from large-scale manufacturing, even though, like their U.S. counterparts, they have recently diversified into real estate and finance.

Members of the upper class tend to create special places in which to live and relax. They often maintain apartments in exclusive city neighborhoods and country estates in secluded communities. They send their children to the most expensive private schools and universities, maintain memberships in the most exclusive social clubs, and fly throughout the world to leisure resorts frequented by members of their own class. Examples of these resorts and enclaves of great wealth include Vail and Aspen in Colorado, Palm Springs in California, Palm Beach in Florida, and large stretches of the New England coastline.

Not Since the Twenties Roared

The very wealthiest Americans—the 145,000 or so taxpayers whose incomes start at $1.6 million and put them in the top 0.1 percent—have pulled away from everyone else in the recent decades, an analysis by *The New York Times* shows.

Growth in Income

The share of the nation's income earned by the taxpayers in the top 0.1 percent has more than doubled since the 1970s, and in the year 2000 exceeded 10 percent, a level last seen in the 1920s.

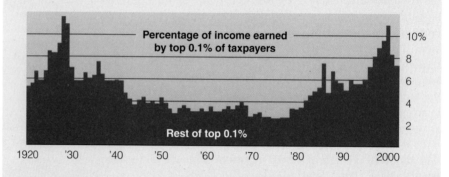

Growth in Wealth

Even after adjusting for inflation, there are five times as many households as there were two decades ago with a net worth of more than $10 million. Not all have high incomes.

Number of households	Households with inflation-adjusted net worth of:			
	All households	$1-5 million	$5-10 million	10 million+
in 1983	84 million	2.2 million	180,500	66,500
in 2001	106 million	4.8 million	729,400	338,400
Percent increase	**+27%**	**+123%**	**+304%**	**+409%**

Figure 11.5 Growth in Income and Wealth, 1920–2000. Source: From The New York Times Poll, *Class Project*, March 14, 2005. Reprinted by permission of the New York Times Agency.

Is There a "Ruling Class?" A question that sociologists continually debate is whether the upper class in America is also the society's ruling class. The hit film *Titanic* presents a vision of the social-class system in which the wealthy are indeed a ruling class; in reality, however, the question of whether great wealth also confers the ability to rule by controlling political actors is open to empirical research and debate. Elite theorists like C. Wright Mills and William Domhoff argue that the upper class not only holds a controlling share of wealth and prestige, which it can pass along to its children, but also maintains a virtual monopoly over power in the United States (Domhoff, 2010). A ruling class, Domhoff states, "is socially cohesive, has its basis in the large corporations and banks, plays a major role in shaping the social and political climate, and

Photographer Barbara Norfleet was studying the lifestyles of the very wealthy in the United States when she caught this patrician gentleman stepping off his yacht on a summer day in Massachusetts.

dominates the federal government through a variety of organizations and methods" (1983, p. 1). Mills (1956) attempted to show that this ruling class produces a *power elite* composed of its most politically active members plus high-level employees in the government and the military. The power elite, these sociologists claim, is the leadership arm of the ruling class.

Other sociologists and political scientists contend that there is no single, cohesive ruling class with an identifiable power elite that carries out its bidding. This is known as the *pluralist* concept (Polsby, 1980; Sullivan, 1998). These researchers agree that there is a readily identifiable upper class in American society. However, its power is not exerted in a unified fashion because there are competing centers of wealth and power within the upper class; moreover, its members' viewpoints on social policy are often opposed to one another. (We return to the debate between the power elite and pluralist theories in Chapter 19.)

In the aftermath of the deep recession that began in 2008 with the crash of the U.S. housing market, and with the revelation of Wall Street scandals and the arrest of financier Bernard Madoff in the largest stock scam in history, one might expect Americans to be more hostile toward the rich, but this is not the case. Surveys show that around six in ten respondents continue to tell interviewers that they would like to be rich, but about the same proportion think it is unlikely that they will become so. Americans also don't believe that rich people are happier than they are. Surveys consistently show, however, that Americans by a slim majority believe professional athletes, celebrities, lawyers, corporate executives, CEOs, senators and congressmen, doctors, and TV anchors are overpaid, too. By the same slight majorities, they also believe that the wealthy have too much power and influence (Bowman, 2009). In 2010, the U.S. Supreme Court vastly increased the power of corporate executives in American politics with its decision in *Citizens United v. Federal Election Commission*, which found that federal agencies do not have the power to limit campaign contributions of corporations. This ruling will have a profound influence on American politics, potentially further weighting the power balance in favor of the wealthiest Americans.

The Middle Classes

Unlike the upper classes, in which family status establishes who is included in the highest stratum and who is not, the middle class includes far more varied combinations of wealth and prestige. As we saw in Table 11.5, subjective measures of social class show that about 45 percent of Americans classify themselves in the middle class, although this figure is lower for African Americans (28 percent). However, the large proportion of Americans who identify themselves as middle class is by no means a homogeneous population. The middle class is often referred to as the "middle classes" to emphasize the existence of strata with differing income, education, and access to wealth.

One group of Americans who usually identify themselves as members of the middle class are highly educated professionals. Most often, they have attended graduate schools and built successful careers as engineers, lawyers, doctors, dentists, stockbrokers, or corporate managers. But their comfortable incomes, often much more than $100,000 a year, may also be derived from family-owned businesses. Members of this class typically live in the suburbs. They join expensive country clubs, are active in community affairs, and spend a good deal of time transporting their children to "enriching" activities.

Another segment of the middle class is composed of people whose income is derived from small businesses, especially stores and community-oriented enterprises (in contrast with large corporations that have offices at many locations). Often referred to as the *petit bourgeoisie* or independent small-business owners, the members of this class may be found among the leaders of local chambers of commerce and other business associations that advocate local economic growth. Erik Olin Wright and his colleagues (1982; Wright, 1997) found that of the 9 percent of Americans who employ other workers, most are members of the petit bourgeoisie and employ fewer than ten people.

The American concept of the middle class is based far more on patterns of consumption and the American dream of shared affluence than on the kinds of economic realities that Marx used to describe social classes. Americans have always believed that they could achieve individual affluence and become part of a great middle class. Thus, in subjective terms, the middle class is the largest single class in American society, but in cultural terms, it is highly diverse because so many different lifestyles are represented within it.

The dominant image of the middle class from the end of World War II until the mid-1970s was of a suburban population living in relatively new private homes (Scott, 1998). The culture of the suburban middle class was thought to be shaped by the experience of frequent changes of residence and long-distance commuting, together with status symbols like "the ranch house with its two car garage, lawn and barbecue, and the nearby church and shopping center" (Schwartz, 1976, p. 327). The suburban middle class was said to be oriented toward family life and suspicious of the offbeat (Fava, 1956; Riesman, 1957). However, empirical research by several sociologists, including Bennett Berger (1961) and Herbert Gans (1976/1967), has shown that suburban communities are far from homogeneous, that many people who think of themselves as members of the working class can be found in them, and that there is no easily identified middle-class suburban culture.

Another change in the nature of the middle class has to do with education. In the 1950s and 1960s, a college diploma was often viewed as a passport to the American middle-class way of life. But we will see later that, although education remains the primary route to upward mobility for people without inherited wealth, the college diploma has lost some of its power to open the door to middle-class prestige.

The most general points that sociologists can make about the middle class are that its members tend to be employed in nonmanual occupations and that they usually have to work hard to afford the material things that the upper-middle class can acquire more easily. Identification with the middle class is likely to be highest among teachers, middle- and lower-level office managers and clerical employees, government bureaucrats, and workers in the uniformed services.

Effects of Economic Insecurity Political candidates attempt to win the support of the middle class because they recognize that large numbers of voters think of themselves as members of that class, rather than as rich or poor. As we have seen, however, the idea of an all-inclusive American middle class is more a myth than a reality for increasing numbers of working Americans. Between the extremes of poverty and wealth are, instead of a unified middle class, two increasingly distinct groups that think of themselves as middle class but face far different life chances and economic realities.

In the more secure segment of the middle class are highly skilled, highly educated professionals and managerial women and men who are doing rather well, although they increasingly experience income stagnation and must find ways to supplement their incomes. Below this segment is a larger stratum of less skilled and less educated people who are experiencing ever-greater economic insecurity resulting from stagnant incomes and declining living standards. Appealing to this growing insecure middle-class stratum is a central goal of political leaders.

Politics aside, it is doubtful that the problems of the middle class can be solved through social policy. Recent figures the Department of Labor released show that people in the middle income quintile are now spending almost half their incomes on three basic necessities of life: housing (approximately 32 percent), utilities (approximately 8 percent), and health care (approximately 7 percent). A generation ago, these expenditures accounted for only one-third of a typical middle-class income. (Spending on food has remained fairly constant.) Also, during the period from 1975 to 2000, the average price of a house rose by nearly 330 percent while the consumer price index increased by over 220 percent. As a result, fewer young people can afford to buy a private home, perhaps the central symbol of membership in the American middle class (Hacker, 1997).

The difficulties Americans are experiencing in realizing the dream of owning their own home are shown in Table 11.7. The proportion of Americans who own their own home has remained between 68 percent and 69 percent in

TABLE 11.7

U.S. Homeownership Rates,* by Race and Ethnicity: 2001–2007

	2001	2002	2003	2004	2005	2007
All households	67.8%	67.9%	68.3%	69.0%	68.9%	67.8%
Whites†	74.3	74.7	75.4	76.0	75.8	75.2
Blacks	47.7	47.4	48.1	49.1	48.2	47.2
Hispanics	47.3	47.0	46.7	48.1	49.5	49.7
Asians/others	54.2	54.5	56.0	58.6	59.2	60.0

*The percentage of households that are homeowners.
†Non-Hispanic.
Source: U.S. Census Bureau, 2005.

the last few years, with a decline between 2004 and 2005, despite an ongoing boom in housing sales and mortgages. The table shows that this decline occurred among white and black households, whereas the rate of home ownership among Asians and Hispanics, which lags far behind that of whites, increased only slightly. These figures suggest that the booming housing market of recent years has primarily benefited families who already owned their own homes, not those who were seeking to buy their first home.

Surveys of how people in the United States view their social class and status consistently show that home ownership is thought of as one of the main criteria of wealth and success. This is evident in Table 11.8. When asked to rate "the

TABLE 11.8

Responses to Survey Question about Symbols of Wealth and Status

What one thing do you think of as a symbol of wealth and status in the United States?	
Money/bank account	22
Freedom/opportunity	2
Have a house/home	26
Family	1
Happiness/contentment/peace of mind	1
"Life in America"	1
High-paying job/successful career	5
Successful	1
Health	1
Comfortable retirement	—
Power	3
Second home/vacation home	1
Vehicles	4
Celebrity	1
Specific items (Rolex, Super Bowl tickets)	1
Consumer goods, in general	1
Own business	1
Education	2
Location/address	2
Charity/generosity/doing for others	—
Other	5
Don't know/not applicable	21

Source: From the New York Times Poll, *How Class Works—Complete Poll Results*, May 15, 2007. Reprinted with permission of the New York Times Agency.

one thing that most symbolizes wealth and status," respondents mentioned many possibilities, but owning a home was the one that was mentioned most frequently (*New York Times* Poll, 2005). Unfortunately, the housing crash of 2008 has accelerated the decline in rates of homeownership, which was beginning as early as 2007 for some population groups (see Table 11.7). By 2009, the overall rate had slipped to 67.4 percent, down from 67.8 percent in 2007 (Aronowitz, 2003; U.S. Census Bureau, 2010).

The Working Class

Forty-four percent of white Americans and almost 57 percent of African Americans identify themselves as members of the working class (NORC, 2000). This class is undergoing the most rapid and difficult changes in America today. Indeed, as with the middle class, those who classify themselves as members of the working class are extremely diverse. There are skilled craftspeople, plumbers, carpenters, machinists, and others who earn wages protected by strong unions, and factory workers whose class interests may or may not be protected by labor unions and whose plants may be threatened with closings or layoffs resulting from automation and the globalization of manufacturing. And some people in the working class work long hours yet cannot earn enough income to lift them very far above the poverty threshold. These struggling working people and their families are sometimes referred to as the "working poor."

The most important characteristic of the working class is employment in skilled, semiskilled, or unskilled manual occupations. Another distinguishing feature of this class is union membership; in fact, labor unions refer to their members as "working people." However, the proportion of workers who are union members is declining. In the mid-1970s, approximately 23 percent of the American labor force belonged to labor unions. Today that proportion has declined to less than 16 percent, largely as a result of the decline in factory employment. One result of this trend is that many workers have less protection against arbitrary changes in their working conditions and earnings (Aronowitz, 1998). Another is that the wages of working people are decreasing, especially in semiskilled and unskilled blue-collar jobs. This controversial issue has been thoroughly documented in the research of Richard Freeman (1994), who has shown that the decreases in union membership and hourly wages account for as much as half of the increase in income inequality referred to in Table 11.1.

There are two major divisions within the American working class. The first is often known as the industrial working class and consists of blue-collar workers who work in large manufacturing industries and are members of industrial unions. The United Automobile Workers, the United Steelworkers, the International Ladies' Garment Workers' Union, the United Mine Workers, and the International Brotherhood of Teamsters are examples of such unions (Aronowitz, 2003; Aronowitz & DiFazio, 1994; Burawoy, 1980). The second major division of the American working class is composed of workers employed in skilled crafts, especially in the construction trades (Halle, 1984; LeMasters, 1975).

Sociologists have found that the industrial working class is more conscious of its class identity and more prone to believe that its fortunes as a class depend on its ability to win in labor–management conflicts. Members of the skilled trades, by contrast, are more likely to be uncertain about their class identity or to think of themselves as members of the middle class. For both segments of the working class, financial worries are a routine fact of life (Halle, 1984). The situation of women in this class is especially problematic because they are often the first to suffer the effects of layoffs and plant closings (Rosen, 1987), an issue that is discussed further in Chapter 13.

These hotel workers, proudly showing off the luxurious accommodations they help maintain, are part of the working class. In fact, far more people are now employed in restaurants and hotels than in metal, automobile, and textile production combined.

More racial and ethnic diversity exists in the working class than in other classes. White workers in the working class are far more likely to work alongside black and Hispanic workers, for example, than are white members of other classes. This situation often brings workers of different races and ethnic backgrounds into competition when jobs are in short supply or an industry is shrinking as a result of automation. The animosity produced by such competition results largely from the fear of losing one's job and skidding downward into the ranks of the poor.

The Poor

Studies of subjective class membership usually underestimate the poor population because many people in this class think of themselves as working people who are underpaid or "down on their luck." This may be particularly true of the NORC measures of subjective social class (see Table 11.9), which use the term "lower class" instead of "poor." In an earlier study by Mary and Robert Jackman (1983), almost 8 percent of Americans classified themselves as poor. In the NORC study, about 5 percent of the sample said that they were in the "lower class." Whichever term one uses, however, these measures greatly underestimate the number of poor people in the United States. Official census statistics show that approximately 12 percent of Americans—about 37 million people—are living in poverty (U.S. Census Bureau, 2009). These figures count only people whose household incomes are below the poverty threshold income of about $420 a week for a family of four. As shown in Table 11.8, African Americans, Hispanics, and children (the majority of whom live in female-headed families) are disproportionately represented among the poor. (See also Figure 11.6.)

Poverty: A Relative Concept Social scientists often point out that poverty is not an absolute concept but must be measured relative to the standards of well-being in particular societies. A family of four in the United States with an annual income of $22,500 is poor by official standards, yet is doing quite well compared with equivalent families in India or Africa. The U.S. family probably has permanent shelter, even if it is in a slum or a run-down neighborhood. It probably receives some form of government income supplement, if only in the form of food stamps and federal medical insurance (Medicaid). The poor family in Africa or India, by contrast, may be living without permanent shelter, forced to beg and sift through scraps to obtain enough food to survive. Although some poor families and individuals in the United States live under similar conditions, most of the U.S. poor are not yet in immediate danger of starvation.

Does this mean that the poor in the United States are better off? In absolute terms (that is, in comparison with the poor in very poor nations), they are not deprived of the necessities of life. In relative terms (that is, in comparison with the more affluent majority in the United States), they are severely deprived. This distinction between absolute and relative deprivation is important in explaining how people feel about being poor. It is no comfort to a child in a poor family in the United States whose parents cannot afford fancy sneakers or a color TV that people are starving in Africa. In short, people feel poor in comparison with others around them who have more.

The Working Poor One of the chief causes of poverty is that people who are working full-time are not being paid wages that provide enough income to raise them above the poverty line. Two-thirds of American children living in poverty have a parent who works full-time for wages that are too low to lift the household out of poverty (U.S. Census Bureau, 2009). Of course, many of these households are one-parent families, but in a study of two-parent families living in poverty, David Ellwood found that 44 percent had at least one member who was working full-time. Low wages, Ellwood states, are a major cause of poverty:

> Work does not always guarantee a route out of poverty. A full-time minimum-wage job (which pays $5.15 per hour) does not even come close to supporting a family of three at the poverty line. Even one full-time job and one half-time job at the minimum wage will not bring a family of four up to the poverty line. (1988, p. 88)

Because so many of the jobs created during the economic expansion of the 1980s and 1990s were low-wage jobs, rates of employment and poverty have risen simultaneously. In 2008, approximately 4.7 million people were in families whose earnings were below the official poverty line but in which there were two working parents. These individuals accounted for 3.1 percent of all Americans living in poverty, an increase from 2.9 such people in working-family households in 2005 (U.S. Census Bureau, 2009). Millions of other workers with full-time jobs did not respond to the census, primarily

TABLE 11.9

Poverty among U.S. Individuals by the Official Poverty Measure, 2008

Characteristic	No. (in 000)	Poverty Rate (%)
All	39,829	13.2
Age		
Under 18	14,068	19.0
18–64	22,105	11.7
65 and over	3,656	9.7
Race/Ethnicity*		
White, not Hispanic	17,027	8.6
Black alone or in combination	9,379	24.7
Hispanic origin⁺	10,987	23.2
Asian alone or in combination*	1,576	11.8
Region of Residence		
Northeast	6,295	11.6
Midwest	8,120	12.4
South	15,862	14.3
West	9,552	13.5

*Federal surveys now allow respondents to report more than one race, which makes possible two basic ways of defining a race group. A group such as Asian may be defined as those who reported Asian and no other race (the single-race concept) or those who reported Asian regardless of whether they also reported another race (the race-alone-or-in-combination concept). This table shows data using the race-alone approach, though the Census Bureau notes, "The use of the single-race population does not imply that it is the preferred method of presenting or analyzing data." The Census Bureau uses a variety of approaches.

Another reason for the increase in poverty is the increase in the number of single-parent, female-headed families. The breakup of marriages or long-term relationships leaves women alone with the responsibility for raising small children and earning the income to do so. Such families often become poor because it is more difficult for a woman to support a family alone than it is for a man. As we will see in Chapter 14, one result of this trend toward the "feminization" of poverty is that a rising proportion of the nation's children are growing up in poor female-headed families and are deprived of the advantages enjoyed by children from more affluent homes.

A common stereotype of the poor in the United States is that they are heavily concentrated in inner-city minority ghettos. This notion is false, as can be seen in Figure 11.7. The poor in large central cities (those living in moderate- or high-poverty neighborhoods) account for just 19 percent of the total poor population, with only about 7 percent concentrated in high-poverty neighborhoods. Fully 29 percent of the poor reside in rural and small-town communities, and 19 percent live in the affluent suburbs of large cities. In short, the poor live everywhere and are not concentrated in a single type of community.

One reason for this pattern is that the poor are an extremely diverse population. A large portion of the poor consists of aged people living on fixed incomes (Social Security or very modest savings or pensions). Other categories of poor people include marginally employed rural workers and part-time miners in communities

because they are working "off the books" or not legally residing in the United States (U.S. Census Bureau, 2006).

from Appalachia to Alaska; migrant farm workers in agricultural areas throughout the United States; chronically unemployed manual workers in industrial cities; disabled workers and their families; and people who

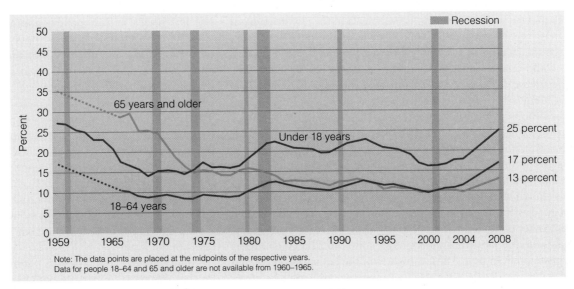

Figure 11.6 Poverty Rates, by Age: 1959–2008. Source: U.S. Census Bureau, 2009.

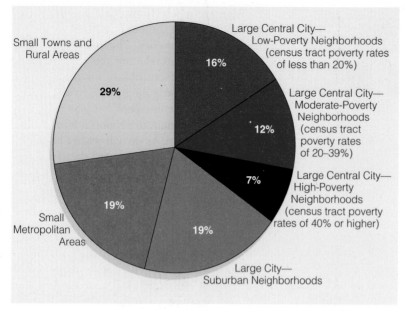

Figure 11.7 **Geographic Distribution of the Poor (percent of total below poverty line).** Figures from the 2000 census indicate that the percentages in this chart have changed somewhat since Elliot conducted his study, but the basic distribution remains the same (the percentages do not add up to 100 because of rounding). Source: Ellwood, 1988; U.S. Census Bureau, 2004.

have been displaced by catastrophes like hurricanes, drought, or arson. For all of these groups, poverty brings enormous problems of insecurity and instability. Many poor people do not even know where they will be living next month or next year. This problem is especially acute for farm families who have been driven from their land.

The Homeless and Indigent Homelessness is one of the most visible effects of severe poverty. In a relatively affluent nation like the United States, homelessness can be defined simply as a social condition in which people do not have regular housing and are forced to sleep in public places, public shelters, or other facilities designed for homeless individuals and families. From a global perspective, the situation is somewhat more complicated. In India, for example, millions of people live on the street, but they are not homeless because they consider the spot they occupy—under a makeshift awning against a building, for example—as home. They may be indigent beggars, new arrivals to the city or town, or people with extremely low-paying work, none of whom can afford more adequate shelter. In Europe and North America, the fact that they live on the street would qualify them as homeless; in their own societies, they may be considered to have homes despite their poverty. In sociological analyses of homelessness, therefore, we must always consider local and regional definitions of the phenomenon. In the following discussion of homelessness, we will draw examples from the United States and similar urban industrial nations (Kornblum & Julian, 2007).

The recession of 2008 brought homelessness back to national attention when individuals and families crowded food lines, soup kitchens, and all available church beds and homeless shelters. But it is extremely difficult to arrive at good estimates of the number of homeless people in the United States or any other urban industrial society. A major part of the problem is that estimates for different cities and states are most often based on counts taken at a given point in time. These measures of the homeless population are called "point prevalence estimates" because they are based on street surveys and visits to shelters at a given moment or point in time—often a

day in midwinter when large numbers of people are in shelters and can be counted (Metropolitan Washington Council of Governments, 2009). By combining all such point prevalence estimates, social scientists come up with estimates that generally range from a very conservative 600,000 to more than 1 million.

Point prevalence counts underestimate the number of people who are homeless for one or more periods during a year, which includes far more adults and children who become homeless because of family disputes, fires, unemployment, and other social and economic disasters. If one considers the number of people who report that they have ever been homeless in a given year, the number of people who experience homelessness is estimated to be approximately 3 million, of whom 38 percent are children and parents.

Two of the leading researchers on homelessness in the United States, Martha Burt and Barbara Cohen, describe the social characteristics of the homeless as follows:

> The homeless population is extremely heterogeneous. The main population is single men, but the fastest growing group appears to be homeless families, generally headed by a single woman. Two constraints most often cited for the failure of people to leave homelessness and become completely self-sufficient are mental illness and drug or alcohol abuse. However, homelessness is not just a problem of mental illness or substance abuse; it is also a consequence of severe poverty and a lack of adequate housing. The effects of poverty and inadequate housing are exacerbated by such problems as mental illness and substance abuse. (Cohen & Burt, 1990, p. 1)

This description is most apt for the homeless encountered in large cities. In rural areas of North America, and in many communities outside the central cities, homelessness is a form of nomadism. Families and individuals who are experiencing periods of homelessness may live for a time in their automobiles or in hidden encampments. Others may go back and forth between different friends' homes. For children, these spells of homelessness usually mean periods of absence from school, which, in turn, leads to school failure and a high likelihood of dropping out. This last point is also true for homeless children in inner-city communities, where school transfers resulting from homelessness are a major problem even if spells of homelessness are not long.

Throughout the world, homelessness touches the most vulnerable populations. Poor people who have inadequate shelter in the first place tend to be those who are driven into homelessness by natural disasters like hurricanes or earthquakes. In Haiti, for example, at least one-tenth of the nation's population was made homeless by a major earthquake in 2010. Recent political disasters in Afghanistan and parts of Africa have created immense numbers of homeless people who populate refugee camps. These homeless populations tend to be made up of people from rural areas who normally make their living in agriculture. On a global scale, they are a small minority of the poor population, but their dire needs and suffering, especially grave among women and children, make them extremely significant from a moral and political standpoint.

Farmers and Farm Families

In analyzing the situation of the poor in America, it is helpful to look at what is occurring in rural areas, where a large proportion of the poor live. Although many rural communities are experiencing renewed economic growth as factories relocate to areas where the costs of energy, space, training, and wages are lower, poverty rates continue to increase in communities based on agricultural production.

Table 11.10 presents some key characteristics of farms in the United States.

Some of the reasons why this is so are revealed in Table 11.10. The large majority of farms earn very little, less than $10,000 a year, and are small, with less than 100 acres. In consequence, fewer than half of all farmers can subsist without working outside their farms. The table also shows that the average age of farmers is increasing. None of the trends indicated in Table 11.10 is new, however. The number of farms in the United States has been decreasing steadily over the past century, although the development of consumer interest in locally grown, organic foods has resulted in a small but perceptible increase in farms on the fringes of major metropolitan regions. Despite this encouraging trend, poverty in rural America, along with problems of substance abuse and ill health, has continued to increase.

This situation repeats a condition that has been part of American social life for more than a century as the nation has been transformed from a rural to an urban society (Dyer, 1996). Although the dream of owning a family farm, or keeping one in the family, has always been an important part of American culture, the rising cost of farm technology, together with the need for ever-larger amounts of land to make farms profitable, has steadily pushed farm families off the land. Very large farms devoted to the

TABLE 11.10

Farm Characteristics

	1997	2002	2007
Average farm size (acres)	431	441	418
Farms by size (percent)			
1 to 99 acres	49.2	51.0	54.4
100 to 499 acres	34.7	33.1	31.0
500 to 999 acres	8.1	7.6	6.8
1,000 to 1,999 acres	4.6	4.7	4.2
2,000 or more acres	3.4	3.7	3.6
Farms by sales (percent)			
Less than $9,999	55.3	59.3	59.8
$10,000 to $49,999	21.4	19.4	18.3
$50,000 to $99,999	7.4	6.6	5.7
$100,000 to $499,999	12.7	11.3	10.9
More than $500,000	3.2	3.3	5.3
Tenure of farmers			
Full owner (farms)	1,384,565	1,428,136	1,522,033
Percent of total	62.5	67.1	69.0
Part owner (farms)	615,902	551,004	542,192
Percent of total	27.8	25.9	24.6
Tenant owner (farms)	215,409	149,842	140,567
Percent of total	9.7	7.0	6.4
Farm organization			
Individuals/family, sole proprietorship (farms)	1,922,590	1,909,598	1,906,335
Percent of total	86.8	89.7	86.5
Family-held corporations (farms)	81,621	66,667	85,837
Percent of total	3.7	3.1	3.9
Partnerships (farms)	185,607	129,593	174,247
Percent of total	8.4	6.1	7.9
Non-family corporations (farms)	8,811	7,085	10,237
Percent of total	0.4	0.3	0.5
Others—cooperative, estate or trust, institutional, etc. (farms)	17,247	16,039	28,136
Percent of total	0.8	0.8	1.3
Characteristics of principal farm operators			
Average operator age (years)	54.0	55.3	57.1
Percent with farming as their primary occupation	47.1	57.5	45.1
Men	2,006,092	1,891,163	1,898,583
Women	209,784	237,819	306,209

Source: U.S. Department of Agriculture, 2010.

production of a narrow range of crops and farm products now dominate North American agriculture. In the United States, more than half of total farm production comes from only about 4 percent of the nation's farms. In consequence, more farmers are employed by large agribusiness operations, and each decade a smaller percentage own family farms (Jackson-Smith & Jensen, 2009).

Larger farms are typically owned by corporations that may also can, freeze, package, or otherwise prepare the crop for sale in supermarkets. These agribusinesses may be highly efficient producers, but they have some negative social consequences for rural counties. Studies by rural sociologists have shown that communities dominated by large farms are characterized by a declining population, a decreasing number of local businesses and civic associations, and an increase in poverty. One study conducted for the U.S. Congress in 200 of the richest agricultural counties in the United States found that the more farm size increases, the more the poverty rate in the county also increases as people are displaced from the land and must search for marginal employment (Office of Technology Assessment, 1986).

As a social class, farmers still take pride in the fact that the United States has the highest level of agricultural production in the world, but this does little to alleviate their difficulties. Moreover, as a class, American farmers are continually at the mercy of a fickle world economy, a condition that they increasingly share with farmers in Europe and other industrialized regions.

One method farmers have traditionally used to hold on to the land they love is for some family members to find work in a nearby town while others continue to work the land. In fact, about half of all farmers farm part-time and work at wage-paying jobs full-time. A growing number of women in farm families work full- or part-time in rural factories. Many also work at home producing knitwear and other apparel on a piecework basis or performing electronic data processing via computer networks. Still others have service jobs.

The strategy of trying to maintain the household's economic position by adding more wage earners is increasingly common in other social classes as well. This trend has far-reaching implications, not just for the way Americans think about social class but for all the society's institutions. For farmers, however, the numbers are discouraging. As an economic class, they are diminishing steadily, especially as young people leave farming for other occupations.

MORE EQUALITY?

A large proportion of the world's people are extremely poor. India, China, Bangladesh, the Philippines, Malaysia, most of sub-Saharan Africa, the mountain regions of Latin America, and many other parts of the world are inhabited by large numbers of people with drastically limited incomes and only the most rudimentary health care, education, and other resources. If poverty continues to increase in those areas and the hopes of impoverished millions for a better life are frustrated, it will become increasingly difficult to maintain peace, achieve slower population growth, and prevent massive environmental damage.

equality of opportunity: Equal opportunity to achieve desired levels of material well-being and prestige.

equality of result: Equality in the actual outcomes of people's attempts to improve their material well-being and prestige.

Poverty is reduced through better health care, more schools, more food and shelter, and more jobs, all of which require economic and social development. The experience of the more affluent nations indicates that such development is likely to produce larger working and middle classes and reduce the rate of poverty. But the persistence of poverty and near-poverty in the United States and other industrialized nations shows how difficult it is to promote greater equality elsewhere in the world.

Two Types of Equality

The presence of millions of poor people in a country as affluent as the United States is a major public policy issue as well as a subject of extensive social research. However, policy debates on this issue are often clouded by problems of definition. Most Americans will say that they believe in equality, citing the claim of the nation's founders that "all men are created equal." When pressed to define equality, however, they often fail to distinguish between **equality of opportunity** (equal opportunity to achieve material well-being and prestige) and **equality of result** (actual equality in levels of material well-being and prestige). Americans may believe that opportunities to succeed should be distributed equally and that the rules that determine who succeeds and who fails should be fair. But their commitment to the ideal of equality falters in the face of their belief that hard work and competence should be rewarded and laziness and incompetence punished. Thus, many Americans believe that poverty is proof of personal failure. They do not stop to ask whether poor Americans have ever been given equality of opportunity, nor do they stop to notice how hard many poor people work (Jencks, 1994; Liebow, 1967). (See the Mapping Social Change box on page 283.)

Sociologists who study inequality in modern societies usually ask how equality of opportunity can be increased and the gaps between the haves and the have-nots decreased. Yet they are also highly aware of how difficult it is to narrow that gap in the United States or any other society. In response to those who believe that government can do nothing effective in its efforts to reduce poverty, sociologists point to the kinds of hard facts presented in Table 11.9. The table shows that poverty among people older than age sixty-five has been reduced dramatically since 1965, largely as a consequence of Social Security legislation and Medicare. Poverty among children, on the other hand, has increased in recent years.

The role the government should play in alleviating poverty is a major issue in American

mapping social change

Poor Counties in the United States

This map shows all of the counties of the United States, grouped by average income. It reveals a great deal about the places where poor people live. For example, along the Mississippi River, especially in Arkansas and Mississippi, we see groups of counties whose residents' average income is in the lowest range. The same is true in parts of the arid Southwest. Why are there such high concentrations of poor residents in these counties?

The Mississippi delta region has rich agricultural land, but most of it is owned by a few white families and agricultural corporations. The land is farmed by African American and poor white workers who earn extremely little. Thus, the delta region has a pattern of agrarian stratification that results in high rates of poverty. In the impoverished counties of the Southwest, the land often contains mineral and water resources that are controlled by a few wealthy individuals and corporations; Chicanos, Native Americans, and poor whites do not share in the wealth created by these resources, and the counties in which they live have low average incomes.

The isolated poor counties of the mountain states in the west often suffer from a newer form of poverty. In these areas, economic growth results from tourism, second-home developments, skiing, and other pursuits of the affluent upper and upper-middle classes. Farms, small-town businesses, and the trailer parks where the resort workers live may be located in a nearby valley, which may be part of a separate county with a lower average income (Jargowsky, 1997).

© Nathan Benn/Woodfin Camp & Associates

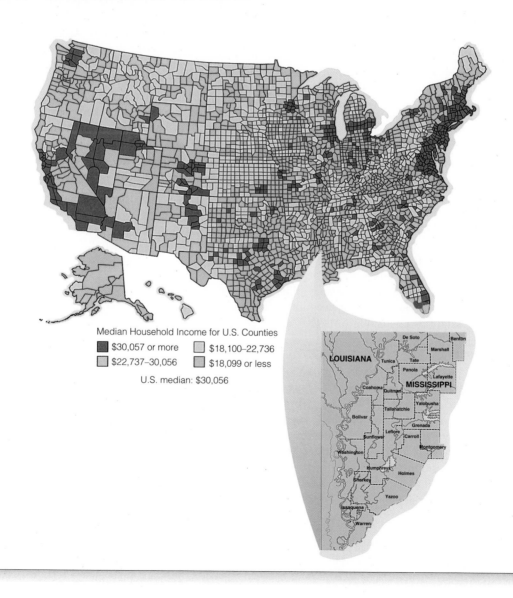

Median Household Income for U.S. Counties

- $30,057 or more
- $22,737–30,056
- $18,100–22,736
- $18,099 or less

U.S. median: $30,056

MORE EQUALITY? **283**

Job seekers wait in line for applications for the apprenticeship program of the Local 3 Elevator Mechanics Union in New York City, a typical scene throughout the United States during the recession that rocked the U.S. economy in 2008.

society today. It is frequently asserted, both in government and in the social sciences, that government has no business engaging in large-scale redistribution of income in the interests of the lower classes. This viewpoint, which is represented in the writings of conservative social scientists like Milton Friedman (1962), George Gilder (1982, 2006), and Charles Murray (1988, 2008b), as well as in current social policies, agrees that public funds must be spent to provide the poor with a "safety net" of programs to prevent suffering, but it also holds that programs to redistribute wealth infringe on individual freedom and the right of private property.

Welfare Reform

The extent to which declines in the income and well-being of poor children in the United States are a consequence of changes in the welfare system is an extremely controversial subject. In 1996, Congress enacted historic changes in the nation's welfare laws, which now require each state to ensure that a rising percentage of its adult aid recipients engage in approved work. The head of each family on government financial assistance ("welfare") is required to find a job within two years after assistance payments begin. Most families may receive benefits for no longer than five years, and states may impose even shorter time limits if they choose to do so. The new law dramatically reduced the welfare population, from about 5 percent of the total U.S. population in 1994 to about 2.6 percent in 2000 (Beneckson et al., 2000).

As noted earlier, when a deep recession slowed economic growth in the United States and elsewhere after 2008, poverty rates rose sharply, especially among children. Most poverty experts agree that, under circumstances of economic recession, the welfare reforms of the 1990s mean that low-income families have fewer options when unemployment deprives them of urgently needed income. In consequence, the percentage of children in poverty is likely to peak at 21 percent in 2010, which means that 27 percent of children—8 million—will likely have at least one parent not working full-time year-round in 2010. Median family income is expected to drop for all families, but especially for single male-headed households (Condon, 2009).

Thinking Critically Is inequality necessary in a nation like the United States? Functionalist sociologists argue that it is: If everyone received the same wages, what incentive would there be to study and work hard for entry into prestigious occupations? Conflict theorists argue that some inequality may create incentives, but too much inequality is a form of punishment for the "have-nots." Where do you stand on this important sociological issue?

Another disturbing fact is that welfare reforms were implemented during a period of sustained economic growth and record high demand for entry-level workers. The downturn in the U.S. economy will likely make it even more difficult for former welfare recipients, most of whom have limited educational credentials, to find work that will sustain their families or offer reasonable income stability.

Is There a War against the Poor?

Many sociologists who study welfare reform are highly critical of the new policies—not because they require people to work rather than receive benefits, but because they seem to increase the number of children living in poverty. Some also argue that the new policies are essentially designed to promote competition for low-wage jobs, which will ensure that the working poor remain poor. Under both past and present policies, sociologist Herbert Gans observes, poverty is so persistent that one can well ask whether it might have positive social functions. With large doses of irony, he goes on to list why he believes that legislators are waging a "war against the poor." He describes several "positive" functions of poverty:

- The existence of poverty ensures that society's dirty work will be done. . . . Society can fill these jobs by paying higher wages than for "clean" work, or it can force people who have no other choice to do the dirty work.
- Because the poor are required to work at low wages, they subsidize a variety of economic activities that benefit the affluent. For example, domestics subsidize the upper-middle and upper classes, making life easier for their employers.
- Poverty creates jobs for occupations and professions that serve or "service" the poor, or protect the rest of society from them. . . . Penology would be minuscule without the poor, as would the police.
- The poor can be identified and punished as alleged or real deviants in order to uphold the legitimacy of conventional norms. (1985, pp. 155–161)

These and other hidden functions of poverty, Gans argues, are far outweighed by its dysfunctions—suffering, violence, and waste of human and material resources. And they could be eliminated by "functional alternatives," such as higher pay for doing dirty work, or by the intentional redistribution of income. He believes welfare payments and tax credits for the poor should be increased so they can qualify for the housing and other subsidies available to better-off Americans (Gans, 1995).

Other sociologists offer proposals that focus on education or local and regional economic development. We saw earlier that success in the job market is directly related to educational attainment. In fact, in today's economy, the lack of a high school diploma puts a job seeker at a tremendous disadvantage. And when the job seeker has limited education and is also African American or Hispanic and a young mother, the disadvantages she faces can be overwhelming. (See Chapters 12 and 13 for more on this subject.) But sociologists like William J. Wilson (1998) point out that many poor people are poor because they happened to be born into households in very poor communities. If public policies were directed toward improving job prospects in these communities, they argue, the gap between the haves and the have-nots would be narrowed as well.

STILL THE LAND OF OPPORTUNITY?

We have seen in this chapter that there is a growing gap between the rich and the poor in the United States and elsewhere in the world. But does this significant increase in inequality also mean that people expect to experience less mobility for themselves and their children? Opinions on this issue vary widely, depending on each respondent's own experience, but the majority of Americans still believe in the American dream of hard work and eventual upward mobility. "They call it the land of opportunity, and I don't think it's changed much," said one respondent in a national survey on issues of class and inequality. The respondent, a homemaker from a declining industrial city in the Northeast, added, "Times are much harder with all the downsizing, but we're still a wonderful country" (NYT Correspondents, 2005, p. 7). Unfortunately, while ever-greater proportions of Americans believe that it is still possible to become rich (see Figure 11.8), the actual trends suggest that opportunities for upward mobility are stagnating at best. In 2005, a full 80 percent of respondents said that it was still possible for a poor person to work hard and become rich, compared to 60 percent who held this opinion in the 1980s. In the future, if current trends toward concentration of wealth in the

A Land of Opportunity?

More than ever, Americans cherish the belief that it is possible to become rich. Three-quarters think the chances of moving up to a higher class are the same as or greater than thirty years ago. Still, more than half thought it unlikely that they would become wealthy. A large majority favors programs to help the poor get ahead.

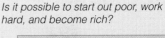

Is it possible to start out poor, work hard, and become rich?

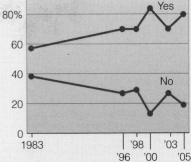

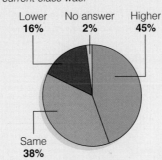

Compared with their social class when growing up, people said their current class was:

Lower **16%**
No answer **2%**
Higher **45%**
Same **38%**

How likely is it that you will ever become financially wealthy?

Already rich **1%**
No answer **1%**
Somewhat **34%**
Very **11%**
Not very **30%**
Not at all **22%**

Figure 11.8 A Land of Opportunity? Source: From The New York Times Poll, *How Class Works—A Nationwide Poll,* May 15, 2005. Reprinted by permission of The New York Times Agency.

hands of a relative few and reduced opportunity to own homes and other highly desired aspects of middle-class life continue, there may be less optimism about prospects for social mobility. On the other hand, Americans' defiant optimism about mobility has stimulated opportunity in the past and may continue to do so in the future.

Figure 11.9 indicates that the American dream of generational mobility is actually more characteristic of other European nations than it is

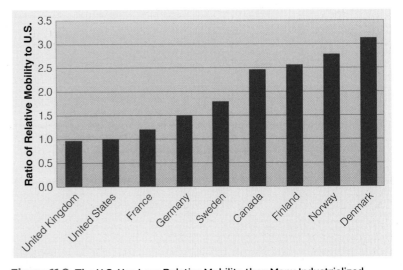

Figure 11.9 The U.S. Has Less Relative Mobility than Many Industrialized Nations. Source: From Sawhill, Isabel and John E. Morton. 2007. *Economic Mobility: Is the American Dream Alive and Well?* Washington, D.C.: Economic Mobility Project, an inititiative of the Pew Charitable Trusts. http://economicmobility.org/reports_and_research/. Reprinted by permission.

visual sociology

The Changing Landscape of Labor

Michael Jacobson-Hardy's photographs contrast the work environment in traditional, basic industries such as paper mills, foundries, textile mills, and shipyards with that in newer, high-technology industries such as computer manufacturing and aircraft production. In addition to examining the physical environment in which industrial production occurs, his images give visibility and voice to the workers themselves—the men and women whose lives and livelihoods have been affected most directly by the social and economic transformations now under way. The photographer explains his approach as follows:

> In 1989 I decided to seek out the effects of capitalism on workers and to record their stories through photography. The division of labor in a factory mirrors society as a whole. I sought out the impact of class oppression on different ethnic, racial, and gender groups. I became enamored with creating dignified portraits of entry-level workers. Images of the rich and famous are plentiful. It is the factory worker who needs visibility and a voice. One worker I met told me, "I've been working in this factory for forty years and nobody's ever asked to take my picture." (Jacobson-Hardy, 1995, p. 43)

The photos shown here are part of an exhibition titled "The Changing Landscape of Labor: Workers and Workplace." Although the focus of the exhibition is New England, the issues addressed are relevant to the United States as a whole, as can be seen in the accompanying comments by workers.

It's a very dirty job [extracting oil from metal shavings]. When you walk downstairs, it's a . . . big cloudy mist of oil. Sometimes you come home and you're oil from head to toe.

The plan is to do away with manufacturing. A lot of good people have been laid off. Both of my parents once worked here and lost their jobs. We can't compete. I like it here. I like the people I work with. Three and a half years ago, we had about 32,000 employees—now, there are about 6,000. Every 3 or 4 months, more people get laid off. It's a slow process.

When I started in the paper industry, it was still . . . a trade. You had to fight to get to be a machine tender. . . . It would take you sometimes 15 or 20 years to get to that position. . . . You had to wait. You had to grow into your job.

of the United States itself. The figure refers to "relative mobility," which is another term for the mobility of children relative to their parents, as opposed to "absolute mobility," which refers to increases in wealth or income due to economic growth. The figure shows that when compared to the same peer group, Germany has 1.5 times more intergenerational mobility than the United States, Canada nearly 2.5 times more, and Denmark 3 times more. Only the United Kingdom has relative intergenerational mobility levels on a par with that found in the United States.

SUMMARY

How is inequality structured in the United States?

- The basic measures of inequality in any society are income, wealth, *occupational prestige,* and *educational attainment.*

- In U.S. society, the distribution of educational attainment and occupational prestige is more nearly equal than the distribution of wealth and income.

- Sociological views of inequality in America have changed as the nation has been transformed from an agrarian society to an urban industrial society and then to a postindustrial society.

- The Jeffersonian view of America envisioned a society in which most families lived on their own farms or ran small commercial or manufacturing enterprises. However, this view did not apply to the larger cities, the southern states, or Native Americans.

- During the Great Depression, the effects of industrialization tended to increase hostility between workers and the owners of businesses.

- In the mid-twentieth century, several important studies of inequality in American communities revealed the existence of a complex social-class system as well as a racial caste system.

- The shift from an economy based on manufacturing to one based on services has resulted in a blurring of class lines and an easing of class conflict between industrial workers and the owners and managers of the means of production. Nevertheless, some sociologists argue that Americans continue to recognize social-class divisions.

- When people are asked what social class they belong to, the largest proportion say that they are members of the middle class. They base their class assignments on *socioeconomic status,* which is derived primarily from occupation but also takes into account family status, education, and earned income.

What kinds of social classes are found in the United States, and how do they compare to classes in other societies?

- Social-class position has important consequences for the daily life of individuals and households. Members of the upper classes tend to have better health and more adequate health care than people in the lower classes. They are also likely to receive more and better education. In politics, the poor and members of the working class generally support the Democratic Party, whereas those in the middle and upper classes support the Republican Party.

- The upper class is estimated at about 1 percent of the U.S. population but controls 40 percent of all personal wealth in the United States. This class may be divided into the wealthiest and most prestigious families, who make up the elite or "high society," and families that have acquired their money more recently.

- Sociologists continually debate whether the upper class in America is also the society's ruling class.

- The middle class, the largest class in American society, is culturally extremely diverse and hence is often referred to as the "middle classes." In the past, it was thought to be associated with a family-oriented, conservative, suburban lifestyle, but recent studies have shown that no easily identified middle-class suburban culture exists.

- The working class, which accounts for 44 percent of white Americans and almost 57 percent of African Americans, is undergoing rapid and difficult changes as production technologies change and industrialization spreads throughout the world. Members of this class are employed in skilled, semiskilled, or unskilled manual occupations, and many are union members. The working class can be divided into industrial workers and those employed in skilled crafts. More racial and ethnic diversity exists in the working class than in other classes.

- Estimates of the proportion of the population living in poverty vary widely, depending on the standard used to define poverty. According to official statistics, about 12 percent of Americans are living in poverty. A significant proportion of the poor have jobs that do not pay enough to support their families. Another large percentage of poor families consists of single-parent families headed by women. Other categories of poor people include aged people with fixed incomes, marginally employed rural workers and part-time miners, chronically unemployed manual workers, and disabled workers and their families. Another group of people in danger of becoming poor are farmers.

How are social classes changing, and what effects do the changes have on members of these classes?

- Policy debates on the issue of poverty are often clouded by problems of definition. Although many Americans believe in *equality of opportunity*, they are less committed to the ideal of *equality of result*.

- Most sociologists agree that it is impossible to achieve a completely egalitarian society; instead, they concentrate on how much present levels of inequality can and should be reduced.

- In 1996, Congress enacted major reforms in the welfare laws, which now require that adult aid recipients engage in approved work. Data on the effects of the reforms indicate that they have had mixed results and may actually have worsened conditions for poor children.

The Kornblum Companion Website

www.cengagebrain.com

Supplement your review of this chapter by going to the companion website to take one of the tutorial quizzes, use the flash cards to master key terms, and check out the many other study aids you'll find there. You'll also find special features, such as GSS data and Web links that will put data and resources at your fingertips to help you with that special project or to do some research on your own.

Inequalities of Race and Ethnicity

FOCUS QUESTIONS

In what way is race a "social construct" rather than a distinctively biological trait of different populations?

What are some of the most common patterns of intergroup relations?

Why do ethnic stratification and inequality persist in societies that are becoming increasingly pluralistic?

How do social scientists explain the existence of racial and ethnic inequalities?

Why are many social problems in the United States associated with race?

When she was only twenty years old and still an undergraduate architecture student at Yale University, Maya Lin had an astounding experience. The young Asian American's design for a monument to commemorate the 58,156 Americans killed in the Vietnam War was chosen from more than 1,500 entries in an international competition. Earlier in the semester, she had submitted the design as a project in her architecture course, but the professor did not like the concept very much. However, Lin had a strong vision of how the polished black granite monument would offer people a chance to reflect on the meaning of the war and remember those who died in it. She believed in her work.

After spending many weeks revising a one-page statement describing the intent of her design, Lin submitted the drawings to the competition's jury, never expecting to win. The rest is history. The Vietnam Veterans Memorial has become the most popular and moving piece of commemorative sculpture in the nation. Visitors run their fingers along the wall and often place bouquets, photographs, poems, or other items below the spot where their loved one's name is inscribed. "I knew people would have a personal experience as they searched for the names of those they knew," Lin (1995) explains, "but I never expected that they would leave things there. That was a surprise to me."

Although the monument commemorates the Vietnam War and does not directly address racism or ethnic conflict, Lin had to deal with a fair amount of racial abuse as the monument was being constructed. Mingled with the praise were letters that said, among other hateful things, that "a gook should not have been chosen to design a Vietnam War memorial." Even at the dedication, which Lin attended with her family, there were racial slights that wounded her deeply. But the popularity of the monument, and the way it brings together Americans of all

Maya Lin.

races to contemplate death and heroism, more than compensate for the hurt. And as described in the Visual Sociology box later in the chapter, the universal appeal of her vision has resulted in other commissions.

THE MEANING OF RACE AND ETHNICITY

How can we explain prejudice and racial discrimination and the inequalities they engender? And how can we measure their effects on individuals and social institutions? Many social scientists focus their research on why some groups in a society have been subordinated and the consequences of that subordination for them and their children. The effects of various forms of subordination, such as slavery, expulsion, and discrimination, on a society's patterns of inequality are at the center of the study of race and ethnicity.

Given the ever-changing diversity of the population, racial and ethnic relations in the United States are always complex and sometimes marked by overt conflict. No sooner does the society appear to be making some progress in combating prejudice and discrimination than new issues and problems appear. After the terrorist attacks on the World Trade Center and the Pentagon, for example, Arab Americans experienced the same kinds of discriminatory searches and questioning, known as racial profiling, that African Americans have struggled against in recent years.

The United States is not unique in the extent to which inequality and hostility among its ethnic and racial groups result in severe social problems. In Canada, conflict between Anglo and French Canadians in Quebec periodically threatens national unity. In Germany, there have been severe riots as native-born Germans have pressed for limits on the nation's acceptance of refugees from eastern European nations that were formerly dominated by the Soviet Union. In what used to be Yugoslavia, hostility among Croats, Serbs, Bosnians, and Albanians produced bloody ethnic wars. More recently, after the U.S. invasion of Iraq, conflict among that nation's three main ethnic and religious groups—ethnic Kurds, Sunni Muslims, and Shiite Muslims—has vastly complicated the development of a new government (Williams & Mousa, 2010).

In view of its importance throughout the world, it is no wonder that the study of racial and ethnic relations has always been a major subfield of sociology. Most societies include minority groups, people who are defined as different according to the majority's perceptions of racial or cultural differences. And in many societies, as the ironic song from the musical *South Pacific* goes, "You've got to be taught to be afraid/of people whose eyes are oddly made/and people whose skin is a different shade./You've got to be carefully taught." Sociologists try to get at the origins of these fears and groundless distinctions that categorize people as different and influence their life chances, often in dramatic ways.

In this chapter, we first look at how race and ethnicity lead to the formation of groups that have a sense of themselves as different from the dominant group. In the second section of the chapter, we analyze patterns of inequality and intergroup relations; in the third section, we discuss cultural consequences, particularly in American society. This is followed by a presentation of social-scientific theories that seek to explain the phenomena of intergroup conflict and accommodation. Rather than discussing each of the major racial and ethnic groups in American society in turn, throughout the chapter we present examples from the experience of blacks, Hispanic Americans, Native Americans, and other ethnic groups that have played an important part in American history.

Race: A Social Concept

Of the millions of species of animals on earth, ours, *Homo sapiens,* is the most widespread. For the past ten millennia, we have been spreading northward and southward and across the oceans to every corner of the globe, but we have not done so as a single people. Rather, throughout our history, we have been divided into innumerable societies, each of which maintains its own culture, thinks of itself as "we," and looks upon all others as "they." Through all the millennia of warfare, migration, and population growth, we have been colliding and competing and learning to cooperate. The realization that we are one great people despite our immense diversity has been slow to evolve. We persist in creating arbitrary divisions based on physical differences that are summed up in the term *race.*

In biology, **race** refers to an inbreeding population that develops distinctive physical characteristics that are hereditary. Such a population therefore has a shared genetic heritage (Graves, 2001; Marks, 1994). But the choice of which physical characteristics to use in classifying people into races is arbitrary. Biologists have used skin color, hair form, blood type, and facial features such as nose shape and eye folds in such efforts. In fact, however, the distribution of these traits overlaps considerably among the so-called races. Human groups have exchanged their genes through mating to such an extent that any attempt to identify "pure" races is bound to be fruitless (Gould, 1981; Graves, 2001). Thus, biologist Joseph Graves points out:

> As far as distinct lineages, throughout history, we have had too much gene flow between so-called races. If sub-Saharan Africans only mated with sub-Saharan Africans and Europeans only mated with Europeans, then there might be unique lineages. But that hasn't occurred, particularly in America. Here, because of our history of chattel slavery, individuals are still classified as black by means of the "rule of hypodescent," whereby one drop of black blood makes one black. However, there is no biological rationale for this rule. (quoted in Villarosa, 2002, p. 5)

Yet doesn't common sense tell us that there are different races? Can't we see that there is a Negro, or "black," race of people with dark skin, tightly curled hair, and broad facial features; a Caucasian, or "white," race of people with pale skin and ample body hair; and a Mongoloid, or "Oriental," race of people with yellowish or reddish skin and deep eye folds that give their eyes a slanted look? Of course these races exist. But they are not a set of distinct populations based on biological differences. The definitions of race used in different societies emerged from the interaction of various populations over long periods of human history. The specific physical characteristics we use to assign people to different races are arbitrary and meaningless—people from the Indian subcontinent tend to have dark skin and straight hair; Africans from Ethiopia have dark skin and narrow facial features; American blacks have skin colors ranging from extremely dark to extremely light; and whites have facial features and hair forms that include those of all the other supposed races. There is no scientifically valid typology of human races; what counts is what people in a society define as meaningful. In short, race is a social concept that varies from one society to another, depending on how the

race: An inbreeding population that develops distinctive physical characteristics that are hereditary.

people of that society feel about the importance of certain physical differences among human beings. In reality, as Edward O. Wilson (1979) has written, human beings are "one great breeding system through which the genes flow and mix in each generation. Because of that flux, mankind viewed over many generations shares a single human nature within which relatively minor hereditary influences recycle through ever-changing patterns, between the sexes and across families and entire populations" (p. 52).

Racism

Throughout human history, many individuals and groups have rejected the idea of a single human people. Tragic mistakes and incalculable suffering have been caused by the application of erroneous ideas about race and racial purity. They are among the most extreme consequences of the attitude known as *racism*.

Racism is an ideology based on the belief that an observable, supposedly inherited trait, such as skin color, is a mark of inferiority that justifies discriminatory treatment of people with that trait. In their classic text on racial and cultural minorities, Simpson and Yinger (1953) highlighted several beliefs that are at the heart of racism. The most common of these is the "doctrine of biologically superior and inferior races" (p. 55). Before World War I, for example, many of the foremost social thinkers in the Western world firmly believed that whites were genetically superior to blacks in intelligence. However, when the U.S. Army administered an IQ test to its recruits, the results showed that performance on such tests was linked to social-class background rather than to race. And when investigators controlled for differences in social class among the test takers, the racial differences in IQ disappeared (Kleinberg, 1935).

Since that time, there have been other attempts to demonstrate innate differences in intelligence among people of different races. The most recent of these efforts is Herrnstein and Murray's controversial study, *The Bell Curve* (1994), which we discussed in Chapter 5. This study argues that because scores on intelligence tests are distributed along a "normal" bell curve, many people will have scores that fall well below the mean, and these individuals will not be capable of performing well in situations requiring reasoning ability or other academic skills. Because the bell curves for African Americans and whites are different, with that of blacks peaking at a somewhat lower mean score, the authors argue that group differences in IQ establish biological limits on ability. This conclusion, in turn, leads them to argue that programs intended to correct differences in educational opportunities, such as Head Start, are doomed to failure.

Critics of these conclusions argue that studies like *The Bell Curve* merely dredge up old academic justifications for the status quo of racial and class inequality (Gould, 1995; Graves, 2001). There is much evidence that IQ tests are biased against members of minority groups and that something as complex and elusive as intelligence cannot be summarized by a single score on a test. The history of efforts to address inequalities in education shows that minority students quickly begin to achieve at the same levels as white students when they have access to high-quality educational programs.

Race Matters Race is not a biologically meaningful concept, but that does not mean it is not a significant aspect of life in societies all over the world. In the United States, as a result both of misunderstandings about race and of overt racism, African Americans, Asian Americans, Native Americans, and other population groups with physical features that set them apart have endured prejudice, discrimination, and economic exploitation. One result is persistent forms of inequality and injustice (West, 1996; Wilgoren, 2003). But there are also more positive outcomes. To survive and to cope with life in a society in which they are considered different and inferior, these groups have developed their own subcultures and a strong sense of themselves as distinct peoples within the larger society. This process occurs among groups that are not considered racially distinct, such as the Irish or Jews, who have struggled against prejudice and discrimination and have also developed strong subcultures.

Social Change and Racism: Confronting Hate Attacks on Arab storekeepers and turban-wearing Sikhs in the aftermath of 9/11, the harassment and beating of Hispanic day laborers, the brutal dragging murder of William Byrd by three white men in Texas, "ethnic cleansing" in Kosovo, continual strife between Palestinians and Jews in Israel . . . where will it strike next? Will it ever end? Perhaps more to the point, what can we as individuals do in the face of the seemingly overwhelming racial and ethnic divisions in our society and the world?

racism: An ideology based on the belief that an observable, supposedly inherited trait is a mark of inferiority that justifies discriminatory treatment of people with that trait.

President Obama's election was a demonstration that the large majority of Americans are anxious to get beyond racial prejudices and stereotypes, but as the president himself has said many times, he has no illusions that the struggle against racism will end quickly just because he is president. On the campaign trail in 2008, then Senator Obama and his wife Michelle experienced a great many insensitive and outright racist comments. Some of them even came from other elected officials, as when a Georgia congressman referred to him and his wife as "uppity," a term that in the past was often used along with the "n" word to refer to African Americans who were trying to improve their situation in society by becoming educated. In Kentucky, another elected official told the press, "I'm going to tell you something: That boy's finger does not need to be on the button." In the end, the Obama victory was the strongest demonstration possible that most Americans intensely desire to bury the nation's legacy of racism and prejudice.

What can we as individuals do when we witness hostility directed against people just because they are different? We may find it difficult to confront racism or ethnic prejudice in people who are our peers and with whom we must interact on a daily basis. Often we look away or pretend to "go along," hiding our feelings of shame and confusion. Another strategy is to confront the problem head-on, as Graves did. It may take courage, but we can tell someone that their behavior is racist or prejudiced. Of course, this could result in the loss of a friend, but perhaps that is not the end of the world. We can befriend others who are not hampered by racism or ethnic bigotry.

It can also help to try to understand the origins of prejudice in people we know and love. Try to see how their attitudes have been affected by social changes such as new forms of competition with recently arrived ethnic or racial groups, or appeals to fear and prejudice by political leaders who stand to benefit from divisions among working people. You won't get very far by giving sociological lectures to those close to you, but by applying your sociological imagination, you may come up with some strategies for changing their minds—or at least giving them an opportunity to see things a bit differently.

Racist beliefs in the innate inferiority of populations that are erroneously thought of as separate races remain one of the major social problems of the modern world. And this tendency to denigrate socially defined racial groups often extends to members of particular ethnic groups as well (Conley, 1999; Duster, 2001).

AP Photo/Jae C. Hong

Ethnic Groups and Minorities

An **ethnic group** is a population that has a sense of group identity based on a distinctive cultural pattern and, usually, shared ancestry, whether actual or assumed. Ethnic groups often have a sense of "peoplehood" that is maintained within a larger society (Kornblum, 1974; Teitelbaum & Weiner, 1995). Their members usually have migrated to a new nation or been conquered by an invading population. In the United States and Canada, a large proportion of the population consists of people who either immigrated themselves or are descended from people who immigrated to the New World or were brought here as slaves. If one follows history back far enough, it appears that all the people living in the Western Hemisphere today can trace their origins to other continents. Even the Native Americans are believed to have crossed the Bering Strait as migrant peoples between 14,000 and 20,000 years ago.

European conquests of the Americas in the sixteenth and seventeenth centuries resulted in people of Iberian (Spanish and Portuguese) origins becoming the dominant population group in Central and South America. People of English, German, Dutch, and French origins became the dominant population groups throughout much of North America. By the late 1770s, almost 80 percent of the population of the thirteen American colonies had been born in England or Ireland or were the children of people who had come from those countries. A century later, because of the influence of slavery and large-scale immigration, the populations of the United States and Canada were far more diverse in terms of national origins, but people from England were still dominant.

According to sociological definitions of the term *dominance,* a dominant group is not necessarily numerically larger than other population groups in a nation. More important in establishing dominance is control or ownership of wealth (farms, banks, manufacturing concerns, and private property of all kinds) and political power—especially control of political institutions like legislatures, courts, the military, and the police. Thus, by the beginning of the twentieth century, white Anglo-Saxon Protestant males (or WASPs, as they are sometimes called) had established dominance over U.S. society even though they no longer constituted a clear majority of the population.

ethnic group: A population that has a sense of group identity based on shared ancestry and distinctive cultural patterns.

Reaching Higher Ground "Do you know the hymn that goes, 'The great trees are bending/The Good Lord is sending/His people up to higher ground'?" The speaker is Polly Heidelberg, a founder of the civil rights movement in Meridian, Mississippi. Among the events that have burned themselves into her memory are the murders of three civil rights workers—James Chaney, Michael Schwerner, and Andrew Goodman—who worked together in Mississippi during the "Freedom Summer" of 1964. Chaney was an African American resident of Meridian; Schwerner and Goodman were volunteers from northern colleges. And Polly Heidelberg was one of the last people the young men saw before they set out for the little village of Philadelphia, Mississippi, where they were murdered and hastily buried in an earthen dam. (Their murder, one of the pivotal moments of the civil rights movement, is the subject of the 1988 film *Mississippi Burning*.) "Miz" Heidelberg, as she is known in Meridian, was Chaney's neighbor, and she feels responsible for "bringing him into the movement." She knew all three men, and even now, when she thinks of them, her eyes cloud over and she prays for their souls.

Richard Avery/Stock Boston Inc.

As she speaks, Heidelberg and I are sitting in a white-owned coffee shop in a white neighborhood of Meridian. One of her granddaughters, a child of about four, has finished eating and is playing quietly on the floor. I am scribbling Heidelberg's words on a notepad while attempting to eat an ample southern breakfast. An interview with Polly Heidelberg is fun. She often slips riddles into her speech to gently tease the listener. She is also a deeply religious and patriotic person, and her speech is filled with allusions to God and country.

Heidelberg tells me about the movement to integrate the armed forces during World War II, about the bitterness of racial strife during the 1950s, and about the exhausting, exhilarating years of the 1960s. There is no end to the movement as she sees it. Civil rights was only one aspect of the struggle for true

Bettmann/Corbis

equality of opportunity. Now there must be more progress toward economic equality.

We have finished our breakfast and the child is becoming restless. Heidelberg asks me, "That line in the hymn that mentions higher ground, what do you think higher ground is?"

Stammering a bit, I begin by saying something about "loftier values, greater spirituality, . . ." but Miz Heidelberg gently interrupts me.

"Sure, you could think about it that way. But to me 'higher ground' is right where we are now. It's you and me and the child having breakfast in this coffee shop here in Meridian, Mississippi."

Between 1820 and 1960, approximately 42 million people immigrated to the United States. Since 1960, an average of at least 300,000 immigrants have arrived each year, with the number increasing to more than 900,000 in the past few years. Since its earliest history, the United States has had one of the most liberal immigration policies of any nation, partly because many of its inhabitants can trace their origins to other countries. And except in the case of the Native Americans, who were expelled from their ancestral lands and forced to resettle on reservations, no area of the nation is seen as the ancestral homeland of a particular people. Unlike many nations, the United States is a multiethnic society where notions of "blood" (descent) do not coincide with claims to ancestral land. In the United States, therefore, there are no separatist movements that claim territory and threaten national unity. However, severe inequalities affect particular populations, especially Native Americans, African Americans, Latinos, and Appalachian whites. Many of these inequalities date back to major periods of settlement and ethnic group formation, which are summarized in the Global Social Change box on pages 296–297.

Minority Groups Immigrant groups that are distinct because of racial features or culture (language, religion, dress, and so on) are often treated badly in the new society and hence develop the consciousness of being a minority group. Louis Wirth (1945), a pioneer in the study of racial and ethnic relations, defined a **minority group** as a set of "people who, because of their physical or cultural characteristics, are singled out from the others in the society in which they live for differential and unequal treatment, and who therefore regard themselves as objects of collective discrimination" (p. 347).

minority group: A population that, because of its members' physical or cultural characteristics, is singled out from others in the society for differential and unequal treatment.

The existence of a minority group in a society, Wirth explained, "implies the existence of a corresponding dominant group with a higher social status and greater privileges. Minority status carries with it exclusion from full participation in the life of the society" (p. 347).

In the United States, the term *minority* often suggests "people of color," meaning African Americans, American Indians, Mexicans (many of whom have the darker coloring of Amerindian ancestry), and Asians. But the term can also be applied to people with lighter skin coloring. In Great Britain, for example, the Irish were a conquered people who were subjected to economic and social discrimination by the English. Not surprisingly, therefore, when the Irish began immigrating to the United States in the nineteenth century, they were treated as an inferior minority by Americans of English ancestry. This attitude was especially prevalent in cities like Boston and New York, to which the Irish came in large numbers.

It should also be noted that the term *minority* does not always imply that the population is numerically inferior to the dominant group. There are counties in some states in the South, and entire cities in the North, in which African Americans constitute a numerical majority, yet they cannot be considered the dominant group because they lack the power, wealth, and prestige of the white population.

Inequality of Wealth Versus Income In their research on income and wealth differences between white and black households, Melvin Oliver and Thomas Shapiro found that if they compared African American households consisting of a married couple and their children with similar white households, the black family had somewhat less income (almost $5,000) but far less wealth: about $46,000 less net worth (the value of all property, equities, and so on, possessed by the household). In other words, income inequality is much less severe than inequality of wealth. The authors found that the cumulative effects of discrimination in employment, housing and mortgage markets, and education go a long way toward explaining this gap. Their comparative lack of wealth creates disadvantages for African American families, who cannot assist their children in buying a home and other aspects of wealth accumulation to nearly the same extent as white families (Oliver & Shapiro, 1995). See the Research Methods box on page 309 for further analysis of how these differences are measured and why these measures are so important.

Ethnic Group Formation Ethnic groups often form in cities or metropolitan regions for reasons of choice as well as necessity. In fact, the identity of a group is often defined through the experience of coping with life in the new society as well as through the group's desire to hold on to its language, food, and other cultural ways (Handlin, 1992; Rodriguez, 1992). In one of the first and best studies of ethnic group formation, W. I. Thomas (1971/1921) described how in the early decades of the twentieth century Italian immigrants settled in small neighborhoods in Manhattan where they could live with people who had come from the same region and spoke the same dialect of Italian. At first, these immigrants identified themselves as Calabrians or Sicilians or Piedmontese, the way they had before leaving their home regions. Thomas found that they identified with those home regions rather than with the nation of Italy. In New York, however, they could speak to each other and appreciate each other's food, and they had reason to join together to compete with other ethnic groups for jobs and homes. They were also called Italians by people outside their neighborhoods. Gradually, therefore, they began to think of themselves as Italians or Italian Americans, but this identity was forged out of their experience in the United States and did not mean the same thing as being an Italian in Italy.

Thinking Critically

We often hear it said that the United States needs an "English-only" policy to ensure that immigrants learn the national language. The history of American immigration and assimilation shows that immigrants quickly learn to speak English when they have opportunities to do so, and that they no longer speak the language of their immigrant parents very well. Do you know native speakers of other languages who are failing to learn English?

Most ethnic populations have had similar experiences that caused them to develop a group consciousness and create organizations to represent their interests. Homosexuals and other populations that are defined in terms of their behavior and cultural traits may become organized into cultural minority groups, but they differ from true ethnic groups because they do not trace their ancestry to another society.

Today, sociologists are interested in knowing whether a similar process of ethnic group formation is occurring in regions where there are large Spanish-speaking populations. They seek to discover whether Mexicans, Puerto Ricans, Colombians, and Dominicans are beginning to think of themselves as Latinos in cities where they are all present in large numbers, or whether their allegiance to their nationality of origin will prevent them from taking on the more inclusive Latino identity. Likewise, researchers are asking

Periods of Migration and Settlement in the United States In 1790, when the republic was newly formed, the vast majority of U.S. citizens and resident foreigners (77 percent) had come from Great Britain and the counties that are now included in Northern Ireland. People from Germany (7.4 percent), Ireland (4.4 percent), the Netherlands (3.3 percent), France (2 percent), Mexico (1 percent), and other countries were present in the population, but with the exception of the Irish, they were also primarily white Protestants. Blacks—involuntary immigrants—and their children actually accounted for 19 percent of the total population, but neither they nor Native Americans were citizens.

1820–1885: The "Old" Northwest European and Asian Migration

By about 1830, the mass migration of different groups into the new nation was under way. The largest inflows of immigrants came from Ireland and Germany, with more than 1 million people from each of these countries coming to the United States to escape political or economic troubles. Smaller waves of immigrants originated in the northern European nations of Sweden, Norway, and Denmark.

Although people from northern Europe made up the most important immigrant groups, far exceeding the continuing influx from Great Britain, 1885 marked the beginning of an inflow of immigrants from China that continued at high rates for the next thirty years. Chinese immigrants settled primarily in the West and contributed immensely to the development of the western states, although they met with sporadic and often violent hostility.

1885–1920: The "Intermediate" Migration from Southern and Eastern Europe and the Beginning of Heavy Immigration from Mexico

From 1880 to 1900, the inflow of immigrants from Italy, Poland, Russia, the Baltic states, and southern Europe (Serbs, Croats, Slovenes, Romanians,

Bulgarians, Greeks, and so on) exceeded the inflow of people from older immigrant groups by about 4 to 1. In 1907 alone, more than 250,000 Italians and 338,000 Poles and other central Europeans were admitted through New York's Ellis Island. Many of these immigrants were Jewish, Roman Catholic, or Orthodox Christian. The newcomers' tendency to live in neighborhoods known as "Little Italy" or "Little Poland" convinced more established Americans that the immigrants would never learn English or become fully American. The mass arrival of physically distinctive groups such as Mexicans and Japanese during this period also aroused fear and hostility among those who disliked the newcomers.

During World War I, thousands of impoverished black southerners were attracted to the cities of the North and the Midwest. This began a trend that would continue throughout the twentieth century whenever wars or economic booms made it possible for blacks to overcome discrimination and get factory jobs.

1921–1959: Immigration by Quota and Refugee Status

Agitation against the "new" immigrants and demands for "Oriental exclusion" led Congress to establish a quota system that drastically altered the inflow of newcomers into the nation after 1920. At first, the quotas were fixed at 3 percent of the number of people from each country who had been counted as U.S. residents in the 1910 census. But because that formula still allowed large numbers of "new" immigrants

---

questions about whether a new Asian ethnicity will be forged out of the experience of various Asian immigrant groups, especially the Koreans and Chinese, in American cities (Alba, 2009; Moore & Pachon, 1985).

WHEN WORLDS COLLIDE: PATTERNS OF INTERGROUP RELATIONS

Throughout history, when different racial and ethnic groups have met and mixed, the most usual outcome has been violence and warfare. In fact, the desire for peaceful and cooperative relations among diverse peoples has emerged only relatively recently. In this section, we explore a continuum of relations between dominant and minority groups that extends from complete intolerance to complete tolerance, as shown in Figure 12.1 on page 298. At one extreme is extermination or genocide; at the other is assimilation.

Genocide and Mass Killing

In a study of the Siane tribe of New Guinea, anthropologist Richard Salisbury (1962) found that the members of this isolated highland tribe believed that anyone from another tribe wanted to kill them. Therefore, the Siane believed that they must kill any member of another tribe they

might encounter. (Fortunately, they excluded the anthropologist from this norm.) We often think of such behavior as primitive, savage, or barbarous. Yet barbarities on a far greater scale have been carried out by supposedly advanced societies. The most extreme of these is **genocide**, state-sponsored mass killing explicitly designed to completely exterminate a population deemed to be racially or ethnically different and threatening to the dominant population.

There have been numerous instances of mass killings and fewer explicit policies of genocide in recent history. Those incidents have been characterized by a degree of severity and a level of efficiency unknown to earlier civilizations. Consider the following:

- The Native American populations of North, Central, and South America were decimated in mass killings by European explorers and settlers between the sixteenth and twentieth centuries. Millions of Native Americans were killed in one-sided wars and through intentional starvation, forced marches, and executions. The population of Native

---

genocide: State-sponsored mass killing explicitly designed to completely exterminate a population deemed to be racially or ethnically different and threatening to the dominant population.

to enter the nation, in 1924 the law was changed so that quotas would be based on the national origins of the foreign-born population as of the 1890 census. This biased the quota system in favor of northern Europeans; in fact, Asians were explicitly excluded.

World War II and the Cold War produced millions of homeless and stateless refugees, especially from Germany, Poland, and the Soviet Union. Almost half a million of these displaced people arrived in the United States between 1948 and 1950; another 2 million came between 1950 and 1957. Their arrival swelled the populations of ethnic enclaves in America's industrial cities.

1960 to the Present: Worldwide Immigration

In the 1960s, national quotas were replaced with a system based on preference for skilled workers and professionals, regardless of country of origin, as well as refugees and people with families already in the United States. Gradually, Congress increased the number of immigrants who may enter the country legally. Today the United States is once again the foremost immigrant-receiving nation in the world. The largest streams of immigrants are from Mexico, other Central American and Caribbean nations, Asia (especially Korea, China,

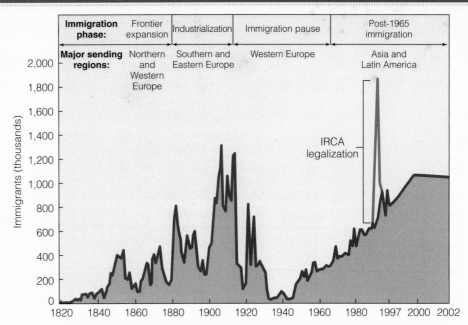

Note: IRCA refers to the amnesty provisions of the Immigration Reform and Control Act of 1986, under which 2.7 million unauthorized foreign residents were transferred to legal immigrant status. Source: INS, 1997, 1999; *Statistical Abstract*, 2005.

India, and Pakistan), and South America. Untold numbers of undocumented, or illegal, immigrants also arrive each year—especially from Mexico, Canada, and the Dominican Republic—but studies show that a high proportion of these newcomers eventually become legal residents or return to their country of origin (Portes & Rumbaut, 1996; Zolberg, 2006).

Americans in North America was reduced from more than 4 million in the eighteenth century to less than 600,000 in the early twentieth century (Thornton, 1987).

- Mass killings were conducted during European colonial conquests: When England, France, Germany, Portugal, and the Netherlands were engaged in fierce competition for colonial dominance of Africa during the nineteenth and early twentieth centuries, millions of native people were exterminated. The introduction of the Gatling machine gun made it possible for small numbers of troops to slaughter thousands of tribal warriors.

- Examples of state-sponsored genocide include the Nazis' systematic killing of six million Jews and 400,000 gypsies during World War II, and the Hutus's extermination of thousands of Tutsis living in Rwanda in the 1990s.

- Genocide is taking place in the Darfur region of Sudan, despite efforts by the United Nations and the African Union to protect vulnerable tribespeople.

Mass executions and other forms of genocide are almost always rationalized by the belief that the people who are being slaughtered are less than human and in fact are dangerous parasites. Thus, the British and Dutch slaughtered members of African tribes like the Hottentots in the belief that they were a lower form of life, a nuisance species unfit even for enslavement. The Nazis rationalized the extermination of Jews and gypsies by the same twisted reasoning (Chalk & Jonassohn, 1990).

Expulsion

In many societies, extended conflicts between racial or ethnic groups have ended in **expulsion**, the forcible removal of one population from territory claimed by the other. Thus, on the earliest map of almost every major American city, there appears a double line drawn at the edges where the streets end. This is the Indian boundary line, and it symbolizes the expulsion of Native Americans from lands that were taken from them in order to create a city in which they would be strangers.

Expulsion was the usual fate of North America's native peoples, who also experienced massacres, deadly forced marches, and other genocidal attacks. These actions were rationalized, in part, by the nineteenth-century doctrine of "manifest destiny," which asserted the inevitability

expulsion: The forcible removal of one population from a territory claimed by another population.

Genocide	Expulsion	Slavery	Segregation	Assimilation
State-sponsored mass killing explicitly designed to completely exterminate a population deemed to be racially or ethnically different and threatening to the dominant population	The forcible removal of a population from a territory claimed by another population.	Ownership of one population by another, which can buy and sell members of the enslaved population and controls every aspect of their lives.	Ecological and institutional separation of races or ethnic groups.	The process by which a minority group blends into the majority population and eventually disappears as a distinct people within the larger society.

Figure 12.1 A Continuum of Intergroup Relations. Note: The placement of slavery to the right of expulsion is not meant to imply that slavery is less severe than expulsion. The reason for the placement is that slaves and slave masters are interdependent populations within a single society, whereas expulsion excludes the subordinated population from any form of membership in the society.

of the westward expansion of Europeans across the territories of the Southwest and West (Davis, 1995). As white settlement expanded westward, Native Americans were continually expelled from their tribal lands. After the 1848 gold rush and the rapid settlement of the West Coast, the pressure of white settlement pushed the Indian tribes into the high plains of the West and Southwest. Between 1865 and the 1890 massacre of Sioux Indians at Wounded Knee, South Dakota, the remaining free tribes in the West were forced to settle on reservations. In the process, the Indians lost more than their ancestral lands. As the famous Sioux chief Black Elk put it: "A people's dream died" (quoted in Brown, 1970, p. 419).

The forced settlement of Native Americans on reservations is only one example of expulsion. Asian immigrants in the American West also suffered as a result of intermittent attempts at expulsion referred to as the Oriental exclusion movement. In an effort to prevent the large-scale importation of Chinese laborers into California and other states, Congress passed the Chinese Exclusion Act of 1882, which excluded Chinese laborers from entry into the United States for ten years. But this legislation did little to relieve the hostility between whites and Asians. Riots directed against Chinese workers were common throughout the West during this period, and Chinese immigrants were actually expelled from some towns. In 1895, for example, a mob killed twenty-eight Chinese immigrants in Rock Springs, Wyoming, and expelled the remaining Chinese population from the area (Lai, 1980).

The most severe example of expulsion directed against Asians in the United States occurred in 1942, after the Japanese attack on Pearl Harbor and the American entry into World War II. On orders from the U.S. government, more than 110,000 West Coast Japanese, 64 percent of whom were U.S. citizens, were ordered to leave their homes and their businesses and were transported to temporary assembly centers (Kitano, 1980). They were then assigned to detention camps in remote areas of California, Arizona, Idaho, Colorado, Utah, and Arkansas. When the U.S. Supreme Court declared the incarceration of an entire ethnic group without a hearing or formal charges to be unconstitutional (*Endo* v. *United States,* 1944), the Japanese were released, but many had lost their homes and all their possessions by then. In 1989, after many years of pressing their claim of human rights violations, Japanese Americans finally succeeded in persuading Congress to create a reparations program for families who had been imprisoned during the war. Although the amounts

paid—generally about $20,000—do not come close to restoring their material or emotional losses, the apology they received from President Reagan, along with the symbolic value of having won their claim, created an important precedent for other groups that have been victims of racism. As Amy Hiratzka, a survivor of the Gila River prison camp, told an interviewer: "I just want to work toward the fact that this doesn't happen to anyone again. I accepted the $20,000 for my daughter. I think of the blacks. I think of the American Indians. Their plight has been just a chronic series of ups and downs and unfair treatment, lack of justice" (quoted in Dymski, 2000, p. 6).

Sociology & Social Justice

THE TRAIL OF TEARS The accompanying artist's rendition of the expulsion of Cherokee Indians from their homes captures some of the tragic aspects of their forced removal to the West. The Cherokees were among the most assimilated of the Native American tribes. They farmed, did business, built communities, and dressed increasingly like their white neighbors. But in his 1829 inaugural address, President Andrew Jackson set a policy to relocate eastern Indians. Between 1830 and 1850, about 100,000 American Indians, Cherokees, Chickasaws, Choctaws, Creeks, and Seminoles living between Michigan, Louisiana, and Florida moved west after the U.S. government coerced treaties or used the U.S. Army against those who resisted. Many were treated brutally. An estimated 3,500 Creeks died in Alabama and on their westward journey. The expulsion came to be known as the Trail of Tears.

The Trail of Tears: A low point in social justice.

Slavery

Somewhat farther along the continuum between genocide and assimilation is slavery. **Slavery** is the ownership of a population, defined by racial, ethnic, or political criteria, by another population that not only can buy and sell members of the enslaved population but also has complete control over their lives. Slavery has been called "the peculiar institution" because, ironically, it has existed in some of the world's greatest civilizations. The socioeconomic systems of ancient Greece and Rome, for example, were based on the labor of slaves. And the great trading cities of late medieval Europe, such as Venice, Genoa, and Florence, developed plantation systems in their Mediterranean colonies that were based on slave labor. In fact, the foremost student of slave systems, Orlando Patterson, makes the following comment:

> There is nothing notably peculiar about the institution of slavery. It has existed from before the dawn of human history right down to the twentieth century, in the most primitive of human societies and in the most civilized. . . . Probably there is no group of people whose ancestors were not at one time slaves or slaveholders. (1982, p. vii)

Figure 12.2 indicates the magnitude of the transatlantic slave trade; the widths of the arrows represent the relative size of each portion of that terrible traffic in humanity. The arrows do not show, however, that for the Americas to acquire 11 million slaves who survived the voyage on the slave ships and the violence and diseases of the New World, approximately 24 million Africans had to be captured and enslaved (Fyfe, 1976; Patterson, 1982, 1991).

The chart demonstrates that the United States imported a proportionately small number of slaves. However, although somewhat less than 10 percent of all slaves were sold in the United States, by 1825, almost 30 percent of the black population in the Western Hemisphere was living in the United States (Patterson, 1982) because of the high rate of natural increase among the American slaves. In Brazil, by contrast, the proportion of slave imports was relatively high, but there was also a high mortality rate among the slaves as a result of disease and frequent slave revolts.

At the time of the first U.S. census in 1790, there were 757,000 blacks in the overall population. By the outbreak of the Civil War, the number had increased to 4.4 million, of whom all but about 10 percent were slaves. In the southern states that fought in the war, one-third of white families owned slaves, the average being nine "chattels" per owner (Farley & Allen, 1987). (The term *chattels* refers to living beings that are considered property, including slaves and, in some cultures, women.)

It can be inferred from the great increase in the U.S. slave population that slaveholders in the United States treated their slaves less badly than did slaveholders in other parts of the Western Hemisphere. This does not mean, however, that slaves in the Americas did not bitterly resent their condition and struggle against it. Slave revolts were extremely common in the Caribbean and elsewhere in the New World. Even when slaves could not rebel openly, they carried out many forms of resistance. Patterson points out that the slave has always striven, against all odds, "for some measure of regularity and predictability in his social life. . . . Because he was considered degraded, he was all the more infused with the yearning

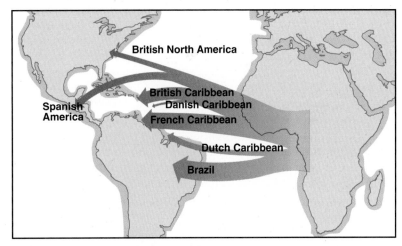

Figure 12.2 **The Transatlantic SlaveTrade.** From the end of the sixteenth century to the early decades of the nineteenth, approximately 11 to 12 million Africans were imported to the New World (Patterson, 1982). The thickness of the arrows shows the approximate volume of the slave trade to each region. Source: Adapted from Curtin, 1969.

slavery: The ownership of one racial, ethnic, or politically defined group by another group that has complete control over the enslaved group.

This historic photo of fugitive slaves during the Civil War was taken at Cumberland, Virginia, on May 14, 1862. The exhausted adults and children are wearing all the clothes they possessed, and many are shoeless. No population group in American history, with the exception of some Native American tribes, began its history as free citizens with such great economic obstacles to overcome as did the former slaves and their children.

Gibson, James F./Library of Congress [LC-DIG-cwpb-01005]

for dignity" (1982, p. 337). Patterson concludes that one of the chief ironies of slavery throughout history has been that "without [it] there would have been no freedmen" (p. 342). In other words, the very idea of freedom developed in large part from the longing of slaves to be free.

Legacies of Slavery Groups that have endured slavery (and this is true of expulsion and genocide as well) typically experience its effects for generations after gaining freedom. Their history of slavery, emancipation, and dealing with slavery's consequences becomes a central aspect of their shared culture. Jews, for example, celebrate the holiday of Passover to commemorate their liberation from Egyptian slavery over twenty-five centuries ago. For African Americans, the memories of slavery and the failures of the U.S. government to make good on promises of land after emancipation in 1863 are far more recent and vivid. Scholars of race relations argue that, just as the Indian reservations are a continuing legacy of expulsion and genocide, so, too, are the nation's areas of segregated rural and urban poverty a continuing legacy of slavery (Conley, 2003).

Does this historical legacy justify such policies as affirmative action, official apologies, or material reparations to the descendants of the enslaved or excluded? In the case of affirmative action, the courts have agreed that it does. And the possibility of some form of reparations is a subject of serious discussion (Robinson, 2003). Yet there is also the question of whether present generations of people should be required to give up money or opportunity to correct the wrongs of the past. These are momentous moral and political questions.

Sociological research, both past and present, provides empirical evidence to inform debates over these issues. It shows that racially distinct groups that have been subjected to the continuing effects of slavery and other extreme treatment do experience long-term disadvantages in their

ability, as a group (with many individual exceptions), to gain wealth and income, educational status, physical and mental health, occupational prestige, and social mobility (Burdman, 2003; Ogbu, 1994). These disadvantages and efforts to address them are discussed more fully in later sections of the chapter.

Segregation

Although African American slaves gained their freedom during the Civil War and became citizens of the United States, this did not mean that they became fully integrated into American society. A long period of segregation followed. **Segregation** is the ecological and institutional separation of races or ethnic groups. The segregation of racially or ethnically distinct peoples within a society may be voluntary, resulting from the desire of a people to live separately and maintain its own culture and institutions. The Amish, the Hutterites, and the Hasidic Jews are examples of groups that voluntarily segregate themselves. But segregation is often involuntary, resulting from laws or other norms that force one people to be separate from others. Involuntary separation may be either **de jure segregation** (created by laws that prohibit certain peoples from interacting with others or place limits on such interactions) or **de facto segregation** (created by unwritten norms that result in segregation, just as if it were "in fact" legally required).

During the 1990s, South Africa underwent a stunning reversal of its system of de jure segregation. Known as *apartheid,* South Africa's racial laws insisted that blacks, whites, and "coloreds" (people of mixed ancestry) remain separate. Intermarriage and integrated schools and communities were forbidden. In 1991, after years of struggle and violence, the white-dominated government agreed to repeal the apartheid laws. A constitutional convention—at which all segments of the population were represented—created a democratic regime in which all citizens' votes count equally. The African majority, led by Nelson Mandela and the African National Congress, assumed power in a relatively peaceful transition. Nevertheless, whites continue to hold most of the nation's wealth, whereas most South African blacks are trapped in dire poverty and illiteracy, just as the former slaves were in the United States after the Civil War.

Jim Crow In the United States, legally sanctioned segregation no longer exists, but this has been true only in recent decades. Before the civil rights movement of the 1960s, de jure segregation was common, especially in the southern states. The system that enforced segregation was supported by **Jim Crow** laws. This term refers to laws

segregation: The ecological and institutional separation of races or ethnic groups.

de jure segregation: Segregation created by formal legal sanctions that prohibit certain groups from interacting with others or place limits on such interactions.

de facto segregation: Segregation created and maintained by unwritten norms.

Jim Crow: The system of formal and informal segregation that existed in the United States from the late 1860s to the early 1970s.

that enforced or condoned segregation, barred blacks from the polls, and the like. (Jim Crow was a nineteenth-century white minstrel who performed in blackface and thereby reinforced black stereotypes.) The system was in effect for about 100 years, from just after the Civil War until the early 1970s. During that period, the so-called color line was applied throughout the United States to limit the places where blacks could live, where they could work and what kinds of jobs they could hold, where they could go to school, and under what conditions they could vote.

At first, the color line was an unwritten set of norms that barred or restricted black participation in many social institutions. By the turn of the twentieth century, however, segregation had become officially sanctioned through legislation and court rulings. This official segregation was rationalized by the "separate but equal" doctrine set forth by the Supreme Court in *Plessy v. Ferguson* (1896). Under this doctrine, separate facilities for people of different races were legal as long as they were of equal quality. In addition, de facto segregation and the existence of a "job ceiling" based on race kept blacks in subordinate jobs and segregated ghettos, as we saw in Chapter 11 in the case of Chicago's "black metropolis."

Only after years of struggle by opponents of de jure segregation did the U.S. Supreme Court finally decide, in the landmark case of *Brown v. Board of Education of Topeka* (1954), that "separate but equal" was inherently unequal. The *Brown* ruling put an end to legally sanctioned segregation of schools, hospitals, public accommodations, and the like. But it took frequent, often violent demonstrations and the mobilization of thousands of citizens in support of civil rights to achieve passage of the Civil Rights Act of 1964. That act mandated an end to segregation in private accommodations, made discrimination in the sale of housing illegal, and initiated a major attack on the job ceiling through the strategy known as *affirmative action*. Despite these judicial and legislative victories, however, de facto segregation remains a fact of life in the United States, especially in large cities (Massey & Denton, 1993).

The Segregation Index Ecological studies of racial segregation show that residential segregation persists in almost all U.S. cities. Table 12.1 is based on a measure known as the *segregation index*. It shows "the minimum percentage of nonwhites who would have to change the block on which they live in order to produce an unsegregated distribution—one in which the percentage of nonwhites living on each block is the same throughout the city (0 on the index)" (Taeuber

TABLE 12.1

Trends in Black-White Segregation in 30 Metropolitan Areas with Largest Black Populations, 1970–2000

Metropolitan Area	1970	1980	1990	2000
Northern areas	(Percentages)			
Boston	81.2%	77.6%	68.2%	65.7%
Buffalo	87.0	79.4	81.8	76.7
Chicago	91.9	87.8	85.8	80.8
Cincinnati	76.8	72.3	75.8	74.8
Cleveland	90.8	87.5	85.1	77.3
Columbus	81.8	71.4	67.3	63.1
Detroit	88.4	86.7	87.6	84.7
Indianapolis	81.7	76.2	74.3	70.7
Kansas City	87.4	78.9	72.6	69.1
Los Angeles–Long Beach	91.0	81.1	73.1	67.5
Milwaukee	90.5	83.9	82.8	82.2
New York	81.0	82.0	82.2	81.8
Newark	81.4	81.6	82.5	80.4
Philadelphia	79.5	78.8	77.2	72.3
Pittsburgh	75.0	72.7	71.0	67.3
St. Louis	84.7	81.3	77.0	74.3
Average	84.5	80.1	77.8	76.6
Southern areas				
Atlanta	82.1%	78.5%	67.8%	65.6%
Baltimore	81.9	74.7	71.4	67.9
Birmingham	37.8	40.8	71.7	72.9
Dallas–Ft. Worth	86.9	77.1	63.1	59.4
Houston	78.1	69.5	66.8	67.5
Memphis	75.9	71.6	69.3	68.7
Miami	85.1	77.8	71.8	73.6
New Orleans	73.1	68.3	68.8	69.3
Norfolk–Virginia Beach	75.7	63.1	50.3	46.2
Tampa–St. Petersburg	79.9	72.6	69.7	64.5
Washington, D.C.	81.1	70.1	66.1	63.1
Average	75.3	68.3	66.5	65.3

Source: Reprinted by permission of the publisher from *American Apartheid: Segregation and the Making of the Underclass* by Douglas S. Massey and Nancy A. Denton, pp. 222. Cambridge, MA: Harvard University Press. Copyright © 1993 by the President and Fellows of Harvard College. Updated with data from the Population Studies Center, University of Michigan, 2006.

& Taeuber, 1965, p. 30). A value of 100 on the index would mean that all nonwhite people live on segregated blocks (that is, the city is 100 percent segregated). In 1970, for example, 81.2 percent of the nonwhite population in Boston would have had to move to another block to produce an unsegregated distribution of whites and nonwhites. By 1990, that proportion had declined to 68.2 percent, clearly an improvement.

Despite such improvements in many metropolitan regions, Table 12.1 shows that most cities, especially in the North, continue to be highly segregated. Although segregation in some regions has fallen below the 70 percent level, high segregation indexes remain a fact of American life. Urban ecologists point out that most changes in segregation patterns are a result of the movement of blacks into newer suburban communities and older white neighborhoods in the cities (Jaynes & Williams, 1989; Massey & Denton, 1993).

Segregation and Poverty One effect of racial and ethnic segregation is to further concentrate poverty. Because African Americans and some Latino groups are at an economic disadvantage and hence are more likely to be poor than other groups in the United States, residential segregation tends to create neighborhoods that are poor as well as segregated. In their studies of the impacts of residential segregation, Douglas Massey and his colleagues note that the growing income and wealth gaps that we discussed in Chapter 11 have been accompanied by increasing spatial separation between classes: "As income inequality rose, so did the degree of class segregation, as affluent and poor families increasingly came to inhabit different social spaces. These trends undermined the socioeconomic well-being of all racial and ethnic groups in the United States" (Massey & Fischer, 2000, p. 670). And although all low-income groups were affected by this widening of class inequalities, African Americans, the most segregated population, were most negatively affected.

An analysis of the 2000 census by sociologist John Logan and others (see Figure 12.3) shows that although the average white or black person in an urban or suburban neighborhood is less likely to live in a racially homogeneous neighborhood than was true a decade ago, Hispanics and Asians are slightly more segregated. Perhaps more important is the fact that a large majority of whites and a smaller majority of blacks still live in segregated neighborhoods. "The average white person," Logan observes, "continues to live in a neighborhood that looks very different from those neighborhoods where the average black, Hispanic, and Asian live." In sum, the United States is more racially and ethnically diverse than ever, but its citizens still either choose or are obliged to live in segregated neighborhoods (Schmitt, 2001).

Other problems of inequality follow from these persistent patterns of racial segregation. It is extremely difficult, for example, to have integrated schools, especially in the primary grades, when children live in segregated neighborhoods. And it is difficult to establish tolerance and understanding among racial and ethnic groups whose members have grown up in segregated neighborhoods and schools.

Assimilation

One of the factors that led to segregation was fear of racial intermarriage; the ideology of white supremacy held that intermarriage would weaken the white race. As late as 1950, thirty states had laws prohibiting such marriages, and even after racist sentiments began to diminish in the 1950s, nineteen states (seventeen of them in the South) maintained such laws until the Supreme Court declared them unconstitutional in 1967 (Holt, 1980). Today, interracial marriages account for about 2 percent of

assimilation: A pattern of intergroup relations in which a minority group is absorbed into the majority population and eventually disappears as a distinct group.

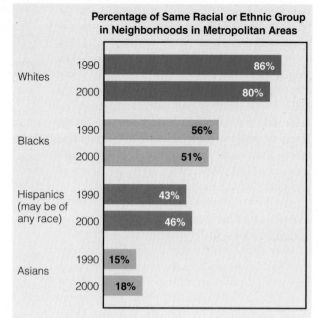

Figure 12.3 Residential Segregation in Metropolitan Areas. Source: John Logan, Lewis Mumford Center for Comparative Urban and Regional Research, State University of New York at Albany; Parisi & Lichter, 2007.

all marriages, but as we see in Figure 12.4, this represents a dramatic rise since 1960. And the popularity of Tiger Woods, Mariah Carey, and other celebrities from multiracial backgrounds suggests that the old norms against interracial families will continue to erode. Figure 12.5 presents more quantitative evidence of the normative shift. Based on extensive survey research, the figure shows that resistance to interracial dating increases as one speaks with representatives of older U.S. generations.

Intermarriage between distinct racial and ethnic groups is an important indicator of **assimilation**, the pattern of intergroup relations in which a minority group is forced or encouraged or voluntarily seeks to blend into the majority population and eventually disappears as a distinct people within the larger society. (In Chapter 3, we saw that assimilation refers to the process by which a culturally distinct group adopts the language, values, and norms of a larger society. Here we are using the term in a broader sense to include social as well as cultural blending.) Needless to say, it makes a great deal of difference whether an ethnic or racial group has been the victim of forced assimilation or has been allowed to absorb the majority culture at its own pace. Many Latin American societies offer examples of peaceful, long-term assimilation of various racial and ethnic groups. For example, as a result of generations of intermarriage, Brazilians distinguish among many shades of skin color and other physical features rather than relying on crude black–white distinctions (Fernandes, 1968).

In the United States, assimilation has had a more troubled history. In an influential treatise on this subject, Milton Gordon (1964) identified three "ideological tendencies" that have affected the treatment of minority groups at various times. These ideologies specify how ethnic or racial groups should change (or resist change) as they seek acceptance in the institutions and culture of American society. They can be summarized as follows:

1. *Anglo-conformity*—the demand that culturally distinct groups give up their own cultures and adopt the norms and values of the dominant Anglo-Saxon culture
2. *The melting pot*—the theory that there would be a biological merging of ethnic and racial groups, resulting in a "new indigenous American type"
3. *Cultural pluralism*—the belief that culturally distinct groups can retain their communities and much of their culture even though they participate in the institutions of the larger society

Because these ideological tendencies have played an important role in intergroup relations in the United States, we will examine them in some detail.

Anglo-Conformity The demand for Anglo-conformity rests on the belief that the persistence of ethnic cultures, ethnic and racial communities, and foreign languages in an English-speaking society should be aggressively discouraged. The catchword of this ideology is *Americanization*, the idea that immigrants and their children must become "100 percent American" by losing all traces of their "foreign" accents, abandoning their ethnic cultures, and marrying nonethnic Americans. This demand is reflected in the movement to make English the official language of the United States, along with opposition to bilingual education. In some cases, discrimination against members of certain ethnic groups is rationalized by the statement that they are not yet fully American. (See the Mapping Social Change box on pages 304–305.)

The Melting Pot In an 1893 paper, a young historian named Frederick Jackson Turner challenged the notion that American culture and institutions had been formed by the nation's original Anglo-Saxon settlers. Turner's essay argued that the major influences on American culture were the experiences of the diverse array of people who met and mixed on the western frontier. Turner held that "in the crucible of

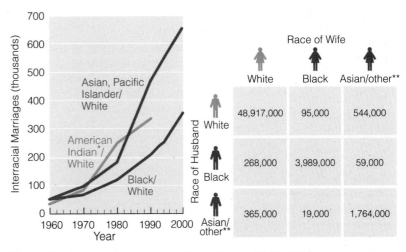

Figure 12.4 **Interracial Marriages, United States, 1960–2000.** *Includes Eskimo and Aleut for 1980–1992. **Race includes Asian, American Indian, and Pacific Islanders; Hispanics may be of any race. Source: Data from *Statistical Abstract*, 2002.

the frontier the immigrants were Americanized, liberated, and fused into a mixed race" (1920/1893, pp. 22–23).

Turner's "crucible" became widely known as the "melting pot," a term taken from a play by Israel Zangwill about a Russian immigrant. David, the play's hero, makes the following declaration:

America is God's crucible, the great Melting Pot where all the races of Europe are melting and re-forming! Here you stand, good folk . . . in your fifty groups, with your fifty languages and histories, and your fifty blood hatreds and rivalries. . . . Germans and Frenchmen, Irishmen and Englishmen, Jews and Russians—into the Crucible with you all! God is making the American. (1909, p. 37)

The melting pot view of assimilation attracted many scholars, artists, and social commentators, but sociologists were not convinced that it was an accurate view of what actually happens to ethnic groups in American society. Their research showed that certain ethnic groups have not been fully assimilated, either through intermarriage or through integration

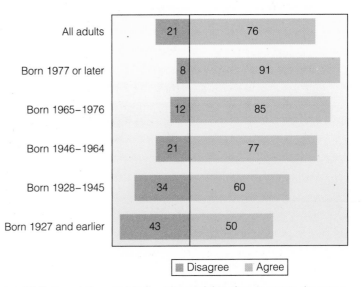

Figure 12.5 **Generations Divided on Interracial Dating.** Percent who agree or disagree that "it's all right for blacks and whites to date each other" by year born. Source: From *Guess Who's Coming to Dinner: 22% of Americans Have a Relative in a Mixed-Race Marriage*, by Pew Social Trends staff. http://pewsocialtrends.org/2006/03/14/guess-whos-coming-to-dinner/. Reprinted by permission of the Pew Research Center.

mapping social change

Receptivity to Immigrants

How welcoming is the United States to new immigrants? We hear a great deal about the increasing presence of immigrants in the "gateway cities"—Los Angeles, New York, Miami, Houston, and Chicago—and somewhat less about the increasing dispersion of new immigrants to smaller metropolitan areas like Charlotte, North Carolina. What kind of reception do immigrants receive in these and other areas of the United States? The question is being addressed by sociologists Gordon F. De Jong and Quynh-Giang Tran (2001), who use data from the General Social Survey to compare how Americans respond to new immigrants in metropolitan and nonmetropolitan regions. The accompanying maps from their research show that there are major differences in opinions about immigrants in different parts of the United States.

The greatest number of immigrants are found in twenty major metropolitan areas, but clearly there are important differences in how well immigrants are received in these areas. Dallas, San Diego, San Francisco, and Tampa appear to be the least receptive, whereas Chicago, Minneapolis, Seattle, and to a somewhat lesser degree, New York, Pittsburgh, and St. Louis are generally warm and welcoming. Within Florida, note the difference between Miami (receptive) and Tampa (less so).

Americans in smaller metropolitan areas, especially in the Midwest, are quite receptive to immigrants, whereas those in small metropolitan areas in the Atlantic states are much less so. And it appears that people in smaller cities and metropolitan regions of Texas and California are somewhat more receptive to immigrants than people in the major metropolitan regions.

The third map looks at nonmetropolitan areas. Here we see that receptivity to immigrants tends to be far lower overall than it is in more historically diverse metropolitan regions. Only nonmetropolitan residents in the industrial Midwest tend to express positive attitudes toward immigrants.

What explains these differences? The sociologists note that, first, much of the reception of immigrants is based on people's previous experience with immigration in their communities. Miamians are familiar with Cuban and other immigrants, whereas immigration to Tampa is more recent and the residents have less experience in accommodating newcomers. A second explanation has to do with border issues: With the exception of Phoenix, most of the metropolitan areas near the nation's borders, where immigration is often clandestine and is associated with hardship and poverty, are relatively negative toward immigrants. Finally, where residents perceive immigrants as economic competitors, especially where there is high unemployment or a high proportion of low-wage jobs for which immigrants often compete, hostility is higher, and this is often the case in nonmetropolitan areas in the South and Southwest. Unfortunately, these data were gathered at a period of high employment. Any prolonged increase in unemployment is likely to engender greater hostility toward immigrants in specific areas, which makes the assimilation of immigrant newcomers more problematic.

Immigrants demonstrate in the hope that Congress will pass legislation easing access to American citizenship. Harsh legislation and enforcement policies in Arizona and other border states—policies that many immigrants view as anti-Latino, are the result of a backlash against the efforts of immigrants to improve their situation.

Sarah Hadley/Alamy

into the nation's major institutions. Instead, there remain distinct patterns of **ethnic stratification**; that is, different groups are valued differently depending on how closely they conform to Anglo-Saxon standards of appearance, behavior, and values. People of Scandinavian or northern European descent, for example, are more readily accepted into the top levels of corporate management than are people of Mediterranean descent, who in turn are more readily accepted than black, Hispanic, and Asian minorities.

Even among groups that have been fully assimilated into U.S. society, there remain many differences that can be traced to their diverse

ethnic stratification: The ranking of ethnic groups in a social hierarchy on the basis of each group's similarity to the dominant group.

20 Major Metropolitan Areas

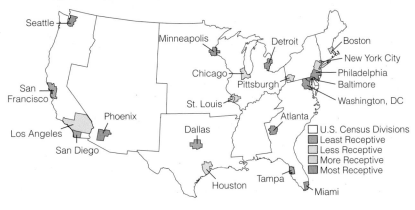

Smaller Metropolitan Areas

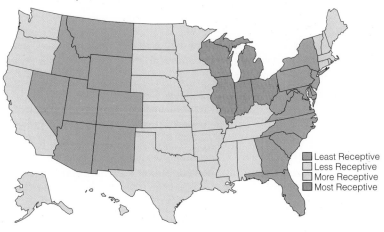

Nonmetropolitan Areas

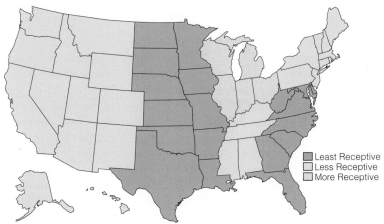

Receptivity Toward Immigrants, by Area. Source: From De Jong & Tran, "Warm Welcome, Cool Welcome: Mapping Receptivity toward Immigrants in the U.S." *Population Today* 29(8):1–4 (2001). Reprinted by permission of the Population Reference Bureau.

backgrounds. In fact, the term **diversity** is often used to refer to the heterogeneous nature of a society made up of numerous different racial, ethnic, religious, and other population groups.

In an early empirical study of assimilation, Ruby Jo Reeves Kennedy (1944) analyzed rates of ethnic intermarriage in New Haven, Connecticut. She found that intergroup marriages increased in frequency between 1870 and 1940, but that while there was a growing tendency for people to marry outside their *ethnic* group, there remained a very strong tendency for people not to marry outside their own *religious*

---

diversity: A term used to refer to the heterogeneous nature of a society made up of numerous different racial, ethnic, religious, and other population groups.

group. From this finding, Kennedy developed the hypothesis of the "triple melting pot," the idea that assimilation occurs first among groups that share the same religion and later among groups with different religions. More recently, in *Beyond the Melting Pot* (1970), Nathan Glazer and Daniel Patrick Moynihan also concluded that ethnic and racial assimilation is far from inevitable. Ethnicity does not disappear as a result of assimilation; ethnic subcultures are continually being created and changed. Thus, *Beyond the Melting Pot* pointed to the emergence and significance of cultural pluralism in American life.

Cultural Pluralism As new waves of immigrants from all over the world have streamed into the United States—Jews fleeing religious intolerance in Russia, Italians and Poles fleeing economic depression in their communities, Central and Latin Americans fleeing dictatorships and poverty, refugees from Asian countries torn by conflict—all have tended to join other members of their nationality group who settled in the United States in earlier decades. Although the influx of new immigrants has led to conflicts between older and newer residents of some communities, in many cases it has resulted in new growth in those communities—in new ethnic businesses such as restaurants and grocery stores, and new social institutions such as churches and social clubs. It has also reinforced pride in ethnic identity as expressed in language and other aspects of ethnic subcultures. This infusion of new energy into ethnic communities confirms Glazer and Moynihan's thesis that the emergence of cultural pluralism is a significant aspect of life in American society.

The recognition that ethnic groups maintain their own communities and subcultures even while some of their members are assimilated into the larger society supported the concept of cultural pluralism. A **pluralistic society** is one in which different ethnic and racial groups are able to maintain their own cultures and lifestyles even as they gain equality in the institutions of the larger society. Michael Waltzer's (1980) comparative research on pluralism in the United States and other societies has shown that although white ethnic groups like the Italians, Jews, and Scandinavians may have the option of maintaining their own subcultures and still be accepted in the larger society, blacks and other racial minorities frequently experience attacks on their subcultures (for example, opposition to bilingual education or African studies courses) and, at the same time, are discriminated against in social institutions. Waltzer concludes that "racism is the great barrier to a fully developed pluralism" (p. 787). In his study of Latinos in the United States, Earl Shorris (1992) essentially agrees with Waltzer. He finds that, despite intense pressure to assimilate, many people from Spanish-speaking nations continue to speak their native language and are using their growing influence in some parts of the country to promote bilingualism. Shorris also notes, however, that Latinos are embracing their own version of the American dream and are not advancing separatist claims.

The problems of pluralism are illustrated by the case of the French-speaking Quebecois minority in Canada. The Quebecois account for almost 28 percent of the Canadian population and constitute the majority in the province of Quebec. This leads to a situation in which a group with its own culture and language seeks protection against pressure to assimilate into the dominant culture and yet demands equal access to the society's political, economic, and cultural institutions. The resulting tensions and hardships have led to the demand that Canada become a bilingual nation, with English and French given equal status. Although the Canadian government has taken firm steps to ensure that French is Quebec's first language, the memory of past discrimination has generated a social movement calling for the creation of an independent French-speaking nation. This movement is especially strong in small towns and rural areas whose entire population is French-speaking. This population strongly backed the separation referendum held in Quebec in 1995. However, the referendum was defeated by an extremely narrow margin, largely because of voters in the more multicultural districts of Montreal, Quebec's largest and most diverse city.

As this example shows, a truly pluralistic society is difficult to achieve. The various ethnic and racial subgroups within a society often feel a strong sense of cultural identity, which they wish to preserve. They also demand equal access to the society's institutions: access to better schools, opportunities to obtain jobs in every field, opportunities to hold important positions—in short, a fair share of the wealth and power available in the society. These desires sometimes conflict, with the result that some groups may be tempted to "go it alone"—that is, form their own cultural and political institutions (their own businesses, newspapers and other media, labor unions, and so on)—or else give up their ethnic identity in order to gain greater access to the society's major institutions.

pluralistic society: A society in which different ethnic and racial groups are able to maintain their own cultures and lifestyles while gaining equality in the institutions of the larger society.

How many ethnic groups can you spot in this photo? Crowded neighborhoods of new immigrants often display signs in the language of people's origin. But the effects of assimilation are powerful, and it is likely that the immigrants' children will prefer to read signs in English when they are adults. This loss of linguistic ability is a mixed blessing for the larger society, which gains new English speakers but loses some of its linguistic diversity.

Thinking Critically

Does going to segregated "ghetto" schools make a difference in a person's life chances? Does living in areas where the majority of young men are imprisoned for some time before they reach full adulthood make a difference in what kinds of life chances these young men will have? Please read the Research Methods box on audit surveys before reaching your conclusions on these questions.

CULTURE AND INTERGROUP RELATIONS

Why do ethnic stratification and inequality persist in societies that are becoming increasingly pluralistic, like those of the United States and Canada? Sociologists have proposed a variety of theories to explain racial and ethnic inequality. Before we explore them, however, we need to understand the cultural basis of ethnic diversity and intergroup hostility—that is, the underlying values and attitudes that shape people's consciousness of other groups and, hence, their behavior toward members of those groups. Chief among these are the tendency to view members of other groups in terms of stereotypes and to use those stereotypes to justify differential attitudes (prejudice) and behaviors (discrimination) toward such individuals.

Stereotypes

People often express the opinion that specific traits of members of certain groups are responsible for their disadvantaged situation. Thus,

in South Africa, it was common for whites to assert that blacks were not ready for full citizenship because "they remain childlike and simple." In the United States, the fact that Hispanics are more likely to be found in low-paying jobs is explained by the assertion that "they don't want to learn English." And the fact that black unemployment rates are generally twice as high as white unemployment rates is explained by the statement that "they don't want to work; they like sports and music, but not hard work, especially in school." These explanations are **stereotypes**, inflexible images of a racial or cultural group that are held without regard to whether they are true.

Sociologist William Helmreich (1982) conducted a study of widely held stereotypes regarding America's major ethnic and racial groups. He found that "every single stereotype discussed turns out to have a reason, or reasons" (p. 242). Those reasons usually stem from earlier patterns of intergroup relations. For example, jokes about stupid Poles stem from a period in the nineteenth century when uneducated Polish peasants immigrated to the United States. The idea that blacks are good at music or sports also has some basis, not because blacks are naturally superior in those areas but because when blacks were barred from other avenues to upward mobility, they were able to succeed in entertainment and sports; as a result, many young blacks have developed their musical and athletic talents more fully than whites have. But even though stereotypes usually have some basis, they never take account of all the facts about a group. As the famous social commentator Walter Lippmann once quipped, "All Indians walk in single file, at least the one I saw did."

Prejudice and Discrimination

The fact that many people hold stereotypical ideas about other groups may be an indication that they are ignorant or prejudiced, but it does not imply that they will actually discriminate against people whom they perceive as different. In a classic study of prejudice, social psychologist Richard LaPiere (1934) traveled throughout the United States with a Chinese couple, stopping at about 250 restaurants and hotels. Only one of the establishments refused them service. Six months later, LaPiere wrote to each establishment and requested reservations for a Chinese couple. More than 90 percent of the managers responded that they had a policy of "nonacceptance of Orientals." This field experiment was replicated for blacks in 1952, with very similar results (Kutner, Wilkins, & Yarrow, 1952; Shibutani & Kwan, 1965).

The purpose of such experiments is to demonstrate the difference between prejudice and discrimination. **Prejudice** is an attitude that prejudges a person, either positively or negatively, on the basis of real or imagined characteristics (stereotypes) of a group to which that person

stereotype: An inflexible image of the members of a particular group that is held without regard to whether it is true.

prejudice: An attitude that prejudges a person on the basis of a real or imagined characteristic of a group to which that person belongs.

belongs. **Discrimination**, on the other hand, refers to actual unfair treatment of people on the basis of their group membership.

The distinction between attitude and behavior is important. Prejudice is an attitude; discrimination is a behavior. As Robert Merton (1948) pointed out, there are people who are prejudiced and who discriminate against members of particular groups. There are also people who are not prejudiced but who discriminate because it is expected of them. With these distinctions in mind, Merton constructed the typology shown in Figure 12.6.

Institutional Discrimination Merton's typology is valuable because it points to the variety of attitudes and behaviors that exists in multicultural and multiracial societies. However, it fails to account for situations in which certain groups are discriminated against regardless of the attitudes and behaviors of individuals. This form of discrimination is part of the "culture" of a social institution. It is practiced by people who are simply conforming to the norms of that institution; hence, it is known as *institutional discrimination.*

At its simplest, **institutional discrimination** is the systematic exclusion of people from equal access to and participation in a particular institution because of their race, religion, or ethnicity. Over time, however, this intentional exclusion leads to another type of discrimination, which has been described as "the interaction of the various spheres of social life to maintain an overall pattern of oppression" (Blauner, 1972, p. 185). This form of institutional discrimination can be quite complex.

The conditions that led to the riot that broke out in South Central Los Angeles in 1992 after the acquittal of the police officers who beat a black motorist, Rodney King, conform very well to what sociologists have observed in similar communities where blacks or other minority groups are trapped in a self-perpetuating set of circumstances caused largely by historical patterns of discrimination (Blauner, 1989; Tuch & Martin, 1997). Blocked educational opportunities result in low skill levels, which, together with job discrimination, limit the incomes of minority group members. Low income and residential discrimination force them to become concentrated in ghettos. The ghettos never receive adequate public services such as transportation, thus making the search for work even more difficult. In those neighborhoods, also, the schools do not stimulate achievement, thereby repeating the pattern in the next generation. The young often grow up bitter and angry and may form violent gangs that engage in vandalism and other activities that the outside world sees as antisocial. At the same time, the police patrol ghettos to the point of harassment, with the result that young blacks are more likely to be arrested—and to be denied jobs because of their arrest records. (See the Research Methods box on page 309.) When these conditions are combined with a precipitating event such as the savage beating of a black person and the acquittal of the police officers responsible for the beating, the results often take the form of violence like that which erupted in South Central Los Angeles.

All of the institutions involved—employers, local governments, schools, real estate agencies, agencies of social control—may claim that they apply consistent standards in making their decisions: They hire the most qualified applicants; they sell to the highest bidder; they apply the law evenhandedly; in short, they do not discriminate. Yet, in adhering to its institutional norms, each perpetuates a situation that was created by past discrimination.

Ethnic and Racial Nationalism

The conditions that contributed to the Los Angeles riot are an example of the pervasive and discouraging effects of institutional discrimination. In the face of such discrimination, racial and ethnic groups frequently organize their members into movements to oppose social inequality. Those social movements often appeal to ethnic or racial nationalism—that is, to the "we feeling" or sense of peoplehood shared by the members of a particular group.

Ethnic (or racial) nationalism is the belief that one's ethnic group constitutes a distinct people whose culture is and should be separate from that of the larger society. Feelings of nationalism among America's ethnic and racial groups have often been strongly affected by nationalist movements outside the United States. American Jews, for example, were deeply influenced by Zionism, a movement that arose in the late nineteenth century with the goal of creating a Jewish homeland in the holy land of Palestine (now Israel). Similarly, Irish Americans have been influenced by the struggle of the Catholic minority in Northern Ireland to gain

	PREJUDICE	
	Yes	**No**
DISCRIMINATION Yes	TRUE BIGOT: does not believe in the American creed and acts accordingly	WEAK LIBERAL: not prejudiced, yet afraid to go against the bigoted crowd
DISCRIMINATION No	CAUTIOUS BIGOT: does not believe in the American creed but is afraid to discriminate	STRONG LIBERAL: not prejudiced and refuses to discriminate

Figure 12.6 Merton's Typology of Prejudice and Discrimination. Source: Figure adapted from "Merton's Typology of Prejudice and Discrimination" [pp. 99–126] from *Discrimination and National Welfare*, edited by Robert M. MacIver. Copyright © 1949 by the Institute for Religious and Social Studies. Reprinted by permission of HarperCollins Publishers.

discrimination: Behavior that treats people unfairly on the basis of their group membership.

institutional discrimination: The systematic exclusion of people from equal participation in a particular institution because of their group membership.

ethnic (or racial) nationalism: The belief that one's own ethnic group constitutes a distinct people whose culture is and should be separate from that of the larger society.

An Audit Survey of Racial Discrimination

Black males are heavily discriminated against in the U.S. labor market. But how can that be proved? One of the best and most convincing methods of proving that employers in a given city practice racial discrimination is an audit survey. Audit surveys are a form of sociological research in which the investigator sends individuals to try to buy a home, to apply for a job, or—as in the classic study of prejudice against Asians in the 1950s—to make hotel reservations for minority and nonminority couples. By sending white and minority applicants to apply for the same opportunities in many different establishments, researchers can observe and calculate any patterns of difference in acceptance and rejection. Audit surveys depend on research assistants, known as "testers," who are trained to represent themselves in a uniform way at each application trial. Recently, Devah Pager, a young sociologist at Princeton University, conducted an extremely important audit survey about racial discrimination in employment. Her research also examined the impact of a criminal background on employment (Pager, 2003; Pager, Bonikowsky, & Western, 2009).

Sociologist Devah Pager.

Pager trained a team of four testers— two black males and two white males of the same age—and sent them to apply for entry-level jobs in a midwestern city. One of each pair of testers had a resumé (fictional) that showed a short prison term for a nonviolent offense, with some positive work experience while in prison. The other had no work experience but also no prison record. The jobs were advertised in a metropolitan daily newspaper. The four testers applied separately for about 350 advertised jobs. The dramatic results are summarized in the accompanying chart. Even among the applicants without a criminal record, the effect of race was very large: In 34 percent of the cases, white applicants were called back for an interview, whereas black applicants— whose resumés listed the same qualifications as those of the white applicants—were called back in only 14 percent of the cases. For those with criminal records, white applicants were called for interviews in 17 percent of the cases, whereas black applicants were called back in only 5 percent of the cases. In other words, a white male applicant for an entry-level job who admits to having a prison record is more likely to be called in for an interview than is a black male with no prison record.

Stigma and Felony Disenfranchisement: The implications of this research cannot be overstated. One in five black males in the United States has a prison record—the majority for minor drug offenses and often the result of inadequate legal representation. The stigma that follows them into the job market is an extremely important problem that can lead to negative consequences such as recidivism. Because at least twelve states also deny voting rights to former prisoners with felony convictions—again, drug use or sale is the most common offense—the result is that almost 2 million U.S. citizens, the majority being African Americans and Hispanic, cannot vote in state and federal elections.

For more detail about Devah Pager's (2003) audit survey, see the original paper, "The Mark of a Criminal Record," in *American Journal of Sociology, 108,* pp. 937–975.

Research That Makes a Difference: Addressing the problems convicted felons have when attempting to reenter the job market, President Bush proposed in his 2003 State of the Union message a $300 million program to provide mentoring and help them find jobs. The director of the White House Office of Faith-Based and Community Initiatives cited Pager's study as one of the sources of information that helped shape the administration's four-year plan. Pager was extremely gratified that her study had some real impact. Over half a million inmates will leave penal institutions in the United States this year, and, she said, "the Administration is finally recognizing that the problems created by our incarceration policies can no longer be ignored." Even if the proposed amount is trivial, she added, the gesture is important (Kroeger, 2004).

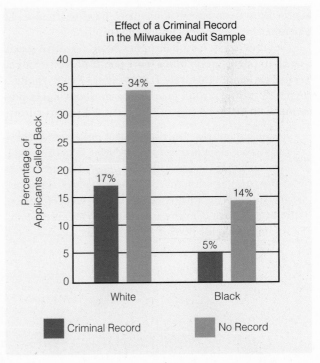

Effect of a Criminal Record in the Milwaukee Audit Sample.
Source: Adapted from "The Mark of a Criminal Record," by Devah Pager, *American Journal of Sociology.* Fig 6, p. 958. Copyright © 2003 University of Chicago Press. Reprinted with permission.

---

independence from England. Nationalist movements often produce a new or renewed ideology of pluralism that replaces the goal of complete assimilation into the larger society. We can see this very clearly among African Americans.

In his famous study *An American Dilemma* (1944), Swedish social scientist Gunnar Myrdal noted that African Americans were in fact "exaggerated Americans." Myrdal believed that blacks had assimilated American values and norms more than any other ethnic or racial group and that they had no distinctive subculture of their own. This theme was repeated in the influential work of E. Franklin Frazier (1957/1939), the leading black sociologist of his generation. Frazier, like Myrdal, believed that African Americans had no distinctive culture. They had the same religions, language, and values as white Americans, and the folkways they had developed (for example, musical forms like jazz and the blues) had become part of mainstream American culture.

Events of the late 1950s and 1960s drastically changed these views of black cultural assimilation. Along with the civil rights movement of the 1960s came a wave of nationalism. Black intellectuals and community leaders began calling not merely for integration but for cultural pluralism. Leaders like Stokely Carmichael of the Student Nonviolent Coordinating Committee (SNCC) and Malcolm X, the charismatic Black Muslim leader, preached that "black is beautiful" and that African Americans should drop the term *Negro* in favor of *black*. In Carmichael's words, "They oppress us because we are black and we are going to use that blackness. . . . Don't be ashamed of your color" (quoted in Bracey, Meier, & Rudwick, 1970, p. 471). In recent years, many black Americans have gone further and adopted the term *African American* in order to emphasize their common heritage and to de-emphasize the racial character of their shared background.

People in the United States are not alone in struggling with issues of nationalism. Now that the cold war between the former Soviet Union and the United States has ended, ethnic and religious nationalism is the primary threat to peace throughout the world. Although North America is relatively free of the extreme nationalism that results in bloodshed, ethnic or racial nationalism has strained the social fabric at various times in North American history. Entire populations, such as African Americans, Puerto Ricans, or French Canadians, have yearned to become entirely separate from the larger culture and society.

At other times, nationalism has been a creative social force, welding a racial or ethnic group into a more politically conscious community able to struggle more effectively for its rights. As the Native American sociologist Russell Thornton notes, there are some highly positive aspects of nationalism or, in the case of Native Americans, tribalism:

> Research shows that American Indians make considerable efforts to reaffirm tribalism in urban areas by living in Indian neighborhoods, by maintaining contacts with reservation areas and extended families, and by creating urban American Indian Centers. Frequent results are a new tribalism for urban American Indians and, sometimes, "bicultural" individuals, that is, American Indians who live successfully in Indian and non-Indian worlds. (1987, p. 239)

A key to Indian survival, Thornton concludes, is the ability to maintain an Indian identity while interacting with non-Indians in an urban environment. Hispanic and African-American sociologists often make the same point. One difficulty in maintaining a bicultural existence is that it requires assertion of the minority group's values, its ways of expression, its history, and its demands for greater equality. The demand for greater equality is often expressed in terms of the need for affirmative action.

Affirmative Action

What separates relatively peaceful multiethnic and multiracial nations from those in which racial, ethnic, or religious hatreds result in war and massacres? One important answer is the rule of law. When minority populations believe that their rights can be protected by government and laws, when they feel that their past grievances will be corrected in time, they will lend their support to the government and laws of their nation. However, in too many countries—even in the United States at times—minorities do not hold these beliefs. As a result, the quest for racial and ethnic equality is a persistent issue in national affairs (Hacker, 1995; Shuck, 2001).

Our society's foremost governmental and economic institutions are besieged with demands for *affirmative action*—that is, policies designed to correct persistent racial and ethnic inequalities in promotion, hiring, and access to other opportunities. Even more forceful in many parts of the country are the demands to do away with affirmative action policies. Many conservatives bitterly oppose affirmative action (Glazer, 1975; Sowell,

Chicago Historical Society

© Peter Turnley/CORBIS

Racial violence and rioting invariably disrupt neighborhoods, often for long periods. During the 1919 Chicago race riots, many African Americans were evicted from their homes in integrated working-class neighborhoods and forced back into the ghetto. During the 1992 Los Angeles riots, so many stores, businesses, and houses were burned that people like this woman felt severe disorientation long after the rioting had ended.

1972), whereas many liberals feel that such policies are necessary if our society is to undo the effects of past discrimination (Ezorsky, 1991; Horowitz, 1979).

At present, the conservative view is in the ascendance on the Supreme Court. Congress has done away with preferential treatment for minority-owned firms in awarding small government contracts. In California, steps have been taken to eliminate affirmative action policies in the state's universities and colleges. It is likely that efforts to eliminate affirmative action will continue and will be increasingly successful. But it is also true, as Martin Kilson points out, that affirmative action has been essential to the advancement of African Americans and, increasingly, Latinos in job markets that were formerly out of bounds to minorities. In 1970, for example, "barely 2 percent of the officers in the armed forces were black; twenty years later, thanks (partly) to affirmative action, 12 percent of officers were African Americans" (Kilson, 1995, p. 470).

We saw earlier that the Supreme Court struck down the doctrine that "separate but equal" facilities and institutions did not violate the constitutional rights of African Americans. It did so largely because social-scientific evidence showed that separate institutions are inherently unequal. Similar arguments have been advanced in affirmative action cases that have reached the Court. If a fire department in a city whose inhabitants are 30 percent black and Hispanic has no firefighters from those minority groups, it can be demonstrated that there is a pattern of discrimination that can be changed only if the department is required to hire a certain number of minority applicants—a quota—within a designated time. However, members of the majority may then feel that they are victims of "reverse discrimination," in which they are being penalized for the wrongs of earlier generations. Thus, difficult choices remain: Should employers mix or replace hiring decisions based on experience and merit with decisions based on race and ethnicity? In essence, the courts have said that they must.

Affirmative action remains an extremely controversial issue in American life. Table 12.2 shows that a majority of Americans support "programs that make special efforts to help minorities get ahead" and that an even greater proportion favor programs that help "people from low-income backgrounds, regardless of their gender or ethnicity, get ahead." But this general support for affirmative action changes when questions deal with the use of federal tax dollars for specific programs. Both whites and blacks tend to favor race-targeted policies that

TABLE 12.2

Blacks' and Whites' Opinions on Racial Policy Issues (in Percentages)

In order to make up for past discrimination, do you favor or oppose programs that make special efforts to help minorities get ahead?[1]

Favor	Oppose	DK/NA
59	30	11

Do you favor or oppose programs that make special efforts to help people get ahead who come from low-income backgrounds, regardless of their gender or ethnicity?[1]

Favor	Oppose	DK/NA
84	10	6

Should federal spending programs that assist blacks be increased, decreased, or kept about the same?[*2]

	Blacks	Whites
Increased	69.5	18.4
Kept the same	28.8	56.5
Decreased	1.7	25.2

Some people say that, because of past discrimination, blacks should be given preference in hiring and promotion. Others say that such preference in hiring and promotion of blacks is wrong because it gives blacks advantages they haven't earned. What about your opinion—are you for or against preferential hiring for blacks?[*2]

	Blacks	Whites
Strongly favor	53.4	6.2
Favor	10.1	7.9
Decreased	1.7	25.2
Oppose	4.9	18.3
Strongly oppose	21.6	67.6

*N = 5,012

[1] Source: From The New York Times Poll, *Class Project*. March 14, 2005. Reprinted by permission of The New York Times Agency.

[2] Source: "Fifty Years after Myrdal: Blacks' Racial Policy Attitudes in the 1990s" by Steven A. Tuch, Lee Sigelman, and Jack K. Martin, in Tuch, Steven A., and Jack K. Martin. *Racial Attitudes in the 1990's*.

offer compensatory programs, such as job training and education, but the majority of whites and a significant minority of blacks oppose employment and educational programs that require preferential treatment in hiring and college admissions decisions (Tuch, Sigelman, & Martin, 1997). The divisions revealed by these attitude data confirm that Americans remain divided along racial lines over how much government should be doing to correct the cumulative impact of slavery and racial discrimination.

It should be noted that affirmative action applies to women as well as to racial and ethnic minorities (Epstein, 1995). The effects of institutional discrimination against women are explored in the next chapter.

THEORIES OF RACIAL AND ETHNIC INEQUALITY

For large numbers of people, a dominant aspect of life in American society is racial or ethnic inequality and, often, hostility. How can we explain the persistence of these patterns of hostility and inequality?

The first thought that comes to mind is that many people are prejudiced against anyone who is different from them in appearance or behavior. This may seem to explain phenomena like segregation and discrimination, but it fails to explain the variety of possible reactions to different groups. Social-psychological theories that focus on prejudice against members of out-groups find the origins of racism and ethnic inequality in individual psychological processes, but other, more sociological theories view prejudice as a symptom of other aspects of intergroup relations.

Social-Psychological Theories

The best-known social-psychological theories of ethnic and racial inequality are the frustration-aggression, projection, and authoritarian personality theories. All of these see the origins of prejudice in individual psychological orientations toward members of out-groups, but they differ in important ways.

Frustration-Aggression The frustration-aggression hypothesis, which is associated with the research of John Dollard, Neil Miller, and Leonard Doob (1939), holds that the origin of prejudice is a buildup of frustration. When that frustration cannot be vented on the real cause, the individual feels a "free-floating" hostility that may be taken out on a convenient target, or **scapegoat**. For example, in the case of workers in eastern Germany who have lost jobs in antiquated factories as a result of unification with western Germany, with its far more modern industries, a convenient scapegoat may be found in gypsies, Turks, or even Jews—all groups that have historically been accused of causing negative conditions for native-born Germans. To justify the hostility directed at the out-group, the prejudiced individual often grasps at additional reasons (usually stereotypes) for hating the "others," such as "Gypsies steal children"; "Jews are usurers"; or "Turks will work for almost nothing."

Projection The concept of projection is also used to explain hostility toward particular ethnic and racial groups. **Projection** is the process whereby we attribute to other people behaviors and feelings that we are unwilling to accept in

ourselves. John Dollard (1937) and Margaret Halsey (1946) applied the concept of projection to white attitudes toward black sexuality. Observers had noted that southern whites frequently claimed that black males were characterized by an uncontrollable and even vicious sexuality. The theory of projection explains this claim as resulting from the white males' attraction to black females, an attraction that was forbidden by strong norms against interracial sexual contact. Thus the white male projected his own forbidden sexuality onto blacks and developed an attitude that excused his own sexual involvement with black women.

The Authoritarian Personality The frustration-aggression and projection explanations of prejudice are general in nature in that anyone can develop pent-up frustrations that will engender hostility under certain conditions, and anyone can project undesirable traits onto others. A more specific explanation of prejudice is the theory of the authoritarian personality, which emerged from attempts to discover whether there is a particular type of person who is likely to display prejudice.

In 1950, a group of social scientists led by Theodor Adorno published an influential study titled *The Authoritarian Personality*. Their research had found consistent correlations between prejudice against Jews and other minorities and a set of traits that characterized what they called the authoritarian personality. Authoritarian individuals, they found, were punished frequently as children. Consequently, such individuals feel an intense anger that they fail to examine (Pettigrew, 1980). They submit completely to people in positions of authority, greatly fear self-analysis or introspection, and have a strong tendency to blame their troubles on people or groups whom they see as inferior to themselves. Unfortunately for Jews, blacks, and other minorities that have been subordinated in the past, the anger and hostility of the authoritarian personality is often directed against them.

Interactionist Theories

Not far removed from these social-psychological theories of intergroup hostility are theories derived from the interactionist perspective in sociology. But instead of locating the origins of intergroup conflict in individual psychological tendencies, interactionists tend to look at how hostility or sympathy toward other groups, or solidarity within a group, is produced through the norms of interaction and the definitions of the situation that evolve within and between

scapegoat: A convenient target for hostility.

projection: The psychological process whereby we attribute to other people behaviors and attitudes that we are unwilling to accept in ourselves.

groups. A few examples illustrate how the interactionist perspective is applied.

From his analyses of interaction in different groups, Georg Simmel concluded that groups often find it convenient to think of nonmembers or outsiders as somehow inferior to members of the group. But why does this familiar in-group–out-group distinction develop? Simmel explained it as arising out of the intensity of interactions within the group, which leads its members to feel that other groups are less important. Once they have identified another group as inferior, it is not a great leap to think of its members as enemies, especially because doing so increases their sense of solidarity (Coser, 1966; Simmel, 1904).

Conflict and hostility between racial or ethnic groups can be overcome by creating situations that require the groups to cooperate to achieve a common goal. Such situations occasionally occur—for example, when people from different ethnic and racial backgrounds compete in sports or when they work side by side in school and in industry. Unfortunately, the integration and friendship found in such settings are not sufficient to overcome the more deep-seated prejudices and fears of many racist individuals. As film director Spike Lee has observed, racists place all black people in one of two categories: "entertainers and niggers." By this, he means that people with racist sentiments tend to exclude famous entertainers and sports figures from their negative feelings about blacks as a group, thereby accentuating their failure to see all blacks as individual human beings.

Functionalist Theories

The difficulty of reducing racism and discrimination through interaction leads to the question of whether racial and ethnic inequalities persist because they function to the advantage of certain groups. One answer was provided by the functionalist theorist Talcott Parsons (1968), who wrote that "the primary historic origin of the modern color problem lies in the relation of Europeans to African slavery" (p. 366). Parsons was not denying that racism is produced through interactions. Rather, he was pointing out that the specific form taken by those interactions—oppression, subordination, domination of blacks by whites—is directly related to the perceived need of white colonialists and traders to use blacks for their own purposes. The whites could abduct, enslave, and sell Africans because their societies had developed technologies (for example, oceangoing ships and navigational instruments) and institutions (for example, markets and trading

corporations) that made them immensely more powerful than the Africans.

From the functionalist perspective, inequalities among ethnic or racial groups exist because they have served important functions for particular societies. Thus, in South Africa, it was functional for the white government to insist on maintaining apartheid because to do otherwise would mean that whites would become a minority group in a black-dominated society. But as world opinion continued to condemn the white regime and blacks continued to build group solidarity and challenge that regime, it became less and less "functional" for the white government to insist on complete apartheid. Indeed, some have speculated that the nation's white rulers released Nelson Mandela and took steps to legalize the formerly banned African National Congress as a way of showing the world that they were willing to begin negotiations toward some form of shared power. Their move incurred the wrath of many South African whites, who feared the consequences of majority black rule under a new constitution. Similarly, the Israeli government's negotiations with the Palestine Liberation Organization enraged many Jewish settlers living in areas that would be placed under the control of Palestinians. This rage resulted in the shocking assassination of former Israeli prime minister Yitzhak Rabin in November 1995, and continues to dominate politics in the Middle East.

In both Israel and South Africa, it appears that the status quo of intergroup relations had become dysfunctional. Change had to come either through peaceful negotiations or through bloodshed. The processes whereby power is redistributed and new social relations develop are often best understood with the help of conflict theories.

Conflict Theories

Conflict theories do a better job than functionalist theories of explaining why groups with less power and privilege, like the South African blacks or the Palestinians in the Middle East, do not accept their place in the status quo. Conflict theories try to explain why these groups often mobilize to change existing intergroup relations (Blauner, 1989). Conflict changes societies, as we see clearly in the case of South Africa, where many years of conflict waged by the disenfranchised blacks led to the creation of a new constitution and majority black rule. But conflict can also destroy societies and nations, as we see all too clearly in the war-torn nations of the former Yugoslavia, where a long history of ethnic conflict has prevented the achievement of a stable, well-functioning society.

Conflict theories trace the origins of racial and ethnic inequality to the conflict between classes in capitalist societies. Marx believed, for example, that American wage earners were unlikely to become highly class conscious because ethnic and racial divisions continually set them against one another, and the resulting strife could be manipulated by the capitalist class. Thus, in American history, we see many examples of black and Mexican workers being brought in as strikebreakers by the owners of mines and mills, especially during the 1920s and 1930s. Strikebreakers from different racial and ethnic groups often absorbed the wrath of workers, anger that might otherwise have been directed at the dominant class. To forge class loyalties despite the divisions created by racial and ethnic differences, Marx believed, it would be necessary for workers to see that they were being manipulated by such strategies.

Internal Colonialism Conflict perspectives on racial inequality also include the theory of **internal colonialism**. According to this theory, many minority groups, especially racial minorities, are essentially colonial peoples within the larger society. Four conditions mark this situation:

1. The "colonial" people did not enter the society voluntarily.
2. The culture of the "colonial" people has been destroyed or transformed into a version of the dominant culture that is considered inferior.
3. The "colonial" population is controlled by the dominant population.
4. Members of the "colonial" people are victims of racism; that is, they are seen as inferior in biological terms and are oppressed both socially and psychologically (Blauner, 1969; Davidson, 1973; Hechter, 1974). This is why African Americans, Native Americans, and Jewish Americans react with such anger to suggestions that they may be different from other populations on measures of intelligence or any other biological trait. They have seen such notions used to rationalize slavery or genocide and hence are not willing to let them pass as harmless speculation or "value-free" science.

Although these characteristics describe colonial peoples everywhere, the theory of internal colonialism asserts that they also apply to subordinated ethnic and racial groups in societies like England, the United States, Canada, and the nations of the former Soviet Union. Michael Hechter (1974) extended the theory to show that societies that have created colonial or "ghettoized" populations within their boundaries also develop a "cultural division of labor" in which the subordinated group is expected to perform types of work that are considered too demeaning to be done by members of the dominant population. The South African institution of *baaskap* was (and in some parts of the nation still is) an example of this phenomenon. It is a set of norms specifying that lower-status, physically exhausting work is appropriate for blacks and higher-status work is appropriate for whites, and that whites should never accept "black" work nor allow themselves to be subordinate to blacks.

Ecological Theories

Are the segregated ghetto communities of black and Hispanic Americans a product of internal colonialism? The answer to this question depends on whether the residents of those communities are able to achieve upward mobility. According to ecological theories of intergroup relations, such mobility should occur naturally in the course of a group's adaptation to the culture and institutions of the larger society. The eventual outcome should be the existence of racially and ethnically integrated communities.

Ecological theories explore the processes by which conflict between racial or ethnic groups develops and is resolved. Along these lines, Robert Park (1914) devised a cyclical model to describe intergroup relations in modern cities. That model consisted of the following stages:

1. *Invasion.* One or more distinct groups begin to move into the territory of an established population.
2. *Resistance.* The established group attempts to defend its territory and institutions against newcomers.
3. *Competition.* Unless the newcomers are driven out, the two populations begin to compete for space and for access to social institutions (housing, jobs, schooling, recreational facilities, and so on); this extends to competition for prestige in the community and power in local governmental institutions.
4. *Accommodation and cooperation.* Eventually, the two groups develop relatively stable patterns of interaction. For example, they arrive at understandings about segregated and shared territories (Suttles, 1967).
5. *Assimilation.* As accommodation and cooperation replace competition and conflict, the groups gradually merge, first in secondary groups and later through cultural assimilation and intermarriage. They become one people. A new group arrives, and the cycle begins again.

Of course, this is an abstract model. It represents what Park and other human ecologists believe are the likely stages in intergroup relations. Korean American sociologist Ilsoo Kim (1981) found that Korean merchants in New York City met resistance both from white merchants and from black residents of ghetto communities when they purchased stores in those communities, but eventually they reached an accommodation with those groups. This finding tends to support Park's model, although the extent to which assimilation will occur remains an open question. On the other hand, critics point out that there is not always a steady progression from one of these stages to the next. Moreover, accommodation, cooperation, and assimilation do not occur in every case. The ecological model fails to explain why and how groups compete for power and under what conditions they eventually come to cooperate. Nevertheless, the model presents a general picture of the stages that culturally distinct groups often (but by no means always) go through over time. (See the Perspectives chart on page 315.)

---

internal colonialism: A theory of racial and ethnic inequality that suggests that some minorities are essentially colonial peoples within the larger society.

A PIECE OF THE PIE

Up to this point, we have explored the patterns of intergroup relations, the cultural basis of those patterns, and some of the theories that have been proposed to explain them. But we have not discussed how groups actually win, or fail to win, a fair share of a society's valued statuses and other rewards—that is, a piece of the social pie. In particular, the persistence of racial inequality in the United States requires more attention.

How can we explain the fact that so many social problems in the United States today are associated with race? As William Julius Wilson (1984) has stated: "Urban crime, drug addiction, out-of-wedlock births, female-headed families, and welfare dependency have risen dramatically in the last several years and the rates reflect a sharply uneven distribution by race" (p. 75). Wilson and others doubt that racial prejudice and discrimination adequately account for the severity of these problems because the period since the early 1970s has seen more anti-discrimination efforts than any other period in American history. The answers are to be found, Wilson argues, in how older patterns of racism and discrimination affect the present situation.

In an important study of this issue, sociologist Stanley Lieberson (1980) asked why the European immigrants who arrived in American cities in the late nineteenth and early twentieth centuries have fared so much better, on the whole, than blacks. Sociologist Marta Tienda asks similar questions about people of Latino

As a result of their unsuccessful resistance to the European invasion, Native Americans were segregated on reservations, many of which lacked adequate resources to permit them to share in the American dream.

ancestry (Tienda & Singer, 1995; Tienda & Wilson, 1992). Wilson, Lieberson, and Tienda all agree that the problem of lagging black mobility is a complex one. For African Americans, the situation can be summarized as follows:

1. African Americans have experienced far more prejudice and discrimination than any immigrant group, partly because they are more easily identified by their physical characteristics. As a result of the legacy of slavery, which labeled blacks as inferior, they have been excluded from full participation in American social institutions far longer than any other group.

2. Black families have higher rates of family breakup than white families. The problems of the black family are not part of the legacy of slavery, however. As Herbert Gutman (1976) has shown, slavery did not destroy black families to the extent that earlier scholars believed it did. Nor did the migration of blacks to the North during

PERSPECTIVES

Theories of Racial and Ethnic Inequality and Prejudice

Theory	Key Variables	Examples
Social-Psychological Theories		
Frustration-Aggression	Prejudice stems from a buildup of frustration and aggressive feelings toward stereotyped victims.	Scapegoating of minority groups such as gypsies or Jews.
Projection	Displaces feelings about ourselves onto others and justifies mistreatment.	Attribution of one's own sexual desires to others.
Interactionist Theories	Intensity of interactions within a group leads in-group members to disparage other groups, but when groups include racially or ethnically distinct others, friendship can supersede hostility.	Groups that need to cooperate (e.g., sports teams) often find ways to reduce racial or ethnic tensions.
Functionalist Theories	Racism and prejudice exist because they bring benefits of some kind to some individuals or group members.	Inequalities derived from prejudice help maintain a supply of low-wage workers.
Conflict Theories	Racial and ethnic inequalities emerge in class conflict under capitalism, but change occurs when subordinated groups resist and rebel.	African Americans were treated as a colonial people until they mobilized to fight back.
Ecological Theories	Helpful in explaining the process of settlement and segregation that often results from prejudice.	People in racial or ethnic ghettos seeking upward mobility.

Maya Lin: A Strong, Clear Vision What measures can a society take to heal the wounds of civil strife and bloodshed? This question is being raised throughout the world as nations seek to diminish intergroup hatreds and cope with the corrosive memories of civil war and genocide. In the United States, one of the strongest examples of successful efforts to address painful memories in a healing fashion is found in the commemorative sculpture of Maya Lin.

Commemoration, in the sense in which it is used to speak about Lin's work, refers to the effort a society makes to officially remember the major events that have marked its history. In the case of the Vietnam Veterans Memorial, the names of all U.S. military personnel who were killed in the war are carved into the shiny black granite of a long, gracefully proportioned wall. People from around the nation come to the wall to remember their loved ones or simply to remember the war and its many consequences for American society. As they stand in contemplation, their own images are reflected by the sheen of the polished stone. Few people leave this monument without being deeply affected by the experience.

Commemorative sculpture for major public sites is never the work of a single person. A committee usually selects from several possible approaches and artists. Thus, in the case of the Vietnam Veterans Memorial, much of the credit for the success of the monument must also go to the committee members who were able to recognize the power of Lin's vision. The same must be said for the committee representing the Southern Poverty Law Center, which commissioned Lin to design the Civil Rights Memorial in Montgomery, Alabama.

Like the Vietnam Veterans Memorial, the Civil Rights Memorial draws viewers into personal meditation. Water from a deep pool on the monument's upper level spills down in even sheets across the face of the wall, on which is a gold-leafed inscription of Martin Luther King's inspiring statement that the fight will continue (in the words of the Bible) "until justice rolls down like waters and righteousness like a mighty stream." An elliptical table, on which are engraved the landmark events in the civil rights era from 1955 to 1968, including the names of forty people who died in the struggle, glistens with a film of water from above. People walk slowly around the monument reading the engraved text, and as they do so many feel compelled to run their fingers across the wet surface in an "act of communion" (Stein, 1994, p. 70).

© William Johnson/Stock, Boston

© Steve Skjold/PhotoEdit

industrialization. A comprehensive review of the status of black Americans (Jaynes & Williams, 1989) notes that "there was no significant increase in male-absent households even after the massive migration to the urban North." Until the 1960s, three-quarters of black households with children younger than age eighteen included both husband and wife. "The dramatic change came only later, and in 1986, 49 percent of black families with children under age 18 were headed by women" (Jaynes & Williams, 1989, p. 528). The report reasons that if black two-parent families remained the norm throughout slavery, the Great Depression, migration, urban disorganization, and ghettoization, then it appears unlikely that there is a single cause for the dramatic decline in two-parent families during the last two decades.

3. Structural changes in the American economy—first the shift from work on farms to work in factories and then the shift away from factory work to high-technology and service occupations—have continually placed blacks at a disadvantage. No sooner had they begun to establish themselves as workers in these economic sectors than they began to suffer the consequences of structural changes in addition to job discrimination.

Further, a great deal of research shows that difficulty in securing adequate employment contributes to the problems young couples experience in forming lasting relationships (Danziger & Gottschalk, 1993; Garfinkel & McLanahan, 1986).

It is significant that, among Hispanic groups in the United States, Puerto Ricans have experienced northward migration and problems of discrimination that are similar to those African Americans in northern cities have experienced. Still a dependent territory of the United States, since World War II, Puerto Rico has sent hundreds of thousands of working people to cities like New York, Philadelphia, and Chicago, which have been rapidly losing manufacturing jobs. The status of Puerto Ricans in the population is similar to that of blacks. There is a growing Puerto Rican middle class and an even more rapidly growing number of Puerto Ricans among the poor. This situation is largely a

Marta Tienda, a leading scholar of poverty among Latino Americans, chose this research area in part because she herself came from a Mexican American family of modest means.

result of changes in the types of jobs available to unskilled blacks and Puerto Ricans (Sandefur & Tienda, 1988).

It is worth noting that, until the Great Recession of 2008, social mobility was more available to blacks today than it was before the civil rights movement. Today, there are millions of middle-class African Americans, and the earnings of black males who have completed college are about 85 to 90 percent of those of whites with comparable education. But although the black middle class is growing, most black workers remain dependent on manual work in industry or lower-level service jobs—types of jobs that are in decreasing supply and in which

unemployment is common. In consequence, the gains in employment equality that blacks made during the last twenty years were almost wiped out in the recession. Figure 12.7 shows that the black–white unemployment gap had narrowed so that the black unemployment rate was only 3.5 percent greater than that of whites in 2007, but shot up again with massive layoffs of hourly workers after 2008.

Another important segment of black workers, male and female, is employed in the public sector. Drastic cuts in public budgets after 2000 drove down their earnings as well. Chronic unemployment is associated with family breakup, alcohol and drug addiction, and depression. These social problems, in turn, severely hamper the ability of individuals to learn the attitudes and skills they need for entry into available jobs (Oliver & Shapiro, 1990; Wilson, 1987).

Nor is the life of middle-class blacks free from the continuing consequences of racism and prejudice. In interviews with a sample of middle-class black Americans in sixteen cities, sociologist Joe R. Feagin found that "they reported hundreds of instances of blatant and subtle bias in restaurants, stores, housing, workplaces, and on the street" (1991, p. A44). And in recent years the number of crimes against minority individuals of all social classes has increased.

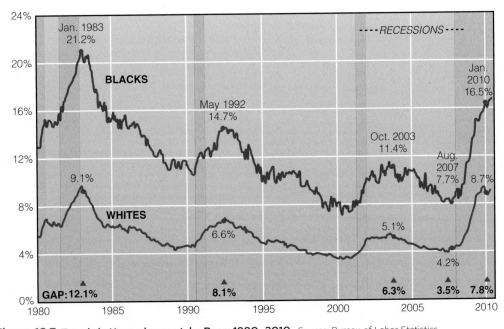

Figure 12.7 **Trends in Unemployment, by Race: 1980–2010.** Source: Bureau of Labor Statistics.

In what way is race a "social construct" rather than a distinctively biological trait of different populations?

- In biology, the term *race* refers to an inbreeding population that develops distinctive physical characteristics that are hereditary. However, the choice of which physical characteristics to consider in classifying people into races is arbitrary. And human groups have exchanged their genes through mating to such an extent that it is impossible to identify "pure" races.

- The social concept of race has emerged from the interactions of various populations over long periods of human history. It varies from one society to another, depending on how the people of that society feel about the importance of certain physical differences among human beings.

- *Racism* is an ideology based on the belief that an observable, supposedly inherited trait is a mark of inferiority that justifies discriminatory treatment of people with that trait.

- *Ethnic groups* are populations that have a sense of group identity based on a distinctive cultural pattern and shared ancestry. They usually have a sense of "peoplehood" that is maintained within a larger society.

- Ethnic and racial populations are often treated as *minority groups*—people who, because of their physical or cultural characteristics, are singled out from others in the society for differential and unequal treatment.

What are some of the most common patterns of intergroup relations?

- Intergroup relations can be placed on a continuum ranging from intolerance to tolerance, or from genocide through assimilation.

- *Genocide* is state-sponsored mass killing explicitly designed to completely exterminate a population deemed to be racially or ethnically different and threatening to the dominant population. It is almost always rationalized by the belief that the people who are being slaughtered are less than human.

- *Expulsion* is the forcible removal of one population from territory claimed by another population. It has taken a variety of forms in American history, including the expulsion of Native Americans from their ancestral lands, the Oriental exclusion movement of the nineteenth century, and the detention of Japanese Americans during World War II.

- *Slavery* is the ownership of a population, defined by racial, ethnic, or political criteria, by another population that has complete control over the enslaved population. Slavery has been called "the peculiar institution" because it has existed in some of the world's greatest civilizations, including the United States.

- Although African American slaves gained their freedom during the Civil War, a long period of segregation followed. *Segregation* is the ecological and institutional separation of races or ethnic groups. It may be either voluntary or involuntary. Involuntary segregation may be either *de jure* (created by laws that prohibit certain groups from interacting with others) or *de facto* (created by unwritten norms).

- *Assimilation* is the pattern of intergroup relations in which a minority group is forced or encouraged or voluntarily seeks to blend into the majority population and eventually disappears as a distinct people within the larger society.

- In the United States, three different views of assimilation have prevailed since the early nineteenth century. They are Anglo-conformity, the demand that culturally distinct groups give up their own cultures and adopt the dominant Anglo-Saxon culture; the melting pot, the theory that there would be a social and biological merging of ethnic and racial groups; and cultural pluralism, the belief that culturally distinct groups can retain their communities and much of their culture and still be integrated into American society.

Why do ethnic stratification and inequality persist in societies that are becoming increasingly pluralistic?

- *Stereotypes* are inflexible images of a racial or cultural group that are held without regard to whether they are true.

- They are often associated with *prejudice,* an attitude that prejudges a person, either positively or negatively, on the basis of characteristics of a group to which that person belongs.

- *Discrimination* refers to actual behavior that treats people unfairly on the basis of their group membership; *institutional discrimination* is the systematic exclusion of people from equal participation in a particular social institution because of their race, religion, or ethnicity.

- Social movements whose purpose is to oppose institutional discrimination are often supported by *ethnic nationalism,* the belief that one's ethnic group constitutes a distinct people whose culture is and should be separate from that of the larger society.

- Policies designed to correct persistent racial and ethnic inequalities in promotion, hiring, and access to other opportunities are referred to as *affirmative action.*

How do social scientists explain the existence of racial and ethnic inequalities?

- Social-psychological theories of ethnic and racial inequality argue that a society's patterns of discrimination stem from individual psychological orientations toward members of out-groups.

- Interactionist explanations go beyond the individual level to see how hostility or sympathy toward other groups is produced by the norms of interaction that evolve within and between groups.

- The functionalist perspective generally seeks patterns of social integration that help maintain stability in a society.

- Conflict theories trace the origins of racial and ethnic inequality to the conflict between classes in capitalist societies.

- The conflict perspective includes the theory of *internal colonialism,* which holds that many minority groups are essentially colonial peoples within the larger society.

- Ecological theories of race relations explore the processes by which conflict between racial or ethnic groups develops and is resolved.

Why are many social problems in the United States associated with race?

- The persistence of racial inequality in the United States is a source of continuing controversy. This complex problem is a result of several factors besides racial prejudice and discrimination. Other factors are high rates of family breakup and the effects of structural changes in the American economy.

- Although social mobility is more available to blacks today than it was before the civil rights movement, most blacks are in insecure working-class jobs or are unemployed.

The Kornblum Companion Website

www.cengagebrain.com

Supplement your review of this chapter by going to the companion website to take one of the tutorial quizzes, use the flash cards to master key terms, and check out the many other study aids you'll find there. You'll also find special features, such as GSS data and Web links that will put data and resources at your fingertips to help you with that special project or to do some research on your own.

13

- Gender and Inequality
- Gender Stratification
- Gender Inequality in Industrial Societies
- The Women's Movement
- Women at Work

Inequalities of Gender

FOCUS QUESTIONS

What are the basic facts about gender inequality in a global scale?

In what ways are gender roles stratified, in the United States and around the world?

How do inequalities of gender operate in different societies to prevent women from realizing their full potential?

How are women mobilizing to challenge inequalities of gender, and in what way does feminist sociology assist in their effort?

What are some problems women encounter in the workplace?

For most people, the word *safari* conjures up images of wild animals—lions, elephants, and rhinos roaming the savannas. I wanted to go and experience all that Kenya had to offer, and yet there was something more that tugged at me. I felt a calling from deep within me to go to East Africa, to go and see the beginnings of all humankind.

The writer of this passage is sociologist Carol Chenault of Calhoun Community College in Decatur, Alabama. She and her husband, Ellis Chenault, traveled to Kenya to observe the way of life of the Maasai people. "I am confident," she says, "that the teaching of culture in college sociology classes for 15 years is a major source of my interest in the Maasai." For her and for the students with whom she now shares her experiences, this sociological voyage was an eye-opening one.

> What I observed was an extremely independent people whose entire lives revolve around the care and herding of their cattle. They truly believe that all of the cattle on earth are their gift from God (Engai). One Maasai warrior asked me if I had any cows. When I told him no, he said, "Good, because we would come get them if you did."

For Chenault, some of the most challenging aspects of life among the Maasai centered on the differences between the roles of men and women. We see in the accompanying photo that Chenault seems to be well accepted by her Maasai hosts,

© Chenault Studios/Decatur, AL

but all of them are males. They could accept a foreign woman visitor and interact with her as an independent person, but that did not mean that their own women could join them in this public greeting. "The Maasai culture has a very traditional division of labor between men and women," she writes.

Women are nurturing, loving mothers who spend a great deal of their time hugging, holding, and kissing their children. However, care of the children is a responsibility that is also shared with the fathers. Early childhood has few responsibilities; time is spent learning the ways of the adults, the language skills and play. The girls gradually begin to learn the tasks of the women, such as care of the animals, making clothing, food preparation, and doing beadwork. Girls are initiated into womanhood through circumcision; however, their attitude toward this is not one of fearful anticipation and bravery. The young women see this as a time when they have to give up their freedom for a strict married life. No longer will they be free to choose their lovers as they once did. Circumcision occurs in the mother's house and does not require the public demonstration of courage and bravery that is required of the males in their circumcision rituals. An elderly Maasai woman told me that it is unclean not to be circumcised.

Maasai girls are betrothed during infancy, or in some cases before they are born. A young couple may be approached and asked if their firstborn is a girl, then may she become the wife of their son. The wedding ceremony is a simple one in which the bride's head is shaved and then adorned with bands of beadwork. Her clothing at the wedding is very colorful in that it is all new and dyed with the familiar red ochre. Both the bride and groom are washed with milk at the time of the wedding. (From Chenault, *Maasai*, 1996, pp. 1–3. Reprinted by permission of the author.)

GENDER AND INEQUALITY

Many women in the nations of Asia and Africa are confronting the cultural norms that keep them in subservient positions and prevent them from asserting their own choices about their futures. The Maasai are unusual in the contemporary world. They are pastoralists who are relatively isolated from the powerful social forces that are rapidly changing other tribes in Kenya and elsewhere in Africa. In consequence, the Maasai insist on the necessity of continuing to adhere to the norms of their own tribal culture. For sociologists, the example of the Maasai is extremely useful because it challenges assumptions about gender and society.

As observant sociologists like Carol Chenault travel to other cultures, they invariably recognize that, despite cultural differences, the differences in the life chances of men and women remain one of the major dimensions of inequality in human societies. This chapter therefore begins by presenting some of the key patterns of gender inequality in the United States and throughout the world. Gender stratification and gender roles are discussed in the second section. The third section of the chapter explores cultural norms and ideologies dealing with gender inequality, especially as they are manifested in industrial societies. The next two sections explore the women's movement and the situation of women in the workplace, where some of the most exciting and difficult changes in patterns of gender inequality are occurring today.

No wonder the subject has become a central theme of global social science. Throughout most of the last century and continuing today, the struggle for female emancipation has been a distinctive characteristic of urban industrial nations of the West (Burum & Margalit, 2002). But even in the West, the notion of equal rights for women has been under attack, especially from extreme conservatives who support the status quo of male superiority and dominance. The Nazi regime in Germany, for example, was bitterly opposed to female emancipation. In the United States as well, certain segments of the population view gender equality as a threat to traditional family values.

Empowerment of Women

The term *empowerment* is often used in discussions of women in developing nations. The term refers to the process whereby relatively powerless people of any gender, race, ethnicity, or social class organize to assert their needs and overcome obstacles to their full participation in the institutions of their societies. The term can also refer to the processes whereby people gain the ability to assert their needs in more personal contexts—in their families, work groups, schools, or communities.

Empowerment of women is a major global issue because such a large proportion of the world's females are trapped in patriarchal societies. Indeed, the extent to which different nations move toward greater parity between male and female holders of elective office (as shown in Figure 13.1) is an international measure of relative progress in dealing with patriarchy. In the United States, which tends to lag behind other Western democracies in electing women to federal office, the empowerment of women is also a major sociological subject because so many women are victims of abuse, violence, poverty, gender discrimination, and other effects of patriarchy.

The United Nations International Conference on Population and Development, held in Cairo, Egypt, in 1994, concluded that "the empowerment of women and improvement of their status are important ends in themselves and are essential for the achievement of sustainable development" (United Nations, 1995).

Sociology & Social Justice

WOMEN AND GLOBAL POVERTY In its Millennium Development Goals, the United Nations called for a halving of world poverty by 2015 (United Nations, 2005). As measured by the number of people living on the equivalent of less than $1 a day in purchasing power, the overall level of poverty in the world's developing and transitional economies decreased by about 6.6 percent from 1990 to 2001, but still

Figure 13.1 Percentage of National Legislative Seats Held by Women, Selected Nations. Source: Data from Inter-Parliamentary Union, www.ipu.org. Graphic Copyright © 2003 India Together. Reprinted by permission of India Together.

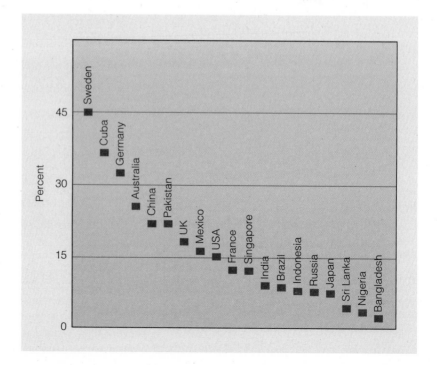

TABLE 13.1

Population Below $1.25 Purchasing Power Parity (PPP) per Day[1]

	Percentage of population living below $1.25 per day	
	1990	2005
Developing countries and transition economies	28	25
Northern Africa and Western Asia	2	4
Sub-Saharan Africa	45	51
Latin America and the Caribbean	11	8
Eastern Asia	33	17
Southern Asia	39	40
Southeastern Asia and Oceania	20	10
Commonwealth of Independent States	0.4	4
Transition countries of southeastern Europe	0.2	2

[1]High-income economies, as defined by the World Bank, are excluded.
Source: UNICEF, *State of the World's Children*, 2009.

TABLE 13.2

Share of Women in Wage Employment in the Nonagricultural Sector

	Share of women in nonagricultural employment	
	1990	2003
World	35.9	39.1
Northern Africa	18.9	21.5
Sub-Saharan Africa	31.5	35.8
Latin America and the Caribbean	38.2	43.5
Eastern Asia	37.9	40.0
Southern Asia	13.4	18.0
Southeastern Asia	36.5	38.6
Western Asia	16.6	20.2
Oceania	28.9	37.3
Commonwealth of Independent States	49.3	50.3
Developed regions	43.5	46.4

Source: *United Nations Millenium Development Goals Indicators*, 2006.

remained above 21 percent, accounting for one-fifth of the world's population. (See Table 13.1.) The majority of these impoverished people are women and children. Note that in certain regions of the world the proportion of the population living in poverty remains very high: In sub-Saharan Africa, for example, over 46 percent of the population is impoverished, and in southern Asia (including India), almost one-third of the population is living in poverty.

To lift women out of poverty, it is essential that they be educated and encouraged to find occupations outside of agriculture, where they and their daughters are most likely to remain impoverished. Table 13.2 shows that in the developed regions of the world, women account for over 46 percent of nonagricultural work, approaching parity with men. In the less

developed regions, however, especially western and southern Asia, women have yet to attain even 20 percent of nonagricultural work. In sub-Saharan Africa, where poverty rates are the highest in the world, women have a larger share (35.8 percent) of nonagricultural employment, but unfortunately this is a small sector of the economies of these nations. Nevertheless, this favorable trend indicates that if and when these economies eventually begin growing, educated women will assume their share of the new opportunities.

Figure 13.2 shows dramatically that women in many parts of the world shoulder a far greater number of domestic burdens than do men. Much of the difference may seem to occur because women take primary responsibility for rearing and nurturing children. One might also argue that, in a developed country like the United States, the differences in men's and women's incomes could be attributed to the fact that so many

Women and children make up 70% of the world's poor.
Women and children account for 80% of the world's refugees.
Women comprise over 2/3 of the world's illiterate population.
Women in developing countries are victims of violence in 1/3 of all families.
Women perform 2/3 of the world's work.
Women produce, process, and market 3/5 of all the world's food.
Women own 70% of all small businesses.
Women in developing countries repay their loans promptly 97% of the time.

YET

Women receive only 10% of the world's income.
Women own less than 1% of the world's property.

Figure 13.2 Essential Facts of Global Gender Inequality.
Source: *United Nations Millenium Development Goals Indicators*, 2006.

global social change

Women and Science in the Islamic World In many nations, including the United States, women lag behind men in the physical and natural sciences. Although they are catching up with men in higher education, many who would otherwise go into research positions drop out of scientific research because there is little social support for their desire to balance work and family responsibilities. The world loses far too much of its human talent through various forms of gender discrimination. The situation in the Islamic world is even more critical, as this excerpt from a speech by a prominent Egyptian scientist demonstrates.

Farkhonda Hassan is a professor of geology at the American University in Cairo and chair of the Commission on Human Development and Local Administration of the Shoura Assembly (Second House of the Egyptian Parliament). Following is an excerpt from an interview with Dr. Farkhonda:

A thousand years ago, the Muslim world made remarkable contributions to science. Muslims introduced new methods of experiment, observation, and measurement. To name but a few: Al-Khwarizmi (born in 780 C.E.) invented algebra (an Arabic word) and the word *algorithm* is derived from his name; Ibn al-Haytham (born in 1039 C.E.) wrote the laws of the reflection and refraction of light and expounded the principles of inertia (long before Isaac Newton formulated his theories). . . . In the words of the science historian George Sarton, "the main, as well as the least obvious, achievement of the Middle Ages was the creation of the experimental spirit, and this was primarily due to the Muslims down to the 12th century."

KHALED DESOUKI/AFP/Getty Images

Farkhonda Hassan (center) listens as Egypt's First Lady Suzanne Mubarak delivers the keynote speech during the Global Summit of Women in Cairo.

Yet today, the number of original research papers published by scientists in Muslim countries is 0.1 percent of the number published by scientists in Europe and the United States. It is to be hoped that this trend is set to change, with many Muslim countries opening new universities and introducing a variety of educational and training programs to improve their capabilities in science and technology. . . .

In many Muslim countries, gender-based discrimination, coupled with social and cultural barriers, limits access and participation of women in higher education. Some people attribute these barriers to the teachings of Islam, but this is false. The teachings of the Holy Prophet of Islam emphasize "the acquiring of knowledge as bounden duties of each Muslim from the cradle to the grave" and that "the quest for knowledge and science is obligatory upon every Muslim man and woman."

. . . Fewer girls than boys are enrolled in high school science curricula because of a bias in the existing education structure that encourages girls to study the arts and humanities. There are various reasons for this related to gender stereotyping, misleading perceptions that science and technology are subjects more suitable for boys, and the failure of curricula to relate science and technology to the everyday life of women. Thus, there is self-inhibition among schoolgirls that affects not only the number of young women entering university to study science and technology subjects, but also results in the reluctance of talented women to introduce their own values and visions into a working world dominated by men.

. . . Women tend preferentially to enroll in the life sciences and chemistry, with far fewer studying physics, mathematics, and engineering. This seems to be more the result of female students choosing these subjects than active discrimination by the education system. This trend is also seen among U.S. and European female students—for example, in the European Union, women constitute 40 percent of natural science undergraduates, 28 percent of mathematics and computer undergraduates, and 20 percent of engineering undergraduates. (From *Science Magazine*. October 6, 2000. Reprinted with permission from AAAS, and Farkhonda Hassan.)

---

women take time out from work and careers to bear and raise children. These familiar (and flawed) arguments raise further questions about the role of gender in social inequality.

The Lenses of Gender

Throughout history, feminist social scientists repeatedly note, three dominant beliefs have prevailed:

- that men and women are fundamentally different in their sexual and psychological natures and that these differences explain the status quo of women's inequality;
- that men are inherently the superior sex, which explains their dominance; and
- that female deference to men and male dominance are natural. (Bem, 1993; Lorber, 2006)

In much of the world and among many Americans, this idea of naturalness remains the justification for the status quo of gender inequality and male dominance. In the twentieth century, this religious explanation was largely replaced by biological reasoning that comes to the same conclusion—that male dominance is a natural phenomenon. In consequence, American political leaders could claim a commitment to equality, but not to gender equality, because of supposed biological differences that make women incapable of many of the pursuits open to men, especially the exercise of power.

The first wave of feminist advocacy established women's basic political rights; it also made the inconsistency between ideology and the treatment of women widely visible for the first time in U.S. history. The second wave, which began in the 1960s, raised social consciousness still further by exposing the sexism inherent in any policy or practice that discriminates on the basis of sex. This second feminist challenge gradually enabled people to see that restricting the number of women in professional schools or paying women less than men for equal work was an illegitimate form of discrimination based on outmoded cultural stereotypes (Lorber, 2006).

Despite the major changes that have taken place over the past century and a half, feminist scholars point out that we still cling to deeply ingrained assumptions about gender. As a result, we "systematically reproduce male power in generation after generation" (Bem, 1993, p. 2). Social psychologist Sandra Bem calls these often unquestioned assumptions "lenses of gender." Her purpose in identifying them and writing about how they shape our lives is to render them visible and hence subject to positive change.

Androcentrism The first of the three lenses of gender she identifies is *androcentrism,* or male-centeredness. This is the "historically crude perception that men are inherently superior to women," which leads to the even more insidious perception that male experience is the natural standard or norm and that "female experience is a sex-specific deviation from that norm."

Gender Polarization The second "lens" that Bem analyzes in her research is *gender polarization.* This is the crude idea that men and women are fundamentally different from each other, which leads to the pervasive use of that perceived difference "as an organizing principle for the social life of the culture." Modes of dress, speech, social roles, "and even ways of expressing emotion and experiencing sexual desire" are subject to gender polarization.

Biological Essentialism The third lens of gender in our culture is that of *biological essentialism,* "which rationalizes and legitimizes both other lenses by treating them as the natural and

inevitable consequences of the intrinsic biological natures of women and men." Of course, Bem does not deny that there are obvious biological differences between men and women, but she and other feminist scholars argue that "these facts have no fixed meaning independent of the way that a culture interprets and uses them, nor any social implications independent of their historical and contemporary context" (Bem, 1993, p. 2).

Feminist social scientists realize that eradication of these lenses of gender would require almost revolutionary changes, especially in the way boys and girls are socialized. But the first step toward this revolution is to open your imagination to a future in which the distortions caused by existing gender lenses are absent.

Gender Socialization and Gender Separation

In Chapter 5, we introduced the concept of gender socialization and discussed its powerful influence on how men and women develop as members of society. We showed, for example, that the values of beauty and youth, especially as they are applied to women by the media, have a strong and often negative influence on how women feel about their bodies and about themselves (Wolf, 1991). Current research on gender inequality is focusing increasing attention on these issues, not only for women but for men as well. Social scientists who are critical of cultures that produce and perpetuate gender

Here we literally see the lenses of gender at work. Note that all the production crew members are male. This 1919 photo of a film star on a studio set is clearly a study in gender polarization.

The "Migrant Madonna" The photo below is perhaps the most famous image of the Great Depression of the 1930s, when unemployment in the United States reached the unimaginable level of over 30 percent. However, a single photo, no matter how powerful its content, never tells an entire story, just as a single sociological study from a particular time period fails to show change over time.

Dorthea Lange/The Library of Congress[LC-DIG-fsa-8b29516]

Known as the "Migrant Madonna," the photo was taken by Dorothea Lange, one of a group of photographers who worked as "sociologists with cameras" to document the suffering and heroism of Americans who faced displacement from their homes because of drought and economic depression. Lange's staff position at the Resettlement Administration (RA), an agency set up to help tenant farmers during the depression, was shaky at best. With no budget for a photographer, Lange had been hired as a clerk-stenographer, and she invoiced her film and travel expenses under the heading "clerical supplies" (Maksel, 2002). Lange found this destitute mother and her children at a camp for migrant farm workers in California. She and her husband had come, with hundreds of other migrant farmers, to pick peas. They found the crop frozen in an unexpected frost, and at the moment the woman, Florence Thompson, agreed to let Lange take her photo, none of the migrants knew where their next meal could be found. When the photo hit the press the next day, it stirred a call of alarm that reached Washington, and the federal government rushed thousands of pounds of relief food to the starving farmers.

Forty years later, Bill Ganzel, a photographer and historian, tracked down Thompson from a story in the local newspaper in Modesto, California, where she had settled with her ten children. He photographed her and three of her daughters—Norma Rydlewski, Katherine McIntosh, and Ruby Sprague—the same daughters that had been included in the original photograph. Thompson told Ganzel that "when Steinbeck wrote in *The Grapes of Wrath* about those people living under the bridge at Bakersfield—at one time we lived under that bridge. It was the same story. Didn't even have a tent then, just a ratty old quilt." The Thompson family got on its feet after the Depression, due largely to New Deal programs pioneered by the administration of Franklin D. Roosevelt.

Bill Ganzel, www.ganzelgroup.com

Florence Thompson and her daughters Norma Rydlewski (in front), Katherine McIntosh, and Ruby Sprague, at Norma's house in Modesto, California, June 1979. Photo by Bill Ganzel from Dust Bowl Descent, University of Nebraska Press, 1984.

inequality point to the many ways in which these inequalities are justified by values, norms, and the workings of major social institutions.

According to many comparative studies and international population reports, a central problem of gender socialization throughout the world is that the traditional separation of men and women in socializing agencies leads to the failure of men to recognize their responsibilities as sexual partners and fathers later in life. As a recent United Nations report states:

> It is clear that women cannot adequately protect their sexual and reproductive health in the context of power imbalances with their male partners. . . . Men must be involved because they share the responsibility of reproductive health, and both men and women do not understand their own and each other's bodies. (Sadik, 1995, p. 23)

In the United States, a major theme in social-scientific research is the way gender socialization, particularly in schools, tends to separate males and females into different social worlds with their own forms of activity and language. The theory that boys and girls are socialized to perform in more or less separate worlds receives a great deal of support in the sociological literature (Fine, 1987; Gilligan, 1982; Lever, 1978; Tannen, 1990). However, we will see shortly that the "separate worlds" thesis also has many limitations (Thorne, 1993, 2000).

A broad version of the gender separation theory as experienced in urban industrial societies is that children absorb ideas about their gender identities from their parents and the media. By the time they are preschoolers, they have strong ideas about what types of dress are appropriate. Boys especially are mortified if they are caught wearing something they feel is inappropriate for their gender. But children's expressions of gender identity are also shaped through interactions in playgroups throughout many years of schooling. From their separate, gender-segregated interactions with peers in school, it is argued, boys and girls develop different values, different ways of bonding with each other, and different types of conflict and ways of resolving conflict. In many ways, they develop separate cultures with their own norms, values, and languages. These differences are thought to persist through life and to exert a lasting influence on how individuals fare in the job market (Kanter, 1977).

In their insightful observation of preadolescent boys' interactions on Little League baseball teams, sociologist Gary Allen Fine and popular writer Stephen King were struck by the depth of male bonding that occurs as the boys undergo the pressure of competition and support each other through their successes and failures on the playing field (Fine, 1987; King, 1994). Fine notes that adult males often want their sons to join teams so that they can "teach their preadolescent boys to see the world from their perspective," which includes teaching what the men consider core values as these values emerge in sports competition. However, "whereas adults focus on hustle, teamwork, sportsmanship, and winning and losing"—as desirable moral behaviors developed through team sports— "the preadolescent transforms these concerns" (Fine, 1987, p. 78). The team members respect the adults' teachings, but they are more concerned with their self-presentation, which includes some of the adult values and some of those that are part of the preadolescent moral code—learning to be tough or fearful under the appropriate conditions; learning to control aggressions, fears, and tears; not being a "rat" (telling on others); and hustling in order to help the team win games. The boys engage in a great deal of sexual talk, much of which expresses their ambivalence about relationships with girls, in which their desire to have sexual experiences is offset by their fears of embarrassment and loss of male friends. The boys' banter with each other also expresses their fear of not being masculine enough. Although they have little or no experience with homosexuality, they use words like *faggot* or *gay* to tease each other or denigrate other boys.

Sociologists who study gender and socialization often point out that the strong same-sex bonds boys form in preadolescence and during their teen years may prove invaluable in the corporate world. In their research on life in American corporations, Rosabeth Moss Kanter and her associates found that sports analogies are commonplace. The male business leader typically thinks of those he works with as members of a team, and as a result of his early socialization experiences, he is more comfortable with male team members than with females (Kanter & Stein, 1980).

Studies of girls' playgroups note that, although boys and girls are increasingly likely to play some games together, the separation of the genders in playgrounds and schoolyards remains a norm (Sutton-Smith & Rosenberg, 1961; Whiting & Edwards, 1988). These studies confirm that girls' games are likely to involve small groups rather than large ones with more complex structures and hierarchies of status. They are interrupted not by arguments over rules or specific plays but by emotional issues, feelings of being slighted, or other hurts— which the girls repair in private conversations

rather than in public protests to umpires or other adults. In addition, girls tend to engage in turn-taking games that emphasize cooperation, whereas boys more often engage in competitive team sports in which they become extremely concerned with establishing hierarchies of skill and success (for example, batting averages, won–lost records) (Lever, 1976, 1978).

Of course, children learn many other social distinctions besides gender from their parents and playmates. Barrie Thorne, who observed children's peer interactions in two primary schools, makes this observation:

> In the lunchrooms and on the playgrounds of the two schools, African-American kids and kids whose main language was Spanish occasionally separated themselves into smaller, ethnically homogeneous groups. . . . I found that students generally separate first by gender and then, if at all, by race or ethnicity. (1993, p. 33)

Problems with the Gender Separation Theory

How fully are men and women socialized into separate social worlds with separate cultures? As mentioned earlier, a great deal of popular and social-scientific writing supports this thesis. The immense success of Deborah Tannen's analysis of men's and women's spoken communication in her popular book *You Just Don't Understand: Women and Men in Conversation* suggests that men and women do feel separate and different in their everyday lives. But recent research on gender differences is critical of the idea of separate gender cultures (Thorne, 2006).

Although there is an undeniable tendency for boys and girls to be socialized in gender-segregated peer groups, they are also raised together in families and often spend a great deal of time together. Boys and girls are increasingly participating in the same types of activities in their schools and communities. In some communities, these changes are controversial, but in many others, teachers, parents, and administrators seek to avoid the worst effects of gender separation (Meier, 1995). Girls' and women's team sports, for example, are undergoing rapid and steady growth. And as they review the literature on separate gender cultures or conduct new research, students of gender often find that the previous emphasis on gender separateness was based on what they refer to as "the Big Man bias"—the tendency to observe only the activities of the most powerful and socially successful individuals. More isolated, less successful children are less likely to be subjects of research, yet their experiences are important because they do not adhere strictly to the norms of gender separation. Finally, as Thorne (1993) argues, generalizations about "girls' culture" are based primarily on research with upper-class white girls, while the experiences of girls from other class, racial, and ethnic backgrounds tend to be neglected.

Thinking Critically

How much difference does gender separation make in the lives of the girls and women you know? Does it seem to contribute to gender inequality? Is it still a cherished aspect of social life? Sociologists warn against generalizing from a very small number of cases, but the concept of gender separation can be quite helpful in understanding the behavior of people around you.

GENDER STRATIFICATION

After class and race, the most important dimensions of inequality in modern societies are gender and age. As stated earlier, gender refers to a set of culturally conditioned traits associated with maleness or femaleness. There are two sexes, male and female; these are biologically determined ascribed statuses. There are also two genders, masculine and feminine; these are socially constructed ways of being a man or a woman. A **gender role** is a set of behaviors considered appropriate for individuals of a particular gender. Controversies over whether women in the armed forces can serve in combat or whether men with children ought to be eligible for family leave from work are examples of issues arising out of the definition of gender roles.

All human societies are stratified by gender, meaning that males and females are channeled into specific statuses and roles. "Be a man"; "She's a real lady"—with these familiar expressions, we let each other know that our behavior is or is not conforming to the role expectations associated with our particular gender. When women's roles are thought to require male direction, as is the case in many households and organizations, the unequal treatment of men and women is directly related to gender roles. The roles assigned to men and women are accorded differing amounts of income, power, or prestige, and these patterns of inequality contribute to the society's system of stratification.

Gender Roles: A Cultural Phenomenon

Until recently, it was assumed that there were two spheres of life, one for women and the other for men. The chief agents of socialization—church, family, and school—worked effectively together to transmit and legitimize the notion that boys would grow up to be leaders in the world outside the home whereas girls would become wives and mothers whose involvement in the world outside the home would be more indirect. Out of this gender-based division of labor, which defined the activities that were appropriate for men and women, grew the notion of differences in men's and women's abilities and personalities. These differences were thought to be natural—an outgrowth of biological and psychological differences between males and females (Epstein, 1985). Behaviors that did not fit these patterns were viewed as deviant and as requiring severe punishment in some cases.

In the twentieth century, evidence from the social sciences called into question the assumption that there are innate biological or psychological reasons for the different roles and temperaments of men and women. Margaret Mead's famous research in New Guinea directly challenged this assumption. Mead was one of the first social scientists to gather evidence to show that gender-specific behavior is learned rather than innate. In her study of gender roles in three tribes, Mead (1950) found that different tribes had different ways of defining male and female behavior. In one tribe, the Mundugumor, men and women were equally aggressive and warlike, traits that westerners usually associate only with men. In a second tribe, the Tchambuli, the men spent their time gossiping about women and worrying about their hairdos, while the women shaved their heads and made rude jokes among themselves. In the third tribe, the Arapesh, both men and women behaved in sympathetic, cooperative ways and spent a great deal of time worrying about how the children were getting along, all behaviors that westerners traditionally associate with women.

Mead's research has been criticized in that she may have been actively looking for gender-role patterns that differed from those that westerners usually associate with males and females. Nevertheless, her study was highly significant because it began a line of inquiry that

gender role: A set of behaviors considered appropriate for an individual of a particular gender.

established that gender roles are not innate. Nor are gender roles wholly determined by a society's relationship to its environment. Although hunting-and-gathering societies sent men out to hunt while women cared for the home, the division of labor in early agrarian societies was less rigid. In early horticultural societies, women had more power than they did in hunting-and-gathering societies or in later feudal societies. Women maintained the grain supply and knew the lore of cultivation, and they were priestesses who could communicate with the harvest and fertility gods (Adler, 1997; Balandier, 1971/1890).

The lesson of this cross-cultural research is that gender roles are heavily influenced by culture. Although the relationship of earlier societies to their natural environment often required that women tend the hearth and home while men went out to hunt for big game, women were also hunting for small game around the encampment and experimenting with new seeds and agricultural techniques. The division of labor by gender was never fixed; it could always be adapted to new conditions. In industrial societies, as the greater strength of males became less important as a result of advances in technology, it made less sense to maintain the earlier divisions of labor. Indeed, modern societies have demanded more involvement of women in a broader range of tasks. Women are now competing with men as military and police officers, engineers, scientists, judges, political leaders, and the like; they may be found in many roles that were assumed to be unsuitable for women less than half a century earlier.

Gendered Organizations Recent research, especially that of feminist sociologists, has expanded our understanding of gender in society by demonstrating that gender stratification is a feature of organizations of all kinds, and not merely an attribute of individuals and their roles (Lorber, 1994; Staggenborg, 1998). The concept of "gendered organizations" encourages researchers to study how assumptions about gender are ingrained in the culture of organizations ranging from the family to corporations and universities. Many law firms, for example, strongly suggest that success is based on long hours of time spent in the office. This assumption distinctly favors male workers whose wives have chosen to be homemakers. And as we will see in Chapter 15, the family is often a highly gendered organization as well.

Especially in the second half of the twentieth century, women have organized in social movements to remove barriers to their employment based on the gendered nature

TABLE 13.3

Percentage of Women in Major Occupational Groups, 1970–2009

Major Occupational Group	1970	2009
Executive, administrative, and managerial	18.5%	46.0%
Professional specialty	44.3	57.5
Technicians and related support	34.4	55.2
Sales occupations	41.3	49.6
Administrative support, including clerical	73.2	79.1
Private household	96.3	91.0
Protective service	6.6	20.6
Other service	61.2	52.1
Farming, forestry, fishing	9.1	21.2
Precision production, craft, and repair	7.3	9.1
Machine operators, assemblers, and inspectors	39.7	36.5
Transportation and material-moving occupations	4.1	11.2
Handlers, equipment cleaners, helpers, and laborers	17.4	20.6
Total	38.0	47.3

Source: Data from Bianchi & Spain, 1986; and U.S. Census Bureau, 2010.

of organizations and the assumption that they could not do certain types of work (for example, policing, fire fighting, or construction). Table 13.3 shows that women have increased their share of employment in many occupational groups that were formerly male "turf." The proportions of female executives and managers have increased dramatically, as have those of technicians and professionals. But older barriers and assumptions continue to stand in the way of equal access to male-dominated occupations such as precision production. Conversely, women continue to be disproportionately represented in "pink-collar" occupational sectors such as clerical and administrative support and domestic service (Goldstein, 2009).

 Thinking Critically

Put yourself in the place of a woman in a poor region of the world. What would your life be like? If you said it would be one of poverty and powerlessness, with limited opportunities for you and your sisters and daughters, you would be on the right track. Of course, you might imagine living the life of a wealthy woman in a poor nation, but that would be counter to the social probabilities. After all, a woman born into a family in a poor region of the world is far more likely to be poor, by "accident of birth," than to be one of the fortunate rich.

Historical Patterns of Gender Stratification

In preindustrial societies, gender often plays a greater role in social stratification than wealth and power. Fewer members of such societies are wealthy or powerful than is true in modern societies, but all are male or female. Cross-cultural research has provided some insight into the development of gender-based stratification in such societies. Among the main findings of that research are the following:

- Preindustrial societies are usually rigidly sex-segregated. Males and females tend to pass through life in cohesive peer groups. Even after marriage, women and men tend to spend more time with their same-sex peers than they do with their spouses.
- As societies increase in size and complexity, women usually become subordinated to men. There have been few, if any, societies in which women as a group controlled the distribution of wealth or the exercise of power (Harris, 1980; Murphy, 2010).

Larger, more complex societies also exhibit distinct patterns of gender-role stratification, but these are more likely to be part of a multidimensional system of stratification in which class, race, ethnicity, gender, and age are intertwined.

The origins of gender inequality in most modern societies can be traced to their feudal periods. Most industrial societies developed out of feudal societies, either as a result of revolutions (in Europe) or through changes brought about by colonialism (in North and South America, Asia, and parts of Africa). Although they are no longer as easily justified, many of the norms that specify separate spheres of activity for males and females, as well as the subordination of women to men, were carried over into modern societies.

In some feudal societies, these norms were far more repressive than those found in modern societies; in others, the subordination of women was disguised as reverence or worship. In European feudal societies, for example, the norms of chivalry and courtly love seemed to elevate women to an exalted status; in reality, however, they reinforced practices that kept women in undervalued roles. These practices resulted in a set of norms that specified that women do not initiate sexual activity, do not engage in warfare or politics but wait for men to resolve conflicts, and do not compete with men in any sphere of life beyond those reserved for women.

The situation of Guinevere in the tales of King Arthur aptly illustrates this point (de Rougemont, 1983; Elias, 1978/1939). Courtly love applied only to women of the nobility. It specified that a woman must be chaste—either a virgin or entirely faithful to her husband at all times. Like Guinevere, she could be worshiped from afar by a noble knight (Lancelot), but she could not be touched or even spoken to without the consent of her male guardian (King Arthur). Figuratively, she was set upon a pedestal. In reality, however, she was imprisoned—kept forever separate from the knightly suitor who worshiped her from a distance and engaged in chivalrous acts in her name. Adultery and the subordination of women were common in the lower orders of feudal society, but the women of the nobility were more likely to be constrained by the norms of chastity.

As feudal societies developed and changed, the norms of courtly love were extended to the new middle classes, and they still exist today in greatly modified forms. They can be seen, for example, in the traditions of courtship that persist in Spanish-speaking nations, in which a man serenades a woman (or hires someone to do it on his behalf) and the woman is never without a chaperone. It should be noted that courtly (or romantic) love eventually elevated the status of women from that of property to that of significant other, a change that had far-reaching implications for the position and influence of women in the family. At the same time, in many Western cultures, norms derived from medieval notions of courtly love continue to justify the notion that there are "good women," whom one reveres and protects, and "bad women," who are available for sexual exploitation. This denial of normal female sexuality is associated with a host of psychological and social problems, including the conflict that some women feel about their sexuality and the inability of some men to relate to their wives as sexual partners.

This medieval illustration depicts the norms of chivalry in action. Notice the women standing on the battlements, aloof from the fray. Their lot is to watch the combat and hope that their hero will not be skewered by his opponent.

Musee Conde, Chantilly, France/Giraudon/The Bridgeman Art Library

GENDER INEQUALITY IN INDUSTRIAL SOCIETIES

In modern industrial societies, age and gender interact to shape people's views of what role behavior is appropriate at any given time. Before puberty, boys and girls in the United States tend to associate in sex-segregated peer groups. Because they model their behavior on what they see in the home and on television, girls spend more of their time playing at domestic roles than boys do; boys meanwhile play at team sports more than girls do. These patterns are changing at different rates in different social classes, but they remain generally accepted norms of behavior. And they have important consequences: Women are more likely to be socialized into the "feminine" roles of mother, teacher, secretary, and so on, whereas men are more likely to be socialized into roles that are considered "masculine," such as those of corporate manager or military leader. It is expected that men will concern themselves with earning and investing while women occupy themselves with human relationships (Rossi, 1980; Witt, 1997).

Childhood socialization explains some of the inequalities and differences between the roles of men and women, but we also need to recognize the impact of social structures. In the United States, for example, it was assumed until fairly recently that boys and girls needed to be segregated in their games. Boys were thought to be much stronger and rougher than girls, and girls were thought to need protection from unfair competition with boys. This widespread belief translated into school rules that did not permit coeducational sports. Those rules, in turn, reinforced the more general belief that girls' roles needed to be segregated from those of boys. Such patterns have significant long-term effects. As sociologist Cynthia Epstein points out, human beings have an immense capacity "to be guided, manipulated, and coerced into assuming social roles, demonstrating behavior, and expressing thoughts that conform to socially accepted values" (1988, p. 240). Through such means, gender roles become so deeply ingrained in many people's consciousness that they feel threatened when women assert their similarities with men and demand equal opportunity and equal treatment in social institutions.

In their adult years, men enjoy more wealth, prestige, and leisure than women do. Working women earn less than men do, and they are often channeled into the less prestigious strata of large organizations. Even as executives, they are often shunted into middle-level positions in which they must do the bidding of men in more powerful positions. Similar patterns are found in all advanced industrial nations.

The Feminization of Poverty

One of the clearest indicators of sexism in a society is the gap in wages earned by men and women in the same occupation. In the United States, women's wages are only about 80 percent of those of men in the same occupations, and the situation is far worse in some fields. Coupled with existing patterns of gender inequality, these income disparities mean that women are more likely to be living at or below the poverty line. Figure 13.3 shows that, between 1979 and 2007, the earnings gap between women and men narrowed for most age groups. The women's-to-men's earnings ratio among thirty-five- to forty-four-year-olds, for example, rose from 58 percent in 1979 to 77 percent in 2007, and the ratio for forty-five- to fifty-four-year-olds increased from 57 percent to 75 percent. The earnings ratio for teenagers and for workers age sixty-five and older fluctuated from 1979 to 2007, but their long-term trend has been essentially flat.

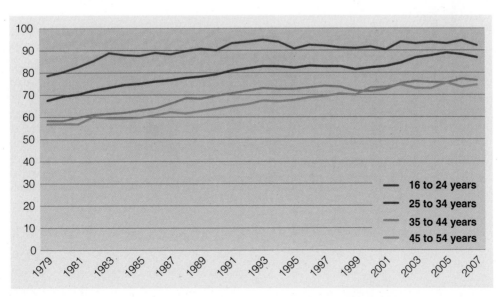

Figure 13.3 Women's weekly earnings as a percent of men's by age, workers aged 16 to 54 years, annual averages, 1979–2007. Source: Bureau of Labor Statistics. http://www.bls.gov/opub/ted/2008/oct/wk4/art03.htm

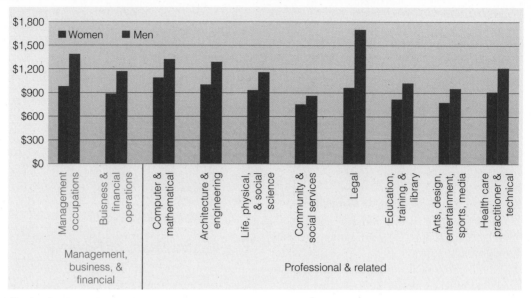

Figure 13.4 Median weekly earnings of women and men in management, professional and related occupational groups, 2008. Source: Bureau of Labor Statistics. http://www.bls.gov/opub/ted/2009/ted_20090807.htm

Encouraging as some of these trends are, when we look at the better-paying management and professional occupations (see Figure 13.4), the gender pay gap is universal across all the major job categories, and in some occupations, like the law and business management, the gap can be more than $500 a week, a substantial difference.

Throughout the developed world, poverty is much more prevalent among women than among men, although as Figure 13.5 shows, there are important differences from one nation to another (Wiepking & Maas, 2005). Only

Australia has a greater proportion of women in poverty than the United States, although poverty rates are comparatively high in Russia as well. In the United States, 36.4 percent of all single women are poor, compared to 22.5 percent of single men, and the risk of being poor is almost 14 percent higher for single women than for single men. In Poland, Belgium, the Netherlands, and Denmark, there are almost no gender differences in poverty rates. These nations all have a strong "safety net" of antipoverty measures that diminish the worst effects of poverty.

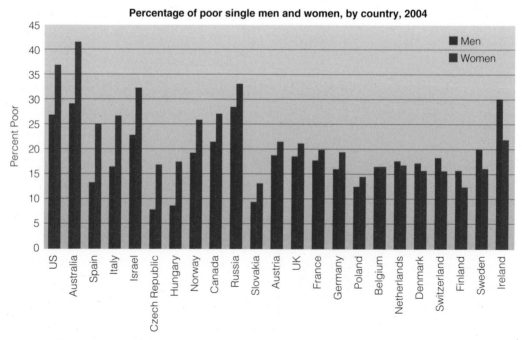

Figure 13.5 Percentage of Poor Single Men and Women, by Country: 2004. Source: Wiepking and Maas, "Gender Differences in Poverty: A Cross-National Study." *European Sociological Review*. 2005; 21: 187–200, Copyright © 2006 by the Oxford University Press. Reprinted by permission.

Men are often compelled by gender norms to behave in stereotypically "masculine" ways in public.

Disparities in employment, with women concentrated in lower-paying jobs, are a major factor in explaining the situation in the United States, but some nations do not have high rates of female labor force participation. In those nations (for example, Italy), female poverty is reduced by the tendency of women to marry men who can support them, and to remain married. In still other nations, social policies play an important role. In "social democratic" nations like Sweden, there is less gender integration in occupations than in the United States. Single women with young children are not obliged to work but are given generous subsidies through income support, child care, education, and other policies (Kammerman & Kahn, 1993). As a result, rates of female poverty are low.

In sum, comparative research indicates that gender inequality and female poverty rates are very much related to the specific workings of other social institutions—particularly those of the economy and government—and that they can be reduced when the society has the resources and will to do so.

Sexism

Gender stratification is reflected in attitudes that reinforce the subordinated status of women. The term *sexism* is used to refer to an ideology that justifies prejudice or discrimination based on sex. It results in the channeling of women into statuses considered appropriate for women and their exclusion from statuses considered appropriate for men. Sexist attitudes also tend to "objectify" women, meaning that they treat women as objects for adornment or sex rather than as individuals worthy of a full measure of respect and equal treatment in social institutions. This can be seen in the case of beautiful women. Such women receive special treatment from both men and women, but their beauty is a mixed blessing. The beautiful woman is often viewed as nothing more than an object for admiration. Being a woman is a master status (see Chapter 4) in that gender tends to outweigh the person's achieved statuses. This is even more painfully true for beautiful women. Someone like Marilyn Monroe is thought of only in terms of her beauty; the person beneath the surface, who was a brilliant comedian and actor, is ignored, a situation that contributed to Monroe's depression and suicide. (See the Research Methods box on pages 334–335.)

The objectification of women can be seen in beauty contests, which came into being in the United States in the summer of 1921, when the first Costume and Beauty Show was held at a bathing beach on the Potomac River. There, the women wore tunic bathing suits and hats, but later that year, a similar contest was held at Atlantic City, New Jersey, that eventually developed into the Miss America Pageant. In that contest, women wore one-piece bathing suits that showed their calves and thighs and created a sensation in the tabloid newspapers. These contests and the publicity they generated made beauty a way for women to gain celebrity and wealth, but they also reflected the dominant male view that the most extraordinary women are those with the most stunning faces and the shapeliest figures (Allen, 1931). Women have been struggling against this view throughout the modern era, a struggle that has frequently been opposed not only by men but also by women who feel threatened by changes in their traditional statuses.

sexism: An ideology that justifies prejudice and discrimination based on sex.

The Beauty Ideal and Women's Self-Image

Recent news articles reveal that employers increasingly seek to hire women whose looks and figures conform to an ideal of beauty, and that major sales corporations like Abercrombie & Fitch and L'Oréal are using standards of female beauty that are possibly discriminatory and may result in charges of racial and age discrimination (Greenhouse, 2003). For feminist sociologists and critics, however, this is old news. For decades, they have been arguing that the images of female beauty presented in the media help maintain gender inequality by promoting an ideal that few women can attain.

In her influential study, *The Beauty Myth*, Naomi Wolf argues that

[t]he more legal and material hindrances women have broken through, the more strictly and heavily and cruelly images of female beauty have come to weigh upon us.... During the past decade, women breached the power structure; meanwhile, eating disorders rose exponentially and cosmetic surgery became the fastest-growing specialty ... pornography became the main media category, ahead of legitimate films and records combined, and thirty-three thousand American women told researchers that they would rather lose ten to fifteen pounds than achieve any other goal.... More women have more money and power and scope and legal recognition than we have ever had before; but in terms of how we feel about ourselves physically, we may actually be worse off than our unliberated grandmothers. (Wolf, 1991, p. 3)

Recent studies of the impact of media images on women confirm Wolf's argument. They also demonstrate some useful methods of exploring these issues. The first of these methods is what is known as *meta-analysis*—that is, analysis of previous studies of a similar subject to determine the degree to which their findings are similar or different. For example, in a meta-analysis of twenty-five studies on the effects of viewing thin images in magazines and television, researchers found overwhelming evidence in all the studies that women felt more negative about their bodies after viewing thin media images than after viewing images of either average-size models, plus-size models, or inanimate objects. The effects were strong for all age groups, but young women below age nineteen felt the strongest negative effects (Groesz, Levine, & Murnen, 2002).

What difference does it make if women tend to feel bad about their bodies after viewing media representations of thin models? Unfortunately, these negative feelings are also associated with eating disorders like anorexia and bulimia, and with lowered self-esteem, which in turn lowers women's confidence and ability to compete in school and in the economy (Clarke, 2002).

Another research method sheds further light on these issues. It is a form of comparative research over time periods that looks at the differences between the bodies of women who are valued in society as models, pinups, or beauty pageant winners and the actual bodies of young women in the larger society. The researchers compared data on the height and weight of Miss America Pageant winners, *Playboy* playmates, fashion models, and young women in general from the 1920s through the 1990s. The results are shown in the accompanying charts. We see that the average height of young women in the general population increased from 63.2 inches in the 1930s to 64.1 inches in the 1990s, but that of models went from over 67 inches to over 69 inches. The changes were even more dramatic for body weight. As the average weight of young women rose from 120.3 pounds to 141.7 pounds over the

The heirs to the Hilton Hotel fortune, Nicky and Paris, are darlings of the media. Impossibly rich and slender, they are shown going out to all the hottest spots with movie stars as their escorts. Research shows that, to the detriment of many women in modern societies, the popular media in the United States and elsewhere are fixated on celebrities who display the idealized thin woman's body.

period studied, that of models actually decreased (Byrd-Bredbenner & Murray, 2003).

As Americans were enjoying the fruits of affluence—driving more, walking less, and responding to advertising that promotes consumption of high-calorie foods—average body weight was increasing. The epidemic of obesity in the United States and other western nations is one consequence (see Chapter 20). Another consequence is the growing "disconnect" between idealized images of the attractive, super-thin woman one sees in the media, and real women who increasingly feel oppressed by these images but respond by trying new diets or by giving up and bingeing. A countertrend, however, is taking hold among some women, who are exercising more, learning how to eat sensibly, and taking control of their own bodies, but unfortunately they are still a minority.

Sexual Harassment

Sexism is also expressed in violence and harassment of women, often intended to "keep women in their place" and maintain male power. In some extremely patriarchal societies, women who have been raped may be beaten or killed by their own husbands, fathers, or brothers, who feel dishonored by the violation of "their" woman. These "honor killings" are merely an extreme example of the violence (physical and emotional) that is routinely directed at women throughout the world.

Two million American women are severely beaten in their homes every year, and 20 percent of visits by women to hospital emergency rooms are caused by battering (Kornblum & Julian, 2007). In popular American culture, especially movies, television, and some popular music, violence against women may not be

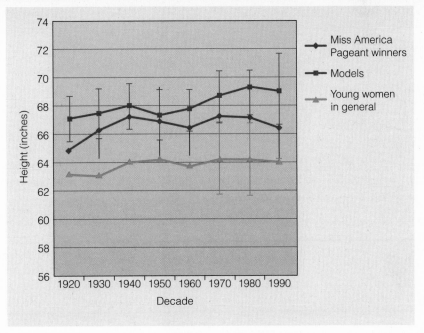

Source: From Byrd-Bredbenner & Murray, *Topics in Clinical Nutrition*, 2003. Copyright 2003 by Lippincott Williams & Wilkins. Reprinted with permission.

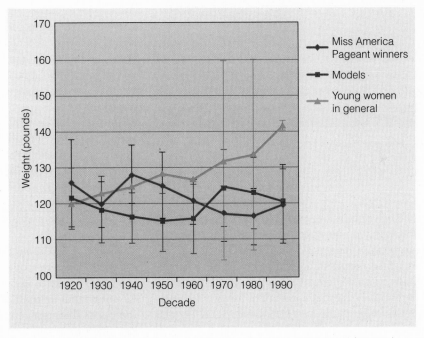

Source: From Byrd-Bredbenner & Murray, *Topics in Clinical Nutrition*, 2003. Copyright 2003 by Lippincott Williams & Wilkins. Reprinted with permission.

condoned but is often presented as a form of thrilling entertainment. Films like *Nightmare on Elm Street* and *The Accused* suggest that the brightest and most independent women are most likely to be victimized. Rap lyrics referring to women as "bitches" and "hos" degrade women and in fact degrade the men who express such sexist sentiments. Examples like these hardly exhaust the overt types of sexism that exist in many areas of social life in the United States. In fact, sexual harassment in the workplace is a continual subject of conflict and social change.

Many definitions of sexual harassment at work rely on the woman's perceptions of the situation. Louise Fitzgerald, one of the foremost researchers in this field, and her colleagues defined sexual harassment as "an unwanted sex-related behavior at work that is appraised by the recipient as offensive, exceeding her resources,

This photo from the 1950s shows the degree of explicit sexism in the airline industry at the time. Although female flight attendants still must conform to dress codes, they have formed unions and have won more stable careers and greater job security.

how one respondent described her perceptions of her work environment:

> I believe there really is sexual harassment. . . . But I also think women abuse it. [pause] Because back there [in my department] they, they comment all the time, you know, about my breasts, or whatever. But I know that they're doing it in the context of a joke, and like I say, it rolls off my back. . . . If I thought they really meant it, then maybe it would bother me.
> —Peggy (quoted in Quinn, 2000, p. 2)

Quinn describes Peggy as "friendly, hardworking, and funny. In her mid-40s, she is divorced, has taken a few college classes, and works a blue-collar job that pays fairly well." The researcher notes that she is typical of the many women who are regularly subjected to sexual harassment by their coworkers. Peggy refuses to publicly label any of their behavior "sexual harassment," even though just talking about it during the interview brings tears to her eyes.

Other women in similar situations resort to counter-joking, also with sexual connotations, and may jokingly insult their male coworkers' sexual prowess as a way of getting back and showing that their taunts and jokes are "not taken personally." But Quinn and others point out that, like the strategy of silence, joking counterattacks by victims allow males to perpetuate a culture of harassment in the workplace. This socially constructed culture of everyday harassment deprives women of control over their own work life and can be corrected only by encouraging women to speak out about their true feelings to their male coworkers and, when necessary, to their supervisors.

Violence in Intimate Relationships Sociologist Claire Renzetti is one of the nation's foremost experts on violence against women. As editor of the influential journal *Violence against Women*, she tracks the most recent research on the subject. She continually seeks ways to help women speak out more effectively against violence in their intimate relations.

Sociologist Claire Renzetti, activist and expert on violence against women.

"If most young women do not believe they are at risk for date rape," Renzetti writes, "it is not surprising . . . that when they are raped—as defined by law— they do not label the event as rape. It

or threatening her well-being" (Fitzgerald, Swan, & Magley, 1997, p. 20). This and similar definitions rely on an appraisal of the victim's perceptions, which makes sense because harassment can include sexist joking, sexual innuendo, and unwanted flirting in addition to more direct sexual contact. But the courts have resisted subjective definitions of harassment and have insisted that definitions of sexual harassment focus more attention on the work environment itself (Sbraga & O'Donohue, 2000). The U.S. Equal Employment Opportunity Commission's (EEOC's) guidelines define sexual harassment as "verbal or physical conduct of a sexual nature that unreasonably interferes with the employee's work or creates an intimidating, hostile, or offensive working environment" (EEOC, 1993, p. 1). This widely accepted legal definition requires the victim to show that harassment made it difficult or impossible for her to do her work as required.

Most researchers in the field of gender relations at work observe that far more sexual harassment occurs than is ever reported (Sbraga & O'Donohue, 2000). Many women are fearful of losing their jobs, of incurring the wrath of supervisors, or simply of "rocking the boat." In a recent study of this aspect of sexual harassment at work, Beth Quinn interviewed women and men in a variety of work environments to find out how women perceive and deal with sexual harassment, especially the common forms of sexual bantering and teasing. She found that women often know that they are being harassed but go along with male coworkers' claims that they should "not take it personally" (Quinn, 2000). Here is is also not surprising that they place most of the blame for the 'bad date' on themselves: It wasn't supposed to happen; therefore, they must have done something that helped make it happen" (1999, p. 227). Renzetti knows that many women rationalize violence against them on dates as a consequence of drinking and drunkenness, but she observes that "even women who were not drunk at the time of the rape, even women who were raped by force or the threat of force, often blamed themselves for the rape." Ashamed and hurt, they are most often "deeply, negatively affected by what happened to them" (1999, p. 227).

Renzetti believes that

> feminists must take the lead in "news making" about violence against—and by—women. We must call press conferences to release our research findings to the public, hold book launches in popular or attention-getting venues and invite the media, write op-ed pieces for local newspapers, and volunteer as consultants or commentators for local radio and television stations—in short, be seen and heard, not after the fact, but proactively. (1999, p. 1235)

Difficult as it may be, by taking a stand, women may ultimately be avoiding the painful next steps that involve formal legal complaints. Unfortunately, in too many situations, women may be required to take such steps if harassment—even of the joking, "don't-take-it-personally" kind—continues. In some law enforcement jurisdictions, the police can intervene in cases of suspected violence against women, but, for the most part, the female victims themselves have the responsibility for opposing harassment. From a feminist viewpoint, the women's movement must continue to strive to educate women about their rights and their need to speak out against sexual harassment.

The Sociology of Masculinities

Throughout much of the world, male dominance over women and children is supported by longstanding cultural norms and values (Connell & Messerschmidt, 2005). Women are typically viewed as subordinate to their fathers and husbands, and marriages are often arranged in ways that may leave women with little choice in the matter. Ideas of masculinity often involve an explicit acceptance of these ideas, such that for a man to accept greater gender equality in his family can be dangerous (Hearn, 2010). Until relatively recently in the Western world, during the twentieth century in particular, the ideals and goals of gender equality were seen as deeply threatening to the stability of society, as they still are in many contemporary societies and in many households that support older ideas of male superiority and dominance.

Although advances in the material and social conditions of women are high on the agenda of world agencies like the United Nations, changes in gender roles that insist on greater equality and women's rights are often seen as socially disruptive, if not revolutionary.

The study of how men behave toward women and toward one another, the power some men crave and compete for, and the ways in which societies create roles for men, are all concerns of the sociology of masculinities. Note that the title of this section emphasizes masculinity in its plural form. Clearly, there is not one single way to be masculine, just as there is surely not one single way to be feminine. Nor are attributes that are typically thought of as male or female found only in men or in women. Research increasingly shows that aspects of the personalities and behavior of members of both sexes incorporate traits that were formerly thought of as exclusively male or female (Kimmel, Hearn, & Connell, 2005). A recent major survey conducted by the Kinsey Institute at the University of Indiana explored this insight.

The Kinsey Institute study found that, contrary to stereotypes about masculinity and male sexuality, men reported that being seen as honorable, self-reliant, and respected was more important to their idea of masculinity than being seen as attractive, sexually active, or successful with women. The study included interviews with more than 27,000 randomly selected men from eight countries (Germany, the United States, the United Kingdom, Spain, Brazil, Mexico, Italy, and France). Regardless of age or nationality, the men more frequently ranked good health, a harmonious family life, and good relationships with their wife or partner as most important to their quality of life. Being seen as a "man of honor" was cited as the most important attribute of masculine identity in Spain, Brazil, Mexico, the United States, and France. "Being in control of your own life" was most important in Germany, the United Kingdom, and Italy. Julia Heiman, director of the Kinsey Institute and an author of the study, commented that "Many meanings, positive and negative, are attached to the term masculinity. To ask a large sample of men what comprises their own sense of masculinity is very useful for both the media and for research.

The State Hermitage Museum, St. Petersburg/Photo by Vladimir Terebenin, Leonard Kheifets, Yuri Molodkovets

Abraham—the patriarch of the Old Testament—famously had two wives: Hagar (Agar), the bondwoman, who bore him a son, Ishmael, the historic progenitor of the Islamic people, and Sarah, the free woman, who bore a son, Isaac, who became the progenitor of the Jews. The history of paternalism and male dominance is evident in many of the central episodes of the Bible, as it is in many other important religious texts. Little wonder, then, that efforts to confront male dominance and enhance gender equality and human rights encounter so much resistance.

These results suggest we should pay attention and ask rather than presume we know."[1]

These findings do not at all challenge the central problem of paternalistic masculinity, whether it is called machismo, toughness, dominance, or anything else, as it exists in so many societies today, but they do suggest that when given choices regarding how to conduct their lives, men have the capacity to move toward greater acceptance of women's rights as an honorable course of action. Unfortunately, sociologists of masculinity also conclude that despite the results of research like the Kinsey Institute study, the situation of male dominance and female subordination will remain an area of conflict and protest for generations to come (Hearn, 2010).

THE WOMEN'S MOVEMENT

In the 1950s, the women's movement was dormant. In the late nineteenth and early twentieth centuries, women had organized social movements to gain full citizenship rights and greater control over reproduction (for example, through family planning and birth control). In the United States, they were known as suffragettes because they campaigned vigorously for woman suffrage—the right of women to vote. But in 1920, after the Nineteenth Amendment to the Constitution extended suffrage to women, the movement faded.

The victory of the suffragettes did not change the patterns of gender stratification, however. For example, in their famous studies of Middletown, Helen and Robert Lynd (1929, 1937) found that women who had worked their way into good factory jobs and professional occupations were severely set back during the Great Depression. Like blacks and members of other minority groups, women were the last to be hired and the first to be fired when times were bad. And although women were hired as factory workers in unprecedented numbers during World War II, they often faced discrimination and harassment and were usually "bumped off" their jobs by returning GIs after the war (Archibald, 1947). Nor did this situation change greatly when women began entering the professions and the corporate world in the 1950s. Women still tended to be limited to roles that were either subordinate to those of men or part of an entirely separate sphere of "women's work."

Early in the twentieth century, women in the suffrage movement were fighting for the right to vote. Their movement often met with organized resistance from men.

[1] For a copy of Julia Heiman's study, "Men Defy Stereotypes in Defining Masculinity," visit www.kinseyinstitute.org/publications/PDF/Sand- Erectile Dys.pdf (from *Psychology & Sociology*, August 26, 2008).

In her review of the origins of the modern women's movement, Jo Freeman (1973) wrote that even in the mid-1960s, the resurgence of the women's movement "caught most social observers by surprise." After all, women were widely believed to have come a long way toward equality with men. Their average educational level was rising steadily, and they were gaining access to the better professions and even to positions in the top ranks of corporate management. Freeman suggested that the movement developed out of an existing network of women's organizations that made many women aware of the inequality still prevailing in American society and encouraged them to organize to demand equal rights. Influential writing by women activists, such as Betty Friedan's *The Feminine Mystique* (1963), and studies of gender inequality by social scientists like Alice Rossi (1964) and Jessie Bernard (1964) played a major part in this growing awareness.

Meanwhile another, less formal network of women was developing. Women who were experiencing sexism within other social movements—the antiwar movement, the civil rights movement, the environmental protection movement, and the labor movement—began to form small "consciousness-raising" groups (Morgan, 1970). By questioning traditional assumptions and providing emotional and material support, these groups attempted to develop a sense of "sisterhood." Their members formed a loosely connected network of small groups that could mobilize resources and develop into a larger and better-organized movement.

The women's movement achieved significant victories during the 1960s and 1970s. In 1963, Congress passed "equal pay for equal work" legislation, and Title VII of the 1964 Civil Rights Act prohibited discrimination against women. In the 1970s, the federal Equal Employment Opportunity Commission began to enforce laws barring gender-based discrimination in employment. Lawsuits filed under the Civil Rights Act and other federal laws forced many employers to pay more attention to women's demands for equality in pay and promotion and an end to sexual harassment on the job. Research shows that men very slowly began to accept roles that had previously been considered "women's work" (Rosen, 1987), and at work, they were more likely to have women as supervisors than was true only a decade earlier. Throughout American society, the attitudes of both men and women were becoming more favorable toward a more equal allocation of political and economic roles (NORC, 1998).

Despite these changes, gender equality is far from complete. Some survey data indicate that there is greater sharing of household tasks, with men doing more of what was once viewed as "women's work" around the home (Gallup Organization, 1997). However, a closer analysis shows that it is premature to assume there have been far-reaching decreases in the sexual division of labor within the average American household. Studies that ask men and women whether the men are doing more housework, and what kinds of work they are actually doing, tend to come up with optimistic figures compared to those obtained in similar surveys a generation ago. But studies based on detailed time-use diaries find far less actual participation by men in housework chores than is reported in the opinion polls. This indicates that men and women may be embarrassed to admit that men are not taking on a fair share of responsibilities for housework and child care, and hence may inflate their estimates in response to pollsters' questions. Time diaries call for detailed information about how time is actually used, and in these diaries, people tend to report their time use more accurately (Press & Townsley, 1998).

The Second Shift

Sociologist Arlie Hochschild (1989) coined the term the *second shift* to describe the extra time working women spend doing household chores after working at a job outside the home. This term emphasizes the expectation that women who work will also perform the bulk of domestic and child care duties. This is an example of the

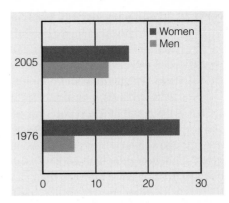

Figure 13.6 The amount of housework done by women has decreased since 1976, while the amount of housework done by men has doubled. In 1976, women did about 26 hours of housework a week; in 2005, they did about 16.5. Men did about six hours a week in 1976, and in 2005, they did about 12.5.
Source: University of Michigan Institute for Social Research (ISR) *Panel Study of Income Dynamics*. Reprinted by permission.

persistence of **patriarchy**, the dominance of men over women. Although women often attempt to influence their male partners to share household chores that were formerly considered "women's work," the men continue to resist despite slow changes—especially in two-career families in which both partners are highly educated. As shown in Figure 13.6, the household division of labor has been changing in favor of greater equality of male and female roles, but there remain significant differences. Men in couples have assumed more responsibility for household chores, but women still do more, and in the realm of child care, as we see in Table 13.4, women typically spend far more time on family child care, especially during the workweek.

patriarchy: The dominance of men over women.

TABLE 13.4
Hours and minutes of child care, parents of infants and of older children, 2003–07

Type of child care and day	Coupled fathers		Coupled mothers		Single mothers	
	With youngest child under age 1	With youngest child aged 1–12	With youngest child under age 1	With youngest child aged 1–12	With youngest child under age 1	With youngest child aged 1–12
Total child care, weekends	5:01	4:13	11:05	7:53	8:56	6:51
Total child care, weekend days	9:31	8:23	11:58	10:31	11:12	9:50
Primary child care, weekdays	1:25	0:53	3:53	1:58	3:13	1:02
Primary child care, weekend days	1:52	1:02	3:19	1:26	2:46	1:18
Total solo child care, weekdays	2:06	2:08	8:08	5:47	8:56	6:51
Total solo child care, weekend days	3:11	3:19	5:50	5:29	11:12	9:50
Sample size, weekdays	489	3,748	617	4,352	116	1,563
Sample size, weekend days	518	3,939	610	4,499	149	1,534

Source: Drago, 2010.

Some sociologists have proposed that, as women gain greater parity with men in occupations, the disparities between men's and women's roles will diminish, and gender roles in the household will become more symmetrical (Bernard, 1982; Willmott & Young, 1971). But as Hochschild (1989) and others report, this remains a speculative hypothesis despite the exceptions one might cite here and there. On the average, data on time budgets show that working women in the United States have at least ten hours a week less leisure time than their husbands or partners because they shoulder far more of the domestic responsibilities (Press & Townsley, 1998). The same disparities do not exist in all industrial nations. France and Holland, for example, incorporate gender equality into national social policies. These nations provide excellent universal child care, family leave, and longer vacations than do lagging nations like the United States, and these policies ease the burdens of dual-earner families in coping with the demands of work and family life (Hacker, 1997).

Outside the home, patterns of gender inequality are also proving to be extremely persistent. The pattern of wage inequality described earlier in the chapter, together with the tendency for women to be segregated in such occupations as secretary, bank teller, and elementary school teacher, has led activists in the women's movement to call for new policies designed to reduce inequality. Increasingly, they have pressed for "comparable worth"—that is, for increases in the salaries paid in traditionally female occupations to levels comparable to those paid in similar, but traditionally male, occupations (Jacobs & Gerson, 2005).

WOMEN AT WORK

In the industrial nations of North America, Europe, and Oceania, the increasing proportion of women who are in the labor force as paid employees is one of the most important aspects of social change in those societies. At midcentury, for example, 34 percent of U.S. women were in the labor force, but that proportion increased steadily in subsequent decades. By 2009, about 60 percent of women considered themselves to be employees, either at work or looking for work (*Statistical Abstract,* 2010). Among women with children over age eighteen, the trends are similar. In 1960, about 20 percent of these older women had jobs or were looking for jobs; by 2009, this figure had almost tripled, to over 57 percent (*Statistical Abstract,* 2010).

In fact, women assumed an important economic role well before industrialization created a sharp distinction between work in the paid labor force and work at home or in the fields. When the United States was an agrarian society, women planted and harvested crops, including extensive household gardens, and were also expected to take responsibility for housework and child rearing, both of which were highly labor intensive in the pre-electricity era. And because women have higher life expectancies than men, widows routinely ran farms or businesses after their husbands died, and many middle-class women gained additional income by taking in lodgers and selling handicrafts or other products of their labor. With the Industrial Revolution came growing demand for factory workers. Although male workers were often preferred, thousands of women and children also swelled the ranks of the new industrial working class. One study has shown that many of the working women in the early years of the Industrial Revolution were immigrants or poor single women from rural backgrounds. Large numbers of African American women were also working for wages in fields and factories and as domestic servants (Hacker, 1997).

World War II marked a significant turning point in women's labor force participation. Although returning servicemen "bumped" women (and minority men) from jobs in factories and offices, thousands of war widows and single women had to continue earning wages whether or not they wanted to. As a result, although women's labor force participation was still far lower in the 1950s than it is today, female employment was actually increasing at a far faster rate than male employment.

At the same time, the 1950s were a time of relative economic prosperity. The middle classes were expanding, and a suburban home became a central feature of the American dream. Popular culture, including the powerful new medium of television, emphasized the ideal norms of the middle-class nuclear family in which the mother was a homemaker and the father worked in the labor force. This "feminine mystique" asserted that women would find fulfillment as wives and mothers. In reality, thousands of women were entering the labor force in a trend that would continue throughout the second half of the century (Epstein, 1988).

Whether they worked outside the home because they wanted to or because necessity forced them to do so, women in the labor force encountered problems of gender segregation and discrimination (Hartmann, 2009). An enduring problem for women in the labor force is segregation into what are known as "pink-collar ghettos." Secretarial and clerical work in particular remain heavily gender segregated, but child care, nursing, and dental assistants also remain largely

Gender Roles among the Maasai When sociologist Carol Chenault and her photographer husband, Ellis, spent time with the Maasai people in 1995, they sometimes felt as if they had wandered back in time to the biblical era. For, as Carol Chenault explains, the Maasai are true pastoralists:

The people are Nilotic, migrating from the Nile regions of North Africa into the Great Rift Valley of Kenya and Tanzania. Myth has it that they lived in a deep crater where life was plagued with drought, famine, and discouragement. The only ray of hope for them was to follow the birds that kept bringing green grass and twigs. Scouts went up and over the steep escarpment and found green, fertile valleys with rivers and streams, and this is where they made their home. The language they speak is Maa; hence the term Maasai, meaning Maa-speaking people. The only transmission of the language is through the spoken word; their language is not written.

As we saw at the beginning of the chapter, gender role distinctions are significant among the Maasai. But unlike western industrial societies (where women are expected to spend more time on personal adornment), attention to beauty is a preoccupation of both males and females among the Maasai:

© Chenault Studios/Decatur, AL

Even the smallest of children are dressed with necklaces and anklets. Ear piercing is done during childhood, well before circumcision. A slit is cut in the earlobe and a green stick is placed in it. Over time this is replaced with successively larger [sticks] until the opening is large enough to be decorated with an earring and leather-beaded strands.

Chenault notes that, despite their relative lack of acculturation to western ways, she saw a Maasai man wearing a 35 mm plastic film canister as an ear adornment, a vivid reminder of the spread of western culture. The Maasai were extremely open and willing to share their culture with the visiting sociologist. "They are a proud people," Chenault concludes, "committed to passing their culture on to their future generations. I think this is the quality that I admire most about them."

These photos provide some clues about social status and gender roles among the Maasai. Notice that the elderly woman with close-cropped hair wears a limited number of personal adornments. Like most elderly women in traditional societies, she seems to feel no need to impress the viewer with her physical appearance. Not so for the woman with the distinctive headpiece and gorgeous array of necklaces. She is dressed in the fashion of the Maasai warrior herdsmen, but only the women wear these headpieces. In another photo, the same woman, the wife of a prominent warrior, is shown with her children. The Maasai women bear children almost as long as they are physically capable of doing so.

The man with a shawl over his shoulders looking quizzically into the camera was one of a group who conducted their own "interview" with the sociologist. They thought it was terrible that the American sociologist was no longer having babies. The women also asked whether American women were circumcised. "When we told them we were not," said Chenault, "they clapped their hands down at their sides, indicating their disapproval."

The Maasai man playing the flute also makes sandals out of old tires. These are called "thousand milers" because the Maasai walk endlessly and need tough, cheap footwear. Making musical instruments and footgear are examples of tasks that are specifically "men's work" among the Maasai.

© Chenault Studios/Decatur, AL

© Chenault Studios/Decatur, AL

© Chenault Studios/Decatur, AL

female occupations despite some increases in the proportions of males in these fields in recent decades. Largely as a result of antidiscrimination laws and women's efforts to break into occupations that previously were more or less closed to them, women have made considerable gains in some areas. Occupations like bus driver, psychologist, and others show significant gains in female employment, but clerical occupations remain a pink-collar ghetto: Clerical work employs one in five working women, a figure that has not changed since the 1950s (Hartmann, 2009).

Throughout the world, new occupations are emerging in professional and technical fields, and as women achieve greater educational parity with

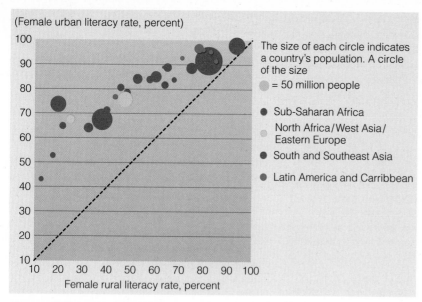

(Female urban literacy rate, percent)

The size of each circle indicates a country's population. A circle of the size

○ = 50 million people

● Sub-Saharan Africa
○ North Africa/West Asia/Eastern Europe
● South and Southeast Asia
● Latin America and Carribbean

Female rural literacy rate, percent

Figure 13.7 Female Literacy Rates in Urban and Rural Populations. Source: From "The March of the Cities," prepared by Patrick Salyer and David Bloom of Harvard University. *Finance and Development* September 2007. International Monetary Fund. Copyright © 2007 by International Monetary Fund. Reproduced with permission of International Monetary Fund in the format Textbook and Other book via Copyright Clearance Center.

men, the comparative statistics show that they are sharing these positions more equally with men. This is the case in Latin America, the Caribbean, and the economically developed regions of the world, but in much of Asia and Africa, where women still lag in education, wide disparities remain.

In administrative and managerial positions, the situation is far less favorable for women. Recent research shows that the United States ranks highest among industrialized nations in promotion of women to managerial positions, and this is in large part the result of affirmative action policies in American business. In contrast to the United States, where women hold 46 percent of such positions, and Sweden (30.5 percent), Japan (8.9 percent) has done extremely little to promote women to managerial roles.

Social scientists believe that this failure to enlist the leadership talent of women is one reason why the Japanese economy continues to stagnate (French, 2003).

Elsewhere in the world, competition for these desirable positions in management is even more intense. Promotion to management is often based not only on education but also on access to social networks and political power. Because women are far less likely than men to have such access, their entry into the managerial ranks is stymied. An exception is found in the island nations of the Caribbean, where women have had access to education and training in business schools longer than in many other regions. Caribbean women often accumulate money and power through their own business activities, which enable them to become owners, managers, and administrators at a greater rate than in the more affluent regions of the world.

Throughout the developing world, which includes over two-thirds of the world's population, female illiteracy remains a critical obstacle to women's advancement in occupations of all kinds, to say nothing of administrative positions. As we see in Figure 13.7, when women are part of the global migration from rural villages to urban towns and cities, they are far more likely to find educational opportunities, as measured by the greater literacy rates among urban women than among rural women. But throughout the world, as demonstrated in Figure 13.8, there are glaring literacy gaps in huge swaths of the developing world.

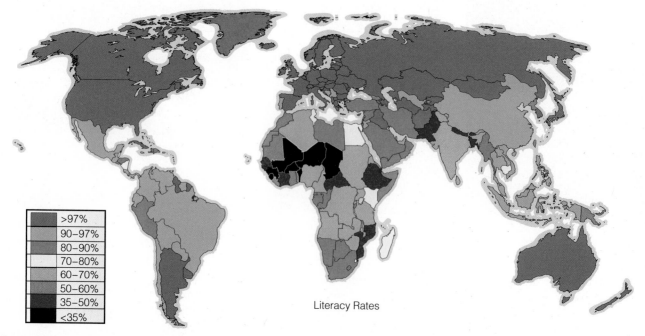

>97%
90–97%
80–90%
70–80%
60–70%
50–60%
35–50%
<35%

Literacy Rates

Figure 13.8 Worldwide Literacy Rates. Source: From United Nations Development Programme, *Human Development Report*, 2007/08, published 2007, reproduced with permission of Palgrave Macmillan.

SUMMARY

What are the basic facts about gender inequality in a global scale?

- Gender inequality stunts women's social and economic development throughout the world. It also impedes the development of entire societies, although it often takes different forms in developed and less developed nations.

- The term *empowerment* is often used in discussions of women in developing nations. The term refers to the process whereby relatively powerless people of any gender, race, ethnicity, or social class organize to assert their needs and overcome obstacles to their full participation in the institutions of their societies.

- Gender socialization in the family and in schools tends to separate males and females into different social worlds with their own forms of activity and language. However, boys and girls are increasingly participating in the same types of activities in their schools and communities, and many teachers, parents, and administrators seek to avoid the worst effects of gender separation.

In what ways are gender roles stratified, in the United States and around the world?

- All human societies are stratified by gender, meaning that males and females are channeled into specific statuses and roles.

- Until recently, it was assumed that there were two separate spheres of life for men and women. Out of this gender-based division of labor grew the notion of differences in men's and women's abilities and personalities. These differences were thought to be based on biological and psychological differences between males and females.

- In the twentieth century, however, evidence from the social sciences established that *gender roles* are not innate but are strongly influenced by culture.

- Preindustrial societies are usually rigidly sex-segregated. As societies increase in size and complexity, women usually become subordinated to men.

- The origins of gender inequality in most modern societies can be traced to their feudal periods. In Europe, for example, the norms of courtly love specified that women do not engage in warfare or politics or compete with men in any sphere of life beyond those reserved for women.

How do inequalities of gender operate in different societies to prevent women from realizing their full potential?

- In modern industrial societies, boys and girls are socialized into "masculine" and "feminine" roles. In their adult years, men enjoy more wealth, prestige, and leisure than do women.

- Gender stratification in modern societies is reflected in attitudes that reinforce the subordinate status of women.

- *Sexism* refers to an ideology that justifies prejudice or discrimination based on sex.

- Sexism is also expressed in violence and harassment of women, often intended to "keep women in their place" and maintain male power.

How are women mobilizing to challenge inequalities of gender, and in what way does feminist sociology assist in their effort?

- The modern women's movement arose in the mid-1960s, emerging from an existing network of women's organizations, together with a less formal network of women in consciousness-raising groups.

- The movement won significant victories during the 1960s and 1970s, and began to change the way men and women think about gender roles.

- Women who work outside the home are also expected to perform the bulk of domestic and child care work. This "second shift" is an example of the persistence of *patriarchy,* the dominance of men over women.

What are some problems women encounter in the workplace?

- Whether they work outside the home because they wanted to or because necessity forced them to do so, women in the labor force encounter problems of gender segregation and discrimination.

- Another problem is the segregation of women into the "pink-collar ghettos" of secretarial and clerical work.

The Kornblum Companion Website

www.cengagebrain.com

Supplement your review of this chapter by going to the companion website to take one of the tutorial quizzes, use the flash cards to master key terms, and check out the many other study aids you'll find there. You'll also find special features, such as GSS data and Web links that will put data and resources at your fingertips to help you with that special project or to do some research on your own.

14

Sources of Age Stratification

Age Stratification and Inequality

Theories of Aging

Death and Dying

Inequalities of Youth and Age

FOCUS QUESTIONS

How do sociologists visualize age stratification?

What patterns of inequality are associated with age stratification?

If everyone ages, becomes old, and eventually dies, why do sociologists need theories of aging?

What do social norms and institutions have to do with death and dying?

Like many of the world's urban industrial nations, the United States has an aging population. Consider this simple fact: In 1960, the number of centenarians, or people older than age 100, in the United States was slightly more than 3,200. Today, it is more than 28,000. Does this momentous change mean that the United States is becoming a society in which to be old is to be respected and appreciated? Most experts on aging would say no. Our culture, as judged by the messages we continually receive on TV or in other media, values youth, vitality, strength, speed, and youthful beauty. No wonder the Irish poet William Butler Yeats wrote in his declining years that an aged person is:

> . . . but a paltry thing, a tattered coat upon a stick, unless soul clap its hands and sing, and louder sing for every tatter in its mortal dress.

Imogen Cunningham.

But how can an elderly person's soul sing in the face of declining health and the gnawing isolation that many elderly people experience? Some answers can be found in the efforts of people with strong sociological imaginations to change our perceptions of the elderly. One of the best examples of this approach may be seen in the photographic work of Imogen Cunningham.

Cunningham, one of the twentieth century's most influential photographers, spent most of her long life in San Francisco. In her later years, she devoted much of her work to photographing people who were more than ninety years old but "still alive and kicking." Her photographs of elderly people, some of which appear in the Visual Sociology box in this chapter, are studies of dignity and courage in the face of advanced age and imminent death.

Ironically, despite the high value placed on youth, young people are another age class that is experiencing severe social stress. This is true in affluent societies like the United States as well as in many developing nations. The specific causes may differ widely from one region to another, but wherever mothers of children and young people are under stress because of poverty and violence, the children suffer as well. Of course, people in other age categories also experience inequalities based on age, but as we will see in this chapter, the young and the aged are particularly subject to age-based inequality.

SOURCES OF AGE STRATIFICATION

In many societies, age determines a great deal about the opportunities open to a person and what kind of life that person leads. Only a little more than fifty years ago, for example,

gerontology: The study of aging and older people.

becoming old in the United States almost automatically meant becoming poor. Today, children are the most impoverished and vulnerable population group in many nations, including the United States. This chapter examines how age stratification—the roles assigned to young people, adults, and the elderly—contributes to inequality. From a discussion of the life course and age structure of societies, the chapter proceeds to a closer examination of the situation of young people. The remainder of the chapter is devoted to the growing sociological subfield known as **gerontology**, the study of aging and the elderly in this and other societies.

The Life Course: Society's Age Structure

All societies divide the human life span into "seasons of life" (Binstock & George, 2006). This is done through cultural norms that define periods of life, such as adulthood and old age, and channel each individual into an **age grade**—a set of statuses and roles based on age. These systems of age grades create predictable social groupings and turning points that everyone in a society can easily recognize. Graduations, communions, weddings, retirements, and funerals are among the ceremonies that are used to mark these turning points.

Age strata are rough divisions of people into layers according to age-related social roles. We speak of infants, preschoolers, elementary school children, teenagers, young adults, and so on; these categories form a series of younger-to-older layers, or strata, in the population. People in different age strata command different amounts of scarce resources like wealth, power, and prestige (Moody, 1998). Numerous laws establish inequalities between youth and adults; they include laws governing the rights to vote, to purchase alcoholic beverages, to incur debt, and the like. In theory, people who lack the rights of adult citizenship will be protected by adults, who are responsible for providing them with adequate food, shelter, and education (and are presumed to have the resources to do so). In practice, however, hundreds of thousands of children and teenagers do not receive the care that is intended to offset their unequal status under the law.

Social scientists often refer to the **life course**, which may be defined as a "pathway along an age-differentiated, socially created sequence of transitions" (Hagestad & Neugarten, 1985, p. 36; see also Cain, 1964; Clausen, 1968; Elder, 1981). The cultural norms that specify the life course and its important transitions create what is thought of as the normal and predictable life cycle (Neugarten, 1996). We expect that we will go to school, find a job, get married, have children, and so on at certain times in our lives, and we consider it somewhat abnormal not to follow this pattern. Social scientists often refer to a ceremony that marks the transition from one phase of life to another as a **rite of passage** (Van Gennep, 1960/1908). The confirmation, the bar mitzvah, the graduation, and the retirement party are examples of rites of passage in modern societies.

In the United States and other Western cultures, the life course is constructed from categories like childhood, adolescence, young adulthood, adulthood, mature adulthood, and old age. But our definitions of these categories lack the stability and uniformity of the age grades found in many traditional societies. For example, French historian Philippe Ariès (1962) showed that, in Western civilization, the concept of childhood as a phase of life with distinct characteristics and needs did not develop until the late seventeenth century. Before that time, children were treated as small adults. They were expected to perform chores and to conform to adult norms to the extent possible. When they reached puberty, they were usually married, often to spouses to whom they had been promised in infancy. Today, many studies of childhood argue that the cultural definitions of childhood and adulthood are again becoming blurred because children are exposed to adult themes in the media, and many children are prosecuted for adult crimes (Applebome, 1998).

Norms regarding gender are closely linked to the life course that each society establishes. Thus, Ariès's study of the emergence of childhood revealed that ideas about the appropriate forms of play and education for boys, and indeed the very concept of boyhood, developed at least a century before the concept of girlhood, emerged. In eighteenth-century European societies, boyhood was conceived of as a time when male children could play among themselves and receive education in the skills they would need as adults. Girls, in contrast, were treated as miniature women who were expected to work alongside their mothers and sisters.

Cohorts and Age Structures

When we think about age, we tend to think in terms of age cohorts. An **age cohort** consists of people of about the same age who are passing through the life course together (Bosworth & Burtless, 1998). We measure our own successes and failures against the standards and experiences of our own cohorts—our schoolmates, our workmates, our senior circle—as we pass through life.

Demographers use the cohort concept in studying how populations change. If we divide populations into five-year cohorts, grouped vertically from age 0 to 100+ and divided into male and female, we can form a *population pyramid,* a useful way of looking at the influence of age on a society. Figure 14.1 compares the age structure of the population of developing nations with that of the more developed ones, including the United States. The base of the population structure of the developing nations is

---

age grade: A set of statuses and roles based on age.

life course: A pathway along an age-differentiated, socially created sequence of transitions.

rite of passage: A ceremony marking the transition to a new stage of a culturally defined life course.

age cohort: A set of people of about the same age who are passing through the life course together.

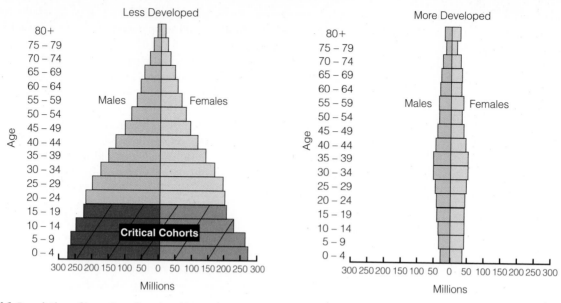

Figure 14.1 **Population of Less Developed and More Developed Nations, by Age and Sex.** Source: Population Reference Bureau, 1998; *Statistical Abstract*, 2005.

wide because of high birthrates. But the structure narrows in each cohort because of high rates of mortality, largely because of lack of access to up-to-date medical care and lack of preventive health care. High mortality rates for people older than age forty result in the pyramidal form that is characteristic of the age structure of developing nations.

The developed nations have a more cylindrical population age structure, a consequence of far lower birthrates and, hence, smaller early-childhood and youth cohorts. For example, the zero to four age cohort is smaller than the five to nine cohort, reflecting the continuing decrease in birthrates in the industrialized and urbanized regions of the world. Note also that death rates are far lower in affluent countries than they are in developing nations, so that in the absence of war or deadly epidemics, the age cohorts pass through the age structure with little attrition until well into middle age, when increasing rates of mortality produce slightly smaller age cohorts.

In developing nations, the age cohorts from birth to age nineteen are termed the "critical" cohorts because the reproductive behavior of these cohorts will largely determine the future size of the world's population. About 2 billion people younger than age twenty are living in less developed nations, of which more than 400 million are already in their early childbearing years (ages fifteen to nineteen) (UNDP, 2003). In some industrialized nations, about 1 percent of teenagers in these cohorts give birth, but in parts of Africa and Asia, the rate can be as high as 24 percent. Overall, in Africa below the Sahara and above the nations of southern Africa, 18 percent of teenagers give birth each year. Were that rate to continue, the region's population would double in twenty-three years. In Latin America, where 8 percent of teenagers give birth each year, the population doubling time would be forty-two years. Concerned about these statistics, in 2000, representatives of more than 180 nations agreed to rapidly increase investments in health care, family planning, and education for young women in the less developed regions of the world in the hope of stabilizing world population growth (United Nations Development Programme, 2008).

The Baby Boom When a population experiences marked fluctuations in fertility, there are bulges in its population pyramid that have

important effects. Perhaps nowhere in the world has this phenomenon been more thoroughly studied than in the United States. The "baby boom" cohorts, which were produced by rapid increases in the birthrate from about 1945 through the early 1960s, have profoundly influenced American society and will continue to do so for the next three decades.

Throughout Europe and North America, the baby boom generations did not have nearly as many children as their parents had. A mean family size of 2.1 children per couple is required for a population to remain constant over time (rapid growth requires a mean number of children closer to 3.0 per family). Since the 1970s, however, the mean number of children per family in industrial societies has been about 1.85 (and much lower in nations like Japan and Germany); as a result, the baby boom has been followed by a relative shortage of children known as the "baby bust" (Himes, 2001; Morgan, 1998). One way to visualize the impact of the baby boom and baby bust is to compare the population pyramids in Figure 14.2. The first of the four pyramids represents the U.S. population in 1900. It is the classic shape for a population, with relatively high birthrates and thus high proportions of people younger than age twenty. The next three pyramids show the large baby boom cohorts born between 1945 (the end of World War II) and the mid-1960s. Note the relatively smaller size of the cohorts just above the baby boomers, which is a result of war and delay of marriage. The 2000 and projected 2030 pyramids show the baby boom cohorts moving through the society's age structure until they are

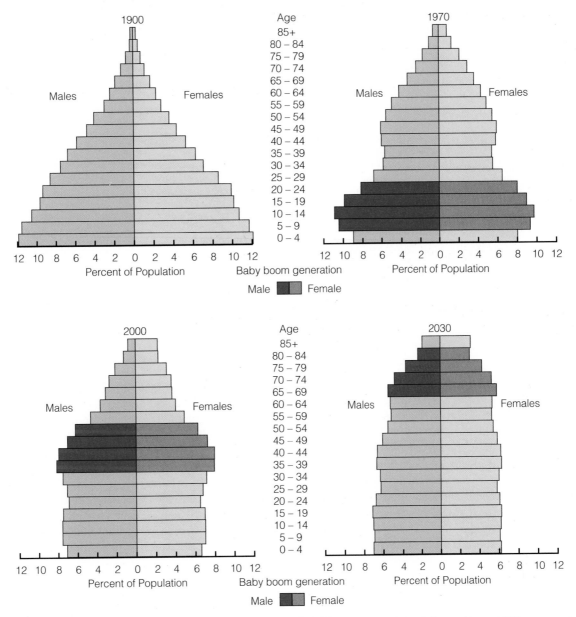

Figure 14.2 U.S. Population, by Age and Sex: 1900, 1970, 2000, and 2030. Source: Population Reference Bureau, 2001.

the most elderly segments of the population. Note also the changing shape of the pyramids as a result of the aging of the population and the relatively low birthrate, which keep the youngest cohorts small compared to those of 1990.

As these cohorts mature, they make new demands on the society's institutions. During the 1960s and 1970s, for example, when the baby boom cohorts passed through their college years, the nation's universities and colleges expanded; the slogan of the day was "Never trust anyone over thirty." Now the baby boom cohorts are moving into their senior years and beginning to draw their Social Security payments and pensions. This generation of "boomers," once hoping to usher in the Age of Aquarius, is now setting its sights more modestly on hopes of retirement and health with insurance coverage

well into older age. But the boomer generation is not the last to have its own experiences and perceptions of the world. Sociologists who study age often speak of more recent cohorts such as Generation X and the Millennials because there are important differences in their experiences and perceptions as well.

In a recent major survey of differences in attitudes and behavior in the U.S. population by age, the Pew Research center identifies the important U.S. population cohorts as the Silents, the Boomers, Gen X, and the Millennials (see Table 14.1). Most of the readers of this text will be in that last cohort, one that has already established its own particular interests and set its own trends, but whose historical experience, and therefore its full adult identity, is still emerging.

The Pew social scientists note that "Generations, like people, have personalities, and Millennials—the American teens and twenty-somethings who are making the passage into adulthood at the start of a new millennium—have begun to forge theirs: confident, self-expressive, liberal, upbeat and open to change." It may be too soon to know how long some of these attitudes will persist, but the data also show (see

TABLE 14.1

What's in a Name?

Generational names are the handiwork of popular culture. Some are drawn from a historic event; others from rapid social or demographic change; others from a big turn in the calendar.

The Millennial generation falls into the third category. The label refers those born after 1980 – the first generation to come of age in the new millennium.

Generation X covers people born from 1965 through 1980. The label long ago overtook the first name affixed to this generation: the baby bust. Xers are often depicted as savvy, entrepreneural loners.

The Baby Boomer label is drawn from the great spike in fertility that begain 1946, right after the end of World War II, and ended almost as abruptly in 1964, around the time the birth control pill went on the market. It's a classic example of a demography-driven name.

The Silent generation describes adults born from 1928 through 1945. Children of the Great Depression and World War II, their "Silent" label refers to their conformist and civic instincts. It also makes for a nice contrast with the noisy ways of the antiestablishment Boomers.

The Greatest Generation (those born before 1928) "Saved the world" when it was young, in the memorable phrase of Ronald Reagan. It's the generation that fought and won World War II.

Generational names are based on the Pew Research Center's recent survey of attitudes of different generations or cohorts in the U.S. population following a set of definitions that are commonly found in popular culture and the media.

Source: From The Pew Research Center. MILLENNIALS: A Portrait of Generation Next: Confident, Connected, Open to Change. February 2010. http://pewsocialtrends.org/assets/pdf/millennials-confident-connected-open-to-change.pdf. Reprinted by permission.

Figure 14.3) that, in contrast to earlier age cohorts, the Millennials are significantly more diverse along racial and ethnic lines. This simple fact has far-reaching implications, as we have seen in the historic election of a nonwhite president who garnered a historic vote from the Millennial generation. Not surprisingly, the Pew study also found that, far more than earlier generations, the Millennials adore new technologies of communication. Perhaps with age, fewer Millennials will actually admit to sleeping with their cell phones (Figure 14.4), but there is little question that this is the most socially networked, in-touch, generation in history.

Aging in Global Perspective As fertility decreases in most regions of the world and fewer babies are born, and as improvements in health care are extended to poorer regions, the proportion of older people in most nations rises accordingly. This "graying" of populations occurs at different rates, and with differing consequences, in developed and less developed regions, as do rates of change in life expectancy. But population aging will affect every man, woman, and child in the world. The steady increase of older age groups will have a direct bearing on relationships within families, inequality across generations, changes in lifestyles, and family solidarity (United Nations Department of Public Information, 2002). In regions with increasing elderly populations and continuing high rates of fertility, especially in Asia and Africa, high proportions of dependent children and increasing proportions of dependent elderly people increase the burden on working adults. We turn, therefore, to a discussion of population aging, or "graying," increased life expectancy and longevity, and their consequences for dependency in the near future.

Europe, North America, and Oceania have "graying" populations, in contrast to regions where the proportions of people age sixty-five and older remain relatively small. These

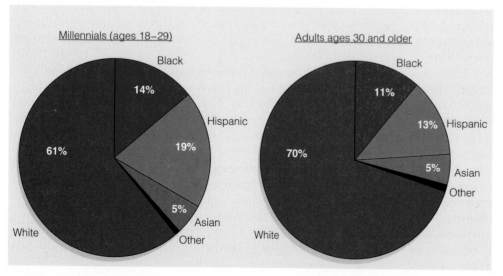

Figure 14.3 Composition of Millennial Generation and Over-30 Population. Source: From The Pew Research Center. MILLENNIALS: A Portrait of Generation Next: Confident, Connected, Open to Change. February 2010. http://pewsocialtrends.org/assets/pdf/millennials-confident-connected-open-to-change.pdf. Reprinted by permission.

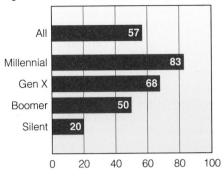

Percent who have ever placed their cell phone on or right next to their bed while sleeping

All	57
Millennial	83
Gen X	68
Boomer	50
Silent	20

Figure 14.4 Do you sleep with your cell phone?

Source: From The Pew Research Center. MILLENNIALS: A Portrait of Generation Next: Confident, Connected, Open to Change. February 2010. http://pewsocialtrends.org/assets/pdf/millennials-confident-connected-open-to-change.pdf. Reprinted by permission.

differences are shown in Figure 14.5. Remember, however, that the figure shows proportions of older people in each region and not absolute numbers. Because Asia has the largest share of the world's population, it also has the highest number of elderly people. Rapidly expanding numbers of very old people represent a social phenomenon without historical precedent. Today, the number of people age sixty or older is estimated to be 629 million. That number is projected to grow to almost 2 billion by 2050, when the population of older people will be larger than that of children under age fifteen for the first time in human history. Fifty-four percent of the world's older people live in Asia. Europe has the next largest share, with 24 percent (United Nations Department of Public Information, 2002; U.S. Department of Commerce, 2001).

The dramatic increase in the proportion of elderly people in the industrialized nations projected for 2025 is largely attributable to the aging of the baby boom cohorts. This increase, often referred to as the "graying of America," is already resulting in greater concern about the needs of the elderly and will augment the influence of the aged on American culture and social institutions. (See the Mapping Social Change box on pages 352–353.)

Demographers and biologists disagree over estimates of how many very elderly people there will be in the U.S. population during the twenty-first century. Most Census Bureau demographers estimate that the population older than age eighty-five will increase from about 3.3 million people (mainly women) in 1990 to about 18.7 million in 2080. But demographers working with the National Institute on Aging argue that there may be no biological limits on life expectancy. With advances in medicine and better prevention of disease, more people may live

into their nineties. If that happens, there could be as many as 70 million people over age eighty-five in the U.S. population late in this century—a change that would have enormous effects on all aspects of social life (Himes, 2001).

Thinking Critically

Many school districts want to forbid students from bringing cell phones to school. Do cell phones interfere with schoolwork? Is the idea of banning them an example of generational conflict? Where do you stand on this issue?

Longevity and Life Expectancy

As people live longer lives—that is, as *longevity* increases—due especially to better nutrition, more medical care, and improved economic and political stability—there are both positive and negative consequences, or risks, for societies and nations. More older people means that more people will live to be grandparents capable of helping their adult children care for their kids. Older people in good health have opportunities to travel and continue to develop their human potential. But greater longevity also means that there will be more "frail" elderly and greater incidence of illnesses such as cancer and Alzheimer's disease, which families and entire societies must learn to cope with. Despite these and other risks, increased life expectancy, as noted in earlier chapters, is a measure of social development because the overwhelming majority of people wish to live longer and more productive lives.

At age sixty-five and beyond, over 50 percent of North American women are widowed, whereas only about 14 percent of men have lost their wives. This difference is due to the fact that the life expectancy of females is at least seven years longer than that of males. (By **life expectancy**, we mean the average number of years a member of a given population can expect to live beyond his or her present age.) Typically, men

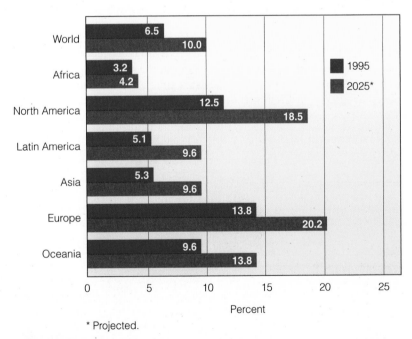

Figure 14.5 Population Age 65 and Older, by World Regions: 1995 and 2025.

Source: Data from United Nations.

* Projected.

life expectancy: The average number of years a member of a given population can expect to live.

mapping social change

The Age Structure of Sun City, Arizona

The aging of the U.S. population is creating major migration streams and new patterns of urban and suburban growth. The accompanying map, which shows the percentage of the U.S. population age sixty-five or older by county, offers a detailed view of the ecology of aging. We are used to thinking of Florida, Arizona, and now Nevada as destinations for retired people, but why are there such high proportions of elderly people in the Great Plains states? In the Dakotas, Nebraska, and Minnesota, cold weather is a severe deterrent to in-migration by the elderly, so what is the reason for the projected growth of this population there? The answer has more to do with the movement of young people out of these states. As young people leave the Great Plains in search of opportunities in California and the rapidly growing states of the Southeast and Southwest, older people constitute a greater proportion of the remaining population.

In many of the plains states, the elderly must be extremely vigorous to survive in the harsh environment and must be highly dedicated to their communities to maintain them in the absence of enough younger people to fill available positions, especially in unpaid volunteer work.

The migration patterns of the elderly also create some unusual inversions of the normal age pyramid in places where they become the most numerous population segment. The age pyramid for Sun City, Arizona, a retirement community in the Sunbelt, is a case in point. In this retirement community, only 0.1 percent of the population is younger than age fifteen and nearly 85 percent is older than age sixty-five. Note also that women outnumber men, as would be expected considering their greater life expectancy.

Population Pyramid for Sun City, Arizona

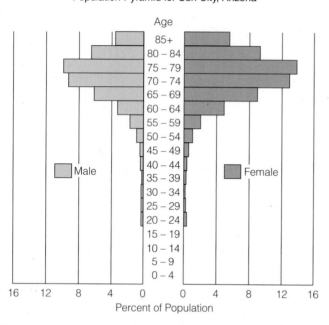

die before their wives do. Older men also tend to marry younger women, making widows the single largest category in the elderly population.

The deep recession that has persisted since 2008 has had perhaps its sharpest effects on the Millennial youth cohort, which is now entering the labor force. From April to July 2009, as new graduates and students looking for summer work began searching for jobs, the number of employed youth ages sixteen to twenty-four increased by 1.6 million, to 19.3 million. In the same period, however, unemployment among youth increased by 1.1 million, about the same as in the summer of 2008 (U.S. Department of Labor, 2009).

Age and Dependency People in the working adult cohorts—that is, those between the ages of eighteen and sixty-four (although many continue to work well after age sixty-four and may start before age eighteen)—contribute disproportionately to the well-being of the young and the elderly. Of course, societies justify this pattern of dependency by recognizing that adults are merely doing in their turn what was done for them as children or will be done for them when they are among the frail elderly. Institutions of modern societies, such as public education and Social Security, ensure that a share of wealth passes to the dependent cohorts.

Jim West/Alamy

Social movements among the elderly are increasing in strength as the older population grows in numbers. Older Americans have taken an active interest in recent efforts to reform the nation's health care system.

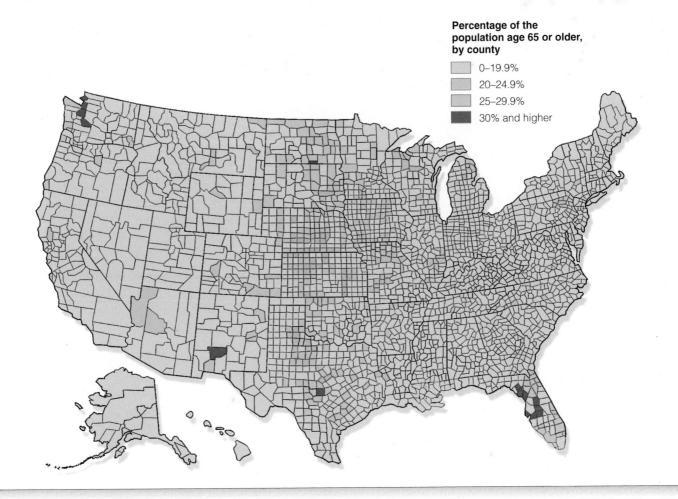

Percentage of the population age 65 or older, by county

- 0–19.9%
- 20–24.9%
- 25–29.9%
- 30% and higher

When a society includes very large numbers of children, as is the case in the developing nations, or increasing numbers of elderly people, as is the case in the older nations of western Europe and North America, working adults may shoulder an increased burden. Figure 14.6 vividly illustrates this dependency problem. It charts the ratio of workers in the U.S. labor force to people receiving Social Security benefits. That ratio has been decreasing sharply with the aging of the U.S. population, so that now there are close to three workers for every retired person drawing on the Social Security fund. The graph also projects the ratio well into the twenty-first century to demonstrate that the dependency situation will continue to worsen as population aging accelerates. According to these projections, by the end of the second decade of this century, more funds will be drawn out of the Social Security system than active workers are paying into it. In consequence, there is a lively debate among legislators about how to modify the system in the face of these changing demographics. Should the retirement age

increase? Should Social Security taxes rise? Should the system be changed so that wealthier people receive lower benefits than less affluent citizens? These are all extremely controversial issues that need to be addressed to ensure that the major gains in well-being made by elderly people since the 1930s are not eroded in coming decades.

AGE STRATIFICATION AND INEQUALITY

In urban industrial societies, there are distinct patterns of stratification in which age defines the roles one plays and the rewards one can expect. We speak of the "age of majority"—the age at which a person crosses the legal boundary between childhood and adulthood. In fact, this age is not always clearly defined. A person can vote at age eighteen but cannot legally consume alcohol until age twenty-one. In addition to the age of majority, other distinct ages mark the passage toward the full rights of adulthood.

Given the opportunity, elderly men and women can have a profound influence on the lives of children in their families and neighborhoods.

Later in life, as people become elderly, they may yield some of their autonomy to their grown children—either willingly through trusts and wills, or unwillingly as they are committed to nursing homes because their care has become too great a burden for their children. Thus, at the early and late extremes of youth and age, the relationship between age and inequalities of power and material resources is most evident.

Age-based inequalities may also affect the life chances of nonelderly adults. As corporations downsize (that is, lay off employees in order to increase profit margins), higher-paid people in their fifties may find that their jobs are combined with other work and offered to younger, less well-paid replacements.

Race and Gender: The Double and Triple Binds of Aging

Inequalities of age are compounded by those of race and gender to produce particular forms of inequality among the aged in industrial societies, especially the United States. If being elderly is a disadvantage, being an elderly woman often places one at a double disadvantage. Being elderly, female, and black or Hispanic places one in triple jeopardy for being isolated, ill, and impoverished (Binstock & George, 2006; Himes, 2001; Holden, 1996). Women of advanced age, especially those older than age eighty-five, are three times more likely than older men to be living in poverty. The situation is worse for elderly black and Hispanic women. Because large numbers of these women either were never married or experienced divorce (a subject to which we return in the next chapter), higher proportions live alone and have lower incomes in their later years than their white counterparts.

At age eighteen, one can join the armed forces without parental permission. At age sixteen, in many states, one is no longer obliged by law to attend school, even though parents still share responsibility for the behavior of school dropouts. Teenagers can drive at age sixteen in many U.S. states, but their parents remain responsible, through insurance systems, for their actions and can be held liable for the consequences. We also make clear distinctions between children in the primary grades and teenagers in high school, and in most homes a child's passage through these age grades is accompanied by various privileges and responsibilities. So although the passage to full adult status may be somewhat vague, it is clear that to be young is to be less equal.

As we have seen, age influences life chances in important ways and helps produce the patterns of inequality that we often take for granted, especially among the young and the old. And age compounded by race and gender often entails further inequality. However, if we study age stratification and inequality from a comparative perspective, we find some surprising differences. In smaller-scale, more traditional agrarian or pastoral societies, age stratification works in different ways than in urban industrial societies. For example, among the Maasai, whom we met in the preceding chapter, there are fixed age grades for males and females that strictly define an individual's roles and opportunities throughout life. Children and adolescents have no individual status in Maasai society. Not until their passage through circumcision and other rituals during puberty do they achieve status as autonomous individuals.

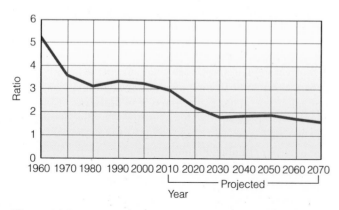

Figure 14.6 **Ratio of U.S. Workers to Social Security Beneficiaries.**
Source: Social Security Administration.

Initiation into adult status occurs for a group of Maasai of roughly similar age, who then constitute an age grade. There are four age grades in the Maasai system, each covering about fifteen years and involving a mandatory set of activities, especially for males. Young initiates engage in military activity. Young elders are married adults with family and economic responsibilities. Elders have political decision-making power, and senior elders have religious ritual power. Each of these age grades has a name in the Maasai language, and passage into it is marked by rituals that convey differing degrees of authority and responsibility (Chenault, 1996). Other smaller-scale societies throughout the world organize age grades according to their own cultural norms, but most, like the Maasai, invest the different age grades with distinct degrees of power and control over the society's wealth.

Throughout the world, the forces of social change unleashed by colonialism, industrialization, urbanization, and population growth have tended to disrupt the formal age grade systems of smaller, more isolated societies. Often the loss of their traditional ways of commemorating the passage from one age to another has been among the more disorienting aspects of social change for people in those societies. The passage into old age is a good example. In most small-scale tribal and peasant societies, the elderly are revered as guardians of moral values, rituals, and religious beliefs. In urban industrial societies, by contrast, they are typically treated as dependents with little status and few rights. The situation of children is similar. In traditional societies, children have well-defined roles. The young girls are typically helpers while the boys are expected to play games that teach them the skills they will need as adult males. But all this changes in urban industrial societies. In those societies, the status of children is extremely fluid and subject to many conflicting ideas and rapid changes in norms (Ariès, 1962).

The Challenge of Youth

In *The Challenge of Youth* (1965), his seminal work on youth in Western societies, Erik Erikson observed that in societies with high levels of youth rebellion, deviance, and even suicide, the problem often lies in the failure of adults to provide young people with clear goals and paths toward a constructive future. Kenneth Keniston (1965), another expert on the problems of growing up in modern societies, adds that the challenges young people face are further complicated by such rapid and far-reaching changes that they are often bewildered about such questions as what careers to pursue, what standards to apply to personal relationships, and how much to devote to oneself and how much to the community and the society. Many young people have parents or other adults in their lives who can help them solve these dilemmas, but many others do not. As we saw in Chapter 8, young people may react to the problems and failures of those around them (for example, alcoholism or divorce) by withdrawing into a world of anomic, alienated youth that explicitly rejects the values of the larger society, sometimes through gang violence.

In this regard, one of the most fascinating unanswered questions in the social sciences is why some children growing up under extremely adverse social conditions appear to be resilient, to be able to succeed in school and in other social settings, while so many others are set back and defeated by the same adverse conditions (Conger, Ge, & Elder, 1994; Garmezy, 1993; Williams & Kornblum, 1994). On a broader sociological level, however, Keniston and others note that changing norms of childhood and youth, on one hand, and the decreasing well-being of young people in industrial nations, on the other, strongly influence the fortunes of youth everywhere.

Youth Cultures and Social Change At the beginning of the twentieth century, when the United States and Canada were still largely agrarian societies (though with booming population growth in their major urban centers), adolescents typically went to work, often in their father's agricultural or industrial workplace. "A son follows his father's trade," was the common expression, and teenage girls were expected to learn to do domestic work in preparation for unpaid careers as homemakers.

AP Photos

Although he lost his skateboard in this jump, this young athlete went on to win in his event class. Skateboarding is part of a youth subculture that attracts young boys and may lead to their participation in extreme forms of the sport as they develop their skills.

then & now

From Adulthood in Miniature to Childhood—and Back? These photos offer versions of ideal behavior among children in the eighteenth century, the mid-twentieth century, and the present. They raise questions about the ever-shifting norms and expectations of childhood. In the eighteenth century, as depicted in the famous Gainsborough portrait known as the *Blue Boy*, the ideal of childhood was one of adulthood in miniature. By age seven or eight, children "were dressed and treated as little adults. They dressed the same, did the same work, and entered into the sexual community of adults. . . . Even in the United States, the age of sexual consent was under 10 in half the states until the end of the 19th century" (Applebome, 1998, p. 3). The idea that children are different and have their own rights to education and protection under the law emerged in the early twentieth century. Today, experts on childhood worry that cultural understandings about childhood have again become blurred. The public's fascination with

Dave Sartin Photography

another "miniature adult," JonBenet Ramsey (the winner of the 1995 Little Miss Colorado beauty pageant, who was murdered in 1996 at the age of six), is only a small part of the story. Perhaps more ominous is the fact that, by 2000, almost all states had passed laws allowing the prosecution and sentencing of fourteen-year-olds as adults.

Do these and many other examples of the blurring of the boundaries of childhood mean that childhood is disappearing? Most experts would disagree. They note that children "usually prove more resilient than grown-ups expect" and that survey research shows "American teenagers to be worldly in ways previous generations were not, but sharing most of the values and sensibilities of earlier times" (Applebome, 1998, p. 3).

Huntington Library / SuperStock

David Deas/DK Stock/Getty Images

Girls from working-class and poor families typically went to work in fields and factories, but the stereotypical image of women's work was the homemaker. Far smaller numbers of boys, and even fewer girls, could be expected to complete high school, and even fewer went on to colleges and universities. Contrast this with the contemporary situation, in which most young men and women continue into postsecondary studies.

One effect of this major change in the social organization of youth is the emergence of an ever-greater variety of subcultures among young people. Because teenagers and young adults spend far more time in educational institutions making themselves ready for an ever-more-demanding labor market, adolescence—the phase of life considered to be relatively free from full adult responsibilities—often lasts well into the young-adult years. And during those years, young people have increased freedom to try out different identities and associate with different kinds of peer groups.

At the beginning of the twentieth century, youth peer groups that developed their own subcultures were common, especially among boys in the cities who spent time in the streets and joined gangs, or among elite young men who joined the secret societies and fraternities of universities and colleges. Today, however, as a result of the increasing racial and ethnic diversity of the North American population and the greater freedom young people have to make choices about friends and activities, the range of youth subcultures is far greater. The existence and behaviors of identifiable groups in any high school—the nerds, skateboarders, surfer dudes, cheerleaders, and on and on—are well known to most teenagers but often far more mysterious to parents, who find it difficult to communicate with their children about sensitive issues of sexuality and substance abuse.

In the 1960s, as the baby boom generation entered its teenage years, the marketing of music, clothes, films, and food to young people became extremely lucrative. Youth marketing became a major industry, and it in turn began to play a significant role in shaping what young people do, how they dress, and how they display their allegiance to different youth subcultures. For example, a popular Canadian website is designed to attract the attention of teenagers and young adults and simultaneously attract businesses that want to market goods and services to them. Its "Trendscan" survey uses sociological research techniques to track trends in youth cultures in Canada and show businesses how to better attract teenage buyers, as the following statement shows:

> Understanding the uniqueness of the 12–13 year old experience versus 14–15, 16–17 and 18–19, male versus female, and the regional variances is essential for developing thoughtful marketing strategies. Further, active brands can profile their specific youth consumers.
>
> Our Early Style Adopter (ESA) Teen is a core component of the Trendscan survey. These youth have been earmarked as the style and opinion leaders within the Canadian youth population. Understanding ESAs can be an essential factor in successfully marketing to a consumer group heavily influenced by "the cool kids." (www.youthculture.com)

Understanding the rapid change occurring in youth cultures and young people's consumer behavior, including their consumption of illegal substances, is the subject of innumerable TV ads and websites, but the intensity of communications about young people is also an indicator that the social changes we have discussed are associated with widespread debate and uncertainty about norms and expectations for children and teenagers in most of the world's nations.

Variations in the Value Placed on Children

Social definitions of childhood—what we expect of children, how we value the child and treat the child—differ immensely throughout the industrial world as well as between modern and traditional societies. In her study of the changing meanings of childhood, sociologist Viviana Selizer shows that in the United States, there occurred a "profound transformation in the economic and sentimental value of children—fourteen years of age or younger—between the 1870s and the 1930s" (1985, p. 3). During that period, child labor laws ended a common practice, at least among poorer families, of requiring children to work for wages. Earlier in the nineteenth century, the incomes of the middle classes had risen enough so that they could keep

their young ones out of the labor force; by the 1930s, this became the norm for all classes.

As a result of this transformation, Selizer points out, children became economically "worthless" but emotionally "priceless"—and this is true today as well. We think of our children as priceless beings to be nurtured, protected, socialized, and gradually introduced to the world of work and careers—not forced into that world abruptly after a brief period of innocent childhood, as was the case in the early years of industrialization. Studies of children's responsibilities in the home find that people justify giving their children chores to do on the grounds that doing so builds character, discipline, responsibility, and so on, not because it contributes to the economic well-being of the family. In fact, in many middle-class homes, the allowance system operates in lieu of any form of wages; the allowance is seen as a form of token money that the child has a right to receive but must also learn to spend responsibly.

But Selizer also shows that this transformation in the value placed on children was surrounded by a great deal of social conflict as the institutions of the larger society, especially courts and schools, were given the responsibility for enforcing the new norms about child protection. Debates over definitions of childhood and the proper roles of children continue today. For example, the home schooling movement has arisen among people who reject the values of the public schools and wish to maintain the absolute authority of parents over their children by educating them in the home.

More important than these conflicts are the class contradictions inherent in the value placed on childhood. Many families are not able to give their children the array of goods and opportunities that are available to children in more affluent families. The worst situations of material deprivation experienced by children and young people, along with their parents, are often hidden from view in racial and ethnic ghettos and class-segregated communities "across the tracks" or in isolated trailer parks (Kozol, 1995). Overall trends in child welfare reveal a widening gap between "priceless" children and children who bear a heavy burden of poverty and deprivation.

Indicators of Well-Being Among Youth

There is a growing consensus among social scientists on the need for a quantitative measure of well-being among youth. In recent years, as ideological debates over social problems like teenage fertility have become more rancorous, this need has become even more pressing (Zill, 1995). Toward this end, a group of eminent sociologists has been working to create a set of

TABLE 14.2
Indicators of Child Well-Being

	Trend Data		
	1990	2000	Percent change 2000–2005
Percent low-birthweight babies	7.0%	7.6%	−8%
Infant mortality rate (deaths per 1,000 live births)	9.2	6.9	0
Child death rate (deaths per 100,000 children ages 1 to 14)	31	22	9
Rate of teen deaths by accident, homicide, and suicide (deaths per 100,000 teens, ages 15 to 19)	71	51	3
Teen birthrate (births per 1,000 females ages 15 to 17)	37	27	36
Percent of teens who are high school dropouts (ages 16 to 19)	10	9	36
Percent of teens not attending school and not working (ages 16 to 19)	10	8	11
Percent of children living with parents who do not have full-time, year-round employment	30	24	−6
Percent of children in poverty (data reflect poverty in the previous year)	20	17	−12
Percent of families with children headed by a single parent	24	28	−3
Teenage violent death rate*	62.8	54.0	NA

*Deaths per 100,000 teenagers ages 15 to 19, 1985 and 1998.

Note: In the "percent change" column, directionality is reversed, so a negative number indicates deterioration and a positive number improvement.

NA: Not applicable.

Source: Annie E. Casey Foundation, 2009.

data books titled *Kids Count.* These data books provide an invaluable and up-to-date array of indicators of trends in child well-being in the United States. They are available for use by local officials, scholars, and those who work with the youth of their communities (Annie E. Casey Foundation, *Kids Count 2000,* www.aecf.org).

Although in many communities most children are healthy and well cared for, the *Kids Count* data indicate that increasing numbers of young people are experiencing poverty and near-poverty conditions. The late Senator Daniel Patrick Moynihan of New York, himself a skilled social scientist, noted that "a child in America is almost twice as likely to be poor as an adult. This is a condition that has never before existed in our history" (1988, p. 5).

Table 14.2 presents some indicators of well-being among American youth. Measures of children in poverty, rates of violent death, infant mortality, teenagers not in school, and births to single teenagers are among the most important ways of looking at the proportion of children and young people whose well-being is questionable and whose daily lives are fraught with risk of further trouble. Perhaps the most controversial of these indicators are those dealing with births to teenagers. In the United States and other urban industrial societies, there is a great deal of debate about increasing rates of teenage childbearing. Table 14.2 shows that, contrary to what many people believe, the rate of births to teenagers decreased from 1990 to 2000. Figure 14.7 demonstrates, however, that U.S. teenagers are much more likely to become pregnant unintentionally than teenagers in other developed countries. They are also more likely to want to become mothers (intentional pregnancies) and to have abortions.

It should be noted that some of the indicators presented in Table 14.2, such as the rate of teenage dropouts, improved between 1990 and 2005. Others, however, such as the percentage of low-birthweight babies, have worsened. These indicators are sensitive to changes in society. The low-birthweight indicator is especially sensitive to changes in the economic situation of low-income households (for example, unemployment and child poverty rates). However, the infant mortality rate—the proportion of infants who die in early infancy—is the most universally recognized measure of how well societies are doing in providing healthy environments for mothers and young children.

Infant Mortality: A Global Measure of Inequality Throughout the world, the infant mortality rate is considered the single most important indicator of the relative well-being of a nation's people (see Chapter 20). It measures the extent and efficacy of a nation's investment in the health of its people. Figure 14.8 shows the encouraging fact that global child mortality declined by almost one-quarter from 1990 to 2006. However, it also highlights the huge gap between industrialized nations and the rest of the world, especially the least developed nations, where child mortality is more than twenty times higher. Note also the extremely high rates of child mortality in many parts of Africa. These are not just a result of poverty, but

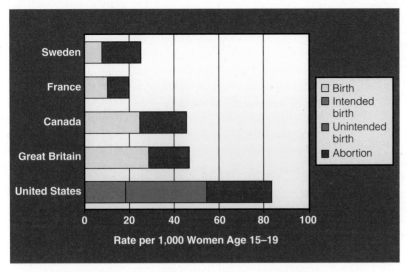

Figure 14.7 Intended and Unintended Pregnancies and Abortions among Teenagers in Selected Western Nations. Source: Data from Alan Guttmacher Institute, 2001.

Age and Inequality

In the United States today, the largest segment of the poor are children. This negative trend is worsened by racial and ethnic disparities in child poverty (see Figure 14.9). African American and Latino children are three times more likely than white children to live in poverty. Even without these disparities, the United States has one of the highest rates of child poverty among the world's highly developed nations. According to the social scientists who compile these data for the Annie E. Casey Foundation,

> [d]espite the enormous wealth in the United States, our child poverty rate is among the highest in the developed world. The gap in the child poverty rate between the United States and other developed countries is partly a product of differences in private-sector income, but differences in governmental efforts to alleviate child poverty greatly accentuate

the disparities. The lack of investment in our children will put us at a competitive disadvantage in the international marketplace of the 21st century.

also a consequence of severe political instability, corrupton, and civil strife, a subject to which we return in Chapter 19.

The United States, one of the world's most affluent nations, actually ranks on the lower end of the scale of infant mortality among the developed nations with which it is often compared. Much of the difference is attributable to the large number of underweight babies born in the United States. Birthweights below 5.5 pounds are a cause of infant death and are associated with higher rates of illness in later years as well as a higher incidence of neurological and developmental handicaps and poor academic performance (Partin & Palloni, 1995). In explaining these trends, sociologists Melissa Partin and Alberto Palloni (1995) point out that although poverty is always associated with low birthweight, a woman's access to prenatal care in her community combined with her decision not to have an abortion are important factors. As more poor minority women choose not to have abortions and find it more difficult to obtain adequate care during pregnancy, the risk of low birthweight increases. Neither side in the abortion debate wishes to see more abortions among poor women. The central implication of these findings is that the rate of low-birthweight babies would be reduced by improved access to health care for young mothers and more education about the need to decrease behaviors such as smoking during pregnancy.

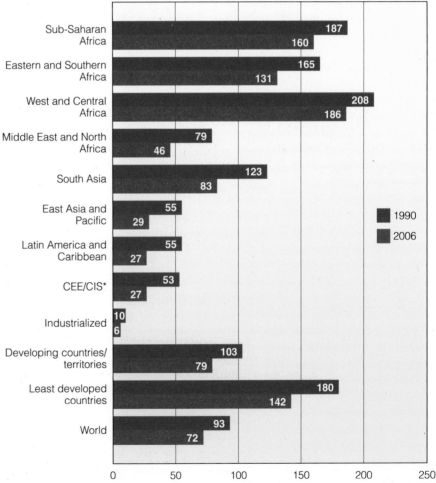

*Central and Eastern Europe and Commonwealth of Independent States.

Figure 14.8 Child Mortality Rates, 1990 and 2006. Source: From UNICEF 2010, *Fact of the Week 74*: The under-five child mortality rate in developing countries dropped to 74 in 2007, from 103 in 1990. http://www.unicef.org/factoftheweek/index_48701.html. Reprinted by permission.

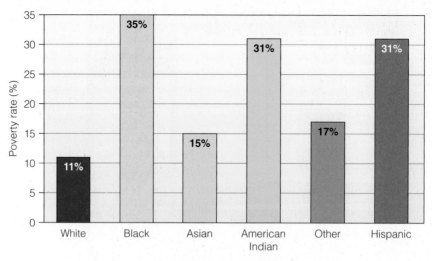

Figure 14.9 Child Poverty Rates by Race/Ethnicity, United States: 2008. Source: From "Who Are America's Poor Children? The Official Story," by Vanessa R. Wight, Michelle Chau, Yumiko Aratani. January 2010, p. 4. National Center for Children in Poverty. Reprinted by permission of the National Center for Children in Poverty, Mailman School of Public Health, Columbia University.

Sociology & Social Justice

AGE AND INEQUALITY At the turn of the twentieth century, the largest segment of the U.S. population living in poverty or near-poverty conditions was the elderly (Preston, 1984). This was also true throughout much of Europe. Until the development of modern systems of employee pensions, Social Security, and basic health insurance (both private insurance systems and public systems like Medicare), people older than age sixty were highly likely to be poor. Although poor and working-class elderly people without substantial wealth or savings might receive assistance from adult children and other family members, even those relatively fortunate individuals often felt deep resentment about having worked all their lives only to end up powerless and dependent. The Great Depression of the 1930s convinced the majority of Americans that something had to be done at the level of the entire society to address problems of poverty in old age. In a great democracy, it did not seem fair that so many had nothing to look forward to after a lifetime of work.

The Social Security Act of 1935 established federal old-age pensions based on employee and employer contributions, as well as unemployment insurance and other antipoverty programs. The results of these programs, and the Medicare system instituted in 1965, have been dramatic. Rates of poverty among the elderly, which were close to 40 percent in the early decades of the twentieth century, declined to about 28 percent in the early 1960s and then to about 14 percent after Medicare went into effect. At present, 12 percent of people age sixty-five and older in the United States live at or below the poverty line, compared to 13.2 percent of the population as a whole (U.S. Census Bureau, 2010).

Among elderly members of minority groups, however, the situation is not nearly so positive. Almost 24 percent of African Americans and about 21 percent of Hispanics older than age sixty-five are living in poverty (U.S. Census Bureau, 2010). These differences often result because elderly people who previously worked at lower-paying jobs, which contributed little or nothing to private pension plans, may be forced to live almost exclusively on their federal Social Security payments, which by themselves are not enough to raise them above the official poverty line. And elderly blacks and Hispanics are especially likely to have had jobs that did not provide pensions.

As more people join the ranks of the elderly, concern about their economic situation becomes an ever-more-powerful theme in American life. Many Americans in their early sixties are subject to mandatory retirement. Once they are forced into retirement, they are limited in what they can earn without experiencing cuts in their Social Security payments and decreases in Medicare support. Matilda White Riley, a leading authority on the sociology of aging, asserts that there is a growing gap between the number of skilled and energetic elderly people in the population and the availability of "meaningful opportunities in work, family, and leisure" (Riley, Kahn, & Foner, 1994). We return to this theme shortly.

Age and Disability As life expectancies have increased throughout the world, so have the numbers of people with major disabilities that seriously impair their ability to function effectively in their daily lives. In the United States, the average life expectancy for people age sixty is nineteen years for women and fifteen years

Robert Frost, shown here in his advanced years, was far more than a poet of rural America. Among other things, he promoted the idea that people need to remain passionate participants in the world around them throughout their lives.

for men. On the average, women who live that many years after their sixtieth birthday can expect to experience major disabilities for five of those years (four years for men). In Germany, another affluent industrial nation, life expectancies for women and men age sixty are twenty-two years and eighteen years, respectively, and the chances of experiencing major disabilities are the same as those in the United States. In Egypt, a developing nation, the comparable figures are thirteen years of life expectancy for women after age sixty and twelve years for men, but because of less advanced health care, the average Egyptian woman can expect to experience six years of disability (four years for men) (Binstock & George, 2006). People who live longer do not automatically suffer major disabilities. As they age and become more frail, however, the experience of living with disability inevitably affects higher proportions of women because of their greater life expectancy.

The problems of health and disability among the elderly are compounded by inequalities of socioeconomic status. The more affluent and educated people are as they enter their sixties, the healthier they tend to be. In their last years, however, the effects of socioeconomic status diminish (Lemme, 2006). People who have worked for decades tending machines in factories, often in polluted and stressful conditions, are more likely to have health problems and to have difficulty obtaining regular medical care than people who have worked in more physically and emotionally favorable conditions. Moreover, data on aging and health show that people from poor and working-class backgrounds are more likely to engage in

psychosocial risk behaviors like smoking and heavy drinking (ninety or more drinks per month). They also tend to experience a higher than average number of **lifetime negative experiences** (for example, the death of a child or spouse, divorce, physical assault), which cause long-term stress. Even when risk behaviors and sources of chronic stress are considered, however, quantitative studies show a strong positive relationship between higher socioeconomic status and better health in elderly people (Himes, 2001; House et al., 1994).

Loss of Social Functions with Age As people age, they experience more medical problems and disabilities—but does this mean that they must inevitably withdraw from social life? Can anyone deny the biological facts of aging and death? Isn't it only natural for the elderly to perform less important roles as their physical capacities diminish? The answer, of course, is that it is not obvious at all. The increasing participation of elderly widows in securities investment clubs throughout the nation is a good example of how elderly people may perform socially important roles even as their physical strength diminishes. It is not uncommon for members of these clubs to impress far younger stockbrokers with their skill as players in the volatile securities market.

---

psychosocial risk behaviors: Behaviors that are detrimental to health, such as smoking and heavy drinking.

lifetime negative experiences: Experiences that cause long-term stress, such as the death of a child or spouse.

Sexual behavior among the elderly is another area in which popular notions about the influence of biology on age roles have been disproved. In the late nineteenth century and well into the twentieth, it was widely believed that people lose their sexual desire and potency after middle age. In fact, however, several studies have shown that the image of the elderly as lacking sexual desire and the ability to enjoy sex is an ageist stereotype. For example, in a pioneering study, Eric Pfeiffer, Adrian Verwoerdt, and Glenn Davis (1972) gathered data from several samples of elderly people—including individuals as old as ninety—that showed conclusively that although sexual interest and activity tended to decline with age, sex remained an important aspect of the subjects' lives. The researchers also found, however, that elderly men are more interested in sex than women of the same age. The explanations for this difference are cultural rather than biological. Elderly men are in short supply. They tend to be married to women with whom they have had a long-standing relationship that includes an active sex life. If they are not married, they are in such great demand that they have less difficulty than women of the same age in finding a sexually compatible partner (Greeley, Michael, & Smith, 1990; Kornblum & Julian, 2007).

Caring for the Elderly: Whose Responsibility? As noted earlier, social scientists who study the situation of the elderly increasingly recognize that longer life spans need to be accompanied by new concepts of social roles in more advanced years. The well-known gerontologists Matilda White Riley and John W. Riley, Jr., believe there is a growing mismatch between the strengths and capacities of older people and their roles in society. As people live longer, they often find themselves living alone with few constructive roles that demand their time and attention. Neglect of the elderly further reduces their mental and physical strength. The Rileys believe that "increasing numbers of competent older people and diminishing role opportunities cannot long coexist" (1989, p. 28). In their view, there is an urgent need for many small-scale programs that create work and volunteer opportunities in which older people can use their skills and feel needed (Riley, Kahn, & Foner, 1994).

Shifts in age statuses can have dysfunctional effects for society as well. A research project of the Carnegie Foundation attempted to assess the impact of an aging society (that is, one in which the median age is increasing) on social roles. For example, whose responsibility is it to pay for long-term care of the elderly? A family may have to pay $65,000 per year or more to provide nursing home care for a disabled elderly mother or father. Should the sons and daughters take on this responsibility? Most of us would say yes, if we are able to do so. The Carnegie study found, however, that an increasing proportion of people say, in effect, "No sir, that's a public responsibility; let them go on Medicaid." In that case, however, we would force the elderly to sell all their assets in order to qualify for government aid because Medicaid is available only to people who have exhausted their own funds before applying to the government for assistance (Pifer & Bronte, 1986).

The question of responsibility for caring for the frail elderly is a controversial policy issue. The conservative position is that family members or close relatives should be required to pay for such care until funds are no longer available. The liberal position is that individuals should not have to be dependent on possibly reluctant family members; moreover, the high costs of long-term care for the elderly are an especially heavy burden for households with low incomes. The elderly themselves, who form a major segment of the U.S. voting population, are increasingly concerned about threats to their economic and social security, especially as represented by efforts to reduce the extent of health coverage under the Medicare system. They also are increasingly sensitive to any slights or discriminatory behavior based on their age.

Ageism The attitude known as **ageism** is similar to sexism (see Chapter 13). The term refers to an ideology that justifies prejudice or discrimination based on age. Ageism limits people's lives in many ways, both subtle and direct. It may label the young as incapable of learning. It labels the elderly as mentally incapable, asexual, or too frail to get around. But people of all ages increasingly reject these notions. In their everyday lives in families and communities, for example, older people continually struggle against the debilitating effects of ageism. "Just because I need help crossing the street doesn't mean I don't know where I'm going," an elderly woman said to community researcher Jennie Keith (1982, p. 198).

Gerontologist Robert Butler observes that "ageism allows the younger generation to see older people as different from themselves; thus they subtly cease to identify with their elders as human beings" (1989, p. 139). Butler, a physician and social scientist, has found that

ageism: An ideology that justifies prejudice and discrimination based on age.

as the proportion of older people in a society increases (as is occurring in the United States and Europe), the prevalence of ageism also increases. The younger generations, he notes, tend to fear that the older, increasingly frail and dependent generations will deprive them of opportunities for advancement. This fear is expressed in demands for reduced spending on Medicare and other programs that assist the elderly, as well as in the belief that the elderly are affluent and do not need social supports.

Stereotyping of the Elderly A form of age-ism that can be corrected through greater socio-logical knowledge about aging is the tendency to view all old people as frail, dependent, inca-pable, and asexual. But as this proportion of the population increases, older people will be found doing many things that are viewed as inappro-priate for their age. A woman in her eighties recently set the age record for solo sailing across the Atlantic. Just before reaching the age of 100, comedian and actor George Burns dedicated a book "to his past five doctors."

Just as other groups in the population reject labels that deprive them of their individuality and self-esteem, elderly people reject ageist labels that relegate them to dependent, unequal status. There will of course be large numbers of dependent older people, but there will also be large numbers of vigorous people of advanced age who need and demand encouragement and support in maintaining their vitality and in con-tinuing to live life to its fullest. Younger people who can apply their sociological imagination to these issues will be highly valued (Kornblum & Julian, 2007).

Elder Abuse

Cynthia, a 93-year-old-woman with diabetes, has lived in the same home for 60 years. Recently, her granddaughter Carol and her boyfriend Kyle moved in with her to provide caregiving assistance in exchange for rent-free housing. Carol convinced Cynthia to add her to her checking account to help her pay bills. Carol is also trying to convince Cynthia to sign over the deed to the house in order to allow Carol to make house payments and generally "run things more smoothly." Neither Carol nor her boyfriend has worked since moving in with Cynthia. Recently, Carol became physically abusive when she was intoxicated and pushed Cynthia down a short flight of stairs. Cynthia will not contact the authorities because she is embarrassed by the situation. She does not want to have her granddaughter arrested. A teller at Cynthia's bank noticed the irregular account activity and made a report to APS.

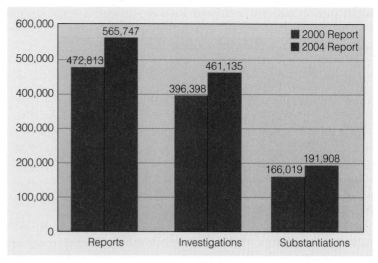

Figure 14.10 Comparison of Total APS Reports, 2004 and 2000 Surveys **(All Ages).** Note: Reports of self-neglect are included in the totals. Source: From *The 2004 Survey of State Adult Protective Services*. National Center on Elder Abuse: Administration on Aging. February 2006, p. 13., Fig. 1. Reprinted by permission.

This is a typical case of elder abuse reported to the national hotline of Adult Protective Services (APS). Elder abuse is defined as the infliction of physical or psychological harm or knowing deprivation of care and services necessary to meet the physical and psychological needs of elderly people. Neglect is refusal to meet one's obligations for the care and welfare of an elderly person, including financial obligations. Self-neglect is defined as inability, due to physical or psychological disability, to perform essential self-care tasks. In Figure 14.10, we see that between 2000 and 2004, the number of fully substantiated reports of elder abuse of all kinds (including self-neglect, which accounts for about 25 percent of all cases) increased by about 24,000 cases. As the U.S. population continues to age rapidly, it is cer-tain that rates of elder abuse will continue to rise dramatically. (This subject is covered in far more detail in social problems and gerontology courses.)

THEORIES OF AGING

Although they have been less far-reaching than the women's move-ment, social movements among the elderly—led by organizations like the Gray Panthers and AARP—have had a significant impact on American society. And as the population continues to age, we can expect to see more evidence of the growing power of the elderly. Consider, for instance, how changes in the consciousness and political activity of elderly people themselves are altering the way sociologists formulate questions about old age.

Disengagement Theory

Until the social movements of the 1960s prompted the elderly to form movements to oppose ageism and fight for their rights as citizens, the most popular social-scientific view of aging was *disengagement theory*. Numerous empirical studies had shown that old people gradually dis-engage from involvement in the lives of younger people and from eco-nomic and political roles that require responsibility and leadership. In a well-known study of aging people in Kansas City, Elaine Cumming and William Henry (1971) presented evidence that, as people grow older, they often gradually withdraw from their earlier roles, and that this process is a mutual one rather than a result of rejection or discrimination by younger people. From a functionalist viewpoint, disengagement is a positive process both for society as a whole (because it opens up roles for younger people) and for the elderly themselves (because it frees them from stressful roles in their waning years).

Activity Theory

The trouble with disengagement theory is that, on one hand, it appears to excuse policymakers' lack of interest in the elderly and, on the other, it is only a partial explanation of what occurs in the social lives of elderly people. An alternative view of the elderly is that they need to be reengaged in new activities. Known as *activity theory*, this view states that the elderly suffer a sense of loneliness and loss when they give up their former roles. They need activities that will serve as outlets for their creativity and energy (Binstock & George, 2006).

Theories of aging are not simply abstract ideas that are taught in schools and universities. The disengagement and activity theories lead to different approaches that often impose definitions of appropriate behavior on people who do not wish to conform to those definitions. Today, gerontologists and elderly activists tend to reject both theories. They see older people demanding opportunities to lead their lives in a variety of ways based on individual habits and preferences developed earlier in life. Elderly people themselves express doubt that activity alone results in successful adjustment to aging or happiness in old age. For example, in her study of a French retirement community, Jennie Keith (1982) made this observation:

> The residents . . . seem to offer support to the gerontologists who have tried to mediate the extreme positions, disengagement vs. activity, by introducing the idea of styles of aging. . . . Some people are happy when they are very active, others are happy when they are relatively inactive. From this point of view, life-long patterns of social participation explain the kinds and levels of activity that are satisfying to different individuals. (p. 59)

In sum, for the elderly as well as for women, social scientists increasingly tend to emphasize individual needs and capabilities. The social movements for gender and age equality also advance the needs of individuals, but in a collective manner, by asserting the needs of entire populations and rejecting preconceived notions of what is best for all women, all youth, all men, or all the elderly.

DEATH AND DYING

As members of the large baby boom generation move through the adult years and approach middle age, their parents enter the ranks of the elderly. This demographic change has led to increased concern about the quality of life of the elderly and about death and the dying process. The questions of how best to prepare to die and whether one can have a "good death" might have seemed strange to Americans of an earlier time, when so many people died as a result of sudden, acute illnesses like pneumonia or tragic accidents in mines and mills. Today, almost two-thirds of the people who die each year experience long struggles with chronic illnesses like cancer and heart disease and are often "sequestered" from the rest of society in hospital intensive-care units. The proportion of deaths occurring in hospitals, nursing homes, and other institutions has risen rapidly since the 1930s, and now exceeds 80 percent of deaths. Sociologists who study death and dying believe that this change is a major reason for poll results showing that most Americans support the right of patients to receive a lethal overdose from their doctors (Horgan, 1997). It is also a reason why so many health care experts and activists are seeking alternatives to hospital care for dying loved ones (Perls & Silver, 1999).

People who are dying often experience a period between life and death known as the *living–dying interval* (Kübler-Ross, 1969). For many dying people and their loved ones, this is an extremely important time. During this time, there may be reconciliations, clarifications, expressions of love and understanding, and, under the best circumstances, a sense of closure and repose for the dying person. But most people die in hospitals. Nurses, doctors, and orderlies may intrude on these emotional scenes to minister to the patient. The hospital is dedicated to prolonging life, even in a dying patient. As the Rileys express it: "The hospital is geared for treatment and cure, it functions according to standards of efficiency and bureaucratic rules, its environment is sterile and unwelcoming to those who would visit patients *in extremis*" (1989, p. 27).

The Hospice Movement

In recent decades, first in Great Britain and later in the United States, a social movement has emerged that is known as the hospice

These elderly men are playing a competitive game of chess in Washington Square Park in Manhattan's Greenwich Village neighborhood. Recent medical research supports sociological theories that emphasize the need to engage in stimulating activities as we age, and that activities that challenge the brain are associated with lower rates of Alzheimer's disease and other forms of senile dementia (*Neurology*, 2002).

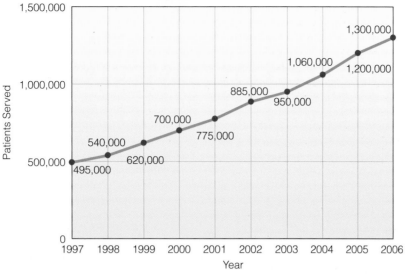

Figure 14.11 **Total Patients Served by Hospice, 1997–2006.** Source: From *NHPCO Facts and Figures: Hospice Care in America* November 2007 Edition, Figure 1, p. 1. © NHPCO November 2007. Reprinted by permission.

movement. The term *hospice* refers to a place, a set of services, or both; increasingly, it refers to a service that can be brought into the dying person's home. The purpose of the hospice is to offer palliative care, which makes the patient as comfortable as possible so that all concerned can use the living–dying interval effectively and humanely. In a hospice, the patient and his or her surrounding social group are viewed as the relevant social unit. Medical personnel and other professionals, such as social workers, are available to help everyone involved.

Figures 14.11 and 14.12 show that the hospice movement has had an astounding impact on how families actually experience the death of loved ones. Approximately 36 percent of all deaths in the United States occur in patients who are in the care of a hospice program, and the total number of patients served in hospice mode,

either in hospice facilities or hospice home care, exceeds 1.3 million. Although about 50 percent of nonhospice patients die in acute care hospitals, 74 of patients who are in hospice care die at home in familiar surroundings (NHPCO, 2007).

Dying is never an easy experience, and the hospice movement cannot make it so. Nevertheless, the hospice movement is providing opportunities for new definitions of the last social role in the life course—the dying role—as well as the roles of those who will remain behind to continue their own life course.

Thinking Critically

How do you think dying in a regular hospital room differs from dying in a hospice setting? In coming up with your answer, remember that in a hospice everyone knows that the patient is in the process of dying.

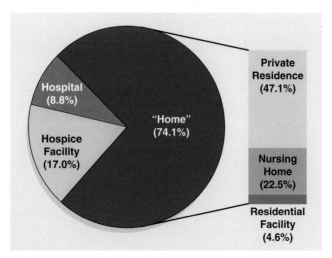

Figure 14.12 **Location of Death.** Source: From *NHPCO Facts and Figures: Hospice Care in America* November 2007 Edition, Figure 3, p. 2. © NHPCO November 2007. Reprinted by permission.

visual sociology

After Ninety Imogen Cunningham's life spanned almost a century, and she spent it making loving photographic portraits of people in the cities where she lived. Born in Scranton, Pennsylvania, at the end of the nineteenth century, she learned photography in its early days, when camera equipment was extremely cumbersome and unpredictable. With the encouragement of her artistic and thoughtful father, Cunningham built her own darkroom and learned every aspect of the fledgling art of photographic portraiture.

In her adult years, Cunningham lived and worked mainly in Seattle and San Francisco, but her career gave her opportunities to take photographs of people all over the world. During the period before World War II, she was one of the most celebrated photographers working for the innovative magazine, *Vanity Fair*. The subjects of her portraits range from world-famous artists and scientists to unknown but singular characters such as the tattooed woman who spent her life in the carnival.

©1975, 2010 The Imogen Cunningham Trust www.ImogenCunningham.com

She said to me, "When you come here nobody knows where you are."

©1975, 2010 The Imogen Cunningham Trust www.ImogenCunningham.com

This woman had been at the carnival all her life, but I found her in the hospital. It looks like lace, doesn't it?

©1975, 2010 The Imogen Cunningham Trust www.ImogenCunningham.com

She is a distinguished radiobiologist who asked me to photograph her. I wasn't taking on commissions anymore, but I did it because she didn't care if she looked old and she didn't hate her face.

At the end of her life, Cunningham became interested in a sociological question: What ideas and activities occupy the minds of very old people? The photos shown here are a representative sample from a collection of stirring portraits she made of people who were, as she put it, "alive and kicking." She photographed many people in their nineties, like Old Norton in his shack and the nursing home resident feeling forgotten and forlorn, for whom the main challenge of life was merely to live through each day. She also photographed people who, like Cunningham herself, had the good fortune to have a passion for work and activity and felt little concern about their physical deterioration because they were absorbed in their creative efforts.

Cunningham died at the age of ninety-three, just as the volume containing these photographs was about to be published. Although she missed the chance to celebrate the appearance of her book about the very old, she herself lived long enough to have the pleasure (denied to many artists) of seeing her work honored during her lifetime. Even from the few photos presented here, we can readily see that her work captures the mystery of the human spirit and demonstrates that extreme age does not extinguish the love of life.

©1975, 2010 The Imogen Cunningham Trust www.ImogenCunningham.com

Old Norton lives by himself in his shack.

SUMMARY

How do sociologists visualize age stratification?

- In many societies, age determines a great deal about the opportunities open to a person and what kind of life that person leads.

- The study of aging and the elderly is termed *gerontology*.

- All societies channel people into *age grades*, or sets of statuses and roles based on age. The transitions among these age grades create a *life course* and are often marked by ceremonies known as *rites of passage*.

- *Age cohorts* are people of about the same age who are passing through life's stages together.

- The baby boom cohorts, which were produced by rapid increases in the birthrate from about 1945 through the early 1960s, have profoundly influenced American society.

- A sizable proportion of the children of the baby boom generation, the "baby boom echo," are members of minority groups.

- By *life expectancy,* we mean the average number of years a member of a given population can expect to live beyond his or her present age.

- As life expectancy in a population increases, the proportion of the population that is dependent on the adult cohorts also increases.

What patterns of inequality are associated with age stratification?

- In urban industrial societies, there are distinct patterns of stratification in which age defines the roles one plays and the rewards one can expect.

- The forces of social change unleashed by colonialism, industrialization, urbanization, and population growth have tended to disrupt the formal age grade systems of smaller, more isolated societies.

- Social definitions of childhood differ immensely throughout the industrial world as well as between modern and traditional societies. As a result of increasing incomes and the passage of child labor laws, children became economically "worthless" but emotionally "priceless." However, there is a growing gap between "priceless" children and children who bear a heavy burden of poverty and deprivation.

- A century ago, the largest segment of the U.S. population living in poverty or near-poverty conditions was the elderly. As a result of programs such as Social Security and Medicare, rates of poverty among the elderly have decreased dramatically. However, the situation is not nearly so positive for elderly members of minority groups.

- As people age, they experience more medical problems and disabilities, but this does not mean that they must inevitably withdraw from social life. Social scientists who study the situation of the elderly

point out that longer life spans need to be accompanied by new concepts of social roles in more advanced years.

- *Ageism* is an ideology that justifies prejudice or discrimination based on age. As the proportion of older people in a society increases, as is occurring in the United States and Europe, the prevalence of ageism also increases.

If everyone ages, becomes old, and eventually dies, why do sociologists need theories of aging?

- As the population as a whole has aged, the impact of the elderly on American society has increased. This is changing the way sociologists view old age. Before the 1970s, the most popular social-scientific view of aging was disengagement theory, the belief that people gradually disengage from their earlier roles as they grow older. An alternative view of the elderly, known as activity theory, states that the elderly need activities that will serve as outlets for their creativity and energy. Today, gerontologists tend to reject both of these theories, seeing older people demanding opportunities to lead their lives in a variety of ways based on individual habits and preferences.

What do social norms and institutions have to do with death and dying?

- The growing proportion of elderly people in the population has led to increased concern about the quality of life of the elderly and about death and the dying process.

- One outcome of this concern is the hospice movement, which attempts to provide dying people and their loved ones with a comfortable, dignified alternative to hospital death.

The Kornblum Companion Website

www.cengagebrain.com

Supplement your review of this chapter by going to the companion website to take one of the tutorial quizzes, use the flash cards to master key terms, and check out the many other study aids you'll find there. You'll also find special features, such as GSS data and Web links that will put data and resources at your fingertips to help you with that special project or to do some research on your own.

15

Families

FOCUS QUESTIONS

How has the family been changing as an institution of modern societies?

How do sociological perspectives explain the causes and consequences of changes in family structure and family relations?

What factors enter into the choice of a mate and the decision to marry?

What factors have contributed to the resilience of the black family?

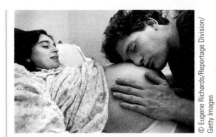

These two photos of "the birth of a family" convey much of the joy of family intimacy: the young couple just before she gives birth and the new father thrilled by his first experience with their child as the mother lays the baby on his chest. In the first photo, we see a couple about to become a family. They will quickly take on a host of new roles with new demands and gratifications that will challenge their love in ways that they can hardly anticipate. In the second photo, our eyes linger on the baby's tiny, perfect fingers and toes and the rapt gaze of the new mother. Is there the slightest hint in the man's somewhat rigid fingers that the crying newborn's vulnerability may tinge his joy with feelings of awkwardness? We cannot know this for sure, but the early days and months of parenthood are filled with many emotions—and love, awkwardness, joy, anxiety, and fatigue are prominent among them.

If we examine these photos from a historical perspective, we will surmise that this is a contemporary couple. There are few clues in the scenes themselves. Her flannel nightgown could be from any decade of the last two centuries. The plain gold band on his left hand is a timeless symbol of matrimony. What reveals them as people of our time is the photographer's choice of them as subjects. No doubt couples in earlier decades behaved in exactly similar ways, but they were not often encouraged to display their joy and intimacy in public. In the early twentieth century, it was more common for photographers to capture the gender-specific aspects of childbirth. The woman's joy at receiving the baby, often with adoring grandmothers or nurses in the background, expressed the heavily gender-specific quality of the childbirth scene. The man was portrayed in a stereotypical fashion, pacing in the hospital corridor, perhaps with one cigarette in hand and another still burning in a nearby ashtray, symbols of his ineffectual nervousness. Other photos of early family moments would focus on the father's first glimpse of the baby through the windows of the hospital nursery. Contemporary approaches to childbirth encourage more direct involvement of the father in the birthing process. They advocate more immediate contact between the new parents and the infant. For thousands of couples, going through the experience of childbirth together provides some of life's most thrilling memories.

As we gaze at these most personal glimpses into family life, we may also be reminded of the mystery of human love and attraction. We come back to these mysteries in the Visual Sociology boxes in this chapter, which capture people in a variety of intimate embraces, all of which convey the power of love and intimacy. Between this intimate chapter opening and the return to intimacy at the chapter's end, however, we will be dealing with many of the challenges to love and intimacy that arise in today's world. The risk in taking this sociological journey into the institutions of family and marriage is that we may forget to what extent love—between a couple, or among siblings, or between parents and children—is a force that can overcome many of the shocks and strains of modern life. On the other hand, we will learn that there are ways in which society makes love of various kinds possible and even inevitable.

THE NATURE OF FAMILIES

We continually hear warnings about the "death of the family" in modern societies. As early as 1934, sociologists like William Fielding Ogburn were writing about the decline of the family. "Prior to modern times," according to Ogburn, "the power and prestige of the family was due to . . . functions it performed. . . . The dilemma of the modern family is caused by the loss of many of these functions in recent times" (Ogburn & Nimkoff, 1934, p. 139). More recent commentators have pointed to the so-called sexual revolution as a threat to this basic social institution. Others insist that, even in the face of trends toward single parenthood, cohabitation without marriage, divorce and remarriage, and smaller family size, the family is "here to stay." Recent surveys of how Americans view the future of marriage and the family confirm this conclusion. The overwhelming majority of Americans value marriage, children, and family. Yet they also believe that their commitments to marriage and family are more voluntary and less obligatory than was the case in previous generations. The same changes are true in all other urban industrial nations. The family as we know it will not disappear, but it will take on new forms and functions, as we will see throughout this chapter (Gerson, 2009; Thornton & Young-DeMarco, 2001).

In this chapter, we will see that there are so many variations in family form, even within a single society, that it has become increasingly difficult to speak of "the family" as a single set of statuses, roles, and norms. Nevertheless, it remains true that in all known societies almost everyone is socialized within a network of family rights and obligations known as *family role relations* (Goode, 1964). In the first section of this chapter, therefore, we examine the family as an institution and show how family role relations have changed in the twentieth century. The second section focuses on how the basic sociological perspectives explain changes in family roles and functions and in the interactions between parents and children. In the third section, we look at the dynamics of family formation—especially mate selection, marriage, divorce, and remarriage.

The Family as an Institution

There is a big difference between a particular family and the family as a social institution. Your family, for example, has a particular structure of statuses (mother, father, sister, brother, grandmother, and so on), depending on who and where its members are. Your immediate family is also part of an extended kinship system (aunts, uncles, cousins, grandparents, and so on) that also varies according to how many branches you and your immediate family actually keep track of. The family as a social institution, however, comprises a set of statuses, roles, norms, and values devoted to achieving important social goals. Those goals include the social control of reproduction, the socialization of new generations, and the "social placement" of children in the institutions of the larger society (colleges, business firms, and so forth).

Social Placement "Social placement" refers to the ways in which parents and relatives work to ensure that their children will achieve the same

or higher social-class positions as their parents. When teachers comment about young children "coming from good families" or when others comment on how "well brought up" a child appears to be, they are referring to the ways in which the parents have developed the child's *cultural capital,* all the knowledge and norms of behavior valued by higher-class people. Reading books to children, listening to classical music with them, teaching them how to be polite, being concerned with how they dress, spending time visiting their schools, pushing them to succeed in school—these are all ways in which parents help place their children socially by developing their cultural capital (Bourdieu, 1993; Coleman, 1994). Extensive research on families and social mobility documents how family background and behavior influence the way children and adolescents are perceived and treated in school. For example, a recent study of primary-school teachers in England found that teachers expect higher achievement from children whose parents have good spoken language skills and seem to read a lot to their children. These expectations, even at a very early age, can help determine how much children enjoy school and how well they learn (Feiler & Webster, 1999).

The Family and Institutional Differentiation These are not the only goals that the family is expected to fulfill. In many societies, especially tribal or peasant societies, the family performs almost all the functions necessary to meet the basic needs of its members. Those needs are listed in Figure 15.1, which compares less differentiated *gemeinschaft* or village-type societies with more differentiated *gesellschaft* societies like that of the United States. It shows that most of the functions that the family traditionally performed are now performed either partly or wholly by other social institutions that are specially adapted to performing those functions.

Figure 15.1 only begins to illustrate the institutional complexity of modern societies. Not only do modern societies have many more separate institutions, but basic social functions are often divided among several institutions. Thus, the institutions that meet protective and social-control needs include governments at all levels, the military, the judiciary, and the police. Replacement needs are met by the family through mating and reproduction. In addition to the family, religious institutions, education, and other cultural institutions meet the need to socialize new generations.

The family's ability to meet the goals society expects it to meet is often complicated by rapid social change. For example, many families in industrial communities throughout the

SOCIAL FUNCTION	TYPE OF SOCIETY	
	Less Differentiated (gemeinschaft)	More Differentiated (gesellschaft)
Communication among members: first through language alone and then through specialized institutions devoted to communication	Family, kin networks	Mass media
Production of goods and services: from the basic items required for survival to items designed to satisfy new, more diverse needs created by increasing affluence	Family groups	Economic institutions
Distribution of goods and services: within societies at first, and later, with the rise of trade and markets, between societies as well	Extended family, local markets	Markets, transportation institutions
Protection and defense: including protection from the elements and predators and extending to defense against human enemies	Family, village, tribe	Armed forces, police, insurance agencies, health care institutions
Replacement of members: both the biological replacement of the deceased and the socialization of newcomers to the society	Family	Family, schools, religious institutions
Social placement: to establish roles and status of children in society as they mature by developing their cultural capital	Family	Family, schools, religious institutions, civic voluntary associations
Control of members: to ensure that the society's institutions continue to function and that conflict is reduced or eliminated	Family	Family, religious institutions, government in all forms

Figure 15.1 **The Family: Institutional Differentiation.** Source: Adapted from Lenski & Lenski, *Human Societies*, 7/e. © 1982 McGraw-Hill Companies. Reprinted by permission of Gerhard Lenski.

United States were quite good at preparing their children for blue-collar manufacturing jobs, but those jobs have been eliminated as a result of automation and the globalization of the economy. For many families, therefore, it is no longer clear how they can equip their offspring to compete in the new job market.

When there are major changes in other institutions in a society, such as economic institutions, families must adapt to those changes. Similarly, when the family changes, other institutions will be affected. As an example, consider the impact of the "sexual revolution" that began in the 1920s. During that decade, often called the Roaring Twenties, young adults began to upset the moral code that had regulated the behavior of couples for generations, at least among people who considered themselves "respectable." According to this traditional set of norms, women were the guardians of morality. Men were more likely to give in to sexual desire, "but girls of respectable families were supposed to have no such temptations" (Allen, 1931, p. 74). After World War I, however, "respectable" young women began to reject the **double standard**, by which they had to adhere to a different and more restrictive moral code from that applied to men. They began smoking cigarettes, wearing short skirts, and drinking bootleg gin in automobiles or at "petting parties." These changes in the behavior of couples produced far-reaching changes in other institutions. Women demanded more education and more opportunities to earn income. This in turn led to demands for new institutions such as coeducational colleges and integrated workplaces, as well as greater access by women to existing institutions such as medicine, law, and science. These changes in women's expectations and behavior, along with the increasing educational and occupational mobility of men after the world wars, contributed immensely to changes in traditional family roles—especially the weakening of the role of the father as the central and most powerful authority figure in the family. The consequences of these fundamental changes in family roles are still felt today.

Changes in the institutions in one area of social life can place tremendous pressure on those in other sectors. This is especially true in the case of the family, as will be evident throughout this chapter. Before we examine

double standard: The belief that women must adhere to a different and more restrictive moral code than that applied to men.

These photos show a conventional nuclear family, with a father, mother, and two children, and an extended family, with three generations of family members, including aunts, uncles, and cousins. Although these may once have been the most common forms of traditional and modern family structures, the numbers of single-parent families and families that share custody of the children after divorce are increasing.

© Paul Conklin/PhotoEdit

© Stephanie Maze/Woodfin Camp & Associates

how families have changed in the twentieth century, however, it is necessary to define some terms that are often used in discussing the family.

Thinking Critically

What are the threats to the family? Do changes in the economy threaten the strength of families? What about marriage between people of the same sex? What is it about love and marriage among same-sex

couples that is so threatening that federal law and many states explicitly define marriage as occurring between a man and a woman, even though thousands of couples who do not fit this definition consider themselves married?

Defining the Family

The family·is a central institution in all human societies, although it may take many different forms (Bianchi & Casper, 2000; Cherlin, 1996). A **family** is a group of people related by blood, marriage, or adoption. Blood relations are often called *consanguineous attachments* (from the Latin *sanguis,* meaning "blood"). Relations between adult people living together according to the norms of marriage or other intimate relationships are called *conjugal relations.* The role relations among people who consider themselves to be related in these ways are termed **kinship.**

The familiar kinship terms—*father, mother, brother, sister, grandfather, grandmother, uncle, aunt, niece, nephew, cousin*—refer to specific sets of role relations that may vary greatly from one culture to another. In many African societies, for example, "mother's brother" is someone to whom the male child becomes closer than he does to his father and from whom he receives more day-to-day socialization than he may from his father. It must be noted that biological or "blood" ties are not necessarily stronger than ties of adoption. Adopted children are usually loved with the same intensity as children raised by their biological parents. And many family units in the United States and other societies include "fictive kin"—people who are so close to members of the family that they are considered kin despite the absence of blood ties (Stack, 1974). Finally, neither blood ties nor marriage nor adoption adequately describes the increasingly common relationship between unmarried people who consider themselves a couple or a family (Medalie & Cole-Kelly, 2002).

The smallest unit of family structure is the **nuclear family.** This term is usually used to refer to a couple and their children, if any. Nowadays one frequently hears the phrase "the traditional nuclear family" used to refer to a married mother and father and their children living together. However, as we will see throughout this chapter, there is no longer a "typical" nuclear family structure. Increasingly, therefore, sociologists use the term *nuclear family* to refer to two or more people related by consanguineous or conjugal ties or by adoption who share a household; it does not require that both husband and wife be present in the household or that there

family: A group of people related by blood, marriage, or adoption.

kinship: The role relations among people who consider themselves to be related by blood, marriage, or adoption.

nuclear family: Two or more people related by blood, marriage, or adoption who share a household.

Kinship Diagrams Kinship diagrams are used in both anthropology and sociology to denote lines of descent among people who are related by blood (children and their parents and siblings) and by marriage. Kinship terms are often confusing because families, especially large ones, can be rather complex social structures. It may help to devote a little time to the kinship diagram shown here.

To understand the chart, one must know the meanings of the symbols used; these are explained in the key to the chart. "Ego" is the person who is taken as the point of reference. You can readily see that Ego has both a family of orientation and

a family of procreation. So does Ego's spouse. Ego's parents become the in-laws of the spouse, and the spouse's parents are Ego's in-laws.

Kinship diagrams like this one provide a visual model of family statuses extending over more than one generation. They are a very useful way of analyzing the social structure of families. We can quickly see the generations in the family, from Ego's grandparents to Ego's children. We can also compare the nuclear and extended families of Ego's parents and cousins.

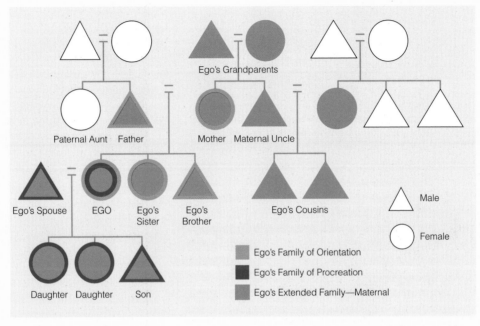

This diagram shows the family relationships of a hypothetical individual, Ego. Ego (a female in this example) was born and socialized in what sociologists call the family of orientation. Ego formed a family of procreation through marriage or cohabitation.

be any specific set of role relations among the members of the household (Benokraitis, 2007).

The nuclear family in which a person is born and socialized is termed the **family of orientation**. The nuclear family a person forms through marriage or cohabitation is known as the **family of procreation**. The relationship between the two is shown in the Research Methods box above.

By studying the chart in the box, we can identify Ego's family of orientation and family of procreation. Like most people, however, Ego also has an **extended family** that includes all the nuclear families of Ego's blood relatives—that is, all of Ego's uncles, aunts, cousins, and grandparents. Ego's spouse also has an extended family, which is not indicated in the chart, nor is it defined as part of Ego's extended family. But relationships with the spouse's extended family are likely to occupy plenty of Ego's time. Indeed, as the chart shows, the marriage bond brings together far more than two individuals.

Most married couples have extensive networks of kin to which they must relate in many ways throughout life.

Variations in Family Structure

For the past several years, sociologists have been demonstrating that the traditional household consisting of two parents and their children is no longer the typical American family. This is not a value judgment but a matter of demographic fact that needs to be carefully assessed. Figure 15.2 shows that the so-called traditional family of mother, father, and children accounts

family of orientation: The nuclear family in which a person is born and raised.

family of procreation: The nuclear family a person forms through marriage or cohabitation.

extended family: An individual's nuclear family plus the nuclear families of his or her blood relatives.

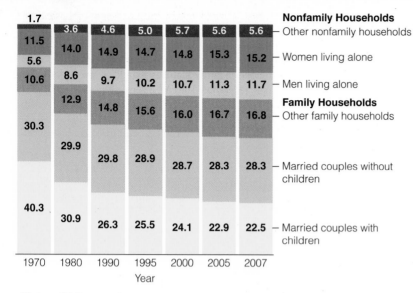

Figure 15.2 Households by Type: Selected Years, 1970–2000 (percent distribution). Source: U.S. Census Bureau.

for less than one-fourth of all households, down from over 40 percent in 1970. From 1970 to 2007, there was also a significant increase in the number of men and women who live alone, outside family households. These are primarily but not exclusively elderly people. Among family households, there has been a significant increase in the proportion of married couples without children (although their children may be grown and living in families of their own). "Other family households" refers primarily to single-parent families with no spouse present but with other relatives, who may be children. This important category increased from 10.6 percent of all households in 1970 to 16.8 percent in 2007.

Do these changes in the composition of American households mean that children are no longer typically being raised in two-parent families? Not at all. Figure 15.3 shows trends in the living arrangements of children in the United States. From this perspective, the two-parent family remains the dominant form, but in 2007, it accounted for 73 percent of families with children, down from 93 percent in the 1950s. The difference

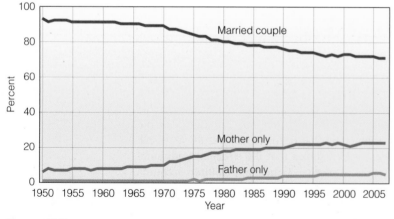

Figure 15.3 Family households with children under 18 by type: 1950 to 2007.
Note: In 2007, it is possible to identify unmarried householders who have a child and live with the second parent of the child. This represents 3.8 percent of family households with children under 18. These households are shown in either mother only or father only in this graph.
Source: U.S. Census Bureau.

is accounted for by the rapid increase since the 1950s of single-mother and, more recently, single-father families.

These trends point to a growing diversity of family types, including far more people living alone and far more women raising children alone. Also, more households are composed of unrelated single people who live together not just because they are friends but because only by doing so can they afford to live away from their families of orientation. Thus, when they discuss family norms and roles, sociologists must be careful not to represent the traditional nuclear family, or even the married couple, as typical. As an institution, the contemporary family comprises a far greater array of household types than ever before.

Sociologists are divided on the meaning and implications of these facts. For some, the trend toward greater diversity of family forms means that the family as an institution is adapting to change. For others, the decline of the family composed of two married parents living with their biological children poses a serious threat to the well-being of children. We will return to this debate over the meaning (and explanation) of changes in family form and function, but first we consider the impact of economic change on the family.

Families and the Economy

The economy exerts extremely strong influences on the family. When family breadwinners are promoted and income increases, there may be more vacations, better food, or a greater chance that the children will be able to go to college, to name only a few possibilities. When income decreases, perhaps because of recession and layoffs or because one of the breadwinners leaves the family, parents and children usually must lower their expectations and spend less. Economic mobility, upward or downward, often means that the family will move to another neighborhood or community, perhaps to another part of the nation, and this uprooting can be extremely difficult for all concerned— especially adolescents, for whom peer relations count so heavily. These examples hardly begin to account for the influence of work and the economy on family life. Think, for example, about how your own family's daily schedules are based on the need to cope with the sometimes conflicting demands of work and family activities.

In the United States and most urban industrial nations, increasing numbers of married women with children are in the labor force

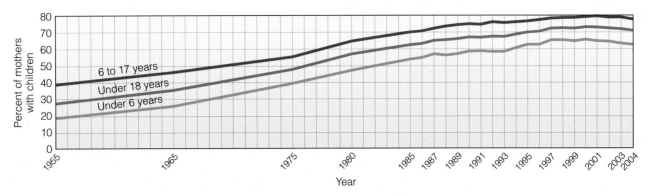

Figure 15.4 Mothers in the Labor Force, 1955-2004.

(Moen, 2001). This striking social change has occurred because families need the income women earn and, in a smaller proportion of families, because women with high-level skills wish to pursue professional careers. Whatever the specific reasons, the trends over the past century have all been in the same direction: toward more families in which both parents work. Figure 15.4 shows that the percentage of women in the U.S. labor force has grown steadily over the last sixty years or more. The figure also reveals that even in the 1950s, which many people think of as a time when women worked in the home as "housewives" and men went out to work at paying jobs, the reality did not match that stereotype. Although women with young children did tend to remain in the home, by the time their children were in school, almost 40 percent of women were working outside the home. By 2004, the vast majority—77.5 percent—of women with school-age children were in the labor force.

Figure 15.5 shows that dual earners are so important in allowing contemporary families to make ends meet that only slightly less than one-fifth (19.6 percent) of all married-couple families conform to the older norm of the traditional family in which only the husband is in the paid labor force.

In families in which both parents work, time pressures on the family increase. As family members work longer hours, role conflicts and stresses within the family also increase. Most couples argue most often about money, but allocation of time within the family is another source of constant tension.

Families and Poverty Economic conditions can produce drastic changes in the number of families that experience deprivation and poverty. The recession following the end of the technology "bubble" and in the aftermath of

the 2001 terrorist attacks reduced working hours and income for thousands of low-wage workers and for thousands of others in the middle classes.

In the mid-1990s, the United States ended the system of Aid to Families with Dependent Children, which allowed single mothers to collect benefits for their children, and replaced it with a variety of state programs funded by the federal government under the heading of Temporary Assistance for Needy Families (TANF). These programs encourage or require single parents to enter the labor force. During the booming economy of the 1990s, most states were able to drastically reduce their welfare rolls and help hundreds of thousands of former welfare recipients enter the labor force. Most of the jobs they found were low-level jobs in the service sector, but overall, the experience of earning incomes and holding steady jobs has been positive. In some states, women have been encouraged to marry the men who fathered their children, men who were banned from official residence in households receiving benefits under the old system (Harden, 2001). Unfortunately, data on labor force participation in recent years have consistently shown that men and women in families with school-age children are also those who are expected to work the longest hours, which adds additional stress to family relations for many couples (Golden, 2010).

The deep recession that gripped the United States and most of the world after the financial collapse in real estate and financial markets in

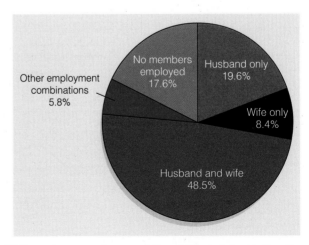

Figure 15.5 Employment Characteristics of Married-Couple Families, 2009.
Source: Bureau of Labor Statistics.

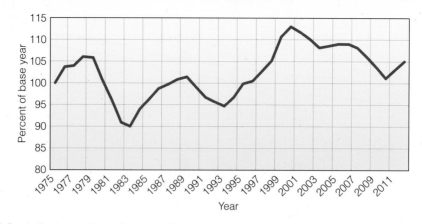

Figure 15.6 Family Economic Well-Being Index, 1975–2008, with Projections for 2009–2012.

Source: From *The 2010 Child Well-being Index*, by Kenneth C. Land and the Foundation for Child Development. June 1, 2010. Figure 3, p. 13. Reprinted by permission of the Foundation for Child Development.

2008 reduced working hours and income for thousands of low-wage workers and for thousands of others in the middle classes. From 2008 to 2009, husbands in U.S. families experienced an 8.5 percent increase in unemployment, some of which was compensated for by an increase in wives working. In families in which the wife became unemployed, however, husbands were still less likely (−2.6 percent) to have been employed in the same period. These figures only begin to describe the effects of recession on families.

Sociologist Kenneth Land has developed a useful index of family economic well-being that is comprised of four key indicators: the poverty rate for all families with children; the proportion of children living in families with at least one parent with full-time employment year-round; median annual income for all families with children; and the proportion of children living in families with some form of health insurance. The starting point is set at 100 for the year 1975, to which subsequent years can be compared.

Figure 15.6 traces changes in this index from 1975 to 2009, with projections extended to 2012. The figure indicates that the family economic well-being index reached a peak of 109 in 2006, before the recession, and a low point of 101 in 2010, with further declines expected for 2011, until a modest improvement sets in. But these findings mean that gains in child well-being over the past thirty-five years will have been wiped out by what is now being called the Great Recession that began in 2008. The proportion of children living in families below the official federal poverty level (see Figure 15.7) indicates that the percentage of children living in poverty is expected to peak at about 21.5 percent in 2010, a level comparable to that recorded in other major recessions (Land, 2010).

In this recession, as in every previous one, children have borne far more than their share of the burdens. Even before the recession hit, children accounted for a disproportionate share of the poor. Although they make up 25 percent of the total population, in 2008, they accounted

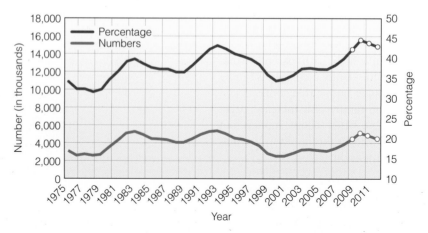

Figure 15.7 Number and percentage of children ages 0–17 living in families below the 100% poverty line, 1975-2008, with projections for 2009–2012. Source: From *The 2010 Child Well-being Index*, by Kenneth C. Land and the Foundation for Child Development. Figure 4, p. 14. Reprinted by permission of the Foundation for Child Development.

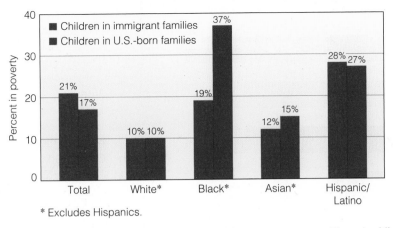

* Excludes Hispanics.

Figure 15.8 Immigrant family background is linked with higher poverty among Hispanic children, but not among black and Asian children. Source: *Population Reference Bureau: Reports on America: Children in Immigrant Families Chart New Path*, February 2009. Figure 5, p. 8. Copyright 2010 Population Reference Bureau. Reprinted by permission.

for 35 percent of the total poor population; 14.1 million children, or 19.0 percent of the U.S. population, were poor according to official poverty guidelines. The poverty rate for children also varies substantially by race and Hispanic origin. Among white citizens in the United States, the rate of child poverty in 2008 was 10.6 percent, whereas it was over three times higher at almost 34 percent among blacks. For Hispanics, the rate was 31 percent, whereas it was 13 percent for Asians (U.S. Census Bureau, 2008). Figure 15.8 shows that immigration status is also linked to the chances that children are living in families below the poverty threshold.

The Great Recession has exposed many of the weaknesses of the welfare reforms of the booming 1990s. Forced to take low-paying jobs or lose their benefits, low-income women with children found, as job losses mounted, that they could not adequately feed their families on the resources available to them from existing programs of the social safety net such as food stamps. In consequence, hunger and "food insecurity" have increased dramatically.

Family Poverty and Food Insecurity

Although they have been widely hailed as a successful new policy, mandatory work programs and deep cuts in other forms of assistance to poor families with children appear to be increasing the problem of food insecurity (Cancian et al., 2003). Figure 15.9 presents a snapshot of food insecurity among U.S. families and households as the Great Recession was taking hold of the U.S. economy but had not yet had its most severe impact. It shows that 14.6 percent of U.S households were experiencing food insecurity, and almost 6 percent felt "very low" food security. "Very low" food security occurs when the food consumed by household members is reduced and their normal eating patterns are disrupted because they lack money and other resources for food (see Figure 15.10). At this writing, the continuing rates of high unemployment—especially among families of the working poor, which were disadvantaged even before the downturn—result in increasing numbers of visits to local food pantries, soup kitchens, and emergency shelters.

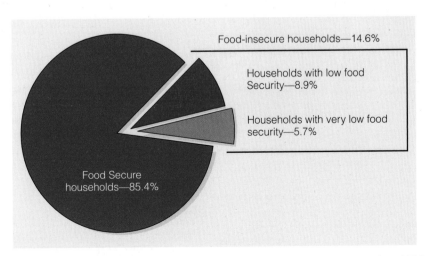

Figure 15.9 U.S. households by food security status, 2008. Source: Nord, Andrews, & Carlson, 2009.

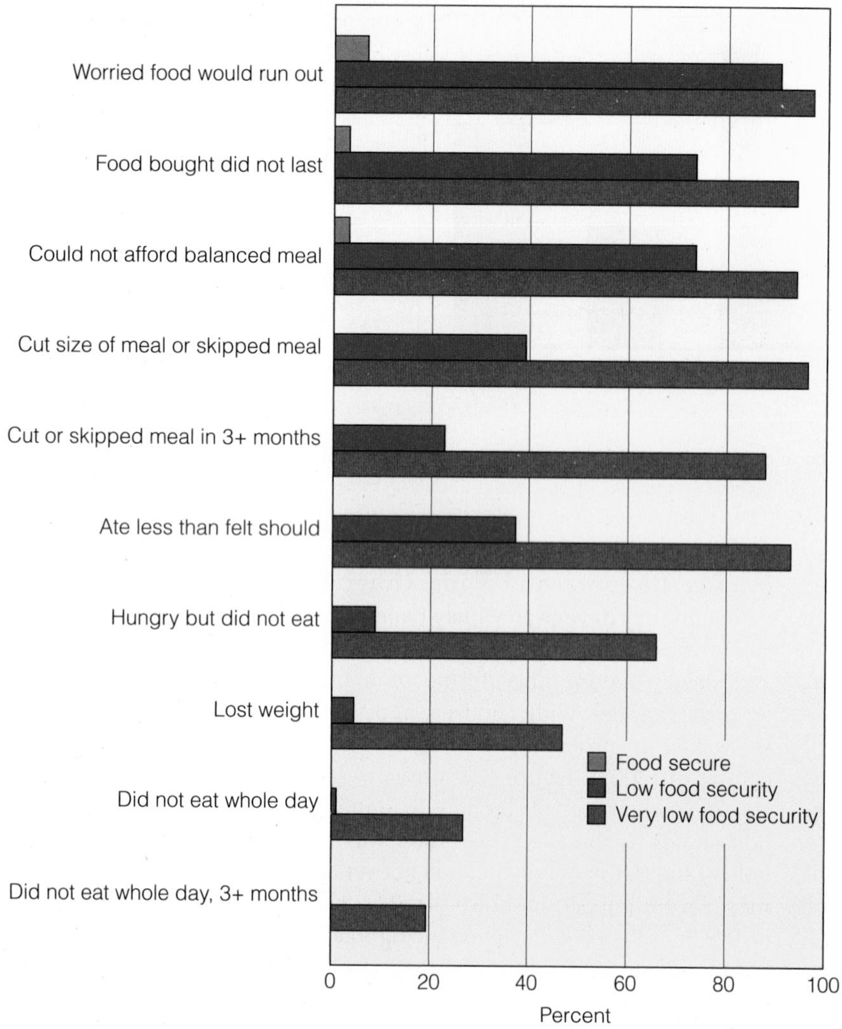
Worried food would run out
Food bought did not last
Could not afford balanced meal
Cut size of meal or skipped meal
Cut or skipped meal in 3+ months
Ate less than felt should
Hungry but did not eat
Lost weight
Did not eat whole day
Did not eat whole day, 3+ months
Food secure
Low food security
Very low food security
0
20
40
60
80
100
Percent

Figure 15.10 Households reporting each indicator of food insecurity, by food security status, 2008. Source: Nord, Andrews, & Carlson, 2009.

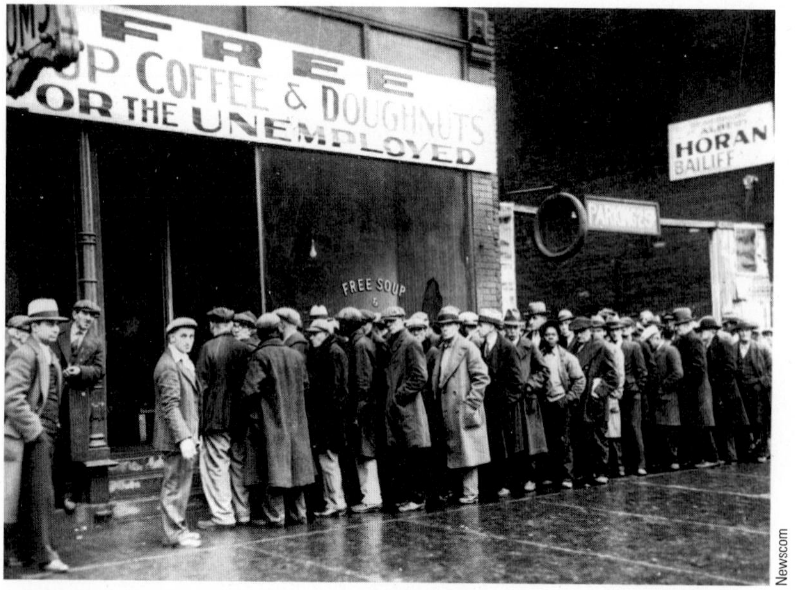
FREE
COFFEE & DOUGHNUTS
FOR THE UNEMPLOYED
FREE SOUP
HORAN
BAILIFF
Newscom

This photo of unemployed men outside a soup kitchen in New York during the Great Depression is both a reminder of how severe that depression was, and an illustration of how different unemployment was in the 1930s than it is today. When unemployed people line up at soup kitchens and food banks in today's difficult economic times, there are far more women and children in the lines.

Sociology & Social Justice

FAIRNESS AND FAMILY POVERTY Is it fair that children from poor working families should disproportionately experience the effects of a recession caused by bankers and high-flying hedge fund managers? Many critics of welfare reform note that the drastic changes in the welfare system passed during the Clinton administration are now causing severe hardship for poor people during a severe recession. They also claim that the "reforms" that weakened the welfare system relied on the use of racialized rhetoric and social-class stereotypes. During the debates over ending the system of Aid to Families with Dependent Children (AFDC), welfare recipients were often portrayed as lazy, dishonest, and pathologically dependent on the government. More conservative policymakers argued that welfare itself had the effect of sustaining poverty rather than alleviating it. Researchers who criticize "welfare reform" as it was actually carried out note that the general public tended to imagine welfare recipients as women of color, even though the majority of welfare recipients had always been white children. Moreover, under the reforms, women with young children were obliged to accept work that did not pay enough to house, feed, and clothe their children. Poor mothers who were forced to take these jobs did reasonably well as long as economic conditions were favorable, and the new laws did offer allowances for day care. However, states with more people of color were likely to impose stricter sanctions on poor mothers and also to be far less generous in their day care allowances. During the recession that began in 2008, the consequences of these biases in thinking about the poor became evident. Clearly, welfare reform has not advanced social justice, at least not during severe economic bad times (Applied Research Center, 2009). Now that many low-paying jobs are gone, the "safety net" of help for poor children and their parents, such as food stamps and emergency assistance, is also under attack by those who whish to further curtail government spending, and poor children of all races are suffering the effects.

Parenting, Stepparenting, and Social Change

It has never been easy to be a parent. Keeping a family intact and raising children who are confident and capable and will become loving, capable parents themselves is a domestic miracle. For better or (sometimes) worse, most people learn to be parents by following the examples set by their own parents when they were children. Much as we may swear that we will improve on the way our parents performed their roles when we are parents ourselves, we often find ourselves acting toward our children much as our parents acted toward us. In fact, many parents seek help from family counselors because they are shocked to find themselves doing and saying the same things to their children that their parents did to them (Scarf, 1995).

Many other adults and children will experience major reorganizations of their families as couples divorce and remarry. When stepfamilies are formed, family members must adapt to new role relations as stepchildren and stepparents, often while maintaining many of their obligations in the original family. Sharon L. Hanna, an expert on stepfamilies, notes that there are more than 20.6 million stepparents in the United States. At present, one in three Americans is involved in some way in a stepfamily situation, and 1,300 new stepfamilies are formed each day (Step Family Foundation, 2010).

"Stepfamilies," sociologists Frank Furstenberg and Andrew Cherlin point out, "are a curious example of an organizational merger; they join two family cultures into a single household" (1991, p. 83). Although this joining of family cultures also occurs in first marriages, in those marriages the couple usually has time to work out differences before children are born. In the case of stepfamilies, the task of working out a mutually acceptable concept of the family may be more stressful than the partners anticipated.

Sociologists point out that positive relationships in any family do not just happen but are worked on consciously through much learning and discussion. These efforts are particularly necessary in stepfamilies (Ganong & Coleman, 1999). Some studies of stepfamilies show that new stepparents often conceive of themselves as "healers," people who can use the force of love to form strong bonds of attachment between two sets of children. Very often, however, the children cling to the fantasy that their original parents will get back together, and the stepparents' efforts may meet with frustration. Stepparents who manage to reorganize the family successfully tend to be diplomatic and sensitive and to avoid competing with biological parents. They gain the support of their spouses and wait for opportunities to gain the acceptance of their stepchildren (Gerson, 2009).

However, resolving the complexity and ambiguity of roles in reorganized families is no simple matter. Many factors come into play when a new family is formed. For example, Furstenberg and Cherlin point out that legal norms "reinforce the second-class status of stepparents" as nonbiological partners (1991, p. 80). A biological father is held responsible for child support even if he has never lived with the child, whereas a stepfather incurs no financial responsibility for a child even if he has lived with that child for many years. "The only legal recourse for stepparents who want to claim rights and responsibilities is to adopt their stepchildren, an act that seems to symbolize a deeper and more permanent tie, more like a biological relationship" (Furstenberg & Cherlin, 1991, p. 80).

Members of all kinds of families experience particular stress in times of economic hardship or when other social changes influence the family. Stress and change in the larger society may cause parents to feel unable to perform their roles as providers of material and emotional support. In the early decades of the twentieth century, for example, the combination of frequent severe recessions, high rates of industrial accidents, and lack of adequate health care had a significant impact on families. Fathers were often forced to leave their families in search of work; infection often claimed the lives of mothers during childbirth. Demographers estimate that, at the beginning of the twentieth century, when divorce rates were still extremely low, almost 25 percent of all children lost at least one parent through death before reaching the age of fifteen (Uhlenberg, 1980).

By midcentury, improvements in health care had reduced this proportion to 5 percent, but divorce rates had increased to 11 percent, and another 6 percent of children were born to unmarried parents. Thus, even in the 1950s—the most stable decade for families and children—demographers estimate that 22 percent of children were being raised in families that had only one parent as a result of death, separation, divorce, or single parenthood (Bumpass, Raley, & Sweet, 1995). Even this relative stability lasted only a decade. By the end of the 1950s, divorce rates had begun their precipitous rise, and so had rates of births to unmarried parents. By 2000, almost 30 percent of births in the United States were to unmarried mothers (National Center for Health Statistics, 2000).

These profound changes in family composition have many causes and many consequences. As families age, children move through the school years, adolescence, and young adulthood while parents pass through adulthood to their senior years. Throughout this period, all families seek to accomplish certain basic socialization goals. As the diversity of families increases, the ways in which they attempt to reach these goals differs, but the goals themselves do not change nearly as much.

The Family Life Cycle

Sociologist Paul Glick, an innovator in the field of family demography and ecology, developed the concept of the *family life cycle,* the idea that families pass through a sequence of five stages:

1. Family formation: first marriage
2. Start of childbearing: birth of first child
3. End of childbearing: birth of last child
4. "Empty nest": marriage of last child
5. "Family dissolution": death of one spouse (Glick & Parke, 1965)

These are typical stages in the life cycle of conventional families. Although other stages could be identified within each of these—such as retirement (the years between leaving the workforce and the death of one spouse) or the "baby" stage (during which the couple is rearing preschool-age children)—the ones that Glick listed are most useful for comparative purposes.

As the "typical" family structure becomes increasingly difficult to identify because of changes in family norms, the stages of family development also vary. In fact, the stages of the family life cycle have become increasingly useful as indicators of change rather than as stages that all or most families can be expected to experience. The Census Bureau estimates, for example, that about 6.4 million heterosexual couples in the United States are cohabiting. This number represents a twelvefold increase since 1970, when an estimated 500,000 cohabiting couples were counted (*Statistical Abstract*, 2010). Cohabiting couples now make up almost 10 percent of all opposite-sex U.S. couples, married and unmarried. (Note that the census figures include only heterosexual couples and therefore vastly underestimate the total number of cohabiting couples.) Although most cohabiting couples will eventually marry, about half will break up or continue to live together without marrying (Farley, 1996).

Age at First Marriage When sociologists look at the median age at which people experience the various stages of the family life cycle, significant trends emerge. Consider age at first marriage: How young or old people are when they marry has many potential effects on children and families that are difficult to judge. People tend to delay marriage during economic recessions. They also delay marriage for longer periods of education and professional training or military service. For all these reasons, and because more young adults are finding it difficult to enter the labor force, the age of first marriage is rising. In 1890, the median age of Americans at first marriage was 26.1 years for men and 22 years for women; in the 1950s and 1960s, it reached the historic lows of 22.8 and 20.3. Then it began rising again slowly, until by 2003, it was 27.1 for men and 25.3 for women (see Figure 15.11). Age at first marriage tends to be somewhat higher for males than for females and is associated with educational attainment; more highly educated people are more likely to delay marriage. Members of minority groups tend to marry later because of the proportionately lower income levels of these populations. People with lower incomes tend to delay marriage or not to marry because they lack the material means to sustain a marital relationship (Farley, 1996).

As it passes through the family life cycle, every family experiences changes in its system of role relations. In analyzing these changes, social scientists often modify Glick's stages in order to focus more sharply on interactions within the family. An example is the set of developmental stages shown in Figure 15.12. There are major emotional challenges at each of these stages. For example, families with adolescent children must adapt to the children's growing independence. This may involve going through a stage of negotiation over such issues as money, cars, and dating. But researchers who study family life note that parents are often confused about how to interact with their adolescent children. They may assume that it is normal for adolescents to leave the family circle and to become enmeshed in their own peer groups, which often get into trouble. However, as Carol Gilligan (1987) has pointed out, adolescents also want the continuing guidance and involvement of adults.

Moving Back Home: Recession's Impact on Family Cycles Social scientists sometimes refer to adults who have had to move back in with their parents as "boomerangers." In a national survey conducted by the Pew Center for Research (Wang & Morin, 2009), researchers found that about 11 percent of all adults age eighteen or older live with their parents in their home and 4 percent of all adults say that they were forced to move back into their parents' home because of the recession. This proportion rises to 10 percent among those ages eighteen to thirty-four.

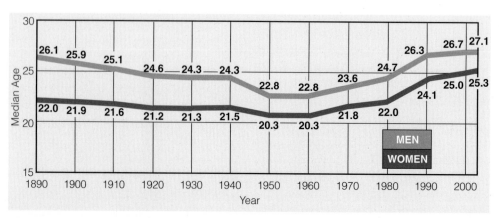

Figure 15.11 **Estimated Median Age at First Marriage, United States, 1890–2000.** Source: Data from U.S. Census Bureau.

Stage	Emotional Process	Required Changes In Family Status
1. Between families: The unattached young adult	Accepting parent–offspring separation	a. Differentiation of self in relation to family of origin b. Development of intimate peer relationships c. Establishment of self in work
2. The joining of families through marriage: The newly married couple	Commitment to new system	a. Formation of marital system b. Realignment of relationships with extended families and friends to include spouse
3. The family with young children	Accepting new generation of members into the system	a. Adjusting marital system to make space for children b. Taking on parenting roles c. Realignment of relationships with extended family to include parenting and grandparenting roles
4. The family with adolescents	Increasing flexibility of family boundaries to include children's independence	a. Shifting of parent–child relationships to permit adolescents to move in and out of system b. Refocus on midlife marital and career issues c. Beginning shift toward concerns for older generation
5. Launching children and moving on	Accepting a multitude of exits from and entries into the family system	a. Renegotiation of marital system as a dyad b. Development of adult-to-adult relationships between grown children and their parents c. Realignment of relationships to include in-laws and grandchildren d. Dealing with disabilities and death of parents (grandparents)
6. The family in later life	Accepting the shifting of generational roles	a. Maintaining own and/or couple functioning and interests in face of physiological decline; exploration of new familial and social role options b. Support for a more central role for middle generation c. Making room in the system for the wisdom and experience of the elderly; supporting the older generation without overfunctioning for them d. Dealing with loss of spouse, siblings, and other peers, and preparation for own death; experiencing life review and integration

Figure 15.12 Stages of the Family Life Cycle. Source: Adapted from McGoldrick & Carter, 1982.

The survey also found that the recession has changed the lives of young adults in other important ways that represent forms of "putting life on hold" (see Figure 15.13). For example, fully 15 percent of single adults younger than age thirty-five say that they have postponed getting married because of the recession, while an additional 14 percent of all young adults delayed having a baby. These trends are evident in Table 15.1. The proportion of adults who postponed their marriage because of the recession increases to 21 percent if the sample of young adults is limited to those ages twenty-five to thirty-four—the age range in which most people get married.

Note, however, that the table shows that the elderly are still almost three times more likely to be living alone than are young people. And social scientists do not think that the trend toward moving in with parents will continue beyond the Great Recession, while they do believe that the trend toward later marriage and singlehood will continue. Does this mean that people will become increasingly isolated? That would be jumping to a conclusion that the empirical facts do not support. Single people have more resources available to them than ever

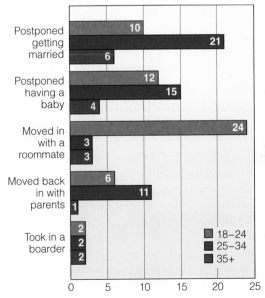

Percent of each group who say they did the following because of the recession...

Postponed getting married: 10 (18–24), 21 (25–34), 6 (35+)
Postponed having a baby: 12 (18–24), 15 (25–34), 4 (35+)
Moved in with a roommate: 24 (18–24), 3 (25–34), 3 (35+)
Moved back in with parents: 6 (18–24), 11 (25–34), 1 (35+)
Took in a boarder: 2 (18–24), 2 (25–34), 2 (35+)

■ 18–24 ■ 25–34 ■ 35+

Note: "Don't know/Refused" responses not shown.

Figure 15.13 Life Interrupted. Source: *Home for the Holidays . . . and Every Other Day Recession Brings Many Young Adults Back to the Nest*, by Wendy Wang and Rich Morin, Pew Research Center, November 24, 2009.

TABLE 15.1

Changes in Share of Adults Who Live Alone: 2007–2009

	2007 %	2009 %	Change %
All	14.0	13.9	−0.1
Age Groups			
18–29	7.9	7.3	−0.6
30–49	9.5	9.5	—
50–64	15.7	15.6	−0.1
65+	30.2	30.1	−0.1

Source: *Home for the Holidays . . . and Every Other Day Recession Brings Many Young Adults Back to the Nest,* by Wendy Wang and Rich Morin, Pew Research Center, November 24, 2009.

before to enable them to engage in friendships, both real and virtual, and in relationships as well. These are issues that will demand research attention well into the future (Klinenberg, 2010; Lamanna & Reidmann, 2009).

In later stages of family life, the parents must be willing to watch their grown children take on the challenges of family formation while they themselves worry about maintaining their marital roles or caring for their own parents. The latter issue is taking on increasing importance in aging societies like the United States. And because women are still expected to be more nurturing and emotionally caring than men, it often falls to women to worry about the question of "where can Mom live?" (Hochschild, 1989). According to Elaine Brody, a leading researcher on this issue: "It's going to be primarily women for a long time. Women can go to work as much as they want, but they still see nurturing as their job." Moreover, "with many more very old people, and fewer children per family, almost every woman is going to have to take care of an aging parent or parent-in-law" (quoted in Lewin, 1989, p. A1).

As if these stages were not stressful enough, consider the complications resulting from divorce, remarriage, and the combining of children of different marriages in a new family. It is increasingly common, for example, for teenagers and young adults to have parents who are dating and marrying, and for children to have four parental figures in their lives instead of two. These and many other changes in the family pose profound problems for sociologists who seek to develop theoretical perspectives on the family as a central institution in contemporary societies, as we will see in the next section.

PERSPECTIVES ON THE FAMILY

High divorce rates do not indicate that the family is about to disappear as an institution in modern societies. Even though many marriages end in divorce and increasing numbers of young adults postpone marriage or decide not to marry at all, most do marry, and most of those who divorce will eventually remarry. Far from disappearing, the family is adapting to new social values and to changes in other institutions, especially economic ones. But one need only listen to an hour or two of talk radio or read a newspaper to find fierce debates about whether it is morally justified for single people to have children or for members of same-sex unions to be legally married or for unmarried adults to live together outside of wedlock, along with many other issues involving "family values." Most sociologists leave the moral debates to people with ideological commitments to one or another concept of the ideal family. Social scientists devote more thought to theoretical issues and to conducting research designed to obtain factual information about the changing family in the United States and elsewhere in the world. In this section, therefore, we review the basic sociological perspectives on family roles and relationships, as well as research that applies those perspectives. (See the Perspectives Chart on page 383.)

The Interactionist Perspective

Interactions within the family cover a wide range of emotions and may take very different forms in different families. Families laugh and play together, work together, argue and bicker, and so on. All of these aspects of family interaction are important, but the arguing and bickering often drive family members apart. Studies of family interaction therefore often focus on the sources of tension and conflict in the family's interactions, but they are equally concerned with understanding family rituals, how family members help make each other feel better, and how they interact as a family with other groups in society.

Observe any normal family with young children, and before long, there will be some whining and complaining of this sort: "Mom, he's teasing me." "Mom, she's taking up too much room." "Dad, you said we were going to stop for ice cream. We want ice cream." The parents may argue with each other about how to deal with these demands, or they may present a unified parental front with a consistent message such as, "You need to stop teasing each other and share the seat. We'll have ice cream after dinner." Such typical family interactions often stem from competition for attention, but they are also caused by the simple fact that children depend so much on their parents for satisfaction of their wants. When children appeal to their parents to address wrongs they feel have been committed by other siblings against them, the underlying cause can be normal sibling rivalry. Skillful parents often develop techniques to diffuse rivalry by devising family games or by singing together, or any number of cooperative techniques that can transform angry interactions into more pleasant ones. But when parents are under stress themselves due to lack of time or resources (especially money), it becomes more difficult for them to develop the interactive skills of more effective parents (Lachs & Boyer, 2003).

The context within which family life occurs can affect family interactions in profound ways. At the lower levels of a society's stratification system, for example, money (or the lack of it) is often a source of conflict between parents or between parents and children. But the

Perspectives on the Family

Perspective	Emphasis	Central Issues	Research Issues
Functionalist	The family is a critical institution in societies but must continually adapt to major social changes.	The family has lost many functions as societies have become more complex, particularly its economic and political roles. Socialization and advancement of children in society remain major functions.	Are societies developing new institutions to support family cohesion and resolve opposing demands of work and home? How well do different families support their children's achievement (e.g., through involvement in schools, aid with schoolwork at home, etc.)?
Conflict	Family cohesion can be threatened by stresses and challenges occurring within the family group and coming from larger social forces in the society.	Conflict within families is often the result of struggles for power among family members and of individuals' efforts to gain more attention and love.	Do abrupt changes in family structure, due to divorce or separation, for example, result in more internal conflict? How do larger social forces, such as loss of work, heavy debt, or the necessity to move the home due to changing economic fortunes, relate to family conflict?
Interactionist	The family is a system of interacting personalities who must also cope with continual changes in family roles and composition as individuals mature.	Many interactions center on individual needs versus those of the family as a group. Sibling rivalry, for example, can stem from competition for a parent's attention.	How do parents learn skills that help them avoid stress and conflict? How do patterns of interaction change and adapt to new family forms after divorce or death in the family?

rich are by no means immune to problems of family interaction. Because they do not have to be concerned about the need to earn a living as adults and because their parents can satisfy any desires they may have, the children of the very rich often develop a sadness that resembles anomie (Wixen, 1979). Their lack of clear goals, which sometimes expresses itself in a compulsion to make extravagant purchases, may create conflict between them and their parents.

To study these situations, either to help a family resolve its problems or to understand the nature of family conflict more thoroughly, the social scientist must interact with and observe the family, either at home, in a laboratory, or in a therapy setting (Minuchin, 1974; Satir, 1972; Whitchurch & Constantine, 1993). For instance, consider the case described by two family therapists, Augustus Napier and Carl Whitaker (1980), in which the parents are deeply troubled by the behavior of their seventeen-year-old daughter, the oldest of their three children. The father is a prominent attorney, the mother a college-educated woman who has devoted herself to homemaking. The parents' definition of the situation that has brought them into therapy is that they are worried about their daughter. She disobeys, is delinquent, and is depressed enough to make them fear that she might commit suicide. The mother and daughter fight bitterly, but usually the mother retreats. The father often attempts to defend the daughter, but eventually he sides with the mother.

After a few sessions, Napier and Whitaker begin to challenge the parents' assumption that it is the daughter who is "the problem." They guess that her depression, anger, and delinquent behavior are symptoms of larger problems in the family. After some gentle probing, they identify a triangle of family conflict in which "two parents are emotionally estranged from each other and in their terrible aloneness they overinvolve their children in their emotional distress" (p. 83). The children blame themselves for their parents' problems and develop a low sense of self-worth. The therapists help the couple face the issues that divide them and that motivate their children to offer themselves as scapegoats or intermediaries who divert attention from the parents' basic problems.

Families like the one just described need to resolve a contradiction inherent in the institution of the family: the need to maintain the *individuality* of each member while providing love and support for him or her within a set of *interdependent* relationships. Many families never succeed in developing ways of encouraging each member to realize his or her full potential within the context of family life. Research shows that the core problem is usually the failure of the adult couple, even in intact families, to understand and develop their own relationship. Such a couple may become what John F. Cuber and Peggy B. Harroff (1980) term "conflict-habituated" or "devitalized." Couples of the first type have evolved ways of expressing their hostility toward each other through elaborate patterns of conflict that persist over many years. In contrast, the devitalized or "empty-shell" marriage may have begun with love and shared interests, but the partners have not grown as a couple and have drifted apart emotionally. Each has the habit of being with the other, a habit that may be strongly supported by the norms of a particular ethnic or religious community. Neither partner is satisfied by the relationship, but neither feels that he or she can do anything to change the situation. Thus, the conflicts that might have produced change are reduced to indifference.

Today, social scientists who study family interaction must deal with family structures that are more complex than the traditional nuclear family. Divorce and remarriage create many situations in which children have numerous sets of parental figures—parents and stepparents, grandparents and surrogate grandparents, and so on. These changes in family form result in new patterns of family interaction. For example, in a study of 2,000 children conducted over a five-year period, sociologist Frank Furstenberg found that 52 percent of children raised by their mothers do not see their fathers at all—partly because some fathers are absent by choice or are denied visitation rights, but also because opportunities for visits decrease as parents remarry or move away (cited in Aspaklaria, 1985). This produces situations in which parents strive to maintain long-distance relationships with their children and occasionally have brief, intense visits with them. Although this specific type of relationship may not be the most desirable, it appears that if parents and children can express love and affection even when the parents are divorced, the children's ability to feel good about themselves and to love others in their turn may not be impaired.

The Conflict Perspective

The conflict perspective on families and family interactions assumes that "social conflict is a basic element of human social life." Conflict exists "within all types of social interaction, and at all levels of social organization. This is as true of the family as it is of any other type of social entity" (Farrington & Chertok, 1993, p. 368). From this perspective, one can observe actual family interactions (the micro level) and ask why conflict occurs, what the issues are, and how they are resolved. At a more macro level, one can ask how conditions of inequality and class conflict influence actual families or the laws and policies governing the family as a central institution in society.

At the micro level of internal family affairs, the conflict perspective points to ways, for example, in which the subordination of women often results in particular forms of powerlessness and anger. The familiar complaint that a parent, particularly a mother, does a lot of nagging is a common one, but it is often a signal that the parent feels relatively powerless to change a situation. In family groups in which power is shared equally and parents have the means to withhold favors or desired rewards from children who do not meet their expectations, nagging is far less frequent. But when the mother is clearly subordinated and does not share equally in the resources the family has at its disposal, conflict in various forms is worsened (Boxer, 2002). Similarly, when problems of alcohol or substance abuse weaken parental authority, all the nagging, cajoling, or punishment meted out by parents

can be ineffective in reaching angry children who feel, often without understanding it, that their own needs for love and attention are not being met. Such problems are often influenced by macro problems of unemployment, stress at work, lack of health insurance, and many other external social influences that conflict theorists focus on in their analysis of family issues.

"Family Values" and Class Conflict As mentioned earlier, Americans are engaged in a major public debate over issues related to the family, including homosexual marriage, freedom to obtain abortions, enforcement of child support and child abuse laws, and a host of other controversies that have a direct bearing on the family as an institution in society. The federal budget battles of the past decade also represent an arena of conflict over family policies, for they centered on the social safety net of welfare agencies, payments to families with dependent children, and many other spending categories that transfer money from more affluent families and individuals to those with fewer resources and, often, great need.

Approximately 8.1 million families in the United States—10.3 percent of all families—are living at or below the official poverty line (*Statistical Abstract,* 2010), the highest level in eleven years. Between 2007 and 2008, at least 1 million children were living in poverty. Poverty affects a far greater proportion of black and Hispanic families than white families, even though most poor families are white.

Sociologists who study poverty argue that most families are poor for the obvious reason that they do not earn or receive enough income. But critics of liberal social-welfare policies argue that self-destructive behaviors of family members themselves, such as dropping out of school, drug use, and out-of-wedlock childbearing, are causes of poverty. Social-welfare policies that are too generous or take the form of entitlements that do not have to be earned are said to encourage behaviors that cause poverty.

Social scientists who study class conflict and poverty policies, notably Herbert Gans (1995) and Christopher Jencks (1993), point out that class conflict helps explain the views of different population groups on family poverty. In the early decades of the twentieth century, they note, a large proportion of impoverished families consisted of elderly couples who did not have Social Security or Medicaid, two programs that have drastically reduced poverty among elderly Americans. Yet at the time that these policies were being debated, many members of the upper classes opposed them, viewing them as undemocratic transfers of wealth to less deserving members of society. The situation today is similar in that members of the upper classes oppose adequate funding for day care and other policies that can help move adult family members from poverty and dependence on public support to employment at decent jobs.

Functionalist Views of the Family

From a functionalist perspective, the family evolves in both form and function in response to changes in the larger social environment. As societies undergo such major changes as industrialization and urbanization, the family must adapt to the effects of those changes. Functionalist theorists like Talcott Parsons and William Goode have called attention to the loss of family functions that occurs as other social institutions like schools, corporations, and social-welfare agencies perform functions that were previously reserved for the family. We have discussed numerous examples of this trend in earlier chapters (for example, the tendency of families to have fewer children as the demand for agricultural labor decreased, the changing composition of households as people were required to seek work away from their families of orientation, and the increasing number of dual-income households). The functionalist explanation of these changes is that as the division of labor becomes more complex and as new, more specialized institutions arise, the family, too, must become more specialized. Thus, modern families no longer perform certain functions that used to be within their domain, but they do play an increasingly vital part in early-childhood socialization, in the emotional lives of their members, and in preparing older children for adult roles in the economic institutions of industrial societies (Cherlin, 1996; Parsons & Bales, 1955).

A good example of empirical studies of how family functioning adapts to social change is research on the effects of child care on infant development. Figure 15.14 shows that there is a wide array of child care arrangements. For very young children, day care is far less common than care by other relatives or care by nonrelatives outside an institution. But by the time children of working parents reach the age of five, day care centers account for fully 75 percent of child care arrangements. An often-cited study (Belsky, 1988; see also Belsky & Steinberg, 1991) offered some evidence that infants who are given over to a nonparent early in life may suffer developmental problems. This finding has been widely challenged. More recent research using more sophisticated methods shows that, with quality care, there are no discernible disadvantages for children of working mothers. Of course, the larger issue is how to provide high-quality care while parents are away working (Eckholm, 1992).

Researchers who approach the issue of child care from a functionalist perspective ask how societies cope with the dilemma faced by families in which both parents work outside the home. In an influential study, Kammerman and Kahn (1981) found that advanced industrial societies are struggling with the question: "Can adults manage productive roles in the labor force at the same time as they fulfill productive roles within the family—at home?" (p. 2). Many European nations answered this question by developing policies to assist working parents, including the development of high-quality child and infant care services (for example, universal day care and school meal programs), paid parental leave, and family allowances that ease the financial burden on families with several children. In the United States, however, there is a great deal of debate about whether such policies are appropriate functions of government or whether they lead to higher taxes, forcing more mothers into the labor force (Pittman, 1993). (See the Global Social Change box on page 386.)

In related studies, sociologist Arlie Hochschild (1989, 1997) asked how adult family members adjust to what she calls the "second

Percentage of children in each age group*

	less than 1	1 year old	2 years old	3 years old	4 years old	5 years old
Children in relative care	24%	24%	19%	21%	18%	15%
Children in nonrelative care	17%	19%	20%	19%	15%	17%
Children in center-based program	7%	11%	19%	41%	65%	75%

*Columns do not add up to 100 because some children participated in more than one type of day care.

Figure 15.14 Child Care Arrangements. Source: U.S. Department of Education, National Center for Educational Statistics.

global social change

Looking to France for Models of Child Care An unintended consequence of efforts to "end welfare as we know it" has been increased pressure on existing child care services. Inadequate to meet the demand before welfare reform, day care and other child care services for working parents in the United States are under even more pressure today. This situation makes it more difficult for low-income parents, who need subsidized day care, to balance work and family obligations. Because early-childhood language and social learning experiences are immensely important to later learning and school achievement, lack of high-quality affordable child care for working parents can lead to higher social and economic costs in the long run.

The shortage of child care services is also a good example of "structural lag." Existing day care institutions (social structures dealing with child care) lag behind the demand created by new norms—that is, the expectation that low-income single mothers must work to receive additional benefits (health care, supplemental income, and the like).

Many American child care experts look to France and other European nations for examples of how to address this problem of structural lag in the provision of day care services. Children are born into poverty in France and the United States at virtually the same rate—25 percent in the United States and 24 percent in France. But by age six, the rate of child poverty in France drops to 7 percent, whereas it hovers around 17 percent in the United States. Social scientists believe the greater improvement in France is largely a result of greater public support of high-quality child care. All French infants are born into a system that guarantees universal health care and financial support for families. The new mother immediately qualifies for fourteen weeks of paid maternity leave, which parents often combine with several weeks of paid vacation, all of which enable French citizens to stay at home with newborns far longer than low-income American parents can. In the United States, there are no requirements for paid maternity leave. Only very recently have parents in companies employing fifty or more workers gained the right to take unpaid maternity leave (a maximum of twelve weeks) and still return to their jobs.

The French system offers numerous benefits that are not available in the United States except for employees in the most family-supportive firms or at the higher levels of the salary scale. Children in France are screened for any health problems in infant child care centers (known as *crèches*) before they enter preschool. These health screenings continue regularly throughout the public school years. When American child care experts toured French child care centers, they noticed "that they had never before seen so many toddlers and preschoolers wearing eyeglasses, and they realized that it was probably because more French children are screened for vision problems earlier than is the case for American youngsters." All child care workers receive extensive professional training in child care and human development. Although new French parents, like their American counterparts, might wish to be at home with their infants, they can at least be confident in the level of care their children are receiving outside the home.

From a critical sociological viewpoint, this comparison indicates that, as a society, France is more committed to raising healthy children than the United States. American families do not value their children less, but they are much more individualistic about such issues—more likely, that is, to believe that child care arrangements are the responsibility of individual families. This makes Americans far more susceptible to claims by employers that universal high-quality child care is too expensive and a detriment to profitability. It also makes many Europeans wonder about the social priorities of Americans (Jacobson, 2001).

---

shift" (see Chapter 13). Only a few decades ago, housework and child care were viewed as primarily the mother's duty, even if she worked outside the home. Today, many working parents cooperate and share domestic roles and responsibilities, but it is often difficult to achieve such cooperation. Hochschild found that "most egalitarian women—those with strong feelings about sharing—did one of two things. They married men who planned to share at home or they actively tried to change their husband's understanding of his role at home" (1989, p. 193). Many of the women Hochschild interviewed were like Adrian Sherman, who "took the risky step of telling her husband 'It's share the second shift or it's divorce.' She staged a 'sharing showdown' and won" (p. 193). Other women do not go so far as to stage a showdown. They may make incremental efforts to change their husbands' attitudes. Still others discuss the issue with their husbands, but the process is not an easy one. More than half of the working women Hochschild interviewed had tried to change domestic roles in one way or another. As Hochschild notes, "If women lived in a culture that presumed active fatherhood, they wouldn't

need to devise personal strategies to bring it about" (p. 193).

In sum, research from all the major sociological perspectives supports the view that the family is a resilient institution; it adapts to changing economic conditions and changing values. As Napier observes:

> Every family is a miniature society, a social order with its own rules, structure, leadership, language, style of living. . . . The hidden rules, the subtle nuances of language, the private rituals and dances that define every family as a unique microculture may not be easy for an outsider to perceive at first glance, but they are there. (Napier & Whitaker, 1980, p. 78)

Thinking Critically With recession hitting families hard, driving more into debt and foreclosure, delaying marriage, and causing young people to move in with their parents even after college, why do sociologists still find the family to be such a resilient institution? When evidence can you cite from reading this chapter in support of family resilience despite social change?

At the macro level, the strength of the family as an institution does not mean the divorce rate will decrease rapidly or families that experience severe stress caused by unemployment, ill health, and the like will have an easier time remaining intact. Whatever the perspective from which they study the family, sociologists recognize that problems such as family violence and family breakup are pervasive and require more effective intervention techniques to protect vulnerable family members and strengthen the family as an institution. We turn, therefore, to examples from contemporary research that show how the basic sociological perspectives on the family can help us understand and intervene more successfully in different situations of family violence.

Family Violence: Applying the Perspectives

Family researchers Jeffrey L. Edleson and Ngoh Tiong Tan (1993) describe cases of family violence that can be analyzed in terms of each of the main sociological perspectives on the family. Their research compares Chinese and North American families in which the wives have experienced incidents of abuse by their husbands.

Spouse Abuse in a Chinese Family Mrs. Lee is referred to a social worker after seeking aid at a police station when her husband struck her in the face. In Chinese and other Asian cultures, the wife is expected to leave her family of orientation to take up residence with her husband and his extended family. Mrs. Lee's husband's family is dominated by her mother-in-law, a powerful widow who treats Mrs. Lee as the person with the lowest status in the family. Mrs. Lee wishes her mother-in-law would be less autocratic and complains about her dictatorial behavior to her own parents and to her husband. Her husband, on the other hand, believes his wife should conform to the ancient norms of Chinese family life and obey his mother. He and his mother consider Mrs. Lee a "traitor" because she has appealed to authorities outside the family.

From a functionalist perspective, the police and social workers are representatives of institutions in the larger society that seek to help families avoid violence. But they cannot simply impose a new form of power on this complex extended family. Instead, they apply an interactionist perspective to the conflict as they understand it. The social worker convenes a meeting of the family, to which Mrs. Lee's older brother is also invited. Together, the group works out a detailed understanding of the roles of all the family members. The social worker explains that Mrs. Lee is not seeking to challenge her mother-in-law. In fact, however, the social worker has demonstrated to the extended family that Mrs. Lee has a support network and resources outside the family—a new notion for a traditional Chinese family. No doubt there will be further conflict, but power relations within the family have been subtly altered as a result of this intervention.

A North American Family Copes with Violence Susan was a regular victim of her husband Bob's violent anger. The beatings began on their honeymoon and ended twenty years later when he struck her with a baseball bat, almost killing her. During those years, their children suffered emotionally from the violent interactions they witnessed and the unhappiness their mother experienced. But Susan loved Bob and struggled to make the marriage work. Now the couple are in court because Bob was arrested after the last assault. He faces a jail sentence, and Susan has abandoned hope of making the marriage work. She understands that she must make a new life for herself. She is seeking opportunities to do so but knows that she also needs protection from possible future assaults.

Unlike Mr. and Mrs. Lee, neither Bob nor Susan lives near an extended-family network. Many Americans do not live in close contact with relatives; on the average, "extended families do not figure as prominently in the lives of Caucasian Americans as they do in other subcultures and in many Asian societies" (Edleson & Tan, 1993, p. 384). Bob and Susan had left their extended families in search of economic opportunity. Susan now has nowhere to turn for help except the police and social welfare agencies. This social safety net will act to protect her so that she and her children can live in peace. From a functionalist perspective, Susan will also have to rely on these agencies to enforce Bob's responsibilities toward his children. The functionalist sociologist argues, therefore, that it is of vital importance that the agencies function effectively in this and similar cases.

Susan needs the help of outside agencies to gain relief from the deadly conflict that has destroyed her family. Yet these very agencies came into being as a result of conflict in the larger society. They emerged after decades of protests and demands by women and social activists for ways to protect innocent family members from their abusers. As we learn almost daily from reports of incidents of spouse and child abuse, however, it is never easy to ensure that the agencies created by society will intervene effectively.

Even if they have the necessary resources, there is no guarantee that they will always act wisely to balance the conflicting values of individual rights and preservation of the family.

Domestic abuse is the most frequently reported issue in American policing (Defina & Wetherbee, 1997). These cases only touch the surface of an immense social problem, but they offer some insight into how sociologists contribute to understanding these extremely complex instances of violence at the micro level. Contemporary law enforcement officers are trained in the basic sociology and psychology of family violence. They understand that the batterer and the domestic victim are often caught in a cycle of repeated hostility and violence. Too often the violent spouse, usually the husband or male partner, expresses contrition after the violent act and is forgiven, and so the cycle resumes. The authorities and the neighbors believe the household is a troubled one but do not believe they can do very much about it (Clement et al., 2009).

In many communities, however, new policies require more stringent approaches to domestic violence. The norms and sanctions are changing to prevent fatalities and trauma to children and spouses. Gradually, the focus is shifting from merely "maintaining the peace" to arresting offenders, protecting victims, and referring battered women to shelters and other community resources available to help victims of domestic violence. Better understanding of the dynamics of violent family interactions, more forceful demands by women and community residents for protection from batterers, and more effective techniques for creating protective organizations are all applications of the basic perspectives of sociology.

Of course, all the training in the world does not explain why so many loving relationships become violent. On the other hand, with so much stress and such great economic, social, and emotional demands placed on husbands and wives, an even better question might be why, despite high rates of divorce and family breakup, so many unions are successful and so many individuals crave love and a stable relationship more than almost any other cherished value. We turn, therefore, to a sociological consideration of romance and marriage.

DYNAMICS OF MATE SELECTION AND MARRIAGE

We may think of mate selection and marriage as matters that affect only the partners themselves, but in reality, the concerns of parents and other family members are never very far from either person's consciousness. As we will see shortly, the values of each partner's extended family often have a significant impact on the mate-selection process.

Marriage as Exchange

People in Western cultures like to think that interpersonal attraction and love are the primary factors in explaining why a couple forms a "serious" relationship and eventually marries. However, although attraction and love are clearly important factors in many marriages, social scientists point out that in all cultures the process of mate selection is carried out according to basic rules of bargaining and exchange (Daniel, 1996). Sociologists and economists who study mate selection and marriage from this exchange perspective ask who controls the marriage contract, what values each family is attempting to maximize in the contract, and how the exchange process is shaped by the society's stratification system.

Among the upper classes of China and Japan before the twentieth century, marriage transactions were controlled by the male elders of the community—with the older women often making the real decisions behind the scenes. In many societies in the Middle East, Asia, and preindustrial Europe, the man's family negotiated a "bride price" with the woman's family. This price usually consisted of valuable goods like jewelry and clothes, but in some cultures it took the form of land and cattle. Throughout much of Hindu India, in contrast, an upper-class bride's family paid a "groom price" to the man's family.

Although such norms appear to be weakening throughout the world, arranged marriages remain the customary pattern of mate selection in many societies. The following account describes factors that are often considered in arranging a marriage in modern India:

> Every Sunday one can peruse the wedding ads in the classifieds. Many people still arrange an alliance in the traditional manner—through family and friend connections. Caste is becoming less important a factor in the selection of a spouse. Replacing caste are income and type of job. The educational level of the bride-to-be is also a consideration, an asset always worth mentioning in the ad. A faculty member at a college for girls has estimated that 80–90 percent of the students there will enter an arranged marriage upon receiving the B.A. Perhaps 10 percent will continue their studies. In this way, educating a daughter is parental investment toward securing an attractive, prosperous groom. (Smith, 1989)

In all of these transactions, the families base their bargaining on considerations of family prestige within the community, the wealth of the two families and their ability to afford or command a given price, the beauty of the bride and the attractiveness of the groom, and so on. Different cultures may evaluate these qualities differently, but in each case, the parties involved think of the coming marriage as an exchange between the two families (Goode, 1964). But do not get the idea that only selfish motives are involved in such marriages. Both families are also committing themselves to a long-standing relationship because they are exchanging their most precious products, their beloved young people. Naturally, they want the best for their children (as this is defined in their culture), and they also want a climate of mutual respect and cooperation in their future interfamily relationships.

Changing Norms of Mate Selection

Endogamy/Exogamy   All cultures have norms that specify whether a person brought up in that culture may marry within or outside the cultural group. Marriage within the group is termed *endogamy*; marriage outside the group and its culture is termed *exogamy*. In the United States, ethnic and religious groups normally put pressure on their members to remain endogamous—that is, to choose mates from their own group. These rules tend to be especially strong for women. Among Orthodox Jews, for example, an infant is considered to have been born into the religion only if the mother is Jewish; children of mixed marriages in which the mother was not born a Jew are not considered Jewish. The conflict between Orthodox and Reform Jews over the status of children born to non-Jewish mothers—even when the mothers have converted to Judaism and the children have been raised as Jews from birth—is an example of conflict over endogamy/exogamy norms. Many African tribes have developed norms of exogamy that encourage young men to find brides in specific villages outside the village of their birth. As noted in Chapter 4, such marriage systems tend to promote strong bonds of kinship among villages and strengthen the social cohesion of the tribe while breaking down the animosity that sometimes arises between villages within a tribe.

Homogamy   Another norm of mate selection is **homogamy**, the tendency to marry a person from a similar social background. The parents of a young woman from a wealthy family, for

Symbols of Marriage   The famous wedding portrait painted by Jan van Eyck in 1434 can be interpreted on several levels. On the surface, it is simply a painting of a bride and groom, people of obvious wealth and social standing, in their bedroom. At a deeper level of meaning, however, the painting is a study in the symbols associated with marriage in Europe during the fifteenth century. The wedding bed was understood as a symbol of marital fidelity, the dog as a symbol of trust; and the couple have removed their shoes to show that they stand on the holy ground of matrimony. These and other symbols in the painting could not be understood, however, without some study of the way people of that time thought about marriage and its relationship to religious belief.

A contemporary artist would use other symbols and perhaps try to convey other emotions in depicting a newly married couple, just as the photographer captures a few very telling symbols in the portraits of the young couple and their new baby at the beginning of this chapter. But the van Eyck painting is considered a great work of art because of the timeless and universal quality of the married couple's expression as they gaze out on an approving society.

The *Marriage of Giovanni Arnolfini and Giovanna Cenani*, by Jan van Eyck, 1434. Panel, 33" x 22 1/2".

© Erich Lessing/Art Resource, NY

endogamy: A norm specifying that a person brought up in a particular culture may marry within the cultural group.

exogamy: A norm specifying that a person brought up in a particular culture may marry outside the cultural group.

homogamy: The tendency to marry a person who is similar to oneself in social background.

example, attempt to increase the chances that she will associate with young men of the same or higher social-class standing. She is encouraged to date boys from "good" families. After graduating from high school, she will be sent to an elite college or university, where the pool of eligible men is likely to include many who share her social-class background. She may surprise her parents, however, and fall in love with someone whose social class, religion, or ethnic background is considerably different from hers. When this happens, however, she will invariably have based her choice on other values that are considered important in the dating and marriage market—values like outstanding talent, good looks, popularity, or sense of humor. She will argue that these values outweigh social class, especially if it seems apparent that the young man will gain upward mobility through his career. Often the couple will marry and not worry about his lower social-class background. On the other hand, "the untalented, homely, poor man may aspire to a bride with highly desirable qualities, but he cannot offer enough to induce either her or her family to choose him, for they can find a groom with more highly valued qualities" (Goode, 1964, p. 33).

Homogamy in mate selection generally helps maintain the separateness of religious groups. Because the Census Bureau does not collect systematic data on religious preferences, it is extremely difficult to obtain accurate data on religious intermarriage. Yet sociologists and religious leaders agree that, although parents continue to encourage their children to marry within their religion, there is a trend away from religious homogamy, particularly among Protestants and Catholics (Scanzoni, 1995). This trend may affect rates of divorce and separation. Research shows that families who maintain an active religious life tend to have lower divorce rates than those who do not participate in religious worship together (National Commission on America's Urban Families, 1993). Similarly, a study by Howard Weinberg (1994) found that shared religion has the strongest effect on the likelihood that couples will be able to overcome a period of separation and achieve a successful reconciliation.

Interracial Marriage The norm of homogamy also applies to interracial marriage. Before 1967, when the U.S. Supreme Court struck them down as unconstitutional, many states had laws prohibiting such marriages. After that decision, the number of marriages between blacks and whites began to increase rapidly, reaching almost 4 million in 2000. Intermarriage of Hispanic and non-Hispanic whites also increased during the same period, from about 500,000 marriages in 1960 to 1.7 million in 2000 (*Statistical Abstract,* 2002). Both of these trends are signs that more people are choosing to marry across racial and ethnic lines, but because most marriages are within racial categories, it is clear that the norm of racial homogamy remains strong in the United States.

Romantic Love

Although exchange criteria and homogamy continue to play significant roles in mate selection, romantic attraction and love have been growing in importance in North American and other Western cultures. Indeed, the conflict between romantic love and the parental requirement of homogamy is one of the great themes of Western literature and drama. "Let me not to the marriage of true minds/Admit impediments," wrote Shakespeare in a famous sonnet. The "star-crossed lovers" from feuding families in *Romeo and Juliet*—like the lovers who have forsaken family fortunes to be together in innumerable stories and plays written since Shakespeare's time—attest to the strong value we place on romantic love as an aspect of intimate relationships between men and women (Cancian, 1994).

In his classic study of worldwide marriage and family patterns, William J. Goode (1964) found that, compared with the mate-selection systems of other cultures, that of the United States "has given love greater prominence. Here, as in all Western societies to some lesser degree, the child is socialized to fall in love" (p. 39). Yet, although it may be taken for granted now that people will form couples on the basis of romantic attachment, major changes in the structure of Western societies had to occur before love as we know it could become such an important value in our lives. In particular, changes in economic institutions as a result of industrialization required workers with more education and greater maturity. These changes, in turn, lengthened the period of socialization, especially in educational institutions. This made it possible for single men and women to remain unattached long enough to gain the emotional maturity they needed if they were to experience love and make more independent decisions in selecting their mates.

However familiar it may seem to us, love remains a mysterious aspect of human relationships. We do not know very much—from a verifiable, scientific standpoint—about this complex emotional state. We do not know fully what it means to "fall in love" or what couples can do to make their love last. But two theories that have stimulated considerable research on this subject are Winch's theory of *complementarity* and Blau's theory of *emotional reciprocity*.

Complementary Needs and Mutual Attraction Robert F. Winch's (1958) theory of complementary needs, based on work by the psychologist Henry A. Murray, holds that people who fall in love tend to be alike in social characteristics such as family prestige, education, and income but different in their psychological needs. Thus, according to Winch, an outgoing person often falls in love with a quiet, shy person. The one gains an appreciative audience, the other an entertaining spokesperson. A person who needs direction is attracted to one who needs to exercise authority; one who is nurturant is attracted to one who needs nurturance; and so on.

Winch and others have found evidence to support this theory, but there are some problems with this research. It is difficult to measure personal needs and the extent to which they are satisfied. Moreover, people also show a variety of patterns in their choices of mates. Some people, in fact, seem to be attracted to each other because of their similarities in looks and behavior rather than because of their differences.

Attraction and Emotional Reciprocity Peter Blau's (1964) theory of emotional reciprocity as a source of love is based on his general theory that relationships usually flourish when people feel satisfied with the exchanges between them. When people feel that they are loved, they are more likely to give love in return. When they feel that they love too much or are not loved enough, they will eventually come to feel exploited or trapped and will seek to end the relationship. In research on 231 dating couples, Zick Rubin (1980; see also Peplau, Hill, & Rubin, 1993) found that, among those who felt this equality of love, 77 percent were still together two years later, but only 45 percent of the unequally involved couples were still seeing each other. As Blau explained it, "Only when two lovers' affection for and commitment to one another expand at roughly the same pace do they mutually tend to reinforce their love" (quoted in Rubin, 1980, p. 284).

Blau's exchange approach confirms some popular notions about love—particularly the ideas that we can love someone who loves us and that inequalities in love can lead to separation. Yet we still know little about the complexities of this emotion and how it translates into the formation of the most basic of all social groups: the married couple. This is ironic because books, movies, and popular songs probably pay more attention to love than to just about any other subject.

Francesca M. Cancian (1994), a leading researcher on love in a changing society, notes that when men and women place a high value on individuality, the price may be "a weakening of close relationships." As people spend time enhancing their own lives, they may become more self-centered and feel less responsible for providing love and nurturance. Cancian's research on loving relations among couples of all kinds has convinced her that people are increasingly seeking a form of love that "combines enduring love with self-development." She regards this kind of relationship as based on interdependencies in which each member of the couple attempts to assist the other in realizing his or her individual potential and simultaneously seeks to strengthen the bond between them. But she finds that, to foster a loving interdependent relationship, couples often need to sacrifice a certain amount of independence and career advancement—not a simple matter in a culture that places a high value on individual achievement.

Marriage and Divorce

More than any other ritual signifying a major change in status, a wedding is a joyous occasion. Two people are legally and symbolically joined before their kin and friends. It is expected that their honeymoon will be pleasant and that they will live happily ever after. However, about 20 percent of first marriages end in annulment or divorce within three years (Cherlin, 1996, 2005). Of course, divorce can occur at any time in the family life cycle, but the early years of family formation are the most difficult for the couple because each partner experiences new stresses that arise from the need to adjust to a complex set of new relationships. As Monica McGoldrick and Elizabeth A. Carter (1982) point out: "Marriage requires that a couple renegotiate numerous personal issues that they have previously defined for themselves or that were defined by their parents, from when to sleep, have sex, or fight, to how to celebrate holidays and where and how to live, work, and spend vacations" (p. 178). For people who were married before, these negotiations can involve former spouses and shared children, resulting in added stress for the new couple.

Most states in the United States now have some form of no-fault divorce that reduces the stigma of divorce by making moral issues like infidelity less relevant than issues of child custody and division of property. Although the growing acceptance of divorce helps account for why divorce rates are so much higher in the United States than in other nations with highly educated populations, Americans today also place a higher value on successful marriage than their parents may have. They also place a higher value on marriage than do people from most European and other English-speaking nations, as measured by the far higher rates of marriage in the United States than in Canada, France, Sweden, Germany, and others (Cherlin, 2005). This means that they often divorce in the expectation of forming another, more satisfying and mutually sustaining relationship (Ahlburg & De Vita, 1992).

Trial Marriage In the 1980s, it was widely believed that the practice of "trial marriage," or cohabitation before marriage, would result in greater marital stability: Couples who lived together before marriage would gain greater mutual understanding and a realistic view of marital commitment, and this would result in a lower divorce rate among such couples after they actually married. By the 1990s, however, it had become evident that these expectations were unfounded; in fact, the divorce rate among couples who had lived together before marriage was actually higher than the rate for couples who had not done so. Within ten years of the wedding, 38 percent of those who had lived together before marriage had divorced, compared to 27 percent of those who had married without cohabiting beforehand.

On the basis of an analysis of data from a federal government survey of more than 13,000 individuals, Larry Bumpass and James Sweet (1989; see also Bumpass & Raley, 1995; Bumpass, Raley, & Sweet, 1995) concluded that couples who cohabit before marriage are generally more willing to accept divorce as a solution to marital problems. They also found that such couples are less likely to be subject to family pressure to continue a marriage that is unhappy or unsatisfactory. In addition, cohabitation has become a predictable part of the family life cycle, not only before marriage but also in the interval between divorce and remarriage. Once again, these changes in families mean that family life has become more variable and uncertain for children. Thus, while some advocate a return to traditional family norms and values, others urge greater tolerance for a variety of family types and accommodation to their needs. This conflict is most evident in discussions of same-sex relationships.

Gay and Lesbian Relationships

In 2003, in a stunning 6–3 decision, the U.S. Supreme Court struck down laws banning homosexual sex in Texas and twelve other states. In its ruling, the Court declared that a Texas law banning what some states have called deviate sex acts, or sodomy, was an unwarranted government violation of the right to privacy. Scholars and gay and lesbian activists said that the decision would make their efforts on behalf of gay rights easier. They also believed that it would add new credibility to their struggle to legitimate same-sex marriages, adoptions, and parental benefits. Critics, both among the minority on the Court and among opponents of gay rights, argued that the ruling would open a floodgate of efforts to legalize forms of sexuality such as prostitution and bestiality. In its ruling, however, the Court recognized that its decision follows legal precedents already established in many other democratic nations.

For sociologists who study these issues, it was clear that the Court was responding to important changes in the norms of sexuality in the American population. As noted elsewhere, attitudes toward same-sex relationships have been becoming more tolerant for the last few decades, so that at least 60 percent of Americans today say that they condone same-sex relationships (Herek, 2002). However, among Christian conservatives, there remains a great deal of opposition to gay rights and to acceptance of gay marriages.

Spain, Canada, the Scandinavian nations, the Netherlands, and Belgium have legalized same-sex marriage. In the United States, the Republican Party used gay marriage as a "wedge issue" in the 2006 congressional elections but abandoned that position after its loss at the polls in 2008.

Current Research on Gay and Lesbian Families It is difficult to know exactly how many gay and lesbian households there are in the United States because respondents still hesitate to reveal this information to census takers. However, the 2000 Census included data on 601,209 gay and lesbian families, of which 304,148 were gay male families and 297,061 were lesbian families. These families were found to be living in 99.3 percent of all U.S. counties, with the greatest concentration found in major metropolitan regions across the nation. These figures must be considered extremely conservative estimates of the total number of same-sex families currently living in the United States because other surveys find a higher number of respondents who say they are in such relationships. The official census figures are likely to be only half the actual number (Smith & Gates, 2001).

There is a substantial body of evidence showing that growing up in a lesbian or gay family does not have a negative influence on child development (Goode, 2001). In fact, in a review

Jamie Bergeron, with her brother, Casey, and parents, Sharon Trinkl, left, and Lynette Bergeron, says she is more open-minded because she has grown up with two mothers.

of this research, sociologists Judith Stacey and Timothy Biblarz (2001) found that some studies indicate that children growing up in lesbian and gay families may be more open-minded about relationships, far less homophobic, and more willing to consider careers that cross traditional gender barriers than children growing up in heterosexual families. This research has been attacked by those who believe that same-sex families should be illegal, but there is no empirical support for this ideological position.

Sources of Marital Instability

Because it generally causes a great deal of emotional stress and pain for everyone involved, to say nothing of its financial costs, divorce is a subject of intense study by social scientists (Amato & Sobolewski, 2001; McManus & DiPrete, 2001). Space does not permit a full treatment of this subject here, but we can discuss some of the major variables associated with divorce.

Age at marriage is one of the leading factors in divorce. It seems that it is best not to marry too young or to wait too long before marrying. Women who marry while still in their teens are twice as likely to divorce as women who tie the knot in their twenties. But those who marry in their thirties are half again as likely to divorce as those who marry in their twenties (Ruggles, 1997). How can we explain these differences?

In a study of age and marital instability, Alan Booth and John N. Edwards (1985) examined data obtained from a national sample of more than 1,700 couples. Among the younger couples, a pattern that the authors identified as inadequate role performance—especially not being attuned to the other partner's sexual needs and not being comfortable with the role of husband or wife—seemed to explain much of the instability in their relationships. Among those who married at an older age, another pattern appeared. These couples were more likely to engage in bitter disputes over the division of labor within the family—that is, over the definition of gender roles. Women who marry at older ages, for example, are more likely to demand that both partners share equally in the housework.

It appears that, until about the middle of the twentieth century, when women were more likely to assume the role of homemaker, marriages had a higher probability of lasting. The demand by women for sharing of household roles—what sociologists often call the demand for a "symmetrical family" in which partners share equally in the roles of provider and homemaker—creates new opportunities for both men and women, but it also places more stress on the family in its early stages. Symmetrical families are not necessarily more dissatisfied with their marriages than nonsymmetrical families, nor do they necessarily encounter more marital problems. It appears, however, that the greater ability of both partners to be economically independent may allow them to consider divorce sooner than would a couple in which the wife is not in the labor force (Bittman, 1993).

Even when a couple is doing well economically, researchers find that the husband's role in the couple's problem-solving efforts is a key predictor of whether the couple will divorce. Couples in which the husband "stonewalls," withdraws from arguments, and fails to make an effort to work things out are more likely to divorce (Brody, 1992).

Several other factors have also been found to be correlated with marital instability and divorce. They include the following:

- The couple met "on the rebound" (that is, after one or both partners had recently experienced a great loss or hurt), and the new relationship may be flawed as a result.
- One of the partners wants to live far away from his or her family of orientation, suggesting that feelings of hostility toward family members may complicate the relationship.
- The spouses' family backgrounds are markedly different in terms of race, religion, education, or social class, which may result in differences in values that can cause conflict between the spouses.
- The couple is dependent on one of the extended families for income, shelter, or emotional support.
- The couple married after an acquaintanceship of less than six months or after an engagement of more than three years.
- Marital patterns in either spouse's extended family were unstable.
- The wife became pregnant before or within the first year of marriage (McGoldrick & Carter, 1982; Bumpass, Raley, & Sweet, 1995).

These correlates of marital instability do not, of course, necessarily mean that marriages that occur under these conditions are doomed to break up. They merely suggest that, based on statistical probabilities, such marriages have a higher chance of ending in divorce.

The Impact of Divorce

At midcentury, it was still considered almost impossible for a divorced man to run for the presidency. Ronald Reagan was the first president who had previously been divorced. Senators Robert Dole and Phil Gramm, candidates in the 1996 presidential election, had each been divorced. Divorce is much more prevalent today. Nevertheless, it remains a significant event in people's lives—as significant as marriage itself. But unlike a wedding, a divorce is not a happy event. Although some divorces turn out well for both partners, most do not.

Research on divorce has shown that many of the most disruptive consequences are due to its economic impact. Women suffer an average decline of about 30 percent in their income in the year after separation, whereas men experience a 10 percent increase (McManus & DiPrete, 2001). In fact, most women who apply for various forms of public support do so because they have recently experienced a drastic decline in income as a result of divorce, separation, or abandonment. In addition, almost 40 percent of divorced mothers (and the children in their custody) move within the first year after divorce, and another 20 percent move after a year, a rate far higher than that for married couples (Cherlin, 1996; Furstenberg & Cherlin, 1991). And as if the breakup of their families is not stressful enough, many children also experience the loss of friends and familiar neighborhoods.

Beyond the material effects of divorce, there are the longer-term effects on family roles and the feelings of family members. In an important longitudinal study, Wallerstein and Blakeslee (1989) tracked sixty families with a total of 131 children for a period of ten to fifteen years after divorce. Both parents and children were interviewed at regular intervals. The data from those interviews show that the turmoil and stress of divorce may continue for a year or more. Many divorced adults continue to feel angry, humiliated, and rejected as much as eighteen months later, and the children of divorced parents tend to exhibit a variety of psychological problems. Moreover, both men and women have a diminished capacity for parenting after divorce. They spend less time with their children, provide less discipline, and are less sensitive to their needs. Even a decade after the divorce, the parents may be chronically disorganized and unable to meet the challenges of parenting. Instead, they come to depend on their children to help them cope with the demands of their own lives, thereby producing an "overburdened child"—one who, in addition to handling the normal stresses of childhood, also must help his or her parent ward off depression (Wallerstein & Blakeslee, 1989; Wallerstein, Lewis, & Blakeslee, 2000).

There is a silver lining to the dark cloud of divorce, however. Because so many adults who are now marrying for the first time come from families that have experienced divorce, they are likely to take more time in selecting their mates in an effort to make sure that their choice is best for both partners; in addition, they try to become economically secure before marrying, thereby eliminating a major source of stress in a new marriage (Cherlin, 2005). Thus, the recent modest downturn in the divorce rate may be expected to continue in the future. In any case, the data on the effects of divorce on adults and children suggest that societies need to do more to ease the stress that young families experience—for example, by providing more day care facilities, establishing more flexible work schedules, and offering opportunities for family leave so that they can care for family members in times of need.

Research on the problems of African American families adds emphasis to the relationship between poverty and problems of family dissolution. Research on black families also confirms the basic need for family-oriented social policies.

THE BLACK FAMILY: A CASE STUDY IN RESILIENCE

The headlines announced a stunning new research finding: "Birth Rate Falls to a 40-Year Low among Unwed Black Women: Education and Contraception Are Seen as Crucial" (Holmes, 1998). The news came from a research project headed by demographer Stephanie J. Ventura of the National Center for Health Statistics. Her research showed that during the 1990s, fertility rates for unmarried black teenagers fell more steeply (31 percent) than those for any other U.S. population group—from 115.5 births per 1,000 unmarried women in 1990 to 79.2 in 2000. This decline appears to be part of a decrease in fertility among all black women—teenagers and adults, married and unmarried, that continued at least until 2008. In fact, births to unmarried women of any race or ethnicity declined during the two decades, but not as rapidly as births to African American teenagers. As Figures 15.15 and 15.16 indicate, fertility rates for unmarried African American (and Hispanic) teenagers remain substantially higher than those for white teenagers, but they are declining at a faster rate. It should also be noted that, although overall rates of teenage childbearing in the United States have been decreasing, they are still the highest rates for any urban industrial nation (Ventura et al., 2001).

In seeking explanations for this significant shift in births to unmarried black women, Ventura and others point to increases in education programs in African American communities, which encourage teenagers and younger adults to stay in school and to use contraception. Donna E. Shalala, former secretary of health and human services, pointed out that increased condom use is associated with greater awareness of the risk of AIDS. "We really did scare people," she commented, and of course AIDS is a matter worthy of fear and great precaution (Holmes, 1998).

Encouraging news about declines in births to unmarried teenagers and young adults is only part of the larger and far more complex picture that describes the strengths and problems of African American families. Poverty rates are higher for black and other minority families than for white families, even though there are many more poor white families. Adding to the confusion is the controversy over "family values." In the case of the black family, the central issue is whether high rates of poverty, crime, and school failure in low-income, segregated black communities are a cause or a consequence of the large number of black single-parent families.

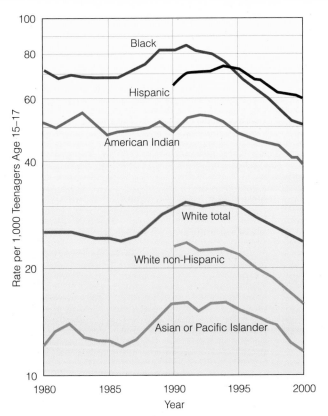

Figure 15.15 Birth Rate for Teenagers Age 15–17, by Race and Hispanic Origin: United States, 1980–2000. Source: Ventura et al., 2000.

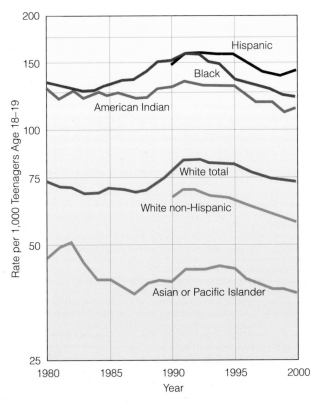

Figure 15.16 Birth Rate for Teenagers Age 18–19, by Race and Hispanic Origin: United States, 1980–2000. Source: Ventura et al., 2000.

This issue is not a new one. More than a half century ago, the eminent African American sociologist E. Franklin Frazier argued in *The Negro Family in the United States* (1957/1939) that the violence of the Middle Passage, in which Africans were captured, shipped in chains, and sold into slavery, had wiped out their previous family ties. Frazier believed that as a result of that experience, the only enduring family form among blacks was the bond between mother and child. He urged black Americans to emulate the white middle-class family. In 1965, his views were endorsed by the famous *Moynihan Report,* which attributed the higher rates of social problems found in poor black communities to inadequate family functioning and the "tangle of pathology" that results from fatherlessness.

Critics of this theory argue that it neglects the effects of both past and present discrimination, which have been particularly severe for black men in America. The extreme difficulty that black men experience in gaining the economic security necessary to sustain a family, it is argued, becomes a formidable obstacle to marriage (Cherlin, 2005). Focusing on the problems of the black family suggests that there is something wrong with poor black people rather than with the larger society (Western, 2007).

Most sociologists agree that the rapid rise in female-headed African American families was attributable mainly to the difficulties that young black men with limited education encounter when seeking jobs. Harriet P. McAdoo, a specialist in the study of disadvantaged women and children, makes the following point:

The tradition used to be that if a woman got pregnant, they were expected to get married, if not forced to get married, to supply income and a name for the child. But if the man has no job, there's no impetus for them to get married. Her parents would have three people to support instead of just two. The boy would become an additional burden on the family. Parents feel: "Why should I force her to marry him? He'll be a drain." (quoted in Cummings, 1983, p. 56)

Other sociologists believe that the problem is multidimensional. As family sociologist Bonnie Thornton Dill points out: "Poverty, unemployment, welfare programs that penalize families: these things explain a major portion of the variance between blacks and whites in the number of female-headed households, not all of it." She notes that changes in the nature of urban poverty that increasingly isolate poor people from quality education and employment opportunities also deprive them of the opportunity to marry. "Teen-age pregnancy among blacks has not increased so much," Dill concludes, "but the rate of marriage among teen-agers has dropped and the number of children born to married couples has dropped. These things are related to perceptions of opportunity" (quoted in Suro, 1992, p. A12).

Even though the data show that the phenomena of teenage pregnancy, female-headed families, and children being raised in poverty are related to social class rather than to race, the fact that more blacks than whites are found in the lower income levels of society means that these problems are proportionally more serious for blacks. In the words of sociologist Joyce Ladner (1973), "Life in the black community has been conditioned by poverty, discrimination, and institutional subordination" (p. 425). Under these conditions, she concludes, the black family has shown surprising resilience. Although a black family may include households headed by single women, the extended family often provides substantial support for its less fortunate members.

Intimate Relationships At the beginning of the chapter, we saw a young couple become a family of three as they shared the intimate experience of childbirth. In closing the chapter, we return to some emotional photos of intimate relationships. All the photos were created by professional photographers trying to capture moments of love and its expression in a diverse social world.

The prom night, the gay lovers expressing their deep devotion and sadness, the rapturous young couple lost in their intimacy—all convey something of the joy of human love. But each of the photos may also make one wonder about the particular story behind it. For example, will the relationship of the lovers at their prom last through the subsequent phases of life as each partner continues to grow and meet new people? If questions like this sound as if they could be from a soap opera, there is a good reason for that. After all, love stories are featured on daytime television because, of all subjects, love is closest to our hearts, our desires, and our doubts. So often, when it seems to be within our grasp, it takes wing. But once we know it at all, we seek it forever.

Phillip Jarrell/Digital Vision/Getty Images

© 1994 Frank Fournier/Contact Press Images

Person with AIDS being comforted by a friend.

Prom night.

© Dan Habib/Impact Visuals

SUMMARY

How has the family been changing as an institution of modern societies?

- In all known societies, almost everyone is socialized within a network of family rights and obligations that are known as family role relations.

- In simple societies, the family performs many other functions as well, but in modern societies, most of the functions that were traditionally performed by the family are performed partly or entirely by other social institutions.

- A *family* is a group of people related by blood, marriage, or adoption, and the role relations among family members are known as *kinship* relations.

- The smallest unit of family structure is the *nuclear family,* consisting of two or more people related by consanguineous ties or by adoption who share a household.

- The nuclear family in which a person is born and socialized is the *family of orientation,* and the nuclear family a person forms through marriage or cohabitation is the *family of procreation.*

- An *extended family* includes an individual's nuclear family plus all the nuclear families of his or her blood relatives.

- The traditional household consisting of two parents and their children is no longer the typical American family. Since the 1940s, there has been a dramatic increase in female-headed single-parent families and in nonfamily households, as well as in the numbers of women and men living alone and in the numbers of unmarried same-sex couples.

- The economy exerts extremely strong influences on the family and can produce drastic changes in the number of families that experience deprivation and poverty.

- These changes are often complicated by divorce, remarriage, and the combining of children of different marriages in stepfamilies.

- The typical stages of the family life cycle are family formation, start of childbearing, end of childbearing, empty nest, and family dissolution (that is, the death of one spouse).

- As it passes through this cycle, every family experiences changes in its system of role relations.

How do sociological perspectives explain the causes and consequences of changes in family structure and family relations?

- The structural context within which family life occurs can affect family interactions in a variety of ways. Problems may arise in connection with the demands placed on the family by institutions of the larger society, or as a result of its position in the society's stratification system.

- A basic contradiction inherent in the institution of the family is the need to maintain the individuality of each member while providing love and support for him or her within a set of interdependent relationships.

- From a conflict perspective, changes in the family as an institution cannot occur without conflict, both within the family and between the family and other institutions. Such conflict is illustrated by public debates over family policies and "family values."

- Functionalist theorists have called attention to the loss of family functions that occurs as other social institutions assume functions that were previously reserved for the family. At the same time, they note that modern families play a vital part in early-childhood socialization, in the emotional lives of their members, and in preparing older children for adult roles.

What factors enter into the choice of a mate and the decision to marry?

- In all cultures, the process of mate selection is carried out according to basic rules of bargaining and exchange.

- In many societies, the customary pattern of mate selection is the arranged marriage, in which the families of the bride and groom negotiate the marriage contract.

- All cultures also have norms that specify whether a person brought up in that culture may marry within or outside the cultural group. Marriage within the group is termed *endogamy*; marriage outside the group is termed *exogamy*.

- In societies in which marriages are based on attraction and love, individuals tend to marry people similar to themselves in social background, a tendency that is referred to as *homogamy*.

- Homogamy generally helps maintain the separateness of religious and racial groups.

- Compared with the mate selection systems of other cultures, that of the United States gives love greater prominence. Yet, from a scientific standpoint, little is known about this complex emotional state. It appears that people who fall in love tend to be alike in social characteristics but different in their psychological needs; however, this is not always the case. There is also considerable evidence that love relationships are more lasting when the partners' affection for each other is roughly equal.

- In the United States and other Western societies, the rate of divorce has risen sharply since World War II.

- In the 1980s, it was widely believed that the practice of cohabitation before marriage would result in greater marital stability, but in fact the divorce rate among couples who lived together before marriage is higher than the rate for couples who have not done so.

- There is substantial evidence showing that growing up in a lesbian or gay family does not have a negative influence on child development.

- Age at marriage has been found to be one of the leading factors in divorce. Marriages that take place when the woman is in her teens or in her thirties are much more likely to end in divorce than marriages that take place when the woman is in her twenties.

- Among other factors that have been found to be correlated with divorce are marked differences in the family backgrounds of the spouses, dependence on either spouse's extended family, patterns of marital instability in either spouse's extended family, and early pregnancy.

- Studies of the impact of divorce have found that the turmoil and distress of divorce may continue for a year or more. Both men and women have a diminished capacity for parenting after divorce and may come to depend on their children to help them cope with the demands of their own lives.

What factors have contributed to the resilience of the black family?

- Even though the data show that the phenomena of teenage pregnancy, female-headed families, and children being raised in poverty are related to social class rather than to race, the fact that more blacks than whites are found in the lower income levels of society means that these problems are proportionally more serious for blacks.

- Under these conditions, the black family has shown surprising resilience. Although within a given black family there may be households headed by single women, the extended family often provides substantial support for its less fortunate members.

The Kornblum Companion Website

www.cengagebrain.com

Supplement your review of this chapter by going to the companion website to take one of the tutorial quizzes, use the flash cards to master key terms, and check out the many other study aids you'll find there. You'll also find special features, such as GSS data and Web links that will put data and resources at your fingertips to help you with that special project or to do some research on your own.

Bill Pugliano/Getty Images

Religion

FOCUS QUESTIONS

What features are common to all religions, no matter what their beliefs or how they are organized?

What are some characteristics of the world's major religions?

Do religions defend the status quo at all times, or have they been involved in lasting social change?

What are the main forms of religious organization?

What are some important trends in religion in the United States today?

The months following the terrorist attacks on the United States saw innumerable acts of cowardly revenge as well as neighborly tolerance and generosity. Vandals in Columbus, Ohio, for example, stole into that city's oldest mosque, drilled holes in the floor, broke water pipes, flooded the building, and tore apart copies of the Koran, the Muslim sacred text. The mosque's president, Mohammed Shareef, said that the sanctuary would not be usable for many months. Non-Muslim groups in the community immediately offered space, money, and other support. A local Jewish congregation and the Rotary Club of Columbus offered to share facilities while the mosque was closed (AP, 2002).

Elsewhere in Ohio, as in many other parts of the United States, there were attacks on Islamic institutions and individual Muslims, but there were also intense efforts by officials, from President Bush to local leaders in hundreds of communities, to reach out to Arab Americans and followers of Islam and assure them that Americans know that Islam is not a religion of terror (Bakalian & Bozorgmehr, 2010).

One of the worst effects of terrorist attacks, however, is to destroy trust among different groups and weaken the guarantees of civil liberties. Despite efforts to allay fears, U.S. government actions against Islamic organizations and against individuals suspected of affiliations with terrorist networks inevitably caused increased fear and sadness in communities like Columbus, where Muslims and non-Muslims live side by side. "The Koran is very clear: Your neighbors are those where you live. America is our neighborhood, so we have to be loyal to America," said Professor Muqtedar Khan of Adrian College in Michigan (quoted in Lampman, 2002, p. 15).

Educated American Muslims often feel most strongly the dilemmas of trying to be loyal citizens while witnessing increased surveillance of their behavior and questions about their trustworthiness. "Rather than treating American Muslims as assets—using their knowledge of the Muslim world for diplomacy and even for intelligence—the government is treating them as suspects," Khan continued (p. 15). The domestic antiterrorism campaign seeks to reassure the majority of Americans that it is safe to travel and lead normal lives. Frequently it has the opposite effect on Muslims who are caught up in searches aimed at finding violent criminals, which many experience as understandable but extremely unsettling.

In 2010, when pleading guilty to ten charges of terrorism and weapons violations, Faisal Shahzad, who had attempted to set off a massive car bomb in New York City's Times Square, was unrepentant and defiant. Originally from Pakistan, he called himself "a Muslim soldier." According to Muslim American journalist Amitabh Pal (2010, June 23), "Shahzad is a disgrace to Pakistani-Americans. . . . They now have to walk around under a pall of suspicion. His actions sully the sacrifice of Mohammad Salman Hamdani, who was killed on Sept. 11, 2001, while trying to rescue people inside the World Trade Center. His deeds also mar the work of Pakistani-American rock star Salman Ahmad, who has made it his life's mission to spread a Muslim message of peace and reconciliation."

After 9/11, non-Muslim Americans flocked to bookstores to buy copies of the Koran and learn more about

Islam and the religious feelings of their Muslim neighbors. But small groups of haters also continued to spread fear and ill will.

All of the world's religions espouse peace and understanding among neighbors. All disavow terrorism. But some aspects of religious belief work against tolerance and, in the minds of zealots, bar the acceptance of differences in belief and behavior. All of these phenomena make the sociology of religion one of the most urgently needed and fascinating subfields within sociology.

RELIGION IN SOCIETY

Religion, one of the oldest human institutions, is also among the most changeable and complex. On the one hand, religion expresses our deepest yearnings for spiritual enlightenment and understanding; on the other, conflicts over religious beliefs and practices have resulted in persecution, wars, and much human suffering, as can be seen, for example, in the violence between Muslims and Jews in the Middle East, between Hindu Indians and Muslim Pakistanis, and between Catholics and Protestants in Northern Ireland. Little wonder, therefore, that the founders of sociology—including Émile Durkheim, Karl Marx, and Max Weber—all wrote extensively about the power of religion and the great changes religion has undergone as societies have evolved.

Defining Religion

Religion is not easy to define. One could begin with a definition that has a concept of God as its core, but many religions do not have a clear concept of God. One could define religion in terms of the emotions of spirituality, oneness with nature, awe, mystery, and many other feelings, but that would not be a very helpful definition because emotions are extremely difficult to capture in words. Taking another tack, one might think in terms of organized religion—churches, congregations, ministers, rabbis, and so on—but clearly the organizational aspect of religion is just one of its many dimensions. It is frustrating

AP Photo/Khalil Hamra

Throughout most of human history, until the past century or two, religion dominated all cultural life, as it still does in many societies. This photo shows Muslim worshippers praying outdoors in Mecca, Saudi Arabia, the most sacred site of the Islamic faith.

to have to work so hard to define something that seems so commonplace, yet without a good working definition of religion, it is impossible to compare different religions or refer to particular aspects of religion.

We can approach a working definition of religion by saying that **religion** is any set of coherent answers to the dilemmas of human existence that makes the world meaningful. From this point of view, religion is how human beings express their feelings about such ultimate concerns as sickness, death, or the meaning of life. An outstanding function of religion is to reduce personal insecurities and anxieties by providing answers to the unknown and incomprehensible (Hunt, 1999). Almost all religions involve their adherents in a system of beliefs and practices that express devotion to the supernatural and foster deep feelings of spirituality. In this sense, we say that religion functions to meet the spiritual needs of individuals.

But religion has also been defined in terms of its social function: It is a system of beliefs and rituals that helps bind people together through shared worship, thereby creating a social group. A **ritual** is a formal pattern of activity that expresses symbolically a set of shared meanings; in the case of religious rituals such as baptism or communion, the shared meanings are sacred. The term *sacred* refers to phenomena that are regarded as extraordinary, transcendent, and outside the everyday course of events—that is, supernatural. That which is sacred may be represented by a wide variety of symbols, such as a god or set of gods; a holy person, such as the Buddha; various revered writings, such as the Bible,

---

religion: Any set of coherent answers to the dilemmas of human existence that makes the world meaningful; a system of beliefs and rituals that helps bind people together into a social group.

ritual: A formal pattern of activity that expresses symbolically a set of shared meanings.

sacred: A term used to describe phenomena that are regarded as extraordinary, transcendent, and outside the everyday course of events.

the Torah, or the Koran; holy objects, such as the cross or the star of David; holy cities, such as Jerusalem or Mecca; and much else. The term *profane* refers to all phenomena that are not sacred (Kurtz, 1995).

Religion as an Institution

Religion is a major social institution because it carries out important social functions and encompasses a great variety of organizations (for example, churches, congregations, charities), each with its own statuses and roles (ministers, priests, rabbis, parishioners, fundraisers) and specific sets of norms and values (the Ten Commandments, the Golden Rule, the Koranic rules). As an institution, religion helps people express their feelings of spirituality and faith. Religion is often said to be a cultural institution because it guides a society's mental life, especially its ideas about morality, goodness, and evil. Of course, religion is not alone in performing these functions, but it remains a powerful source of moral precepts (Dudley, 2002; Wolf, 2007).

Religion also confers legitimacy on a society's norms and values. Families seek the "blessing of holy matrimony" in wedding ceremonies. Baptisms, bar mitzvahs, confirmations, and other religious ceremonies mark the passage of children through their developmental stages and are occasions for statements about proper behavior and good conduct. Swearing on a Bible is common in courtrooms and on other occasions when norms of truth and fairness are being enforced. The political institutions of society also often look to religion for legitimation. The monarchs of eighteenth-century Europe invoked the will of God in their activities and sometimes claimed to rule by "divine right." The Pledge of Allegiance includes the phrase "under God" to reinforce the feeling that the destiny and unity of the nation and the values of liberty and justice are human efforts to carry out an even higher purpose.

Church and State Most Western nations are characterized by separation of church and state. For the most part, this means that these societies, including the United States, prohibit the establishment of an official state religion. The First Amendment to the U.S. Constitution states: "Congress shall make no law respecting an establishment of religion, or prohibiting the free exercise thereof; or abridging the freedom of speech, or of the press; or the right of the people peaceably to assemble, and to petition the government for a redress of grievances."

But there are ambiguities in this law. Although the government clearly is prohibited from doing anything that might seem to favor one religion, such as Christianity, over another, neither can it prohibit the voluntary or free exercise of prayer or other forms of religion. Does this mean, for example, that children cannot pray in school? No, but it does mean that anything that even hints of mandatory prayer in school is contrary to the separation doctrine. Controversies over issues like school prayer are an indication of even more basic conflicts about exactly what separation of church and state means (Whittle, 2003). We return to this issue later in the chapter.

Religion in the Public Square In the United States and Canada, there is a great deal of discussion of "religion in the public square." This phrase refers to the claim that it is not a violation of the separation of church and state for people with strong religious views, and even for paid religious leaders, to argue for or against specific policies, such as prayer in schools or gay marriage, based on one's religious beliefs (Dunn, 2009). But when, if ever, does this view of religion and public life begin to cross the boundary between church and state? Clearly, citizens, whether or not they are religious, have every right to express their opinions about moral and spiritual issues and to use these as a basis for supporting one political candidate or party or another. But what about pastors, rabbis, and priests? Some advocates of religion in the public square stop short of advocating that religious leaders take political stands. They argue that when a church leader speaks in support of a political party, he or she compromises the church's neutrality, and that when pastors take political positions on behalf of their congregations, they assume, without necessarily knowing, that all their parishioners hold the same opinions.

> **Thinking Critically** Should religious values be considered when evaluating a candidate who is running for election? Should religious monuments be built on public lands? Can one believe that religion has a place in the "public square" and still oppose placing religious symbols in public places?

Many nations do not separate church and state; instead, they have chosen an official state religion, which usually conforms to the beliefs of the vast majority of the citizens. Israel, where Judaism is the official state religion, other nations of the Middle East, where Islam is the official state religion, and Greece, where Orthodox Christianity is the state religion, are examples of societies that do not practice

---

profane: A term used to describe phenomena that are not considered sacred.

separation of religious and governmental institutions. Other religions in these societies are, in principle, protected under the law, but a particular religion is chosen as the national religion and its clerics are granted powers that those of other religions do not have. Nations with some form of state religion often experience conflict when religious minorities fear that their rights will not be protected equally under the law, as occurred, for example, in parts of the former Yugoslavia in the 1990s.

Theocracy A more extreme form of church involvement in a society's politics is *theocracy*. In Iran, for example, religious leaders actually have the power to control elected officials, especially through their control of the judicial branch of government and of the media. Although the voting population has voiced its desire for more secular government, the hold of the Islamic leaders over civil politics remains quite strong (Kazemipur & Rezaei, 2003). In the ancient world, theocracies were far more common, and many of the societies described in the Old Testament, including Israel, were theocracies. With the secular revolutions of the modern age, from the American and French Revolutions of the eighteenth and nineteenth centuries to the Russian Revolution in the early twentieth, it appeared that secular states that excluded clerics from official power would become the norm. But events in the Muslim world and elsewhere, in which religious leaders have taken power from secular ones, cast doubt on the theory that secularization is inevitable. Even in the United States, which is far from being a theocracy, the growing power of fundamentalist Christian religious leaders over elected political leaders shows that the separation of church and state is not complete.

The Power of Faith

Until comparatively recent times, religion dominated the cultural life of human societies. Activities that are now performed by other cultural institutions, particularly education, art, and the media, used to be the province of religious leaders and organizations. In hunting-and-gathering bands and in many tribal societies, the holy person, or shaman, was also the teacher and communicator of the society's beliefs and values. In early agrarian societies, the priesthood was a powerful force; only the priests were literate and, hence, able to interpret and preserve the society's sacred texts, which represented the culture's most strongly held values and norms. For example, in ancient Egypt, where the pharaoh was worshipped as a god, his organization of regional and local priests controlled the entire society.

Today, religion continues to play an important part in the lives of people throughout the world, even though the influence of organized religions is diminishing in many societies. In the United States, for example, the Gallup Poll routinely asks Americans whether they believe in God. In 1944, 96 percent of the population said they were believers; at the beginning of the twenty-first century, the proportion of believers remains over 90 percent (National Opinion Research Center, 2006). The strength of religious attitudes and the influence of some religions can also be seen in the conflict over abortion, which plays such a prominent role in American politics, and in the controversies generated by Christian fundamentalists who believe in the literal interpretation of the Bible and may therefore deny the validity of evolutionary theory. Outside the United States, the Islamic world is torn by religious strife between liberals and fundamentalists, and Northern Ireland and Lebanon remain deeply divided largely because of conflict between Protestants and Catholics or between Christians and Muslims.

At the same time that religion is a source of division and conflict, however, it can also be a force for healing social problems and moving masses of people toward greater insight into their common humanity. This occurs at the micro level of interaction—for example, in groups like Alcoholics Anonymous, in which spirituality is an essential part of the recovery program. At the macro level, the power of faith can be seen in impoverished rural and urban communities throughout Latin America. In those communities, Catholic church leaders, parish priests, and lay parishioners have embraced the ideals of a "social gospel" that seeks the liberation of believers from poverty and oppression. These "base Christian communities," as they are often called, have become a powerful force in the movement for social justice and other far-reaching changes in their societies (Tabb, 1986).

Secularization and Its Limits Despite the great power of spirituality, since medieval times, the traditional dominance of religion in many spheres of life has been greatly reduced. The process by which this has occurred is termed **secularization**. This process, according to Robert A. Nisbet (1970), "results in . . . respect for values of utility rather than of sacredness alone, control of the environment rather than passive

---

secularization: A process in which the dominance of religion over other institutions is reduced.

These two pictures of cathedrals provide a graphic illustration of secularization. In the medieval town, the cathedral dominated the landscape; it was the highest, and symbolically the most important, building. In the modern city, even the great cathedral is dwarfed by the greater scale and importance of the commercial buildings that surround it.

submission to it, and, in some ways most importantly, concern with man's present welfare on this earth rather than his supposed immortal relation to the gods" (p. 388). Secularization usually accompanies the increasing differentiation of cultural institutions—that is, the separation of other institutions from religion. In Europe during the Middle Ages, for example, there were no schools separate from the church. The state, too, was thought of as encompassed by the church or at least as legitimated by the official state religion and church organization. Laws and courts were guided by religious doctrine, and clerical law could often be as important as civil law—indeed, to be tried as a heretic often meant torture and death. Churches engaged in large-scale economic activity, owned much land and property, and often mounted their own armies.

The Renaissance, the Enlightenment, and the revolutions of the eighteenth and nineteenth centuries all speeded the process of differentiation in which schools, science, laws, courts, and other institutions gained independence from religious control. However, this process has not occurred at the same rate throughout the world. For example, the removal of education from the control of religious institutions has occurred more slowly in some societies than in others. In eastern Europe, all education was controlled by the state until recently. In most countries of western Europe and the Americas, there are religious schools, but these are separate from and overshadowed by the state-run educational system (Kurtz, 1995).

The emergence of cultural institutions like the secular public school and the weakening of the influence of religion on government do not result in complete secularization. People who are free to determine their own religious beliefs and practices may attend church less often or not at all, but total secularization does not occur (Finke & Stark, 1992). Moreover, in almost every society that has experienced secularization, one can find examples of religious revival. Indeed, modern communication technologies, especially television, have contributed immensely to the revival of interest in religion, as witnessed by the popularity of "televangelists" like Oral Roberts and Pat Robertson (Stark & Bader, 2008).

Faith and Tolerance: Uneasy Together

People who believe that theirs is the "one true faith" are found in every religion. For such true believers, it can be extremely difficult, if not impossible, to practice tolerance and accept other beliefs and other religions. Thus, the Taliban's destruction of ancient Buddhist statues in Afghanistan in 2001 was widely viewed as an act of extreme religious intolerance. So were some extremely anti-Islamic statements made by a few Christian and Jewish religious leaders in the United States after the events of September 11. But is Islam more or less intolerant than Christianity, Judaism, Hinduism, or any other major religion?

Adherents of all the major religions have at times acted as if theirs were the one true faith, and they have used this claim as a basis for persecuting those who deviated from their beliefs, both within their own religion and in other religions. In our own time, the assassination of peacemaker Yitzhak Rabin by fanatical Jewish fundamentalists is an example of extreme intolerance within a religion. In Christianity, for centuries before the Enlightenment, it was common practice to harass and kill people who practiced pagan religions and to attack believers in other monotheistic faiths (particularly Judaism and Islam), who were called "infidels" or nonbelievers. During the same historical period, far more tolerance for other religions was practiced in the Islamic world. Jews expelled from Spain during the Catholic Inquisition in the 1490s, for example, were often given safe havens in Islamic societies. So from a historical perspective, we see that all major religions pass through phases during which they are more or less tolerant of other religions.

The separation of church and state, a hallmark of American culture and society as well as many other democracies, negates the idea that there is only one true faith. Instead, it promotes the idea that "you may believe what you like, and so may I, and neither can impose belief on the other." Thus, Thomas Cahill, a scholar of religion and an expert on religious tolerance, writes:

> Each of the great religions creates, almost from its inception, a colorful spectrum of voices that range from pacifist to terrorist. But each religion, because of its metaphorical ambiguity and intellectual subtlety, holds within it marvelous potential for development and adaptation. This development will be full of zigzags and may sometimes seem as slow as the development of the universe, but it runs—almost inevitably, it seems—from exclusivist militancy to inclusive peace. (Cahill, 2002, p. 4)

© Reuters/Corbis

This Afghan statue of the Buddha, carved into a mountainside over 1,500 years ago, was destroyed by the Taliban, the intensely fundamentalist Islamic government that was deposed in 2001 by a United Nations–sanctioned invasion. The Taliban ruled Afghanistan according to its own interpretations of Islamic and tribal Pashtun laws. Islamic leaders throughout the world condemned the destruction as "un-Islamic" and "an act of cultural genocide against humanity."

VARIETIES OF RELIGIOUS BELIEF

Religious sentiments and behavior persist even in highly secularized societies like the United States. As sociologist Robert Wuthnow (1988) points out:

> The assumption that religion in modern societies would gradually diminish in importance or else become less capable of influencing public life was once widely accepted. That assumption has now become a matter of dispute. . . . Modern religion is resilient and yet subject to cultural influences; it does not merely survive or decline, but adapts to its environment in complex ways. (p. 474)

Many of the adaptations that religions make to social change become evident when one examines the varieties of religious belief in the world today, and especially when one studies the major world religions.

Major World Religions

In a 1913 essay, Max Weber commented that "by 'world religions' we understand the five religions or religiously determined systems of life-regulation which have known how to gather multitudes of confessors around them" (1958/1913, p. 267). Among these Weber included, in addition to Christianity, "the Confucian, Hinduist, Buddhist, and Islamic religious ethics." He added that, despite its small population of adherents, Judaism should also be considered a world religion because of its influence on Christianity and Islam as well as on Western ethics and values even outside the religious sphere of life.

In discussing religion, sociologists often refer to the "Islamic world" of the Middle East, the "Roman Catholic world" of Latin America and southern Europe, the "Hindu world" of the Indian subcontinent, and the "Buddhist world" of the Far East. The United States, northern Europe, and Australia are among the societies in which Protestantism is strongest. There are also, of course, the nations of eastern Europe and the former Soviet Union, where until recently communism as a civil religion was the only legitimate belief

NORTHERN AMERICA		(296)
Roman Catholic	75	(25%)
Protestant	121	(41%)
Other Christian	58	(20%)
Jewish	6	(2%)
(All other)	36	(12%)

LATIN AMERICA		(490)
Roman Catholic	409	(83%)
Protestant	35	(7%)
Other Christian	12	(2%)
(All other)	34	(7%)

EUROPE		(728)
Roman Catholic	269	(37%)
Protestant	80	(11%)
Other Christian	21	(3%)
(All other)	370	(51%)

AFRICA		(748)
Muslim	309	(41%)
Christian	361	(48%)
Ethnic Religionist	70	(9%)
(All other)	8	(1%)

THE WORLD		(5,804)
Christian	1,955	(37%)
Muslim	1,126	(19%)
Hindu	793	(14%)
Buddhist	325	(6%)
(All other)	1,605	(28%)

ASIA		(3,513)
Christian	303	(9%)
Muslim	778	(22%)
Hindu	707	(20%)
Buddhist	322	(9%)
(All other)	1,403	(40%)

OCEANIA		(29)
Christian	24	(82%)
(All other)	5	(17%)

Note: Percentages may not add up to 100 because of rounding.
Source: Data from *Britannica Book of the Year*, 2002.

MAJOR RELIGIONS OF THE WORLD, RANKED BY SIZE (MILLIONS OF MEMBERS)

1. Christianity	2,000	7. primal-indigenous	190	15. Shinto	4
2. Islam	1,300	8. Sikhism	23	16. Cao Dai	3
3. Hinduism	900	9. Yoruba religion	20	17. Tenrikyo	2.4
4. Secular/Nonreligious/ Agnostic/Atheist	850	10. Juche	19	18. Neo-Paganism	1
5. Buddhism	360	11. Spiritism	14	19. Unitarian-Universalism	0.8
6. Chinese traditional religion	225	12. Judaism	14	20. Scientology	0.75
		13. Baha'i	6	21. Rastafarianism	0.7
		14. Jainism	4	22. Zoroastrianism	0.15

Figure 16.1 Estimated Religious Membership, by Continent (in millions).

system (although millions of people resisted the state's efforts to eradicate traditional religious faiths) (Robertson, 1985).

Figure 16.1 shows the distribution of the world's major religions. Note that in many parts of the world, particularly in Asia and Africa, large numbers of people practice other religions. Many are indigenous peoples whose religious practices may be influenced by the major world religions but continue to be based on beliefs and ways of worship that are unique to that culture. For example, in her sociological portrait of Mama Lola, a Haitian priestess, Karen McCarthy Brown notes that many Haitians practice voodoo, which is derived from African religious beliefs. Some of Mama Lola's followers may worship in Catholic churches yet continue to practice the traditional rituals of voodoo as well (Brown, 1991, 1992).

Classification of Religious Beliefs

The religions practiced throughout the world today vary from belief in magic and supernatural spirits to complicated ideas of God and saints, as well as secular religions in which there is faith but not God (Bowker, 1997). With such a wide range of religious beliefs and practices to consider, it is useful to classify them in a systematic way. One often-used system classifies religions according to their central belief. In this scheme, the multiplicity of religious forms is reduced to a more manageable list consisting of five major types: simple supernaturalism, animism, theism, abstract ideals, and civil religion. (See the For Review chart on page 407.)

In this section, we describe each type briefly. Be warned, though, that not all religions fit neatly into these basic categories.

Simple Supernaturalism In less complex and rather isolated societies, people may believe in a great force or spirit, but they may not have a well-defined concept of God or a set of rituals involving God. Studies by anthropologists have found that some isolated peoples (for example, South Pacific island cultures and Eskimo tribes) believe strongly in the power of a supernatural force but do not attempt to embody that force in a visualized conception of God. In this form of religion, called **simple supernaturalism**, there is no discontinuity between the world of the senses and the supernatural; all natural phenomena are part of a single force. Consider these remarks by an Inuit Eskimo:

> When I was small I knew a man who came from the polar bears. He had a low voice and was big. That man knew when he was a cub and his bear mother was bringing him to the land from the ocean. He remembered it. (quoted in Steltzer, 1982, p. 111)

Animism More common among hunting-and-gathering societies is a form of religion termed **animism**, in which all forms of life and all aspects of the earth are inhabited by gods or supernatural powers. Most of the indigenous peoples of

simple supernaturalism: A form of religion in which people may believe in a great force or spirit but do not have a well-defined concept of God or a set of rituals involving God.

animism: A form of religion in which all forms of life and all aspects of the earth are inhabited by gods or supernatural powers.

Forms of Religion

Form	Description	Example
Simple Supernaturalism	A form of religion in which there is no discontinuity between the world of the senses and the supernatural; all natural phenomena are part of a single force	Some Inuit Eskimo cultures
Animism	A form of religion in which all forms of life and all aspects of the earth are inhabited by gods or supernatural powers	Native American culture; some African tribal cultures
Theism	A form of religion in which gods are conceived of as separate from humans and from other living things on the earth, although the gods are in some way responsible for the creation of humans and for their fate	
Polytheism	A form of theism in which there are numerous gods, all of whom occupy themselves with some aspect of the universe and of human life	The pantheon of gods of the ancient Greeks and Romans
Monotheism	A form of theism that is centered on belief in a single all-powerful God who determines human fate and can be addressed through prayer	Christianity, Islam, Judaism
Abstract Ideals	A form of religion that is centered on an abstract ideal of spirituality and human behavior	Buddhism, Confucianism
Civil Religion	A collection of beliefs, and rituals for communicating those beliefs, that exists outside religious institutions	Marxism-Leninism; some versions of humanism

the Western Hemisphere were animists, and so were many of the tribal peoples of Africa before the European conquests. Europeans almost invariably branded American Indians "heathens and barbarians" because, among other things, the Indians believed that "people journeyed into supernatural realms and returned, animals conversed with each other and humans, and the spirits of rocks and trees had to be placated" (Jennings, 1975, p. 48). The same can be said of European attitudes toward African religions. Determined to subjugate nature and make the earth yield more wealth for new populations, the Europeans could not appreciate the meanings of animism for people who lived more closely in touch with nature.

Yet if one takes some time to read about the perceptions of animistic religions, it becomes clear that they contain much wisdom for our beleaguered planet. In one beautifully written account, an Oglala Sioux medicine man, Black Elk, speaks "the story of all life that is holy and is good to tell, and of us two-leggeds sharing in it with the four-leggeds and the wings of the air and all green things; for these are children of one mother and their father is one Spirit" (Neihardt, 1959/1932, p. 1). Black Elk's prayer continues:

Grandfather, Great Spirit, lean close to the earth that you may hear the voice I send. You towards where the sun goes down, behold me; Thunder Beings, behold me! You where the White Giant lives in power, behold me! You where the sun shines continually, whence come the day-break

star and the day, behold me! You in the depths of the heavens, an eagle of power, behold! And you, Mother Earth, the only Mother, you who have shown mercy to your children! (p. 5)

Traces of animism can also be seen in the religious beliefs of the ancient Egyptians, Greeks, and Romans. The Greeks, for example, spoke of naiads inhabiting rivers and springs, and of dryads inhabiting forests. These varieties of nymphs were believed to be part of the natural environment in which they dwelled, but sometimes they took on semihuman qualities. They thus bridged the gap between a quasi-animistic religion and the more familiar theistic systems that evolved in Greece and Rome.

Theism Religions whose central belief is **theism** usually conceive of gods as separate from humans and from other living things on the earth, although these gods are in some way responsible for the creation of humans and for their fate. Many ancient religions were **polytheistic**, meaning that they included numerous gods, all of whom occupied themselves with some aspect of the universe and of human life. In the religion of the ancient Greeks, warfare was the concern of Ares; music, healing, and prophecy were the domain of Apollo; his sister Artemis was concerned with hunting; Poseidon was the god of seafaring; Athena was the goddess of handicrafts and intellectual pursuits; and so on. A similar division of concerns and attributes could be found among the gods of the Romans and, later, among the gods of the Celtic tribes of Gaul and Britain.

The ancient Hebrews were among the first of the world's peoples to evolve a **monotheistic** religion—one centered on belief in a single

theism: A belief system that conceives of a god or gods as separate from humans and from other living things on the earth.

polytheistic: A term used to describe a theistic belief system that includes numerous gods.

monotheistic: A term used to describe a theistic belief system centered on belief in a single all-powerful God.

Among the Quechua-speaking Indian peoples of the Andes, Christianity and older religious practices often coexist. Shown here is a ritual in which guinea pigs, believed to possess magical powers, are sacrificed for the well-being of the community.

all-powerful God who determines human fate and can be addressed through prayer. This belief is expressed in the central creed of the Jews: "Hear O Israel, the Lord our God, the Lord is One." Jewish monotheism, based on the central idea of a covenant between God and the Jewish people (as represented in the written laws of the Ten Commandments, for example), helped stimulate the codification of religious law and ritual, so that the Jews became known as "the people of the book." As they traveled and settled throughout the Middle East, the Jews were able to take their religion with them and hold on to the purity of their beliefs and practices (Johnson, 1987; Kurtz, 1995).

Christianity and Islam are also monotheistic religions. The Roman Catholic version of Christianity envisions God as embodied in a Holy Trinity consisting of God the Father, Christ the Son, and the Holy Spirit of God, which has the ability to inspire the human spirit. The fundamental beliefs of Islam are similar in many respects to those of Judaism and Christianity. Islam is a monotheistic religion centering on the worship of one God, Allah, according to the teachings of the Koran as given by Allah to Mohammed, the great prophet of the Muslim faith. In his early preachings, Mohammed appears to have believed that the followers of Jesus and the believers in Judaism would recognize him as God's messenger and realize that Allah was the same as the God they worshiped. The fundamental aim of Islam is to serve God as he demands to be served in the Koran.

Another basically monotheistic religion, Hinduism, is difficult to categorize. On the one hand, it incorporates the strong idea of an all-powerful God who is everywhere yet is "unsearchable"; on the other hand, it conceives of a God who can be represented variously as the Creator (Brahma), the Preserver (Vishnu), and the Destroyer (Shiva). Each of these personifications takes several forms in Hindu ritual and art. Of all the great world religions, Hinduism teaches most forcefully that all religions are roughly equal "paths to the same summit."

Abstract Ideals In China, Japan, and other societies of the Far East, the dominant religions are centered not on devotion to a god or gods but on an abstract ideal of spirituality and human behavior. The central belief of Buddhism, perhaps the most important of these religions, is embodied in these thoughts of Siddhartha Gautama, the Buddha:

> Life is a Journey
> Death is a return to the Earth
> The universe is like an inn
> The passing years are like dust

Like all the world's major religions, Buddhism has many branches. The ideal that unifies them all, however, is the teaching that worship is not a matter of prayer to God but a quest for the experience of godliness within oneself through meditation and awareness.

Another important religion based on abstract ideals is Confucianism, which is derived from the teachings of the philosopher Confucius (551–479 B.C.). The sayings of Confucius are still revered throughout much of the Far East, especially among the Chinese, although the formal study of Confucius's thought has been banned since the communist revolution of the late 1940s and early 1950s. The central belief of Confucianism is that one must learn and practice the wisdom of the ancients. "He that is really good," Confucius taught, "can never be unhappy. He that is really wise can never be perplexed. He that is really brave can never be afraid" (quoted in McNeill, 1963, p. 231).

In Confucianism, the central goal of the individual is to become a good ruler or a good and loyal follower and thus to carry out the *tao* of his or her position. *Tao* is an untranslatable word that refers to the practice of virtues that make a person excellent at his or her discipline. As is evident even in this brief description, Confucianism is a set of ideals and sayings that tend toward conservatism and acceptance of the status quo, although the wise ruler should be able to improve society for those in lesser positions. Little wonder that, under communism, this ancient and highly popular set of moral principles and teachings was banned in favor of what sociologists call a civil religion.

Civil Religion In the last few decades, some social scientists, notably Robert Bellah (1970), have expanded the definition of religion to

Sister Gertrude Morgan, Religious Folk Artist As we experience other societies and cultures, it is remarkable how much of our time is spent looking at and thinking about the expression of spiritual feelings in their art, literature, and architecture. Anyone traveling through the capitals of Europe or Asia or any other continent will find that the greatest examples of a culture's achievements are often inspired by religious feelings and institutions. Michelangelo's breathtaking sculptures and frescos in the Vatican's Sistine Chapel, the great mosques of Istanbul, the prayer wheels on Tibetan mountain crags, the processions of carved saints in a Mexican Indian village festival, and the whimsically spiritual work of American religious folk artists are only a few of the thousands of examples of expressions of faith that form the heart of so many of the world's cultures.

Sister Gertrude Morgan is considered one of the most distinctive religious artists in the United States. Born in 1900 to a devoutly Christian African American family in LaFayette, Alabama, Morgan experienced a series of religious visions that shaped her life and led her, at the age of twenty-three, to begin drawing and painting. In an unschooled but inspiring and highly personal style, she portrays the religious life of her society in the form of a spiritual "autobiography." Her work is an outstanding example of how religious feelings and thoughts find expression in artistic works by people who did not necessarily have artistic training and did not necessarily think of themselves as artists.

The Barefoot Prophetesses, by Sister Gertrude Morgan, New Orleans, Louisiana, ©1971. Watercolor on paper, 11½" x 16".

Collection of the Museum of American Folk Art, New York Gift of Mr. and Mrs. Edwin C. Braman. 1983.14.4.

Morgan, for example, regarded her art as a form of private communication between herself and her God. The images and the words she wrote on them came pouring out of her in response to the experience of religious inspiration. But she never exhibited her work or had any intention of selling it. In the 1970s, art lovers in New Orleans, her adopted hometown, discovered her work and began featuring it in exhibitions. Much against her will, Morgan became something of an artistic celebrity. This acclaim disturbed her; she felt that it distracted her from her devotion and dedication to good works. As a result, she stopped producing art, believing that she had been so ordered by God.

After her death in 1980, Morgan's art became even more highly acclaimed. In 1982, it was prominently featured in a historic exhibition titled "Black Folk Art in America" at Washington's Corcoran Gallery. Its simple beauty and evidence of saintly devotion continue to inspire. Morgan's story is not dissimilar to that of many other folk artists whose religious vision seems unquenchable.

Sister Morgan is but one of many well-known religious folk artists. Somewhere in your area, undiscovered folk artists may be creating similar works, and it is likely that even if their work is largely secular, some of it expresses their deepest spiritual feelings.

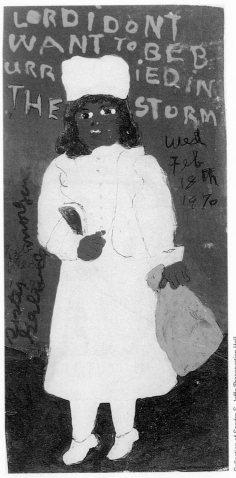

Lord I don't want to be buried in the storm, by Sister Gertrude Morgan, 1970. Acrylic and ink on cardboard, 1,912 cm x 912 cm.

Collection of Sandra S. Jaffe/Preservation Hall

---

include **civil religion.** A civil religion is a collection of beliefs, and rituals for communicating those beliefs, that exists outside religious institutions (Swift, 1998). Often, as in the former Soviet communist societies, they are attached to the institutions of the state. Marxism-Leninism can be thought of as a civil religion, symbolized by the reverence once paid to Lenin's tomb. Central to communism as a civil religion is the idea that private property is evil, but property held in common by all members of the society (be it the work group, the community, or the entire nation) is good. The struggle against private property results in the creation of the socialist personality, which values all human lives and devalues excessive emphasis on individual success, especially success measured by the accumulation of property. Although the communist regimes of the Soviet Union and eastern Europe have fallen, there are millions of people in those nations and in China who were socialized to believe in these principles.

In the United States, certain aspects of patriotic feeling are sometimes said to amount to a civil religion: reverence for the flag, the Constitution, the Declaration of Independence, and other symbols of America

civil religion: A collection of beliefs and rituals that exist outside religious institutions.

is cited as an example. Thus, most major public events, be they commencements, political rallies, or Super Bowl games, begin with civil-religious rituals such as singing the national anthem or reciting the Pledge of Allegiance, in which a nonsectarian God is invoked to protect the nation's unity ("one nation under God").

Although there is no doubt that Lenin's image and the American flag may be viewed as sacred in some contexts, neither communism nor American patriotism can compete with the major world religions in the power of their central ideals and their spirituality. In consequence, sociologists tend to concentrate on religions in the traditional sense—that is, on the enactment of rituals that represent the place of sacred beliefs in human life. In the remainder of this chapter, we discuss the structure of religious institutions and the processes by which new ones arise.

Does one have to belong to a particular religion to have a fulfilling spiritual life? Does one have to belong to the same religion or congregation as one's parents? Among the people you know, for example, how do you think these questions might be answered?

RELIGION AND SOCIAL CHANGE

Now that the nations behind the former iron curtain are enjoying new freedoms, the role of religion in bringing about social change is an important aspect of life in those societies. Indeed, in many parts of the world, religion is one of the primary forces opposing or supporting change. However, it is not always a simple matter to predict whether religion will encourage change or hinder it. In Israel, for example, highly Orthodox Jews, although they account for a small minority of the electorate, often hold the votes necessary to keep the ruling party in power. The orthodox political parties favor continued Jewish settlement on the West Bank of the Jordan River and oppose the creation of a Palestinian state near Israel's borders.

In the United States, the Catholic Church plays an active role in seeking social change. The Church strongly opposes women's right to obtain abortions legally. Instead, it supports a return to the traditional view of abortion as a crime, which is based on the belief that humans must submit to the will of God and not use their technological skills to achieve power over life and death. In this instance, therefore, although the Church is promoting social change, the change

represents a return to an earlier moral standard. Throughout much of Latin America, in contrast, the Catholic Church is fighting in support of the masses of urban and rural poor who seek social justice and equitable economic development. In Brazil, for example, the typical Catholic priest or nun favors the political left, which seeks a more egalitarian distribution of wealth and income and is highly critical of the rich.

We could add many other examples of the role played by religion and religious organizations in social change. However, in attempting to generalize about the relationship between religion and social change, it is useful to return briefly to classic sociological theories.

The pioneering European sociologists, particularly Karl Marx and Max Weber, noted the prominent role of religion in social change. But they wondered whether the influence of religious faith is a determining force in social change or whether religious sentiments and the activities of religious organizations are an outgrowth of changes in more basic economic and political institutions.

Marx, as we have seen, believed that economic institutions are fundamental to all societies and that they are the source of social change. In his view, religion and other cultural institutions are shaped by economic and political institutions; they are a "superstructure" that simply reflects the values of those institutions—of markets, firms, the government, the military, and so on. The function of cultural institutions, especially religion, is to instill in the masses the values of the dominant class. In this sense, they can be said to shape the consciousness of a people, but they do so in such a way as to justify existing patterns of economic exploitation and the existing class structure. Religion, in Marx's words, is "the opium of the people" because it eases suffering through prayer and ritual and deludes the masses into accepting their situation as divinely ordained rather than organizing to change the social system.

For Marx, the French Revolution represented a major example of what would happen to religious institutions in a time of revolutionary change. The French Revolution eventually established democracy and capitalism as dominant systems in many Western nations in the eighteenth and nineteenth centuries. It stripped power and privilege from the landowning aristocracy and the church officials who used their religious authority to justify the status quo of rule by kings and their allies in the aristocracy.

For Weber, on the other hand, religion can be the cause of major social change rather than the outcome or reflection of changes in other institutions. Weber set forth this thesis in one of his most famous works, *The Protestant Ethic and the Spirit of Capitalism* (1974/1904). Noting that the rise of Protestantism in Europe had coincided with the emergence of capitalism, Weber hypothesized that the Protestant Reformation had brought about a significant change in cultural values and that this was responsible for the more successful development of capitalist economic systems in Protestant regions.

As Weber explained it, Protestantism instilled in its followers certain values that were conducive to business enterprise, resulting in the accumulation of wealth. Because the early Protestants believed that wealth was not supposed to be spent on luxuries or "the pleasures of the flesh," the only alternative was to invest it in new or existing business enterprises—in other words, to contribute to the rapid economic growth characteristic of capitalist systems (Kurtz, 1995). This view was reinforced by the belief—also part of the Protestant ethic—that a person who worked hard was likely to be among those predestined for salvation.

Some have questioned the validity of Weber's thesis regarding Protestantism and early capitalism, but there can be no doubt that religious institutions are capable of assuming a major role in shaping modern societies. Throughout the Islamic world, there are currents of orthodoxy and reform that threaten to cause both civil and international wars. And as noted earlier, in many Latin American communities, the Catholic Church leads the movement for large-scale social change. So Marx was wrong in his claim that religion functions largely to maintain the existing values of more basic social institutions. On the contrary, religion can often lead to new ways of organizing societies—to new political and economic institutions as well as whole new lifestyles.

The Marxian view of religion is still relevant to those who are critical of the influence of religious institutions. These critics assert that religion, along with other cultural institutions such as education, reaffirms and perpetuates inequalities of wealth, prestige, and power. When the poor are encouraged to pray for a better life, for example, they are further oppressed by a religion that prevents them from realizing that they need to marshal their own power to challenge the status quo. Thus, there is still some question about whether (and how much) religious institutions change society in any fundamental way.

To make informed judgments about this and related issues, we need to have a better understanding of the nature of religious organizations and how they function in modern societies. In the next section, therefore, we describe the main types of religious organizations. We then focus on trends in religious belief and practice in the United States.

STRUCTURE AND CHANGE IN MODERN RELIGIONS

Religion today is a highly structured institution, with numerous statuses and roles within a variety of organizations as well as many kinds of smaller, less bureaucratic groups. This was not always so. The religions of tribal peoples were not highly institutionalized; that is, there were no separate organizations like churches or interfaith councils or youth fellowships. It is true that the occupational status of a holy person or priest might exist.

Thus, most Native American peoples had spiritual leaders who specialized in the rituals and symbols through which the members of the tribe could address the Great Spirit and the sacred spirits of their ancestors. Even in societies that had spiritual leaders or priests, however, religious practice was intertwined with tribal and family life. There was no concept of the church as a separate institution specializing in religious rituals.

Religion as a fully differentiated institution developed in agrarian societies, and formal religious organizations first appeared in such societies. As we saw in Chapter 4, agrarian societies produce enough surplus food to support a class of priests and other specialists in religious rituals. In those less complex societies, religion was incorporated into village and family life; it had not yet become differentiated into a recognized, separate institution with its own statuses and roles (Warner, 1998). Over time, however, the development of religious institutions resulted in a wide variety of organizations devoted to religious practice. Today, those organizations include the church, the sect, and the denomination.

Forms of Religious Organization

Churches and Sects A **church** may be defined as a religious organization that has strong ties to the larger society. Often in its history it has enjoyed the loyalty of most of the society's members; indeed, it may have been linked with the state itself (Weber, 1963/1922). An example is the Church of England, or Anglican Church.

church: A religious organization that has strong ties to the larger society.

A **sect**, by contrast, is an organization that rejects the religious beliefs or practices of established churches. Whereas the church distributes the benefits of religious participation to anyone who enters the sanctuary and stays to follow the service, the sect limits the benefits of membership (that is, salvation, fellowship, common prayer) to those who qualify on narrower grounds of membership and belief (McGuire, 1987; Weber, 1963/1922).

Sects require strong commitment on the part of their members and usually are formed when a small group of church members splits off to form a rival organization. The sect may not completely reject the beliefs and rituals of the church from which it arose, but it changes them enough to be considered a separate organization. Most "storefront churches" are actually sects that have developed their own particular interpretations of religious ritual.

An important difference between churches and sects is that churches draw their adherents from a large social environment (that is, from a large pool of possible members), whereas the size of the population from which a sect draws its members tends to be small (Iannaccone, Stark, & Finke, 1998). Also, churches make relatively limited demands on their members, whereas sects make heavy claims on their members' time, money, and emotional commitment. As Robert Wuthnow (1988) writes: "Churches attempt to regulate or fulfill a few of the activities or needs of large numbers of people; sects attempt to regulate or fulfill many of the activities or needs of small numbers of people" (p. 495).

Denominations A third type of religious organization is the **denomination**. Unlike a sect, a denomination is on good terms with the religious institution from which it developed, but it must compete with other denominations for members. An example of a denomination is the United Methodist Church, a Protestant denomination that must compete for members with other Protestant denominations such as Presbyterians, Episcopalians, and Baptists. Denominations sometimes evolve from sects. This occurs when the sect is successful in recruiting new members and grows in both size and organizational complexity (Wuthnow, 1994).

In a recent study of the strength of attachments to denominations (for example, Lutheranism, Congregationalism, or Presbyterianism), sociologist Nancy Ammerman (2000) and her colleagues interviewed people about the importance of denominational differences in Christian congregations. They visited more than 500 churches, which included ninety-one different denominational and other religious groups, including fifty-one congregations that were nondenominational or interdenominational and another twenty-two that were part of more informal networks.

Although many people change congregations one or more times during their lives and some churches are less attached to their parent denomination than others, Ammerman found that denominational differences are hardly about to disappear. Her main finding was that:

> A congregation's sense of identification with a particular denominational tradition is closely tied to how many of its members grew up in the tradition. Relatedly, individuals who themselves have "switched" are less likely to say that denomination is important to them. Which is the chicken and which the egg is hard to say. But congregations full of "switchers" are much less likely to report that denomination is an important part of the way they "do church" (49 percent versus 69 percent of congregations with few switchers).
>
> The phenomenon of "switching" is relevant to a lot of congregations. Indeed, nearly three-quarters of the white Protestant congregations we studied reported that half

This photograph of a religious festival in the province of Chiapas in Mexico clearly shows the social organization of the village church. The male officials of the church (visible in the background) march with banners and candles while the women bear the saint through the village.

Courtesy of Susan Kornblum

sect: A religious organization that rejects the beliefs and practices of existing churches; usually formed when a group leaves the church to form a rival organization.

denomination: A religious organization that is on good terms with the institution from which it developed but must compete with other denominations for members.

or more of their members grew up in another denomination. African-American Protestants are much less denominationally mobile. Only 11 percent of the ones in which we interviewed people had similar numbers of switchers in their pews. Similarly, almost all the Catholic parishes contain mostly people who grew up Catholic. (Ammerman, 2000, p. 3)

Cults Still another type of religious body, the **cult**, differs in significant ways from the organizations just described. Cults are usually entirely new religions whose members hold beliefs and engage in rituals that differ from those of existing religions. Some cults have developed out of existing religions. This occurred in the case of early Christianity, which began as a cult of Jews who believed that Jesus of Nazareth was the Messiah and who practiced rituals that were often quite different from those of Judaism. Cults may also be developed by people who were not previously involved in a church or sect, such as individuals who become active in pagan cults like those based on ancient forms of witchcraft (Barrett, 1996). Most major religions began as seemingly insignificant cults, but new cults are formed every day throughout the world, and few of them last long enough to become recognized religions. The Mormons, who began in 1830 as a cult with only six members, grew into a large religious organization with more than 60,000 members by 1850. Today, the Mormon Church is one of the fastest-growing Christian churches in the world (Stark, 2005).

In a study of trends in church, sect, and cult membership, sociologists Roger Finke and Rodney Stark found that, contrary to common assumptions, the rate at which cults are formed is no higher today than it was in earlier historical periods. Moreover, compared to some of the earlier cults, contemporary cults often have less success in gaining members:

> No cult movement of the sixties or early seventies seems ever to have attracted more than a few thousand American members, and most of even the well-publicized groups counted their true membership in the hundreds, not the thousands. For example, after more than thirty years of missionizing, it is estimated that there have never been more than 5,000 followers of the Unification Church (the Moonies) in the United States, and some of them are from abroad. (Finke & Stark, 1992, p. 241; see also Melton, 1989)

In a similar example, Barry Kosmin and his colleagues (1991) estimated, on the basis of the largest sample survey of religious beliefs ever conducted in the United States, that actual membership in the Church of Scientology very likely does not exceed 45,000, despite the cult's claims that its domestic membership is more than 450,000. On the basis of findings such as these, Finke and Stark conclude that cults have always interested some relatively small proportion of U.S. citizens and still do. Despite sensational media accounts of the activities of cults like the Branch Davidians, the rate of cult formation in the United States has not increased appreciably in recent decades, nor are cults more successful in gaining members than they were before.

One reason cults and sects do not always grow in numbers or influence is that established religions often absorb them through the process known as *co-optation*. This term is not limited to religious institutions; we encountered it in Chapter 9 in connection with social movements. It refers to any process whereby an organization deals with potentially threatening individuals or groups by incorporating them into its own organizational structure (Selznick, 1966). For example, over the centuries, the Roman Catholic Church has been particularly successful at co-opting regional Catholic sects by including their leaders in the panoply of lesser saints, thereby allowing people to worship a holy person of their own culture while remaining true to the world Church.

Religious Interaction and Change

Sects and cults are a major source of change in religious organizations. People who are not satisfied with more established churches and denominations, or are otherwise alienated from society, often form or join a cult or sect (Barrett, 1996). One of the most convincing explanations of the emergence of sects was suggested by H. Richard Niebuhr (1929), borrowing from Max Weber's (1922) pioneering analysis of churches and sects. According to Weber, churches tend to justify the presence of inequality and stratification because they must appeal to people of all classes. Sects, on the other hand, may be led by charismatic individuals who appeal to people who have felt the sting of inequality. Niebuhr agreed with Weber that class conflict is a primary cause of sect formation. But he observed that as a sect becomes more successful and better organized, it becomes more like a church and begins to justify existing systems of stratification. This creates the conditions in which new sects may emerge.

cult: A new religion.

Another motivation for the formation of sects or cults is dissatisfaction with the interactions that occur in more established organizations. In church rituals, for example, prayer is often led by a priest or other religious professional and is relatively restrained, whereas in sects and cults, communication between God and the individual is more direct and typically allows the individual to express deep emotions. The differing styles of interaction in different types of religious organizations can be illustrated by the contrast between the hierarchy of statuses and roles that characterizes the Catholic Church (with its pope, cardinals, bishops, priests, and other well-defined statuses) and the seemingly greater equality and looser structure of a cult like Krishna Consciousness or the Unification Church.

People who are attracted to cults are often influenced by a charismatic leader who inspires them to new and very personal achievements, such as ecstatic experiences, a sense of salvation, or a release from physical or psychological suffering. Some become cult members simply because they are lonely; others are born to cult members and are socialized into the cult.

Some, but by no means all, cults are extremely authoritarian and punitive. Their leaders may demand that members cut themselves off entirely from family and friends and sacrifice everything for the sake of the cult. The leaders may also insist that they themselves are above the moral teachings to which their followers must adhere. Under these conditions of isolation and submission to a dominant authority, cult members may be driven to incredible extremes of behavior—even mass suicide, as occurred in the case of the followers of Heaven's Gate in 1997. However, not all cults are so dangerous or so easily condemned, and there is an ongoing conflict between norms that protect the right of individuals to belong to cults and efforts to protect people from the harm that can occur when cult leaders place themselves above morality and the law.

Sociological Insights into Waco and Wounded Knee The confrontation that took place in Waco, Texas, on February 28, 1993, between the Branch Davidians and agents of the Bureau of Alcohol, Tobacco, and Firearms (BATF) revealed the need for greater understanding of alternative religious groups (Foster et al., 1998). Among sociologists who study religion and social change, there is consensus that the term *cult* can be dangerous when it is applied as a pejorative label to all religious groups that are outside the mainstream of religious organizations.

Nancy Ammerman, one of the foremost authorities on Protestant religious groups in the United States, notes that "one of the primary interpretive lenses through which the general public views groups such as the Branch Davidians is the lens supplied by 'cult awareness' groups and 'exit counselors.' Namely, most people think members of a group like the Branch Davidians must have been 'brainwashed' into joining" (Foster et al., 1998, p. 27). Analyses of the Waco tragedy now place greater emphasis on the deep misunderstandings that existed between the government agencies, primarily the FBI and the BATF, that laid siege to the Branch Davidian community, and the members of the community—especially its leader, David Koresh. These misunderstandings appear to have been deepened by hysterical press and television coverage during the episodes of violence and siege (Bates, 1999).

Sociologist Joel Martin observes that the Branch Davidians were not newcomers but an established alternative religious group that had lived in the Waco area since the mid-1930s. The group did believe in an unconventional variant of Seventh-day Adventist millenarianism and that Armageddon and the end

of time were close at hand. They also had some unorthodox and disturbing sexual practices, but these charges, and the charge that they were stockpiling weapons, could have been addressed by quietly apprehending the group's leader for questioning rather than engaging in the large-scale operation that resulted in an armed standoff and the eventual death of the entire group in a blaze of fire and bullets.

Many sociologists point to the similarities between the Waco disaster and the tragedy at Wounded Knee, South Dakota, in 1890. In both cases (and many others like them), a new religion attracted men and women with visions of a new heaven and earth. As they sought to live in accord with these visions, they withdrew into their own community, altering their behavior in ways that may have seemed strange and even threatening to others. People outside the community began to spread rumors about them. Government authorities got nervous and planned to move against the group. Journalists congregated, anticipating a big story. When a confrontation took place, gunfire erupted and a few of the authorities were killed. In the end, the millenarian community was destroyed. After the tragedy, officials, scholars, historians, and journalists interpreted what happened. Many concluded that the community's members had been caught up in a "messiah craze." Thus, the dead were blamed for their own deaths.

Such was the case with the Lakota Indian Ghost Dancers who were killed at Wounded Knee in 1890. In the Ghost Dance religion, some

The Ghost Dance religion spread rapidly among American Indians in the 1890s and was based on the belief that through the dance, all Native Americans, living and dead, would be reunited on an earth restored to its original natural state.

National Anthropological Archives

Big Foot, leader of the Sioux, lies frozen on the battlefield of Wounded Knee, South Dakota. Photograph ca. 1890.

has gradually declined, partly as a result of upward mobility and the tendency of more affluent and educated people to be less active in religious institutions. Meanwhile, the proportion of the population who identify themselves as Catholics has been growing, primarily as a result of the large numbers of Hispanic immigrants who have entered the United States in recent years. Also, since the mid-1960s, the percentage of people who express no religious preference has increased. Thus, it is possible that in coming years less than half of all Americans will be Protestant (Ammerman, 2005; Finke & Stark, 2005).

Membership in a religious organization is different from identification with a religious faith, and this is reflected in statistics on Christian church membership in the United States. Because data on membership are obtained from the organizations themselves, there are significant differences in who is included. (Some churches, for example, count all baptized infants, whereas others count only people above a certain age who are enrolled as members.) The Mapping Social Change box on page 416 illustrates the geographic distribution of reported church membership in the United States.

Indians began to believe they might be invulnerable to white attackers. In a sense, they courted death at the hands of white men. Many believed that if the Ghost Dance did not protect them, they would at least achieve speedy transport to a happier existence than the one they led among the conquering white people (Hackett, 1995). The situation of the Branch Davidians killed outside Waco a century later was similar. The issue is not whether either group was right or wrong but how people in a position to bring understanding to these tragic situations can help prevent similar events from occurring over and over again.

Trends in Religion in the United States

In studying religion in the United States, sociologists are unable to use census data because the Census Bureau does not collect information on religious affiliation. However, some statistics are available from smaller sample surveys such as those conducted by the National Opinion Research Center (NORC). The available data indicate that slightly more than half of Americans older than age eighteen identify themselves as Protestant, one-quarter are Roman Catholic, and 1 percent identify themselves as Jewish. All other religions combined account for about 3 percent of the population. Fifteen percent say that they have no religious affiliation; they constitute the third-largest category of religious preference in the United States (NORC, 2008). (See Figure 16.2.)

The religious preferences of Americans have been changing slowly but steadily over the last two generations. The Protestant majority

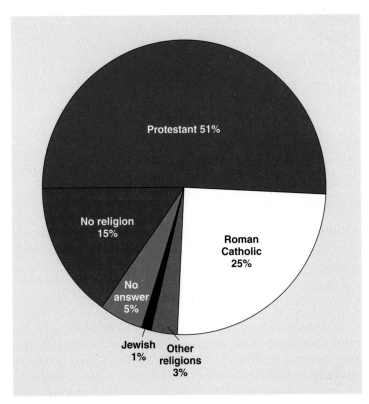

Figure 16.2 Percentage Distribution of Religious Preference in the United States (persons 18 years of age and older). Source: Data from *Statistical Abstract*, 2005.

Religious Pluralism in the "Bible Belt"

The wide swath of states in the South and Southwest where Baptist Church congregations predominate is often known as the "Bible Belt." This term refers to the strong religious convictions of many of the region's residents and to the powerful influence of Baptist churches in thousands of small communities. On the accompanying map, the Bible Belt stands out clearly. But maps showing the distribution of religious beliefs often mask the changes occurring in regions that appear more homogeneous than they are. Migration, urbanization, and differences in the growth of specific religious groups can lead to more religious pluralism (or less, depending on the region), even though a map shows a static distribution.

When we look closely at communities in west Texas, at the western end of the area of heavy Baptist church membership, we find higher levels of Catholic church membership. This is largely a result of the growing presence of Latino residents.

Wherever there are changing patterns of church membership, there is also likely to be more interfaith communication and more frequent discussion of the need for religious tolerance.

Similar evidence of pluralism can be found in other regions of the United States. In the heavily Catholic Northeast, for example, the map of Christian congregations does not indicate non-Christian, particularly Jewish, congregations, which are concentrated in the larger cities of this region. Catholic–Jewish interfaith dialogues are strongest in these urban centers. In sum, although ecological maps like this one are useful, they only begin to suggest the important sociological facts that will be revealed by further research.

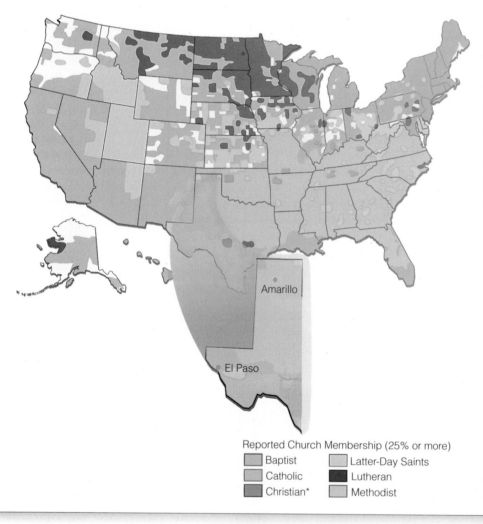

Amarillo

El Paso

Reported Church Membership (25% or more)

- Baptist
- Catholic
- Christian*
- Latter-Day Saints
- Lutheran
- Methodist

*Includes Christian Church (Disciples of Christ), Christian Churches/Churches of Christ, and Churches of Christ. Blank areas indicate counties where no group has 25 percent or more of the adherents in that county.
Source: Churches and Church Membership in the United States, 1980, by Bernard Quinn et al., and www.adherents.com, 2003.

As can readily be seen, various regions of the country differ with respect to the church membership of most of their residents. Protestants outnumber members of all other religions in the South, for example; in the West and Southwest, the Roman Catholic Church is dominant.

There are also significant differences in church membership by size of place: Protestants are most likely to live in rural places, small cities, and suburbs of smaller cities, whereas native-born Catholics are found in medium-sized cities and suburbs of larger cities. Much of the rapid growth of Catholic congregations in the United States is accounted for by immigrants to major cities such as Miami, Los Angeles, and New York (Niebuhr, 1995). Jews, who originally migrated to the nation's largest cities, are most numerous in major metropolitan regions such as New York, Los Angeles, Chicago, and Miami (NORC, 2008).

Religion has always had an especially important place in the lives of African Americans, and a major proportion of the black population is deeply religious. Many of the functions of religion as a central cultural institution—meeting the spiritual needs of individuals, binding together the members of social groups, and sometimes engendering social change—are illustrated by the history of the black church in America and by the internal diversity and social change that characterize the powerful Southern Baptists in the United States.

The Black Church in America

The respected black sociologist E. Franklin Frazier was a firm believer in the principles of the Chicago school of ecological research and an astute student of race relations. He specialized in two of the most basic social institutions—the family and the church—as they affected and were affected by the situation of black Americans.

Frazier's (1966) study of the history of the black church in America emphasized the role of religion in the social organization of blacks ever since they were imported from Africa to the New World to be sold as slaves. Few aspects of the slaves' African culture were able to survive the experience of abduction and slavery, but one of those that did was dancing, which Frazier refers to as the most elemental form of religious expression. Although the slaves were encouraged to dance, the religious beliefs expressed in their dances were not permitted to flourish in America.

With the coming of Baptist and Methodist missionaries, the slaves found a form of religion in which they could express their most intense feelings. The religious services in which they participated helped bind them together into close-knit groups, replacing the bonds of kinship and tribal membership they had lost when they were sold into slavery. However, the freed slaves resented the subordinate status of blacks in the white-dominated churches. They left those churches and established all-black churches.

Out of these black churches, Frazier noted, other forms of black organizational activity emerged. The churches began to play an important role in the organization of the black community in such areas as economic cooperation and the building of educational institutions. The church also became the center of political life for black Americans, the arena in which they learned how to compete for power and position.

The large-scale urbanization of blacks that began in World War I transformed the black church. In particular, the church became more secular, placing less emphasis on salvation and

Internationally acclaimed singer Aretha Franklin attributes her success to the vitality and inspiration of the black church in the United States.

turning its attention to the serious problems of blacks in the here and now. Changes in the class structure of the black community have also been reflected in the black church, with many middle-class blacks shifting to the Presbyterian, Episcopalian, and Congregational churches. A countertrend may be seen in the emergence of "storefront churches" (that is, sects) that satisfy the spiritual needs of those who prefer a more intimate and expressive form of worship.

At the same time, blacks have become more fully integrated into American society. This has naturally affected the organization of black communities and the black church. Today, the church is no longer the dominant institution for blacks that it once was (Patterson, 1999). Nevertheless, as Frazier pointed out, the black church has left its imprint on most aspects of African American life. And it is clear that the black church (including the Black Muslims and other Islamic sects) provides much of the leadership and many of the rank-and-file supporters of movements for political and economic justice among minority groups in the United States.

Social Change and the Southern Baptists

On the conservative side of cultural and political life in the United States, the Southern Baptists have emerged as the most powerful and vocal critics of many contemporary social trends, including secular education, evolution, same-sex marriage, and many other issues. Nancy Ammerman, a leading sociologist of religion, views issues of religion and social change as central to understanding the split between fundamentalist and moderate Southern Baptists (Ammerman, 2007). She writes:

> The disruption against which fundamentalists struggle is often labeled "modernity." Whatever else that label may mean, the "modern" world is one in which change is a fact of life, in which people of multiple cultures live side by side, and in which religious rules have been largely relegated to a private sphere of influence. (1990, p. 150)

To assess differences among Baptists in the way they view social change and diversity, Ammerman surveyed more than 900 Baptist leaders in congregations throughout the United States. Her survey sample included men who were pastors and church deacons and women who were presidents of their local Baptist women's organizations. After examining their responses to key questions about their beliefs, Ammerman developed categories (or "theological parties") with labels corresponding to the terms respondents often used to describe themselves (for example, *fundamentalist, conservative, moderate*). She also developed a scale for measuring individuals' degree of opposition to modernity. The questions on which the scale is based and the method of scoring responses are discussed in the Research Methods box on page 419. Table 16.1 shows that, "among those who disapprove of pluralism, fundamentalists outnumber moderates nearly ten to one" (Ammerman, 1990, p. 151). But the data in the table and the box also show that Baptists classified as "conservative," whose views fall somewhere between those of the fundamentalists and moderates, are the most numerous category and are quite divided in their views of pluralism.

When she examined how respondents felt about modernity in relation to such variables as education, occupation, and rural versus urban residence, Ammerman found some surprising results. Although people often think of individuals who hold strong fundamentalist beliefs and reject modernity as uneducated people living in remote farming hamlets, and of individuals who approve of modernity as educated middle-class urban dwellers, Ammerman's results challenge this conventional wisdom:

> Those among Southern Baptists who were the most skeptical of change, choice, and diversity were blue collar workers, those with middle incomes, people who had moved from farm to city, and those who had been to college, but did not have a degree. And combinations were important: people who had moved from farm to city *and* had some college or people with middle income *and* blue collar households, for instance. These were people who knew exactly what the modern world was all about, and they were less enthusiastic about embracing it than

were any other Southern Baptists. And the less enthusiasm they had for modern attitudes, the more likely they were to adopt fundamentalist beliefs and identity. (1990, p. 155)

Eventually, Ammerman believes, these differences in outlook may lead to an irremediable schism in which moderates may form a new Baptist denomination. She is not convinced that religious fundamentalism poses dire threats. The rise of Christian fundamentalism does, in her view, challenge widespread beliefs about the inevitable progression of modern societies toward ever-greater "differentiation, individualism, and rationalism as 'the way things are'" (1997, p. 202). Modern frameworks for understanding the social world, she argues, emphasize the "individualized 'meaning system' that would be carved out of differentiation and pluralism." But we live in a world

> where organizational boundaries are more fluid, where mergers and out-sourcing and flextime and telecommuting are as common as the time clock and the stockholder corporation. We live in a world where rationality and the scientific method are valued, but also critiqued, where multiple sources of wisdom are finding a voice. We live in a world where people are both more rooted in particularistic ethnic and religious communities that refused to melt and more aware of the larger world and the choices that have brought them to their current practices. (1997, p. 202)

A clear implication of Ammerman's brilliant empirical research, and the conclusions she draws from her findings, is that fundamentalist communities are the result of unquestioned religious socialization of some people and the rational choices of others. These congregations may

TABLE 16.1

Percentage in Different Theological Parties, by Responses to Pluralism

Theological Party	Strong Approval of Pluralism	Moderate Approval of Pluralism	Disapproval of Pluralism	Total
Self-Identified Moderate	21%	5%	3%	9%
Moderate Conservative	16	8	3	9
Conservative	38	55	43	48
Fundamentalist Conservative	18	24	27	23
Self-Identified Fundamentalist	8	9	23	11
Total	101%	101%	99%	100%
(Number of Cases)	(256)	(500)	(187)	(944)

Note: Difference statistically significant at $p < .001$. Some percentages do not total 100 due to rounding.
Source: Ammerman, 1990.

Scales and Composite Scores When analyzing questionnaire responses, sociologists often create *scales* in order to arrive at a composite score for a particular variable. Nancy Ammerman created scales to measure the strength of a person's fundamentalist beliefs and opposition to modernity. (See the accompanying tables.) A scale is a set of statements that fit together conceptually in that they deal with aspects or examples of the same attitude or behavior. For each statement, the respondent may be asked to choose among responses ranging, for instance, from 1 ("Strongly disagree") to 5 ("Strongly agree"), with "Unsure" in the middle. The statements in a particular scale (for example, fundamentalist beliefs) are scattered throughout the questionnaire to avoid *response set,* the tendency to answer all such items similarly without giving them much thought; response set is a frequent problem when similar items are grouped together. To ensure that the statements in a scale actually measure the behavior or attitude the researcher is studying, great care must be taken in their selection and wording.

Once the questionnaires have been completed and the data entered into a computer, the researcher can arrive at a simple composite score by adding up the scores on all the items in a scale. Then the researcher can either use each respondent's composite score as a continuous variable or establish cutoff points that will create categories for analysis, as Ammerman did. In her measurement of opposition to modernity, for example (see table), Ammerman decided that those who agreed or strongly agreed with all the items (that is, whose scores ranged from 5 to 9) could be classified as "strongly agreeing with pluralism." Those who agreed with most of the items but not all of them (that is, scored from 10 to 14) were classified as showing "moderate approval." Those who scored 15 or above—in other words, those who disagreed as often as they agreed—were classified as "disapproving."

Fundamentalist Beliefs

Items	Scoring
The Scriptures are the inerrant Word of God, accurate in every detail.	Strongly agree = 5
God recorded in the Bible everything He wants us to know.	Strongly agree = 5
The Genesis creation stories are there more to tell us about God's involvement than to give us a precise "how and when."	Strongly disagree = 5
The Bible clearly teaches a premillennial view of history and the future.	Strongly agree = 5
It is important that Christians avoid worldly practices such as drinking and dancing.	Strongly agree = 5

Distribution	
Range	Percent
5–10	5
11–15	14
16–20	47
21–25	34

Opposition to Modernity

Items	Scoring
I like living in a community with lots of different kinds of people.	Very untrue = 5
Public schools are needed to teach children to get along with lots of different kinds of people.	Strongly disagree = 5
I sometimes learn about God from friends in other faiths.	Very untrue = 5
One of the most important things children can learn is how to deal creatively with change.	Strongly disagree = 5
Children today need to be exposed to a variety of educational and cultural offerings so they can make informed choices.	Strongly disagree = 5

Distribution	
Range	Percent
5–9	29
10–14	52
15–19	17
20–25	2

present severe challenges to religious pluralism, but in the long run, it is in pluralistic societies that they will eventually find a secure, though limited, place (Ammerman, 2007).

Unofficial Religion

Religion in the United States is increasingly voluntaristic. People are less likely to practice a religion simply because their parents did so. They are more likely than ever to join a religious group that appeals to their desire for membership in a community of like-minded peers and that upholds their particular moral standards (Swift, 1998). People in the United States and Canada are also more likely to engage in what sociologists term *unofficial religion*—a set of beliefs and practices that are "not accepted, recognized, or controlled by official religious groups" (McGuire, 1987, p. 89). Sometimes called folk or popular religion because it is practiced by ordinary people rather than by religious professionals in formal organizations, unofficial religion takes many forms. It is engaged in by people who purchase religious books and magazines, follow religious programs on television, make religious pilgrimages, or practice astrology, faith healing, transcendental meditation, occult arts, and the like—and who may belong to organized churches at the same time. Contrary to older sociological theories that viewed these practices as holdovers from rural folk cultures, research has shown that urban societies continually produce their own versions of popular religion (Fischer, 1987/1976).

Muslims and Hindus in the United States

Largely as a consequence of recent waves of immigration from Asia and the Near East, membership in Islamic and Hindu religious groups is growing rapidly. By 2000, there were about 5 million Muslims in the United States and Canada, and more than 1,000 Islamic centers with about a half million Muslims attending mosque services as official members. The largest numbers of Muslims are found in Chicago, Washington, New York, Toronto, and southern California. The largest ethnic groups among North American Muslims are African American and Indo-Pakistani, with Arab Americans ranked third. The oldest continuously functioning mosque in the United States is located in Cedar Rapids, Iowa, and is known as the Mother Mosque of America.

Most Muslims in the United States belong to the Sunni sect, but membership in the more fundamentalist and militant Shiite sect is also increasing. Muslims, particularly those of the first immigrant generation, tend to hold views on moral issues that are similar to those of conservative Jews and Christians, and to oppose abortion, pornography, and extramarital sex. First-generation Muslim immigrants, like other first-generation immigrants, tend to be concerned about whether their children will marry within the faith and continue to share their Islamic beliefs and values (Swift, 1998).

Immigration from Pakistan, India, and nations of Africa and the Caribbean that were formerly part of the British Empire has caused a rapid increase in the number of Hindus in North America in the past decade or so. An estimated 400,000 Hindus now reside in the United States, most of them concentrated in southern California, New York, and Washington, D.C. Most Hindus worship at home, but Hindu temples have recently been established in several major U.S. cities. A far smaller proportion of practicing Hindus in the United States are native-born Americans who are attracted to the religion's values of inner peace and enlightenment and have studied with Indian gurus like Bagwan Shree Rajneesh (Swift, 1998).

Pagans

Another trend, though not strictly a result of immigration, nonetheless reflects the influence of other cultures on contemporary religious practices. Sociologists studying religion in the United States note the small but significant increase in rates of participation in pagan religions, witchcraft, or Wicca. Wicca is particularly attractive to feminists who reject what they perceive as the patriarchal tendencies of Judaism and Christianity. Pagan holy days and rituals are associated with, but not limited to, the solstices and equinoxes. Pagans believe that women participate in the divine nature of the Goddess, the source of human life and of all nature. Many pagans are reluctant to reveal their religious beliefs and practices because of the history of Christian suppression of paganism (Adler, 1997).

Religiosity

Neither church membership figures nor self-reports of religious identification are accurate indicators of the aspect of religious behavior known as **religiosity**. This term refers to the depth of a person's religious feelings and how those feelings are translated into religious behavior. Responses to questions about whether one believes in God and how strongly, whether one believes in a life after death, whether one's religious beliefs provide guidance in making major decisions, and the frequency of one's church or temple attendance all can be used to measure religiosity (Smith et al., 2002).

When sociologists study religiosity as opposed to church membership, some important results emerge. Although a common stereotype of Americans describes them as materialistic, individualistic, pleasure seeking (hedonistic), and secularized, data on their religious faith and behavior do not confirm the stereotype. Compared with citizens of other urban-industrial nations, Americans are more religious. In fact, the World Values Survey of the University of Michigan reports that:

> Fully 44 percent of Americans attend church once a week, excluding funerals and christenings, compared with 27 percent of Britons, 21 percent of the French, 4 percent of Swedes, and 3 percent of Japanese. Not only do they go to church, but 53 percent of Americans say religion is very important in their lives, compared with just 16 percent of the British, 14 percent of the French, and 13 percent of Germans. (Reese, 1998)

religiosity: The depth of a person's religious feelings.

When Pope Benedict assumed the papacy in 2005, he made it known that a goal of his tenure as pope would be to revitalize the Church in Europe. This has proven to be an elusive goal, a subject to which we will return. But even in the United States, where church attendance is high compared to attendance in Europe, the figures mask many variations. Close examinations of church attendance statistics have found that a sizable proportion of Americans who say they attend church or synagogue regularly actually exaggerate their estimates. Analysis of their responses suggests that the figures for American church attendance are inflated by 50 percent or more in some instances (Hadaway & Marler, 1998). But other sociologists note that successful congregations have a core group of regular attenders who tend to be most active in all of the congregation's activities. Those on the margins of this core group tend to be those who also overestimate their attendance (Ploch & Hastings, 1995; Reese, 1998).

There is far less disagreement over what the poll data reveal about other aspects of religiosity, particularly about beliefs as opposed to behavior. About 90 percent of Americans believe in the existence of God, and 86 percent express certainty in a life after death (NORC, 2008). Adults in the Pacific states, which tend to have the lowest rates of church membership, do not differ very much on these measures from those in other regions. The lower rates of church membership on the West Coast are correlated with the greater spatial mobility and lower median age of the populations of those states. People who move often tend to sever their attachments not only to churches but to all organizations in the community. Yet people who do not belong to organized churches can nevertheless hold deeply cherished religious beliefs.

Data on church attendance do not support the notion that the United States is becoming an increasingly secular society. Although Catholics have the highest weekly attendance rates and Jews the lowest, those rates can change considerably from one year to the next, for reasons that are not fully understood. Nevertheless, it is clear that the data do not bear out the secularization hypothesis. It should also be noted that increases in immigration during the 1980s and 1990s are significantly increasing the diversity of religious practice in the United States. As noted earlier, Muslims, Hindus, Buddhists, and others are well represented among the new immigrants (Esposito, 2003). In a society that prides itself on religious freedom, it is also no surprise that so many varied and different beliefs and religious practices flourish.

New findings indicate that many Americans are perfectly capable of trying out or embracing multiple faiths or versions of spirituality (Ammerman, 2007).

New findings on religious beliefs and practices in the United States also show that the religious life of an increasing proportion of Americans does not fit neatly into conventional categories. A 2009 poll by the Pew Research Center's Forum on Religion and Public Life found that large numbers of Americans engage in multiple religious practices, mixing elements of diverse traditions. About one-third of Americans (35 percent) say that they attend religious services at more than one place, and most of these (24 percent of the public overall) indicate that they sometimes attend religious services of a faith different from their own. Aside from when they are traveling and special events like weddings and funerals, three out of ten Protestants attend services outside their own denomination, and one-fifth of Catholics say that they sometimes attend non-Catholic services. Many also blend Christianity with Eastern or New Age beliefs such as reincarnation, astrology, and the presence of spiritual energy in physical objects. And sizeable minorities of all major U.S. religious groups say that they have experienced supernatural phenomena such as being in touch with the dead or with ghosts. These trends are shown in Table 16.2 and Figure 16.3. Note that in the figure we see a robust increase over the past fifteen years in the proportion of Americans who say that they have had a personal religious or mystical experience.

TABLE 16.2

Supernatural Experiences and Beliefs

	Total	Christians
% who have . . .	%	%
Been in touch with the dead	29	29
Had ghostly experience	18	17
Consulted psychic	15	15
% believe in . . .		
Spiritual energy in trees, etc.	26	23
Astrology	25	23
Reincarnation	24	22
Yoga as spiritual practice	23	21
Evil eye (i.e., casting of curses)	16	17
Sample size	2,003	1.589

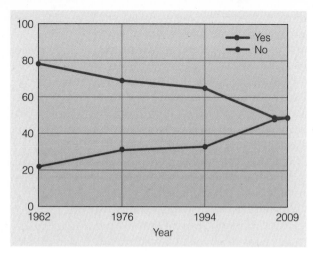

Figure 16.3 Percentage of Respondents Who Have Ever Had a Religious or Mystical Experience

Religious Pluralism

Differences in the distribution of religious belief are important because they show how religion helps account for cultural differences among populations within a society. Thus we use the expression "Bible Belt" to refer to the most strongly Protestant areas of the nation—the South and Midwest. The large Catholic populations in some cities make them quite different in cultural terms from the Protestant-dominated suburbs. And the large Jewish populations in New York and Miami are reflected in the cultural life of those cities—in joint Hanukkah/Christmas celebrations in offices and schools, for example, and in the incorporation of Jewish holidays into the school calendar.

The United States is an example of a society in which the state protects religious pluralism. The First Amendment to the Constitution guarantees freedom of religion, which in turn allows different religions to exist and to compete for adherents if they wish to do so. Religious pluralism is also protected in most European nations today, although before the Protestant Reformation of the sixteenth century, many European states made Roman Catholicism the official state religion and attempted to force all citizens to convert to Catholicism. And in the early 1990s, the religious diversity and tolerance of the former Yugoslavia dissolved into a genocidal religious and nationalist war.

In societies in which religious pluralism is protected, one usually can observe the continual formation of new religious organizations. New religious groups tend to arise either out of schisms within existing organizations or as a result of the teaching of charismatic leaders who attract people to new religious movements. Many new religious movements are also stimulated by severe inequalities and the widening gap between rich and poor. Whether it originates in schisms within existing religious organizations, as a result of the emergence of charismatic religious leaders, or in response to severe inequalities, religious pluralism presents many personal questions and challenges. Two examples of these challenges are found in movements for social change within the Catholic Church and the rise of fundamentalism in the United States and elsewhere.

Social Change and the Catholic Church It is a striking fact that, throughout the world, one finds deeply devout individuals involved in movements for social change. Religious workers were killed in El Salvador because they advocated greater democracy and equality. In Poland

before the fall of the Soviet empire, the Catholic Church was a leader in the movement to democratize that nation. American bishops speak out against poverty and inequality in the United States. The civil rights movement has tended to be led by members of the clergy and bases many of its appeals on biblical principles of justice and equity. Yet, on the other hand, we could cite many examples of religious intolerance and an apparently high correlation between devotion to religious faith and resistance to social change.

Support for and resistance to social change are relative ideas—relative, that is, to the position of their advocates. If one supports a woman's right to choose whether to have an abortion, the social change in question is tolerance for abortion rights (which does not necessarily mean that one favors abortion). If one is opposed to abortion rights, the change one seeks is a return to a situation in which abortion is illegal. As noted earlier in the chapter, by opposing abortion rights, the Catholic Church stands for social change in a conservative direction.

In religious life, the same institution can stand for conservative social change in one place and for more radical change in another. For example, the conservative Episcopalian

Pope Benedict XVI addresses the people of Ireland, apologizing for the sex abuse scandal in the Catholic Church. The televised appearance was watched by Catholics from Boston to Berlin. The church is only beginning to come to terms with decades of child abuse in its parishes and schools. The scandals first emerged in Canada and Australia in the 1980s, followed by Ireland, the United States, and recently Benedict's German homeland.

(Anglican) Church in Northern Ireland favors the continuation of British colonial rule there, whereas the Episcopalian Church in the United States supports liberal causes such as abortion rights and environmental action. The relationship between the Catholic Church and social change is even more complex. In the United States, the church is accused by some of its own leaders of having too conservative a view of social change. In Brazil, by contrast, the church is accused of being too liberal. How can a single institution be perceived so differently in two societies? Part of the answer is that a church that is so large and embraces so many separate organizations can never be free from internal politics (Espin, 2009).

Sociology & Social Justice **THE CHURCH IN CRISIS** Many people inside and outside the Catholic Church view the crisis the Church is experiencing through the lens of their own ideas about social justice. Conservative Catholics tend to view the sexual abuse scandal as having been caused by a small number of homosexual priests who deserve to be excluded from the priesthood. More liberal Catholics and those who feel persecuted because of their sexual preferences tend to regard the Church as an institution that must reform itself by changing some of its basic norms, particularly the norm of priestly celibacy and the ban on ordination of women. Parents and their children, however, especially those who have been directly affected by sexual abuse by Church officials, tend to blame those in positions of authority within the Church because of the many proven instances of "cover-up" of priestly sexual abuse. Their view, one with which the greater public largely sympathizes, is that this is an issue of social justice that requires forceful action to right a history of past wrongs and conspiracies within different Catholic Church jurisdictions.

Clearly, the sexual abuse scandal the Catholic Church has been facing for the past decade or more represents by far the worst crisis the Church has faced since the Protestant Reformation (Cahill, Garvey, & Kennedy, 2006; Greeley 2004). At this writing, the abuse of children within the Church, a scandal that seemed for a while to have been largely confined to the United States, has been revealed to have also taken place in many parts of Europe and elsewhere in the world. Worse yet, efforts were made to transfer guilty priests to new parishes while covering up the details of their sins. In the United States, accusations of sexual abuse of children by priests, and long-standing patterns of cover-up of such abuses by prominent bishops in Boston, New York, and elsewhere, have caused inordinate conflict among parishioners and church leaders. Gradually, however, through new appointments of bishops and progress in addressing legal claims by injured families, the Church is beginning to address the conflict and trauma among its faithful. Whether this means that the Church will reconsider its traditions of priestly celibacy or its staunch opposition to the legitimation of homosexuality remains an open question (Cahill, Garvey, & Kennedy, 2006; Scheper-Hughes & Devine, 2003). The crisis in the Church over sexual abuse demonstrates that issues of sexuality among priests and laypeople continue to produce diverse attitudes and behaviors. Although no one involved condones abuse of children in any form, open discussion of how the Church deals with sexuality remains highly controversial.

Other major religions and Christian denominations in the United States are also facing conflict and controversy over issues of sexuality among church leaders and parishioners. The Episcopal Church in America is divided over the appointment of an openly gay bishop to be the leader of the church in New Hampshire. Although the majority of Episcopalians does not oppose this move, a strong and vocal minority does, and this conflict threatens to divide the church into two separate organizations if calmer voices do not prevail (Goodstein, 2003).

Fundamentalism Religious *fundamentalists* are believers (and their leaders) who are devoted to the strict observance of ritual and doctrine. They hold deep convictions about right and wrong in matters of faith and lifestyle, with little tolerance for differences in belief and practice, and they are fiercely opposed to astrology, magic, unorthodox conceptions of religion, and any form of civil religion. In all the world religions, there are divisions between liberal and conservative approaches to the religious norms that guide daily life. Among fundamentalists, however, there is a belief in the

> basic, intrinsic, essential, inerrant truth about humanity and deity; that this essential truth is fundamentally opposed by forces of evil which must be vigorously fought; that this truth must be followed today according to the fundamental, unchangeable practices of the past; and that those who believe and follow these fundamental teachings have a special relationship with the deity. (Hunsberger, 1995, p. 9)

Clearly, these beliefs, passionately held as they are, present enormous challenges for societies in which there is considerable religious diversity. (The consequences of Islamic fundamentalism, particularly for the lives of women, are presented in the Global Social Change box on pages 424–425.) In the United States, fundamentalist believers may be found in every major religion, but fundamentalism is especially strong among some Protestant churches and sects.

Historically, the influence of religious fundamentalism on the culture and politics of American life reached its height in 1919, when the Eighteenth Amendment to the U.S. Constitution prohibited the manufacture, sale, or transportation (and hence the consumption) of "intoxicating liquors," regardless of whether one supported the fundamentalist view of alcohol consumption. The repeal of Prohibition in 1933 represented a defeat for religious fundamentalism in American culture and politics.

Another famous episode involving fundamentalism was the so-called "monkey trial" of 1925, in which John T. Scopes, a biology teacher in Tennessee, was charged with teaching the then-forbidden theory of evolution, which many religious fundamentalists oppose because it contradicts the account of creation presented in the Bible. The prominent lecturer and religious fundamentalist (and former presidential candidate) William Jennings Bryan prosecuted the controversial case, which received

global social change

Women in Islam. More than one billion people adhere to the Islamic faith. Most of them live in the Arab states of the Middle East and North Africa, but Muslims may be found in growing numbers in every corner of the globe, including the United States.

Relations between the Islamic world and Western societies are colored by a great deal of misinformation and bias. An example from the social sciences is the theory that social changes do not affect the Islamic world in the same way that they do, or did, affect the Western nations. In particular, it is argued that because of the subordinate status of women in the Islamic religion, the movements for gender equality, limiting of family size, and greater equality in politics that have brought about such dramatic change in the West will not occur in the Islamic world (Caldwell & Caldwell, 1988). This hypothesis attributes the slow pace of demographic change in many Islamic nations to Orthodox Islam's traditional values, such as polygyny, which permits men to father many children with as many as four wives. The tradition of female submission in Orthodox Islam is also said to account for the persistence of gender inequalities in those nations. Islamic sociologists point out, however, that there are significant differences among the various Islamic societies and that some have long-standing norms supporting birth control and education for women.

When we examine empirical data about the Islamic nations, many ironies become apparent. There are nations where traditional Islamic norms are quite powerful and women cannot hope for educational or occupational equality with men, and there are others where these differences are diminishing rapidly. For example, in Saudi Arabia, patriarchal norms are in effect and the subordination of women is strictly enforced. Women are not allowed to drive, yet that nation is the world's foremost producer of petroleum. As shown in the accompanying table, in Saudi Arabia, only 65 percent of girls of primary school age are enrolled in school, among the lowest rates in the Arab world. Extremely poor, peripheral Islamic nations like Somalia (with only 13 percent female primary school attendance) are even less advanced, but in these nations, it is usually the challenge of day-to-day survival rather than religious orthodoxy that prevents children from attending school. In Iraq, on the other hand, orthodoxy is less powerful and women's average level of education is among the highest in the Islamic world.

Many other differences among the Arab nations in the heart of the Islamic world show up clearly in the table. (Note that Iran is not included in this list because, although it is an Islamic nation, its culture is Persian, not Arab.) There are major differences among these nations in basic social and demographic indicators. All have high fertility rates, but in more urbanized Arab nations like Kuwait—where education for women is more easily available and polygyny is waning—rates of fertility and child mortality are closer to those of other more developed nations. In more agrarian societies like Egypt, which also tend to have larger populations, fertility remains extremely high, as does child mortality. Sociologists from these societies find, however, that Islamic teachings themselves do not account for these disparities so much as economic conditions such as poverty and concentration of the population in rural villages and farms (Obermeyer, 1992).

© Tony Savino/The Image Works

worldwide attention. Despite a strong defense by Scopes's lawyer, Clarence Darrow, Scopes was convicted and fined $100. (Although courts later reversed the conviction on a technicality, it would be more than thirty years before the Tennessee legislature repealed the law that prohibited teaching evolution theory in state-supported schools.) Despite the conviction, the Scopes trial is seen as a turning point for the intellectual opponents of fundamentalism. Although fundamentalism continued to pose a challenge to mainline Baptist, Methodist, and Presbyterian denominations, after the Scopes trial it lost influence—until recently—as a religious ideology (Simpson, 1983; Swift, 1998).

Sociologists of religion tend to view the rise of fundamentalist religious movements as part of a larger effort by people in modern societies to make moral and spiritual sense of their lives. In the Islamic world, fundamentalism is also associated with the failure of elites to deliver on their promises of well-being for the masses of people in their societies (Hussein, 1994). In more affluent nations like the United States, however, the appeal of fundamentalism seems to be more often based on rejection of moral relativism, racial integration, and cultural pluralism, and on the desire to find moral absolutes that can guide people through complex and ambiguous social environments.

Demographic and Socioeconomic Indicators in Arab Countries

	Population (millions)	Gross National Product per Capita (US$, 1989)	Percent Urban	Percent of Age Group Enrolled in Primary School*		Total Fertility Rate 1985–1990	Infant Mortality Rate 1985–1990	Life Expectancy 1985–1990 (years)	
				Male	Female			Male	Female
Algeria	24.9	2,230	51.7	105	87	5.4	74	63	65
Bahrain	0.5	6,340	82.9	111	110	4.1	16	69	73
Egypt	52.4	640	46.7	100	79	4.5	65	58	60
Iraq	18.9	—	71.3	105	91	6.4	69	63	65
Jordan	4.0	1,640	68.0	98	99	6.2	44	64	68
Kuwait	2.0	16,150	95.6	95	92	3.9	18	71	75
Lebanon	2.7	—	83.7	105	95	3.8	48	63	67
Libya	4.5	5,310	70.2	128	119	6.9	82	59	63
Mauretania	2.0	500	46.8	61	42	6.5	127	44	48
Morocco	25.1	880	48.0	85	56	4.8	82	59	63
Oman	1.5	5,220	10.6	103	92	7.2	40	62	66
Qatar	0.4	15,500	89.4	121	121	5.6	31	67	72
Saudi Arabia	14.1	6,020	77.3	78	65	7.2	71	62	65
Somalia	7.5	170	36.4	26	13	6.6	132	43	47
Sudan	25.2	480	22.0	59	41	6.4	108	49	51
Syria	12.5	980	50.4	115	104	6.8	48	63	67
Tunisia	8.1	1,260	54.3	126	107	4.1	52	65	66
United Arab Emirates	1.6	18,430	77.8	98	100	4.8	26	69	73
Yemen, YAR (North)	9.2	640	25.0	141	40	8.0	120	50	50
Yemen, PDRY (South)	2.5	430	43.3	96	35	6.7	120	49	52

*Enrollment ratios are expressed as a percentage of the total population of primary school age. For countries with universal education, the gross enrollment ratio may exceed 100 percent because some pupils are above or below the official primary school age.

Sources: Obermeyer, 1992; UNDP, 2000.

This fact is reflected in the remarks of Robert Bellah, a well-known commentator on American culture. For several centuries, he writes, Americans "have been embarked on a great effort to increase our freedom, wealth, and power. For over a hundred years, a large part of the American people, the middle class, has imagined that the virtual meaning of life lies in the acquisition of ever-increasing status, income, and authority, from which genuine freedom is supposed to come" (Bellah et al., 1985, p. 284). Yet many Americans seem uneasy about their lives despite their material comfort. They seem to yearn for spiritual values without necessarily wanting to return to traditional religious practices. They adhere to the values of individualism, but they also long for the stronger sense of community and commitment that one finds in religious congregations.

Bellah predicts that Americans will continue to seek self-actualization as individuals, but that they will increasingly express their desire for community attachments and higher values, either in traditional religions or in civil-religious practice. As we saw earlier in the chapter, however, research on the challenges faced by Southern Baptists suggests that the divisions between more secular and more fundamentalist believers will continue to strain the institutions of pluralistic societies like the United States.

The Changing Face of Islam Can American Muslim women and men like those shown in these photos play a role in transforming the Islamic world? That might seem like a far-fetched question to many, but if one studies the sociology of religious and ethnic groups in the United States, it becomes less so. After all, Jews in the United States greatly aided in creating the modern state of Israel after World War II, and American Catholics, who often take a relatively liberal stance in disputes about issues such as birth control or the use of local languages in the mass, have influenced Vatican policies on these issues over the years. It is quite likely, then, that American Islamic women who seek to reconcile the demands of their religion with those of living as modern women with careers and families will have some influence on debates about these issues elsewhere in the Islamic world.

AP Photos

Here, for example, is a statement about women in Islam from the Islamic Center of Southern California, which represents the desire of some Muslims in the United States to address the critical question of the future of women in Islam:

Contrary to widespread erroneous belief,
Contrary to widespread negative stereotyping, and
Contrary to regrettable practices in some Islamic societies where anti-Islamic culture traditions have won over Islamic teachings and where women are subdued (and men even more so).

This information has been written with the objective of briefing you on the true Islamic teachings regarding women laid down by the Quran and prophet Mohammad over 14 centuries ago.

- Islam declared women and men equal.
- Islam condemned pre-Islamic practices degrading and oppressing women.
- The same injunctions and prohibitions of Islam equally apply to both sexes.
- Islam gave women the right of inheritance and the right of individual independent ownership unhampered by father, husband, brother, son, or anyone else.
- Islam gave women the right to accept or reject a marriage proposal free from pressure, and by mutual agreement to specify in the marriage contract that she has the right to divorce (if she misses that option she has the right to seek court divorce if she deems the marriage to have failed beyond repair).
- Islam does not require a woman to change her name at marriage.

- Islam protects the family and condemns the betrayal of marital fidelity. It recognizes only one type of family: husband and wife united by authentic marriage contract.

(Published by Islamic Center of Southern California.)

American Muslims share basic religious beliefs with Muslims throughout the world, but compared with the leaders of Islam in Saudi Arabia, where an especially patriarchal form of Islam prevails, they are far more liberal and tolerant of others and more receptive to the claims of women for equal treatment.

For better or worse, the events following the terrorist attacks on the United States have brought American Muslims to the forefront of public scrutiny, and as these photos suggest, they are a diverse people who share the American dream of success, security, and freedom.

© Steve Rubin/The Image Works

© Steve Rubin/The Image Works

SUMMARY

What features are common to all religions, no matter what their beliefs or how they are organized?

- *Religion* is among the oldest and most changeable and complex of human social institutions. It has been defined as any set of coherent answers to the dilemmas of human existence that makes the world meaningful.

- It has also been defined as a system of beliefs and rituals that helps bind people together into a social group.

- *Rituals* are formal patterns of activity that express a set of shared meanings; in the case of religious rituals, the shared meanings are *sacred,* pertaining to phenomena that are regarded as extraordinary, transcendent, and outside the everyday course of events.

- Most Western nations practice some form of separation of church and state, but many other nations have chosen an official state religion. A few nations are theocracies, in which religious leaders control the government.

- Until comparatively recent times, religion dominated the cultural life of human societies. Since medieval times, however, the traditional dominance of religion over other institutions has been reduced by a process termed *secularization.* This process is never complete; religion continues to play an important role in the contemporary world.

- Adherents of all the major religions have at times acted as if theirs were the one true faith, and they have used this claim as a basis for persecuting those who deviated from their beliefs.

What are some characteristics of the world's major religions?

- In simpler and rather isolated societies, people may believe in a great force or spirit, but they do not have a well-defined concept of God or a set of rituals involving God. This form of religion is called *simple supernaturalism.*

- More common among hunting-and-gathering societies is *animism,* in which all forms of life and all aspects of the earth are inhabited by gods or supernatural powers.

- *Theism,* in contrast, comprises belief systems that conceive of a god or gods as separate from humans and from other living things on the earth.

- Many ancient religions were *polytheistic,* meaning that they included numerous gods.

- The ancient Hebrews were among the first of the world's peoples to evolve a *monotheistic* religion, one centered on belief in a single all-powerful God.

- In China, Japan, and other societies of the Far East, the dominant religions are centered not on devotion to a god or gods but on an abstract ideal of spirituality and human behavior.

- Some social scientists have expanded the definition of religion to include *civil religions,* or collections of beliefs and rituals that exist outside religious institutions.

Do religions defend the status quo at all times, or have they been involved in lasting social change?

- A major controversy in the study of religious institutions has to do with the role they play in social change.

- Karl Marx believed that cultural institutions like religion are shaped by economic and political institutions and that they function to instill in the masses the values of the dominant class.

- Max Weber, on the other hand, argued that religion can cause major social change by instilling certain values in the members of a society, in turn causing changes in other institutions.

What are the main forms of religious organization?

- Religion today is a highly structured institution, with numerous statuses and roles within a variety of organizations as well as many kinds of smaller, less bureaucratic groups.

- The three main types of religious organizations are the church, the sect, and the denomination.

- A *church* is a religious organization that has strong ties to the larger society and has at one time or another enjoyed the loyalty of most of the society's members.

- A *sect* rejects the religious beliefs or practices of an established church and usually is formed when a group of church members splits off to form a rival organization.

- A third type of religious organization is the *denomination,* which is on good terms with the religious institution from which it developed but must compete with other denominations for members.

- A *cult* is an entirely new religion. Along with sects, cults are a major source of change in religious organizations.

- People who are not satisfied with more established churches and denominations may form or join a cult or sect. New religious movements arise when a "religious innovator" attracts followers. This is particularly likely when traditional religions fail to meet the needs of their members or when a society is undergoing rapid secularization.

What are some important trends in religion in the United States today?

- The black church has left its imprint on most aspects of African American life and provides much of the leadership and many of the rank-and-file supporters of movements for political and economic justice.

- Membership in a religious organization is quite different from identification with a religious faith, and this is reflected in trends in religion in the United States. Among those trends are a growing tendency to practice unofficial or "folk" religion and an emphasis on religiosity as opposed to church membership.

- *Religiosity* refers to the depth of a person's religious feelings and how those feelings are translated into behavior.

- Studies of religiosity find high percentages of Americans believing in the existence of God and in a life after death.

- In societies characterized by religious pluralism, one usually can observe the continual formation of new religious organizations.

The Kornblum Companion Website

www.cengagebrain.com

Supplement your review of this chapter by going to the companion website to take one of the tutorial quizzes, use the flash cards to master key terms, and check out the many other study aids you'll find there. You'll also find special features, such as GSS data and Web links that will put data and resources at your fingertips to help you with that special project or to do some research on your own.

Education

Education for a Changing World

Attainment, Achievement, and Equality

The Structure of Educational Institutions

FOCUS QUESTIONS

In what ways is education a central institution for addressing the learning needs of people in the developed and developing nations?

How do issues of educational equality differ between developed and developing regions of the world, and in what ways are they similar?

What are some of the key structural aspects of educational institutions, and how do educational reform movements in the United States address structural issues?

Wendy Kopp was an idealistic American college senior, about to graduate and filled with the desire to make the world a better place. In her senior thesis, she proposed a plan to develop a corps of new teachers who would go into communities where too many children were failing. That was in 1989. Today, Kopp is president of Teach for America (TFA), one of the most successful teacher recruitment efforts for underserved communities in the nation's history.

Wendy Kopp.

Courtesy of Jonas Chartock

Only days after her graduation, Kopp secured a seed grant and began working tirelessly to put her thesis into practice. Her central idea was that young people like herself would respond with enthusiasm to the chance to be trained for teaching assignments in poor urban and rural schools. She recruited some influential educators and foundation and business executives to her cause. Under her leadership, they raised the first $2.5 million to begin Teach for America as a nonprofit corporation. The organization received 2,500 applications, from which it selected the first 500 recruits for its new training institute. School districts in six states were prepared to hire recruits who successfully completed the training.

Teach for America has grown rapidly. In 2008, the organization saw its largest applicant pool ever, with approximately 35,000 students applying for a little over 4,000 positions in thirty-five regions across the country. Congress's passage of the Consolidated Appropriations Act of 2010 earmarked $21 million to help TFA reach its goal of growing to more than 8,000 corps members within the year (Johnson, 2008).

Teach for America veterans continue to work for educational equality and social justice. Many have gone on to start their own schools, become principals, develop innovative teaching methods and programs, or become doctors and lawyers who practice in disadvantaged communities. Wendy Kopp remains president of the original corporation, which is linked to the federal AmeriCorps program, whose recruits receive modest incomes for two years of national service.

All this success came after an enormous amount of work and struggle. Teach for America had—and still has—critics who believe that putting underprepared teachers in classrooms does no one a service. Some professional educators fault the program for promoting what they say is the dangerous myth that "anybody can teach." The 1994 attack on a wide range of social programs in the Republicans' Contract with America would have ended AmeriCorps and severely stunted the growth of Teach for America. But Congress was not about to scuttle a program that was so appealing to the nation's young graduates and so needed in disadvantaged regions. In fact, the program's measurable success has made Teach for America more secure. Surveys show that 77 percent of principals in the schools where the recruits teach feel that the recruits do better than the first-year teachers usually assigned to them. Eighty-five percent of recruits complete their full two-year commitment, a rate far higher than those most school systems in poor communities report for their entering teachers (Chaddock, 1999).

Teach for America has won high marks from educators and policymakers. In a 2009 speech, President Obama cited TFA as a shining example of young people's desire to serve their country. "I've seen a rising generation of young people work and volunteer and turn out in record numbers," he said. "They have become a generation of activists possessed with that most American of ideas—that people who love their country can change it."

Wendy Kopp has been honored with many awards, but no doubt she is proudest of the testimony of Teach for America recruits like Jonas Chartock:

> There is not another job that would bring me back to work excited day in and day out as teaching does. I enjoy my life as the 23-year-old chief executive officer of the nonprofit organization known as Room 104B. I have a responsibility matched by few. The potential for "growth" and "increasing shareholder value" and "opening new markets" in my industry cannot be quantified.

EDUCATION FOR A CHANGING WORLD

Teach for America is one among many current initiatives in North America and throughout the world that address the need for more effective education. Poorer regions of the world face a dire need for universal access to schools; in the more developed regions, serious issues of educational equity need to be addressed. And amid the clamor for school reform in the United States is concern that new standards and new pressures for higher performance will not be matched by the necessary resources of money and talent. Although critics of public schooling abound, there is little consensus about how to achieve better student performance (DeCoker, 2010). But why are schools and education so often at the center of controversy and debate? To begin answering this question, we turn to basic concepts in educational sociology.

Efforts to change the goals and methods of education are not new. One of the most famous trials in history took place in ancient Greece when Socrates was accused of corrupting the morals of Athenian youth with his innovative ideas and educational methods. Today, the fact that proposals for educational reforms are receiving so much hopeful attention is evidence of the importance of education in modern societies. There is great concern with the need to improve education in order to train new generations of workers. At many points in this chapter, we will see that there is a widespread desire to reform schools without significantly increasing school funding. However, although critics of the public schools abound, little consensus exists about how to achieve better student performance (Sadker & Sadker, 2005).

Education may be defined as the process by which a society transmits knowledge, values, norms, and ideologies and in so doing prepares young people for adult roles and adults for new roles; in other words, it transmits the society's culture to the next generation. Education thus is a form of socialization that is carried out by institutions outside the family, such as schools, colleges, preschools, and adult education centers. Each of these is an educational institution because it encompasses a set of statuses and roles designed to carry out specific educational functions—it is devoted to transmitting a specific body of knowledge, values, and norms of behavior. A particular school or college may be referred to as an *institution* in everyday language, but in sociological terms, it is an *organization* that exemplifies an educational institution. Thus, El Centro College in Dallas, Texas, is an organization that exemplifies the institution of higher education. The high school you attended was also an organization, but its curriculum and norms of conduct were those of the institution known as secondary education.

Educational institutions have a huge effect on communities in the United States and other modern societies. Upwardly mobile couples often base their choice of a place to live on the quality of the public schools in the neighborhood. Every neighborhood has at least an elementary school, and every large city has one or more high schools and at least one community college or four-year college. Cities usually also have a variety of school administrations—public and private—and some owe their existence, growth, and development to the presence of a college or university (Huggins & Johnston, 2009).

Education and Social Change

If education is so desirable, why are there so many controversies about how best to educate new generations? Educational institutions, like those of religion, seem to be always under intense scrutiny and ideological attack. And the level of controversy over education and schools seems to be particularly high in democratic nations like the United States. Here are some basic sociological explanations for why education and schooling draw so much fire:

1. *High stakes*—Everyone wants what they feel is the best education for their children. Educational success counts a great deal in explaining later patterns of success and failure, not only for individuals but for entire social groups and communities (Kingston et al., 2003).

---

education: The process by which a society transmits knowledge, values, norms, and ideologies and, in so doing, prepares young people for adult roles and adults for new roles.

2. *Class and cultural diversity*—Schools, especially public schools, bring together extremely diverse groups of students, parents, and educators, whose values and attitudes about what should be learned and how it should be learned are often in conflict (Arnot, 2002).

3. *Citizen involvement*—In democratic societies, people expect to speak freely about their beliefs and values and exert political influence to create changes in the institutions they pay for and are involved in; this is particularly true of schools and other educational institutions (Bowles & Gintis, 2000).

4. *Rapid change in knowledge*—Our expectations about what education can and must achieve for the society are always changing as scientific and humanistic knowledge changes. School curricula therefore are always open to revision, which leads to further ideological debate about how best to teach and study, and the value of particular levels of education (Moore & Muller, 2002).

These and other, more specific issues—such as how people think reading should be taught, or whether they believe teachers should lead children in prayer, or whether the schools in a community are failing—account for much of the daily debates about education and schooling in America and elsewhere. How can one go beyond these debates and gain a more sophisticated understanding of educational quality? Social scientists, and especially sociologists, conduct systematic research on a host of educational questions. Sociological theory and research cannot resolve debates among people with widely differing values, but it can contribute in extremely important ways to our factual knowledge about how students, teachers, schools, and other educational organizations are actually working or not working in this and other societies. Application of the major sociological perspectives to education and schooling provides an overview of major areas of research on education.

Changing Functions of Education and Schooling To sociologists, the most common educational institution, the school, is a specialized structure with a special function: preparing children for active participation in adult activities. In simpler societies, however, education of the young is everybody's task in the village or settlement. Children learn through direct demonstration around the hearth or in pastures or fields. For example, among the Maasai, whom we encountered in Chapter 13, boys learn to hunt from their uncles and fathers; girls learn their roles by helping their mothers perform their tasks. However, in urban industrial societies with a great deal of division of labor and highly developed social institutions, schools of all kinds, including universities, have become highly specialized institutions designed to teach young people the skills and values the society wishes to foster. The typical school has a clearly defined authority system and set of rules. In fact, sociologists often cite schools as examples of bureaucratic organizations (Ballantine, 1997; Mulkey, 1993).

From this functionalist perspective, another primary function of schools is to help assimilate the children of immigrants into U.S. society. Through language instruction and socialization of students into "the American way of life," schools have been active agents in the assimilation of waves of immigrants over the past 150 years of U.S. history. But debates over bilingual education and interpretations of American history (for example, the treatment of Native Americans by the U.S. government) in school curricula are indications that the assimilation function is also subject to controversy.

Manifest and Latent Functions of Education Sociologists summarize the explicit, or manifest, functions of education in the United States

and most other highly developed nations as follows:

- Formal education transmits the culture of a society to new generations.
- It prepares future generations for appropriate occupational and citizenship roles.
- Educational institutions continually evaluate and select competent individuals.
- Education transmits the requisite social skills for functioning in society.

In addition to these manifest functions, however, there are many latent functions that are implicit in the way we actually go about organizing the education of our children. A latent function of schools is to help reproduce the existing class structure of societies, for example. Parents with knowledge and sufficient wealth can send their children to better public schools by moving to suburban communities that pay teachers more and recruit more highly qualified teachers than are available in central cities, where there are higher concentrations of students from poor backgrounds with less preparation for school. A latent function of institutions of higher education is to bring young men and women from similar class and cultural backgrounds together in an informal "marriage market." Of course, there are many exceptions to these latent functions; people from different class backgrounds may find each other in college, for example, but the basic trends remain in effect, often despite educators' efforts to foster greater equality of access and educational outcomes.

Interaction in Schools A more interactionist viewpoint sees the school as a set of behaviors; that is, the central feature of the school is not its bureaucratic structure but the kinds of interactions and patterns of socialization that occur there. From this perspective, we need to examine all the roles involved and see how they interact—how they mesh or fail to mesh. For example, homework is assigned, but who besides the student is involved in completing the assignment? Are the parents involved? In what ways—as helpers or merely as disciplinarians? Do the parents ever ask what the teacher did with a homework assignment? The interactionist perspective thus insists on examining all the factors involved in learning, often in an effort to determine how they may be strengthened or challenged so as to create more effective learning environments (Torres & Mitchell, 1998).

The Hidden Curriculum Interactionist sociologists ask questions about what actually goes on in the daily life of schools and classrooms.

Their research focuses on the interactions among students, between students and their teachers, or among teachers and administrators. Their research on these questions usually helps answer questions asked by those with other sociological perspectives as well. For example, observation of classroom interactions might reveal that teachers are unknowingly following what is sometimes called a "hidden curriculum" that communicates different expectations for girls and boys, with boys expected to do better at math and science and girls expected to do better at analyzing characters and emotions, common gender-stereotypical roles (Lynch, 1990).

In their research on global educational issues, interactionist sociologists ask questions about how schools work to reach as many of a nation's children as possible, how well students actually learn in those schools, and how the schools function with regard to other major social institutions. Are the schools secular or religious institutions, or both? How free are educators to decide on what is to be taught? In more formal terms, how differentiated are the schools from the society's other social institutions? This question is the familiar one of separation of church and school, or of political party and school. Even in relatively secular nations such as the United States, where separation of church and school is the law, there are continual battles over issues like prayer in school and the teaching of "creationism" or "intelligent design" (the idea that God set in motion the mechanisms of evolution). In less secular societies, such as many Islamic nations, religious institutions are entirely responsible for running the schools, in which case little value may be placed on scientific objectivity or freedom of inquiry.

The Conflict Perspective: Mobility versus Reproduction
Conflict theorists view education in modern societies as serving to justify the status quo of class inequality. They ask whether the education produces genuine upward mobility for new generations or simply reproduces the existing class structure by giving students from less advantaged backgrounds mediocre education and schools (Arrow, Bowles, & Durlauf, 2000; Bowles & Gintis, 2000; Castells et al., 1998). Their emphasis on the reproduction of inequality through education challenges the more popular view that education is the main route to social mobility and that it can offset inequalities in family background (Bell, 1973, 1999/1973). When sociologists analyze the impact of educational institutions on society, they generally conclude that the benefits of education are unequally distributed and tend to

This photo of Tibetan refugee children in a classroom in northern India illustrates the similarity of classroom organization in many parts of the world.

reproduce the existing stratification system (Aronowitz & Giroux, 1985; Sadovnik & Semel, 2010).

As noted earlier, sociologists with a conflict perspective on social research often ask whether schools are actually involved in creating equal opportunities for social and group advancement or whether they are in fact working to justify and reproduce existing patterns of inequality. Are they preparing all students, regardless of class, race, ethnicity, or gender, to rise in the social-class hierarchy, for example, or are they preparing students to accept their positions in society, especially when those positions are less advantaged?

The different perspectives generate many specific questions and diverse research projects. This is especially true when one takes a global perspective because educational issues vary widely across different regions of the world. The Perspectives chart on page 434 summarizes the researchable questions these basic sociological perspectives raise in developing and developed regions. Note that the question of who actually goes to school is a critical one. In most urban industrial societies, over 95 percent of all children attend school, but homeschooling is becoming more popular, and differential school dropout rates mean that questions about who goes to school at different levels of education are just as relevant in the developed nations as they are in the developing regions of the world.

Thinking Critically

If one of the major functions of education is to prepare young people to live as well-functioning citizens in their societies, don't students benefit from attending schools that reflect the nation's ethnic, racial, and class diversity? If you are a parent, how do you think about these issues for your children, or how will you think about them if and when you become a parent?

Who Goes to School?

The idea that all children should be educated is a product of the American and French Revolutions of the late eighteenth century. In the European monarchies, the suggestion that the children of peasants and workers should be educated would have been considered laughable. In those societies, children went to work with adults at an early age, and adolescence was not recognized as a distinct stage of development. Formal schooling, which was generally reserved for the children of the elite, typically lasted three or four years, after which the young person entered a profession.

Global Perspectives on Education

| Perspective | Typical Issues Raised | |
	Developed Regions	Developing Regions
Functionalism	How can schools be better organized to carry out their mission?	Are schools devoted to elite or mass education? How independent are the schools from other social institutions, such as mosques and political parties?
Conflict Theory	Do schools promote equal opportunity and mobility, or do they reproduce the status quo? of inequality? Do they foster free inquiry or stifle it?	Whose children actually get to go to school? Do schools allow freedom of thought and critical thinking?
Interactionism	How does learning take place in classroom interactions? What else is being communicated and taught in addition to the lessons? How do peer interactions work to assist or impede learning?	Questions are essentially the same, based on observation of classroom interactions, but perhaps with more emphasis on how traditional cultures are dealt with in school and classroom interaction.

Even after the creation of republics in France and the United States and the beginning of a movement for universal education, the development of a comprehensive system of schools took many generations. In the early history of the United States, the children of slaves, Native Americans, the poor, and many immigrant groups, as well as almost all female children, were excluded from educational institutions. The norm of segregated education for racial minorities persisted into the twentieth century and was not overturned until 1954, in the Supreme Court's famous ruling in *Brown v. Board of Education of Topeka*. Even after that decision, it took years of civil rights activism to ensure that African Americans could attend public schools with whites. Thus, although the idea of universal education in a democracy arose early, it took many generations of conflict and struggle to transform it into a strong social norm (Cremin, 1980).

The idea of mass education based on the model created in the United States and other Western nations has spread throughout the world. Mass education differs from elite education, which is designed to prepare a small number of privileged individuals (generally sons of upper-class families) to run the institutions of society (for example, the military, the clergy, the law). Mass education focuses instead on the socialization of all young people for membership in the society. It is seen as a way for young people to become citizens of a modern nation-state. Mass education establishes an increasingly standardized curriculum and tries to link mastery of that curriculum with personal and national development (Benavot et al., 1991; Meyer, Ramirez, & Soysal, 1992).

Figure 17.1 indicates that, by 1980, all the nations of the world had adopted the basic model of a mass educational system (although many had not extended educational opportunities to the majority of young citizens). Note the sharp upsurge in the 1950s. This was the decade when large numbers of colonial nations in Africa and Asia became independent. Commitment to a mass educational system became a hallmark of modernity for these and other new nations. It was also a time when agencies like the World

Bank, UNESCO (United Nations Educational, Scientific, and Cultural Organization), and the U.S. Agency for International Development began to actively encourage and provide financial support for mass educational institutions.

Sociology & Social Justice

GLOBAL INEQUALITY IN ACCESS TO EDUCATION The continuing gap between rich and poor nations in access to education is a global issue of social justice. The United Nations (2010) warns that hundreds of millions of children throughout the world are denied the right to quality basic education. They are therefore growing up ill equipped to make decent lives for themselves. In *The State of the World's Children 1999*, the United Nations Children's Fund (UNICEF) warned that "hundreds of millions of

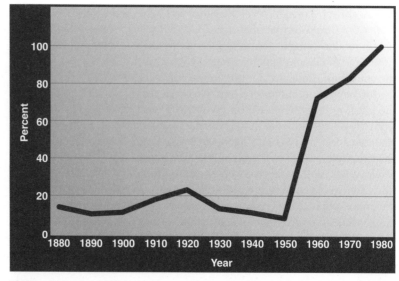

Figure 17.1 Percentage of Nations Developing Systems of Mass Education, by Decade. Source: Adapted from Meyer, Ramirez, & Soysal, 1992.

global social change

Poverty and Illiteracy There is a strong correlation between illiteracy and poverty throughout the world (see the map on pages 436–437). The poor nations of the Sahel, such as Mali, Chad, Niger, Ethiopia, and the Sudan, suffer from some of the highest rates of illiteracy in the world. High rates of illiteracy are also evident in much of southern Asia, especially on the Indian subcontinent. On the other hand, the correlation is not absolute: Illiteracy rates are quite low in poor nations such as Mexico and Cuba, where the ideology of social development places strong emphasis on educating the mass of citizens to the fullest extent possible.

Sociological theories of modernization stress the need for populations to become literate so that their members will be better-informed voters, more highly skilled workers, more careful parents, and generally better able to realize their human potential. Reductions in the level of ignorance yield improvements in every aspect of a nation's social and civic life.

This line of argument is clearly supported by the correlation between poverty and illiteracy in India (see the accompanying chart). Overall, the chart shows that as illiteracy decreases, the amount of money a household is able to spend each month increases. The chart also shows that illiteracy and the poverty associated with it are much more prevalent among women than among men, with rural women showing the highest rates of illiteracy and poverty (about 90 percent illiterate in the high-poverty category). That men are far more likely than women to become literate in India is a reflection of the immense gap in prestige between the sexes in that society; men are considered far more worthy of education than women.

Research on the effects of literacy on vital measures of social change such as reduced fertility clearly shows the importance of educating women. Data from research conducted in Thailand and other Southeast Asian nations suggest that, until women gain access to at least minimal educational opportunities, fertility rates in those nations will remain high. This research confirms the hypothesis that "demographic change is unlikely if the movement towards mass schooling is confined largely to males" (London, 1992, p. 306).

These relationships among gender, illiteracy, and social indicators like poverty and fertility offer a warning that investments in literacy alone are necessary but not sufficient to accelerate social change in a population. Investments in literacy must be accompanied by strategies to reach the most impoverished and discriminated against segments of the population (such as rural women in many societies). Such strategies, in turn, are difficult to develop in a society in which the powerful may fear their effects. In nations where poor rural women are offered more education, for example, the women often begin to demand greater equality. It takes farsighted leadership to actively promote such strategies.

Another consequence of increasing literacy and education in poor societies is that some of the "best and brightest" are attracted by opportunities in richer societies. This phenomenon is known as the "brain drain" and is a significant problem for developing nations (Alam, 1988; O'Leary, 2006). Relatively well-educated people, often with professional training, are an important source of migration from third world nations to the wealthier nations of North America and western Europe. Indeed, immigration policies in North America actively encourage immigration by well-trained individuals and establish quotas that discourage less educated people from seeking entry.

To address the problems created by the brain drain phenomenon, the poorer nations have developed policies requiring that people trained at public expense (for example, through scholarships and training programs) must serve in their own nation until it has recouped its investment in their education. Such policies are not popular with educated people in those nations, especially if they are aware of more attractive opportunities outside their own society. But if the full benefits of literacy and education are to be applied to national development, such policies are necessary.

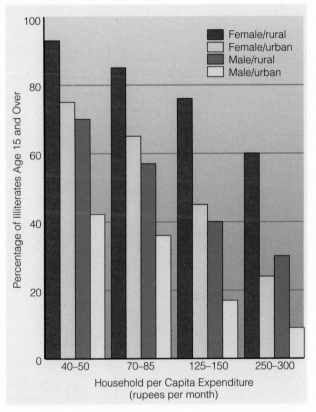

Source: Data from third quinquennial survey on employment and unemployment, Ministry of Planning, New Delhi, India.

children throughout the world are denied the right to quality basic education, and are therefore growing up ill-equipped to make decent lives for themselves." According to the research documents on which the report is based, nearly one billion people—one-sixth of humanity—are classified as functionally illiterate, and two-thirds of these adults are female (United Nations, 2010). (For more on the consequences of this situation, see the Global Social Change box above.)

Children who grow up without basic education find it harder not only to sustain themselves and their families but also to make their way as adults in society with a spirit of tolerance and understanding. Girls who are excluded from school and remain illiterate are even more subject to the dictates of patriarchal figures than those who have received some education and can voice their opposition, at least among themselves. At present, at least 90 million children worldwide are denied any schooling at even

Worldwide Illiteracy Rates. Source: Adapted from Peters, 1990; United Nations, 1997.

the first-grade level, and 232 million have no access at all to secondary education. Throughout much of Africa, for example, annual school fees for families are typically under $30 per child, but for too many families earning below poverty wages, even this sum is out of reach. The UNICEF researchers calculated that, in order to provide basic education for all children worldwide, governments would need to spend only an additional $7 billion per year over the next ten years, which is less than one-tenth of the world's annual military spending.

Turning their attention to educational issues in the developed nations, the UNICEF researchers' findings echo those of critical sociologists. They found that academic underachievement, especially in math and science, is a problem in industrialized as well

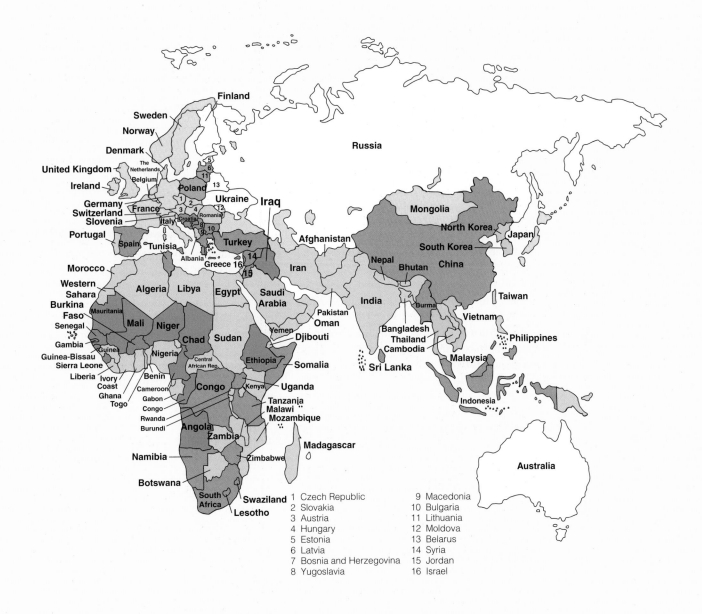

1	Czech Republic	9	Macedonia
2	Slovakia	10	Bulgaria
3	Austria	11	Lithuania
4	Hungary	12	Moldova
5	Estonia	13	Belarus
6	Latvia	14	Syria
7	Bosnia and Herzegovina	15	Jordan
8	Yugoslavia	16	Israel

as developing nations, and that violence is a major problem in schools around the world. The gap between educational haves and have-nots is widening, the research showed, partly as a result of the influence of new learning technologies that are unequally distributed. Most important, the study signaled the long-term negative effects of ignorance and illiteracy caused by the educational exclusion of so many children in the world's poor nations (United Nations, 2001).

In the more affluent industrial nations, extremely high proportions of students who complete secondary school go on to universities and colleges or some kind of professional or occupational training. The demand for greater access to higher education was pioneered in the United States and Canada. An extension of mass education in the United States was the expansion of public institutions of postsecondary education (and the parallel expansion of private colleges and universities, often with federal aid and research funding). This expansion was fueled by the post–World War II baby boom and the massive increase in the college-age population in the 1960s. During those years, there was a parallel boom in employment for elementary school teachers and then for high school teachers and college professors. But after the boom came the bust: The birthrate fell sharply beginning in the late 1950s, and by the mid-1970s, college enrollments also began to decline. These changes had dramatic effects on primary schools, but at the college level, they were partially offset by an unprecedented countertrend: the immense increase in the number of older students seeking higher education. Today, large numbers of adults are returning to college. Unlike the typical student of earlier years, adult students are in the labor force, are married and living with their spouses, are going to school part-time, and are seeking skills and knowledge to enhance their careers (Sadker & Sadker, 2005; Torres & Mitchell, 1998).

The return of so many adults to educational institutions has led many social scientists to describe a future in which education will be a lifelong process in which people of all ages will move in and out of educational institutions. Nevertheless, the most rapidly growing area of education is preschool programs, an important trend that is discussed later in the chapter.

Schools and Adolescent Society

A key feature of education is the fact that schools structure the lives of children and adolescents. This is particularly true at the high school level. In 1961 James Coleman published a famous study of the effect of schools on adolescents and youth, *The Adolescent Society*. Its main point was that schools help create a social world for adolescents that is separate from adult society. According to Coleman, this is an almost inevitable result of the growing complexity of industrial societies, in which, as we saw earlier, functions that were formerly performed by the family are increasingly shifted to other institutions, especially educational institutions. Yet schools cannot provide the same kind of support and individual attention that the family can. As a result, according to Coleman, students are pushed into associations primarily with others of the same age. These small groups of age-mates come to "constitute a small society, one that has most of its important interaction within itself, and maintains only a few threads of connection with the outside adult society" (1961, p. 3).

Coleman's research was not unique. Other social scientists have analyzed what has come to be known as the "youth culture" in terms of changes in the structure of American society. Briefly, what has happened is that the rising level of expectations regarding educational attainment has placed more and more demands on students. This pressure often is not matched with a clear means of meeting the new demands. This mismatch between ends and means can lead to the emergence of deviant individuals and groups, as we saw in our discussion of Merton's typology of deviance in Chapter 8. The resulting frustration often leads teenage peer groups to become extremely ambivalent and even cynical about adult expectations. They develop fierce loyalty and conformity to the peer group, strictly observing group norms and not tolerating any deviance from those norms, which themselves may deviate from those of their teachers and parents.

The risk for society in the development of isolated adolescent cultures is that teenagers will fail to learn how they can become involved in shaping their own society. Great cultural gaps between youth and their parents are often a signal that the older generation has not offered young people enough opportunities to work with adults for constructive social change (Gilligan, cited in Norman, 1997; Knecht et al., 2010).

Education and Citizenship

Another important aspect of education in the United States is the relationship between education and citizenship. Throughout its history, this nation has emphasized public education as a means of transmitting democratic values, creating equality of opportunity, and preparing new generations of citizens to function in society. In addition, the schools have been expected to help shape society itself. During the 1950s, for example, efforts to combat racial segregation focused on the schools. Later, when the Soviet Union launched the first orbiting satellite, American schools and colleges came under intense pressure and were offered many incentives to improve their science and mathematics programs so that the nation would not fall behind the Soviet Union in scientific and technological capabilities.

Education is often viewed as a tool for solving social problems, especially social inequality. The schools, it is thought, can transform young people from vastly different backgrounds into competent, upwardly mobile adults. Yet these goals seem almost impossible to attain (Cahill, 1992). In recent years, in fact, public education has been at the center of numerous controversies arising from the gap between the ideal and the reality. Part of the problem is that different groups in society have different expectations. Some believe that students need better preparation for careers in a technologically advanced society, others that children should be taught basic job-related skills, still others that education should not only prepare children to compete in society but also help them maintain their cultural identity (and, in the case of Hispanic children, their language). On the other hand, policymakers concerned with education emphasize the need to increase the level of student achievement and to involve parents in their children's education (Wilson, 1993).

Some reformers and critics have called attention to the need to link formal schooling with programs designed to address social problems. Sociologist Charles Moskos, for example, is a leader in the movement to create a system of voluntary national service. National service, as Moskos defines it, would entail "the full-time undertaking of public duties by young people—whether as citizen soldiers or civilian servers—who are paid subsistence wages" and serve for at least a year (1988, p. 1; Moskos, 2006). In return for this period of service, the volunteers would receive assistance in paying for college or other educational expenses.

Advocates of national service and school-to-work programs believe that education does not have to be confined to formal schooling. In devising strategies to provide opportunities for young people to serve their society, they emphasize the educational value of citizenship experiences gained outside the classroom. AmeriCorps and Teach for America are examples of such programs, but they are minuscule compared to the need.

Education and Military Service

Each year thousands of young men and women, most with high school diplomas or the equivalent, join the nation's armed services. The armed forces include more than 800,000 active-duty personnel, some serving in Afghanistan or Iraq, others stationed elsewhere abroad or on military bases in the United States. What effect does military service have on how much education and training an individual eventually acquires? This is an important question, if only because it deals with such a large segment of the nation's youthful population and affects their earnings over a lifetime. But because a disproportionate number of minority group members and people from lower-income backgrounds serve in the armed forces, the issue is of even greater importance for them. Joining the military is often one of the only routes into an independent adulthood for people from less advantaged backgrounds.

For veterans returning to civilian life after World War II, the Roosevelt administration created the Servicemen's Readjustment Act of 1944, which was enacted by Congress and became known as the GI Bill. One of the most lauded legislative acts of the twentieth century, the GI Bill helped hundreds of thousands of ex-GIs attend college or professional training programs (Humes, 2006). The education benefits were even more successful among veterans of the Vietnam War. Some changes made between 1990 and 2000 made some of the benefits contingent on the individual soldier's decision to opt into the plan by making monthly payments that the government would match. In his research on this change, sociologist of education Jay Teachman found that those who would stand most to benefit from education assistance after service were least likely to have been able to afford the payments. They were thus less likely to be able to draw on GI Bill funds to help with their education after their military service. In 2008, newly elected Senator Webb of Virginia, himself a veteran, championed legislation to improve GI Bill benefits. Veterans are now eligible for greatly expanded benefits, or the full cost of any public college in their state. The new bill also provides a housing allowance and a $1,000-a-year stipend for books, among other benefits (Teachman, 2007).

ATTAINMENT, ACHIEVEMENT, AND EQUALITY

Although some educational reformers focus on the need to expand learning opportunities through nonschool service experiences, most scholars and administrators seek improvements in educational institutions themselves. Their efforts often focus on *educational attainment*, or the number of years of schooling that students receive, and *educational achievement*, or the amount of learning that actually takes place. Both aspects of education are closely linked to economic inequality and social mobility (see Chapter 11).

Educational Attainment

In any discussion of education as a major social institution, the concept of **educational attainment** (number of years of school completed) holds a central place. Educational attainment is

---

educational attainment: The number of years of school an individual has completed.

TABLE 17.1

Educational Attainment of U.S. Population, 1940–2008

Year	Median Years of School Completed
2008	13.1
2000	12.9
1990	12.9
1980	12.5
1970	12.2
1960	10.6
1950	9.3
1940	8.6

Source: Based on data from U.S. Census Bureau, 2008.

correlated with income, occupation, prestige, attitudes and opinions, and much else. It is essential, therefore, for social scientists to understand the impact of recent trends in school enrollment on the educational attainment of the population as a whole and of various subgroups of the population.

Table 17.1 shows the median number of years of school completed by the population as a whole in the decades since 1940. Table 17.2 presents data on educational attainment by race, ethnicity, and sex. Both tables apply to the population age twenty-five and older. It is immediately clear that the average American today has much more education than the average American of the early 1940s. It is also clear that whites are more likely to complete high school than are blacks or Hispanics, and considerably more likely to attend college. The data indicate the increasingly high value placed on education, especially college education. People older than age seventy-five, who were born in the early decades of the twentieth century, have, on the average, considerably lower levels of educational attainment than their children and grandchildren (Farley, 1996; *Statistical Abstract*, 2002).

Barriers to Educational Attainment

Tracking and Inequality The rise of mass education gave the middle and lower classes greater opportunities for upward social mobility through educational attainment. As early as the 1920s, however, many schools began to use "tracking" systems in which higher-achieving students were placed in accelerated classes while others were shunted into vocational and other types of less challenging classes. Today, tracking remains a major problem in public schools. Parents of "gifted" children seek educational challenges for their sons and daughters and do not want them to be held back by slower learners. However, tracking systems can make average students feel less valued, and there is a danger that gifted but alienated students will be labeled as nonachievers.

Figure 17.2 is based on a national study of more than 14,000 eighth grade students in public schools. It shows clearly that white and Asian students are far more likely than black and Hispanic students to be tracked into high-ability groups. (Ability grouping is synonymous with tracking.) Some social scientists argue that these differences are a result of ability differences among racial and class groups; others assert that they are a consequence of race and class bias (Wolfe, 1998). Because far more Hispanic and black students are in the lower socioeconomic classes, there is little question that social factors outweigh biological ones in explaining these results. In any case, tracking separates children and is increasingly viewed as leading to educational inequalities. A current trend in educational practice, therefore, is to institute detracking

TABLE 17.2

Educational Attainment, by Race, Ethnicity, and Sex: 1947–2003

Year	Percentage High School Graduate						Percentage College Graduate					
	White		Black		Hispanic		White		Black		Hispanic	
	M	F	M	F	M	F	M	F	M	F	M	F
2003	84.5%	85.7%	79.6%	80.3%	56.3%	57.8%	29.4%	25.9%	16.7%	17.8%	11.2%	11.6%
2000	88	88	77	78	56	56	31	25	14	17	11	11
1991	80	80	67	67	51	51	25	19	11	12	10	9
1980	71	70	51	51	46	44	22	14	8	8	10	6
1970	57	58	32	35	NA	NA	15	9	5	4	NA	NA
1962	47	50	23	26	NA	NA	12	7	4	4	NA	NA
1947	33	37	13	15	NA	NA	7	5	2	3	NA	NA

NA = Data not available.

Source: U.S. Census Bureau, 1992; *Statistical Abstract*, 2005.

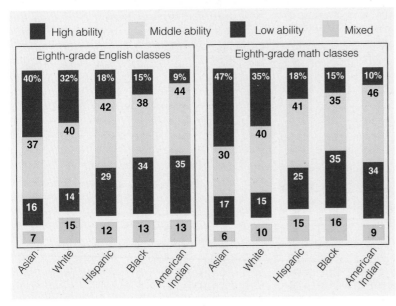

Figure 17.2 Ability Grouping ("Tracking") in the U.S. Public Schools. Source: Adapted from Mansnerus, 1992.

aware of tracking when it exists and that it does not contribute to the academic confidence of those in lower tracks.

Research on the lasting effects of tracking gives strong support to the critical "reproduction" theories of schooling. Students in lower tracks tend to end up in lower-paid and lower-prestige occupations once they finish their schooling, as we see clearly in Table 17.3. In this longitudinal study, which compared student experience with tracking with careers later in life, the researchers found that almost 41 percent of those in the academic track were in professional, technical, or managerial careers later in life, as opposed to less than 20 percent of those in vocational-commercial tracks. Tracking does not automatically translate into reproduction of the class system, but these data indicate that it contributes to maintaining the status quo of class inequality.

programs and to provide highly gifted students with additional challenges through cooperative education (in which they have opportunities to teach others) and through after-school programs (Betts, 1998; Mansnerus, 1992).

Tracking most often begins in junior high school or middle school, where students do not stay in a single classroom but move from one subject to another. It persists into high school, where it is often made explicit in tracks such as the common, academic, general, and vocational-commercial curricula of many secondary school systems. Teachers and administrators often support tracking systems because they believe that students learn more effectively and develop greater self-esteem when they are grouped with academically equivalent students. These claims may be legitimate, but most of the available research indicates that students are intensely

Dropping Out The educational attainment of various subgroups of the population is virtually identical up to the age of thirteen (Barrow & Rouse, 2006; Farley, 1996). The picture begins to change in the early secondary school years, however. Until recently, the dropout rate for black and Hispanic students in high school was almost twice the rate for white students. In 1970, for instance, 10.8 percent of white students dropped out without finishing high school, whereas the proportion was 22.2 percent among black students. By 2007, the situation had improved remarkably as a result of strenuous efforts by students, parents, and teachers across the United States. In that year, 7.3 percent of white students and 7.2 percent of black students dropped out (*Statistical Abstract*, 2010).

International comparisons show that with a national secondary school completion rate just below 80 percent, the United States ranks well below other nations in Europe and Asia with which it is often compared, such as Japan, Germany, and France. In the early twentieth century, the United States led the world in the rate at which its teenagers completed high school, but as Figure 17.3 demonstrates, it has slipped to the middle of the rankings (Peterson, 2003). Such comparisons lead some political and educational leaders to argue that greater school accountability, particularly

TABLE 17.3

Cross-Tabulation of Track Placement by Occupation of Respondent

Occupation	Academic		General		Vocational-Commercial	
	N	Percent	N	Percent	N	Percent
Service	48	10.0	137	20.5	41	19.1
Operator, laborer	70	14.6	165	24.7	39	18.1
Precision production and craft	24	5.0	65	9.7	29	13.5
Administrative, sales	141	29.5	147	22.0	64	29.8
Professional, technical, managerial	195	40.8	153	22.9	42	19.5

Note: Data used in this table are from the 1991 National Longitudinal Survey of Youth, a nationally representative sample of 12,686 men and women first interviewed in 1979, when they were between ages fourteen and twenty-two. Slightly more than 1,900 cases were randomly selected and analyzed.

Source: Broussard & Joseph, 1998.

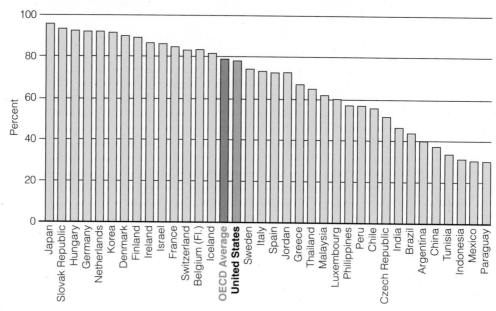

Figure 17.3 **International Secondary School Completion Rates.** Source: Organization for Economic Cooperation and Development, 2000.

through the use of standardized testing of all students annually, is required to make the United States more competitive. As we turn to issues of educational achievement, however, we will see that equating school success with test scores can lead to additional and unanticipated problems.

The main reason for dropping out of school is poor academic performance, but there are other reasons as well. Students often drop out because of the demands of work and family roles; many are married, or unmarried and pregnant, and/or working at regular jobs. Whatever the reason, the effects of dropping out can be serious.

Dropouts have less chance of joining the labor force than high school graduates; whatever jobs they find tend to be low-paying ones. Since the 1970s, the income advantage from completing college has increased for all individuals, regardless of gender or minority status. But black–white differentials are found at each level of attainment. White college graduates, for example, earn at least 20 percent more than black college graduates of the same age.

High-Stakes Testing and Dropout Rates Research has shown that states that base their funding of school districts on how well schools do on standardized tests actually increase, rather than decrease, high school dropout rates. This is an unintended consequence of such "high-stakes testing." Mandated by the Bush administration's "No Child Left Behind" education bill of 2001, the move to increase school accountability through increased use of uniform tests, and then rating schools on how well their students do on these tests, has been a growing trend since the early

1980s. Unfortunately, new research shows that although such tests do result in small gains in student performance, they also encourage school systems to engage in "creative" measurement of school achievement.

In 2003, Texas audited sixteen middle schools and high schools and found that, of 5,500 teenagers who had left school in the 2000–2001 school year, about 3,000, or 55 percent, should have been reported as dropouts. The research showed that the rankings of fourteen of the sixteen schools should be lowered and that Houston's school system should be ranked unacceptable. Houston as a whole reported a 1.5 percent annual dropout rate, although education experts estimate that the true percentage of students who quit before graduation is nearer 40 percent. The city contended that it may be guilty of shoddy record keeping, but not of fraud (Schemo, 2003).

But Houston is far from alone in using statistical sleight of hand to inflate graduation rates (and deflate dropout rates). Many other school districts whose funding is now based on national test results have also been found to be taking steps to make their graduation rates appear better than they are.

The congressionally mandated 1999 National Academy of Sciences report *High Stakes: Testing for Tracking, Promotion and Graduation* cites five other studies that draw similar conclusions (Orfield & Wald, 2001).

Degree Inflation The trend toward increasingly higher levels of educational attainment has had an unexpected effect known as "degree inflation" (Kealy, 2008). Employers have always paid attention to the educational credentials of potential employees, but today they require much more education than in the past. For example, early in the twentieth century, a person could get a teaching job with just a high school diploma; now a bachelor's or master's degree is usually required. The same is true of social work. And secretaries, who formerly could get by without a high school diploma, now are often required to have at least some college education or, in some cases, a college degree.

Degree inflation is discouraging to some students and prevents them from continuing their education. It also adds to the expense of education (both directly and indirectly in terms of lost income) and hence prevents less advantaged students from undertaking advanced studies. Degree inflation also increases the amount of time that must be devoted to formal education. It therefore raises questions about the meaning of educational achievement—that is, the value of the time spent attaining educational credentials.

Educational Achievement

International differences in levels of educational attainment are viewed as a sign that public education is not meeting the expectations of society in terms of the quantity of education provided to citizens. There is also controversy over the quality of education, or educational achievement as reflected in scores on standardized tests like the Scholastic Aptitude Test (SAT). **Educational achievement** refers to how much students actually learn, measured by mastery of reading, writing, and mathematical skills. It is widely believed that the average level of educational achievement has declined drastically in recent decades. In a historic 1983 report titled *A Nation at Risk*, the National Commission on Excellence in Education pointed to the decline in the average test scores of high school students since the mid-1960s, and stated that the schools have failed to maintain high educational standards. In recent years, standardized scores in math have risen somewhat, but verbal scores (reading and vocabulary) have not (Jencks & Phillips, 1999). Many observers attribute the decline in test scores to a variety of social conditions such as too much television viewing and changing values related to family life.

Critics of excessive reliance on standardized tests might agree with prominent educational reformer Ted Sizer, who points out that "none of the major tests used in American elementary and secondary education correlates well with long-term success or failure. SAT scores, for example, suggest likely grades in the freshman year at college; they do not predict much thereafter" (1995, p. 58). For better or worse, however, the SAT and other standardized tests continue to be used as primary measures of individual and school performance.

Functional Incompetency Another problem related to educational achievement is the high rate of "functional incompetency" among Americans. Many people are unable to read, write, keep a family budget, and the like. Although more and more people are obtaining a college education, many others are being left behind—particularly members of the lower social classes, people for whom English is a second language, and people with learning disabilities. As increasing amounts of education are required for better jobs, this cleavage between educational haves and have-nots becomes an increasingly dangerous trend. And recent research shows that because people with less education have more stressful lives, with longer periods of unemployment and other risks to their health, mortality rates for people ages twenty-five to sixty-four are much higher for those with less education. In other words, people with less than twelve years of education are at greater risk of dying than are those with twelve or more years of education (Battle, 2003).

International Comparisons Recent research on school achievement offers even more disturbing evidence of deficiencies in the educational achievement of U.S. students, beginning at an early age. Tests of achievement by thousands of schoolchildren in comparable U.S. and Asian cities—Minneapolis, Chicago, Sendai (Japan), Beijing (China), and Taipei (Taiwan)—found that the mathematics scores of American first graders were lower than those of Asian first graders. In some American schools, first graders' scores were similar to those of Asian students; by the fifth grade, however, all the American students had fallen far behind. In computation, for example, only 2.2 percent of children in Beijing scored as low as the mean score for U.S. fifth graders. On a test consisting of word problems, only 10 percent of fifth graders in Beijing scored as low as the average U.S. student. Gaps also appeared in reading skills, although vast differences in written languages make it somewhat difficult to compare achievement in those skills (H. W. Stevenson, 1992, 1998).

educational achievement: How much the student actually learns, measured by mastery of reading, writing, and mathematical skills.

TABLE 17.4

Average Mathematics Scale Scores of Eighth-Grade Students, by Country: 2003

Country	Average Score	Country	Average Score
International	466	Romania	475
Singapore	605	Norway	461
Korea, Republic of	589	Moldova, Republic of	460
Hong Kong SAR[1]	586	Cyprus	459
Japan	570	(Macedonia, Republic of)	435
Belgium-Flemish	537	Lebanon	433
Netherlands	536	Jordan	424
Estonia	531	Iran, Islamic Republic of	411
Hungary	529	Indonesia	411
Malaysia	508	Tunisia	410
Latvia	508	Egypt	406
Russian Federation	508	Bahrain	401
Slovak Republic	508	Palestinian National Authority	390
Australia	505	Chile	387
(United States)	504	(Morocco)	387
Lithuania	502	Philippines	378
Sweden	499	Botswana	366
Scotland	498	Saudi Arabia	332
(Israel)	496	Ghana	276
New Zealand	494	South Africa	264
Slovenia	493		
Italy	484		
Armenia	478		
Serbia	477		
Bulgaria	476		

▢ Average is higher than the U.S. average.

▢ Average is not measurably different from the United States.

▢ Average is lower than the U.S. average.

[1]Hong Kong is a special administrative region (SAR) of the People's Republic of China.

Note: Countries are ordered by 2003 average score. The test for significance between the United States and the international average was adjusted to account for the U.S. contribution to the international average. The tests for significance take into account the standard error for the reported difference. Thus, a small difference between the United States and one country may be significant, whereas a large difference between the United States and another country may not be significant. Parentheses indicate countries that did not meet international sampling or other guidelines in 2003. Countries were required to sample students in the upper of the two grades that contained the largest number of thirteen-year-olds. In the United States and most countries, this corresponds to the eighth grade.

Source: International Association for the Evaluation of Educational Achievement (IEA). Trends in International Mathematics and Science Study (TIMSS), 2003.

Recent data from standardized math tests administered to students globally show that, by the eighth grade, U.S. students score higher than the world average but still lag significantly behind students in many modern Asian and northern European nations. Note in Table 17.4 that there is no statistical difference between Malaysia, whose eighth graders scored an average of 508 on the test, and the United States, with an average of 504. Note also that the margin of error in the averages is plus or minus 4 points. In other words, differences that are within that range may be due to random error and not the result of real differences in students' math ability.

Education and Social Mobility

Studies by educational sociologists have consistently found a high correlation between social class and educational attainment and achievement. As we see in Table 17.5, there has been a marked erosion of the incomes of people without college degrees over the past twenty-five years. Only college graduates have experienced significant increases in earnings. This fact indicates that social stratification in the United States is increasing, rather than declining as a result of greater equality of opportunity.

Changes in the nature of work and global competition at all skill levels place ever-greater pressure on educational institutions to produce better results. Nevertheless, there is evidence that the promise of equal education for all remains far from being fulfilled. Educational institutions have been subjected to considerable criticism by observers who believe that they hinder, rather than enhance, social mobility (Fullan, 1993).

As noted earlier, critical sociologists often argue that the schools actually reproduce existing patterns of stratification and inequality rather than promote social mobility. The point here is that working-class and poor children do not automatically inherit their lower-class position through the failure of education to reach them or through their own failure to accept the value of education. The influence of specific family and school environments has been demonstrated numerous times in empirical research on how inequality is reinforced by school experiences, especially in schools in low-income neighborhoods (MacLeod, 1995/1987; Torres & Mitchell, 1998).

Inequality in the Classroom Two contemporary studies of the school experiences of students from lower social classes offer strong support for "reproduction theory," the idea that the educational experiences of children of a given social class serve to keep them in that class once their schooling is completed.

Paul Willis's research in an industrial community in England draws on the experience of a group of boys from low-income homes who thought of themselves as "the lads." They delighted in making fun of higher achievers in their school, whom they called "the earholes." In many different ways, however, the lads indicated that they believed their teachers were pushing them into low-prestige futures. Willis interpreted the hostility and alienation of these students as a way of resisting the social forces they encountered in school and in their communities that were channeling them into working-class careers (Willis, 1983).

TABLE 17.5

Average Earnings, by Level of Education: 1979 and 2005

	1979	2005
Without a High School Education		
Males	$27,690	$19,676
Females	18,302	14,680
High School Graduates		
Males	34,978	25,811
Females	23,041	18,679
College Graduates		
Males	55,068	46,914
Females	29,919	30,022

Note: In constant 1998 dollars.

Source: Data from National Center for Education Statistics, Bureau of Labor Statistics and U.S. Census Bureau.

In a study that confirms some aspects of Willis's research and challenges others, Jay MacLeod (1995/1987) spent a great deal of time observing and interacting with two groups of students from working-class or poor backgrounds in an American high school in a poor community. MacLeod's white subjects, who called themselves the "Hallway Hangers," were very much like "the lads." In school, they were angry and alienated. They believed that there was no point in raising their educational aspirations above those of their parents and older siblings. Their school experiences did nothing to change their resigned outlook. In contrast, a black peer group, "the Brothers," had taken on the higher aspirations of an African American version of the American dream of hard work and eventual success. Although objectively, they were not doing much better in school than the Hallway Hangers, in MacLeod's view, their greater optimism and resolve call into question the apparent inevitability of social-class reproduction in the public schools. Although MacLeod does not deny the existence of educational forces that tend to reproduce inequalities from one generation to another, he believes that there is always the possibility of "agency"—that is, determined efforts by individuals and groups to challenge the status quo of inequality, even if this challenge often produces very mixed results (MacLeod, 1995/1987).

These studies are based on close observation of students in inner-city communities in England and the United States. They document the processes whereby class inequalities are reinforced in school as children from working-class and poor backgrounds are shown in subtle and not-so-subtle ways that they need not have high expectations for their educational future. If these lower expectations are not countered at home in the family environment, and if the students are not exceptionally gifted in school, the resulting cycle of school failure and lowered expectations for mobility through education is difficult to break.

In a classic study of how this cycle is established, Ray C. Rist (1973) found that "the system of public education in the United States is specifically designed to aid in the perpetuation of the social and economic inequalities found within the society" (p. 2), despite the widespread belief that education increases social mobility. He reached the following conclusion:

> Schooling has basically served to instill the values of an expanding industrial society and to fit the aspirations and motivations of individuals to the labor market at approximately the same level as that of their parents. Thus it is that some children find themselves slotted toward becoming workers and others toward becoming the managers of those workers. (1973, p. 2)

More recent studies of the reproduction of inequality in schools and classrooms often point to problems stemming from cultural differences among students and teachers. This new emphasis is a reflection of the growing impact of immigration and cultural diversity in schools throughout urban industrial nations (Castells et al., 1998). One of the primary issues in studies of culture and education is bilingualism, as shown, for example, in recent empirical research on language and culture among Latino students.

Language and Inequality A study by Lourdes Diaz Soto (1997; Soto & Malewski, 2001) traces how the recent retreat from bilingual education has affected children from homes in which Spanish is the primary language. Although they were always controversial, programs that offered instruction in the student's native language as well as in English are being cut back in U.S. schools, largely in response to demands for "English-only" instruction and what is often known as "total immersion" in a single-language environment. Diaz Soto gathered data from participant observation and interviews of teachers, students, and parents in a small industrial city in Pennsylvania where a bilingual program had been replaced with an English-only curriculum. Her research documents the frustration and disappointment felt by Spanish-speaking students and their parents as opportunities to enhance the students' ability to speak, read, and write Spanish are withheld. Most painful

is the message that only English is permitted in the schools. Diaz Soto's research shows that, although this message may be motivated by the desire to have all students master one language (English), it has the negative effect of communicating to Latino students and their parents that their culture is devalued and that their voices will be silenced.

The Spanish-speaking parents argued that there was a need for earlier opportunities for students to learn a second language in elementary school so that children whose home language was not English could see that their language and culture were valued and would be taught in school. Diaz Soto, like many other educational researchers, notes the waste of human resources as children's abilities in languages other than English are lost. She points out that elsewhere in the world, "second- and multiple-language learning is common: Danish schools' educational system includes compulsory second-language learning at age eleven; Swedish schools initiate second-language learning in the lower grades; France is initiating language learning for children under five; three-year-olds in northern parts of Germany are experimenting with second-language learning" (1997, p. 94). But in the United States, despite the growing emphasis on global relations and global social change, language education is extremely weak and "bilingual education continues to be suspect." Unfortunately, students who feel that their language and culture are devalued in school, be they Hispanic, Native American, Asian American, or others, often withdraw from participation and fail to develop their full potential.

Important as language and other cultural issues may be in explaining school achievement, economic class may be even more significant. It remains true that students from poor families are likely to attend schools with far fewer resources, less favorable student–teacher ratios, and more staff turnover than the schools attended by middle- and upper-class students (Torres & Mitchell, 1998). These class differences are also striking at the level of higher education.

Inequality in Higher Education Inequality in higher education is primarily a matter of access—that is, ability to pay. Ability to pay is unequally distributed among various groups in society, and as a result, students from poor, working-class, and lower-middle-class families, as well as members of racial minority groups, are most likely to rely on public colleges and universities. Social-scientific evidence indicates that such inequalities are alleviated by federal aid to students from families with low or moderate income. It is likely that, without

Pell grants, work-study funds, and student loan programs, the proportion of students from such families would be far lower than it currently is. Even with such aid, the proportion of low-income students in colleges and universities is lower than that of students from high-income families (Fuller & Elmore, 1996).

Sandra Baum (1987) examined the educational careers of 2,000 students who were high school seniors in 1980. She found that college attendance rates were higher for students from high-income families than for students from low-income families (60 percent versus 46 percent). However, she found that achievement in school seemed to account for an even larger difference: Students from low-income families who scored high on achievement tests had far higher rates of college attendance than low-scoring students from affluent homes. Thus, achievement in the primary school grades seems to be at least as important as family income in explaining a student's success in higher education—but remember that, in the lower grades, students from more affluent homes tend to achieve more than students from less advantaged homes.

Community Colleges: Dead End or Route to Mobility? Community colleges play an ever more important role in U.S. education as a transition both to occupations and to higher education. Often highly innovative, especially in new areas of education such as online learning and cooperative education, they offer high school graduates a route to higher education in their own community. Community college students often juggle many more work and family responsibilities than is typical of students who go directly from high school to universities or four-year colleges. They also tend to be more diverse in race, ethnicity, and social class than students at state universities and private institutions of higher education (Ishitani & McKitrick, 2010). But this diversity also presents challenges for community colleges, especially as they adjust their curricula to the demands of a changing labor market.

Baum and other researchers have found that, despite the diversity of community college student bodies, low-income students tend to be overrepresented in two-year colleges and are more likely to drop out before completing a degree than students who enter four-year colleges. On the basis of findings such as these, most educational researchers agree that without college assistance for needy students, differences in educational attainment and achievement would be much greater. And because public institutions of higher education remain the primary route to college degrees

for low-income students, increased support for such institutions also tends to diminish inequality in access to higher education (Torres & Mitchell, 1998). The severe economic downturn beginning in 2008 has led to cuts in community college staffs, and increases in tuition costs threaten these gains (Lewin, 2010).

Education for Equality

The question of whether and in what ways education leads to social mobility remains open. Some social scientists view education as an investment like any other: The amount invested is reflected in the future payoff. This is known as *human capital theory*. In this view, differences in payoffs (jobs and social position) are justified by differences in investment (hard work in school and investment in a college education). However, critics of the educational system point out that the resources required to make such "investments" are not equally available to all members of society.

The question of whether the society as a whole, or only its more affluent members, will invest in its "human capital" through such means as preschool programs and student loans is a major public policy issue. This is especially true as technological advances coupled with degree inflation increase the demand for educated people (Castells et al., 1998; Reich, 1992). Empirical evidence of the effectiveness of such investments may be seen in the results of numerous evaluations of Head Start and other early-education and prekindergarten educational programs (Kagan & Neuman, 1998). One of these evaluations, conducted by the High-Scope Educational Research Foundation and completed in the early 1980s, thoroughly demonstrated the lasting positive effects of a high-quality preschool program.

The subject of the study was a preschool program at the Perry Elementary School, located in a low-income black neighborhood of Ypsilanti, Michigan, a small industrial city on the outskirts of Detroit. The researchers randomly selected 123 children—three-year-old boys and girls—and assigned them either to the preschool group or to another group that would not go to the preschool program. The latter children, like most children at that time, would not be enrolled in school for another two years.

About twice a year for the next twenty years, the researchers traced the experiences of the experimental (preschool) and control (no-preschool) groups. The data in Table 17.6 show that the preschool experience produced important differences between the two groups. Members of the preschool group achieved more

TABLE 17.6

Major Findings at Age Nineteen in the Perry Preschool Study

Category	Number Responding*	Preschool Group	No-Preschool Group
Employed	121	59%	32%
High school graduation (or its equivalent)	121	67	49
College or vocational training	121	38	21
Ever detained or arrested	121	31	51
Females only: teen pregnancies, per 100**	49	64	117
Functional competence (APL Survey: possible score 40)	109	24.6	21.8
Percent of years in special education	112	16	28

*Total N = 123.

**Includes all pregnancies.

Source: Data from Berrueta-Clement et al., 1984.

in school, found better jobs, had fewer arrests, and had fewer illegitimate children; in short, they experienced more success and fewer problems than members of the control group. These findings confirm what other, less well-known studies had shown. For example, pioneering studies by Benjamin Bloom (1976) and by Piaget and Inhelder (1969) had indicated that as much as 50 percent of variance in intellectual development takes place before the age of four.

The most significant contribution of studies like the HighScope research is their documentation of the actual improvements that a good preschool program can produce in disadvantaged children. The HighScope study was also able to translate its findings into dollar figures. One year in the Perry program cost about $1,350 per child. The researchers showed that the "total social benefit" from the program (measured by higher tax revenues, lower welfare payments, and lower crime costs) was equivalent to $6,866 per person.

This research became a classic in the international social-scientific study of education because it demonstrated the undeniable benefits of preschool education to society (Peyton, 2005). It also appeared during a period when funding for public preschool programs was in danger of being drastically reduced. The findings of the HighScope study gave congressional advocates of preschool education the evidence they needed to justify continued federal funding. In many instances, however, social scientists know what needs to be done to improve educational achievement and increase equality of educational opportunity but are frustrated by the resistance to change that is inherent in large-scale bureaucratic systems such as the public school systems of many states and municipalities.

THE STRUCTURE OF EDUCATIONAL INSTITUTIONS

A significant barrier to educational reform is the bureaucratic nature of school systems. We noted earlier that sociologists view the school as a specialized structure with a special socializing function and that it is also a good example of a bureaucratic organization. As any student knows, there is a clearly defined status hierarchy in most schools. At the top of the hierarchy in primary and secondary schools is the principal, followed by the assistant principal and/or administrative assistants, the counselors, the teachers, and the students. Although the principal holds the highest position in the system, his or her influence on students usually is indirect. The teachers, on the other hand, are in daily command of the classroom and therefore have the greatest impact on the students. In this section, we discuss several aspects of the structure of educational institutions and attempts to change those institutions.

Schools as Bureaucracies

As the size and complexity of the American educational system have increased, so has the tendency of educational institutions to become bureaucratized (Torres & Mitchell, 1998). The one-room schoolhouse is a thing of the past; today's schools have large administrative staffs and numerous specialists such as guidance counselors and special education teachers. Teachers themselves specialize in particular subject areas or grade levels. Schools are also characterized by a hierarchy of authority. The number of levels in the hierarchy varies, depending on the nature of the school system. In large cities, for example, there may be as many as seven levels between the superintendent and school personnel, making it difficult for the superintendent to control the way policies are carried out. Similarly, in any given school, it may be difficult for the principal to determine what actually happens in the classroom.

School bureaucracies are often criticized for being "top heavy" with administrators who do not teach and whose regulations seem to stifle creativity at the classroom level. What accounts, then, for the prominence of these complex administrations and their influence

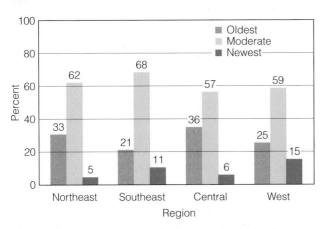

Figure 17.4 Percentage Distribution of Public Schools According to School Condition, by Region. Source: U.S. Department of Education, 1999.

on the conduct of schooling? In attempting to answer this question, we need to consider several aspects of what makes a successful school system in a city, town, or rural area.

First, in addition to everything else they are, schools are a collection of buildings, sports fields, laboratories, and real estate. They require capital funds for building and maintenance, planning for new facilities, budgeting and accounting for fiscal oversight, and much more. As Figure 17.4 shows, in some parts of the nation, overly high percentages of students are going to schools that are old and need to be refurbished or replaced. The average age of schools in the United States is forty-two years. Teachers, students, and parents want to know that buildings are safe and designed to facilitate learning, but they do not have the time or ability to take care of these needs themselves. School administrators must work with political leaders and parents' groups to get the physical work of the district done, and they often labor under extremely difficult conditions of budgetary constraint and ideological conflict. Numerous studies conducted since the national federal study of school buildings conditions was published in 1999 have found demonstrable negative effects on learning and student self-esteem that are correlated with poor physical conditions of the school and its facilities (Fine, 2002).

A second social force that tends to expand educational bureaucracies is their need to respond to so many different public needs and values. Children with learning disabilities need special programs; sports teams demand budgets and leagues; federal and state authorities demand compliance with new requirements for testing, record keeping, school security, student health standards, and much more. Every demand on a school system that administrators are expected to fulfill adds to their workload and produces a need for more administrative personnel. Some school administrations become large because political leaders want to place their allies in jobs (this is known as *patronage*) or because of *cronyism* within the administration (appointing people to jobs as a favor), but much of the growth of school administrations is caused by demands that citizens and their representatives place on the school system.

The Classroom In most modern school systems, the primary school student is in the charge of one teacher, who instructs in almost all academic subjects. But as the student advances through the educational structure, the primary school model (which evolved from the one-room school with a single teacher) is replaced by a "departmental" structure in which the student is taught by several different teachers, each specializing in a particular subject. The latter structure is derived largely from that of the nineteenth-century English boarding school. But these two basic structures are frequently modified by alternative approaches such as the "open" primary school classroom, in which students are grouped according to their level of achievement in certain basic skills and work in these skill groups at their own pace. The various groups in the open classroom are given small group or individual instruction by one or more teachers rather than being expected to progress at the same pace in every subject.

Open classrooms have not been found to produce consistent improvements in student performance, but they have improved the school attendance rates of students from working-class and minority backgrounds. Students in open classrooms tend to express greater satisfaction with school and more commitment to classwork. The less stratified authority structure of the open classroom and the greater amount of cooperation that occurs in such settings may help students enjoy school more and, in the long run, cause them to have a more positive attitude toward learning.

The Teacher's Role Given the undeniable importance of classroom experience, sociologists have done a considerable amount of research on what goes on in the classroom. They often start from the premise that, along with the influence of peers, students' experiences in the classroom are of central importance to their later development. It is rare, however, that social scientists have the opportunity to trace the influence of teachers over students' lifetimes. A classic exception to this limitation is a study (Pedersen & Faucher, 1978) that examined the impact of a single first-grade teacher on her students' subsequent adult status. As they examined the IQ scores of pupils in the school's first-grade classes, the researchers found a clear association between changes in a pupil's family background, first-grade teacher, and self-concept. Further examination of the school's records revealed a startling fact: The IQs of pupils in one particular teacher's first-grade class were significantly more likely to increase in subsequent years than those of pupils in first-grade classes taught by other teachers. And pupils who had been members of that class were more than twice as likely to achieve high status as adults as pupils who had been members of other first-grade classes (see Table 17.7).

What was so special about Miss A, as the outstanding teacher was labeled? First, she was still remembered by her students when they were interviewed twenty-five years after they had been in her class. More than three-quarters of those students rated her as very good or excellent as a teacher. "It did not matter what background or abilities the beginning pupil had; there was no way that the pupil was not going to read by the end of grade one." Miss A left her pupils with a "profound impression of the importance of schooling, and how one should stick to it" and "gave

then & now

Pigskin Mania Football emerged in more or less its present form in the 1890s, as a popular sport among wealthy young men at elite eastern universities like Yale and Harvard. In North America, football has become not just a sport but a generator of enormous revenues and a cultural form that many people almost feel they could not live without. The growth in popularity of the sport was intimately associated with the powerful influence of the mass media, which can broadcast games to ever-larger audiences (Oriard, 2009).

But the growth of media institutions, in turn, depends on the revenues made possible by the huge audiences attracted to major games. Thus, there is a symbiotic relationship between television and the educational institutions that can afford to support football teams (Kriegel, 1993; Lawton, 1984). More than 50 million fans attend one or more football games in a given year, and no one attending a college or university in the United States needs to be told that football games are central events in the life of many campuses. In fact, for many men and an increasing number of women, football has come to resemble a civil religion. There are fans whose lives become subordinated to the rituals of preparation for the Saturday game. Legendary coaches like Alabama's Bear Bryant or Penn State's Joe Paterno assume a status that is well below that of a god but far above that of an ordinary mortal (C. R. Wilson, 1987). The rivalries that emerge between regional powerhouse teams and are played out each year become material for local legends. Many will admit to having prayed for a successful third-down conversion.

The metaphors inherent in football—success, hard work, strength, speed, organization, strategy—all make the sport a study in cultural ritual that mirrors the values of the larger society in which it has flourished (Brokaw, 1994). These cultural qualities of the game also make it extremely popular to television viewers and enable broadcasters to make large payments to the universities; these, in turn, help support other, less profitable sports. The popularity of football also enhances the university's prestige and its ability to compete for students and faculty.

Bettmann/CORBIS

© Mark Lewis

---

extra hours to the children who were slow learners." In nonacademic matters, too, Miss A was unusual:

> When children forgot their lunches, she would give them some of her own, and she invariably stayed after hours to help children. Not only did her pupils remember her, but she apparently could remember each former pupil by name even after an interval of 20 years. She adjusted to new math and reading methods, but her success was summarized by a former colleague this way: "How did she teach? With a lot of love!" (Pedersen & Faucher, 1978, pp. 19–20)

TABLE 17.7
Adult Status, by First-Grade Teacher

Adult Status	Miss A	Miss B	Miss C	Others
High	64%	31%	10%	39%
Medium	36	38	45	22
Low	0	31	45	39
Total	100%	100%	100%	100%
(N) Mean adult status	(14) 7.0	(16) 4.8	(11) 4.3	(18) 4.6

(N) = Number of students who could be located and interviewed twenty-five years later.

Note: "Adult status" was determined from interviews that included questions on occupational status and work history, highest grade completed, rent paid and number of rooms, and related indicators of social position.

Source: Table 7A, "Adult Status by First-Grade Teacher," from Eigil Pedersen and Thérèse Annette Faucher, with William W. Eaton, "A New Perspective on the Effects of First-Grade Teachers on Children's Subsequent Adult Status," *Harvard Educational Review*, 48:1 (February 1978), p. 16. Copyright © by the President and Fellows of Harvard College. All rights reserved. For more information, please visit www.harvardeducationalreview.org.

In summing up their findings, Pedersen and Faucher stated that their data "suggest that an effective first-grade teacher can influence social mobility" (p. 29). The surprising results of this study have important implications. Good teachers clearly can make a big difference in children's lives, a fact that gives increased urgency to the need to improve the quality of primary school teaching. The reforms carried out by educational leaders suggest that when good teaching is combined with high levels of parental involvement, the results can be even more dramatic.

Because the role of the teacher is to change the learner in some way, the teacher–student relationship is an important part of education. Sociologists have pointed out that this relationship is asymmetrical or unbalanced, with the teacher being in a position of authority and the student having little choice but to passively absorb the information the teacher provides. In other words, conventional classrooms provide few opportunities for students to become actively involved in the learning process. On the other hand, students often develop strategies for undercutting the teacher's authority: mentally withdrawing, interrupting, and the like (Darling-Hammond, 1997; Rose, 1995). Much current research assumes that students and teachers influence each other instead of assuming that the influence is always in a single direction.

School Peer Groups and Violence The shootings at Columbine High School in Colorado and in other seemingly safe suburban schools in Oregon, Arkansas, Michigan, and Kentucky, along with the outbreaks of gang violence in many urban schools in the late 1990s, were generally seen as symptoms of widespread school violence and student alienation. These tragic episodes received sensational television coverage and scared millions of parents, but they are not representative of the actual data on school violence. As Figure 17.5 shows, rates of serious school violence have remained relatively stable over the past few decades. The most common form of crime that high school students experience is theft (about 40 percent). Rates of vandalism and threats with or without a weapon actually decreased during the 1990s. Unfortunately, however, rates of victimization in U.S. schools have begun to rise again in the past few years (U.S. Department of Education, 2006).

Public reactions and social policies often are determined not by facts but by perceptions of social conditions. The perception that schools were becoming unsafe, even in communities assumed to be immune from severe peer violence, resulted in much new legislation intended to make schools safer. Perhaps a more positive outcome, however, is that educators, parents, and students have become more aware of the causes of school violence. Much more emphasis has been placed on the need to understand school peer groups (and identifiable cliques) and their interactions. Teachers are being taught how to recognize bullying, how to protect the rights and safety of gay and lesbian students, and how to communicate better with students about their feelings and experiences (Hazler, 2000). Much remains to be done to create school environments that are free of violence and where students can speak to adults about their problems with peers without feeling that they are "snitches." As the diversity of racial and ethnic groups in the schools continues to increase, the challenge of combating school violence will become more difficult. But the statistics show progress, even if perceptions cause some parents to seek alternatives to their local public school or, in some cases, withdraw their children from school.

Changing the System

The bureaucratic organization of school systems makes them highly resistant to change. The basic classroom unit has remained

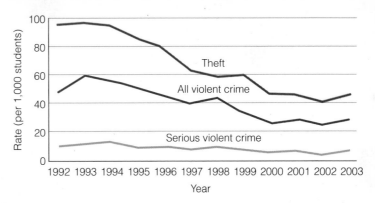

Figure 17.5 Rates of Victimization in U.S. Schools, 1992–2003. Source: DeVoe et al., 2005.

essentially unchanged since the days of the one-room school, as has the structure in which one teacher is in charge of a roomful of students. But despite the difficulties, there are endless demands for school reform and changes in the way students and teachers are evaluated. A full treatment of controversies over educational policy is offered in most courses on the sociology of education. Here we discuss three current controversies that have a direct bearing on whether schools can deliver on their promise of equality of opportunity. They concern class size, school choice, and desegregation.

Comprehensive School Reform *Raise the standards. Create a level playing field. End social promotion. Use vouchers for school choice.* These are only a few of the most popular slogans people use in debates over education. There is no end to the controversies over what should and should not be accomplished in the nation's schools and colleges. Most of these calls for reform seek to address the needs of schools that are experiencing high levels of failure and poor performance. These schools also tend to be those receiving disproportionately lower shares of public support per student.

Because measures such as universal achievement testing do not address the overall needs of failing schools, some school reformers have developed approaches to comprehensive school reform, which affects all the groups that have a stake (sometimes called stakeholders) in the quality of learning in schools. Among the most prominent of these approaches is the one developed by the pioneering educator and Yale psychiatrist James Comer. His school reform model has inspired many others that are being used to develop new schools and revamp old school systems. We therefore can consider the Comer model as a case study of current efforts at comprehensive school reform.

School districts in Prince George's County (Maryland), Chicago, Miami, New Haven (Connecticut), Dallas, and many other areas are undertaking major efforts at educational reform. Nowhere are their goals quite as ambitious as those of the Prince George's County school district. The Comer-inspired reform efforts involve schoolchildren in collaborative planning with their parents, teachers, and principals. "What I've tried to do here," Comer explains, "is to shape the system, the school, so that it becomes the advocate and support for the kid, and a believer in the kid in the same way that my parents were" (quoted in Schorr, 1988, p. 232).

Comer's research and experience convinced him that children from neighborhoods undergoing severe strain—because of poverty, racism, unemployment, violence, drug addiction, and related social problems— enter school with major deficits. They are often "underdeveloped in their social, emotional, linguistic, and cognitive growth." As a result, they often withdraw, act up, and do not learn. Of course, not all such students have these problems. Those who do, however, are labeled as "slow learners" or as having "behavior problems" and consequently fall even further behind their classmates. Comer's solution appears simple on the surface: Change the "climate of demoralized schools by paying much more attention to child development and to basic management of the school" (quoted in Schorr, 1988, p. 233).

Of course, the process of changing school climates and management is much more difficult than this simple formula suggests. The Comer model of institutional change calls for the formation of a School Planning and Management Team in each school; the team is directed by the principal and has from twelve to fourteen members, including teachers, aides, and parents. A second team is made up of helping professionals—the school psychologist, social workers, special education teachers, and counselors; it acts as an advisory group. The purpose of the teams is to ally the parents with the school so that, in Comer's words, "you reduce the dissonance between home and school and you give the kid a long-term supporter for education at home" (quoted in Schorr, 1988, p. 233; see also Comer, 1997; O'Neil, 1997).

Inspired by the research of Comer and others, large cities such as Chicago and New York are instituting more opportunities for parents to get involved in school issues and are seeking to reduce the size of schools so that teachers, students, and administrators feel that they are part of a "learning community" in which they all know one another and are concerned with one another's well-being. A prominent example that builds on the Comer model and extends it to early-childhood education is the Harlem Children's Zone (HCZ), a project started by educational innovator Geoffrey Canada. The HCZ project has developed a comprehensive system of programs covering nearly 100 blocks of Central Harlem in New York City; its goal is to keep children on track through college and into the job market. While on the presidential campaign trail in 2008, Barack Obama announced a plan to replicate the HCZ in twenty cities across the United States. His Race to the Top Initiative, which began to go into effect in 2009, encourages local institutional reforms, encourages comprehensive approaches like those based on the Comer model and the Harlem Children's Zone, and allocates funds to states that can present evidence of consensus among teachers, parents, and school administrators that educational reforms will be made that address the needs of children from early ages through secondary school. The federal initiative also emphasizes reforms like the creation of smaller schools, often through charter arrangements whereby new schools are started by innovators within the existing public school system.

In many school districts, however, the pace of reform is slowed by bitter disagreements about how to fund the schools in an era of reduced public spending. There are also controversies over issues such as sex education, tolerance for homosexual students, distribution of condoms to prevent AIDS and teenage pregnancy, bilingual education, and school prayer. These debates can divide communities and prevent structural changes like the move toward smaller schools and more local responsibility for educational decisions (McFadden, 1998; Rose, 1995).

Class Size Although the basic organization of the classroom has not changed much over time, the ratio of students to teachers varies widely. Higher-quality education is associated with lower student–teacher ratios—that is, fewer students per teacher. The more students there are in the classroom, the more difficult it is for the teacher to spend time with individual students. One reason parents who can afford to do so often send their children to private schools is so their children will receive more attention and direct instruction from their teachers.

In public schools, classrooms in lower-income communities tend to have the highest student–teacher ratios. Classrooms in suburban schools in affluent communities have among the lowest student–teacher ratios in public education. This is one reason that so many families seek to move to those communities (Ballantine, 1997). Extensive studies in Tennessee and in various urban school districts provide quantitative support for the hypothesis that student performance is indirectly correlated with class size: The lower the number of students per teacher, the higher average school performance will be in terms of grades and performance on standardized tests (Gursky, 1998).

School Choice One of the most controversial issues confronting educational institutions today is that of school choice. In fact, "school choice" involves several types of choices. As educational sociologist Carol H. Weiss notes:

> Choice plans include magnet schools: public schools with special emphases and/or facilities that draw students from across a district, such as technology schools or music and art schools. Choice plans include charter schools, authorized by a number of state legislatures to be free of most local school district regulations; students apply for admission. (quoted in Fuller & Elmore, 1996, p. vii)

Among conservative school reformers, school choice often means plans to allow students to choose private schools as well as public ones by giving parents vouchers to pay for tuition and leaving them free to decide where and how they will use the vouchers. These "free-market" school choice plans have been popular among conservatives since the 1950s, when economist Milton Friedman first suggested this approach.

At this writing, a great deal of research is being conducted to evaluate one or another of the school choice initiatives. Note, however, that the range of choice is so wide that no single study can demonstrate the overall effect of school choice on all the desirable outcomes of education (for example, knowledge, mobility, equality, morality, citizenship). And there are so many actual school choice behaviors—including moving to better school districts, homeschooling children, vouchers or education grants (such as the Pell grants in higher education), and the choice among parochial, private, and public schools—that researchers are hard-pressed to respond to the challenges presented by these changes in educational institutions (Fuller & Ellmore, 1996). But most liberal critics of school choice plans are afraid that, although greater choice is attractive to parents in poor school districts as well as to those in more affluent ones, the overall consequence of increasing choices among different types of schools will be to increase the segregation of students by class and race. And as such segregation increases, so will rates of failure in schools that have been abandoned by better students and more aware parents (Torres & Mitchell, 1998). Indeed, from this critical perspective, increasing racial segregation in public and private education can already be viewed as an outcome of school choice.

Early results of social-scientific evaluations of student achievement provide little support for school choice and cash vouchers as superior to public schools. In a three-year study of Washington, D.C. primary schools, 1,300 students from low-income families were chosen by lot from about 20,000 applicants for cash payments that enabled them to choose a religious or secular private school and to attend that school for at least four years, beginning in the fourth grade. After three years at these schools, the students who received vouchers performed no better on standardized tests than a control group of students who remained in the public schools (Wolf, Howell, & Peterson, 2000).

The Homeschooling Movement Between 1.2 and 1.5 million children are now being taught at home. Because approximately 44 million children are in public schools at all levels and another 10 million are enrolled in private schools, the 3 percent being schooled at home represent a small minority. On the other hand, this number has grown rapidly in the past decade. Its advocates come from all social classes and backgrounds, but a disproportionate 96 percent of homeschooled children are white, and most are Protestant (McDowell & Ray, 2000).

Most homeschooling families are motivated by the desire to teach their children more moral precepts and religious values than they would be taught in public or private schools. They usually want to keep their children away from negative peer influences, materialistic values, and the possibility of violence and intimidation. Parents who wish to educate their children outside the public schools have traditionally sought private schools, often parochial schools that combine secular and religious curricula not available in the secular public schools. Some critics of homeschooling question whether children learn well outside schools, but the limited number of studies that have addressed this question indicate that homeschooled students score on the average significantly higher than children in public schools on standardized tests (Ray, 2000).

Critics of homeschooling in Europe—where it is also becoming an increasingly popular alternative to public education—note that it is part of a general trend in many societies to value private rights over public rights and private goods over public ones. We can think of education as something that the individual gains, which eventually makes the individual more competitive in the job market or a more moral person—that is, a person with more private goods to share with selected others as he or she sees fit. But we can also think of education as serving the common good of a vibrant society to which children and parents all contribute through learning; in this view, education is a public rather than a private good. By rejecting the schools, homeschoolers are withdrawing the resources of their children and families from the community and thus denying this common good (Lubienski, 2000).

Homeschooling families argue that they and their children should have the right to contribute to the common good in their congregations and peer groups. They should not be obliged to send their children to a school if they do not choose to. Courts have tended to rule that as long as parents can demonstrate to educators in their communities that they are not depriving their children of the right to high-quality education, they can school them at home. Similar rulings exist in the United Kingdom and other European nations, with the exception of Germany, where judges have tended to rule against homeschooling largely on the ground that all children need to contribute to the learning community of the public schools.

The future of homeschooling may have lasting implications for the strength of public schooling in democratic societies. An estimated 3 million children were being homeschooled by 2010, and the annual rate of increase was about 7 percent (higher in some states, such as South Carolina). Whether this rate continues to increase depends largely on what happens in the schools as well.

The Retreat from School Desegregation Perhaps the most significant change in educational institutions in the late twentieth century was desegregation. Efforts to desegregate public schools began in 1954, with the Supreme Court's ruling in *Brown v. Board of Education of Topeka*, when the Court held that segregation had a negative effect on black students even if their school facilities were "separate but equal" to those of white students. In the first few years after this landmark decision, most of the states in which schools had been legally segregated instituted desegregation programs. However, in some states, particularly in the Deep South, desegregation orders were resisted, and by 1967, only 26 percent of the nation's school districts were desegregated (Orfield & Eaton, 1996).

Further progress toward desegregated schools was made after the passage of the Civil Rights Act of 1964 and the Elementary and Secondary Education Act of 1965; by 1973, de jure (legally sanctioned) segregation had been almost entirely eliminated. However, de facto (in-practice) segregation remained common in many parts of the nation. This pattern was created by two factors: (1) school districts' traditional policy of requiring students to attend schools in their own neighborhoods, and (2) "white flight," the tendency of whites to move to the suburbs to take advantage of suburban housing opportunities and, in some cases, to avoid sending their children to central city schools with large percentages of black students. On the basis of an extensive study of the impact of desegregation programs, Coleman (1976) concluded that "policies of school desegregation which focus wholly on within-district segregation . . . are increasing, rather than reducing or reversing, the tendency for our large metropolitan areas to consist of black central cities and white suburbs" (p. 12).

Coleman's findings had a significant influence on the policy debates that took place throughout the 1970s and 1980s over what could be done

The Toughest Job You'll Ever Love In late 2001 and early 2002, the terrorist attacks on the United States and the sinking U.S. economy created a wave of interest in teaching among Americans from all social backgrounds. This was extremely good news to U.S. educators, who face a potential shortage of teachers as retirements among older teachers thin the teaching ranks. There are now about 2.8 million teachers, but in the next ten years another

Participants in Teach for America use their leadership skills to help students fulfill their potential.

A group of TFA teacher trainees.

2.4 million will need to be recruited. When the economy is doing well, it is far more difficult to recruit teachers, but now the states are finding it relatively easy to find new recruits. In time, as the economy heats up again, it will be more difficult to recruit new teachers, but programs like Teach for America will be there to fill the gaps, especially by recruiting recent college graduates.

The recruiting slogan for teachers, "The Toughest Job You'll Ever Love," is a reminder that, in addition to job security and a refuge from the harsh world of business, teaching offers opportunities for personal growth and a sense of having helped others that is more difficult to find in other professions. An ad in the New York City subways says, "No one goes back ten years later to thank a middle manager" and, "Your spreadsheets won't ever grow up to be doctors and lawyers" (Goodnough, 2002). These testimonials by Teach for America teachers do not ensure that they will stay in teaching or become professional educators, but they certainly show that young Americans are still able to find outlets and true challenges for their idealism.

TFA teacher Orin Gutlerner with his fifth- and sixth-grade students at Clark Street Elementary School in Henderson, North Carolina.

to increase racial equality in public education, especially the issue of whether schoolchildren should be bused across city lines. Educational policymakers are now seeking ways of promoting racial integration, not just by means of busing plans but also through changes in school curriculums and instructional methods.

When education critic Jonathan Kozol visited public schools in central cities and suburban communities, he was shocked by the stark evidence of continuing racial and class segregation he found:

> What startled me most was the remarkable degree of racial segregation that persisted almost everywhere. . . . Moreover, in most cities, influential people that I met showed little inclination to address this matter and were sometimes even puzzled when I brought it up. Many people seemed to view the segregation issue as "a past injustice" that had been sufficiently addressed. . . . I was given the distinct impression that my inquiries about this matter were not welcome. (1995, pp. 2–3)

The children Kozol interviewed understood quite well the way class and racial segregation affected their own educational opportunities. Samantha, an eighth grader in the impoverished schools of East St. Louis, Illinois, described to Kozol the great differences in resources between the schools she attended and those in the nearby white suburbs—lack of adequate textbooks, no computers, fewer after-school programs, and the like. Her mother had tried to get her into one of the better schools but had been told that she had to stay "in her jurisdiction." "Is it a matter of

race?" Kozol asked Samantha. He notes that, in answering him, the girl chose her words slowly and with great care: "Well, the two things, race and money, go so close together what's the difference? I live here, they live there, and they don't want me in their school" (pp. 30–31).

Clearly, the process of racial integration still has a long way to go. Although school choice plans may increase racial segregation in schools, residential segregation has a much greater impact on school segregation. As sociologist John Yinger points out, "segregation in housing obviously leads to segregation in schools" (1995, p. 142). In his view, the failure to ensure freedom of residential choice for African Americans and other people of color remains a major obstacle to equality of education through desegregation.

Samantha's comment in the interview just quoted is a reminder of the central place schools occupy in American communities. The schools can bring different groups together in the pursuit of common goals, or they can separate children and families according to class, race, or religion. The schools, especially public schools, community colleges, state colleges, and state universities, are explicitly designed to build the capabilities of communities and entire cities or states while developing competent citizens. The same can be said about specific aspects of school programs, such as school and college sports.

Thinking Critically

The No Child Left Behind Act attempts to address the underachievement of minority students in poor communities by instituting a system of standardized testing and requiring schools that do not measure up on these tests to hire more qualified teachers. Do you think that recruiting more qualified teachers for failing schools might influence parents to send their children to those schools, thereby helping to decrease segregation?

SUMMARY

In what ways is education a central institution for addressing the learning needs of people in the developed and developing nations?

- *Education* is the process by which a society transmits knowledge, values, norms, and ideologies, and in so doing, prepares young people for adult roles and adults for new roles. It is accomplished by specific institutions outside the family, especially schools and colleges.

- Schools are often cited as examples of bureaucratic organizations because they tend to be characterized by a clearly defined authority system and set of rules.

- A more interactionist viewpoint sees the school as a distinctive set of interactions and patterns of socialization.

- Conflict theorists view schools as institutions whose purpose is to maintain social-class divisions and reproduce the society's existing stratification system.

- Hundreds of millions of children throughout the world are denied the right to quality basic education.

- In the United States, the post–World War II baby boom caused a bulge in elementary school enrollments beginning in 1952, and an expansion of the college-age population in the 1960s. These trends were reversed in the 1960s and 1970s; at the same time, however, more and more adults have sought additional education.

- A key feature of education is the fact that schools structure the lives of children and adolescents. They help create a social world for adolescents that is separate from adult society.

- Another important aspect of education in the United States is the relationship between education and citizenship. Education is also viewed as a tool for solving social problems, especially social inequality.

How do issues of educational equality differ between developed and developing regions of the world, and in what ways are they similar?

- *Educational attainment* refers to the number of years of school a person has completed. It is correlated with income, occupation, prestige, and attitudes and opinions.

- The average American today has much more education than the average American of the early 1940s.

- Research on the effects of tracking has found that students in lower tracks tend to end up in lower-paid and lower-prestige occupations once they finish their schooling.

- Social-scientific research has shown that states that base their funding of school districts on how well schools do on standardized tests actually increase, rather than decrease, high school dropout rates.

- One effect of higher levels of educational attainment is "degree inflation," in which employers require more education of potential employees.

- *Educational achievement* refers to how much students actually learn, measured by mastery of reading, writing, and mathematical skills. The decline in average SAT scores is viewed as a sign that average levels of educational achievement have declined.

- Cross-cultural research has shown that these deficiencies are apparent from an early age.

- Educational institutions have been criticized by observers who believe that they hinder, rather than enhance, social mobility. Higher levels of educational attainment provide the credentials required for better jobs, and students who are able to obtain those credentials usually come from the middle and upper classes.

- A factor that has been shown to affect students' school careers is teacher expectations regarding students, which are affected by the teacher's knowledge of the student's family background. Other factors include language and social class.

- Inequality in higher education is primarily a matter of access—that is, ability to pay.

What are some of the key structural aspects of educational institutions, and how do educational reform movements in the United States address structural issues?

- The American educational system is highly bureaucratized, a fact that acts as a major barrier to educational change.

- Among the factors that tend to expand educational bureaucracies are the need to maintain buildings and other real estate and the need to respond to many different public needs and values.

- Recent outbreaks of violence in schools have led to greater emphasis on efforts to understand school peer groups, recognize bullying, protect the rights of gay students, and communicate better with students about their feelings and experiences.

- Most calls for school reform rely on single reforms such as testing at every grade level. Proposals for comprehensive school reform seek to address a wide array of issues.

- The number of children schooled at home has grown rapidly in the past decade. Homeschooling families are motivated to teach their children more moral precepts and more religious values than would be the case in public or private schools.

- When educational bureaucracies have been forced to change, as in the case of school desegregation, they have often proved to be very adaptable.

The Kornblum Companion Website

www.cengagebrain.com

Supplement your review of this chapter by going to the companion website to take one of the tutorial quizzes, use the flash cards to master key terms, and check out the many other study aids you'll find there. You'll also find special features, such as GSS data and Web links that will put data and resources at your fingertips to help you with that special project or to do some research on your own.

Economic Institutions

FOCUS QUESTIONS

How do markets work, and why do they play such a powerful role in shaping the growth of societies?

What are the basic approaches that different nations employ to try to balance economic growth with social justice?

How are states and economic institutions interrelated in major political economic systems?

What are the dominant dimensions of conflict and change among workers, managers, and corporations?

What are some important sociological theories about workers and the workplace?

What findings of sociological research help explain how people find work and jobs?

An American businessman, E. L. Winthrop, was on a vacation trip to Mexico. While there, he traveled to a remote and rather quaint village somewhere in the province of Oaxaca. As he wandered through the dusty streets of that little pueblo, which had neither electricity nor running water, he came upon an Indian man squatting on the earthen-floored porch of his palm hut. The Indian was weaving baskets, but these were not ordinary baskets. Winthrop had never seen such colors and such fine detail as the Indian was weaving into the little containers, using only natural fibers and dyes from the nearby jungle. Winthrop immediately decided that he must buy one no matter what the price.

When Winthrop learned that the man was selling his wares for only 50 centavos, perhaps a tenth of what he had expected to pay, Winthrop could not help thinking that here was a fine opportunity to start a thriving business. The baskets would make ideal containers for gift candy; he could sell them at home for three or four times the price. And even counting shipping costs and other expenses, both Winthrop and the Indian would surely make out very well.

"I've got big business for you, my friend," he announced to the surprised basket weaver. "Do you think you can make me one thousand of those little baskets?"

Winthrop explained that the Indian could enlist the help of his family and kin in the project, and that they all stood to gain tremendously at the prices he could pay.

That might be true enough, the man admitted, but who then would look after the corn and beans and goats? Yes, he could see that he and his kin would have plenty of money to buy these things as well as many other items that they did not yet possess, but, he explained, "Of the corn that others may or may not grow, I cannot be sure to feast upon."

Not easily discouraged, Winthrop proceeded to show the basket weaver how wealthy he and his entire village could become even if they continued to sell the baskets for only 50 centavos. The Indian seemed interested, which spurred Winthrop on to more detailed descriptions of the basket assembly line they could begin together in the dusty little town. But after it was all explained, the Indian was still unmoved. The thousands of pesos that could be earned in this way were more than he could reckon with, because he had always used the few pesos from his basket selling only to supplement the modest needs of his family. There was another reason, too:

> Besides, señor, there's still another thing which perhaps you don't know. You see, my good lordy and caballero, I've to make these canastitas my own way and with my song in them and with bits of my soul woven into them. If I were to make them in great numbers there would no longer be my soul in each, or my songs. Each would look like the other with no difference whatever and such a thing would slowly eat up my heart. Each has to be another song which I hear in the morning when the sun rises and when the birds begin to chirp and the butterflies come and sit down on my baskets so that I may see a new beauty, because, you see, the butterflies like my baskets and the pretty colors on them, that's why they come and sit down, and I can make my canastitas after them. And now, señor jefecito, if you will kindly excuse me, I have wasted much time already, although it was a pleasure and a great honor to hear the talk of such a distinguished caballero like you. But I'm afraid I've to attend to my work now, for day after tomorrow is market day in town and I got to take my baskets there. Thank you, señor, for your visit. Adios. (Traven, 1966, p. 72)

SOCIOLOGY AND ECONOMICS

This simple story captures the essence of what Max Weber meant when he observed that as societies grow in size, complexity, and rationality, there is less enchantment in human life. The Indian could never imagine himself as a captain of industry, the manager of a workforce of Indians who would buy their corn in a store with a portion of their wages. But neither could Winthrop appreciate the full meaning of the Indian's refusal and the metaphor of his song. The basket weaver feared that mass production would rob his baskets of their poetry and their maker of his soul. The businessman was used to dealing with markets and with the logic of profit and loss. He believed that he would be creating a "win-win" situation. He would profit, and he would be helping the Indian earn a better living for himself and his family. After all, the Indian was engaged in selling his baskets in the market. We will see in the Visual Sociology box at the end of this chapter that in the modern world there is a constant struggle to come out ahead in markets yet at the same time to maintain one's sense of creativity—one's soul. Nor is this the only dilemma forced upon us by modern economic institutions (Pozo & Pozo, 2007).

In this chapter, we turn to an analysis of the major economic institutions of modern industrial societies. As we do so, we should keep in mind examples from nonindustrial societies like that of the basket weaver to appreciate the enormous impact of economic and social change in the past two centuries. Such examples will also show that there is not just one path toward economic modernization, nor only one blueprint for how the economic institutions of a modern society should work (Dobbin, 2004).

In any society, economic institutions specialize in the production of goods and services. How we survive in modern societies depends on whether we have jobs and income, on the nature of markets, on the public policies that involve governments in economic affairs, on worker–management relations, and so on. Thus, it is little wonder that sociologists have dedicated much research to these questions. But if that is true, what distinguishes sociology from economics? In our review of the sociology of economic institutions, we begin by distinguishing between these two social-scientific fields.

Nobel Prize–winning economist Paul Samuelson (1998) defines *economics* as the study of how people and social groups choose to employ scarce resources to produce various commodities and distribute them for consumption among various people and groups in society. Sociologists are also concerned with how individuals and societies make choices involving scarce resources such as time, talent, and money. But sociologists do not assume that scarcity or supply and demand are the only reasons for making such choices (Dobbin, 2004; Elster, 1997). Sociologists are deeply concerned with showing how the norms of different cultures affect economic choices. For example, we cannot assume that people with equal amounts of money or talent are free to choose the same values or activities. We often find that women who might have chosen to pursue careers in business management have been channeled into such occupations as teaching or social work. Until recently, American culture defined the latter occupations as appropriate for women, whereas managerial roles were viewed as appropriate for men. Women therefore did not have the same choices open to them even when they had equal amounts of scarce resources like talent.

It would not be accurate to stress only the differences between sociology and economics, however. Increasingly, some sociologists (and economists) use the central theory of economics to study all situations in which people choose to allocate scarce resources to satisfy competing ends (Hechter & Kanazawa, 1996; Smelser & Swedberg, 2005). By "central theory," we mean the central proposition of economics: *People will attempt to maximize their pleasure or profit in any situation and will also try to minimize their loss or pain.* Applying this idea to nonbusiness situations, another Nobel Prize–winning economist and sociologist, Gary S. Becker, has conducted research on racial discrimination, marriage, and household choices. In this research, he applies an economic "rational-choice" approach to the study of the behavior of people in groups and organizations that do not produce goods or services. Becker points out that even the choice of a mate or the decision to have a child can be looked upon as an economic decision because people who make these decisions intuitively calculate the "price" of their actions in terms of opportunities gained or lost (Becker, 1997). Having a child means giving up other scarce resources, such as free time or income that could be used in other ways. And we saw in Chapter 15 how intuitive calculations enter into the choice of a marriage partner.

Following the lead of rational-choice theorists like Becker, sociologist Michael Hechter (1987) has studied the larger social consequences of the individual propensity to maximize pleasure and avoid pain. He concludes, among other things, that the larger the number of individuals in an organization, the more the organization will be required to invest in inspectors, managers, security personnel, and the like to control its members' propensity to "do their own thing." The same point was made over a century ago by Karl Marx: "All combined labour on a large scale requires . . . a directing authority. . . . A single violin player is his own conductor, an orchestra requires a separate one" (1962/1867, pp. 330–31).

Although it is tempting to apply principles of rational choice or exchange theory to noneconomic institutions like the family and the church, in this chapter we focus specifically on behavior in economic institutions such as businesses and markets and on the social implications of changes in these institutions. We will see that sociologists are continually comparing the ways in which such institutions are organized in different societies. The comparative study of how these differences relate to other features of a society's culture and social structure is very much a part of the sociological perspective on economic institutions.

Sociologists who study economic institutions are also interested in how markets for goods and services change as people learn to use new technologies and to divide up their labor in increasingly specialized ways (Drucker, 1992). They also look at how competition for profits may lead to illegal activities and cause governments to try to regulate economic institutions. In addition, sociologists attempt to show how the labor market is organized and how professions develop. Thus, markets and the division of labor, the interactions between government and economic institutions, and the nature of jobs and professions are the major subjects of sociological research on economic institutions (Wallerstein, 1999). In the rest of this chapter, we discuss each of these subjects in turn.

GLOBAL MARKETS AND THE DIVISION OF LABOR

A hallmark of an industrial society is the production of commodities and services to be exchanged in markets. In London, Hong Kong, New York, and other major cities throughout the world are markets for gold, national currencies, stocks and bonds, and commodities ranging from pork bellies to concentrated frozen orange juice. However, not all societies have economies dominated by industry and markets. Subsistence economies, in which people live in small villages like that of the Mexican basket weaver, do not have highly developed markets. In such economies, people seek to produce enough food and other materials to enable them to meet their own needs, raise their children, and maintain their cultural traditions. The basic unit of production is the family rather than the business firm as is the case in a capitalist economy. But there are few, if any, purely subsistence economies in the world today. Most subsistence economies must engage in some trade in international markets to obtain goods such as medicine or tools. These markets are dominated by the larger firms and entrepreneurs of the global capitalist economy, as we saw in the case of the basket weaver and Mr. Winthrop.

Socialist economies like those of Cuba or China also buy and sell products in international markets. Within their societies, however, they usually try to limit the influence of markets. Prices for essential goods, especially foodstuffs, are not entirely subject to the laws of supply and demand; often they are prevented by law from rising too high or falling too low. Many types of goods and services, including firms themselves, cannot be sold to private buyers because they are said to belong to all the people of the society. We will return to the conceptual differences among these types of economies later in the chapter. For now, remember that few contemporary societies conform to one or another of the basic economic types. Most nations encompass capitalist markets, subsistence economies, and elements of socialism as well. Nevertheless, the market is an increasingly dominant economic institution throughout the world (Stehr & Adolf, 2010).

The Nature of Markets

Markets are economic institutions that regulate exchange behavior. In a market, different values or prices are established for particular goods and services, values that vary according to changing levels of supply and demand and are usually expressed in terms of a common measure of exchange, or currency. A market is not the same thing as a marketplace. As an economic institution, a market governs exchanges of particular goods and services throughout a society. This is what we mean when we speak of the "housing market," for example. A marketplace, on the other hand, is an actual location where buyers and sellers make exchanges. Buyers and sellers of jewelry, for instance, like to be able to

---

market: An economic institution that regulates exchange behavior through the establishment of different values for particular goods and services.

gather in a single place to examine the goods to be exchanged. The same is true for many other goods, such as clothing and automobiles.

Market transactions are governed by agreements or contracts in which a seller agrees to supply a particular item and a buyer agrees to pay for it. Exchanges based on contracts are a significant factor in the development of modern societies. As social theorist Talcott Parsons pointed out, the use of contracts makes impersonal relations possible. Contracts neutralize the relevance of the other roles of the participants, such as kinship and other personal relationships, that govern exchanges in nonmarket situations. In contractual relations, for example, the fact that people are friends or kin does not, in principle, change the terms of their agreement and the need to repay debts (Parsons, 1991).

Among hunting-and-gathering peoples and in relatively isolated agrarian societies before the twentieth century, markets in the modern sense of the term did not exist. Think of the Mexican Indian basket weaver who could not imagine buying corn and other foods. He was used to growing foodstuffs himself and exchanging them with members of his extended family. The idea of buying them with currency was foreign to him. In social-scientific terms, a society cannot be said to have a fully developed market economy if many of the commodities it produces are not exchanged for a common currency at prices determined by supply and demand.

The spread of markets into nonmarket societies has been accelerated by political conquest and colonialism as well as by the desire among tribal and peasant peoples to obtain the goods produced by industrial societies. To illustrate this point and to show what happens as a smaller-scale society becomes integrated into world markets, consider the case of the Tiv.

Markets in a Subsistence Economy In precolonial times, the Tiv, a small West African tribal society living in what is now Nigeria, were a good example of a *subsistence economy,* one in which producers try to meet the needs of their immediate and extended families and do not produce goods or services for export beyond the family, village, or tribe. Among the Tiv, the land was divided up into garden plots on which each family grew its own food. When the harvest was completed, the remaining vegetation was slashed and burned. The ashes served to fertilize the soil, but sometimes a plot was allowed to lie fallow for a few seasons. This meant that there was always a lot of discussion about which plots were to be used by which families in any given season. Never was a garden plot bought

or sold—the idea that land could be bought and sold was not part of Tiv culture. The land belonged to the tribe as a whole. Individual families were granted the right to use the land, but there was no such thing as private property in the form of land that could be bought and sold.

Nor were the products of the land thought to be the property of the grower. Food was distributed according to need, and a person who accepted food was expected to return the gift at some future time. Only a few goods, mainly items of dress worn on ceremonial occasions, were traded for currency—in this case, short brass bars. The Tiv also accepted brass bars from foreign traders in exchange for carvings or woven cloths. But they would have laughed at the idea of trading the bars for food. They would have shared their food with strangers without any expectation of payment.

The conquest of Tivland by the British in the early twentieth century brought immense changes to this nonmarket society. The British colonial administrators were shocked when they observed Tiv weddings, in which it appeared to them that young women were being traded for ceremonial goods. In reality, these exchanges were not market transactions; the items that were given in exchange for a bride were a symbol of respect for the bride's family.

The British also noted that the Tiv seemed always to be at war, either with neighboring tribes or among themselves. To them, this was a further sign that they needed to "civilize" the Tiv; in fact, however, the warfare was an outcome of the colonial system of taxation. Each tribal chief was required to pay a certain sum, either in British currency or in a commodity such as pigs. Because the Tiv had no currency and had to struggle to produce the required commodities, they were forced to cultivate more land. Their conflicts with neighboring clans or tribes were caused by the need to acquire more land. Thus the wars that the British observed were an almost immediate result of their own policies.

When the British succeeded in ending the warfare that had broken out among the Tiv, the only remaining source of cash to pay the taxes was wage labor on the colonial plantations. The Tiv therefore sent their sons and daughters to work for the British, and on their own land they began to grow crops that they could sell. Before long, they were part of the British market economy, selling their crops and labor for cash with which they could buy manufactured goods. Eventually, like thousands of other tribal societies, they were engulfed by the larger market system of the more powerful European nations (Bohannon & Bohannon, 1953, 1962).

In this idealized depiction of British colonial rule in Africa, white administrators confer in the foreground while happy natives work in the background. In reality, the "blessings" of colonialism were extremely mixed. As we see in the case of the Tiv, the experience of becoming part of the world market system made it difficult to maintain a small-scale society and culture.

Markets and the World Economic System

In the late fifteenth and early sixteenth centuries, according to sociologist Immanuel Wallerstein (1999), a "European world-economy" came into existence. The new economy was a kind of social system that the world had not known before. It was based on economic relationships, not on political empires; in fact, it encompassed empires, city-states, and the emerging nation-states.

Great empires had been a feature of the world scene for at least 5,000 years before the dawn of the modern era. But the empires of China, India, Africa, the Mediterranean, and the Middle East were primarily political rather than economic systems. Wallerstein argues that because the great empires dominated vast areas inhabited by peoples who lacked military and political power, they were able to establish a flow of economic resources from the outlying, or peripheral, areas to the imperial centers or "core" areas of the empire, such as Rome and all of Italy during the Roman Empire. The means used were taxation, tribute (payments for protection by the imperial army), and trade policies in which the outlying societies were forced to produce certain goods for the imperial merchants.

But this system—exemplified most clearly in the case of the Roman Empire—required a huge military and civil bureaucracy, which absorbed much of the imperial profit. Local rebellions and wars continually increased the expense of maintaining imperial rule. Political empires thus can be viewed as a primitive means of economic domination. "It is the social achievement of the modern world," Wallerstein (1974) comments, "to have invented the technology that makes it possible to increase the flow of the surplus from the lower strata to the upper strata, from the periphery to the center, from the majority to the minority" (p. 16) without the need for military conquest.

What technologies made the new world system possible? They were not limited to tools of trade, such as the compass or the oceangoing sailing vessel, or to tools of domination like the Gatling machine gun. They also included organizational techniques for bringing land, labor, and local currencies into the larger market economy: ways of enclosing and dividing up land in order to charge rent for its use; financial and accounting systems that led to the creation of new economic institutions like banks; and many others.

Although the European colonial powers (and the United States) often used political and military force to bring isolated societies into their markets, in the twentieth century they allowed their former colonies to gain independence, yet still maintained economic control over them. This occurred because the economies of the colonial societies had become dependent on the technologies and markets controlled by the Western powers. Today, former colonies are struggling to develop independent economic systems, but their ability to compete effectively in world markets is limited by the increasing power of **multinational corporations,** or *multinationals,* economic enterprises that have headquarters in one country and conduct business activities in one or more other countries (Barnet, 1994).

Multinationals are not a new phenomenon. Trading firms like the Hudson's Bay Company and the Dutch East India Company were chartered by major colonial powers and granted monopolies over the right to trade with native populations for furs, spices, metals, gems, and other valued commodities. Thus exploitation of the resources of colonial territories has been directed by multinational corporations for over two centuries (Hannah, 2006). Modern multinationals do not generally have monopolies granted by the state, yet these powerful firms, based primarily in the United States, Europe, and Japan, are transforming the world economy by "exporting" manufacturing jobs from nations in which workers earn high wages to nations in which they earn far less. This process, which is particularly evident in the shoe, garment, electronics, textile, and automobile industries, has accelerated the growth of industrial working classes in the former colonies while greatly reducing the number of industrial jobs in the developed nations.

Thinking Critically We often hear economists assert that government should not interfere with the workings of free markets by passing laws that regulate them because the markets will take care of any problems themselves if given the opportunity. Don't consumers need some basic level of protection from unscrupulous marketers who lie about their products? What are your views on the question of free markets versus government regulation, and what are the implications of your views for consumers?

Changes in worldwide production patterns challenge Wallerstein's thesis that there is a world economic order dominated by the former colonial powers. In particular, industrial nations like the United States and Japan are increasingly losing manufacturing jobs while newly industrializing nations like Brazil and Indonesia are gaining them. As the Global Social Change box on page 465 shows, however, the economic survival of millions of rural workers continues to be heavily influenced by changing markets in the highly developed nations.

Economic Globalization and Deregulation

Economic globalization can be broadly defined as a worldwide shrinkage of economic distances (costs of doing business) between nations. It includes two closely related processes: the globalization of production and trade, and the globalization of flows of finance (funds) and capital (tools, equipment, and services). These aspects of globalization have been greatly facilitated in recent decades by three important changes:

1. *Innovations and advances in transportation* (for example, jet travel and air freight) *and communications* (computers and the Internet), which allow extremely rapid exchanges of people, funds, and capital between regional markets.

2. *Global economic liberalization*—especially the reduction of national tariffs that may create barriers to the free flow of trade and investment funds—are encouraged through global trade institutions, particularly the General Agreement on Tariffs and Trade (GATT) and the World Trade Organization (WTO) in the case of world trade in goods and services, and the International Monetary Fund (IMF) in the case of global finance and capital flows.

3. *Use of market incentives* that encourage people to rely on supply and demand rather than government-enforced quotas, regulations, or treaties to regulate economic behavior—to speed up the development of production and trade everywhere in the world (Rajan, 2001).

According to its advocates, especially in the business sector, the modern market economy encourages constant change and innovation. Some of that change, however, involves risk to jobs and incomes. If one region produces steel at a higher cost than another region, for example, a market free of protective tariffs will make the lower-cost steel more competitive and create plant closings and layoffs in the region where labor costs (wages and benefits) are higher. Advocates of worldwide free trade and rapid globalization believe that nations that lose out in market competition will find ways of adapting and changing their economies so that workers in uncompetitive industries will eventually find work in more dynamic sectors of the economy.

multinational corporation: An economic enterprise that has headquarters in one country and conducts business activities in one or more other countries.

Globalization's opponents, including many trade union members, environmentalists, and critics of unregulated market forces, argue for "fair trade" rather than free trade. By that they mean trade that does not eliminate protections against exploitation and environmental pollution in the rush to deregulate all markets. For example, if steelworkers in the United States are able to form unions and bargain for higher wages and decent benefits, should they be vulnerable to competition from companies in nations where workers are forbidden or systematically discouraged from joining unions and are severely exploited and paid extremely low wages? Fair trade would insist that certain standards, such as bans on child labor, the right to form unions, and regulations on pollution, be applied to all trade treaties and that nations or regions that do not have such regulations be barred from markets in nations that do have these protections (Wayne, 2001).

Globalization and Third World Agriculture

Subsidies for the agricultural products of rich farming nations, especially the United States and Europe, present some of the most pressing problems of globalization and economic development. Although Western leaders sing the praises of "free trade" and claim that global supply and demand, not political considerations, should determine the international flow of goods and services, their actions go against their ideologies. This is particularly true in the case of politically motivated subsidies for domestic agriculture. Growers of corn, cotton, sugar, tobacco, and many other crops in the United States and other affluent nations are routinely granted subsidies, often in the form of guaranteed minimum prices, that encourage more production than would otherwise be economically

feasible. The surpluses that these subsidies generate, in turn, flood the world market with agricultural products that squeeze out the farm products of poor nations and add to the suffering of impoverished rural people throughout the world.

Cotton, for example, can be produced quite cheaply and in large quantities in the vast semidesert regions of Africa, where the average person lives on less than a dollar a day. But American cotton farmers have a good deal of political influence over southern and southwestern congressional representatives and use this influence to win guaranteed crop prices and other subsidies that encourage them to produce bumper crops that they would not be able to sell easily without the subsidies. In 2004, U.S. cotton farmers received over $4 billion in subsidies from the federal government. As U.S. cotton floods the world market, prices paid to other farmers decrease, but the American farmers are guaranteed their price due to the subsidies. This is having a dire effect on cotton-producing countries in Africa, such as Mali, where one-third of the population is dependent on cotton for their income. In the 2004–2005 growing season, Malian cotton farmers received just 18 cents per kilogram (2.2 pounds) for their cotton, whereas U.S. farmers received 80 cents per kilogram.

In 2010, Brazil, acting on behalf of many other cotton-producing nations, threatened the U.S. government with a suit in the International Court and also said that it would impose retaliatory trade measures unless the United States agreed to reduce subsidies to its domestic cotton producers, who form a powerful lobbying group in Congress. To settle the dispute, the United States announced that it would provide $150 million to Brazil to compensate Brazilian cotton producers. It did nothing to address the needs of

Cotton farmers like Bakary Diarra face crisis due to unfair competition and trade rules that work against their interests.

Adrian Arbib/Alamy

Addictive Substances and World Markets In 1983, when sociologist Edmundo Morales returned to his birthplace in northern Peru to conduct research on land reform, he found that the peasants of the Andes villages had become dependent on the worldwide traffic in cocaine. Their efforts to gain cooperative ownership of the land they farmed had faded. Instead, most of the able-bodied men in the villages were busy packing for the long trip down the eastern face of the

Andes. Most of them would work for months in coca-processing laboratories hidden in the jungle. Some would earn cash income by selling food, especially pork and beef, to the managers of the cocaine factories. The demand for cocaine in the United States and other societies had changed their lives.

Morales was dismayed to see so much urgently needed food being shipped from the impoverished highlands to the jungles. He felt that the enormous growth of the cocaine market had led his people down a false path. They were gaining cash income, perhaps, but in the bargain, they were becoming increasingly addicted to cocaine and were neglecting the need for more basic economic progress in their own villages (Morales, 1986, 1989).

Morales was able to use his knowledge of the culture and his training as a sociologist to identify the connections between the plight of the Andean peasants and the world market for addictive substances. He knew that

cocaine was simply another in a long line of substances for which huge markets had emerged in the richer nations, with drastic results for the less developed world. Foremost among those substances was sugar. In the thirteenth century, sugar was a luxury available only to royalty. But as its use as a source of energy and as a basis for alcoholic beverages—especially rum—became more widespread, the demand for sugar production became one of the main causes of European expansion into the tropical regions of the world. Moreover, sugar production required that huge amounts of land be devoted to growing sugarcane, and many cane cutters were needed to harvest the crop.

Thus, along with the expansion of sugar production went slavery, first in parts of North Africa and the Azores and later in the Caribbean islands and Latin America (Bennett, 1954; Mintz, 1985). And sugar was only the beginning. The cultivation of wheat, tea, and coffee; the extraction of iron, tin, gold, silver, and copper; and the widespread cutting of primeval forests were all spurred by the emergence of mass markets in the developed nations of the West. Andean coca for cocaine and Indochinese poppies for heroin were latecomers in a series of crops that had changed the earth and all its peoples (Capra, 2000).

Some scholars argue that economic development in the world's commercial capitals actually produced *underdevelopment*—dependency and poverty and overpopulation—in huge parts of Africa, Asia, and Latin America. But Morales questions the notion of "development of underdevelopment." He believes that, despite their tendency to become addicted to cocaine, the villagers of northern Peru are becoming more sophisticated in their market dealings and their understanding of politics. When they turn their attention once again to land reform, as eventually they must, their experiences in recent years will have made them less culturally isolated and more politically effective.

Edmundo Morales took this photo of a Peruvian highlander dipping into his sack of coca leaves. Andean Indians have traditionally chewed coca leaves to increase energy and reduce hunger. In the small bottle he carries is a form of lime paste, which he mixes with the leaves in his mouth to strengthen the effects, a process similar to what cocaine manufacturers do on a far larger scale with the leaves they buy from Indian cultivators.

poorer nations like Mali, which will have to wait until a broader agreement is negotiated through the World Trade Organization. The result is that Mali loses revenues, its population sinks further into poverty, and third world leaders become increasingly cynical about the benefits of free trade that is not in fact free.

Globalization and Culture Globalization and free trade may also threaten the values of people in nations where cultural products like music, movies, fast food, and fashions seem to promote deviance from strict religious and gender norms. Thus, orthodox religious leaders in many nations continually protest against free markets that bring American or European movies that question religious authority, or fashions that reveal women's bodies, or music and foods that depart from local traditions and norms.

ECONOMICS AND THE STATE

The earliest social scientists were deeply interested in understanding the full importance of modern economic institutions. In fact, since the eighteenth century, almost all attempts to understand large-scale social change have dealt with the question of how economic institutions operate. But the age-old effort to understand economic institutions has not been merely an academic exercise. The fate of societies throughout the world has been and continues to be strongly influenced by theories about how economic institutions operate, or fail to do so, in nation-states with different types of political institutions. The major economic ideologies thus are also political ideologies, and economics is often called *political economics* (Heilbroner, 1995).

Political-Economic Ideologies

In this section, we review three economic ideologies—mercantilism, capitalism, and socialism—and some variations of them. (See the For Review chart on page 466.)

Political-Economic Ideologies

Ideology	Description	Example
Mercantilism	An economic philosophy based on the belief that the wealth of a nation can be measured by its holdings of gold or other precious metals and that the state should control trade	European nations under feudalism
Laissez-Faire Capitalism	An economic philosophy based on the belief that the wealth of a nation can be measured by its capacity to produce goods and services and that these can be maximized by free trade	England at the time of the Industrial Revolution
Socialism	An economic philosophy based on the concept of public ownership of property and sharing of profits, together with the belief that economic decisions should be controlled by the workers	The Soviet Union before 1989; China and Cuba today
Democratic Socialism	An economic philosophy based on the belief that private property may exist at the same time that large corporations are owned by the state and run for the benefit of all citizens	Holland and the Scandinavian nations
Welfare Capitalism	An economic philosophy in which markets determine what goods will be produced and how, but the government regulates economic competition	The United States

Mercantilism The economic philosophy known as **mercantilism**, which was prevalent in the sixteenth and seventeenth centuries, held that a nation's wealth could be measured by the amount of gold or other precious metals possessed by the royal court. The best economic system, therefore, was one that increased the nation's exports and thereby increased the court's holdings of gold.

The mercantilist theory had important consequences for economic institutions. For example, the guilds, or associations of tradespeople, that had arisen in medieval times were protected by the monarch, to whom they paid tribute. The guilds controlled their members and determined what they produced. In this way, they were able to produce goods cheaply, and those goods were better able to compete in world markets, thereby increasing exports and bringing in more wealth for the court. But the workers were not free to seek the best jobs and wages available. Rather, the guilds required them to work at assigned tasks for assigned wages, and the guildmasters fixed the price of work (wages), entry into jobs, and all working conditions. Land in mercantilist systems was not subject to market norms either. As in feudalism, land was thought of not as a commodity that can be bought and sold—that is, a commodity subject to the market forces of supply and demand—but as a hereditary right derived from feudal grants.

Laissez-Faire Capitalism The ideology of **laissez-faire capitalism** attacked the mercantilist view that the wealth of nations could be measured in gold and that the state should dominate trade and production in order to amass more wealth. The laissez-faire economists believed that a society's real wealth could be measured only by its capacity to produce goods and services—that is, by its resources of land, labor, and machinery. And those resources, including land itself, could best be regulated by free trade in world markets (Halevy, 1955; Smith, 1965/1776).

The ideology of laissez-faire capitalism also sought to free workers from the restrictions that had been imposed by the feudal system and maintained under mercantilism. Thus, it is no coincidence that the first statement of modern economic principles, Adam Smith's *The Wealth of Nations*, was published in 1776. Revolution was in the air—not only political revolution but the Industrial Revolution as well. And some of the most revolutionary ideas came from the pens of people like Adam Smith, Jeremy Bentham, and John Stuart Mill, who understood the potential for social change contained in the new capitalist institutions of private property and free markets.

Private property (as opposed to communal ownership) is not merely the possession of objects but rather a set of rights and obligations that specify what their owner can and cannot do with them. The laissez-faire economists believed that the owners of property should be free to do almost anything they liked with their property in their efforts to gain profit. Indeed, the quest for profit would provide the best incentive to produce new and cheaper products, and therefore required free markets in which producers could compete to provide better products at lower prices.

These economic institutions are familiar to us today, but the founders of laissez-faire capitalism had to struggle to win acceptance for them.

mercantilism: An economic philosophy based on the belief that the wealth of a nation can be measured by its holdings of gold or other precious metals and that the state should control trade.

laissez-faire capitalism: An economic philosophy based on the belief that the wealth of a nation can be measured by its capacity to produce goods and services (that is, its resources of land, labor, and machinery) and that these can be maximized by free trade.

No wonder they thought of themselves as radicals. In fact, their economic and political beliefs were so opposed to the rule of monarchs and to feudal institutions like guilds that they could readily be seen as revolutionary. They believed that the state should leave economic institutions alone (which is what *laissez-faire* means). In their view, there is a natural economic order, a system of private property and competitive enterprise, that functions best when individuals are free to pursue their own interests through free trade and unregulated production.

Smith opens *The Wealth of Nations* by describing how wealth and prosperity are created through a self-regulating system of markets in which participants enter into contracts to buy and sell without outside influences or constraints. This classical model of capitalism and free markets hinges on three primary characteristics:

1. *Freedom:* the right to produce and exchange products, labor, and capital.
2. *Self-interest:* the right to pursue one's own business and appeal to the self-interest of others.
3. *Competition:* the right to compete in the production and exchange of goods and services.

Smith's observations of markets convinced him that these three characteristics would produce a "natural harmony" among workers who contribute their labor power, landlords who rent land and other property, and capitalists who buy and build the machines and equipment and develop new enterprises to produce goods and services. When free to buy and sell their products or services in unfettered markets, the self-interest of millions of individuals creates an orderly and wealthy society without the need for central direction from the state (government). As Smith wrote:

> It is not from the benevolence of the butcher, the brewer, or the baker, that we expect our dinner, but from their regard to their own interest. We address ourselves, not to their humanity, but to their self-love. . . . Every individual. . . [who] . . . employs capital . . . and . . . labours . . . neither intends to promote the public interest, nor knows how much he is promoting it. . . . He is . . . led by an invisible hand to promote an end which was no part of his intention. . . . By pursuing his own interest he frequently promotes that of the society.

This doctrine of enlightened self-interest, often called "the invisible hand," is the central idea of laissez-faire or classical economics. As we will see, it is an idea that is continually invoked

by those who wish to make markets the primary institution regulating economic and other transactions within and across the world's nations (Skousen, 2001).

Socialism As an economic and political philosophy, **socialism** began as an attack on the concepts of private property and personal profit. These aspects of capitalism, socialists believed, should be replaced by public ownership of property and sharing of profits. As we have noted in earlier chapters, this attack on capitalism was motivated largely by horror at the atrocious living conditions caused by the Industrial Revolution. The early socialists thought of economics as the "dismal science" because it seemed to excuse a system in which a few people were made rich at the expense of the masses of workers. They detested the laissez-faire economists' defense of low wages and wondered how workers could benefit from the Industrial Revolution instead of becoming "wage slaves." They proposed the creation of smaller-scale, more self-sufficient communities that would produce modern goods but would do so within a cooperative framework.

Karl Marx viewed these ideas as utopian dreams. He taught that the socialist state must be controlled by the working class, led by their own trade unions and political parties, which would do away with markets, wage labor, land rent, and private ownership of the means of production. These aspects of capitalism would be replaced by socialist economic institutions in which the workers themselves would determine what should be produced and how it should be distributed.

Marx never completed his blueprint of how an actual socialist society might function. That chore was left to the political and intellectual leaders of the communist revolutions—Lenin, Leon Trotsky, Rosa Luxemburg—and, finally, to authoritarian leaders like Joseph Stalin, Mao Ze-dong, and Fidel Castro. They believed that all markets and all private industry must be eliminated and replaced by state-controlled economic planning, collective farms, and worker control over industrial decision making. Unless capitalist economic institutions were completely rooted

out, they believed, small-scale production and market dealings would create a new bourgeois class.

In the socialist system as it evolved under the communist regimes of the Soviet Union and China, centralized planning agencies and the single legal party, the Communists, would have the authority to set goals and organize the activities of the worker collectives, or soviets. Party members and state planners would also devise wage plans that would balance the need to reward skilled workers against the need to prevent the huge income inequalities found in capitalist societies. In Soviet-style societies, markets were not permitted to regulate demand and supply; this vital economic function was supposedly performed by government agencies. Societies that are managed in this fashion are said to have *command economies*. The state commands economic institutions to supply a specific amount (a quota) of each product and to sell it at a particular price. (Note that not all command economies are dominated by communist parties. Germany under the Nazis and Italy under the fascists were also command economies [Smelser & Swedberg, 2005].)

With the collapse of the Soviet political empire, the economies of Russia and the other nations of the former Soviet Union (as well as its satellites in eastern Europe) are attempting to make a transition to capitalist economic institutions. It has become evident that command economies provided these nations with overdeveloped industrial infrastructures and limited capability to compete in world markets for goods and services. Indeed, it has been said that the command economic system of the former Soviet Union gave it "the most impressive nineteenth century industrial infrastructure in the world, 75 years too late" (Ryan, 1992, p. 22).

The Soviet system was notorious for the inefficiency of its economic planning and industrial production. Under the command system, factory managers must continually hoard supplies, or raid supplies destined for similar factories in other regions, in order to meet their production quotas. Or they must trade favors with other factory managers to obtain supplies, a system that creates a hidden level of exchange based on bribes and favoritism. And because there are no free markets, goods that are desired by the public often are not available simply because the planners have not ordered them. In fact, in the Soviet-style economies, clandestine markets operated outside the control of the authorities and supplied goods and services to those with the means to pay for them. These underground markets thereby increased inequality in those societies and generated

These children worked in a Rhode Island textile mill in the 1890s. Children were preferred as workers in this industry because it was believed that their nimble fingers helped maintain the pace of production.

Library of Congress Prints and Photographs Division

socialism: An economic philosophy based on the concept of public ownership of property and sharing of profits, together with the belief that economic decisions should be controlled by the workers.

greater public disillusionment with command economies and one-party communist rule.

The massive social upheavals of 1989 and 1990 brought an end to the communist command economic system, but they also initiated a period of great economic and social uncertainty in the former Soviet-dominated nations. At present, efforts to privatize their economies and use market forces to regulate the supply of goods and services, including labor, have begun to produce economic growth in some of those nations. Rapid changes have also produced new forms of insecurity and corruption that have caused many people there to wish for the relative stability and security of the communist past (Ash, 1999; Dasgupta, 2004). These problems are made even worse by the collapse of the political system in some nations of the former Soviet empire, which has enabled organized criminal gangs known as the *mafiya* (not related to the Mafia) to flourish. Without the rule of law established by a legitimate system of government, it is extremely difficult to modernize an economy and prevent severe abuses of power through violence and terror. (The role of legitimacy is discussed further in Chapter 19.) In recent elections in Russia, Slovakia, and the former East Germany, many people voted for former Communist Party officials who wish to slow down the conversion to a capitalist economy (Tagliabue, 1998).

State Capitalism China, Vietnam, Cuba, and other nations that once followed the Soviet model of a socialist state with a command economy are now considered to be moving rapidly toward a system that can be termed *state capitalism* (Satya, Resnick, & Wolff, 2008). Under state capitalism, the central government (the state) may allow individuals to accumulate a certain amount of wealth and even invest that wealth in new businesses. But the state reserves ultimate control over individuals and their businesses and can force mergers or otherwise ensure that a business does nothing to interfere with the political power of the state. Prosperous individuals or groups may be allowed to run entire industrial sectors of the economy. Some competition among firms may be allowed, and markets for goods and services may be allowed to regulate commerce, but the state can intervene whenever it considers business practices to conflict with its interests. It is considered illegitimate for workers to form trade unions that are independent of the state. With unions kept completely under state control, workers are not free to strike for higher wages.

Much of mainland China's astounding economic growth over the past two or more decades is attributable to its system of state capitalism. China has allowed foreign producers to help construct factories that are ultimately run by Chinese managers, and the government and its business allies have kept workers' wages as low as possible in order to compete effectively in global markets. Labor unions do exist, but they represent the interests of the managers rather than those of the workers. In 2010, however, a series of suicides by workers protesting harsh labor conditions galvanized unrest in the nation's huge manufacturing sector. Strikes broke out in plants that produced parts for Japanese and U.S. corporations like Honda and KFC. The new workers' movement has demonstrated not only that disenfranchised sections of society are willing to show their determination to win advancements in wages and working conditions, but also the greater need for the Chinese government to address the distribution of wealth (Chan & Unger, 2009).

Democratic Socialism A far less radical version of socialism than that attempted in the Soviet bloc is known as **democratic socialism**. This economic philosophy is practiced in the Scandinavian nations, especially Sweden, Denmark, and Norway, as well as in Holland and to a lesser extent in Germany, France, and Italy. It holds that the institution of private property must continue to exist because people want it to, and that competitive markets are needed because they are efficient ways of regulating production and distribution. But large corporations should be owned by the nation or, if they are in private hands, required to be run for the benefit of all citizens, not just for the benefit of their stockholders. In addition, economic decisions should be made democratically.

Democratic socialists look to societies like Sweden and Holland for examples of their economic philosophy in practice (Harrington, 1973). In Sweden, for instance, workers can invest their pension benefits in their firm and thereby gain a controlling interest in it. This process is intended to result in socialist ownership of major economic organizations.

The United States has a long history of conflicted relations between workers and owners of capital. Cooperative systems in which authority

Chinese workers assemble a car at the Honda plant in the southern city of Guangzhou. Workers like these walked off their jobs in 2010, as part of a wave of worker unrest and strikes that led the Chinese government to agree to begin raising workers' wages. China's leaders are fearful that worker unrest could spread and threaten their tight control over the nation's system of state capitalism.

China Newsphoto/Reuters/Landov

democratic socialism: An economic philosophy based on the belief that private property may exist at the same time that large corporations are owned by the state and run for the benefit of all citizens.

and even ownership are shared are developing, but more slowly than in the social democracies of Europe. Social scientists who study the U.S. economy note that in recent years there has been a reaction among owners of capital against the principles of democratic socialism, especially those that stress cooperation between labor and management and the right of workers to organize unions and engage in collective bargaining (Freeman, 2000). Employers have also shown diminished commitment to welfare capitalism, the political-economic model that has had a great deal of influence in the noncommunist modern economies until recently (Dobbin, 2004). President Obama and many of his supporters in the Democratic Party avoid the term *democratic socialism* because the term *socialism* in any form remains a toxic label for U.S. politicians, but in fact, his efforts to extend health care benefits to a far greater numbers of Americans qualifies as a policy favored by democratic socialists. Although critics consider this and similar policies to be a form of socialism, health care reform was debated endlessly as part of the democratic process, both before and after the 2008 election (Goldberg, 2010).

Welfare Capitalism Emerging to some extent as a response to the challenge posed by the Russian Revolution (which called attention to many of the excesses of uncontrolled or laissez-faire capitalism), **welfare capitalism** represented a new way of looking at relationships between governmental and economic institutions. Welfare capitalism affirms the role of markets in determining what goods and services will be produced and how, and it also affirms the role of government in regulating economic competition (for example, by attempting to prevent the control of markets by one or a few firms).

Welfare capitalism also stresses the role that governments have always played in building the roads, bridges, canals, ports, and other facilities that make trade and industry possible. Expanding on this role, the theory of welfare capitalism asserts that the state should also invest in the society's human resources—that is, in the education of new generations and the provision of a minimum level of health care. Welfare capitalism also guarantees the right of workers to form unions to reach collective agreements with the owners and managers of firms regarding wages and working conditions. It creates social-welfare institutions like Social Security and unemployment insurance. And to stimulate production and build confidence in times of economic depression, welfare capitalism asserts that governments must borrow funds to finance large-scale public works projects like the construction of the American interstate highway system during the 1950s.

The theory of welfare capitalism is associated with the writings of John Maynard Keynes, Joan Robinson, John Kenneth Galbraith, and James Tobin, all of whom contributed to the revision of laissez-faire economic theory. Welfare capitalism dominated American economic policy from World War II until the 1970s, when a succession of economic crises—inflation, energy shortages, and unemployment—turned the thoughts of many Americans once again toward laissez-faire capitalism. Today, however, the increasingly evident gap between the haves and the have-nots in U.S. society has created renewed interest in enhancing the institutions of the welfare state—unemployment insurance, health insurance, low-income housing, public education, and others. At the same time, fear of governmental control over people's lives, and its costs in the form of higher taxes, has generated a revolt against the "welfare state" in many segments of the population in the United States and other industrial nations. This trend has resulted in conservative electoral victories and unprecedented reductions in social spending. What is not clear is whether the public's desire for relative security from economic stress will outweigh the desire for lower taxes and reductions in the scope of government bureaucracies.

The United States: A Postindustrial Society?

Although many people in the United States see the failure of socialism in the Soviet bloc as a victory for capitalism, sociologists and poll takers have called attention to a growing sense of unease about the U.S. economy. As demographer Valerie Kincade Oppenheimer observes: "Since about 1970, the average real earnings position of young men has deteriorated considerably. This has been true not only for high school dropouts but also for high school and college graduates" (1994, p. 331). (Figure 18.1 shows the value of continuing one's education through college and beyond.) The declining incomes of U.S. workers over the past quarter century have contributed to the growth in dual-earner households.

The disparity between the incomes of people with professional and technologically sophisticated jobs and those of people with less intellectually demanding jobs is likely to increase as economic growth centers on computers,

welfare capitalism: An economic philosophy in which markets determine what goods will be produced and how, but the government regulates economic competition.

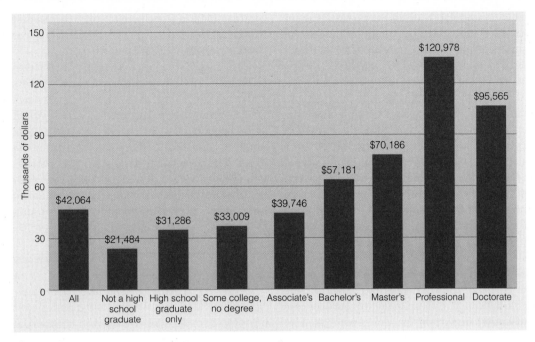

Figure 18.1 Mean Earnings, by Highest Degree Earned.

telecommunications, and biological technologies. Jobs that require lower levels of education and training may be more onerous, but the number of people who can perform them is relatively large. The basic economic laws of supply and demand therefore drive up the wages of highly educated workers while those of less-educated workers are driven downward.

In an influential study, *The Coming of Post-Industrial Society,* sociologist Daniel Bell (1973, 1999/1973) reviewed changes in American economic institutions indicating that we are undergoing a transition from an industrial to a postindustrial society. A postindustrial society "emphasizes the centrality of theoretical knowledge as the axis around which new technology, economic growth, and the stratification of society will be organized" (p. 116). As societies undergo the transition to postindustrial economic institutions, an "intellectual technology" based on information arises alongside machine technology. Industrial production does not disappear, but it becomes less important. New industries devoted to providing knowledge and information become the primary sources of economic growth (Block, 1990; Sabel, 1995).

Changes in the distribution of jobs in the goods-producing and service-producing sectors of the American economy clearly show the growing importance of services, especially those employing professional and technical workers. For Bell, this change is inevitable. Industries decline, but new ones appear. A century ago, for example, over 70 percent of Americans were employed in agriculture and related occupations. Now this figure is below 3 percent. Similar transformations are taking place in manufacturing today, and they can be made less painful if the society is willing to invest in education and the retraining of workers in declining industries like mining, steel production, or manufacturing for military purposes (Smelser & Swedberg, 2005).

Other social scientists view these changes less optimistically. Some note that declines in industrial production have pushed thousands of skilled workers into lower-paying jobs in the service sector (for example, in the fast-food industry or as janitors or security guards), which helps explain the decline in average wages (Aronowitz, 1996; Brass, 2009). Others are concerned that the gains made by blacks and women in access to better-paying industrial jobs since World War II are being erased. William Julius Wilson (1998), for example, has observed that access to control over the means of production is increasingly based on educational criteria and thus threatens to solidify the subordinate position of the black lower class.

European sociologists have criticized the theory of postindustrial society on even broader grounds, saying that it tends to promote belief in the inevitability of an increasingly impoverished working class with even more reason to demand a more equal distribution of wealth (Dobbin, 2004; Touraine, 1998). In the United States, the theory of postindustrial society has been criticized on the ground that even though manufacturing accounts for a declining share of the labor force, the society still depends on manufactured goods of all kinds. Critics argue that the United States cannot relinquish its manufacturing base simply because a theory says it is outmoded (W. J. Wilson, 1998).

A Changing Social Contract? The massive changes brought about by economic globalization, technological advances, and the growth of service industries are increasing the pressure on employers to keep their costs down in order to be competitive. This focus on cost savings results, as we have seen, in declining real wages and widening economic inequalities. In the social democracies of Europe, especially Scandinavia, the impact of these changes is eased by social-welfare policies. In the United States, the welfare state and its "safety net" of economic and social policies (for example, Medicare, the minimum wage, unemployment insurance) are considered important by those who need to be protected from the harshest consequences of sudden economic change. Others, however, attack the policies as wasteful and costly government intervention in the economy.

In his analysis of postindustrial society, Daniel Bell (1973, 1999/1973) assumed that the "social contract" in the United States and other advanced nations would be maintained. In return for hard work, saving, and efforts to improve their skills, citizens could expect modest increases in their living standards, adequate health care, protection from crime, decent public education for their children, and protection of their rights. However, the Reagan administration of the 1980s and the conservative Congress of the 1990s challenged the assumption that government would provide these benefits if they were not provided by the economy itself. This alteration in the basic social contract that has guided social policy for the past half century has had far-reaching consequences (Blau, 1993; Galston, 2008).

The consequences of a fraying social safety net—with fewer protections for workers and children and more cash-strapped local governments—have been especially evident since 2008, as the nation faced its worst economic crisis since the Great Depression of the 1930s. Clearly, it is far easier to build and sustain the institutions of welfare capitalism when the economy is growing than it is when massive unemployment makes life uncertain for millions of citizens.

Sociology & Social Justice

WHEN WORK DISAPPEARS Figure 18.2 tells the story of employment and unemployment in the United States over the past two decades. During the 1990s, the United States enjoyed its longest period of sustained economic growth in the twentieth century. Unemployment dropped to lows not seen since the post–World War II boom years. Worker productivity increased steadily. In the financial markets, there were record increases in the prices of securities (stocks and bonds) and mutual funds, and because a large proportion of U.S. employees have their pension funds invested in these securities, there was a widespread sense of well-being during most of the decade. This positive outlook was not shared by workers in the lower ranks of labor, who continued to experience erosion of their spending ability. However, in relation to the major European and Asian economies, where economic crisis and chronic unemployment were serious problems, the U.S. economy assumed a position of almost undisputed dominance.

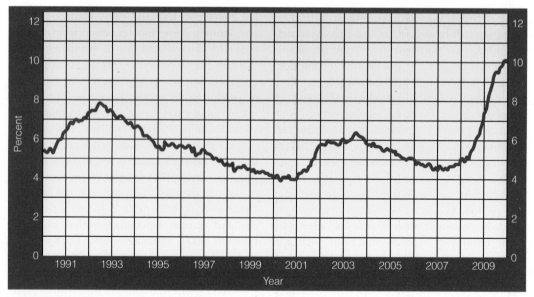

Figure 18.2 Unemployment Rates, 1991–2009. Source: Bureau of Labor Statistics, 2010.

Even during this booming economy, many social scientists warned that the United States was becoming ever more indebted to China and other rapidly growing economic powers. Americans were borrowing against the value of their homes, counting on ever-increasing real estate prices. A recession loomed in the near future, but few were ready for the abrupt crash that came in 2008. When the bottom fell out of financial and real estate markets throughout the world, unemployment began to soar, as is evident in the figure, which shows the unemployment rate hovering around a low of about 4.6 percent in 2007–2008 and then shooting up to 10 percent in 2010. At this writing, the U.S. economy is still sputtering, and employment growth is painfully slow. Millions of people remain jobless.

Figure 18.3 demonstrates that the pain of unemployment is not distributed fairly through the U.S. population. Its toll falls most heavily on people who are young and just entering the labor force, and on minority workers. For decades, the unemployment rate for African Americans has been more or less double that of whites, with Latinos faring somewhat better but not nearly on a par with whites. Why is this so? Part of the answer is the "last hired, first fired" situation of minority workers. It takes a very strong economy to put high proportions of minority workers in jobs, but when a recession hits, these workers are usually the first to become unemployed. The graph also reveals that minority men are especially hard hit in a severe recession like the one that began in 2008. Note that unemployment of minority women was almost 14 percent in April 2010, but that rate was far lower than the overall black unemployment rate of 15.7 percent.

Clearly, the difference is accounted for in the statistics for black males, and indeed, though not shown in the graph, the 2010 unemployment rate for black males was almost 18 percent (Bureau of Labor Statistics, 2010), a level almost comparable to that recorded in the Great Depression of the twentieth century. In consequence, when Congress failed to renew unemployment benefits in 2010, with Republicans arguing that nothing should add to the nation's growing deficit, the burden of that failure fell most heavily on those who were least equipped to cope without the aid of such social safety net policies. The tension between Democrats' emphasis on welfare capitalism and Republicans' emphasis on the free market and fiscal austerity places severe limits on social justice. For sociologists, it also raises the question of how workers and their employers will act collectively once the severe recession begins to ease.

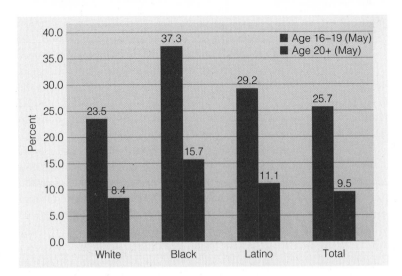

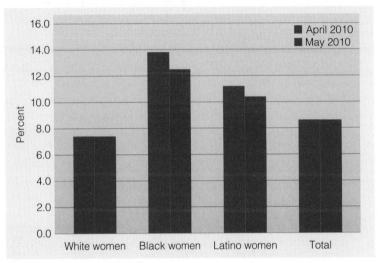

Figure 18.3 National Unemployment, by Race and Age, as of May 7, 2010.
Source: Bureau of Labor Statistics, 2010.

WORKERS, MANAGERS, AND CORPORATIONS

Even in the midst of overall prosperity during the 1990s and the early years of the new century, sociologists noted that certain social problems in U.S. economic institutions remained unresolved. Feminist sociologists pointed to the persistence of discrimination against women in the workplace. Others warned of worker alienation, while still others focused on the increasing power and dominance of corporations over the lives of individuals (Chang, 2003; Dobbin, 2004; Odaka, 1998).

Gender and Workplace Diversity

One of the most outstanding results of affirmative action efforts over the past generation has been the entry of women into jobs and professions that were formerly closed to them. Although affirmative action to correct racial discrimination and inequality has drawn fire from critics of quotas and admissions policies in education, conflicts arising in the workplace have also been fierce on occasion. In 1998, for example, Texaco Inc. agreed to settle a gender discrimination claim by a group of its female employees. The women were working at all levels of the company but were paid less than their male counterparts for the same work. Though not admitting its guilt, Texaco agreed to pay $3.1 million in back pay to the 186 women in the lawsuit brought on their behalf by the U.S. Department of Labor. This settlement sent a clear signal to other major domestic and global corporations that the push for gender equality in workplaces continues (Stevenson, 1999). In fact, in the same year, the Supreme Court ruled against the Virginia Military Institute's exclusion of women and, in so doing, made it clear that affirmative action to achieve gender equality is justified on economic grounds (Rapoport et al., 2002).

Feminist sociologists note that as important as these victories may be, they should not convey a sense of uniform progress toward greater gender equality and tolerance of diversity in the workplace (Bradley, 1998; Epstein, 1995). Major corporations like Texaco often adopt far more conciliatory positions on these issues than do smaller corporations, which are less likely to be in the public spotlight. For example, the Walt Disney Company recognized the growing diversity of its labor force by conceding the right of cohabiting adults, no matter what their gender, to receive the same health and other employee benefits as heterosexual married couples. But for every company like Disney, there are scores of others in which the forms of inequality that produce differences in pay for women and members of minority groups persist.

Globalization of the activities of major corporations has an immense impact on gender relations outside the United States. On the Mexican border, for example, there are hundreds of so-called *maquiladora* industries. These factories do production work of all kinds, often on a contract basis, for larger corporations. Women and men employed in these factories often experience conditions and hours of work that would be illegal in the United States. In Mexico, however, workers are often thankful to have the work and feel that they must endure the hardships. As *maquiladora* workers begin to protest and seek to join labor unions, they find themselves enmeshed in class conflict and labor struggles that may be entirely new to them. The experiences of earning their own wages and, especially, taking on leadership roles at work often motivate female workers to challenge traditional male-dominated family and community relations as well. Thus, experiences in the workplace produce a movement for more egalitarian gender relations in other institutions. This trend is not limited to Mexico. Sociologists who study economic globalization and gender relations are finding many instances of these changes throughout the world (Cowie, 1998; Dasgupta, 2004).

Increases in the demand for part-time workers, the trend toward contracting for services by freelance workers (who may be highly skilled professionals), decreases in the willingness of employers to pay health care benefits, and efforts to prevent employees from joining labor unions are only a few of the major trends in the economies of industrialized nations that are causing workplace conflict and employee alienation. As we see next, alienation is not a new problem, but it takes on new aspects as the nature of work and economic institutions changes.

Employee Mentoring versus Worker Alienation

Many sociologists use the concept of alienation to explain the gap between workers and managers. Karl Marx first applied the term *alienation* to the situation of workers under early capitalism. A worker in a factory performed only a fraction of the work that went into a product and therefore could feel little sense of ownership of the final product. And because the worker typically could not control the work process very much, he

---

alienation: The feeling of being powerless to control one's own destiny; a worker's feeling of powerlessness caused by inability to control the work process.

---

or she came to feel like a mere cog in a giant machine, a feeling that produced a sense of not being able to control one's own actions on the job (Fromm, 1961; Marx, 1961/1844). Today, the term *alienation* is often used to describe the feeling of being powerless to control one's own destiny (Burawoy, 1996; Freeman, 2000). At work, people may feel alienated because their labor is divided up into activities that are meaningless to them. For example, in a classic study of autoworkers, oil workers, and others in different factory settings, Robert Blauner (1964) found that workers who perform highly repetitive tasks on an assembly line are more alienated than workers in groups whose tasks involve teamwork.

In a study of organizational techniques that have been used successfully in Japan, Ezra Vogel (1979) pointed out that Japanese firms tend to avoid worker alienation by strengthening the role of the small group:

> The essential building block of a company is not a man with a particular role assignment and his secretary and assistants. The essential building block of the organization is the section. . . . The lowly section, within its sphere, does not await executive orders but takes the initiative. . . . For this system to work effectively leading section personnel need to know and to identify with company purposes to a higher degree than persons in an American firm. They achieve this through long experience and years of discussion with others at all levels. (pp. 143–145)

The irony of this finding and others like it is that, even before World War II, American and European industrial sociologists had demonstrated the benefits of team approaches to work (McCord, 1991). The famous studies by Elton Mayo, discussed in Chapter 2, were designed to find out how workers at any level of an organization could be made to feel less alienation and a greater sense of ownership of their jobs. This research demonstrated the importance of primary-group relationships at work. But the function of the manager is to motivate employees to work more efficiently and at higher levels, a goal that is not easily achieved (Hickins, 1998).

Large companies are often extremely concerned about worker alienation. Because they invest heavily in personnel training, losing employees to other businesses or having their productivity decrease because of feelings of discontent can result in lower profits. Employee mentoring programs are a popular approach to preventing alienation. Mentoring usually matches an experienced employee with a newcomer; it is also used to groom employees for advancement in the organization.

In essence, a mentor is a personal teacher and helper. The experienced employee shows the newcomer how to do the work more effectively, how to avoid "burnout" (the sense that one has no more energy to devote to the job), how to deal with the quirks and foibles of certain supervisors, and much more. The mentoring approach is widely used in business, education, health care, and government—and, of course, some programs are more effective than others.

Sociologists or people with sociological training are often asked to evaluate mentoring programs. For example, in one large metropolitan hospital, evaluators studied more than 600 professional, administrative, and clerical employees who were involved in a mentoring program. They found a strong negative correlation between time spent in mentoring and measures of employee alienation. This means that the more mentoring an individual received, the less alienation he or she felt. The evaluation also suggested ways to measure the effectiveness of mentoring aside from merely looking at whether the employee had a mentor or how much time the mentor and the employee spent together. Overall, organizations that institute mentoring programs and attempt to improve them through ongoing monitoring and evaluation are extremely pleased with the results (Koberg, Chappell, & Ringer, 1997).

Mentoring is especially important for professional workers, whose success often depends on social skills, and those in service occupations that require a great deal of interpersonal skill. But throughout the world, for workers in manufacturing and service occupations, especially in the middle and lower ranks, any improvements in wages, benefits, and working conditions are most influenced by whether or not they are represented by labor unions.

Labor Unions in Global Context

Labor unions promote and defend the interests of workers in the private and public sectors of the economy. They negotiate with owners and managers on behalf of the workers for pay and benefits. In the United States, approximately 16 million of the 130 million workers in the labor force are union members. The highest proportion of union members are in police and other protective services, and the lowest in farming, but unions represent workers in thousands of job settings and are especially important in construction, manufacturing, hospitals, county and state governments, retail trades, airlines, and many other major sectors of the economy (*Statistical Abstract,* 2010).

As noted earlier, in China (and other nations dominated by a communist party), free labor unions are forbidden, and those that do exist are controlled by the party (MacLeod, 2010). When workers begin to organize for their own benefit, they must first create trade unions. As a first step toward real unions, workers seek the right to bargain freely with the owners and managers of capital, and to enter into labor-management contracts that specify hours of work and set wage standards. Very often, the owners of companies—Wal-Mart, for example—fight hard against this demand. They often believe that unions, which protect workers' rights, hinder their ability to freely hire and fire and thereby are an infringement on the basic rights of capitalists to control the means of production that they own.

Collective Bargaining The union–management negotiating process occurs only after management has agreed to sign a union contract and to engage in *collective bargaining* with the union on issues of pay, benefits, and job conditions. These contracts and the negotiations they lead to are by far the most important reform of laissez-faire capitalism in the past two centuries. Under laissez-faire capitalism, workers can only bargain as individuals with their employer. The owner or manager reserves the right to set wages, establish working conditions, and decide on whether or not to pay benefits. A highly valued worker might get more pay and benefits than others, but most will be told, if they ask for improvements in their situation, to accept the status quo or quit. The employees as individuals do not have any real power to influence these decisions except by finding another job. But if they are represented by a union that has a collective bargaining contract with the employer, this contract will allow union leadership to bargain with the boss for improvements in wages and working conditions. Union contracts also provide for some measure of justice and due process at work by establishing a grievance process whereby workers who feel that they have been unfairly disciplined or subjected to unsafe conditions can appeal to the union to represent them in a grievance against the company or agency. Nonunion workplaces sometimes have a grievance system put in place by company owners, but these systems invariably give almost all the decision-making power to management (Freeman & Rogers, 2002).

What Unions Achieve for Their Members
Workers represented by labor unions earn more pay and receive better insurance and pension benefits than workers in nonunion jobs. These

TABLE 18.1

U.S. Union Wage and Benefit Premiums, 2001

	Hourly Pay			Total Compensation
	Wages	Insurance	Pension	
All workers				
Union	$21.40	$2.48	$1.52	$27.80
Nonunion	16.67	1.14	0.51	19.98
Union premium				
Dollars	$4.73	$1.34	$1.01	$7.82
Percentage	28.4%	117.5%	198.0%	39.1%
Blue collar				
Union	$21.10	$2.66	$1.70	$28.07
Nonunion	13.72	1.12	0.37	16.93
Union premium				
Dollars	$7.38	$1.54	$1.33	$11.14
Percentage	53.8%	137.5%	359.5%	65.8%

Source: Mishel, Bernstein, & Boushey, 2003.

facts are clearly demonstrated in Table 18.1. The wage premium, or gain, for union membership is 28.4 percent, and almost 40 percent if one considers all elements of hourly compensation (including insurance and pensions). The U.S. Department of Labor recently reported that a typical union job pays $640 a week compared to $478 a week for nonunion workers. This difference means an extra $8,400 a year for a union member.

Labor unions also bargain for vacations, holidays, sick pay, family leave, and hours of employment, including overtime pay. Improvements like overtime pay, pay for working on holidays, compensation for injuries suffered on the job, and many more improvements in the conditions of employees can be credited to the efforts of the labor movement and dedicated workers over many years of struggle with the owners of capital and with high-level managers in the public sector. In 2003, for example, the Bush administration issued new regulations, through the U.S. Department of Labor, that defined many thousands of workers as part of management or as contingent workers (not full-time employees), and therefore not subject to the regulations that require overtime pay after eight hours of work in a day or forty hours in a week. It claimed this would lead to improvements in the wages of many part-time workers and give parents more flexibility to spend more time with their children. Unions and their members opposed this directive, calling it an attack on the incomes of their members and all workers in the U.S. economy (Swindell, Allen, & Graham-Silverman, 2003).

What Unions Achieve for Society as a Whole Sociologists and historians credit the labor movement (along with its allies in Congress) for spearheading many of the most significant reforms of American capitalism, including the forty-hour workweek, Social Security, health and welfare plans, job safety, generous pensions, and a host of other benefits that many Americans now take for granted and many others still only dream about (Bronfenbrenner, 2003). With some exceptions, unions have been in the front ranks of other social movements that have brought forth positive changes for minorities, women, the elderly, gays and lesbians,

and many others. In the case of the civil rights movement, for example, internationally respected union leaders like A. Philip Randolph, president of the Brotherhood of Sleeping Car Porters, were instrumental in bringing about many of the key rallies and events leading up to the great 1963 March on Washington at which Dr. Martin Luther King, Jr., gave his "I Have a Dream" speech.

Unions and "Free Riders" Unions play a significant role in improving wages and working conditions and reducing inequality even for employees who are not union members and, often without knowing it, are "free riding" on the efforts of union members. Because of the efforts of unionized workers in many industries, employers seek to pay something close to the "prevailing wage" in their region and industry. These wages are often set by union workers, who may have to make sacrifices, and even go on strike, to win gains in wages and benefits from their employers. Although nonunion workers in these regions and industries rarely get the same raises and improved benefits, those they do get are often the result of "free riding" on the efforts of unionized workers (Wallerstein & Moene, 2003).

Declines in U.S. Unions Despite all their positive achievements, unions in the United States have been losing membership and strength since they reached their peak numbers in the 1950s. In 2002, about 13.2 percent of U.S. workers were union members, although the percentage varies widely from the major industrial states of the Midwest, where union densities (the proportion of workers in unions) are over 20 percent, to those of the South, where union densities are generally lower than 9 percent. But union densities have been declining in other Western nations as well, as shown in Table 18.2.

The table also shows, however, that with the exception of France, the United States has the lowest union density among the nations listed. The Scandinavian nations, where social inequalities are lower than anywhere else on earth, show the highest union densities, and unions in these nations have played a major role in decreasing inequality. But if unions are so good for workers, why is their strength declining?

Deindustrialization and Globalization of Labor Much of the reduction in union density in the United States and other Western nations is due to deindustrialization and the transformation of the economy from one dominated by manufacturing to one in which services predominate, although industrial production remains in

TABLE 18.2

Union Density Trends, Selected Countries, 1960–1994

Country	1960	1974	1980	1990	1994	Change
Sweden	70.7%	71.9%	78.3%	82.4%	88.6%	16.7%
Denmark	59.6	63.8	77.5	74.5	79.5	15.7
Austria	57.8	51.7	50.8	45.2	39.4	−12.3
Norway	51.6	51.1	54.1	53.1	52.6	1.5
Australia	47.9	47.5	46.4	42.6	35.0	−12.5
Belgium	40.7	49.9	56.6	56.7	58.8	8.9
United Kingdom	44.3	50.4	52.8	40.9	33.8	−16.6
Netherlands	39.4	35.1	32.6	22.3	22.3	−12.8
Switzerland	37.0	29.2	30.6	26.3	23.3	−5.9
Germany	34.2	32.9	33.6	29.9	27.4	−5.5
Finland	31.9	63.4	70.0	72.5	78.9	15.5
Japan	31.0	33.2	29.9	24.5	23.2	−10.0
United States	28.0	24.2	20.5	14.6	13.9	−10.3
Canada	27.5	31.3	33.2	31.2	30.8	−0.5
Italy	22.4	42.7	44.4	33.6	33.2	−9.5
France	19.2	21.0	17.1	9.2	8.8	−12.2%

Note: Union density figures represent the proportion of workers who are in unions among all workers, so that union membership can be growing, as it is slowly in the United States, but the labor force can be growing at a faster rate, which results in lowering union density figures.

Source: Ebbinghaus & Visser, 2000.

a highly automated form. Manufacturing plants in major industries like steel, mining, textiles, and garment manufacturing have been moved to places where it is much more difficult, even impossible, for workers to form unions (Scruggs & Lange, 2002). Automation has also eliminated hundreds of thousands of union jobs in the United States and Europe. Automation increases the productivity of the remaining workers, and as such, it is of great value to every nation's economy, but it is always an open question whether the new jobs created in the economy will be as high paying and secure as those that are lost to globalization and automation.

Antiunionism What about those jobs that cannot be "exported" to low-wage nations and regions? Are they more likely to be unionized because the owners of capital cannot send them away? The answer is no, in large part because of the strenuous efforts of employers and their allies among conservative political leaders to oppose union representation. Compared to those in Europe, where there is greater acceptance of labor unions and collective bargaining, employers in the United States are extremely antiunion. Although major unionized employers like the auto companies have long recognized their workers' rights to bargain collectively—through the United Auto Workers and other unions—many U.S. employers, including major ones like General Electric, IBM, and Wal-Mart, have never willingly accepted unions and actively engage in efforts to prevent their workers from joining unions. In smaller nonunion companies, owners often feel that unionization of their workers deprives them of their right to run their businesses the way they wish and forces them into labor agreements that drive up the cost of making their products or delivering their services. When faced with union efforts to organize their workers, the majority threaten to move their production facilities (where feasible) to lower-wage regions of the world rather than agree to pay union wages and benefits. In situations where the owners cannot globalize their production by moving it to low-wage regions, unions are somewhat more successful in winning contracts, but in the United States, many employers simply threaten to move to antiunion regions of the country (Bronfenbrenner, 2000; Milkman & Voss, 2004).

 Thinking Critically

Opponents of labor unions often label them as representing "special interests," but unions seek to represent the broad economic and social needs of people as workers, whether they are mounting unionization drives against Wal-Mart, for example, or lobbying in Congress for an increase in the minimum wage. Do you think these activities make unions equivalent to the U.S. Chamber of Commerce or the lobbying associations of the health care industry?

False Consciousness Another important reason nonunion workers seem complacent about their lack of bargaining power at work, their lack of adequate health care benefits, and their overall passivity is that they often identify with their bosses because they want approval, rather than with other workers in the same situation who are seeking union representation. And workers in low-wage jobs, where there is a great deal of turnover of personnel, are typically reluctant to confront their supervisors with demands for better treatment, pay, or benefits. They know that if they begin inquiring about their rights or their need to have a union, they may be fired. Although these responses are understandable under the circumstances of their employment, they are also a form of *false consciousness;* that is, as workers, they are aware of their lower status and inequality, but they do not think there is much they can effectively do about it.

To make matters worse, nonunion workers tend to believe that they have rights at work that they actually do not have. In a survey of what American workers think about how much voice they should have and what rights they have in their workplaces, economist Richard Freeman and sociologist Joel Rogers found that "employees vastly exaggerate their workplace rights" (Freeman & Rogers, 2002, p. 118). Here is a list of questions they asked about this issue. How would you answer each of them?

Is it legal or illegal for your employer to:
Fire you for no reason?
Move you to a lower-paying job because you wanted to form a union?
Refuse to let you take time off from work to take care of your sick baby?
Fire you for refusing to do hazardous work?
Permanently replace you if you went on strike?
Avoid hiring minorities for "good business reasons"?

Figure 18.4 indicates that an average of 74 percent of American workers believed that three or more acts that are actually legal are against the law. Eighty-three percent believed, wrongly, that it is illegal to fire a worker for no reason; 83 percent thought it is illegal to fire a worker for refusing hazardous work. In fact, these acts are perfectly legal and are fundamental principles of laissez-faire capitalism. Extremely high proportions of workers also believe, wrongly, that illegal acts by employers, such as firing or moving someone to a lower-paying job for trying to form a union, avoiding hiring minorities for "good business reasons," or refusing family leave for a sick child (all of which are illegal according to U.S. labor laws) are legal (Freeman & Rogers, 2002).

Can the Labor Movement Rejuvenate Itself? Antiunionism, globalization and deindustrialization, and worker misunderstandings and false consciousness are some of the main reasons for the decline in union strength in the United States, but it is also true that as a social movement, labor is confronted with a negative climate due to the prevalence of far more conservative thinking in U.S. society today than in the early twentieth century. The Bush administration was opposed to doing anything that would help unions organize new members. But this was not a new situation, for many other administrations have been unsympathetic to organized labor; moreover, the movement's original victories were won in an extremely antiunion social environment during the first decades of the twentieth century. The prospects for a rejuvenated labor movement, therefore, will depend on aggressive and creative leadership in the unions (Bronfenbrenner, 2003).

One currently popular strategy is for unions to open up their membership and the benefits they can supply to workers in nonunion workplaces. This "open source" unionism can bring lower medical costs, better insurance plans, and many more benefits, short of actual union contracts, which still need to be won in local representation elections. But such a strategy can eventually convince more and more

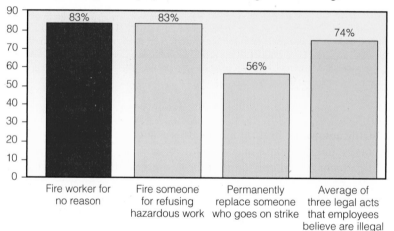

Percentage of Employees Who Believe Legal Acts are Illegal

- Fire worker for no reason: 83%
- Fire someone for refusing hazardous work: 83%
- Permanently replace someone who goes on strike: 56%
- Average of three legal acts that employees believe are illegal: 74%

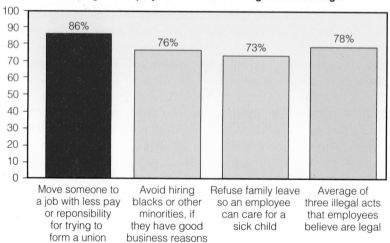

Percentage of Employees Who Believe Illegal Acts are Legal

- Move someone to a job with less pay or reponsibility for trying to form a union: 86%
- Avoid hiring blacks or other minorities, if they have good business reasons: 76%
- Refuse family leave so an employee can care for a sick child: 73%
- Average of three illegal acts that employees believe are legal: 78%

Figure 18.4 Employee Beliefs about the Legality of Selected Actions by Employers. Source: Freeman & Rogers, 2002.

workers to take on the responsibilities of helping to organize actual union membership drives in their workplaces. Successful modern labor unions like the Service Employees International Union (SEIU) are currently using such strategies to take on major antiunion employers like Wal-Mart (Milkman & Voss, 2004).

Sociological Perspectives on the Workplace

Industrial sociology is concerned with the social organization of work and the types of interactions that occur in the workplace. Like sociologists in other fields, industrial sociologists use the functionalist, conflict, and interactionist perspectives in their research. The results of that research have been used to support various approaches to labor–management relations.

Human Relations in Industry: A Functionalist Perspective The *human relations* perspective on management is associated with the research of Elton Mayo and his colleagues at Western Electric's Hawthorne plant and in aircraft and metal production plants. Their goal was to use experimental methods and observation of workers and managers on the job in an attempt to understand how the factory's formal organization and goals are affected by patterns of informal organization within the workplace. This was in sharp contrast with earlier approaches to labor–management relations, especially the *scientific management* approach developed early in the twentieth century by Frederick W. Taylor.

Scientific management (or Taylorism, as it is often called) was one of the earliest attempts to apply objective standards to management practices. After rising through the ranks from laborer to chief engineer in a large steel company, Taylor turned his attention to the study of worker productivity. He noticed that workers tend to adhere to informal norms that require them to limit their output. For example, he found that the rate of production in a machine shop was only about one-third of what might normally be expected. Taylor decided to use the authority of management to speed up the workers.

He fired stubborn men, hired "green hands" who did not know the norms of the experienced workers, and experimented with ways of breaking down the labor of each worker into its components. Every job, he claimed, could be scientifically studied to determine how it could be performed most efficiently. Such "time and motion studies," combined with piecework payment systems that induced workers to produce more because they were paid for each unit produced above a set number, became the hallmarks of scientific management (Miller & Form, 1964). Taylor's principles were quickly incorporated into the managerial practices of American businesses, but workers and their unions often resisted them (Braverman, 1974).

Mayo's experiments were intended to determine what conditions would foster the highest rates of worker productivity. His observations convinced him that increased productivity could be obtained by emphasizing teamwork among workers and managers rather than through pay incentives or changes in such variables as lighting, temperature, and rest periods (Bendix, 1974). We saw in Chapter 2 that Mayo's research showed that it was the attention given to the workers that mattered, not the various experimental conditions. But Mayo drew another, more important conclusion from the Hawthorne experiments:

> The major experimental change was introduced when those in charge sought to hold the situation humanly steady . . . by getting the cooperation of the workers. What actually happened was that six individuals became a team and the team gave itself wholeheartedly and spontaneously to cooperating in the experiment. (Mayo, 1945, pp. 72–73)

Although Mayo's research focused on the interactions between workers and managers, the human relations approach that grew out of that research can be said to represent the functionalist perspective on the workplace because it stresses the function of managerial efforts in increasing worker productivity. The functionalist perspective is also illustrated by William F. Whyte's classic study of the restaurant industry. In a small restaurant, the organization's structure is simple: "There is little division of labor. The owner and employees serve together as cooks, countermen, and dishwashers." But when the restaurant expands, various supervisory and production occupations are added to its role structure. According to Whyte, this magnifies old problems and creates new ones:

> In a large and busy restaurant a waitress may take orders from fifty to one hundred customers a day (and perhaps several times at each meal) in addition to the orders (much less frequent) she receives from her supervisor. When we add to this the problem of adjusting to service pantry workers, bartenders, and perhaps checkers, we can readily see the possibilities of emotional tension—and, in our study, we did see a number of girls break down and cry under the strain. (1948, p. 304)

Whyte discovered that tension and stress could be reduced, and customers served more happily, if the restaurant was organized so that lower-status employees were not required to give orders directly to higher-status ones. For example, waiters and waitresses should not give orders directly to cooks but should place their orders with a pantry worker or use a system of written orders. By changing the organization of statuses and roles in the restaurant (and, thus, the way they functioned as a social system), it would be possible to increase work satisfaction and output.

Conflict at Work The human relations approach just discussed seeks to improve cooperation between workers and managers in order to achieve the organization's goals. Industrial sociologists who take a conflict perspective believe that this approach automatically condones the goals of managers and fails to consider more basic causes of worker–management conflict such as class conflict (Touraine, 1998). This results in continual fine-tuning of organizational structures rather than in more thorough reforms.

In an influential study of a midwestern metal products factory, Michael Burawoy (1980) found that even after changes were made on the basis of the human relations approach, the workers continued to limit their output. The workers called this "making out" and saw it as a way of maximizing two conflicting values: their pay and their enjoyment of social relations at work. But instead of wondering why the workers refused to produce more, Burawoy asked why they worked as hard as they did. He concluded that the workers were actually playing into the hands of the managers and owners because neither they nor the unions ever questioned management's authority to control basic production decisions. The norms of the shop-floor culture, of which "making out" is an example, caused the workers to feel that they were resisting the managers' control. As a result, they did not feel a need to engage in more direct challenges to the capitalist system itself (Burawoy, 1996).

A fundamental problem for conflict sociologists is that class conflict must be shown to exist; it cannot be assumed to exist simply because Marxian theory defines workers and business owners as opposing classes. In fact, there is considerable evidence that despite their misfortunes, workers do not view their interests as automatically opposed to those of owners and managers. As workers and supervisory personnel lost their jobs in the severe recession of the early 1990s, they often expressed discontent with the nation's leadership and with company managers. Generally, however, the anguish of the unemployed tends to be turned inward, taking the form of depression rather than militant class consciousness. Class consciousness may be popularized in the lyrics of a Bruce Springsteen song—"They're closing down the textile mill/ across the railroad track/Foreman says these jobs are going boys/and they ain't coming back/ to your hometown"—but it has not produced much overt class conflict or new social movements (Bensman & Lynch, 1987).

The Debt Crisis: Consumers in Trouble

Where workers and their families are currently experiencing the greatest stress, and voicing the most anger politically, is in their relationships with banks and financial institutions (Surowiecki, 2010). Because so many people have lost their homes to foreclosure, seen the value of their pension savings and 401k plans diminish due to no fault of their own, and remain in deep debt to banks and credit card agencies, this is entirely understandable. For the overall economy, the fears workers and their families express about their economic future are translated into weak consumer demand. About two-thirds of domestic production in the United States is purchased by U.S. consumers, a fact that helps explain the intense pressure on those consumers to go to the mall or order merchandise from mail-order catalogs. But credit card and other forms of consumer debt, which

fuel the economy in good times, can become a major problem if the economy weakens and bankruptcies rise. This is especially true for people employed in part-time service work or in manufacturing, where workers are most vulnerable to plant closings and layoffs.

In 2005, Congress passed the Bankruptcy Abuse Prevention and Consumer Protection Act in response to heavy lobbying by major credit card companies and banks, which sought to stiffen bankruptcy laws and make it far more difficult for consumers to write off their debts by declaring bankruptcy. In consequence, the rate of personal bankruptcies increased by over 50 percent as consumers in debt hurried to file for bankruptcy before the new law took effect. (The act did nothing to make it more difficult for consumers to obtain credit cards.)

As Figure in 18.5 shows, nonbusiness (personal) bankruptcies soared to over 650,000 in 2006, and then fell to just over 100,000 the next year as the new law made it far more difficult to declare bankruptcy and still protect family assets. But the ensuing financial crisis has brought the volume of household financial failure to levels similar to those experienced before the new legislation, but with far greater consequences in loss of homes and savings.

Although family financial failure generates a great deal of anger and despair, much of it can be explained by financial ignorance among borrowing consumers. A recent study by Annamaria Lusardi, who studies how much people really understand about personal finance, found that a majority of borrowers did not know the terms of their mortgage or the interest rate they were paying. Yet the financial industry was (and still is in many instances) offering more and more complex loans and financing schemes. "It's like we've opened a faucet and told people they can draw as much water as they want, and it's up to them to decide when they've had enough. But we haven't given people the tools to deduce how much is too much" (quoted in Surowiecki, 2010, p. 23). The financial reforms that Congress passed in 2010 created a consumer financial protection agency, but it remains to be seen how strong its enforcement powers will be. In the meantime, many consumers are increasingly wary of lending agencies and justifiably dubious about the honesty of professional bankers—an attitude that raises further questions about the future of professions in the U.S. economy.

Professions: An Interactionist Perspective

Factory workers and other low-status employees are not the only subjects of research by sociologists who study work and employment. Because

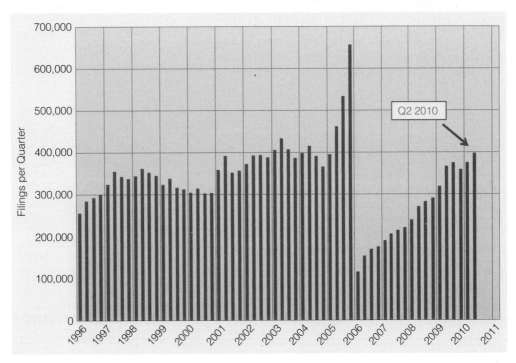

Figure 18.5 Nonbusiness U.S. Bankruptcy Filings, by Quarter. Source: www.calculatedriskblog.com.

they are a growing segment of modern economic institutions, professionals have also been studied extensively. A **profession** is an occupation with a body of recognized knowledge and a developed intellectual technique. Its knowledge is transmitted by a formal educational process, and entry into the profession is regulated by testing procedures. A code of professional ethics governs each profession and regulates relations with colleagues, clients, and the public. Most professions are licensed by the state (Cant & Sharma, 1998; Montagna, 1977). This definition could be applied to a variety of occupations. For example, doctors and attorneys have long been considered professionals, but what about nurses and stockbrokers? Sociologists have addressed this question in their research on the "professionalization" of occupations—that is, on the way different occupations attempt to achieve the status of professions.

According to Everett C. Hughes (1959), a profession is a set of role relations between "experts" and "clients," in which the professional is an expert who offers knowledge and judgments to clients. The professional often must assume the burden of the client's "guilty knowledge": The lawyer must keep secret the client's transgressions, and the physician must hide any knowledge of the sexual behavior or drug use of famous and not-so-famous patients. Professions thus "rest on some bargain about receiving, guarding, and giving out communications" (Hughes, 1959, p. 449; see also Abbott, 1993).

Interactionist sociologists pay special attention to the role relations that develop in various professions. They are particularly interested in the processes of *professional socialization*—learning the profession's formal and informal norms. For example, in a well-known study of professional socialization in medical schools, Howard Becker and his colleagues (1961) found that, in the later years of their training, medical students' attention is increasingly devoted to learning the informal norms of medical practice. The formal science they learn in class is supplemented by the practical knowledge of the working doctor, and the student's skill at interacting with higher-status colleagues becomes extremely important to achieving the status of a professional physician. In a highly acclaimed study on the same subject, Charles Bosk (1979) showed that resident surgeons in a teaching hospital are forgiven if they make mistakes that the older surgeons believe to be "normal" aspects of the learning process, but not when their mistakes are repeated or are thought to be caused by carelessness. The point of these and similar studies is that role performance is learned through interaction. Role expectations vary from one profession to another, as do the ways in which members of the profession experience these expectations (Abbott, 1988).

Our discussion of professions can serve as a reminder of a basic characteristic of economic institutions, one that we noted early in the chapter: Economic institutions are concerned with allocating scarce resources to satisfy competing ends. Thus, just as the guilds of feudal societies controlled access to skilled trades in medieval times, professional associations like the American Medical Association attempt to control access to the professions. Efforts by nation-states to regulate markets and trade, and efforts by managers to increase the productivity of workers, are all concerned with achieving the most advantageous use of scarce resources. This fundamental fact is at the heart of the sociological study of economic institutions.

profession: An occupation with a body of knowledge and a developed intellectual technique that are transmitted by a formal educational process and testing procedures.

HOW PEOPLE FIND JOBS

How do people find their jobs? Do they rely on the want ads? Family contacts? College placement services? From a strictly economic viewpoint, the question is answered by relying on market mechanisms. Employers seek workers at given wages for particular kinds of work. Potential employees seek employers who need their skills and will pay the highest rate for their services. The supply and demand for particular services will explain changes in the prevailing wages for those services. But how do people actually learn about available jobs, and how do they obtain them once they know about them? Market models of job-seeking behavior often assume that there is a perfect flow of information. In fact, however, sociological research uncovers some surprising gaps in the flow of information and reveals that both employers and employees actually behave in ways that are not the most rational from an economic viewpoint.

Mark Granovetter (1995/1974), one of the nation's foremost economic sociologists, is largely responsible for discovering what is now known as "the strength of weak ties." This discovery turns out to be a key to explaining how a large proportion of people actually go about finding their jobs. By "weak ties," Granovetter refers to people who are part of our social networks even if we are not always aware that they are. They are acquaintances or friends of acquaintances. It turns out that these weak ties, as opposed to stronger ties of kinship, can be essential in the success of job searches. Here is how Granovetter formulated his research problem and collected the data for his pioneering research.

While a graduate student in sociology at Harvard, Granovetter developed an interest in chains of contact and flows of information through them. His mentor, the economic sociologist Harrison White (1970), had made some important discoveries about how information in markets flows through chains of acquaintance, and Granovetter wanted to extend this research to labor market and job-seeking behavior. He decided to interview a sample of prospective employees to determine whether they also relied on chains of acquaintance within social networks to help them find jobs. To reduce the possible variance among different occupations, and because he could not afford to interview more than about 280 respondents, Granovetter decided to limit his sample to professional, technical, and managerial workers. He conducted the interviews in the Boston suburb of Newton, obtaining the names of eligible male respondents through a commercial directory service that lists residence, occupation, and related information for each entry. (In subsequent research, he broadened his study to include women.) In collecting the data, he combined in-person interviews with career histories and basic demographic information—age, education, race or ethnicity, place of birth, and so on.

The first surprises in Granovetter's research came as he looked at the data about which job-finding methods worked for which types of job seekers. Examples of these findings from his original survey are presented in Tables 18.3 and 18.4. In no case do formal job search methods—that is, the use of an impersonal intermediary (newspaper ad, placement service, employment agency, professional association, and the like)—account for the majority of job placements. Direct application, in which the applicant shows up at the employer's office without an appointment, and personal contacts through social networks are more important than formal searches, especially at the middle levels of the salary scale. Table 18.3 shows that 154 of 275 respondents (56 percent) found their jobs through personal

TABLE 18.3

Level of Income of Respondent in Present Job, by Job-Finding Method Used, in Percentages

| Income | Method Used | | | | Total |
	Formal Means	Personal Contacts	Direct Application	Other	
Less than $10,000	28.0%	22.7%	50.0%	5.3%	27.6%
$10,000–14,999	42.0%	31.8%	30.8%	26.3%	33.1%
$15,000–24,999	24.0%	31.2%	15.4%	52.6%	28.4%
$25,000 or more	6.0%	14.3%	3.8%	15.8%	10.9%
N	50	154	52	19	275

Source: Adapted from *Getting A Job*, 2nd edition, by M. Granovetter, copyright © 1995/1974. Used with permission of the University of Chicago Press.

TABLE 18.4

Origin of Job, by Job-Finding Method of Respondent, in Percentages

Origin of Job	Method Used				Total
	Formal Means	Personal Contacts	Direct Application	Other	
Direct replacement	47.1%	40.5%	58.0%	38.9%	44.9%
Added on	31.4%	15.7%	18.0%	27.8%	19.9%
Newly created	21.6%	43.8%	24.0%	33.3%	35.3%
N	51	153	50	18	272

Note: Percentages may not add up to 100 because of rounding.

Source: Adapted from Granovetter, 1995. *Getting A Job*, 2nd edition. Copyright 1995 University of Chicago Press. Used with permission of the University of Chicago Press.

contacts, clearly the most effective of the three methods of job seeking. Table 18.4 shows that personal contacts are particularly important when the position being filled is newly created. Indeed, the lengthy interviews Granovetter conducted with his respondents revealed a pattern in which a personal contact mentioned a job that another contact in a certain company had mentioned as just being created—thus providing the job seeker with "insider information" in seeking the position.

Many readers may think that these findings are not actually surprising. Don't we often hear it said that "it's not what you know, but who you know" that explains how people succeed in getting jobs? Indeed, this is a time-honored cliché—but Granovetter explored its meaning more fully. After all, "who you know" should include family members and close kin. Yet family and close kin contacts and referrals accounted for a minority of the personal referrals that paid off in terms of obtaining a job. Far more important were links between friends and acquaintances. The following three cases from Granovetter's original study illustrate how weak ties actually work:

> George C. was working as a technician for an electrical firm, with a salary of about $8,000 and little apparent chance for advancement. While courting his future wife, he met her downstairs neighbor, the manager of a candy shop, a concession leased from a national chain. After they were married, Mr. C. continued to see him when visiting his mother-in-law. The neighbor finally talked him into entering a trainee program for the chain, and arranged an interview for him. Within three years, Mr. C. was earning nearly $30,000 in this business.

> Herman D. was the owner of a fruit and vegetable store, which he sold (at age 45) because of ill health. He took a vacation; meanwhile his brother, a business executive, attended a meeting where a colleague mentioned that he was looking for someone to do inventory management.

Mr. D. had done similar work before buying his store, and his brother therefore suggested him. He was hired several days later.

> Gerald F. was a salesman for a wholesale liquor distributor. A friend who was a doctor asked him if he would be interested in managing a nursing home, and if so, to put together a résumé. One of the references Mr. F. used for the résumé was his wife's cousin, owner of a fashionable antique shop. When the nursing home job didn't come through, the wife's cousin, now aware that Mr. F. was considering changing jobs, offered him a job as business manager of his shop, which he accepted. (Granovetter, 1995/1974, p. 49)

These are not the family ties that one often associates with "who you know." In fact, another surprising finding of Granovetter's research is that it takes on average about two or more contacts in the chain of network associations to get the job applicant to appear as a likely candidate for a prospective employer, so it is not the direct contact but the contact through network chains that makes the difference in a successful job search.

One implication of this research is that people who are on the margins of society because of their race, gender, or social class are less likely to have the network resources to draw upon when looking for a job. Granovetter used this finding to argue in favor of affirmative action in hiring. African Americans and other groups that have experienced systematic exclusion from particular types of jobs and careers have plenty of personal contacts and social networks, Granovetter explains. However, because they are underrepresented in the existing structures of employment, these contacts do not produce chains of referral into these careers. Once a core of formerly excluded people are included, largely through intentional efforts like affirmative action, the normal network system of weak ties will work for them as it does for others.

Crafts Traditions in Modern Economies

As these photos show, crafts traditions are flourishing. The Indian basket maker whom we met at the beginning of the chapter is not alone. In fact, in industrial nations like England and the United States, many people are engrossed in using their creative powers to make traditional and modern craft objects of all kinds. Millions of people throughout the industrial world are facing the same problem as the Indian basket weaver: how to market their goods while maintaining their pride in their skills and their love of the process.

Contrary to the dire predictions of many social critics and early sociologists, such as Karl Marx and Max Weber, who lived more than a century ago during the earlier phases of capitalism, the enchantment people find in creating arts and crafts has not withered in the face of mass production and mass marketing. On the contrary, the easy availability of mass-produced goods seems to have made people even more eager to buy unique objects made by people who are carrying out enduring cultural traditions. And global forms of marketing like the World Wide Web are proving to be extremely powerful institutions for disseminating information about craftspeople and their products.

Although craftspeople will always face an uphill struggle to make a living from their work, particularly in traditional societies, we see proof that the struggle can pay off. Early in the era of industrial capitalism, William Morris, a man with keen sociological insights and a passion for arts and crafts, started what became known as the arts and crafts movement. Morris, a contemporary of Marx, was deeply concerned that industrial expansion and urbanization would wipe out craft traditions throughout the world. If he were alive today, he would still be concerned about the loss of traditional skills and techniques for producing baskets and other objects of everyday life, but he would be extremely gratified to see the burgeoning activity in this area of life—which we can observe by simply browsing the Web.

William Morris.

SUMMARY

How do markets work, and why do they play such a powerful role in shaping the growth of societies?

- *Economics* is the study of how individuals and societies choose to employ scarce resources to produce various commodities and distribute them among various groups in the society.

- Sociologists are also concerned with how individuals and societies make such choices, but much of their research focuses on how cultural norms affect those choices.

- The main subjects of sociological research in this area are markets and the division of labor, the interactions between government and economic institutions, and the nature of jobs and professions.

What are the basic approaches that different nations employ to try to balance economic growth with social justice?

- A hallmark of an industrial society is the production of commodities and services to be exchanged in markets. A *market* is an economic institution that regulates exchange behavior.

- In a market, different values are established for particular goods and services; those values are usually expressed in terms of a common measure of exchange, or currency. Market transactions are governed by agreements, or contracts, in which a seller agrees to supply a particular item and a buyer agrees to pay for it.

- The spread of markets throughout the world began in the late fifteenth century as a result of the development of new *technologies* that facilitated trade.

- Today, world markets are dominated by *multinational corporations,* economic enterprises that have headquarters in one country and conduct business activities in one or more other countries.

- An important modern trend is economic globalization, a worldwide shrinkage of economic distances between nations.

How are states and economic institutions interrelated in major political economic systems?

- The fate of societies throughout the world has been strongly influenced by the economic ideologies of mercantilism, capitalism, and socialism.

- *Mercantilism* held that the wealth of a nation could be measured by its holdings of gold, so the best economic system was one that increased the nation's exports and thereby increased its holdings of gold.

- *Laissez-faire capitalism,* on the other hand, argued that a society's wealth could be measured only by its capacity to produce goods and services—that is, its resources of land, labor, and machinery. The major institutions of capitalism are private property and free markets.

- *Socialism* arose out of the belief that private property and personal profit should be replaced by public ownership of property and sharing of profits. According to Marx, the socialist state would be controlled by the workers, who would determine what should be produced and how it should be distributed.

- Soviet-style socialist societies were characterized until recently by command economies, in which the state commands economic institutions to supply a specific amount of each product and to sell it at a particular price.

- *Democratic socialism* holds that private property must continue to exist but that large corporations should be owned by the nation or, if they are in private hands, required to be run for the benefit of all citizens.

- In *welfare capitalism,* markets determine what goods and services will be produced and how, but the government regulates economic competition. Welfare capitalism also asserts that the state should invest in the society's human resources through policies promoting education, health care, and social welfare.

- In the United States, the transition from an economy based on the production of goods to one based on the provision of services has resulted in the displacement of thousands of skilled workers into lower-paying jobs in the service sector. These trends appear to signal a change in the basic social contract in which individuals who work hard can expect increases in their living standards.

What are the dominant dimensions of conflict and change among workers, managers, and corporations?

- Although much progress has been made, gender discrimination in the workplace has not been eliminated.

- The concept of *alienation* refers to the feeling of being powerless to control one's own destiny. At work, people may feel alienated because their labor is divided up into activities that are meaningless to them.

- Mentoring programs have been found to raise employee morale and prevent burnout.

- Labor unions promote and defend the interests of workers in the private and public sectors of the economy. Workers represented by labor unions earn more pay and receive better insurance and pension benefits than workers in nonunion jobs.

- Unions in the United States have been losing membership and strength since they reached their peak numbers in the 1950s. Much of the reduction in union density is due to deindustrialization and the transformation of the economy from one dominated by manufacturing to one in which services predominate. Other factors are antiunion attitudes among employers and "false consciousness" (identification with the employer's interests) among workers.

What are some important sociological theories about workers and the workplace?

- The scientific management approach to labor–management relations attempted to increase productivity by determining how each job could be performed most efficiently and by using piecework payment systems to induce workers to produce more.

- Efforts to determine what conditions would result in the highest rates of worker productivity led to the recognition that cooperation between workers and managers is an important ingredient in worker satisfaction and output. A result of this recognition was the human relations approach to management, which seeks to improve cooperation between workers and managers.

- Conflict theorists believe that the human relations approach fails to consider the basic causes of worker–management conflict. They study how social class and status both at work and outside the workplace influence relations between workers and managers.

- Interactionist theorists have devoted considerable study to professionalization, or the way in which occupations attempt to gain the status of *professions,* and to the processes of professional socialization (that is, learning the formal and informal norms of the profession).

What findings of sociological research help explain how people find work and jobs?

- Recent research has found that formal job search methods—that is, the use of an impersonal intermediary (newspaper ad, placement service, employment agency, professional association, and the like)— do not account for the majority of job placements. Direct application, in which the applicant shows up at the employer's office without an appointment, and personal contacts through social networks are more important than formal searches, especially at the middle levels of the salary scale.

The Kornblum Companion Website

www.cengagebrain.com

Supplement your review of this chapter by going to the companion website to take one of the tutorial quizzes, use the flash cards to master key terms, and check out the many other study aids you'll find there. You'll also find special features, such as GSS data and Web links that will put data and resources at your fingertips to help you with that special project or to do some research on your own.

19

The Nature
of Politics
and Political
Institutions

The Nation-State
in Crisis

Political
Institutions:
A Global View

Perspectives
on Political
Institutions

Military
Institutions

Politics and Political Institutions

FOCUS QUESTIONS

How are a society's political institutions shaped by its ideas of power and legitimate authority?

Why has globalization created crises for nation-states?

What are the main characteristics of different types of political regimes in the world today?

How do the basic sociological perspectives contribute to our understanding of political sociology?

What are the particular political challenges of the modern military in a democratic society?

W e are living in a particularly trying period of global politics, in which unstable governments, rife with corruption and ethnic animosities, fail to realize the ideals of democratic self-government and the rule of law. Even the most stable of democratic nations, including the United States, find their electorates deeply divided and mistrustful of their elected leaders. But the most immediate political crises are found in nations where the threat of terrorism has brought war and deadly unrest. In Iraq and Afghanistan, in particular, troops from the United States and other nations are attempting to defeat the Taliban, al Qaeda, and related terrorist organizations. To do so requires that the populations of these war-torn nations can feel secure and believe in their leaders' genuine desire to establish the rule of law and legitimate political leadership. At this writing, however, corrupt or extremely close election results in both Iraq and Afghanistan have cast grave doubts on the ability of the central government to actually govern with the consent of the voters.

Because of its invasion and occupation of Iraq, most world leaders consider the United States to be responsible for the current situation. The "war on terror" initiated by the Bush administration after the terrorist attacks of September 11, 2001, led to major commitments of U.S. military and civilian personnel in Iraq, Afghanistan, and elsewhere in the world, but especially in the Middle East. The Obama administration has begun withdrawing forces from Iraq, under conditions of grave political uncertainty there, while it has increased U.S. and international troop commitments in Afghanistan.

The American troops and their allies are working day and night to help the Iraqis build democratic institutions based on a constitution that the people feel is legitimate. However, this experiment in building democracy during a military occupation is full of risks to all parties. The creation of stable democratic political institutions in Iraq (and Afghanistan as well) must occur despite high

SHAH MARAI/AFP/Getty Images

levels of tensions and mistrust, not only between those nations' citizens and the Americans in their midst, but also among long-established religious, ethnic, and regional factions.

As the world watches, the stakes of this political experiment increase. Social scientists and experts on democracy note that it is not easy to manufacture the conditions for liberal democracy. No quick fix replaces the hard work of building trust in laws, establishing checks and balances, and encouraging civil debate. Recent attempts to impose democracy in countries such as Cambodia, Bosnia, and Angola have failed dismally (Carnegie Endowment for International Peace, 2003; Zakaria, 2003). Once a major effort at building democracy is under way, as it is in Iraq, much depends on the hard, slow work of getting groups and leaders together to create the specific ruling institutions that will make the experiment work for Iraq's people. That will be the true test of whether the political experiment is a success or a failure. In Afghanistan, the United States and its allies have scaled back their hopes for building democratic political institutions that would replace tribal and feudal forms of rule. Instead, they are recruiting local leaders and local police forces who can bring stability and security in areas terrorized by the Taliban. At this writing, the outcome of these efforts is impossible to predict.

It is appropriate to ask what right the United States has to be conducting military and civilian operations in other nations in the first place. This is where the role of international peacekeeping institutions, particularly the United Nations, is of central importance. In 2010, the United Nations Security Council passed a resolution that reaffirmed earlier resolutions commissioning military operations against the Taliban by a force consisting of troops provided by the United States, Britain, and other nations: "Supporting international efforts to root out terrorism, in keeping with the Charter of the United Nations, . . . Recognizing the urgency of the security and political situation in Afghanistan in light of the most recent developments, particularly in Kabul, condemning the Taliban for allowing Afghanistan to be used as a base for the export of terrorism by the Al-Qaida network and other terrorist groups and for providing safe haven to Osama Bin Laden, Al-Qaida, and others associated with them, and in this context supporting the efforts of the Afghan people to replace the Taliban regime." In other words, the community of world nations is on record as opposing any right of a sovereign nation, Afghanistan in this case, to shelter and allow the export of international terrorism. Despite such efforts, it is extremely difficult to ensure security and establish the rule of law in a land where the central government cannot adequately govern and where loyalties to local warlords and to one's tribe are stronger than attachments to the nation and its political institutions.

THE NATURE OF POLITICS AND POLITICAL INSTITUTIONS

"Politics," Max Weber liked to tell his students, "is the slow boring of hard boards." By this, he meant that political change is almost never achieved easily. Creating new political institutions or changing old ones, even just changing the leadership of existing institutions, usually requires years of effort. There will be endless meetings, ideological debates, fund-raising, negotiations, and campaigning. At times, however, especially during times of revolutionary social change, the pace of political change is fast and furious.

We are living through a period of severe crisis and dramatic change in world politics. It is not at all clear that the democratic nation-state, the rule of law, and guarantees of the rights of minorities can be established in many parts of the world (Kaplan, 2001). In the view of many political sociologists, the twentieth century ended with the world in a state of political unrest and nationalistic conflict much like that at the end of the nineteenth century (Abu-Lugod, 1996; Feldman, 2004). This uncertainty in world and national politics makes the comparative study of political institutions a major growth area in sociology.

Well before the invasion of Iraq in 2003, the fall of the Berlin Wall in 1989 symbolized the beginning of a new era in world politics. The cold war between the former Soviet Union and the Western capitalist democracies had dominated the politics of many parts of the world for more than forty years. Although many nations, including the United States, still feel the effects of the arms race that accompanied the competition between the superpowers, with the end of the cold war has come an end to the nuclear "balance of terror" and to the suppression of political expression within the former Soviet empire. Nationality groups in the former Soviet republics are asserting their desire for independence or, at least, for protection of their rights within a multiethnic state.

As we have seen in Iraq since the American occupation began, the removal of an authoritarian regime may leave a nation without an effective governing authority to restrain populations who yearn to form their own independent nation or who covet the lands and resources of their neighbors. This has frequently been the case as well in regions of the world that were formerly dominated by authoritarian communist parties, as can be seen in Chechnya and other former "satellite republics" of the Soviet Union where ethnic strife threatens to create chaos and civil war (Fukuyama, 2006). Political upheavals in the former Yugoslavia, Africa, the Middle East, and India and Pakistan; the struggle to transform South Africa into a truly democratic state; and the rise of militias and other violently antigovernment hate groups in the United States—all these changes and others promise to make the comparative study of politics a central area of social-scientific research for years to come.

We can gain much insight into what the nationality groups in eastern Europe, the black majority in South Africa, and others who are trying to effect political change are experiencing by looking more carefully at politics and political institutions and the place they occupy in national life. This section of the chapter reviews some of the essential concepts of politics, with special emphasis on power, authority, and legitimacy.

The second section explores the evolving nature of the nation-state, and the third section describes the political institutions that are typical of modern nations and compares the political regimes of states and territories throughout the world. The fourth section explores the major theories that have been proposed to explain why political institutions operate as they do and how they change or resist change. In the final section, we look at the role of the military in the political life of contemporary nations.

Sociology & Social Justice

BASIC HUMAN RIGHTS Before looking more closely at the key features of politics and political institutions, it is important to understand the relationship of political institutions to broad issues of human rights. The unprecedented and systematic human rights abuses committed during World War II, including the Nazi genocide of Jews, Roma (gypsies), and other groups, paved the way for the subsequent Nürnberg trials, in which many of those responsible for the atrocities were tried and sentenced. Those events also signaled the need to hold future perpetrators of such atrocities accountable for their action. As the United Nations (UN) was being created by nations exhausted and demoralized by two World Wars, world leaders agreed that the new international body needed to be clear about where it would stand on issues of human rights. Following is the preamble to the United Nations General Assembly's Universal Declaration of Human Rights resolution 217 A (III), issued on December 10, 1948 (Reprinted by permission of the United Nations):

> The General Assembly recognizes that the inherent dignity and the equal and inalienable rights of all members of the human family is the foundation of freedom, justice and peace in the world, human rights should be protected by the rule of law, friendly relations between nations must be fostered, the peoples of the UN have affirmed their faith in human rights, the dignity and the worth of the human person, the equal rights of men and women and are determined to promote social progress, better standards of life and larger freedom and have promised to promote human rights and a common understanding of these rights.

Table 19.1 presents a more detailed listing of the actual tenets of the declaration. The original draft, written in more legal language than this summary, was created by a committee that was chaired by Eleanor Roosevelt and included members from Lebanon, China, France, and Canada. And although the declaration does not suggest that the United Nations actually has the power to enforce these basic human rights, it does commit the community of nations to continual efforts to do so, whenever possible, through diplomacy and other peaceful means.

What is the relationship between human rights and social justice? A quick reading of the thirty statements of human rights should make it clear that the two are the same in many instances. For example, the ideas that "everyone has the right to be treated equally under the law" and that "the law is the same for everyone" are basic to any understanding of the rule of law. Similarly, the idea that "no one has the right to hurt you or to torture you" establishes a universal criterion for what we mean by security as a human right.

We are far from realizing the goals of the declaration. But does that mean we are incapable of doing so? In the era of kings and queens, there was widespread agreement that average people, "the common folk," could never govern themselves as well as the royalty could. During the American Civil War, there was equally widespread speculation that freed slaves could never be competent citizens, but today those falsehoods have been disproved many times over. We do great wrong by misunderstanding our capabilities for self-rule, but we also deny the dream of social justice by failing to understand the enormous difficulties involved in creating just and well-managed political institutions.

TABLE 19.1

Summary of the Universal Declaration of Human Rights

1. Everyone is free and we should all be treated in the same way.
2. Everyone is equal despite differences in skin color, sex, religion, language, and the like.
3. Everyone has the right to life and to live in freedom and safety.
4. No one has the right to treat you as a slave nor should you make anyone your slave.
5. No one has the right to hurt you or to torture you.
6. Everyone has the right to be treated equally by the law.
7. The law is the same for everyone, it should be applied in the same way to all.
8. Everyone has the right to ask for legal help when their rights are not respected.
9. No one has the right to imprison you unjustly or expel you from your own country.
10. Everyone has the right to a fair and public trial.
11. Everyone should be considered innocent until guilt is proved.
12. Everyone has the right to ask for help if someone tries to harm you, but no one can enter your home, open your letters, or bother you or your family without a good reason.
13. Everyone has the right to travel as they wish.
14. Everyone has the right to go to another country and ask for protection if they are being persecuted or are in danger of being persecuted.
15. Everyone has the right to belong to a country. No one has the right to prevent you from belonging to another country if you wish to.
16. Everyone has the right to marry and have a family.
17. Everyone has the right to own property and possessions.
18. Everyone has the right to practice and observe all aspects of their own religion and change their religion if they want to.
19. Everyone has the right to say what they think and to give and receive information.
20. Everyone has the right to take part in meetings and to join associations in a peaceful way.
21. Everyone has the right to help choose and take part in the government of their country.
22. Everyone has the right to social security and to opportunities to develop their skills.
23. Everyone has the right to work for a fair wage in a safe environment and to join a trade union.
24. Everyone has the right to rest and leisure.
25. Everyone has the right to an adequate standard of living and medical help if they are ill.
26. Everyone has the right to go to school.
27. Everyone has the right to share in their community's cultural life.
28. Everyone must respect the "social order" that is necessary for all these rights to be available.
29. Everyone must respect the rights of others, the community, and public property.
30. No one has the right to take away any of the rights in this declaration.

Eleanor Roosevelt holding a poster of the Universal Declaration of Human Rights.

Politics, Power, and Authority

In any society, Harold Lasswell (1936) argued, politics determines "who gets what, when, and how." Different societies develop their own political institutions, but everywhere the basis of politics is competition for power. **Power** is the ability to control the behavior of others, even against their will. To be powerful is to be able to have your way even when others resist (Mills, 1959). The criminal's power may come through a gun and the threat of injury, but many people wield power over others without any threat of violence, merely through the agreement of the governed. We call such power *authority*. **Authority** is institutionalized power—power whose exercise is governed by the norms and statuses of

organizations. Those norms and statuses specify who can have authority, how much authority is attached to different statuses, and the conditions under which that authority can be exercised.

A **political institution** is a set of norms and statuses that specializes in the exercise of power and authority. The complex set of political institutions—judicial, executive, and legislative—that operate throughout a society form the **state** (see Figure 19.1). This chapter will concentrate

power: The ability to control the behavior of others, even against their will.

authority: Power whose exercise is governed by the norms and statuses of organizations.

political institution: A set of norms and statuses pertaining to the exercise of power and authority.

state: The set of political institutions operating in a particular society.

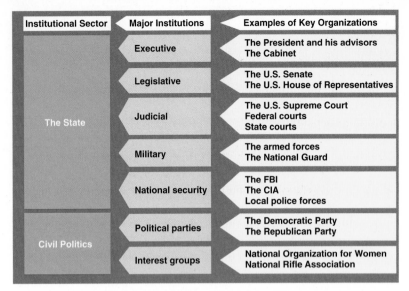

Institutional Sector	Major Institutions	Examples of Key Organizations
The State	Executive	The President and his advisors The Cabinet
	Legislative	The U.S. Senate The U.S. House of Representatives
	Judicial	The U.S. Supreme Court Federal courts State courts
	Military	The armed forces The National Guard
	National security	The FBI The CIA Local police forces
Civil Politics	Political parties	The Democratic Party The Republican Party
	Interest groups	National Organization for Women National Rifle Association

Figure 19.1 Political and Military Institutions of the United States.

primarily on these explicitly political institutions, but keep in mind that in modern societies many institutions that are active in politics are not part of the state. Labor unions, for example, are economic institutions because they represent their members in bargaining with business owners and managers, but they play an active role in politics when they support particular political candidates or when they lobby for government-funded benefits for workers and their families (Fitzgerald, 2006).

When we look at the ways in which conflicts over scarce resources such as wealth, power, and prestige occur, we are looking at politics. From this perspective, there can be a politics of the family, in which its members vie for attention, respect, use of the family car, money, and other resources. There can be a politics of the school, in which teachers compete for benefits such as smaller class size or better students. And the politics of government at all levels of society determines how the society's resources are allocated among various groups or classes. This view of politics, which looks at competition for power and at conflict over the use of power in a variety of settings, was reflected in the works of Karl Marx and Max Weber, two pioneering sociologists who devoted much of their work to the analysis of politics and change in political institutions. According to Marx, the way power is distributed in a society's institutions is a feature of its system of stratification. People and groups that have power in economic institutions are most likely to have power in political institutions as well. However, much research in political sociology is devoted to the analysis of competition for power in political parties and electoral campaigns (Lipset, 1998). A central problem of political sociology is whether the electoral process can sustain people's belief in legitimate authority in a society, despite inequalities in power and wealth (Archer, 2010).

Legitimacy and Authority

A basic dilemma that every political institution must solve is how to exercise legitimate authority—that is, how to govern with the consent and goodwill of the governed. "You can't sit on bayonets," goes an old political

legitimacy: The ability of a society to engender and maintain the belief that the existing political institutions are the most appropriate for that society.

traditional authority: Authority that is hereditary and is legitimated by traditional values, particularly people's idea of the sacred.

expression. This is a way of saying that although a state can exercise its power through the use of police force or coercion, the use of force will not be sufficient to govern a society in the long run. When the Soviet Union abandoned the use of force to control its satellite states in eastern Europe, the governments of those nations (Poland, Czechoslovakia, East Germany, Romania) were soon toppled by popular revolts. The communist parties in those nations had governed for too long without the consent of the governed—that is, without legitimacy.

But why do people consent to be governed without the use of force? For political sociologists since Weber's time, the answer to this question begins with the concept of legitimacy. As defined by political sociologist Seymour Martin Lipset (1981), **legitimacy** is the capacity of a society "to engender and maintain the belief that the existing political institutions are the most appropriate for the society" (p. 64). In other words, legitimacy results from citizens' belief in the norms that specify how power is to be exercised in their society. Even if they disagree with some aspects of their political institutions or dislike their current leaders, they still hold to an underlying belief in their political system.

Traditional, Charismatic, and Legal Authority Weber (1947) recognized that shared beliefs about legitimate authority may differ widely in different groups or societies. These differences result in three basic types of authority—traditional, charismatic, and legal—that have important effects on the way societies cope with social change. (See the For Review chart on page 493.)

1. *Traditional authority.* In tribal and feudal societies, Weber observed, **traditional authority** prevails. The rulers usually attain authority through heredity, by succeeding to power within a ruling family or clan. Traditional authority is legitimated by the people's idea of the sacred, and traditional leaders are thought to derive their authority from God. The absolute monarchs of Europe, for example, were believed to rule by "divine right." Traditional leaders usually embrace all the traditional values of the people they rule. This gives immense scope to their authority but makes it difficult for them and their systems of government to adapt to social changes that challenge those values. Thus the absolute monarchs of Europe had difficulty adapting their methods of governing to the growing power of new social classes like the

proletariat and the bourgeoisie. The traditional chiefs of African societies today find it difficult to retain their authority as people in those societies gain more education and wealth and abandon their traditional village and agrarian way of life.

2. *Charismatic authority.* **Charismatic authority**, Weber's second type, finds its legitimation in people's belief that their leader has God-given powers to lead them in new directions. A charismatic leader, such as Jesus or Joan of Arc, comes to power not through hereditary succession but through a personal calling, often claimed to be inspired by supernatural powers. The charismatic leader usually develops a new social movement that challenges older traditions. Charismatic leaders may appear in all areas of life, "as prophets in religion, demagogues in politics, and heroes in battle. Charismatic authority generally functions as a revolutionary force, rejecting the traditional values and rebelling against the established order" (Blau & Scott, 1962, pp. 30–31).

Charismatic authority tends to bring about important social changes but is usually unstable because it depends on the influence of a single leader. Moreover, as noted in Chapter 9, charisma tends to become institutionalized as the charismatic leader's followers attempt to build the values and norms of their movement into an organization with a structure of statuses and roles and its own traditions. This process occurred in the early history of Christianity as the disciples built organizations that embodied the spirit and teachings of Christ.

3. *Legal authority.* Weber's third type of authority, **legal authority**, is legitimated by people's belief in the supremacy of the law. This type of authority "assumes the existence of a formally established body of social norms designed to organize conduct for the rational pursuit of specified goals. In such a system obedience is owed not to a person—whether a traditional chief or a charismatic leader—but to a set of impersonal principles" (Blau & Scott, 1962, p. 31). The constitutions of governments establish legal forms of authority. All formal organizations, not only modern nation-states, but also factories, schools, military regiments, and so on, have legal forms of authority.

Ideally, legal forms of authority adhere to the principle of "government of laws, not of individuals." We may be loyal to a leader, but that leader,

Weber's Three Types of Authority

Type	Definition	Example
Traditional Authority	Authority that is hereditary and is legitimated by traditional values, particularly people's idea of the sacred.	Tribal chiefs; absolute monarchs
Charismatic Authority	Authority that comes to an individual through a personal calling, often claimed to be inspired by supernatural powers, and is legitimated by people's belief that the leader does indeed have God-given powers.	Joan of Arc; Mahatma Gandhi
Legal Authority	Authority that is legitimated by people's belief in the supremacy of the law; obedience is owed not to a person but to a set of impersonal principles.	Presidents; prime ministers

in turn, must adhere to the laws or regulations that establish the rights and obligations of the leader's office. In practice, individual leaders often abuse their office by using their authority for private gain, as we have seen in cases of government corruption ranging from Watergate to the Enron scandal. But legal systems and the rule of law cannot maintain their legitimacy for long if official wrongdoing is not punished (Breiner, 1996; Spohn, 2010).

Of Weber's three basic types, legal systems of authority are the most adaptable to social change. In legal systems, when the governed and their leaders face a crisis, they can modify the laws to adapt to new conditions. Such modifications do not always occur easily or without political conflict, however, especially when momentous changes are under way.

Political Culture Because the stability of political institutions depends so directly on the beliefs of citizens, we can readily see that political institutions are supported by cultural norms, values, and symbols like the Statue of Liberty. These are commonly referred to as the society's *political culture*. When we look at other societies, it is clear that their political cultures differ significantly from ours. Americans justify a political system based on competitive elections by invoking the values of citizen participation in politics and equality of political opportunity. Chinese Communist Party leaders justify their single-party political system by downplaying the value of citizen participation in politics and asserting the need for firm leaders and a centralized state that will hold the society together and reduce inequalities. Because of widespread skepticism about the ability of most people to participate in the political process, political sociologists believe it will be difficult to promote democracy and the rule of law in many parts of the world (Spohn, 2010).

Nationalism Perhaps the strongest and most dangerous political force in the world today is nationalism. **Nationalism** is the belief of a people that they have the right and the duty to constitute themselves as a nation-state

charismatic authority: Authority that comes to an individual through a personal calling, often claimed to be inspired by supernatural powers, and is legitimated by people's belief that the leader does indeed have God-given powers.

legal authority: Authority that is legitimated by people's belief in the supremacy of the law; obedience is owed not to a person but to a set of impersonal principles.

nationalism: The belief of a people that they have the right and the duty to constitute themselves as a nation-state.

The Lincoln Memorial is a shrine not only to a great president but also to the essential values of American political culture. It evokes feelings of reverence for the ideas of national unity and the rule of law.

(Hechter, 2000). Religion, language, and a history of immigration or oppression are among the shared experiences that can cause people to feel that they ought to have their own state. The situation of the French-speaking population of Quebec is a good example. Their language and their history of discrimination in English-dominated workplaces set the Quebecois apart from English-speaking Canadians. In 1995, the people of Quebec narrowly defeated a separatist referendum, but it is still possible that nationalist sentiments in Quebec may result in the formation of a new nation on the North American continent.

In the case of Quebec, where political institutions and the rule of law are well established, such a change would not necessarily be traumatic. Elsewhere in the world, however, the rise of ethnic hostility and nationalism threaten to produce chaos. As former secretary of state

nation-state: The largest territory within which a society's political institutions can operate without having to face challenges to their sovereignty.

Warren Christopher noted in a statement before the House Foreign Relations Committee, "If we don't find some way that the different ethnic groups can live together in a country, how many countries will we have? We'll have 5,000 countries rather than the one hundred plus we have now" (quoted in Blinder & Crossette, 1993, p. A1). Although the numbers may be exaggerated, few political sociologists doubt that new nations will be created as a consequence of nationalist conflicts in coming years.

Crises of Legitimacy Societies occasionally undergo political upheavals like those that are occurring at this writing in Nigeria, Zimbabwe, the former Yugoslavia, the former Soviet Union, and Columbia. When these periods of unrest and instability involve enough of the citizens to such a degree that they challenge the legitimacy of the nation's political institutions, sociologists call the situation a *crisis of legitimacy*. For example, the Declaration of Independence was written during a crisis in which the American colonists challenged the legitimacy of British rule. In modern South Africa, where the black majority now rules with the cooperation and participation of the white minority, it still remains to be seen whether political legitimacy can be maintained in the face of extremes of wealth and poverty (Hunter-Gault, 2010).

In sum, as we scan the globe today, it is evident that the world's political ecology includes many states that are extremely stable and many others in which political instability threatens to develop into a crisis of legitimacy.

THE NATION-STATE IN CRISIS

The most important political territory in world affairs is the **nation-state**, the largest territory within which a society's political institutions can operate without having to face challenges to their sovereignty. Throughout the world, however, we see nation-states in crisis. In Africa, there has been persistent civil strife in many nations, particularly Somalia, Sierra Leone, Liberia, Angola, and Nigeria. In India, the world's most populous democracy, strife between Hindus and Muslims continues to threaten peace and political stability. In Indonesia, economic crisis and rampant corruption led to the fall of the Suharto regime. At this writing, the stability of Iraq and Afghanistan's national governments, and their ability to control religious militias and the armies of tribal "strongmen" in order to achieve national unity and legitimacy, is still very much in question. Nor is the

United States immune from terrorism directed against the legitimacy of the nation—as we saw in the case of the Oklahoma City bombing and in the threats of terrorism by armed militia groups that challenge the state's legitimacy, detest the United Nations and other agencies of world government, and claim that the U.S. federal government's protection of abortion clinics or the rights of homosexuals and other minorities justifies armed resistance (Horowitz, 1999).

Fortunately, one can also point to improvements in the stability of some nations that have faced severe threats to their existence and to social peace within their borders. Northern Ireland, Chile, and Mali are examples of nations that have experienced marked improvements in national legitimacy in the recent past. On balance, however, the world has begun a new millennium facing serious challenges to the stability and viability of many of its nations.

We noted in Chapter 4 that nation-states claim a legitimate monopoly over the use of force within their borders. By this, we mean that within a nation-state's borders, only organizations designated by the state (that is, the government) may use force. The war in Lebanon between Israel and the armed Hezbollah party in 2006 was a clear demonstration of why nation-states must be able to exercise this monopoly. Although it legitimately claimed to be a Lebanese political party, Hezbollah also maintained a militia along the border with Israel that was not subject to control by Lebanon's central government; this situation proved to be extremely dangerous both for Israel and for the future of Lebanon itself.

But if the citizens believe that their government is legitimate, why is force necessary at all? The answer is that even when citizens believe their government to be legitimate, force (or, more often, the threat of force) may be needed to maintain order and ensure compliance with the law. For example, citizens may agree in principle that some of their income should be taken by the state through taxes to cover its expenses, but some individuals or groups (such as the Freemen in the United States) may attempt to avoid paying taxes.

In many parts of the world, bribery and corruption among public officials are commonplace. When elites do not observe the rule of law, it is difficult to convince less powerful and less wealthy citizens to do so. Conflicts arising from this issue take the form of scandals such as the Enron bankruptcy, the lobbying abuses that rocked the Republican Party in 2006, or the collusion between politicians and corrupt businesspeople in Russia. The point is that those charged with enforcing the laws should be beyond reproach. Otherwise the legitimacy of government and the laws is diminished in the minds of citizens—and in their behavior.

People who believe that the state uses its power in an illegitimate manner are not allowed to form private armies to resist the state. They may challenge the legitimacy of a particular law in the courts, where rational arguments and interpretations of law replace violent conflict. Nevertheless, all the institutions of the state, including the courts, are ultimately based on the state's monopoly over the use of force (Conversi, 2010). In societies in which the state is weak or particular groups deny its legitimacy, as in Afghanistan and Iraq, it is common to find bandits, guerrilla armies, or terrorist groups.

Citizenship and the Rule of Law

The term *citizenship* refers to the status of membership in a nation-state. Like all statuses, citizenship is associated with a specific set of rights and obligations. As political sociologist Reinhard Bendix put it: "A core element of nation-building is the codification of the rights and duties of all adults who are classified as citizens" (1969, p. 89). In other words, the roles of citizens in the society's political life must be made clear to all.

In feudal societies, most of the members of the society did not participate in political life. The needs of various groups in the population were represented, if at all, by the edicts of powerful landholders, generals, and clergymen. In modern nation-states, as a result of major social movements, this form of representation has been replaced by a form based on citizen participation, in which representatives elected by the citizens are entitled to vote on important public issues. But as societies adopted the principle of citizen participation, conflicts arose over the question of who would be included among a society's citizens. This is illustrated by the violent struggle over citizenship and political participation for blacks in South Africa, led by Nelson Mandela and the African National Congress, that finally resulted in a new constitution and majority rule.

Rights of Citizenship The rights of citizenship include much more than the right to vote. T. H. Marshall (1964) defined those rights as follows:

- *Civil rights* such as "liberty of person, freedom of speech, thought and faith, the right to own property and to conclude valid contracts, and the right to justice."

---

citizenship: The status of membership in a nation-state.

- *Political rights* such as the right to vote and the right of access to public office.
- *Social rights* ranging from "the right to a modicum of economic welfare and security to the right to share fully in the social heritage and to live the life of a civilized being according to the standards prevailing in the society." (pp. 71–72)

This list does not necessarily imply that these rights actually exist in a given society or that they are shared equally by all of the society's citizens. In the United States, for example, before 1920, women were considered citizens but were denied the right to vote. No society in which these rights are denied to citizens can be said to have established the rule of law within its boundaries.

Thinking Critically

What social institutions are required to establish the rule of law in any decent society, and how must these institutions actually function? These questions arise throughout this chapter and others. What would it be like to live in a society without the rule of law? In such a society, how would you ensure that you and your loved ones were safe? Do we want to live in a society where it is necessary for each person to create their own "law" with guns and other weapons?

Participation in Local Politics The question of who is entitled to full participation in politics is a key issue at the local level as well as at the national level. At the national level, participation is vital to a group's position in society, as can be seen in the struggles by women and blacks to win the right to vote. However, in the cities, towns, and communities in which daily life is lived, the same guarantee of full participation plays a role in social mobility.

W. E. B. Du Bois's *The Philadelphia Negro*, which we discussed in Chapter 1, includes this poignant petition by black community leaders to the city's mayor:

> We are here to state to your excellency that the colored citizens of Philadelphia are penetrated with feelings of inexpressible grief at the manner in which they have thus far been overlooked and ignored by the Republican party in this city. . . . We are therefore here, sir, to earnestly beseech of you as a faithful Republican and our worthy chief executive, to use your potent influence as well as the good offices of your municipal government, if not inconsistent with the public weal, to procure for the colored people of this city a share at least, of the public work and the recognition which they now ask for and feel to be justly due to them. (1967/1899, p. 374)

At the turn of the twentieth century, Philadelphia, like most American cities, was dominated by leaders who were among the town's richest and most powerful citizens. Immigrant workers and the poor were effectively excluded from political participation. The Great Depression drastically altered this pattern. Fearing that the misery created by the depression would lead to unrest among the lower classes that could result in social disorder and even revolution, upper- and middle-class Americans began to look more favorably on the call for the creation of new social-welfare institutions. Such institutions would ease the plight of the poor and the unemployed without fundamentally changing capitalist economic institutions. At the same time, the poor and the working class began to exercise the right to vote in increasing numbers.

In the century and a half since the Civil War, the extension of full rights of citizenship and political participation has helped the United States avoid crises of legitimacy and civil strife. As we will see in the next section, however, by no means do all nations experience the same political developments, nor are they necessarily striving to extend the rights of participation in democratic political institutions to all citizens. Nor have all nations been as fortunate as the United States and others that have managed to solve their political crises without civil strife (Spohn, 2010).

POLITICAL INSTITUTIONS: A GLOBAL VIEW

The central problem of contemporary politics, according to Seymour Martin Lipset, is this: "How can a society incorporate continuous conflict among its members and social groups and yet maintain social cohesion and legitimacy of state authority?" (1979, p. 108). This question has taken on particular urgency since World War II, when the colonial empires of Europe crumbled in the face of nationalistic movements and hundreds of new nations were created. In comparing the political institutions of the world's nations, it is helpful to begin

These new citizens at a swearing-in ceremony have had to demonstrate knowledge of basic English and awareness of the rights and responsibilities of U.S. citizens.

David R. Frazier Photolibrary, Inc./Alamy

by outlining the nature of the political institutions one expects to find in any modern state and noting the problems associated with their development.

Any political leader in the contemporary world who speaks of the "rule of law" or spreading the benefits of democracy or making a commitment to nation building outside his or her nation's own borders must have a firm grasp of the full range of institutions and political processes that make freedom possible in the modern nation-state. As we have seen with tragic and dramatic consequences in Iraq and Afghanistan, as well as in Palestine, the conduct of elections alone does not make a nation a democracy or its citizens free. Freedom in a nation-state usually includes a voice for citizens in national affairs, justice before the law, freedom of speech and religion, and domestic peace, among other cherished values. Free elections are a hallmark of democracies and are an important way of giving citizens a voice in their affairs, but elections alone, in the absence of other essential political institutions, will not bring about either democracy or stability. We will return to this issue later in the chapter, but it is important to remember it as we analyze different institutions of modern political systems, not all of which are necessarily democratic.

The Public versus the Electorate

Those who study political institutions in North America and throughout the world make an important distinction between a nation's public and its electorate. The *public* refers—among other things—to the population of a nation that forms opinions about current events. These opinions are often reported in public opinion polls. The *electorate*, on the other hand, refers to the nation's population of eligible voters. Just because the public feels one way or another about a given issue does not mean that the electorate will vote on the basis of those feelings. In the United States, for example, the practice of *gerrymandering*, the drawing of electoral district boundaries to include voters from certain racial, ethnic, or class backgrounds, may limit the overall influence of large categories of voters. Other laws, many derived from the Constitution itself, preserve an imbalance of power in which some regions have more influence than others. There are two U.S. senators from each state, independent of the state's population, a situation that discriminates against the more heavily urbanized states and gives an advantage to states with small populations. In other nations, similar practices limit the equitable

functioning of democratic systems. Corruption, including paying voters for their support, barring some voters from the polls in order to affect the results, bribing politicians to support certain laws, and many other corrupt practices often make supposedly democratic regimes far less open and democratic than they appear to be in their constitutions. Corruption has proven to be a fatal flaw of many fledgling democracies in Africa and the Middle East, but as we have seen in recent lobbying scandals in the United States, it also threatens the rule of law in the most advanced democracies (Fukuyama 2004; Lebovics, 2006).

Characteristics of Modern Political Systems

A Rational Administration Modern nations are governed by elected and appointed officials whose authority is defined by laws. To the extent possible, they are expected to use their authority for the good of all citizens rather than for their own benefit or for that of particular groups. They are forbidden to use their authority in illegitimate ways—that is, in ways that violate the rights of the citizens.

Modern political institutions usually specify some form of separation of powers. The authors of the U.S. Constitution, for example, were careful to create a system of checks and balances among the nation's legislative, executive, and judicial institutions so that abuses of authority by one could be remedied by the others. Thus, when agents of the Nixon campaign organization were caught breaking into the Democratic Party headquarters during the 1972 presidential election campaign, and the president and some of his advisers were found to have covered up their role in this and other illegal activities, the impeachment proceedings initiated by the legislative and judicial branches of the government led to the president's resignation.

In many other nations, abuses of state authority are far more common. The power of rulers is unchecked by other political institutions, and although citizens may question the legitimacy of their rule, they are powerless to prevent the rulers' use of coercion in violation of their rights. Such states are often ruled by a **demagogue**, a leader who uses personal charisma and political symbols to manipulate public opinion. Demagogues appeal to the fears of citizens and essentially trick them into giving up their

---

demagogue: A leader who uses personal charisma and political symbols to manipulate public opinion.

rights of political participation. Hitler's ability to sway the suffering German masses made him the outstanding example of demagoguery in the twentieth century.

A Party System Organizations of people who join together to gain legitimate control of state authority—that is, of the government—are **political parties**. Parties may be based on ideologies, or they may simply represent competing groups with the same basic values. Many American political sociologists assert that nations must make certain that other political parties are able to compete with the ruling party (Archer, 2010; Janowitz, 1968). Failure to protect the existence of an opposition party or parties leads to **oligarchy**—rule by a few people who seek to stay in office indefinitely rather than for limited terms of office.

Parties that seek legitimate power and accept the rule of other legitimate parties form a "loyal opposition" that monitors the actions of the ruling party, prevents official corruption, and sustains the hopes of people whose needs are not adequately met by the ruling party. Revolutionary political parties, it should be noted, do not view the state as legitimate. They therefore do not agree to seek authority through legitimate procedures like elections. For this reason, they tend to be banned by most governments. On the other hand, many modern nations, including the United States, have banned or repressed nonrevolutionary communist parties, largely because of their opposition to private property and their sympathy for the communist-dominated regime of the former Soviet Union (Pierson & Skocpol, 2007).

Institutions for Maintaining Order We have seen that states control the use of force within their borders. They also seek to protect their territories against attacks by other states. For these purposes, most states maintain police forces and armies. Sometimes, however, the state's leaders have difficulty controlling these institutions, and military factions often seize power in what is called a *coup d'état* (or simply *coup*).

A coup usually results in the establishment of an oligarchy in which the state is ruled by a small elite that includes powerful members of the military. The *junta* is a special type of oligarchy in which military generals rule, usually with the consent of the most powerful members of the nation's nonmilitary elite. Their rule is commonly opposed by members of the intellectual professions, especially journalists, professors, writers, and artists. The dissent expressed by these individuals often leads to censorship and further repression of nonmilitary political institutions. Juntas have appeared in all parts of the world, especially where military institutions are strong and institutions that stand for democracy and the rule of law are weaker, as is the case in many developing nations.

Regimes that accept no limits to their power and seek to exert their rule at all levels of society, including the neighborhood and the family, are known as **totalitarian regimes**; Nazi Germany and the Soviet Union under Stalin are examples. Such regimes cannot exercise total power without the cooperation of the military. It would be incorrect, however, to attribute the existence of oligarchies and totalitarian regimes to the power of the military alone. On many occasions, military leaders have led coups that deposed the state's existing rulers and then turned over state power to nonmilitary institutions.

Nevertheless, in modern societies, military institutions have become extremely powerful. Even before the end of World War II, Harold Lasswell (1941) warned that nations might be moving toward a system of "garrison states—a world in which the specialists in violence are the most powerful group in society" (p. 457). This theme was echoed by Dwight D. Eisenhower, one of the nation's celebrated soldier-presidents. As he was leaving office, Eisenhower warned that the worldwide arms buildup, together with the increasing sophistication of modern weapons, was producing an "industrial-military complex." By this he meant that the military and suppliers of military equipment were gaining undue influence over other institutions of the state. (We discuss military institutions more fully in the final section of the chapter.)

Secular Governmental Institutions It is extremely difficult for minorities in a nation-state, be they religious minorities, ethnic or racial groups, or others, to feel that their rights are protected and honored if the institutions of government are dominated by members of the majority group. This is the case in Israel, where the Arab minority feels that, because Israel is a Jewish state, the institutions of government are biased against them. It is even more true in other nations of the Middle East, such as Saudi Arabia, Jordan, and Iran. These governments

---

political party: An organization of people who join together to gain legitimate control of state authority.

oligarchy: Rule by a few people who stay in office indefinitely rather than for limited terms.

totalitarian regime: A regime that accepts no limits to its power and seeks to exert its rule at all levels of society.

may be theocracies like Saudi Arabia, where religious leaders dominate the courts and the agencies of government, or democracies like Iran, where separation of church and state has made some progress in recent years but clerics still dominate many of the vital institutions of government and the press. As the world witnessed in the case of the Taliban regime in Afghanistan before 2002, a state ruled by religious tyrants can ensure law and order among its people, but lacking legitimacy among minority groups and the educated, it has great difficulty promoting economic and social development.

Abdullahi An-Na'im, a professor at Emory University in Atlanta and a political exile from the repressive regime in the African nation of Sudan, is an expert on political reform in the Islamic world. "It takes time," he observes, "and certain conditions of stability and prosperity, for societies to relax their defenses and actively allow civic engagement. It took the Americans 200 years, including a horrendous civil war and a long civil rights movement, and still you have far to go" (2002, p. 20). He also points out that every Muslim country in the world today was either colonized by Western powers or controlled by Western powers the way Iran and Saudi Arabia were, and that "colonialism was not in the business of promoting democratic values or institutions." The long struggle for separation of church and state and democratization of Muslim societies will never be imposed by Western power, he observes; it will have to come from democratic political movements in the Islamic world itself. There are already many precedents to draw on, despite the more visible influence of antidemocratic fundamentalists. This can also be said of many other nations in Asia, Africa, and Latin America: Democratic reforms will depend on the efforts of leaders within those regions as well (Reardon, 2002).

Democratic Political Systems

In contrast to oligarchies and totalitarian regimes, democratic societies offer all of their citizens the right to participate in public decision making. Broadly defined, **democracy** means rule by the nation's citizens (the Greek *demos*, from which the word *democracy* is derived, means "people"). In practice, democratic political institutions can take many different forms as long as the following conditions are met:

1. The political culture legitimizes the democratic system and its institutions.
2. One set of political leaders holds office.
3. One or more sets of leaders who do not hold office act as a legitimate opposition (Lipset, 1995, 2000).

Abdullahi An-Na'im on the Emory University campus. He sees a need for internal discourse "that has to be done by Muslims."

The two most familiar forms of democratic political rule are the British and American systems. In the British *parliamentary* system, elections are held in which the party that wins a majority of seats in the legislature "forms a government," meaning that the leader of the party becomes the head of the government and appoints other party members to major offices. Once formed, the government generally serves for a specified length of time. If no party gains a majority of the legislative seats, a coalition may be formed in which smaller parties with only a few seats can bargain for positions in the government. Such a system encourages the formation of smaller parties.

In the American *representative* system, political parties attempt to win elections at the local, state, and national levels of government. In this system, the president is elected directly, and the party whose candidate is elected president need not have a majority of the seats in the legislature. Success therefore depends on the election of candidates to national office. In the United States, two political parties, the Republicans and the Democrats, have developed the resources and support necessary to achieve this goal.

Representative versus Direct Democracy Efforts to create direct democratic institutions, whereby everyone in the electorate votes to decide on an issue, periodically arise in representative democracies. Many Americans become impatient with representative democracy and its legislative deliberations and wish for more immediate solutions to high taxes, open immigration, or whatever problems are most troubling to them. California's 2003 gubernatorial recall election, in which movie star

democracy: A political system in which all citizens have the right to participate in public decision making.

Arnold Schwarzenegger was in the political spotlight, is a highly instructive case in the problems of direct democracy as applied to a large and complex society. California is larger than many nations of the world and boasts the world's fifth most important economy (Kolbert, 2003).

Twenty-five years ago, Californians voted on Proposition 13, an anti–property tax measure that was placed on the statewide ballot after voters signed enough petitions to qualify it as a referendum issue. The measure passed, and voters seemed so pleased with their experience in direct democracy that an average of sixty-four initiatives have been filed in California each year since then. A more recent example was California's Proposition 8, which barred same-sex marriages in that state. The referendum before that one, Proposition 49, initiated about $500 million a year of new expenditures but provided no new revenues for them, thus helping to drive up the state deficit. That referendum was financed in part by Arnold Schwarzenegger. Sociologist Mark Baldassare, research director of the Public Policy Institute of California, finds that many of the problems of California's economy can be traced directly to changes in the way revenues are collected that resulted from Proposition 13. The state's education ratings, for example, once the pride of the nation, have declined sharply, and there is a growing crisis in housing and basic infrastructure, also largely due to the inability of local governments to fund needed rehabilitation projects. So although it may feel good to reject an unpopular governor or pass one's favorite piece of legislation immediately, the state has suffered from its rush to pass referendums. Direct democracy works well in small communities where town hall meetings and local citizen involvement are long-standing traditions, but as we see in the case of California, it is far from the best way to ensure political stability and legitimacy in a large polity (Baldassare, 2003).

Small Parties in Representative Systems A disadvantage of representative democratic systems like that of the United States is that smaller parties, which have no hope of electing candidates to national office or even to offices in most statewide elections, are effectively barred from

Is the tea party a political party or a political movement, or both? Although popular political figures seek its support, the tea party has remained a rather loosely organized conservative social movement rather than a registered political party with an official leadership, but this could change as the next presidential elections near.

AP Photo/Paul Sakuma

sharing power with the larger parties, as they often do in parliamentary systems. Even when they do win a few seats in a state legislature or in Congress, they are not asked to form coalitions that would place their leaders in high cabinet positions. But small parties do exist in American politics, and they often exert an extremely important influence on the outcomes of elections.

The Reform Party created by Ross Perot and the Green Party headed in the 2000 presidential election by Ralph Nader are examples of smaller parties that have garnered enough votes on the national level to alter the outcomes of presidential elections. Votes for Perot in 1992, many taken away from the Republican column, helped defeat George Bush; similarly, votes for Nader in 2000, most of which ordinarily would have gone to Al Gore, helped elect George W. Bush. In close elections, smaller parties can play a decisive role in the outcome of national elections. Ironically, however, their impact is often to swing hotly contested elections toward the candidate who is least in favor of their ideas. Does this mean that smaller parties have no positive effects? On the contrary, the existence of smaller parties, which represent specific sets of issues such as greater concern for the environment and consumer affairs, or reform of government, often push the major parties to adopt platforms that incorporate their demands. They also tend to enlist people in active political life who would otherwise be uninvolved or apathetic. So even though the third parties or smaller, narrow-issue parties in a representative system like that of the United States normally do not share power after elections, they can have an enormous impact on the course of political history (Cohen, 2000).

One-Party Systems Suppose that a nation has only one political party. Does this mean that it is undemocratic? Not in principle. A one-party state or region can be democratic as long as the citizens are free to form other parties, as long as the party's leaders can be replaced through democratic processes, and as long as the leaders do not violate the rights of citizens. But how likely are these conditions to be met if there are no opposing parties? The evidence suggests that the odds are against democracy in one-party states and regions, as we see in communist one-party states and many noncommunist one-party dictatorships (Ishiyama & Quinn, 2006).

A Typology of Regimes

Social scientists use analyses of political institutions to develop typologies of political regimes. One such typology, which is widely used, was developed by J. Denis and Ian Derbyshire (1996). They classify national regimes into these categories:

1. *Liberal democracies.* These regimes are marked by multiparty elections, competitive parties, separation of powers, and guarantees of the rights of minorities and individuals. Examples are the United States, Canada, Great Britain, France, Brazil, and Japan.

2. *Emergent democracies.* These have constitutions that specify all or most of the institutions and processes of liberal democracies but are marked by problems in fully establishing democratic processes caused by one-party dominance, insurgencies, corruption, and so forth. They are often viewed as "liberal democracies on trial." Examples are Chile, Ivory Coast, Mali, Haiti, Morocco, Tunisia, and the Philippines.

3. *Communist regimes.* These are run by a "revolutionary dictatorship" and a single communist party that in principle is serving the interests of the working class. There are very limited guarantees of individual or minority rights. Political command over economic institutions is widespread but subject to market experimentation. China, North Korea, and Cuba are examples of existing communist regimes.

4. *Nationalistic socialist regimes.* These are similar to communist regimes, with a single socialist party; however, they are more inclined to promote the interests of one national group over others and to allow private commerce. There is little or no protection of individual or minority rights. Examples are Iraq, Libya, Tanzania, and Syria.

5. *Authoritarian nationalist regimes.* The extreme nationalism of these regimes leads to intolerance and the exclusion of other races and creeds, often in the most brutal or genocidal fashion as in Nazi Germany or contemporary Zimbabwe.

6. *Military regimes.* These are ruled by a military elite or junta, usually with extremely limited protection of citizens' rights and no free elections. Current examples are Sierra Leone, Sudan, and Myanmar, although many emergent democracies and some liberal democracies are plagued by problems of civilian control over the military.

7. *Islamic nationalist regimes.* These are ruled by nationalistic political regimes devoted to fundamentalist Islam. Afghanistan under the Taliban and Iran are examples of this type of regime.

8. *Absolutist regimes.* These are usually ruled by an absolute monarch who passes power to successors through a hereditary line. Constitutional forms of government, popular assemblies, judiciary rules that counter the executive power, and political parties are banned. Sultanates, emirates, and traditional monarchies such as Jordan, Saudi Arabia, and Swaziland are examples of this type of regime.

Most of the world's people live in liberal or emergent democracies, although China, the world's most populous nation with more than one billion people, is ruled by a communist regime. In research that charts trends in world political regimes since the mid-twentieth century, T. R. Gurr (2002) shows that the numbers of democracies and of states in possible transition to democracy have been increasing steadily, with the greatest progress being made since 1990 (see Figure 19.2). Meanwhile, autocracies—authoritarian states, including all the types of nondemocratic regimes in Denis and Derbyshire's classification—have been decreasing in number since the 1960s, and have done so especially rapidly since the fall of the Soviet Union in the late 1980s. These are encouraging trends, suggesting that the world's peoples will strive for some kind of democratic self-rule when given the opportunity. In recent years, however, a far more disturbing trend has emerged in world politics, a trend often referred to as "the rise of failed states."

Failed States and Political Violence A major change in world politics in the new millennium is the rise of failed states, which often contribute to problems of regional violence and international terrorism. Studies by the U.S. government, the United Nations, and the Carnegie Endowment for International Peace have warned that almost two billion people throughout the world live in nations whose governments are in danger of collapse. *Failed states* are nations whose governments are unable to maintain domestic peace or prevent the predations of drug barons, tribal warlords, or independent militias, and whose domestic tranquility is continually threatened by ethnic or religious strife, guerilla movements, and extreme corruption and criminality. Although it is relatively simple to define what we mean by a failed state, it is far more difficult to arrive

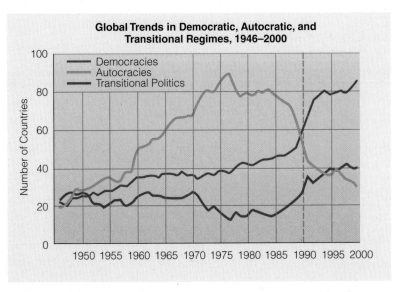

Figure 19.2 **Global Trends in Democratic, Autocratic, and Transitional Regimes, 1946–2000.** Source: From Ted Gurr, "Attaining Peace in Divided Societies: Five Principles of Emerging Doctrine," 2:27, *International Journal of World Peace*, 2002. Reprinted by permission of the author.

TABLE 19.2

Key Political, Economic, Military, and Social Indicators of Instability

Demographic pressures—overpopulation, rapid urban growth, resource depletion

Refugees and displaced people

Group grievance—the presence of minorities with serious grievances

Human flight

Uneven development—particularly the growth of a very rich elite and a large mass of extremely poor citizens

Economic decline

Delegitimization of state—the sense that pervasive corruption and graft make national leaders unworthy of their positions

Decline of public services

Lack of protection of human rights

Security apparatus—military and police institutions that are not professional, are graft-ridden and liable to engage in power struggles

Factionalized elites

External intervention—threats from outside the nation, from neighboring states or ethnic groups

Source: *Foreign Policy*, 2005.

at designations for failure in actual nations. In consequence, social scientists have developed a number of ways to measure state failure. One of the most respected of these was developed by the Fund for Peace and the journal, *Foreign Policy*. In developing their index of state failure, the *Foreign Policy* researchers ranked the severity of a variety of problems that contribute to state failure, as indicated in Table 19.2. Each of these variables can be analyzed and assigned a quantitative score. When this is done, the research yields a composite score for every nation in the world. The top ten states at risk of state failure are listed in Table 19.3 and include current hot spots such as Somalia, Iraq, and Pakistan. The authors note, however, that the index includes others whose instability is less widely acknowledged, including Bangladesh, Guatemala, Egypt, Saudi Arabia, and Russia.

TABLE 19.3

Top Ten Nations Ranked on the Failed States Index

Nation	Index Score
Somalia	114.7
Zimbabwe	114
Sudan	112.4
Chad	112.2
Congo	108.7
Iraq	108.6
Afghanistan	108.2
Central African Republic	105.4
Guinea	104.6
Pakistan	104.1

Source: *Foreign Policy*, 2010.

Even more threatening to prospects of world peace is the fragility of states like North Korea (thirteenth on the list), Pakistan (thirty-fourth), and Iran (fifty-seventh)–which possess nuclear weapons or are seeking to develop them and have relatively high rankings on the failed state index. The *Foreign Policy* researchers end their analysis by asking,

> What are the clearest early warning signs of a failing state? Among the 12 indicators we use, two consistently rank near the top. Uneven development is high in almost all the states in the index, suggesting that inequality within states—and not merely poverty—increases instability. Criminalization or delegitimization of the state, which occurs when state institutions are regarded as corrupt, illegal, or ineffective, also figured prominently. Facing this condition, people often shift their allegiances to other leaders—opposition parties, warlords, ethnic nationalists, clergy, or rebel forces. Demographic factors, especially population pressures stemming from refugees, internally displaced populations, and environmental degradation, are also found in most at-risk countries, as are consistent human rights violations. Identifying the signs of state failure is easier than crafting solutions, but pinpointing where state collapse is likely is a necessary first step. (*Foreign Policy*, 2005)

PERSPECTIVES ON POLITICAL INSTITUTIONS

Our description of democratic political institutions leaves open the question we posed earlier: How can a society accommodate competition or conflict among its members and social groups and yet maintain social cohesion and legitimacy of state authority (Archer, 2010)? One answer is that a society can resolve conflicts through democratic processes. But this leads us to ask what conditions allow democratic institutions to form and, once formed, what ensures that they actually function to reduce the inequalities that engender conflict. The broadest test of democratic institutions is not whether they are embodied in formal organizations like legislatures and courts but whether they are able to address the problems of inequality and injustice in a society.

At least three schools of sociological thought attempt to address these questions. The first, derived from the functionalist perspective, asserts that democratic political institutions can develop and operate only when certain "structural prerequisites," such as a large middle class, exist in a society. The second school of thought, often referred to as the **power elite model**, is based on the conflict perspective. It is highly critical of the functionalist view, supporting its criticism with evidence of the ways in which so-called democratic political institutions actually operate to favor the affluent. A third position, known as the **pluralist model**, asserts that the existence of ruling elites does not mean that a society is undemocratic, as long as there are divisions within the elite and new groups are able to seek power and bargain for policies that favor their interests.

Structural Prerequisites of Democracy and the Rule of Law

Seymour Martin Lipset (1981, 1994) argues that democratic political institutions are relatively rare because if they are to exist and function well, the society must have attained a high level of economic and cultural development. To prove his theory, Lipset surveyed data on elections, civil rights, freedom of the press, and party systems in forty-eight nation-states, and subsequently in many other countries (Lipset, 1994). He found that the presence of these institutions was correlated with a nation's level of economic development, its degree of urbanization, the literacy of its citizens, and the degree to which its culture values equality and tolerates dissent. The Research Methods box on page 504 presents Lipset's findings in more detail.

In this cross-cultural research and in his research on democracy in the United States, Lipset attempted to show that the growth of a large middle class is essential to democracy. The middle-class population tends to be highly literate and, hence, able to make decisions about complex political and social issues. Moreover, middle-class citizens believe that they have a stake in their society and its political institutions. Accordingly, they often support policies that would reduce the class and status cleavages—the distinctions between the haves and the have-nots—that produce social conflict.

Sociologists who study voting behavior tend to support Lipset's thesis that the stability of democratic institutions rests on structural features that diminish conflict in a society. But their research has revealed something less than complete stability. Morris Janowitz (1978) made this observation, which is even more true today:

> Since 1952, there has been an increase in the magnitude of shifts in voting patterns from one national election to the next. . . . Increasingly

important segments of the electorate are prepared to change their preference for president and also to engage in ticket splitting [voting for candidates of different parties]. (p. 102)

These shifts in voting behavior may represent new alignments of voters that could lead to major changes in the nation's public policies. The important point, however, is that although these realignments influence which parties and political leaders gain or lose power, they do not affect the process of democratic competition itself.

The Power Elite Model

However important elections are to the functioning of democratic institutions, there are strong arguments against the idea that they significantly affect the way a society is governed. Some social scientists find, for example, that political decisions are controlled by an elite of rich and powerful individuals (Zweigenhaft & Domhoff, 1998). This "power elite" tolerates the formal organizations and procedures of democracy (elections, legislatures, courts, and so on) because it essentially owns them and can make sure that they act in its interests no matter what the outcome of elections may be. C. Wright Mills, the chief proponent of this point of view, has described the power elite as follows:

> The power elite is composed of men whose positions enable them to transcend the ordinary environments of ordinary men and women. They are in positions to make decisions having major consequences. . . . They are in command of the major hierarchies and organizations of modern society. They rule the big corporations. They run the machinery of the state. . . . They direct the military establishment. They occupy the strategic command posts of the social structure. . . . The power elite are not solitary rulers. . . . Immediately below the elite are the professional politicians of the middle levels of power, in the Congress and in the pressure groups, as well as among the new and old upper classes of town and city and regions. Mingling with them in curious ways . . . are those professional celebrities who live by being continually displayed. (1956, p. 4)

power elite model: A theory stating that political decisions are controlled by an elite of rich and powerful individuals even in societies with democratic political institutions.

pluralist model: A theory stating that no single group controls political decisions; instead, a plurality of interest groups influence those decisions through a process of coalition building and bargaining.

Diagramming Social Change

The accompanying chart is an example of how a diagram can effectively communicate ideas about the directions and influences of social change. This diagram presents the social structures from which democratic political institutions arise (listed on the left) and some possible further consequences of the emergence of those political institutions (on the right) (Archer, 2010).

The arrows in the diagram indicate either conditions that contribute to democracy or possible consequences of the emergence of democratic institutions. As in most social-scientific analysis, the author, political scientist Seymour Martin Lipset, is careful not to make direct assertions about causality. He does not state, for example, that an open class system (measured by upward and downward social mobility) inevitably leads to the emergence of democratic regimes. Instead, social mobility and the other aspects of social structure listed on the left are "necessary but not sufficient" preconditions for the emergence of democracy. That is, they must exist to some degree in a society, but they are not sufficient to "cause" democracy. For that to happen, other conditions must also be present, such as the leadership of people who can sacrifice their desire for personal power in the interest of creating democratic political institutions. George Washington, for example, is revered as a pioneer of democracy because, when he had the chance to become a monarch, he chose to relinquish personal power and lead the newly independent colonies toward the creation of a constitutional monarchy.

Note that the arrows on the right side of the diagram suggest that democratic rule is associated with the emergence of social phenomena that run counter to democracy. Bureaucracy, for example, is a common consequence of democratic regimes. Laws passed by democratic processes tend to spawn administrations designed to deliver services, provide protection, enforce regulations, and much more. These bureaucracies, in turn, may stifle democracy by replacing the will of the majority with rules, regulations, and procedures. Note, however, that these excesses can be corrected by the actions of democratic institutions (for example, further legislation). The arrows merely indicate that democratic rule has certain counterdemocratic consequences.

In an expansion of the comparative research on which this diagram is based, Lipset makes the following comment:

> Not long ago, the overwhelming majority of the members of the United Nations had authoritarian systems. As of the end of 1993, over half, 107 out of 186 countries, have competitive elections and various guarantees of political and individual rights—that is, more than twice the number two decades earlier in 1970. (1994, p. 1)

But Lipset, a veteran of many decades of close observation of democratic rule in nations throughout the world, is cautious about predicting a trend toward the emergence of democratic regimes. Too many nations, like Nigeria and Russia, have turned toward more authoritarian rule after promising attempts to develop democratic political institutions. Above all, Lipset warns:

> [If nations] can take the high road to economic development they can keep their political houses in order. The opposite is true as well. Governments that defy the elementary laws of supply and demand will fail to develop and will not institutionalize genuinely democratic systems. (1994, p. 1)

Conditions and Consequences of Democracy

CONDITIONS — INITIAL POSSIBLE CONSEQUENCE — ADDITIONAL CONSEQUENCES

Conditions: Open class system; Economic wealth; Equalitarian value system; Capitalist economy; Literacy; High participation in voluntary organizations

DEMOCRACY

Additional Consequences: Open class system; Equalitarian value system; Political apathy; Bureaucracy; Mass society; Literacy

Source: From Lipset, Seymour Martin. *Political Man: The Social Bases of Politics.* pp. 61. © 1959, 1960, 1981 by Seymour Martin Lipset. Reprinted with permission of the Johns Hopkins University Press.

When it was first published, Mills's *The Power Elite* created a stir among sociologists. Mills challenged the assumption that societies that have democratic political institutions are in fact democratic. He asserted instead that party politics and elections are little more than rituals. Power is exercised by a ruling elite of immensely powerful military, business, and political leaders that can put its members into positions of authority whenever it wishes to do so. Mills's claim was significant for another reason as well: Although Mills appreciated Marx's views on the role of class conflict in social change, he did not believe that the working class could win power without joining forces with the middle class. He therefore attempted to demonstrate the existence of a ruling elite to an educated public, which would then, he hoped, be able to see through the rituals of political life and make changes through legitimate means.

Other sociologists have questioned the power elite thesis on methodological grounds. For example, Talcott Parsons (1960) noted that the power elite was supposed to act behind the scenes rather than publicly. Therefore, its actions could not be observed, and the power elite thesis could not be either proved or disproved. The power elite thesis therefore was not scientifically sound, according to Parsons.

Numerous adherents of the power elite thesis have attempted to show that a ruling elite does indeed exist and that its activities can be observed. One of the best known of these researchers is Floyd Hunter (1953), whose classic studies of Atlanta's "community power structure" attempted to show that no more than forty powerful men were considered to have the ability to make decisions on important issues facing the city and its people. Most of these men were conservative, cost-conscious business leaders. The Hunter study stimulated many attempts to find similar power structures in other cities and in the nation as a whole (Domhoff, 1983; Zweigenhaft & Domhoff, 1998). Although the term *power structure* has found its way into the language of politics, these studies have been strongly criticized for basing their conclusions on what people say about who has power rather than on observations of what people with power actually do (Oppenheimer, 1999). On the other hand, recent corporate scandals—especially the Enron and Minerals Management Services cases—lend support to Mills's original power elite thesis because of the cozy relationships that exist between many energy and media corporate leaders and core members of the Bush administration (see the Global Social Change box on page 506).

The Pluralist Model

The power elite thesis has not gone unchallenged in the research literature. In another famous study of politics—this one in New Haven, Connecticut—Robert Dahl (1961) found that different individuals played key roles in different types of decisions. No single group was responsible for all of the decisions that might affect the city's future. No power elite ruled the city, Dahl argued. Instead, several elites interacted in various ways on decisions that affected them. In situations in which the interests of numerous groups were involved, a plurality of decision makers engaged in a process of coalition building and bargaining.

The pluralist model calls attention to the activities of interest groups at all levels of society

(Reidy, 2001). **Interest groups** are not political parties; they are specialized organizations that attempt to influence elected and appointed officials on specific issues. These attempts range from **lobbying**—the process whereby interest groups seek to persuade legislators to vote in their favor on particular bills—to making contributions to parties and candidates who will support their goals. Trade unions seeking legislation that would limit imports, organizations for the handicapped seeking regulations that would give them access to public buildings, and ethnic groups seeking to influence U.S. policy toward their country of origin are among the many kinds of interest groups that are active in the United States today (Oberschall, 1996).

In recent decades, as the number of organized interest groups has grown, so has the complexity of the bargaining that takes place between them and the officials they want to influence. Social scientists who study politics from the pluralist perspective frequently wonder whether the activities of these groups threaten the ability of elected officials to govern effectively. For example, a study of a federal economic development project in Oakland, California, found that the personnel in the agency responsible for such projects had to learn to deal with supporting and opposing interest group leaders and elected officials throughout the life of the project. It took six years of constant negotiation on many unanticipated issues—affirmative action in employment, environmental concerns, design specifications—before a firm could win a contract to build an airplane hangar (Pressman & Wildavsky, 1984).

Politics and Social Interaction

From the interactionist perspective, political institutions, like all others, are "socially constructed" in the course of human interaction. They do not simply come into being as structures of norms, statuses, and roles and then continue to function without change. Instead, they are continually being shaped and reshaped through interaction. This view closely parallels the popular notion that democratic political institutions must be continually challenged if they are to live up to their ideals. Otherwise they will become oligarchies that rule for the benefit of small cliques rather than for the mass of citizens (Oberschall, 1996).

---

interest group: An organization that attempts to influence elected and appointed officials regarding a specific issue or set of issues.

lobbying: The process whereby interest groups seek to persuade legislators to vote in their favor on particular bills.

global social change

The Power Grid Why is it that so many recent political scandals have involved the energy industry, and especially giant global corporations like British Petroleum (BP) or Enron? Any sociological analysis of this question begins with the relationships that exist between energy companies and political leaders who need funds for their campaigns. Another closely related phenomenon is the "cozy relationships" that develop between governmental regulators and oil executives, which became evident once again in the British Petroleum oil spill in the Gulf of Mexico in 2010. Analyses of both these scandals exposed the darker side of politics on the global stage.

This article from the *London Financial Times*, published as the Enron bankruptcy scandal was beginning to unfold, focuses on the close relationships between Enron executives and Texas political leaders:

> Enron was, politically speaking, the best connected company in America. It donated millions to both main political parties but was closer to Republicans, especially Mr. Bush, who, when he was governor of Texas, referred to Mr. Lay [Enron's Chief Executive Officer] as "Kenny Boy." Half a dozen senior officials in the administration either worked directly for the company or held substantial amounts of equity in it (Barker et al., 2002)

The Minerals Management Services (MMS) scandal, which began in 2008 and was involved in the 2010 Gulf oil disaster, exposed a related but somewhat different aspect of the influence of energy corporations on political institutions. Responsible for managing energy production in federal waters, including inspections of all deep-water drilling rigs in the Gulf of Mexico, personnel of the agency were found to have had improper dealings and corrupt relationships with oil company executives. Instead of enforcing regulations, some MMS officials attended lavish parties with oil executives. The Obama administration was too slow in replacing MMS officials, and in consequence, MMS's failures to enforce federal regulations continued and played a part in the lax enforcement of the BP Deepwater Horizon drilling operation.

In both the Enron and MMS cases, the underlying problem is the influence of money on politics. Elected officials often receive large campaign contributions from energy company executives. Lobbyists for the energy companies seek out regulators whose allegiance may be to the same elected officials. Favors are traded, but the public's need for protection gets short shrift, and corruption, especially by officials involved in the energy sector, is once again a global news story.

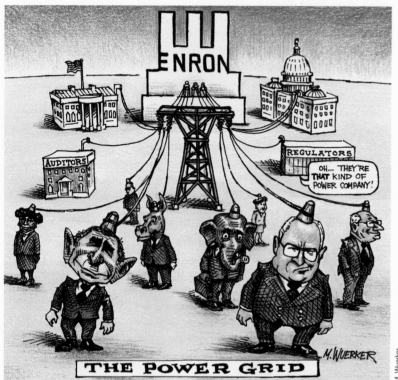

THE POWER GRID — M. WUERKER

The question of how social interaction affects a society's political institutions was a central concern of the leading philosophers of ancient Greece. In his *Politics,* for example, Aristotle "held that humans were made for life in society just as bees were made for life in the hive" (Bernard, 1973, p. 30). Because humans are "political animals," Aristotle believed, their constant discussions of political issues and their ability to form coalitions allow them to arrive at a consensus regarding what a good society is and how it should be governed. Unfortunately, he concluded, existing political systems were flawed. The divisions and gaps in interaction created by wealth prevented consensus and created conflict. Moreover, people in occupations like farming did not have enough time to examine all the sides of an issue and work toward agreement with their fellow citizens. In a good society, therefore, all the citizens must be free to

devote their full attention to political affairs. (In practice, Greek democracy required an economy based on the labor of slaves.)

Nearly 2,000 years later, the Italian political adviser Niccolò Machiavelli (1469–1527) again took up the relationship between human interaction and political institutions. Machiavelli believed that political institutions had to be based on the recognition that all human beings are capable of evil as well as good. In his *Discourses,* he made this observation:

> All those who have written upon civil institutions demonstrate . . . that whosoever desires to found a state and give it laws, must start by assuming that all men are bad and ever ready to display their vicious nature, whenever they may find an occasion for it. (1950/1513, Vol. 1, p. 3)

Some people, he admitted, are merciful, faithful, humane, and sincere, but even though there are virtuous people in every society, political leaders must anticipate the worst possible behavior. In his most famous work, *The Prince*, Machiavelli suggested that the wise ruler or prince would be a master of astuteness—the ability to "read" the intentions of allies and opponents in the tiniest of gestures and reactions—and would also have mastered the skills of diplomacy. He used the following comparison to make this point:

> A prince . . . must imitate the fox and the lion, for the lion cannot protect himself from the traps, and the fox cannot defend himself from wolves. One must therefore be a fox to recognize traps, and a lion to frighten wolves. Those that wish to be only lions do not understand this. (1950/1513, Vol. 1, p. 66)

In the five centuries since they were originally formulated, these ideas have had a powerful influence on political thinkers. The authors of the U.S. Constitution, for example, attempted to avoid situations in which "foxes" or "lions" could take advantage of the weaknesses of others. They anticipated the more self-serving aspects of political interaction rather than simply assuming that a new society would bring about cooperation among citizens: "If men were angels, no government would be necessary" (*The Federalist*, No. 51). This is why, as noted earlier in the chapter, they planned a government in which each of the major branches would be able to check any abuse of power by the other branches.

Political Interaction in the "Democratic Experiment"

In the 1830s, Alexis de Tocqueville, a young French aristocrat, visited the world's first experiment in national democracy. Although the United States was by no means a true democracy—it condoned slavery, indentured servants, and war against Native Americans—European social thinkers took a keen interest in the model of citizen rule through democratic institutions that was being established in the new nation. Tocqueville was stunned by the "tumult" of American political interactions, which "must be seen in order to be understood." In America, he wrote, political activity was in evidence everywhere:

> A confused clamor is heard on every side; and a thousand simultaneous voices demand the immediate satisfaction of their social wants. Everything is in motion around you; here, the people of one quarter of a town are met to decide upon the building of a church; there, the election of a representative is going on; a little further, the delegates of a district are traveling in a hurry to the town in order to consult upon

> some local improvements; or, in another place, the labourers of a village quit their plows to deliberate upon the project of a road or a public school. (1980/1835, p. 78)

Tocqueville expressed surprise at the fact that although they were denied the vote, in the United States, "even the women frequently attend public meetings, and listen to political harangues as recreation after their household labours" (p. 79). In fact, the culture of America was so suffused with political issues that he believed Americans could not converse except to discuss issues. The typical American, he observed, "speaks to you as if he were addressing a meeting" (p. 79).

Tocqueville was concerned that democratic societies might encourage the rise of demagogues who could manipulate the masses, but his observations persuaded him that the immense number of competing groups in the American political system would prevent this from happening. And while he worried about the possibility of political conflict tearing the young nation apart, he became convinced that Americans' love of liberty and belief in the legitimacy of their political institutions would carry them through any crisis that might arise.

The Media and Political Communication

Television has brought about vast changes in the politics of democratic societies. The televised debate between Vice President Richard Nixon and John F. Kennedy in 1960 marked the first time an entire nation could watch candidates for the nation's highest office as they presented themselves and their views to a mass audience. Most viewers believed that Kennedy came across as the more attractive candidate, and he subsequently won the election. From then on, all election campaigns in the United States, except those for smaller local offices, increasingly relied on the media to reach voters with political messages. Campaigns had always made extensive use of newspapers and other print media, but after 1960, television and radio became by far the dominant media of political communication.

Table 19.4 demonstrates the overwhelming importance of television viewing as compared to other forms of media consumption in the United States. To understand the influence of the major networks (that is, ABC, CBS, NBC, Fox, and TBS), one needs only to see that the average adult in the United States spends 865 hours per year watching network broadcasts. Radio is also an extremely important medium, but these comparative figures suggest that newspapers command a relatively insignificant proportion of media attention. In fact, however, individuals

TABLE 19.4

Average Media Consumption, United States (People Age 18 and Over)

	Hours per Person
Total	3,496
Television	1,613
Broadcast TV	639
Network-affiliated stations	577
Independent stations	60
Cable networks	973
Basic cable networks	824
Premium cable networks	149
Broadcast and satellite radio	782
Recorded music	177
Newspapers	171
Pure-play Internet services	189
Consumer magazines	125
Consumer books	109
Video games	85
Home video	61
Box office	12
Pure-play mobile services	15
Educational books	8

Source: *Statistical Abstract*, 2010.

who influence other people's opinions of events and political leaders tend to spend far more time reading newspapers and magazines than the average citizen does. Sociologists refer to such individuals, who are well informed and like to reach independent judgments about current events, as *opinion leaders*. Although they may be a minority in any community, their views can help sway the opinions and votes of others. In consequence, political campaigns attempt to get their positions and candidates covered in the print media, but as the actual election approaches, they spend millions of dollars on television advertising designed to reach the mass of voters, many of whom are undecided until just before election day.

Today, television dominates most major political campaigns, and its influence has stimulated the movement for campaign finance reform. Candidates who can afford to buy television advertising have advantages that other candidates simply cannot match. Thus, in recent elections in the United States, billionaires like Ross Perot, Michael Bloomberg, Jon Corzine, Steve Forbes, and many others have relied on personal funds to buy enough television advertising to make their names household words among voters who might never have heard of them before the campaign. For nonbillionaires, the greatest amount of campaign time is now spent not meeting crowds of voters but meeting crowds of rich donors. As we have seen in the case of Enron, this imbalance of influence in American politics has stimulated a powerful campaign finance reform movement.

In many European nations, political candidates and parties receive free television time and there are structured debates that present candidates and issues to voters. Candidates cannot simply purchase as much media advertising as they can afford. The United States probably will not adopt a similar system in the near future, but sociological research and polling data show that most U.S. voters want significant reforms in the current system of campaign finance (Duffy et al., 2002).

For those out of power, political communication and the ability to gain the attention of the mass media are major concerns. Sociologist Todd Gitlin (1993) reminds us how important this aspect of politics was to the student radicals who challenged U.S. policies in Vietnam in the 1960s and 1970s. They were only one among many social movements of that period that discovered the great power of television in a society with mass publics (see Chapter 9). Through skillful use of television coverage of staged confrontations, sit-ins, and demonstrations, it was possible for a relatively small group of activists to mobilize larger numbers of supporters. The labor movement had used similar tactics in earlier decades, but by the 1960s, it had become easier to communicate to mass audiences and sway public opinion—at least in a society in which the press, including the television news, is relatively free from governmental interference.

In a world in which the techniques of communication are increasingly sophisticated, it becomes ever more important to analyze political communications and to read between the lines of political rhetoric. Political jargon often masks deeds that leaders would rather not admit to. George Orwell, a master of political commentary, made this point in his essay "Politics and the English Language":

> Political speech and writing are largely the defense of the indefensible. . . . Defenseless villages are bombarded from the air, the inhabitants driven out into the countryside, the cattle machine-gunned, the huts set on fire with incendiary bullets: this is called *pacification*. Millions of peasants are robbed of their farms and sent trudging along the roads with no more than they can carry: this is called *transfer of population* or *rectification of frontiers*. People are imprisoned for years without trial, or shot in the back of the neck or sent to die of scurvy in Arctic lumber camps: this is called *elimination of unreliable elements*. Such phraseology is needed if one wants to name things without calling up mental pictures of them. (1950, p. 136)

If Orwell sounds highly pessimistic about the conduct of politics, remember that he was writing in a time of even greater political cruelty and chaos than our own. In his lifetime, he had seen nations with the most advanced constitutions commit the most brutal acts of war and repression. The point of his essay is that to be politically objective, to seek the true meanings of political acts and the consequences of political beliefs, we must "start at the verbal end":

> The great enemy of clear language is insincerity. When there is a gap between one's real and one's declared aims, one turns as it were instinctively to long words and exhausted idioms, like a cuttlefish squirting out ink. . . . If you simplify your English . . . when you make a stupid remark its stupidity will be obvious, even to yourself. Political language . . . is designed to make lies sound truthful and murder respectable, and to give an appearance of solidity to pure wind. One cannot change this all in a moment, but one can at least change one's own habits, and from time to time one can even, if one jeers loudly enough, send some worn-out and useless phrase . . . into the dustbin where it belongs. (p. 140)

George Orwell, the author of *Animal Farm* and *1984*, addressed the abuse of language for political purposes in his essay "Politics and the English Language."

AP Photo

then & now

Television Reshapes the Campaign Trail In 1948, Harry Truman campaigned across the United States, making speeches from a special car at the back of a train. This was one of the last "whistle stop" campaigns in American presidential politics. The older style of campaign, in which the candidate attempted to meet as many voters as possible, was accompanied by a great deal of "back room" politics—the candidate met with local political leaders, who bargained for favors in return for bringing out their supporters on election day. These dealings, of course, were not visible to the people who gathered at the "whistle stops."

Bettmann/CORBIS

Today, the power of television brings the candidate and the campaign message into the homes of millions of voters, diminishing the importance of town-by-town campaigning. But television demands camera skills that many politicians do not necessarily possess. Television has also vastly increased the importance of fund-raising to pay the high costs of television advertising. The influence of local political party leaders has declined, but "back room" politics and the influence of wealthy and powerful contributors have yet to disappear.

Ed Andrieski/AP Photo

MILITARY INSTITUTIONS

The problem of civilian control of the military remains the most important issue in many of the world's nation-states. At this writing, for example, Nigeria, the largest emergent democratic regime in Africa, is experiencing a difficult transition from military absolutism to constitutional democracy. After years of harsh military dictatorship, extreme political and economic corruption, and suppression of the media, universities, and the courts by a military "strongman," two major parties are competing peacefully for electoral support. The press is thriving again, and exiled intellectuals and artists are returning (Gbadamosi & Adewoye, 2010). Nigeria, however, is not an isolated example of the problem of social control over the military. Throughout the world, the problem of how to maintain the allegiance of the military often looms as the largest threat to democracy and to the legitimacy of governments. These situations raise a major question in political sociology: How can states control their military institutions? Answers are to be found in knowledge about the nature of the modern military as a social institution as well as in the political culture of different nations.

President Obama's abrupt firing of Stanley McChrystal, the general heading U.S. operations in Afghanistan, for derogatory remarks made to a *Rolling Stone* reporter, once again raised the key question about military institutions in all societies: How can the military remain under the command of elected civilian leaders? A democratic society cannot remain democratic for long if the military usurps the authority granted to it by the civilian institutions of the nation-state. This is a common occurrence—taking the form of coups led by military "strongmen"—in nations in which the military is not adequately controlled by other institutions.

The problem of civilian control of the military was recognized by Tocqueville, who noted that in aristocratic nations the government could control the military because the military leaders were aristocrats and therefore were part of the court. "The officer is noble, the soldier is a serf," he observed. "The one is naturally called upon to command, the other to obey" (1980/1835, p. 128). Tocqueville formed the hypothesis that greater equality would make a nation less likely to go to war because its citizens would have more to lose in wars and more power to prevent them. Yet he also feared the ambitions of military leaders in democratic nations. "In democratic armies," he wrote, "all the soldiers may become officers, which . . . immeasurably extends the bounds of military ambition" (p. 129).

Morris Janowitz (1960), a highly innovative political sociologist, responded to Tocqueville's concerns in his analysis of the history of military institutions in democratic nations. He argued that the military has often played a crucial role in establishing and protecting democratic institutions. In the Greek city-states and the Roman Republic, as well as the democracies that emerged from the revolutions of the late eighteenth century, citizens were obligated to serve in the armed forces. The military institution could function only through the enlistment of "citizen soldiers." Because the citizen soldier was committed to democratic institutions and was serving because such service was a requirement of citizenship, the military would be unlikely to take over the functions of democratic political institutions. Soldiers would presumably place the values of their society above the demands of the military.

When Janowitz reviewed the impact of new methods of warfare on the military, he concluded that because of the increasing sophistication of military technology, the military in many modern nations had become staffed by "professional soldiers" for whom service in the armed forces is a career. In addition, the shift to a volunteer military, rather than one that depends on citizen soldiers, may threaten the control of the military by civilian institutions (Abbott, 2009; Janowitz, 1960).

The Economic Role of the Military

Even in the United States, where the threat of a coup is not considered great, the military has so much influence on the economy that social control of the military becomes difficult. The function of the armed forces is, of course, the defense of the nation-state; in the United States, however, the military also serves an important economic function as a producer of jobs and revenue. The economies of states like Alaska, Connecticut, Maryland, California, and Virginia

President Barack Obama is greeted by General Ray Odierno, the top U.S. commander in Iraq, as he arrives in Baghdad. Presidential visits to overseas war zones are occasions not only to demonstrate the President's concern for fighting forces, but also to remind military leaders and the public that the President is the Commander in Chief of all the Armed Forces.

AP Photo/Charles Dharapak

became highly dependent on the wealth gained from defense contracts during the decades of the cold war and the arms race. In addition, the private companies that vie for military contracts became a major source of employment, often providing some of the most highly paid and secure jobs in their regions. Throughout the United States, the influence of the military "pork barrel"—federal spending on defense contracts and the resulting employment—often makes it difficult for legislators to support cuts in military budgets and programs.

The growth of military production in different parts of the United States also adds to the potential for armed conflict elsewhere in the world. Faced with declining domestic sales, U.S. arms makers seek to supply advanced weapons to buyers in other nations. Indeed, weapons account for a significant portion of U.S. exports; the United States is the world's leading exporter of conventional weapons (Boese, 1998).

Military Socialization

Social control of the military is partially explained by the professional socialization of military personnel. This point was vividly illustrated by Tom Wolfe in *The Right Stuff* (1980). "There were many pilots in their thirties," Wolfe wrote, "who to the consternation of their wives, children, mothers, fathers, and employers, volunteered to go active in the reserves and fly in combat in the Korean War." This was in vivid contrast to the attitude of the foot soldiers in that war, whose morale during some periods of the prolonged conflict "was so bad it actually reached the point where officers were prodding men forward with gun barrels and bayonets" (p. 32).

This contrast between the "fighter jocks," socialized to want to prove they had "the right stuff" and thus win promotion to higher ranks, and the far more cautious draftees and recruits on the ground, raises a series of questions: What motivates a person to face death in war? Is it desire for glory and advancement? Is it fear of punishment? Is it the pressure of collective action? From an interactionist perspective, this contrast results from the ways in which people are socialized in military institutions, coupled with their definitions of the situation. The fighter jocks define the situation as one in which they must act as they do—it is the job for which they have volunteered. The draftees do not define the situation in the same way; they are more interested in their own survival and in adhering to group norms like "Never volunteer."

In his classic study of socialization in a military academy, Sanford Dornbusch (1955) showed that traditional military socialization

processes are designed to develop a high level of motivation and commitment to the institution. Those processes include the following:

- *Suppression of previous statuses.* Through haircuts, uniforms, and the like, the recruit is deprived of visible clues to his or her previous social status.
- *Learning of new norms and rules.* At the official level, the recruit is taught obedience to the rules of the military; through informal socialization, he or she is taught the culture of the military institution.
- *Development of solidarity.* Both informal socialization and harsh discipline build solidarity and lasting friendships among recruits; they learn to depend on one another.
- *The bureaucratic spirit.* The recruit is taught unquestioning acceptance of tradition and custom; orders are taken and given from morning to night.

The controversy that greeted President Bill Clinton's order that the military cease discriminating against homosexuals is an indication of the sensitivity that surrounds issues of military socialization. Gay men and women have been serving in the armed forces for decades but have had to deny their sexual preferences in military examinations and screening procedures. Objections to openly admitting homosexuals into the military centered on the possibility that gay sexuality would be a divisive influence and a detriment to fighting morale. Gay activists in and outside the military answered that there is no reason why the norms of military conduct would not apply to openly gay soldiers just as they always have to gay and nongay military personnel.

Despite his personal opposition to the president's order, General Colin L. Powell, then chairman of the Joint Chiefs of Staff, stated that "we know that there will be changes and . . . you and the Commander in Chief [the President] can always count on us for faithful support and execution of your decisions" (quoted in Schmitt, 1993, p. A1). General Powell was referring not only to the issue of acceptance of homosexuals in the military but also to the issue of cuts in military budgets. His statement indicates the control the constitutional rule of law provides over the powerful institution of the military.

Social Change and Military Institutions

Like all major social institutions, the military has undergone many changes, some of which can be seen in the transformation of American military institutions in the past two centuries.

visual sociology

Building Democracy in Occupied Nations It is one thing to conquer a nation or society and quite another to institute a government that its people believe is legitimate. In the nineteenth century, the colonial powers, including the United States, found it easy to conquer people all over the less industrialized and militarized world, but ruling the colonized peoples in ways they believed to be fair and representative of their own wishes proved almost impossible for any length of time. The photo of an American soldier on a tank delivering food in Somalia is a graphic reminder that even humanitarian missions like the American-led, U.N.-sanctioned mission to restore order in Somalia and feed its starving people can end in disaster when there is no legitimate national government to maintain order (Dobbins et al., 2003).

The photo of the chief of the U.S.-led occupation of Iraq, Ambassador L. Paul Bremer, is also a reminder of how difficult it has proven to be to bring stability and democratic political institutions to Iraq. This photo was taken just after the terrorist bombing of the United Nations headquarters in Baghdad, which led to added controversy over whether the occupying coalition (primarily the United States and Great Britain) had "enough boots on the ground" to offer security to Iraqis who were cooperating with the goals of the occupation.

occupying nation's growing desire to extricate itself from its costly responsibilities in the conquered nation often results in a rush to create a government and hold elections. In this regard, the Democracy and the Rule of Law Project at the Carnegie Endowment for International Peace notes that rushing elections reinforces the divisions that already exist in a society. As a nation rapidly "decompresses," tribal and religious loyalties fill the vacuum. Citizens attach themselves to what is familiar—often the most belligerent separatists.

Dmitri Kessel/Time Life Pictures/Getty Images

When the British ended their rule in Iraq in 1921, they created the new nation's first king, Faisal. But aside from his friendships with the British, Faisal, whose parents came from Mecca in what is now Saudi Arabia, had never lived in Iraq and was a Sunni Muslim in a land dominated by the Shia Muslims who make up about 60 percent of the population. After years of political unrest, Britain decided to reassert its control over Iraq and its valuable oil fields after World War II.

Unfortunately, the key lesson of Iraq's history, as well as that of so many former colonial regions, is that people prefer bad rule of their own kind to good rule by somebody whom they see as foreign. And the

AP Photo/Jerome Delay

© Stephanie Sinclair/Corbis

In the era of the citizen soldier, people expected to be called into service only in emergency situations and to be trained by a small cadre of professional officers. Today, in contrast, the permanent armed forces have more than 1.5 million uniformed members (*Statistical Abstract,* 2005). In response to the war on terrorism and the projected need for more—and more sophisticated—military hardware and training, for fiscal year 2011, President Obama requested $548.9 billion for the Department of Defense base budget. This was $15 billion more than the $533.7 billion Department of Defense base budget request for fiscal year 2010. The world's only superpower now spends more on defense than all of its allies combined.

Today, the requirements of highly technological forms of warfare and the need to maintain a complex set of military organizations throughout

the world have changed our concept of a military career. The military as a modern social institution requires professionally trained officers who commit themselves to spending much of their working lives in the military. Thus, as the military has evolved, so has its need to recruit and train career officers in specialized military academies.

Over time, the institutions that train professional soldiers have had to adapt to certain changes. For example, as the larger society has accepted the demand for gender equality and women have gained access to occupations that were previously reserved for men, the military has had to redefine its historical perception that women soldiers are better equipped for desk duty than for service in combat. Although few women have fought in ground warfare to date, women in the military are now being trained for combat roles. The idea of women engaging in ground combat remains controversial among military traditionalists—primarily males—but the increasing automation of air and ground combat makes it more reasonable for women to assume the same military roles as men (Browne, 2010), as some are doing in Afghanistan at this writing. Yet the massive military buildup for the war against terrorism is having important consequences for other institutions for many years to come. Education, Social Security, health care, and many other social-welfare activities of the federal and state governments will be under greater stress because of shifts in spending toward defense and domestic security (Alvarez, 2009).

Thinking Critically In a speech given when he left office in 1961, President Dwight D. Eisenhower warned U.S. citizens to beware of the rise of what he called the "military-industrial complex," a permanent and ever-growing war machine that had the potential to threaten the other institutions of American democracy. How do you feel about this situation? Is the growth of the military and its demands on public funds vital to protecting the United States from terrorists? Do you think military spending is justified, or is it "out of control"?

SUMMARY

How are a society's political institutions shaped by its ideas of power and legitimate authority?

- Politics determines "who gets what, when, and how." The basis of politics is competition for *power,* or the ability to control the behavior of others, even against their will.

- *Authority* is institutionalized power, or power whose exercise is governed by the norms and statuses of organizations.

- Sets of norms and statuses that specialize in the exercise of power and authority are *political institutions,* and the set of political institutions that operate in a particular society forms the *state.*

- Although a state can exercise its power through the use of force, the use of force will not be enough to govern a society in the long run. When people consent to be governed without the use of force, the state is said to be legitimate.

- *Legitimacy* is a society's ability to engender and maintain the belief that the existing political institutions are the most appropriate for that society. Legitimacy is the basis of a society's political culture—the cultural norms, values, and symbols that support and justify its political institutions.

- According to Max Weber, the political cultures of different societies give rise to three different types of authority: *traditional, charismatic,* and *legal.*

- Perhaps the strongest and most dangerous political force in the world today is *nationalism,* the belief of a people that they have the right and the duty to constitute themselves as a nation-state.

- The most important political territory in world affairs is the *nation-state,* the largest territory within which a society's political institutions can operate without having to face challenges to their sovereignty.

Why has globalization created crises for nation-states?

- Throughout the world, there are nation-states in crisis. Even when citizens believe that their government is legitimate, force may be needed to maintain order and ensure compliance with the law.

- *Citizenship* is the status of membership in a nation-state.

- The rights of citizenship include civil rights (for example, freedom of speech, thought, and faith), political rights (the right to vote), and social rights (the right to a certain level of economic welfare and security).

- The question of who is entitled to full participation in politics is a key issue at the local level as well as at the national level.

What are the main characteristics of different types of political regimes in the world today?

- Modern nations are governed by elected and appointed officials whose authority is defined by laws. To prevent abuses of authority, modern political institutions usually specify some form of separation of powers in which abuses by one institution can be remedied by others.

- *Political parties* are organizations of people who join together to gain legitimate control of state authority. Parties that accept the rule of other legitimate parties form a "loyal opposition" that monitors the actions of the ruling party and prevents the emergence of *oligarchy,* or rule by a few people who stay in office indefinitely. Regimes that accept no limits to their power and seek to exert their rule at all levels of society are known as *totalitarian regimes.*

- It is extremely difficult for minorities in a nation-state to feel that their rights are protected and honored if the institutions of government are dominated by members of the majority group.

- *Democracy* means rule by the nation's citizens: Citizens have the right to participate in public decision making, and those who govern do so with the explicit consent of the governed.

- In the British parliamentary system, elections are held in which the party that wins a majority of the seats in the legislature "forms a government": The leader of the party becomes the head of government and appoints other party members to major offices.

- In the American representative system, the party whose candidate is elected president need not have a majority of seats in the legislature.

- Although third parties in a representative system normally do not share power after elections, they can have an enormous impact on the course of political history.

- National regimes can be classified into several categories: liberal democracies, emergent democracies, communist regimes, nationalistic socialist regimes, authoritarian nationalist regimes, military regimes, Islamic nationalist regimes, and absolutist regimes. Most of the world's people live in liberal or emergent democracies.

How do the basic sociological perspectives contribute to our understanding of political sociology?

- Functionalist theorists assert that certain "structural prerequisites" must exist in a society for democratic political institutions to develop and operate. Among these are high levels of economic development, urbanization, and literacy, as well as a culture that tolerates dissent.

- The *power elite model* holds that the presence of democratic institutions does not mean that a society is democratic; political decisions are actually controlled by an elite of rich and powerful individuals.

- This view is challenged by the *pluralist model,* which holds that political decisions are influenced by a variety of *interest groups* through a process of coalition building and bargaining.

- Many political thinkers have been concerned with the relationship between social interaction and political institutions. Among the most influential was Machiavelli, who argued that political institutions must be based on the recognition that human beings are capable of evil as well as good. This recognition played a major part in the planning of the government of the United States, in which each branch of the government is able to check abuses of power by the other branches.

- Today, television dominates most major political campaigns, and its influence has stimulated the movement for campaign finance reform.

- In modern nation-states, political communication is a major concern for those out of power as well as those in power. Analyzing political communications and reading between the lines of political rhetoric is important, particularly in societies characterized by mass publics and sophisticated communication techniques.

What are the particular political challenges of the modern military in a democratic society?

- A major question in political sociology is how states can control their military institutions. A society cannot remain democratic for long if the military usurps the authority granted to it by civilian institutions.

- One factor that contributes to this problem is the fact that the military is staffed by professional soldiers for whom service in the armed forces is a career.

- Another factor is the immense influence of the military on the economy. On the other hand, military socialization instills norms and values that may contribute to social control of the military.

The Kornblum Companion Website

www.cengagebrain.com

Supplement your review of this chapter by going to the companion website to take one of the tutorial quizzes, use the flash cards to master key terms, and check out the many other study aids you'll find there. You'll also find special features, such as GSS data and Web links that will put data and resources at your fingertips to help you with that special project or to do some research on your own.

Rachel Epstein/PhotoEdit

20

Health and
Medicine in Global
Perspective

Issues of Health
Care in the United
States

Medicine and
Technology

Health and Medicine

FOCUS QUESTIONS

What are some key health care issues in the developing nations?

What factors make it difficult to control health care costs in the United States?

How have modern science and technologies transformed medical institutions?

What is social epidemiology, and how does it contribute to our understanding of health and medicine?

Médecins Sans Frontières (MSF), known also as Doctors without Borders, is a nongovernmental organization that recruits doctors and medical professionals to work in medical crisis situations all over the world. At present, their volunteers can be found on the medical front lines in most of the world's war-torn regions, from Somalia to Kyrgyzstan, or providing medical relief in areas devastated by natural disasters, such as Port-au-Prince and other cities and villages of Haiti, the poorest nation in the western hemisphere.

When MSF's La Trinité Hospital was destroyed in the January 12, 2010, earthquake, Haiti lost its only specialized treatment unit for victims of severe burns. MSF quickly put up a temporary burn unit. Dr. Rémy

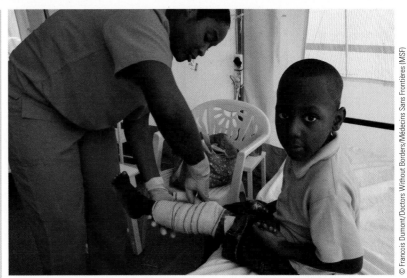

Walderson is attended to at the burn unit at MSF's Saint-Louis Hospital in Port-au-Prince. For more than a month, his parents have been keeping watch at the bedside of their five-year-old son. They take turns at the hospital while caring for their other three children. The family lives in a tent set up in front of their house, which was badly damaged by the earthquake that hit the island on January 12, 2010.

Zilliox, a plastic surgeon and burn specialist at Edouard Herriot Hospital in Lyon, France, describes his experience with Haitian burn victims:

I started working in the burn unit just after it opened in late March. At that time, we had two or three admissions of varying severity every day. The leading cause of burns is related to the very precarious conditions in which hundreds of thousands of people live. Because they lack housing, all aspects of family life take place in just one, often very cramped space.

Family members sleep, play, and cook in that same area. Women and children are often burned during domestic accidents—a pot of boiling water or oil tips over, or a candle sets fire to a blanket. We see so many children that we had to set up a separate tent. The unit includes three tents and thirty beds for severe burn patients, both children and adults.

Men are most often burned when handling flammable products, primarily fuel containers. Because there are fuel shortages, people store fuel, which increases the risk of accident. I've also seen electrical burns, which are particularly serious. Electrical wires that trail on the ground are very dangerous. I had to amputate the feet or hands of four patients who were burned by electricity. Two of them did not survive their burns.

Before the earthquake, La Trinité hospital treated patients who were burned in domestic accidents or by electrocution.

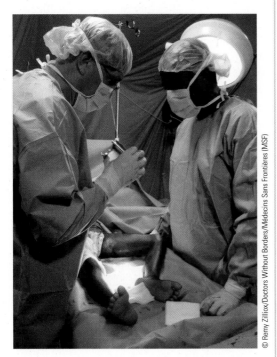

Doctors tend to a young patient in the new burn unit in MSF's Saint Louis Hospital in Port-au-Prince.

Electrical facilities were already in poor shape, especially after the 2007 hurricane. The causes haven't changed much, but burns are increasingly frequent and severe now because more people are living in even more precarious conditions.

A burn patient generally requires very regular care for three weeks to a month. Dressings must be changed every other day until the burns are fully healed. . . . Daily treatment is complicated, and you must have the equipment to perform skin grafts on severe burn patients. We perform three or four grafts every day . . . The skin that is to be grafted is processed in a machine that expands its surface. This makes it possible to take as little skin as possible from the patient, because the procedure can be performed using only the patient's own skin. Appropriate conditions are required to carry out the operation, and you must have a blood bank for transfusions (Remy Zilliox, *Voice from the Field Haiti: Treating Severe Burn Victims After the Earthquake* from Medecin Sans Frontiers, May 6, 2010. Reprinted with permission.)

With the help of funds from hundreds of thousands of concerned donors throughout the world, a burn unit has now been installed in MSF's new Saint-Louis Hospital in Port-au-Prince.

HEALTH AND MEDICINE IN GLOBAL PERSPECTIVE

The selfless work of MSF brings modern technologies of medicine and health care to nations and regions that are in desperate need of their services. MSF attempts to deliver health care and medical services in areas where existing institutions are overwhelmed by crisis or are nonexistent, as is often the case, for example, in refugee camps where thousands of people live in dire conditions. MSF's activities also highlight a number of compelling moral and sociological issues concerning health and medicine. Not surprisingly, these same issues are raised in one way or another in almost every nation of the world, whether rich or poor.

First, health care is distinct from health. "Health care" refers to medical services, but not to the health of individuals or populations. Health care is only one determinant of health. The health of an individual is highly associated with personal income; for a nation, the overall health of the population is associated with the degree of poverty and economic inequality in the society. In Haiti, for example, poverty is the leading cause of illness, but in addition, people who cannot pay have limited access to decent health care services. That is why nongovernmental agencies like MSF are playing such a vital role there.

Health itself may be viewed as a characteristic of individuals, who may be relatively healthy or unhealthy, but as just noted, health can also be an attribute of entire societies. When social scientists and medical professionals speak of the health of populations or nations, or of regions and cities within them, they are referring to "public health." As we have seen in earlier chapters, social scientists use basic measures of public health, such as infant mortality rates, maternal death rates, and life expectancy, as global indicators of public health. These measures of public health also serve as indicators of global inequality and social injustice (Sen, 2001).

A second critical perspective on medicine and health care is equality of access to medical care. How people in a society do or do not gain access to the services provided by these institutions goes a long way toward determining the quality of life of different groups in society. When we compare nations from a global viewpoint, issues of access to medical care, and the quality of that care, are important dimensions of analysis. In Haiti, there are not enough Haitian doctors, partly because many migrate to places where they can earn an adequate living and work in modern hospitals and clinics. But it is also expensive to train medical personnel, and very poor nations like Haiti lack the means to train a sufficient number of health care professionals.

Third, health care itself is a major institutional sector of society composed of a variety of institutions, including hospitals, clinics of physicians and medical professionals, pharmaceutical and insurance companies, and medical research centers in universities and private corporations. Coordinating the activities of these health care institutions is a huge task for any nation. Making them economically viable, encouraging innovation, and holding them to appropriate moral and scientific standards are also major challenges. The furious debates raging in the United States about health care reform provide ample evidence of the difficulty of making health care both affordable and accessible (Starr, 2010). In Haiti and other extremely poor nations, local health care institutions are so overwhelmed with patients that they must call upon global voluntary health providers like MSF to supplement existing health care services or provide specialty care like that provided by the burn unit at Saint-Louis Hospital. When disaster strikes, as it did in the 2010 earthquake in Port-au-Prince, this need becomes even more urgent.

Should poor nations count on medical assistance from more affluent nations and from international organizations like the United Nations? Should people in the more affluent nations shoulder the responsibility for helping to provide this aid? These questions raise even larger ethical and sociological issues. What responsibility do we have for the misfortunes of others? Even in the affluent nations like the United States, there are people who are desperate for adequate health care. Why should scarce resources be devoted to international aid, medical or otherwise?

TABLE 20.1

Global Estimates of Migrant Populations

Category of Migrant	Population Estimates
Regular immigrants	Annual flow of ~ 2.4 million
International Students	~ 2.1 Million
Migrant workers	~ 86 Million
Refugees	16 million
Asylum seekers or refugee claimants	650,000
Temporary—recreational or business travel	900 million per year
Trafficked (across international borders)	Estimated 800,000 per year
Internally displaced	51 million (includes those displaced by natural disasters and conflict)

Source: Adapted from Gushulak, B., Weekers, J., & MacPherson, D.W. 2009.

One can begin to formulate answers to these questions from a pragmatic viewpoint that does not make assertions about ethics or morality, but rather, sees advancing global health as being in the best interests of all individuals and nations in a rapidly globalizing social environment. Without invoking ideas of global ethics or morality, there are strong arguments for supporting international assistance for health care. And when we look at issues of global health from a social justice viewpoint, the issues of responsibilities and rights in a global setting become even more evident.

A Pragmatic View of Global Health

It makes sense to invest in developing global health care systems and improving the health of people throughout the world because the boundaries between nations and regions are no longer as rigid as in the past. No nation in the contemporary world can exist without allowing a certain amount of movement of foreign people into and out of its territory. If people in the affluent nations don't want to be exposed to deadly illnesses like dengue fever, AIDS, or other potentially deadly epidemics, efforts must be made to treat and eradicate these diseases where they are most prevalent, which is often in the less developed regions of the world. Table 20.1 documents the annual flow of migrants of different types throughout the world. It shows that 2.1 million foreign students, many of whom come to institutions of higher learning in the United States, contribute immensely to the economies of the receiving nations. Nine hundred million recreational and business migrants move around the world

each year. Do we want to exclude these important streams of foreign visitors? Would it even be possible? The World Health Organization has concluded that only through global approaches to monitoring and fighting diseases can nations protect their populations against influenza and similar pandemics that travel along with international migrants (Gushulak, Weekers, & MacPherson, 2009).

The Importance of a Safe Water Supply

Contaminated water is the leading cause of infant death throughout the world. Creating a safe water supply is therefore one of the most important ways of improving global public health. The most essential step in

Desmond Kwande/AFP/Getty Images

Only about 40 percent of Africa's population has access to a supply of safe drinking water.

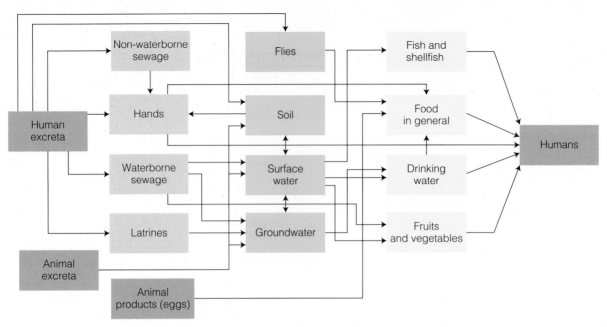

Figure 20.1 Contaminants of Water Supplies. Source: From "Safer Water, Better Health: Costs, benefits and sustainability of interventions to protect and promote health," by Annette Prüss-Üstün, Robert Bos, Fiona Gore, Jamie Bartram. World Health Organization, Geneva, 2008. © World Health Organization 2008. All rights reserved.

developing a supply of clean water anywhere is the careful separation of human and animal waste, and other forms of sewage, from the water supply used for drinking and irrigation. This relationship is shown in Figure 20.1. Anyone who is involved in development work needs to understand this basic diagram. Figure 20.2 shows how many serious diseases are spread through faulty water systems.

Improvements in a nation's water supply can become a precursor to economic and social development, which has positive benefits not only in the poorer nations but throughout the world. The World Health Organization estimates that "An important share of the total burden of disease worldwide—around 10 percent—could be prevented by improvements related to drinking-water, sanitation, hygiene and water resource management" (Prüss-Üstün et al., 2008, p. 7). Indeed, much of the global economic growth that has occurred in the last few decades—which has benefited all the world's nations—is attributable to the rapid development of the economies of Asia, especially those of India and China, Indonesia, Malaysia, and Vietnam, where major advances have occurred in making clean water available to urban and rural populations. China and India have both made huge investments in clean water systems and in the past decade have ensured that over 70 percent of their populations have access to safe drinking water. More remains to be accomplished, but the progress in the past decade has been an outstanding aspect of economic growth in both nations.

Africa still lags behind other continents in this vital area of development, often because so many African nations are so politically instable that international aid projects cannot proceed. South Africa, however, is one of the continent's more promising nations. Hosting the 2010 soccer World Cup was a symbol of that African nation's progress toward development, but world viewers also saw how much remained to be accomplished to address problems of poverty and public health in the impoverished townships. Improvements in the

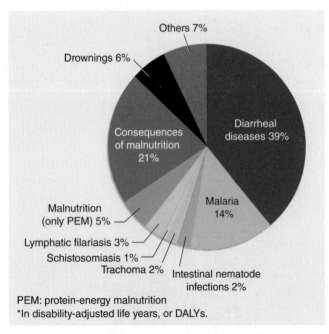

Figure 20.2 Diseases Contributing to the Water-, Sanitation-, and Hygiene-Related Disease Burden*. Source: From "Safer Water, Better Health: Costs, benefits and sustainability of interventions to protect and promote health," by Annette Prüss-Üstün, Robert Bos, Fiona Gore, Jamie Bartram. World Health Organization, Geneva, 2008. © World Health Organization 2008. All rights reserved.

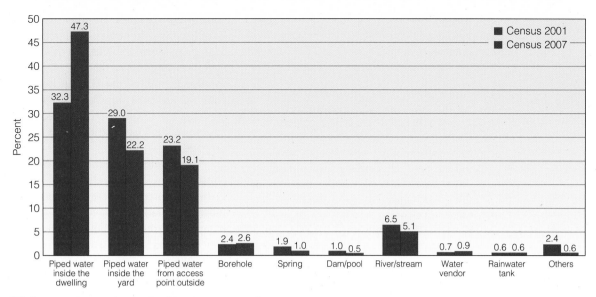

Figure 20.3 **Improvements in South Africa's Water Supply.** Source: From Viljoen, F.C. 2010. The World Health Organization's water safety plan is much more than just an integrated drinking water quality management plan. *Science and Technology*, 61, 173–179. Reprinted by permission.

nation's water supply system, however, are an even more important indicator of development than the shiny new soccer stadiums. We see in Figure 20.3 that from 2001 to 2007, the nation achieved a 15 percent increase in the number of homes with inside drinking water, but the graph also shows that a significant proportion of the nation's population still draws water from streams and rivers and other potentially unsafe sources.

ISSUES OF HEALTH CARE IN THE UNITED STATES

In the history of the Western, developed nations, the separation of drinking water from sewage and groundwater runoff was the single most important improvement contributing to public health and economic development. Much of this task was accomplished more than a century ago, although problems continue to appear even in the most affluent nations. A focus on water issues in the developing regions of the world helps us understand how development projects can account for vast improvements in people's quality of life and thereby help untold millions of children have a better chance of realizing their full human potential.

Sociology &
Social Justice

HEALTH IN THE DEVELOPED WORLD During the contentious debates leading up to the passage of health care reform in 2010, the social justice perspective figured prominently on both sides (Rawles, 2001).

Opponents of the Obama administration's proposals were particularly angry about the proposal to require individuals to have some form of health insurance, either public or private (a plan that had been proven to work in Massachusetts). They argued that the requirement was an infringement of their liberties, and pursued lawsuits attempting to bar this aspect of the law from taking effect. This argument is based, however, on an idea of individual rights rather than on what is just and fair for the society as a whole. The more liberal argument for universal access to health care begins with the belief that health care should be an equal right of all people. A "right" implies that the government guarantees something to everyone. Rights come in two categories: individual freedoms and population-based entitlements. Some conservatives and libertarians support only the former category, whereas modern liberals support both. Entitlements require that the government either appropriate money for a service or demand that another entity pay for the service—for example, education or health care (Bodenheimer, 2008).

In his September 2009 address to a joint session of Congress, President Obama clearly set forward a social justice argument for health care as a right. He quoted a letter from Senator Ted Kennedy, written earlier in the year but delivered to the president after the senator's death, stating that health care "is above all a moral issue; that at stake are not just the details of policy, but fundamental principles of social justice and the character of our country." But the president also took into consideration moral arguments from religious groups and agreed that the health care bill would not include federal funding of abortion (Wallis, 2009).

The social justice argument for health care reform in the United States is also based, quite often, on comparisons between the United States, where 46 million people had no health insurance as of 2009, to European and other advanced, industrialized nations. In a careful examination of health care systems in these nations, T. R. Reid (2009) concluded that:

> those Americans who die or go broke because they happened to get sick represent a fundamental moral decision our country has made. Despite all the rights and privileges and entitlements that Americans enjoy today, U.S. citizens have never decided to provide medical care for everybody

who needs it. In the world's richest nation, we tolerate a health-care system that leads to large numbers of avoidable deaths and bankruptcies among our fellow citizens. . . .

Like most international observers, Reid notes that all the other developed countries on earth have made a different moral decision:

> All the other wealthy, technologically advanced, industrialized democracies guarantee medical care to anyone who gets sick. Countries that are just as committed as we are to equal opportunity, individual liberty, and the free market have concluded that everybody has a right to health care—and they provide it. (Reid, 2009)

Although Reid's statement is true overall, there is no single system that other nations have adopted for delivering universal health care. England and France have universal coverage, in which the government pays almost all health care costs and individuals visit doctors, clinics, and hospitals that are essentially run by the state. Some nations, like the United States, rely on an insurance-based system in which individuals and families can select one of a variety of health care plans that allow them varying levels of choice in the selection of doctors. The United States relies on an employer-based system in which the majority (52 percent) of its population receives their benefits while working. Sudden unemployment can mean a painful loss of health care coverage or require them to purchase it on the open market at an extremely high cost for most moderate- or low-income families. The elderly and retired citizens, and the very poor, qualify for Medicare or Medicaid coverage, which is funded by tax revenue and represents, especially in the case of Medicare, a successful version of a public health care plan for the elderly.

Many people, including Senator Kennedy, hoped that the Medicare system could be extended to cover all Americans, but this proposal was defeated by conservatives in both political parties. Starkly divided along ideological lines, Congress passed a compromise reform plan that promises to add 32 million people to those with health care insurance, in part by insisting that all Americans be covered by some form of health insurance by 2014. Few legislators, and perhaps fewer American voters, were entirely pleased with the reforms, but the bill does culminate many decades of effort to bring the U.S. health care system in line with those of the other highly developed nations to which we often compare ourselves. It remains to be seen in the next few years, as the reforms are implemented and Congress attempts to continue the work of making health care more accessible and controlling its costs, whether the efforts will have been worth the pain and political sacrifice of so many elected officials. At this writing, however, it is safe to say that the measures passed in 2010 to reform the U.S. health care system count as the most significant legislation passed by Congress since the creation of the Medicare system in 1965 (see the Research Methods box on page 523).

Proponents of health care reform in the United States did not rely exclusively on social justice arguments. At least as important was the more pragmatic argument that reform would benefit everyone by finally addressing the problem of rising health care expenses (Marmor, Oberlander, & White, 2009). An analysis by the Congressional Budget Office (see Figure 20.4) shows that as a percentage of the total U.S. gross domestic product (GDP), health care spending, especially that of employers and private citizens (not Medicare or Medicaid), has risen steadily from 1960, when it accounted for slightly less than 5 percent, to the present, when it accounts for over 14 percent. The Organization for Economic Cooperation and Development (OECD), an international organization that helps governments tackle the economic, social, and governance challenges of a globalized economy, reports a far lower average: 8.9 percent of its members' GDP.

A paradox of the U.S. health care system is that higher costs do not bring higher quality of care. The United States spends 5 percent more of its GDP on health care than does France, the nation with the second-highest level of health care spending among the thirty wealthy countries in the OECD. But the United States has little to show for these far greater expenses. For example, it has fewer physicians per capita than do most other OECD countries: 2.43 per 1,000 population versus an OECD average of 3.1. Austria, Belgium, Iceland, Ireland, the Netherlands, and Norway all spend at least one-third less of their GDP on health than does the United States yet average almost four doctors per 1,000 population. Only four OECD countries have fewer acute-care hospital beds per capita than does the United States, which has 2.7 per 1,000 population versus an OECD average of 3.8. Japan has 8.2 acute-care beds per 1,000 population, despite spending half as much of its GDP on health care as the United States does.

Comparative research also shows that the United States does not achieve significantly better health through its vastly higher health care spending. Americans' life expectancy at birth, considered the best general measure of a population's health, has increased by 8.2 years since 1960. In contrast, life expectancy has risen more

A Social Justice Typology of Health Care Systems

Uwe Reinhardt (2008) of Princeton University is among the leading U.S. social scientists who study health care systems. In his typology of health care systems throughout the developed world, he clearly describes how each system is financed and bears the costs of care, and some of the related social justice issues. In his writings, he emphasizes the problem of "moral hazard," which arises frequently in all discussions of social justice politics and policies. If people are offered an entitlement such as unemployment insurance or, in this case, free or low-cost medical care, there is the hazard that they are likely to abuse the privileges and overuse the system.

In the accompanying table, cell A represents pure "socialized medicine," such as the Veterans Administration (VA) health system. In that system, government performs all the basic health care functions. Cells A through F represent "social insurance" systems. In these, government performs the financing and risk-pooling functions, and the insured individual's contribution to that risk pool is based on her or his ability to pay. Under social insurance systems, health care can be purchased under two distinct arrangements: (1) a single-payer approach (cells A, B, C), such as Medicare, the provincial Canadian health plans, or Taiwan's single-payer national health insurance system; or (2) a multiple-payer system (cells D, E, F), such as the private Medicare Advantage Plans or Medicaid managed-care plans in the United States, or the Statutory Health Insurance System in Germany, under which over 200 independent, nonprofit sickness funds compete for enrollees mainly based on the quality of their services. Under either arrangement, however, the delivery side can embrace all forms of ownership and control. Government manages only the financing and risk-pooling functions and sometimes the purchasing function as well.

In health care systems that rely mainly on private, not-for-profit insurers (cells G, H, I) or on for-profit insurers (cells J, K, L), the individual's contribution to risk pools typically is not based on ability to pay, but rather, is either a per capita levy, if insurance premiums are community rated, or a so-called "actuarially fair" premium that is based on the individual's health status and set to come close to the insurer's expected outlays for that individual's health care in the coming period. In the eyes of Europeans and Canadians, the per capita basis and the actuarially fair approach to setting premiums violate the principle of social solidarity. Many Americans, however, seem to find them ethically acceptable.

Finally, the complete or partial lack of insurance (cells M, N, O) approximates a genuinely free market in health care because it avoids the "moral hazard" inherent in health insurance. "Moral hazard" refers to the potential for overuse of health care, because at the point of using health care, an insured person pays much less than the true full cost of producing that care. Although some may consider this arrangement ideal, few modern societies embrace it. First, it fails to reap the benefits of protection against the financial inroads of illness. Second, it violates widely shared principles of fairness.

| Ownership of Providers | The Financing of Health Care | | | | Insurance |
| | Social Insurance (Financed by ability-to-pay) | | Private Insurance (Actuarially financed) | | No Health Insurance (Out-of-pocket) |
	Single Payer	Multiple Carriers	Non-Profit	For-Profit	
Government	A	D	G	J	M
Private but non-profit	B	E	H	K	N
Private and Commercial	C	F	I	L	O

Source: From Reinhardt, U.W. "The True Cost of Care: The complicated relationship between the market and government health programs." From *America Magazine: The National Catholic Weekly*, September 8, 2008, pp. 10–13. Reprinted by permission.

in most other OECD countries. In Canada, life expectancy has risen by 9.4 years, and it has risen by over 10 years in both Germany and France. Life expectancy rose by almost 15 years in Japan during the same period. Infant mortality is also a commonly used comparative measure of the quality of a health care system. In 2006, 6.7 infants died per 1,000 live births in the United States—a sharp decline from 26 per 1,000 in 1960. But the infant mortality rate is lower in every other OECD country except Turkey and Mexico. The average rate for all OECD countries is 4.9 deaths per 1,000 live births.

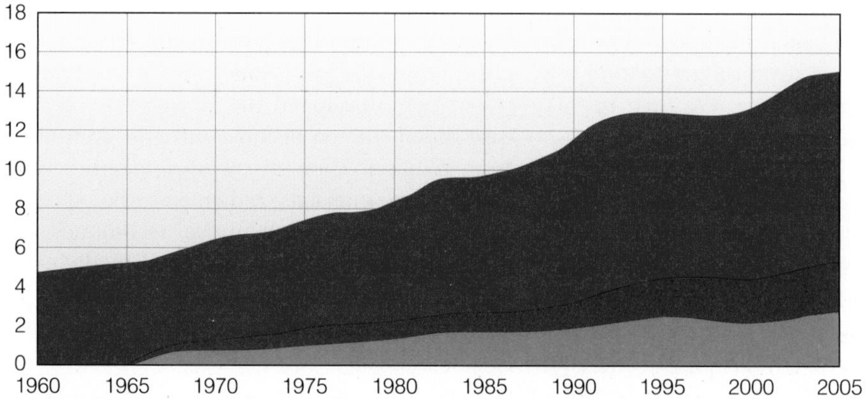

Medical sociologists have been arguing for decades that the United States needs to work harder to control medical costs, and the Obama administration made cost control a central argument in promoting its reform plans (Bloom, 2002). During his campaign, Obama argued that bringing rising health costs under control is vital to the long-term health of the U.S. economy. Critics have argued that it is not clear how the health care bill that was finally passed in

Figure 20.4 U.S. Health Care Costs as a Percentage of GDP. Source: Congressional Budget Office.

2010 will achieve these cost controls. Part of the administration's answer lies in increasing levels of preventive medicine. The administration has continued to develop the details of its reform program and issued new rules requiring health insurance companies to provide free coverage for dozens of screenings, laboratory tests, and other types of preventive care. The theory behind this policy is that, with more preventive measures to decrease costly behaviors like smoking and the unhealthful eating patterns that cause obesity, and by increasing incentives for people to get more medical tests done, early detection of serious illnesses such as diabetes, colon cancer, and heart disease will have the effect of reducing costs. The sicker a patient is when first treated, the higher the costs of care. Kathleen Sebelius, Secretary of Health and Human Services, notes that 100,000 deaths could be averted each year if doctors and patients made more effective use of just five services: colorectal and breast cancer screening, flu vaccines, counseling on smoking cessation, and aspirin therapy to prevent heart disease (Pear, 2010).

Another costly aspect of the U.S. health care system is its extremely high rate of use of expensive medical technologies. Critics of the Canadian and European health care systems often argue that patients in those nations must wait longer for advanced medical tests like MRIs or CT scans. However, research has shown that U.S. physicians often order these expensive tests because patients expect them to be done, rather than because they are absolutely necessary. In 2007, there were twenty-six MRI machines per 1 million population in the United States, compared to an average of fewer than ten in the OECD nations. This gap in high-tech equipment is shrinking, although the rate at which U.S. physicians order such tests remains extremely high relative to the rate in other nations (Bartlett, 2009).

The subject of medical technologies is a complex one that warrants a broader consideration of how medicine has changed in the modern era as scientific knowledge has transformed it from a mystical art to a highly sophisticated human service. We turn to this subject in the next section.

Thinking Critically

Do you think the free market can eventually provide universal health care for people in the United States? If so, how do you think this could happen? How would free markets accommodate the needs of the elderly, veterans, or the poor?

MEDICINE AND TECHNOLOGY

Throughout most of human history, limitations on food production, together with lack of medical knowledge, have placed limits on the size of populations. Dreadful diseases like the bubonic plague have actually reduced populations. In England, the plague, known as the Black Death, was responsible for a drastic drop in the population in 1348 and for the lack of population growth in the seventeenth century. In 1625 alone, 35,417 residents of London died of the plague. Smallpox and dysentery have had similar, though less dramatic, effects (Davis, 1992; Wrigley, 1969).

Until relatively recently, physicians were powerless either to check the progress of disease or to prolong life. In fact, they often did more harm than good—that is, their remedies were more harmful than the illnesses they were intended to cure. As Lewis Thomas (1979) has stated:

> Bleeding, purging, cupping, the administration of infusions of every known plant, solutions of every known metal, every conceivable diet including total fasting, most of these based on the weirdest imaginings about the cause of the disease, concocted out of nothing but thin air—this was the heritage of medicine up until a little over a century ago. (p. 133)

Thomas's point is that, before the nineteenth century, when scientists finally began to understand the nature of disease, physicians based their treatments on folklore and superstition. In fact, with few exceptions, the practice of healing, like many other aspects of science, was closely linked to religion. In ancient Greece, people who suffered from chronic illnesses and physical impairments would journey to the temple of Asclepius, the god of healing, in search of a cure. In medieval times, pilgrims flocked to the cathedral at Lourdes in France (as many still do today) in the belief that they would thereby be cured of blindness, paralysis, or leprosy. Not until Louis Pasteur, Robert Koch, and other researchers developed the germ theory of disease did medicine become fully differentiated from religion. Their discoveries, together with progress in internal medicine, pathology, the use of anesthesia, and surgical techniques, led to the twentieth-century concept of medicine as a scientific discipline (Cockerham, 2004).

During the nineteenth century, scientific research resulted in the discovery of the causes of many diseases, but at first this progress led physicians to do less for their patients rather than more: They began to allow the body's natural healing processes to work and ceased to engage in damaging procedures like bloodletting. At the same time, they made major strides toward improving public health practices. They learned about hygiene, sterilization, and other basic principles of public health, especially the need to separate drinking water from wastewater. These innovations, which occurred before the development of more sophisticated drugs and medical technologies, contributed to a demographic revolution that is still under way in some parts of the world. Suddenly rates of infant mortality decreased dramatically, births began to outnumber deaths, and life expectancy increased. This change resulted not from the highly sophisticated techniques of modern medicine, but largely from the application of simple sanitation techniques and sterilization procedures. In fact, these simple technologies have had such a marked effect on infant survival that the rate of infant mortality in a society is often used as a quick measure of its social and economic development (see Figure 20.5).

In sum, as medical science progressed toward greater understanding of the nature of disease and its prevention, new public health and maternal care practices contributed to rapid population growth. In the second half of the nineteenth century, such discoveries as antiseptics and anesthesia made possible other life-prolonging medical treatments. In analyzing the effects of these technologies, sociologists ask how people in different social classes gain access to them and how they can be more equitably distributed among the members of a society. The ways in which medical technologies have been institutionalized in hospitals and the medical profession are a central focus of sociological research on these questions.

The Hospital: From Poorhouse to Healing Institution

In the twentieth century, the nature of medicine changed dramatically as scientific investigation expanded our knowledge of the causes and cures of disease. That knowledge led to the development of a vast array of technologies for the prevention and cure of many known illnesses, as well as the long-term care of terminally ill patients. Because the more complex of these technologies are applied in a hospital setting, it is worthwhile to consider the development of the hospital as the major social institution for the delivery of health care.

Historically, hospitals evolved through several stages, beginning as religious centers and eventually developing into centers of medical technology (Cockerham, 2004). The first hospitals were associated with the rise of Christianity; they were community centers for the care of the sick and the poor, providing not only limited medical care but also food, shelter, and prayer. During the Renaissance, hospitals were removed from the jurisdiction of the church and became public facilities. Because they offered food and shelter to the poor regardless of their health, they soon became crowded with invalids, the aged, orphans, and the mentally ill. The third phase in the development of hospitals began in

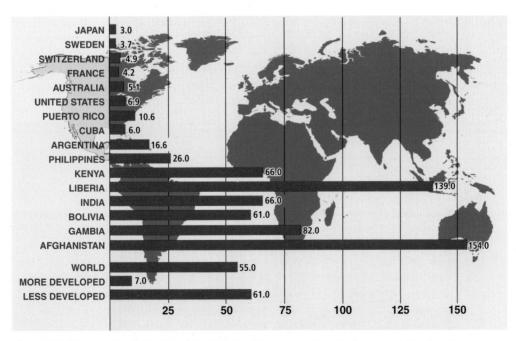

Figure 20.5 **Infant Mortality Rates for Selected Countries.** Note: The infant mortality rate is the number of deaths of infants younger than 1 year of age per 1,000 live births in a given year. Source: Data from Population Reference Bureau, 2003.

the seventeenth century, when physicians gained influence over the care of patients in hospitals. Gradually, the nonmedical tasks of hospitals disappeared, and the hospital took on its present role as an institution for medical care and research.

The modern hospital began to emerge at the end of the nineteenth century as a result of the development of the science of medicine. Especially important were advances in bacteriology and increased knowledge of human physiology, along with the use of ether as an anesthetic. Because the new medical technologies were more complex and often more expensive than earlier forms of treatment, they were centralized in hospitals so that many physicians could use them. Physicians also began to refer patients of all social classes to hospitals, and those patients paid for the services provided to them there. In the United States, the number of hospitals grew rapidly in the twentieth century—from a few hundred at the turn of the century to more than 6,500 in the 1990s.

Today, hospitals play an important role in the control of medical practice and access to medical care. For example, doctors who want to practice in a particular hospital must be accepted by the hospital's medical board. Patients who want high-quality care in private hospitals must be able to pay the fees charged by those hospitals or have the necessary insurance coverage. And hospitals have a monopoly on advanced medical technologies, a fact that has had a major impact on the American health care system.

Medicine and Social Change

Today the technologies available for the diagnosis and treatment of serious illnesses are often described as "miracles of modern medicine." The magnetic resonance imaging (MRI) scanner, for example, allows hospital technicians to observe a patient's internal organs without the use of X-rays; renal dialysis is used to prevent patients from dying of kidney failure; open-heart surgery is practically a routine operation, with the patient kept alive during the process by an external heart–lung machine; other surgical procedures involve the use of laser beams and fiber optics to perform delicate operations. All of these technologies require the use of extremely expensive equipment and highly trained personnel.

The development of increasingly sophisticated and costly medical technologies, together with the practice of requiring patients (or their insurance companies) to pay for hospital services, has led to a crisis in American medical care. The high cost of medical care has become a

major public issue, as has the fact that some groups in the population are unable to obtain adequate care. (Over 40 percent of Americans have inadequate medical insurance, or none at all.) Some critics claim that the American health care system is suffering from *hypertrophy,* by which they mean that it has expanded to a size and complexity at which it has become dysfunctional. In their view, excessive emphasis on technological progress has created a situation in which the needs of the patient are subordinated to those of the providers of health care.

According to Paul Starr (1982, 1995), the problems of the American health care system stem from the way medical institutions evolved. As medical knowledge increased and technological advances were made, physicians developed narrow specialties and hospitals invested in specialized equipment. The physicians referred their patients to the hospitals for sophisticated medical testing and treatment. At the same time, the institution of health insurance emerged in response to demands for a more equitable distribution of health care. Insurance companies or the government began to pay for the services provided by hospitals. Physicians and hospitals became highly interdependent, so much so that they began to "assert their long-run collective interests over their short-run individual interests" (Starr, 1982, p. 230). Their collective interests involve continued investment in complex technologies, with the result that medical care is becoming more and more expensive. The high cost of medical care makes it more difficult for the poor, the elderly, and other groups to afford high-quality care and heroic life-preserving measures. In the opening years of the twenty-first century, efforts to make health care less costly while preserving the insurance systems

This painting by Jan Beerblock (1739–1806) depicts the sick wards in Saint-Janshospitaal in Bruges, Belgium, a bleak forerunner of modern hospital organization.

that provide medical coverage for the majority of Americans are proving to be one of the most controversial and rancorous issues in U.S. political life.

Medical Sociology

Starr's 1982 study of the evolution of health care institutions had a major impact on the health care reform movement of the 1990s. It is an excellent example of *medical sociology.* This relatively new field of study has emerged in response to the development of medicine as a major institution of modern societies. Many sociologists are employed by health care institutions, and some medical schools have established faculty positions for sociologists. When they work in medical research, sociologists often conduct studies that add to our knowledge of epidemiology, especially social epidemiology.

Social Epidemiology

Social epidemiology is a new and important subfield of epidemiology that draws heavily on research in medical sociology. *Epidemiology* is the study of how diseases originate and spread in human populations and how societies can develop better public health systems to attack the causes of illness. Epidemiologists trace the outbreaks of new diseases like Ebola or severe acute respiratory syndrome (SARS), and they also track the spread and decline of epidemics like AIDS, malaria, and many others. Epidemiologists are trained in biology and some aspects of medicine, and they are particularly well trained in biostatistics.

Social epidemiology looks for the fundamental causes of health problems in basic social and cultural conditions that are often the precursors of health problems. The policies they advocate, based on their research, often address social issues like poverty, racism, gender bias, and the stresses working people encounter on their jobs (Burris, 2002).

AIDS and Social Epidemiology In the United States, the Centers for Disease Control and Prevention (CDC), with headquarters in Atlanta, Georgia, is a pioneer in social epidemiology and the use of medical–sociological research. The CDC is the lead federal agency for protecting the health and safety of Americans both at home and abroad. It is expected to provide credible information to enhance health decisions and to promote health through strong partnerships with community groups of all kinds. For example, the CDC is responsible for developing programs to detect and combat possible terrorist use of bioweapons. And since the emergence of AIDS in the 1980s, the CDC has taken a major leadership role in tracking the course of the AIDS epidemic in the United States and elsewhere in the world.

Since the start of the AIDS epidemic, the CDC has been the leading agency helping to prevent the spread of HIV (the virus that causes AIDS) as well as providing surveillance. According to the CDC,

> At the end of 2006, an estimated 1,106,400 persons . . . in the United States were living with HIV infection, with 21% undiagnosed. In 2008, CDC estimated that approximately 56,300 people were newly infected with HIV in 2006 (the most recent year that data are available). Over half (53%) of these new infections occurred in gay and bisexual men. Black/African American men and women were also strongly affected and were estimated to have an incidence rate that was 7 times as high as the incidence rate among whites. (Hall et al., 2008)

The largest proportion of HIV/AIDS diagnoses is seen in men who have sex with men, followed by adults and adolescents infected through heterosexual contact. Among women, by far the most common form of AIDS transmission is through heterosexual sex. Women account for over one-fourth of new AIDS cases in the United States. African Americans, who make up approximately 12 percent of the U.S. population, account for half of the estimated number of HIV/AIDS cases diagnosed (see Figure 20.6 on the next page).

The CDC suggests that heterosexual contact followed by injection drug use is the leading cause of HIV infection among African American women. It cautions people not to think of race and ethnicity as risk factors in themselves, because they are not, but people from certain racial and ethnic groups have to endure many challenges that are associated with the risk of HIV infection, such as poverty, denial and discrimination, partners at risk, and substance abuse. Currently, and due in part to the research conducted by the CDC, people are living longer with HIV/AIDS than they have in the past.

The CDC employs a number of strategies to prevent the spread of HIV and AIDS; such strategies include monitoring the epidemic to target prevention and care activities, researching the effectiveness of prevention methods, funding local prevention efforts for high-risk communities, and fostering linkages with care and treatment programs. The CDC also provides counseling, testing, community outreach services, and even treatment and prevention for other sexually transmitted diseases. It also

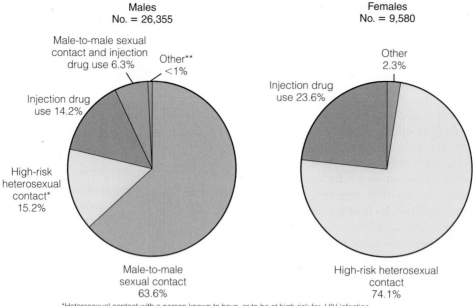

Transmission categories of adults and adolescents with HIV/AIDS diagnosed during 2007

Males
No. = 26,355

Females
No. = 9,580

Male-to-male sexual contact and injection drug use 6.3%

Other** <1%

Injection drug use 14.2%

High-risk heterosexual contact* 15.2%

Male-to-male sexual contact 63.6%

Other 2.3%

Injection drug use 23.6%

High-risk heterosexual contact 74.1%

*Heterosexual contact with a person known to have, or to be at high risk for, HIV infection.
** Includes hemophilia, blood transfusion, perinatal exposure, and risk not reported or not identified.

Sex of adults and adolescents with HIV/AIDS diagnosed during 2007

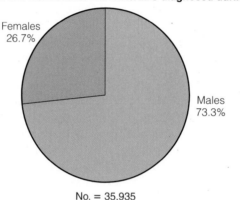

Females 26.7%

Males 73.3%

No. = 35,935

Figure 20.6 Transmission Categories and Estimated Numbers of AIDS Cases, United States, 2007. Source: Centers for Disease Control, 2010.

encourages creative approaches to communicating with the public about sexual behavior and HIV/AIDS.

The Social Epidemiology of Obesity Another serious threat to public health in the United States and other affluent nations is obesity. A certain amount of body fat is essential to good health—for stored energy, heat insulation, shock absorption, and other functions. Women as a rule have more body fat than men, but men with more than 25 percent and women with more than 30 percent body fat are obese, according to most public health experts. Measuring how much fat one carries is not a simple matter, however, because of other aspects of body mass such as height and bone density.

By calculating a person's body mass index (BMI) to control for height, and studying the distribution of BMI in different populations, epidemiologists can determine the proportions of a population that are overweight or obese. By these measures, approximately 25 percent of Americans are obese, and rates of obesity are rising rapidly in Europe as well, although the problem is by far the most acute in the United States.

Aside from considerations of appearance, what is so bad about obesity? Epidemiologists know, for example, that Type II diabetes is

associated with excess body weight and the consumption of high levels of sugar and fatty foods. Cardiovascular disease is also closely related to obesity. When one considers all health-related problems, obesity costs almost $100 billion a year in excess health care costs, making it second only to smoking as a public health problem. And it is also a social problem in that it is closely related to other aspects of inequality and to the way high-fat foods are marketed in the U.S. food industry (Stephen, 2003).

The image of people in the United States stuffing themselves on supersized starchy meals and sugary drinks is a common stereotype, but unfortunately, like many other stereotypes, it is based on an important core of fact. In *Fast Food Nation,* his popular critique of the fast-food industry, Eric Schlosser (2001) points out that fast food is now served at restaurants, stadiums, airports, zoos, schools, and universities; on cruise ships, trains, and airplanes; at supermarkets and gas stations; and even in hospital cafeterias. Americans now spend more money on fast food—over $110 billion a year—than they do on higher education.

Social epidemiologists have shown that the increases we are seeing in obesity are fundamentally related to inequalities of class and race. Poor people, who are also more likely to be members of minority groups, are less likely to have access to healthy diets and more likely to have fast food stores in their neighborhoods, where sugars and fatty foods become a staple of the family diet. People in low-income urban communities or more isolated rural areas have far fewer choices in available foods. "They don't have big supermarkets in their areas," notes sociologist William Julius Wilson, an expert on the lifestyles of the poor, and people on low incomes eat large amounts of more inexpensive but delicious starchy-fat foods like fried potatoes and macaroni with cheese, which are extremely high in calories. As Figure 20.7 shows, people whose household income is under $10,000 are especially prone to obesity, an effect that minority status only increases (Barboza, 2000).

In addition to low income, the number of hours parents spend at work—the hours that children spend watching television and sitting around eating during the day—is highly correlated with children's body fat. And these practices are more common in neighborhoods where parents are afraid to let their children go outside for exercise. Cuts in school budgets for athletics and physical education and more emphasis on cramming for high-stakes standardized tests are other afflictions of the poor that are implicated in the obesity epidemic.

So are the practices of the fast food industry, which markets its high-fat and high-sugar content foods to children and adults with particular success in communities of the working poor, where harried parents have less time to prepare more nutritious dinners and less knowledge about the dangers of high-fat diets. In response to criticism and to lawsuits by obese consumers, the major fast food chains have begun to offer more vegetables and salads and less fat-saturated foods (Carpenter & Bartley, 1994; Gale, 2003).

Chronic Illnesses and the Challenge to World Health

Obesity is a major risk factor for adult diabetes, a disease that, if left untreated, often results in heart disease, kidney failure, blindness, and untimely death. World health officials worry that as the global population becomes

Race, Income, and Obesity

A study done by the Centers for Disease Control and Prevention shows that obesity rates in the United States have a direct correlation to race, ethnicity, and income level.

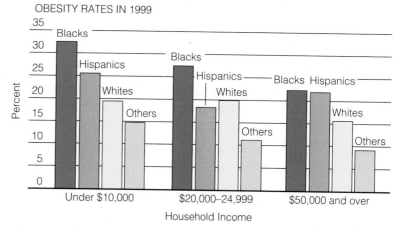

OBESITY RATES IN 1999

Figure 20.7 Race, Income, and Obesity. Source: Barboza, 2000.

somewhat more affluent and can afford the tasty fast foods that Americans and Europeans consume, rates of diabetes will continue to increase rapidly. But the diabetes epidemic reveals an even larger trend in world health. As nations become more affluent, chronic diseases, especially heart disease and cancer, begin to replace infectious diseases like influenza and tuberculosis as primary causes of death. These trends are shown in Table 20.2 on the next page, which also demonstrates that as automobile use increased in the United States early in the twentieth century, motor vehicle accidents became as great a health risk factor as major infectious diseases.

Chronic diseases often require the most advanced and costly medical techniques, which helps account for the rapidly rising share of medical care in national budgets around the world. Of course, death eventually

Bill and Melinda Gates at an AIDS conference in Toronto in 2006. The Bill and Melinda Gates Foundation has made combating the spread of AIDS one of its highest priorities—an example of how fortunes based on technology are taking the lead in social philanthropy.

TABLE 20.2

Death Rates, by Cause of Death, United States, 1900–2002 (per 100,000 Population)

Year	Tuberculosis, All Forms	Malignant Neoplasms (Cancer)	Major Cardiovascular Diseases	Influenza and Pneumonia	Motor Vehicle Accidents
1900	194.4	64.0	345.2	202.2	NA
1910	153.8	76.2	371.9	155.9	1.8
1920	113.1	83.4	364.9	207.3	10.3
1930	71.1	97.4	414.4	102.5	26.7
1940	45.9	120.3	485.7	70.3	26.2
1950	22.5	139.8	510.8	31.3	23.1
1960	6.1	149.2	521.8	37.3	21.3
1970	2.6	162.8	496.0	30.9	26.9
1980	0.9	183.9	436.4	24.1	23.5
1990	0.7	203.2	368.3	32.0	18.8
2000	0.3	200.5	340.4	24.3	15.2
2001	0.3	194.4	323.9	21.8	15.4
2002	0.3	193.8	318.3	22.9	15.5

NA = Information not available.

Source: National Center for Health Statistics.

claims all humans, so some causes of death, such as heart failure, will always be extremely important. There is a great difference, however, between cardiovascular disease that has its onset in earlier adult years, often due to stress, smoking, and poor health practices, and heart failure among the elderly. Thus, the table indicates that since the 1970s, when public health advocates began showing people how to use diet and exercise to combat cardiovascular disease, rates of deadly heart disease have declined significantly. Take note also of the trend in motor vehicle accidents, which peaked in the 1980s and then, with the advent of campaigns to increase seat belt use and decrease drunken driving, began to decrease. These are the consequences of public policies and thoughtful educational campaigns. Social scientists who have studied the behavioral aspects of chronic diseases and accidents that result in enormous medical expenses have made major contributions to our understanding of adverse health behaviors and how to address them through public health programs (Cockerham, 2004).

SUMMARY

What are some key health care issues in the developing nations?

- The health of an individual is highly associated with personal income; for a nation, the overall health of the population is associated with the degree of poverty and economic inequality in the society.

- Because the boundaries between nations and regions are no longer as rigid as they were in the past, it makes sense to invest in developing global health care systems and improving the health of people throughout the world.

- Creating a safe water supply is one of the most important ways of improving global public health. The most essential step is the careful separation of human and animal waste, and other forms of sewage, from the water supply used for drinking and irrigation.

What factors make it difficult to control health care costs in the United States?

- The health of Americans could be improved by investing in more preventive measures to decrease costly behaviors like smoking and unhealthy eating patterns, and by increasing incentives for early detection of serious illnesses such as diabetes, colon cancer, and heart disease.

- A key factor in the high cost of health care in the United States is the extremely high rate of use of expensive medical technologies.

How have modern science and technologies transformed medical institutions?

- The complex interactions between technology and other aspects of the social order are illustrated by the case of medical technology.

- Until relatively recently, physicians were powerless either to check the progress of disease or to prolong life. Scientific research during the nineteenth century led to the discovery of the causes of many diseases, but in the twentieth century, a vast array of technologies was developed both for the prevention and cure of illnesses and for the long-term care of terminally ill patients.

- The technologies used in the diagnosis and treatment of serious illnesses require extremely expensive equipment and highly trained personnel. This has caused health care to become very expensive, and as a result, some groups in the population are unable to obtain adequate care.

- Some critics claim that extreme emphasis on technological progress has created a situation in which the needs of the patient are subordinated to those of the providers of health care.

What is social epidemiology, and how does it contribute to our understanding of health and medicine?

- Social epidemiology looks for the fundamental causes of health problems in basic social and cultural conditions that are often the precursors of health problems. The policies they advocate often address social issues like poverty, racism, gender bias, and job stress.

- The Centers for Disease Control and Prevention is a pioneer in the use of epidemiological studies to track the AIDS epidemic, increasing rates of obesity, and other serious health conditions.

- As nations become more affluent, chronic diseases, especially heart disease and cancer, begin to replace infectious diseases like influenza and tuberculosis as primary causes of death.

The Kornblum Companion Website

www.cengagebrain.com

Supplement your review of this chapter by going to the companion website to take one of the tutorial quizzes, use the flash cards to master key terms, and check out the many other study aids you'll find there. You'll also find special features, such as GSS data and Web links that will put data and resources at your fingertips to help you with that special project or to do some research on your own.

Brasil2/iStockphoto.com

Population, Urbanization, and the Environment

FOCUS QUESTIONS

How do sociologists measure and predict population changes, and how does the growth of cities affect population?

What changes can be seen over time in urban areas?

Are urban communities doomed to become anonymous, faceless places?

In what ways can cities be described as "engines of social change?"

The peasant families with many children were asked to come sit in front of the gathering. The huge crowd watched as hundreds of parents and children filed toward the front of the throng of people seated in a grain field, sweating in the afternoon sun. They were waiting to see and hear Chandrababu Naidu, one of India's most powerful political leaders.

When the families were settled in the front, Naidu began speaking, but his words were not designed to make the large families happy. He criticized them for having too many children. They could not provide adequate education and health care for so many children, he warned, and the government could not afford to help them either. He urged the parents to immediately make appointments for voluntary sterilization. The parents seemed embarrassed. Many people in the crowd could not suppress their nervous laughter. They knew that the subject was one of great importance, but it was also one that they preferred to whisper about in the privacy of their homes.

Summoning all his courage, a peasant farmer named Nara Singh, a father of four with another baby on the way, stood up in front of the crowd, took the microphone, and firmly but with great conviction insisted that he needed even more children to help work his small farm.

The politician listened to him respectfully, but when the farmer was finished, he asked the crowd, "Is this man on the right path?" Very few of the listeners raised their hands in agreement with the farmer. He had won their respect for speaking his mind, but they knew that he represented the old, and now discredited, idea that having as many children as possible was desirable for rural people.

After years of public discussion on radio and television and in sweaty gatherings like this one, many of the people in the crowd understood that India needed to control its population growth. The nation had added more than 180 million new mouths to feed in the previous decade alone. At current growth rates, India's population, which recently passed the billion mark, will surpass China's by the middle of the century, making India the world's most populous nation. But the idea of vasectomies for men or tubal ligations for women still struck many as a drastic measure.

Those in the crowd with more education and the ability to read the newspapers also knew that the politician's policies were controversial. Their state had the highest rate of voluntary sterilizations in India, and it appeared that the policy had had some success in decelerating the region's population growth rate. But other states of India had shown even more success by instituting programs to dramatically increase female literacy, health care, and contraception. In their state, men and women with no more than two children who underwent voluntary sterilization were given a small cash payment and preference in the distribution of small farm lots or building plots. These were powerful incentives, but they were not making nearly as much progress in raising literacy levels, and low literacy rates could be an obstacle to economic development over the long term.

A combination of tactics, including education, health care, and incentives for sterilization, would be even more effective, some of the educated listeners argued. But the need for urgent measures was clear even to the least educated in the crowd. Growing populations were threatening to exhaust available supplies of wood for cooking fires. Water resources were being stretched to their limits. The landless were being forced into the nearby cities, where they often had to resort to begging and life in the streets. The political leaders of their state were making sterilization attractive enough to the small landholders that it was beginning to become a common practice. Clearly, the peasants, and especially the women, wanted to control their fertility, which was a good sign for the future of their society and its natural environment (Dugger, 2001).

PEOPLE, RESOURCES, AND URBAN GROWTH

This story about the reproductive dilemmas poor people face in India only begins to touch on the many issues that population growth raises throughout the world. The struggle to make ends meet, the degradation of the natural environment, and massive migrations to crowded cities are not limited to the developing regions of the world. More affluent nations have their own versions of these dilemmas.

The human population continues to spread throughout the world. Although most of the world's people live in rural villages, in modern societies they typically live in cities or metropolitan areas rather than on farms or in small towns. In fact, 90 percent of Americans live within twenty-five miles of a city center (*Statistical Abstract*, 2010). This means that they live in the human-built environment of cities and suburbs; relatively few live in the "country" environment of rural areas. It also means, among many other things, that settlement occurs in unstable places that invite disaster.

The study of population growth and the growth of cities as centers of demographic and social change has long been a central field in sociology (Kleniewsky, 2006). In this chapter, therefore, we see how increasing population size and other important changes in populations are connected with the rise of cities and metropolitan areas throughout the world. At the end of the chapter, we turn to the study of natural and social disasters.

The Population Explosion

For many decades, sociologists have been conducting research on the impacts of population growth, on how such growth places stresses on the earth's natural resources, and on the rapid growth of cities throughout the world. This chapter highlights some of the main issues addressed by this research. First, however, we must note that the most basic facts about how many people there are on the earth and how fast their numbers are growing (and will grow in the future) are surprisingly complicated. In Chapter 4, we described the extremely rapid growth of the world's population in the past three centuries. If you refer back to Figure 4.2, you can readily see that the *rate* of population growth increased dramatically in the twentieth century, giving rise to the often-used term *population explosion*. The world's population is currently estimated at 6.85 billion, and according to United Nations forecasts, it is likely to increase to

about 8.8 billion by 2025 (O'Neill & Balk, 2001). This would represent growth over a thirty-year period that is roughly equal to the entire world population in 1965 (Mitchell, 1998).

At present, about 90 million new people are added to the world's population each year. Most forecasts predict that this rapid growth will begin to decline by midcentury, but by then there could be as many as 10.7 billion people on a very crowded planet. Demographers point out, however, that crop production, famine and war, contraception, economic development, and many other variables affect population. The potential effects of these variables make it extremely difficult to predict trends in population growth. It is certain, though, that rapid population growth, especially in the poorer regions of the world, will continue in coming decades and will pose severe challenges to human survival and well-being (United Nations Population Fund, 2005). The pressures of population growth on resources of food, space, and water have produced changes in economic and social arrangements throughout history, but never before have those changes occurred as rapidly or on as great a scale as they are occurring today (Cohen, 1998; Smil, 2001).

Catastrophists versus Cornucopians Is there a danger that the earth will become overpopulated? The debate over population growth is not new. For almost two centuries, social scientists have been seeking to determine whether human populations will grow beyond the earth's capacity to support them. *Catastrophists* believe that it is possible that rapid population growth may lead to increasing social and environmental disasters: famines, wars driven by people desperate to expand into others' lands, and extreme depletion of water and soil resources (Brown, 1995; Ehrlich & Ehrlich, 2004). The far more optimistic *cornucopians* believe that there are endless possibilities for humans to control populations and devise solutions to environmental and social problems. Many argue that population growth is necessary for economic growth and is not an obstacle to development (Simon, 1996; Singer, 1999).

The earliest and most forceful theory of overpopulation appeared in Thomas Malthus's *An Essay on the Principle of Population* (1927–1928/1798). Malthus attempted to show that population size normally increases far more rapidly than the food and energy resources needed to keep people alive. Couples will have as many children as they can afford to feed, and their children will do the same. This will cause populations to grow *geometrically* (2, 4, 8, 16, 32, . . .). Meanwhile, available food supplies will increase *arithmetically* (2, 3, 4, 5, 6, . . .) as farms are expanded and crop yields increased. As a result, population growth will always threaten to outstrip food supplies. The resulting poverty, famine, disease, war, and mass migrations will act as natural checks on rapid population growth.

History has proved Malthus wrong on at least two counts. To begin with, we are not biologically driven to multiply beyond the capacity of the environment to support our offspring; Malthus himself later recognized that people could limit their reproduction through delay of marriage or celibacy. The second fault of Malthusian theory is its failure to recognize that technological and institutional change could expand available resources rapidly enough to keep up with population growth. This has occurred in the more affluent regions of the world, where improvements in the quality of life have tended to outstrip population growth. Improvements in agricultural technology have also increased the yield of crops in some of the less developed parts of the world, such as India. But rates of population growth and exhaustion of environmental resources (firewood, water, grazing land) are highest in the poorest nations (Furedi, 1997). Will reductions in population growth rates and increases in available resources also occur there? Or will the fears of the catastrophists prove correct in the long run? The theory known as the *demographic transition* provides a framework for studying this question, but before we explore this theory we need to know more about measuring population change.

British clergyman Thomas Malthus's predictions were dubbed "Malthus's Dismal Theorem" by his contemporaries. His theory predicted that any form of increased food supply or aid to the poor would only lead to bad behavior, illegitimacy, and debauchery. Time has shown his predictions to be false.

Rates of Population Change Populations change as a consequence of births, deaths, out-migration, and in-migration. The most basic measures of population change are *crude rates,* or the number of events of a given type (for example, births or deaths) that occur in a year divided by the midyear population (Bogue, 1969). Thus, the **crude birthrate** (CBR) is the number of births occurring during a year in a given population divided by the midyear population, and the **crude death rate** (CDR) is the number of deaths occurring during a year divided by the midyear population. These fractions are usually expressed as a rate per thousand people. They are "crude" because they compare the total number of births or deaths with the total midyear population, when in fact not all members of the population are equally likely to give birth or to die.

The **rate of reproductive change** is the difference between the CBR and the CDR for a given population. It is a measure of the natural increase of the population; that is, it measures increases caused by the excess of births over deaths and disregards in- and out-migration. At present, the rate of reproductive change in several nations is zero or less, meaning that there is no natural population growth. Germany, for example, had a CBR of 9 and a CDR of 10 in 2001, for an annual rate of increase of 20.1. In Greece, the CBR was 10 and the CDR was 10, for an annual rate of increase of 0. In the United States, the rate of population growth is about 0.6 percent, representing an increase of about 1.65 million people per year. These rates are in dramatic contrast with the annual growth rates of countries like Liberia and Nicaragua, which are above 3 percent. Rates of population growth in selected countries are listed in Table 21.1.

The last column in Table 21.1 shows the time it would take for the population of each nation to double if existing growth rates continued without change. A growth rate just under 1 percent, such as China has achieved, will double the population in seventy-nine years, but a growth rate of almost 3 percent, such as we see in Nigeria, doubles the population

---

crude birthrate: The number of births occurring during a year in a given population, divided by the midyear population.

crude death rate: The number of deaths occurring during a year in a given population, divided by the midyear population.

rate of reproductive change: The difference between the crude birthrate and the crude death rate for a given population.

TABLE 21.1

Population Growth and Doubling Time, Selected Countries

Country	Population (millions)	Annual Growth Rate* (percent)	Doubling Time* (years)
India	1,002.1	1.8%	39
China	1,264.5	0.9	79
Brazil	170.1	1.5	45
Bangladesh	128.1	1.8	38
Nigeria	123.3	2.8	24
Pakistan	150.6	2.8	25
Indonesia	212.2	1.6	44
Russia	145.2	20.6	—
Mexico	99.6	2.0	36
United States	275.6	0.6	120

*Annual rate of natural increase.

**Number of years in which the population will double at current rate of increase.

Source: UNFPA, 2009.

in twenty-four years. In the United States, in contrast, an annual growth rate of 0.6 percent doubles the population in 120 years.

We can easily see from Table 21.2 that an annual rate of population growth of only 1 percent will lead to an increase of almost 270 percent in a century, and growth rates of 2 percent or more, such as we find in many of the poorer nations of the world, would yield populations of over 16 billion within a century. But current estimates for the end of this century, based on declining fertility rates, indicate that the world's population is likely to cease its rapid growth and level off somewhat below 10 billion (Hale, 2010). The reasons for these changes are

TABLE 21.2

Relationship of Population Growth per Year and per Century

Population Growth per Year (percent)	Population Growth per Century (percent)
1%	270%
2	724
3	1,922

Source: Data from Worldwatch Institute.

demographic transition: A set of major changes in birth and death rates that have occurred most completely in urban industrial nations in the past 200 years.

summarized in the phenomenon known as the demographic transition.

The Demographic Transition The **demographic transition** is a set of major changes in birth and death rates that have occurred most completely in urban industrial nations in the last 200 years. We saw in Chapter 4 that the rapid increase in the world population in the past 150 years resulted largely from rapid declines in death rates. Beginning in the second half of the eighteenth century and continuing until the first half of the twentieth century, death rates in the countries of northern and western Europe declined significantly. Improvements in public health practices were one factor in that decline. Even more important were higher agricultural yields because of technological changes in farming methods, as well as improvements in the distribution of food as a result of better transportation, which made cheaper food available to more people (Bacci, 1997; Vining, 1985). At the same time, however, birthrates in those countries remained high. The resulting gap between birth and death rates produced huge increases in population. It appeared that the gloomy predictions of Malthus and others would be borne out.

In the second half of the nineteenth century, however, birthrates began to decline as couples delayed marriage and childbearing. As a result of lower birthrates, the gap between birth and death rates narrowed and population growth slowed. This occurred at different

Demographic Transition and Development

Stage of Population Growth	Type of Society	Main Developmental Features
High Growth Potential	Most types of preindustrial societies	High death rates due to infant mortality and low life expectancy; high rates of fertility; relatively low rates of increase
Transitional Growth	Most societies in early stages of urban and industrial development where basic public health measures are being introduced (e.g., safe drinking water systems)	Decreasing mortality rates; continuing high rates of fertility; high rates of female illiteracy and limited protection of women's rights
Incipient Decline	Societies in more advanced stages of urban and industrial development where people are delaying marriage and are more likely to use birth control	Decreasing rates of mortality and fertility, with low or even negative rates of increase; high rates of female literacy and high levels of medical care for women in their reproductive years

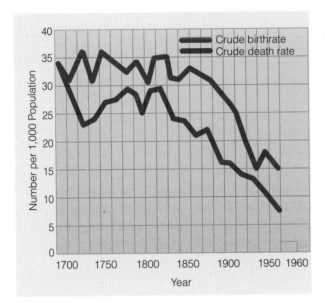

Figure 21.1 The Demographic Transition: Sweden, 1691–1963. The peaks in birth and death rates in the early 1800s were a result of social unrest and war. The drop in deaths and simultaneous rise in births in the early 1700s were a result of peace, good crops, and the absence of plagues. Source: Judah Matras, *Population and Societies*, Englewood, Cliffs, NJ: Prentice Hall, ©1973. Permission of Armand Éditeur.

In the second, or transitional growth, stage of the demographic transition, the population not only grows rapidly but also undergoes changes in its age composition. Because people now live longer, the proportion of elderly people increases slightly. There is also a marked increase in the proportion of people younger than age twenty as a result of significant decreases in infant and child mortality. This is the stage in which many less developed countries find themselves today: Death rates dropped in the twentieth century because of improved medical care and public health measures, as well as increased agricultural production. Yet in these societies, birthrates have remained high, causing phenomenal increases in population, especially in the younger, more dependent age groups.

No population has entered the third stage of the demographic transition without limiting its birthrate. This can be achieved by encouraging couples to marry later and postpone childbearing or by preventing pregnancies or births through various birth-control techniques. In advanced industrial societies, many couples use both approaches, making their own decisions about whether and when to have children. In other societies, such as China and India, the state has attempted to limit population growth by promoting birth control through educational programs or, in the case of China, imposing penalties on couples who have more than a prescribed number of children. Such measures have had some transitory success, but often at a high political cost to national leaders. In China, for example, the strict one-child rule is not being enforced rigorously in areas undergoing rapid industrial and commercial development, where labor is in demand (Faison, 1997). We will see in more detail in the next chapter that guarantees of education for women, protection of their reproductive rights, and guarantees of their safety and right to vote are all highly correlated with decreases in fertility rates (Dasgupta, 2000).

Note that economic and social development is essential if the demographic transition is to occur. Death rates cannot decrease, or food supplies increase, without progress in social institutions such as public health, medicine, and transportation. People in more developed societies tend to limit their family size because they seek economic advancement and wish to delay marriage and childbearing until they can support a family.

In many highly industrialized nations, on the other hand, population growth rates have fallen below the rate necessary to maintain the

times in different countries (Figure 21.1 graphs the demographic transition in Sweden), but the general pattern was the same in each case and consisted of three stages:

1. A stage of high birthrates and death rates (the *high growth potential* stage)
2. A stage of declining death rates (the *transitional growth* stage)
3. A stage of declining birthrates (called the stage of *incipient decline* because it is possible for population growth rates to decrease at this point)

These three stages are summarized in the For Review chart above.

population at the existing level. If a fertility rate of two children or less per couple became the norm for an entire population, the growth rate would slow to zero or even a negative rate. The United States, New Zealand, Japan, Australia, and Canada all have total fertility rates of 2.1 or less (Population Reference Bureau, 2001; United Nations Population Fund, 2005). Similar low rates are appearing in many European nations. This development is highly significant; if it continues over a generation or more, it could result in negative rates of natural increase and, possibly, other consequences such as slower economic development and labor shortages. Continued low fertility could also result in increased immigration from countries with high birthrates to countries with lower birthrates.

A related issue is the rapid aging of populations in Europe and the United States as the proportion of elderly people increases while the proportion of children declines. This is somewhat less of a problem in the United States than in Europe because of continuing high rates of immigration, but as we saw in Chapter 14, the entry of the large baby boom cohort into its retirement years represents a major challenge for social policy in the United States (Koretz, 2003).

A full exposition of the processes and politics of population control is not possible here. Suffice it to say, most sociologists would agree with the demographer Philip Hauser (1957) that low population growth rates are caused primarily by a combination of delay of marriage, celibacy, and use of modern birth-control techniques. This trend is beginning to occur in many less developed nations (Allen, 2007; Ofosu-Amaah, 1998). Figure 21.2 illustrates the demographic transition in Singapore, a nation in Southeast Asia that has undergone this important population change more recently. Note that the decline in deaths, which took almost 150 years in Western industrial nations, took less than 50 years in Singapore. The decline in births began in the late 1950s, spurred by rapid economic development, the new desire for smaller families, and widespread availability of contraceptives. Whereas this phase of declining births unfolded over more than 100 years in Western nations like Sweden, this phase occurred in only about 20 years (between about 1955 and 1975) in Singapore. Although world population growth remains a serious problem, the demographic transition in formerly poor nations like this one shows that population control efforts along with

economic development can have dramatic effects (Davis, 1991; Lesthaeghe, 2010).

WOMEN'S REPRODUCTIVE RIGHTS The world's population increases by 70 to 80 million a year, partly because 200 million women lack access to family planning. Thus, central to any discussions of population growth and control is the right of women to choose the number and spacing of their children and have access to reproductive health services, including contraception. This rights-based approach to population issues is at the heart of the Programme of Action adopted at the Cairo Conference on Women and Globalization in 1995. This document forcefully stated the right of women to control their own reproduction and the responsibility of nations to protect women from gender-based violence.

Now these goals are at risk. In some countries, fertility levels have not stabilized as expected. When low birthrates are accompanied by increased life expectancy and an aging population, policy makers often become increasingly concerned about dependency ratios, productivity, and economic prosperity. As a result, a number of governments have reversed their previous policies to encourage higher birthrates. Yet even within those countries, unequal access to education, health care, housing, water, sanitation, and employment persists. Addressing such social injustice needs to be an urgent priority to ensure that all people can lead lives of dignity, meaning, and respect.

In the world's poor nations, increases in political instability and civil strife have resulted

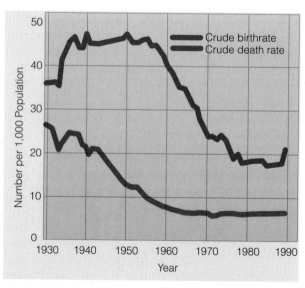

Figure 21.2 Crude Birth and Death Rates: Singapore, 1930–1988. Source: Davis, 1991.

in epidemics of violence against women, as in the case of widespread rape in the Congo.

While governments in Singapore, Hong Kong, Korea, Russia, Italy, Japan, and elsewhere urge women to have more children, two out of three women in poor countries lack access to modern contraception, and funding for contraceptives has stagnated since 2001, even though demand is expected to rise by over 25 percent between 2000 and 2015 (UNFPA, 2008). Poor sexual and reproductive health accounts for one-third of the global burden of disease for women between the ages of fifteen and forty-four. Moreover, many governments still fail to recognize the critical importance of women's empowerment, education, sexual and reproductive health, and rights to poverty elimination and sustainable development (Greer, 2009).

Life Expectancy in Global Context *Life expectancy*—usually defined as the number of years one can expect to live—is one of the most important indicators of how well a society cares for its population. Highly developed societies with advanced systems of health care and well-developed public health systems (sewage, drinking water, shelter, hospitals) have resulted in average life expectancies well up in the seventies (see Table 21.3). Societies that are in the earlier stages of industrialization and urbanization, such as Brazil and China, have lower average life expectancies, but impoverished, war-torn societies like Angola and Zimbabwe have life expectancies that are extremely low and may be declining, largely due to extremely high rates of infant mortality.

Infant mortality and life expectancy are closely related indicators of how well societies care for their populations. Newborn infants are extremely susceptible to infections and diseases, often the result of failures in basic sanitation and unclean drinking water. When poor societies make improvements in their public health systems, even before making more lasting improvements in health care and hospitals, life expectancy increases because far fewer infants die in the first year or so after birth. This is the period when death rates are falling but birthrates remain high, so populations can increase quite rapidly in this phase of the demographic transition. This is the situation in much of India at present.

Healthy Life Expectancy As we have seen in earlier chapters, it is one thing to live a long life, and quite another to live a long and healthy life. Standard measures of life expectancy, such as those given in Table 21.3 on the next page, are based only on measures of mortality. Recently the World Health Organization has developed a more sophisticated measure that takes into account how long the average person in a nation can expect to live without severe and disabling illness. This *disability-adjusted life expectancy (DALE)* summarizes the average number of years a person can expect to live in "full health." According to this measure, Japan has the world's best healthy life expectancy at 74.5 years, and Sierra Leone, which has suffered years of civil war and rampant corruption, has the world's lowest at 26 years. Under this system of measurement, the United States falls to twenty-fourth in the world with a disability-adjusted life expectancy for newborn babies of 70 years.

The WHO cites various reasons for why the United States ranks relatively low among wealthy nations. Those reasons include the following:

- In the United States, some groups, such as Native Americans, rural African Americans, and the inner-city poor, have extremely poor health, which is more characteristic of a poor developing country than of a rich industrialized one.
- The proportion of young and middle-aged people suffering death or disability due to HIV is higher in the United States than in most other advanced countries. HIV/AIDS cut three months from the healthy life expectancy of male American babies born in 1999, and one month from that of female babies.
- The United States has one of the highest rates of cancers relating to tobacco use, especially lung cancer. Tobacco use also causes chronic lung disease.
- The United States has a high rate of coronary heart disease; this rate has dropped in recent years but remains high.
- The United States has fairly high levels of violence, especially homicides, compared to other industrial countries (Murray, 2000).

 Thinking Critically

When we see the United States ranking low on social indicators like life expectancy and infant mortality compared to other highly developed nations, we must realize that the United States has always been a far more diverse society, with greater extremes of wealth and poverty and a relatively large immigrant population. But does this diversity make the nation's poor rankings acceptable?

Environmental Impacts

The growth of the human population in the past two centuries has begun to exert major stresses on the earth's resources of arable soil, clean air,

TABLE 21.3

Infant Mortality and Life Expectancy for Selected Countries, 2006

Country	Infant Mortality[1]	Life Expectancy[2]	Country	Infant Mortality[1]	Life Expectancy[2]
Albania	20.8	77.4	Japan	3.2	81.2
Angola	185.4	38.6	Kenya	59.3	48.9
Australia	4.6	80.5	Korea, South	6.2	77.0
Austria	4.6	79.1	Mexico	20.3	75.4
Bangladesh	60.8	62.5	Mozambique	129.2	39.8
Brazil	28.6	72.0	New Zealand	5.8	78.8
Canada	4.7	80.2	Nigeria	97.1	47.1
Chile	8.6	76.8	Norway	3.7	79.5
China	23.1	72.6	Pakistan	70.5	63.4
Costa Rica	9.7	77.0	Panama	16.4	75.2
Cyprus	7.0	77.8	Peru	30.9	69.8
Czech Republic	3.9	76.2	Poland	7.2	75.0
Denmark	4.5	77.8	Portugal	5.0	77.7
Ecuador	22.9	76.4	Russia	15.1	67.1
Egypt	31.3	71.3	Slovakia	7.3	74.7
Finland	3.5	78.5	South Africa	60.7	42.7
France	4.2	79.7	Spain	4.4	79.7
Germany	4.1	78.8	Sri Lanka	14.0	73.4
Greece	5.4	79.2	Sweden	2.8	80.5
Guatemala	30.9	69.4	Switzerland	4.3	80.5
Hungary	8.4	72.7	Syria	28.6	70.3
India	54.6	64.7	United Kingdom	5.1	78.5
Iran	40.3	70.3	United States	6.4	77.8
Ireland	5.3	77.7	Venezuela	21.5	74.5
Israel	6.9	79.5	Zimbabwe	51.7	39.3
Italy	5.8	79.8			

[1] Infant deaths per 1,000 live births.
[2] Life expectancy at birth, in years, both sexes.
Source: U.S. Census Bureau, International Database.

and freshwater for drinking and irrigation, as well as on its sources of fuels. Three thousand years ago, when the total population was less than 100 million, the impact of human activities like cooking and mining was negligible. Today, however, the combined impact of more than 6 billion people is enormous, and it is especially severe because of the extremely high use of fossil fuels and high levels of consumption of steel and other durable goods in the richest nations of the world.

The United States, for example, with less than 5 percent of the earth's population, consumes more than 20 percent of its energy resources and, in consequence, puts far more than its share of pollutants and earth-warming gases into the atmosphere. Much of this consumption and pollution is directly related to the high rate of automobile ownership and use in the United States, where there are about 134 million private autos on the road, or 488 per 1,000 people. In Germany, there are 508 autos per 1,000 people, perhaps the highest rate of auto ownership in the world, but the total number of autos (42 million) is far lower than in the United States (*Statistical Abstract*, 2010). Figure 21.3 shows that, by 2010, there will be more than 1 billion automobiles operating globally, with

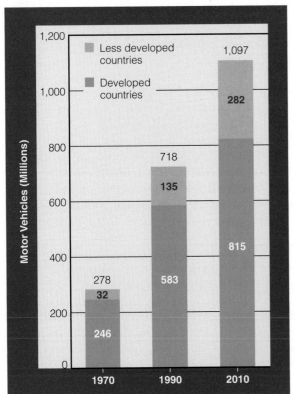

Figure 21.3 Increase in Motor Vehicles, 1970–2010.
Source: Livernash & Rodenberg, 1998.

about 26 percent in the developing nations. Low incomes in those nations prevent ever-higher rates of auto acquisition, but ownership of less expensive two- and three-wheeled motor vehicles is increasing rapidly. In India, for example, motorcycle ownership is increasing by about 17 percent annually. All growth in the use of internal combustion engines contributes to the production of carbon dioxide, the most prevalent "greenhouse gas" associated with the growing problem of global warming. Overall, however, industrialized nations produce 90 percent of the world's hazardous wastes (Livernash & Rodenburg, 1998).

Increases in population in poor regions of the world place additional pressures on critical resources, especially soil for crops, water for irrigation, forests for lumber and firewood, and fish species in nearby lakes and oceans. As noted earlier, the catastrophist view of these resources, especially land and arable soils, is that they will be exhausted and incapable of supporting rapid increases in population. But this gloomy view has proven, so far, to be overstated. According to one analysis, "despite the unprecedented population growth in the second half of the twentieth century, the world's farmers produce enough food to feed the current population and will probably be able to feed several billion more people in the next century" (Livernash & Rodenberg, 1998, p. 19).

On the other hand, most of the available soil resources in the world are being used, and each year a small proportion of those resources is lost to urban expansion and degradation through overuse. Still, improvements in irrigation, fertilization, and crop rotation continue to produce increased yields (Boserup, 1965; Dasgupta, 2000). The same is not true, however, of world marine harvests, which have been increasing steadily and have resulted in widespread overfishing of wild fish stocks such as salmon, herring, and others. In Figure 21.4, we see that China catches an increasing proportion of the world's wild fish, and that elsewhere in the world, catches of marine fish are declining steadily. In theory, therefore, agriculture and fishing can continue to feed growing populations, but in practice this may not be true for all regions of the world, particularly those most subject to dire poverty and political unrest.

In coming decades, many of the answers to the problems of poverty, resource depletion, and political unrest currently experienced in poor regions of the world will emerge from the rapidly growing cities of those regions. Rapid urbanization can worsen social conditions, but over time, the growth of cities usually signifies that societies are moving into stages of development marked by greater emphasis on education, invention, and economic modernity.

Population Growth and Urbanization

Studies of the demographic transition in Europe (Bacci, 1997; Laslett, 1972, 1983) have concluded that its specific course in any given society depends on a complex combination of factors: higher age at marriage and fewer couples marrying, use of birth-control techniques, increased education, migration to other countries, and rural–urban migration. In each case, however, the bulk of the population growth was absorbed by the cities. Thus, rapid urbanization is partially an outgrowth of the demographic transition.

Urbanization refers to the proportion of the total population concentrated in urban settlements. Although urbanization and the growth of cities have occurred together, it is important to distinguish between the two. Cities can continue to grow even after most of the population is urbanized and the society's dominant institutions (for example, government agencies, major markets, newspapers, television networks) are located in its urban centers.

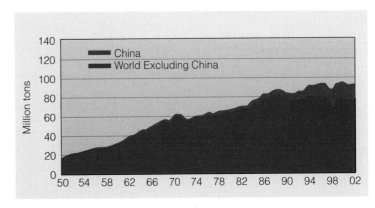

Figure 21.4 World Capture Fisheries Production. Note: Excludes marine production from aquaculture. Source: Food and Agriculture Organization, 2004.

urbanization: A process in which an increasing proportion of a total population becomes concentrated in urban settlements.

Urbanization contributes to the lower birthrates that are characteristic of the third stage of the demographic transition. As Louis Wirth pointed out, "The decline in the birthrate generally may be regarded as one of the most significant signs of the urbanization of the Western world" (1968/1938, p. 59). In the city, a variety of factors lead to postponement of marriage and childbearing. For one thing, living space is limited. For another, newcomers to the city must find jobs before they can even think about marrying, and often they lack the ties to family and kin groups that might encourage them to marry and have children.

Despite declining birthrates in many non-Western countries, the populations of some cities in Asia and Africa are increasing faster than those of rural areas. The reason seems to be that rapid change in rural areas, especially the mechanization of agriculture, pushes rural people to the cities, where they often live in shantytowns and villagelike settlements. In those settlements, the rural tradition of large families is not quickly altered. In addition, because they have greater access to health care during pregnancy and childbirth, the infant death rate may be lower among the migrants than in rural villages. However, when these populations become part of the urban economy, they too begin to limit their family size (Beauregard, 2009; Gugler, 1997). (See the Global Social Change box on page 544.)

Latino Migration to Small Towns and Cities

Previous waves of immigrants to the United States tended to arrive in what are known as the gateway cities, especially New York, Los Angeles, Miami, Chicago, Houston, and a few others. But the most recent trends find Latino, especially Mexican, immigrants, both legal and illegal, moving directly to small towns and cities throughout the United States. In Figure 21.5, we see that Hispanic populations are increasing most rapidly not in the major gateway cities but in rural areas of the South, the Southwest, and the mountain states. These trends are creating new, and in some cases disturbing, social forces in communities that have little experience in dealing with new racial and ethnic groups.

"We really haven't had this sort of rapid demographic change in 100 years," said Jeffrey S. Passel, a demographer at the Pew Hispanic Center in Washington, D.C. In the last fifteen years, the number of Hispanics, both documented and undocumented, shot up from about 560,000 to over 2.4 million, according to census data released in 2006 (Swarns, 2006). Latino immigrants are drawn to small towns and cities in the South and other more rural parts of the nation because these are areas that have been losing white and African American populations for many years. As a consequence of these population losses, these counties seek out low-wage workers and find them among Mexican immigrants. This sudden shift is causing a good deal of local conflict, as well as incidents of neighborliness, in different communities. Although they frequently encounter hostility and bigotry, the Mexicans, like other immigrants before them, are reacting to the new opportunities available to them in these communities by beginning to limit their family size, learn English, and establish economic and cultural footholds as Mexican Americans.

In sum, urbanization is closely linked with rapid increases in population, yet the nature of life in cities tends to limit the size of urban families. Cities grow primarily as a result of migration, but new migrants do not find it easy to form families. Thus, in his research on the changing populations of Western industrial cities, Hauser (1957) showed that birthrates were lowest in the areas of the city that had the highest proportions of new migrants. Mounting evidence from archaeology also suggests that child sacrifice in ancient cities may have helped control urban populations (Browne, 1987). The eventual result of large-scale migration to cities may be a slowdown in population growth: As an increasing proportion of a society's population lives in cities, the rate of growth of the population as a whole tends to drop.

The Growth of Cities

In Chapter 4, we noted that cities became possible when agricultural populations began to produce enough extra food to support people who were not directly engaged in agriculture, such as priests, warriors, and artisans. Changes in the technology of food production made it possible for increasingly larger populations to be supported by the same number of agricultural workers. This has been a central factor in the evolution of cities, but as we will see shortly, the most dramatic increases in urban populations have occurred only in the past 200 years.

The Urban Revolution The increasing tendency of people throughout the world to live in cities has been referred to as the *urban revolution*. The extent of this "revolution" can be grasped by comparing a few figures. In 1800, only 3 percent of the world's people lived in cities with populations of over 5,000, and of this

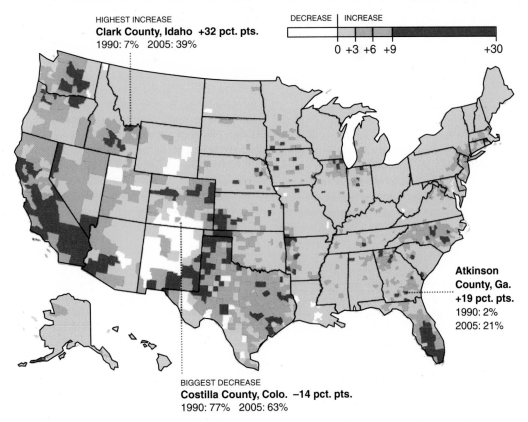

HIGHEST INCREASE
Clark County, Idaho +32 pct. pts.
1990: 7%   2005: 39%

DECREASE | INCREASE

0 +3 +6 +9   +30

**Atkinson
County, Ga.
+19 pct. pts.**
1990: 2%
2005: 21%

BIGGEST DECREASE
Costilla County, Colo. −14 pct. pts.
1990: 77%   2005: 63%

Figure 21.5 Tracking Growth: Percentage Increase in Number of Hispanics in Each County, from **1990.** Source: "In Georgia, Newest Immigrants Unsettle an Old Sense of Place," by Rachel L. Swarns, *The New York Times*, August 4, 2006. Copyright 2006. The New York Times.

proportion, a mere 2.4 percent lived in cities with populations of over 20,000. Between 1800 and 1970, a period during which the world's population increased fourfold, the percentage of people living in cities with 5,000 or more inhabitants increased elevenfold, whereas that of people living in cities with 100,000 or more inhabitants increased almost fourteenfold. By 1970, fully one-third of the world's population lived in cities (Gugler, 1997; United Nations Population Fund, 2005).

These data indicate not only that increasing percentages of the world's population are living in cities but also that the cities themselves are larger than ever before. The growth of cities in this century has given rise to the concept of the **metropolitan area,** in which a central city is surrounded by several smaller cities and suburbs that are closely related to it both socially and economically. Most people in the United States live in large metropolitan areas.

Large-scale urbanization is a relatively recent development in human history. As Kingsley Davis (1955) pointed out:

Compared to most other aspects of society— such as language, religion, stratification, or the family—cities appeared only yesterday, and urbanization, meaning that a sizable proportion

of the population lives in cities, has developed only in the last few moments of man's existence. (p. 429)

Although there were a few cities as early as 4000 B.C.E., they were very small and were supported by large rural populations. The famous cities of ancient times were minuscule by modern standards: Babylon covered roughly 3.2 square miles, Ur some 220 acres (Davis, 1955).

Preindustrial cities like Ur and Nineveh, early Athens and Rome, and the ancient Mayan cities were vastly different from the cities we know today. They did not grow around a core of office buildings and retail outlets the way industrial cities do. Instead, they were built around temples or other ceremonial buildings (for example, Notre Dame in Paris). Close to the temple, one could find the palaces of the rulers and the courtyards of the royal families. Parade grounds and public shrines made the core of the ancient city a spacious place where the city's population could meet on special occasions. On the outskirts of these ancient cities, one found not the rich, as in contemporary cities with

metropolitan area: A central city surrounded by several smaller cities and suburbs that are closely related to it both socially and economically.

global social change

The Power of Rural-Urban Migration

Immigration and migration are social forces that continually bring about change of all kinds. In eastern Europe, for example, the animosities among nationality groups that periodically erupt in violence can often be traced to earlier periods of immigration and resettlement of populations resulting from war and conquest. In the United States as well, immigration continually produces social change, especially in the major cities that serve as the gateways for newcomers.

Migration, which includes both immigration from a foreign nation and movement within national boundaries, is, along with birth and death rates, a key factor affecting population size. People have migrated from one region or country to another throughout history, compelled to seek new homes because they have exhausted their food supplies or been driven out by invaders or because they believe they will find a better life in the new land. Although people have migrated from rural areas to the city throughout history, rural-urban migration has become especially pronounced in the last two centuries.

The forces that impel residents of rural areas to migrate to cities are not fully understood. There is continual debate over whether rural people are "pushed" into cities by conditions beyond their control or "pulled" to cities by the attractions of city life. Of course, both push and pull factors affect rural-urban migration, but it is not clear exactly how they interact. It appears, however, that "it is the push of existing rural circumstances which suggests to the rural resident that things might be better in the urban area" (Breese, 1966, p. 80).

Several factors may be responsible for pushing people out of rural areas. Chief among these is overpopulation, which reduces the amount of food and work available per rural resident. Others include lack of opportunities to obtain farmland (e.g., as a result of *primogeniture,* in which only the oldest son inherits land) and the seasonal nature of employment in agriculture. When rural people who are experiencing these conditions become aware of higher living standards in urban places, they may develop a sense of relative deprivation (see Chapter 9). "They view with great interest the reported higher income, access to education, and other rumored facilities of the urban area" (Breese, 1966, pp. 80–81). The feeling of relative deprivation is intensified by improvements in communication and transportation that provide rural people with more feedback about the advantages of life in the city.

A significant pull factor is the presence of relatives and friends in the city. These individuals can be called on for help when the rural migrant arrives in search of a new home and a job. This factor produces what is known as *chain migration,* a pattern in which a network of friends and relatives is transferred from the village to the city over time. The result is the formation of a small, homogeneous community within the city (Massey, Durand, & Malone, 2003).

Chain migration operates across national boundaries as well as between rural and urban areas in the same country. (For example, a large proportion of the immigrants to American cities in the twentieth century were rural people from European and Asian countries.) Once a few hardy souls establish a foothold, their families and friends can join them. In this way, small ethnic communities are formed within the city—like the Chinatowns of New York and San Francisco, the Slovenian community in Cleveland, or the Mexican community in Chicago. Recently this pattern has been established in suburban portions of metropolitan areas as well as in central cities. Thus, the suburbs of metropolitan areas in the South and West have become home to large numbers of immigrants from Vietnam (Portes & Rumbaut, 1996).

At this writing, debate rages over a proposed Arizona state law that encourages police to check the legal immigration status of anyone they have stopped for traffic violations or any other reason. This controversial measure, which may not be approved by the courts, is a reflection of the intense feelings aroused by illegal border crossings in states along the Mexican border, but it is also infuriating Latino populations in the United States who feel that they are being unfairly targeted (Archibold, 2010b).

A U.S. Border Patrol agent marches a group of illegal immigrants across a portion of the Yuha Desert after they were found hiding in the brush near El Centro, California.

well-developed transportation networks, but the poor, who lived in hovels and were often pushed from one location to another according to the whims of those with more wealth and power (Gugler, 1997; Sjoberg, 1968).

A variety of factors limited the size of cities. Among them were farming methods that did not produce enough surplus food to feed many city dwellers, the lack of efficient means of transporting goods over long distances, inadequate technology for transporting water in great quantities, and the lack of scientific medicine. Not until about 1800 did large-scale urbanization become possible.

The speed with which urbanization has changed the size and layout of cities is remarkable: "Before 1850 no society could be described as predominantly urbanized, and by 1900 only one—Great Britain—could be so regarded. Today . . . all industrial nations are highly urbanized, and in the world as a whole the process of urbanization is accelerating rapidly" (Davis, 1968, p. 33). In 2008, for the first time in history, more than half of the world's population lived in towns and cities. By 2030, this number is expected to swell to almost 5 billion, with urban growth concentrated in Africa and Asia (UNFPA, 2008). Although megacities have captured much public attention, most of the new growth will occur in smaller towns and cities, which have fewer resources to respond to the magnitude of the change. (Africa

MARWAN IBRAHIM/AFP/Getty Images

U.S. soldiers on patrol with Iraqi forces during a massive sweep against jihadists in the Nineveh province of which Mosul is the capital, June 2008.

now has the fastest rate of urbanization, and about 70 percent of the population of Latin America lives in urban areas—a rate of urbanization comparable to that of North America and Europe.)

As shown in Table 21.4 on the next page, many metropolitan regions in older advanced nations like the United States (especially Greater New York), Great Britain (Greater London), and Germany (Rhein-Ruhr) reached their peak of growth decades ago and are now growing slowly if at all. In many parts of Latin America and Asia, in contrast, metropolitan regions are experiencing explosive growth and are attracting vast populations of urban newcomers, many of whom are living as squatters on vacant land.

Rapid urbanization occurring throughout the world brings together diverse groups of people in cities that often are not prepared to absorb them. The problems caused by such urbanization are immense. They include housing, educating, and caring for the health of newcomers; preventing gang violence and intergroup hatred; and many other difficult tasks. Moreover, as the world becomes ever more urbanized, populations become increasingly interdependent.

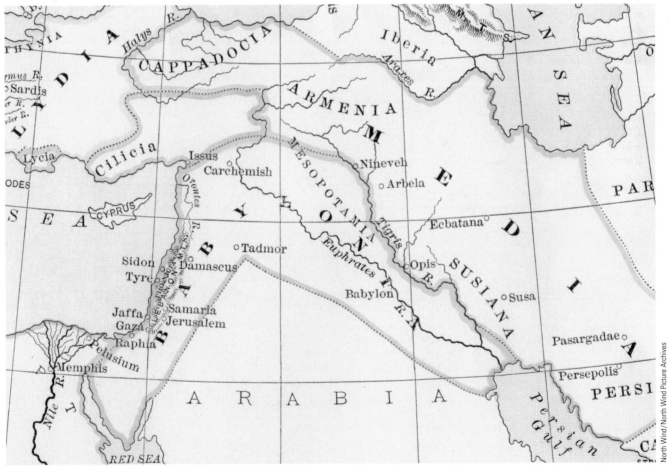

North Wind/North Wind Picture Archives

When the United States and other nations invaded and occupied Iraq in the first and second Gulf Wars, they were also discovering one of the world's oldest centers of urban civilization. The cities of Nineveh and Babel, shown on the accompanying map of the world's Bronze Age cities, date to before 2000 B.C.E. The Bronze Age refers to a period of time, varying in different parts of the world between 3000 B.C.E. and about 1500 B.C.E., during which metallurgy had advanced to the point of making bronze—an alloy of tin and copper—from natural ores, but not yet to the point of systematic production of iron (which occurred during the Iron Age). The Bronze Age was more advanced than the Stone Age, in which artifacts and tools were largely made from carved stone.

Rank in 2000	Country	Urbanized Area	Thousands of Inhabitants		
			1950	1975	2000
1	Mexico	Mexico City	3,050	11,610	25,820
2	Japan	Tokyo–Yokohama	6,736	17,668	24,172
3	Brazil	São Paulo	2,760	10,290	23,970
4	India	Calcutta	4,520	8,250	16,530
5	India	Bombay	2,950	7,170	16,000
6	United States	New York–N.E. New Jersey	12,410	15,940	15,780
7	Republic of Korea	Seoul	1,113	6,950	13,770
8	Brazil	Rio de Janeiro	3,480	8,150	13,260
9	China	Shanghai	10,420	11,590	13,260
10	Indonesia	Jakarta	1,820	5,530	13,250
11	India	Delhi	1,410	4,630	13,240
12	Argentina	Buenos Aires	5,251	9,290	13,180
13	Pakistan	Karachi	1,040	4,030	12,000
14	Iran	Tehran	1,126	4,267	11,329
15	Bangladesh	Dhaka	430	2,350	11,160
16	Egypt	Cairo–Ginza–Imbaba	3,500	6,250	11,130
17	Iraq	Baghdad	579	3,830	11,125
18	Japan	Osaka–Kobe	3,828	8,649	11,109
19	Philippines	Manila	1,570	5,040	11,070
20	United States	Los Angeles–Long Beach	4,070	8,960	10,990
21	Thailand	Bangkok–Thonburi	1,400	4,050	10,710
22	United Kingdom	London	10,369	10,310	10,510
23	Russia	Moscow	4,841	7,600	10,400
24	China	Beijing	6,740	8,910	10,360
25	Germany	Rhein–Ruhr	6,853	9,311	9,151
26	Peru	Lima–Callao	1,050	3,700	9,140
27	France	Paris	5,525	8,620	8,720
28	Nigeria	Lagos	360	2,100	8,340
29	Italy	Milan	3,637	6,150	8,150
30	India	Madras	1,420	3,770	8,150

Note: The urban places listed in this table are areas of dense contiguous settlement whose boundaries do not necessarily coincide with administrative boundaries; they usually include one or more cities and an urbanized fringe.

Source: UNDP, 2002.

Urban populations are supported, for example, by worldwide agricultural production, not just by the produce grown in the surrounding countryside. In the same way, the problems of one major city or one large urbanizing region can no longer be thought of as isolated from the problems of the older, more affluent urbanized regions.

© Peter Menzel/Stock, Boston

Urban sociologists have always looked at the relationship between the planned and unplanned aspects of urban growth and the social groups that represent each aspect. Usually the poor and the immigrants represent unplanned additions to the city, whereas the rich and the powerful guide the processes of urban planning and organization to suit their interests.

living. In an urban society, not everyone lives in the cities, but no one can escape the pervasive influence of urban centers.

The concept of an urban society may become clearer if we look at a society that has not become fully urbanized, such as India. Until the early twentieth century, the rate of city growth in India was relatively slow. Beginning in the 1920s, however, the populations of India's cities increased dramatically, and by 1960, seven Indian cities had populations surpassing 1 million. Yet India has not become an urban society. It remains, in the words of Noel P. Gist (1968), "a land of villagers." Because of persistently high birthrates and declining death rates in the villages, India's rural population is growing almost as fast as its urban population. Moreover, although the growth of the cities has affected village life, especially through the efforts of city-trained teachers, health officials, administrators, and storekeepers, it has had little effect on the social structure of rural India. Thus, as a whole, Indian society exhibits the extremes of rural isolation and urban dynamism, with all the chaos and poverty that are characteristic of societies undergoing major social change (World Bank, 2005).

Throughout the twentieth century, sociologists devoted considerable study to urbanization and the changes that accompany it. The ways in which the growth of cities and metropolitan regions alter the surface of the planet are part of the sociological study of urbanization. So are questions about how cities change our experience of community and our relationships with others. Finally, sociologists continue to focus on the changing patterns of inequality and conflict that occur in metropolitan regions. The remainder of this chapter explores these issues in greater detail.

Urban Societies Urbanization produces *urban societies.* By this we mean not only that cities are the cultural and institutional centers of a society but also that urban life has a pervasive influence on the entire society (Durkheim, 1964/1893; Weber, 1962/1921; Wirth, 1968/1938). Today, the United States is spanned by interstate highways that link the nation's rapidly growing urban and suburban places, and carry traffic through rural areas at high speeds. Waterways, forests, hills, and valleys are channeled, cut, and bulldozed to make way for expanding settlements. Once considered far from the urban scene, national parks and forests now receive millions of visitors from the metropolitan centers. And in an urban society, more and more people, even those living in isolated rural communities, share in the mass culture of that society—the television and radio programs, the movies, the books and magazines, all of which stress themes that appeal to people who are familiar with metropolitan

THE URBAN LANDSCAPE

Urban Expansion

The effects of urbanization began to be felt in American society in the mid-nineteenth century. As growing numbers of people settled in the West, waves of immigrants from Ireland, Germany, Italy, and many other parts of central and southern Europe streamed into the cities of the East. In the 1840s and 1850s, for example, approximately 1.35 million Irish immigrants arrived in the United States, and in just twelve years, from 1880 to 1892, more than 1.7 million Germans arrived (Bogue, 1985). In other chapters, we have more to say about the impact of the great migrations from China and Korea and Latin America, the importation of slaves from Africa, and the large numbers of people of all races and ethnic groups who continue to arrive in U.S. cities. The point here is that, for 150 years, North American cities have been a preferred destination of people from all over the world; as a result, they have received numerous waves of newcomers since their period of explosive growth in the nineteenth century.

The science of sociology found early supporters in the United States and Canada partly because the cities in those nations were growing so rapidly. It often appeared that North American cities would be unable to absorb all the newcomers who were arriving in such large numbers.

then & now

Street Urchins and Trailer Kids A century ago, as millions of newcomers from throughout the world crowded into U.S. cities in search of opportunity, rapid social change produced some of the world's most notorious urban slums. Jacob Riis, a photojournalist and amateur sociologist, brought the harsh living conditions in those slums to the attention of the world in his

© Bettmann/Corbis

famous photos of "how the other half lives." Children who spent much of their time on the streets and survived by their wits were called "street urchins" (or, as Riis called them, "street Arabs").

In today's world, the central city is no longer viewed as a place to find new opportunities. Mobile families and individuals travel long distances, seeking manufacturing and service jobs outside major city centers. They often find what they hope will be temporary lodging in trailer parks. Poor and lacking health insurance and other forms of social security, residents of trailer parks tend to be stigmatized by those who are more

fortunate. They and their children are sometimes called "trailer trash," a term that reveals the deep prejudices that continue to divide Americans, even in a society that is far more affluent than the one described by Riis.

© Kirk Condyles

Presociological thinkers like Frederick Law Olmsted, the founder of the movement to build parks and recreation areas in cities, and Jacob Riis, an advocate of slum reform, urged the nation's leaders to invest in improving the urban environment, building parks and beaches, and making better housing available to all (Cranz, 1982; Kornblum & Lawler, 1999). As we saw in Chapter 1, these reform efforts were greatly aided by sociologists who conducted empirical research on the social conditions in cities. In the early twentieth century, many sociologists lived in cities that were characterized by rapid population growth and serious social problems. It seemed logical to use empirical research to construct theories about how cities grow and change in response to major social forces as well as more controlled urban planning.

The founders of the Chicago school of sociology, Robert Park and Ernest Burgess, attempted to develop a dynamic model of the city, one that would account not only for the expansion of cities in terms of population and territory but also for the patterns of settlement and land use within cities. They identified several factors that influence the physical form of cities. Among them are "transportation and communication, tramways and telephones, newspapers and advertising, steel construction and elevators—all things, in fact, which tend to bring about at once a greater mobility and a greater concentration of the urban populations" (Park, 1967/1925, p. 2). The important role of transportation is described in one of Park's essays:

> The extent to which . . . an increase of population in one part of the city is reflected in every other depends very largely upon the character of the local transportation system. Every extension and multiplication of the means of transportation connecting the periphery of the city with the center tends to bring more people to the central business district, and to bring them there oftener. This increases the congestion at the center; it increases, eventually,

the height of office buildings and the values of the land on which those buildings stand. The influence of land values at the business center radiates from that point to every part of the city. (1967/1926, pp. 57–58)

The Concentric Zone Model Park and Burgess based their model of urban growth on the concept of *natural areas*—that is, areas in which the population is relatively homogeneous and land is used in similar ways without deliberate planning. In Park's words:

> Every great city has its racial colonies, like the Chinatowns of San Francisco and New York, the Little Sicily of Chicago. . . . Most cities have their segregated vice districts . . . their rendezvous for criminals of various sorts. Every large city has its occupational suburbs, like the Stockyards in Chicago, and its residential enclaves, like Brookline in Boston. (1967/1925, p. 10)

Park and Burgess saw urban expansion as occurring through a series of "invasions" of successive zones or areas surrounding the center of the city. For example, migrants from rural areas and other societies "invaded" areas where housing was cheap. Those areas tended to be close to the places where they worked. In turn, people who could afford better housing and the cost of commuting "invaded" areas farther from the business district, and these became the Brooklines, Gold Coasts, and Greenwich Villages of their respective cities.

Park and Burgess's model, which has come to be known as the *concentric zone model,* is portrayed in Figure 21.6 (Figure 21.7 applies the model to Chicago). Because the model was originally based on studies of Chicago, its center is labeled "Loop," the term that is commonly applied to that city's central commercial zone. Surrounding the central zone is a "zone in transition," an area that is being invaded by business and light manufacturing. The third zone is inhabited by workers who do not want to live in the factory or business district but, at the same time, need to live reasonably close to where they work. The fourth, or residential, zone consists of higher-class apartment buildings and single-family homes. And the outermost ring, outside the city limits, is the suburban or commuters' zone; its residents live within a thirty- to sixty-minute ride of the central business district (Burgess, 1925).

Studies by Park, Burgess, and other Chicago school sociologists showed how new groups of immigrants tended to become concentrated in segregated areas within inner-city zones, where they encountered suspicion, discrimination, and hostility from ethnic groups that had arrived earlier. Over time, however, each group was able to adjust to life in the city and to find a place for itself in the urban economy. Eventually many of the immigrants were assimilated into the institutions of American society and moved to desegregated areas in outer zones; the ghettos they left behind were promptly occupied by new waves of immigrants (Kasarda, 1989).

Note that each zone is continually expanding outward. Thus, Burgess wrote, "If this chart is applied to Chicago, all four of these zones were in its early history included in the circumference of the inner zone, the present business district. The present boundaries of the [zone in transition] were not many years ago those of the zone now inhabited by independent wage-earners" (1925, p. 50). Burgess also pointed out that "neither Chicago nor any other city fits perfectly into this [model]. Complications are introduced by the lake front, the Chicago River, railroad lines, historical factors in the location of industry, the relative degree of the resistance of communities to invasion, etc." (pp. 50–51).

The Park and Burgess model of growth in zones and natural areas of the city can still be used to describe patterns of growth in cities that were built around a central business district and continue to attract large numbers of immigrants. But this model is biased toward the commercial and industrial cities of North America, which have tended to form around business centers rather than around palaces or cathedrals, as is the case in so many other parts of the world. Moreover, it fails to account for other patterns of

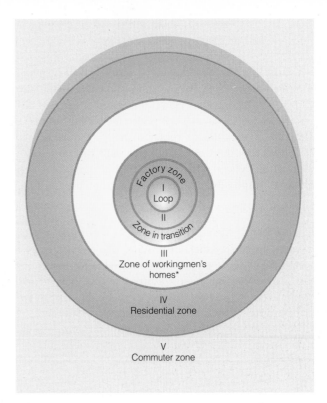

Figure 21.6 The Concentric Zone Model. *Workers' homes ("workingmen" was the accepted terminology of the time). Source: Park & Burgess, 1967/1925.

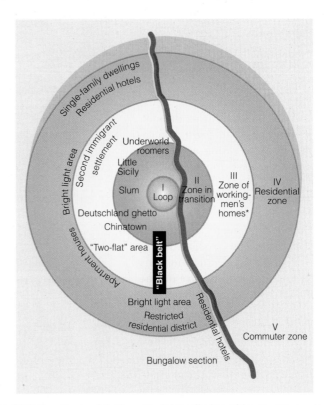

Figure 21.7 The Concentric Zone Model Applied to Chicago. *Workers' homes ("workingmen" was the accepted terminology of the time). Source: Park & Burgess, 1967/1925.

urbanization, such as the rise of satellite cities and the rapid urbanization that occurs along commercial transportation corridors.

Satellite Cities Outside the city of Detroit lies the town of River Rouge, long famous as a center of steel and automobile production. Outside the city of Toronto lies Hamilton, also a smoky manufacturing town where much of Canada's steel is produced. And outside Chicago is Gary, Indiana, another major center of heavy industrial production. Outside New York City, there are many other *satellite cities,* some devoted to heavy industries whose needs for space, rail and water service, and energy make it impossible to locate them in the central business districts.

Other satellite cities are devoted to less environmentally stressful industries. Fort Worth, Texas, for example, is becoming a major center for white-collar industries that are leaving older central-city locations in search of cheaper space and a well-educated labor force. Still others, like Tysons Corner, Virginia, near Washington, D.C., have grown around large shopping centers and mall complexes. In the second half of the twentieth century, the growth of these and other satellite cities was accelerated by public investment in the interstate and metropolitan highway systems, a point to which we will return shortly in discussing the rise of metropolitan urban systems (Baldassare, 1986; Clay, 1994; Fishman, 1987).

Strip Development The growth of satellite cities specializing in the production of a particular commodity or product is typical of the period of rapid industrial growth that occurred before World War II. A more current model of urbanization is known as *strip development;* it is shown in schematic form in Figure 21.8. Strip I represents a typical nineteenth- or early twentieth-century farming area bisected by a road and a stream. In time the bridge over the stream and the path along the stream became a road that intersected the original road (Strip II), creating an intersection that stimulated the growth of a village with a mill; families living in the village depended on wages from the mill. As the village grew into a town, population growth and increasing automobile traffic created the need for wider roads and bypass roads, as shown in Strips III and IV. Specialization as an "automobile-convenient" strip, with the development of more motels, a drive-in theater, and drive-in businesses, can be seen in Strip V. At this time, the residential functions along the strip began to disappear and the area assumed an increasingly commercial character.

Strip VI shows the strip's present stage of development. A spur of the limited-access highway system creates new growth around a cloverleaf (including a new community college situated between the artificial lake and the shopping center). Note that along the original main road are office buildings and a modern

industrial park rather than the smaller retail businesses that once defined the town center. Indeed, as the strip has developed and surrounding roads have been enlarged to form part of the metropolitan highway system, the town center itself has largely disappeared (Clay, 1980). The strip development model thus describes the incorporation of smaller communities and towns into a larger metropolitan area.

Metropolitan Areas

Megalopolis After 1920, new metropolitan areas developed largely as a result of the increasing use of automobiles and the construction of a network of highways covering the entire nation (Flink, 1988). The shift to automobile travel brought former satellite cities within commuting distance of the major industrial centers, thereby adding to the size of those metropolitan areas. In the South and Southwest, new metropolitan areas developed. In recent years, these have become the fastest-growing urban areas in the nation (Exum & Messina, 2004).

Since World War II, sociologists have been studying an increasingly important urban phenomenon: the emergence of large multinuclear

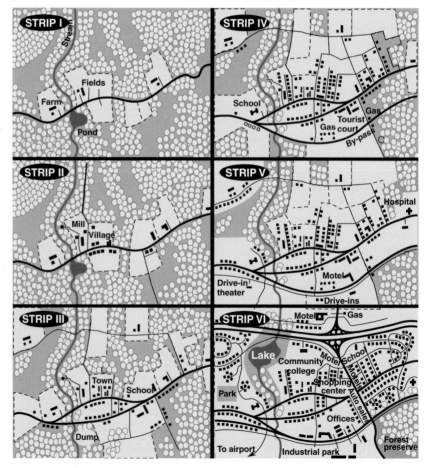

Figure 21.8 Strip Development and Sprawl. Source: Clay, 1980.

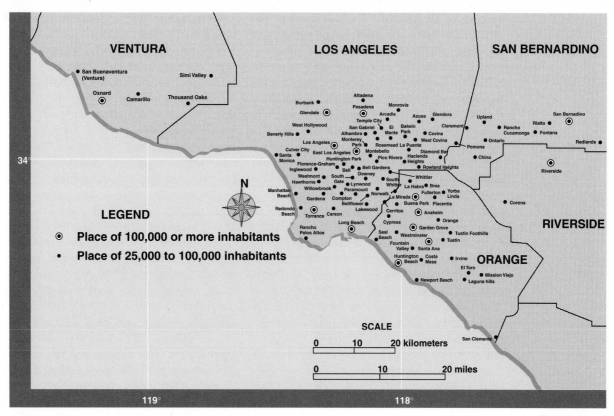

Figure 21.9 **The Los Angeles–Anaheim–Riverside Megalopolis.** Source: U.S. Census Bureau.

urban systems. The term *megalopolis* is used to describe these vast complexes, whose total population exceeds 25 million. Jean Gottmann (1978) pointed out that a megalopolis is not "simply an overgrown metropolitan area"; rather, it is a system of cities distributed along "a major axis of traffic and communication" (p. 56). Gottmann identified six megalopolises: the American northeastern megalopolis, the Great Lakes megalopolis, the Tokaido megalopolis in Japan, the megalopolis in England (the London area), the megalopolis of northwestern Europe (extending from Amsterdam to the Ruhr), and the Urban Constellation in China (centered on Shanghai). Four others are developing rapidly: the Rio de Janeiro–São Paulo complex in Brazil, the Milan–Turin–Genoa triangle in northern Italy, the Valley of Mexico, and the urban swath extending from San Diego to the San Francisco Bay area (Gioioso, 2010).

A megalopolis is characterized by an "intertwined web of relationships between a variety of distinct urban centers . . . expressed partly in a physical infrastructure consisting of highways, railways, waterways, telephone lines, pipelines, water supply, and sewage systems criss-crossing the whole area, and partly in more fluid networks, such as the flows of traffic, the movement of people and goods, the flows of telephone calls [and] of mail" (Gottmann, 1978, p. 57). Despite their interdependence, however,

"the sizes and specializations of the various . . . components [of a megalopolis] are extremely varied, as demonstrated by the diverse characteristics of the cities, towns, villages, suburban and rural areas that form the vast system" (p. 57). Therefore, a megalopolis can best be described as a huge social and economic mosaic.

Los Angeles provides a good example of the development of a megalopolis (see Figure 21.9). Between 1960 and 1970, the population of Los Angeles increased by more than 2 million, double the growth of Chicago and more than that of New York and San Francisco put together (Smith, 1968). Today, the Los Angeles metropolitan area continues to grow, although at a somewhat slower rate. As a result of its extraordinary growth, the region must struggle to control the effects of air pollution from more than 3 million automobiles. It must also struggle to furnish adequate supplies of water for its residents. In a review of conflicts between Angeleno leaders and the residents of smaller towns near water sources, sociologist John Walton (1992) has shown that control over water and other natural resources, especially by powerful groups that dominate the regional real estate markets, is a key to understanding the history not only of the Los Angeles metropolitan region but of the entire desert West as well.

Another problem affecting many of the world's megalopolitan regions is the possibility of major disasters. Many are located in disaster-prone areas—either on coastal shorelines, as in the case of the Boston–Washington corridor, or worse, along major fault lines of earthquake activity, as in the case of Los Angeles (Davis, 1998).

Decentralization One effect of the growth of megalopolitan areas is *decentralization,* in which outlying areas become more important at the

> **megalopolis:** A complex of cities distributed along a major axis of traffic and communication, with a total population exceeding 25 million.

expense of the central city. This trend is not new. In the 1960s and 1970s, large numbers of middle-income city dwellers moved to suburban areas while the poor remained in the central cities. Business and industry also moved to the suburbs, creating widespread speculation that vibrant central cities would become a thing of the past. Over the past generation, however, the central cities of some metropolitan areas—such as New York, Philadelphia, Chicago, and Baltimore—have shown renewed vitality. Far from decaying, they have become major financial and cultural centers serving the needs of huge populations. On the other hand, medium-sized cities like Gary, Indiana, and Paterson, New Jersey, have suffered because their central business districts have little to offer suburban dwellers in the way of financial services like banking and insurance or cultural attractions like theaters and symphony orchestras (Kasarda, 1988).

An important feature of megalopolitan areas—especially as they become more decentralized—is their diversity. These huge urban regions include many different kinds of communities: ethnic communities in both central cities and suburbs, middle-class "bedroom" suburbs, industrial towns, areas devoted to truck gardening or dairy farming, "second home" communities (for example, beachfront areas), and so on. Each meets the economic and cultural needs of a specific urban population. Sociologists have devoted considerable study to these urban communities, and we discuss their findings in the next section.

URBAN COMMUNITIES

"The city," wrote Robert Park, is more than a set of "social conveniences—streets, buildings, electric lights, tramways, and telephones, etc.; something more, also, than a mere constellation of . . . courts, hospitals, schools, police, and civil functionaries of various sorts. The city is, rather, a state of mind, a body of customs and traditions. . . . It is a product of nature, and particularly of human nature" (1967/1925, p. 1).

The connection between the city and human nature has been a recurrent theme in literature throughout history. Many literary images of the city are negative. In the Bible, for example, the cities of Sodom and Gomorrah are symbols of the worst aspects of human nature. The poet Juvenal complained that Rome produced ulcers and insomnia and subjected its residents to burglars and dishonest landlords. American literature also contains many negative images of the city. Thomas Jefferson, for instance, wrote, "I view great American cities as pestilential to the morals, health, and the liberties of man." And Henry David Thoreau escaped from the city to Walden Pond in an effort to "rediscover his soul" (Fischer, 1987/1976).

Social scientists who study cities have devoted a great deal of attention to the tension between "community" and "individualism" as it relates to life in cities (Lofland, 1998). Country dwellers have been thought of as "happily ensconced in warm, humanly rich and supportive social relationships: the family, neighborhood, town," whereas city dwellers are "strangers to all, including themselves. They are lonely, not emotionally touching or being touched by others, and consequently set psychically adrift" (Fischer, 1987/1976, p. 19). On the other hand, country dwellers are sometimes viewed as "stifled by conventionality, repressed by the intrusion and social control of narrow-minded kin, neighbors, and townsmen," whereas city dwellers are "free to develop individual abilities, express personal styles, and satisfy private needs" (p. 20). These views of the city are obviously contradictory, and much research has been devoted to the question of how urban life affects individuals and communities. In this section, we look at some of the findings of that research and the theories of urban life that have been proposed on the basis of those findings.

The Decline-of-Community Thesis

Early studies of the nature and effects of urban life were dominated by efforts to evaluate the differences between rural and urban societies. They tended to reach rather gloomy conclusions (Beall, Guha-Khasnobis, & Kanbur, 2010). We have already noted (see Chapter 4) that Ferdinand Tönnies described the process of urbanization as a shift from *gemeinschaft* (a community based on kinship ties) to *gesellschaft* (a society based on common interests). Émile Durkheim reached a similar conclusion: Small rural communities are held together by ties based on shared ideas and common experiences, whereas urban societies are held together by ties based on the interdependence of people who perform specialized tasks. Both Tönnies and Durkheim believed that urban life weakens kinship ties and produces impersonal social relationships.

In a 1905 essay, "The Metropolis and Mental Life," Georg Simmel focused on the effects of urban life on the minds and personalities of individuals. According to Simmel, cities bombard their residents with sensory stimuli: "Horns blare, signs flash, solicitors tug

In urban societies, where there are large numbers of people who are strangers to one another, sociologists have noted a propensity to wear identifying "uniforms" that indicate the kind of person one wishes to know and interact with.

at coattails, polltakers telephone, newspaper headlines try to catch the eye, strange-looking and strange-behaving persons distract attention" (quoted in Fischer, 1976, p. 30). The urban dweller is forced to adapt to this profusion of stimuli, which Stanley Milgram (1970) termed *psychic overload.* The usual way of adapting is to become calculating and emotionally distant. Hence the image of the city dweller as aloof, brusque, and impersonal in his or her dealings with others.

This view of the effects of urban life found further expression in the work of Louis Wirth, especially in his essay "Urbanism as a Way of Life" (1968/1938). Wirth began by defining the city as "a relatively large, dense, and permanent settlement of socially heterogeneous individuals" (p. 28). He then attempted to show how these characteristics of cities produce psychological stress and social disorganization. The primary psychological effect of urban life, according to Wirth, is a weakening of the individual's bonds to other people. Without such bonds, the individual must deal with the crises of life alone, and often the result is mental illness. In other cases, the city dweller, again because of the absence of close ties to friends or kin, lacks the restraints that might prevent him or her from engaging in antisocial behaviors.

Wirth linked social disorganization to the diversity that is characteristic of cities. Unlike rural residents, city dwellers work in one place, live in another, and relax in yet another. They divide their social lives among coworkers, neighbors, friends, and kin. Their jobs, lifestyles, and interests are extremely varied. As a result, no single group—be it the family, the friendship group, or the neighborhood—controls their lives. In Wirth's view, this absence of social control produces anomie or normlessness (see Chapter 8). Urban dwellers often do not agree on the norms that should govern their lives, and hence they are likely to either challenge existing norms or ignore them. Consequently, instead of being controlled by the norms of primary groups, the lives of city dwellers are controlled by impersonal agencies such as banks and police forces.

One consequence of the impersonality of urban life, some argue, is greater callousness among city dwellers. An often-cited example is the case of Kitty Genovese, discussed in Chapter 6. After that episode, in which a young woman was murdered while thirty-eight residents of nearby apartments—who heard her cries—did nothing, many commentators called attention to the callous character of city dwellers. Subsequent research on bystander apathy revealed that the presence of many other people tends to diffuse the sense of responsibility. We are less likely to take action if we have reason to believe that someone else will do so (Hunt, 1985; Latané & Darley, 1970). In this sense, the bigger the city and the more one is surrounded by strangers, the more likely it is that such behavior will occur—but this does not mean that city dwellers are alienated from one another when they are among people they know, or that when alone they would not help someone in trouble.

On the basis of many years of research on urban social interaction, together with an extensive review of the literature, urban sociologist Lyn Lofland (1998) concludes that city dwellers adhere to a set of norms that govern behavior among strangers in the city. These norms stress civility along with inattention, concern along with avoidance of any appearance of prying. To some observers, the urban dweller's behavior might seem hostile or unconcerned, but according to Lofland, this is largely a misunderstanding. Moreover, although urbanites may feel less responsibility for others when surrounded by strangers, this condition of city life is hardly evidence for the decline-of-community thesis.

The Persistence of Urban Communities

The idea that urbanization leads to the decline of community has been criticized on several grounds. Rural life is nowhere near as pleasant as some urban sociologists have assumed it to be; evidence of this is the almost magical attraction that cities often have for rural people. On the other hand, urban social disorganization is not as extensive as the early urban sociologists believed. Many city dwellers maintain stable, intimate relationships with kin, neighbors, and coworkers. Moreover, urban life is not necessarily stressful or anomic.

Subcultural Theory A more recent view of urban life sees the city as a "mosaic of social worlds" or intimate social groups. Numerous studies have shown that the typical urban dweller does not resemble the isolated, anomic individual portrayed by Simmel and Wirth. In fact, communities of all kinds can be found in cities. Many urban dwellers, for example, are members of ethnic communities that have not become fully assimilated into the melting pot of American society and are unlikely to do so. They may be the children or grandchildren of immigrants who formed ethnic enclaves within large cities in the late nineteenth and early twentieth centuries, or they may be recent immigrants themselves, trying to build a new life in a strange culture.

But group ties among urban dwellers are not based solely on ethnicity. They may be based on kinship, occupation, lifestyle, or similar personal attributes (Fischer, 1987/1976; Hummin, 1990; Lofland, 1998). Thus many cities contain communities of college students, elderly people, homosexuals, artists and musicians, wealthy socialites, and so on. Although the members of any given group do not always live in the same neighborhood, they are in close touch with one another much of the time. Their sense of community is based not so much on place of residence as on the ability to come together by telephone, in special meeting places like churches or synagogues, or even in restaurants and bars (Duneier, 1992; Fischer, 1987/1976; Oldenburg, 1997).

An example of this point of view, known as *subcultural theory,* is Illsoo Kim's (1981) detailed study of the Korean community in New York City. This growing ethnic community has developed since the passage of the Immigration Act of 1965, which eliminated the nationality quotas established earlier in the twentieth century. Largely because of population pressure in South Korea (which has led to overcrowded cities and high rates of unemployment), more than 750,000 Koreans have immigrated to the United States since 1970, with at least 100,000 finding new homes in the city of New York.

Like all immigrants, the Koreans have had to create a new way of life. They have had to find new ways of making a living, and they have had to adapt to a new culture. The first problem has been solved mainly by opening small businesses, the second by establishing neighborhoods where the immigrants can maintain their own culture while they and their children learn the values and norms of Western culture. Kim explains the Koreans' inclination to open small businesses (mainly grocery stores) in this passage:

> Old-timers frequently tell newcomers that "running a jangsa (commercial business) is the fastest way to get ahead in America." The language barrier partly explains this inclination.... This is an insuperable barrier to most Korean immigrants; it deprives them of many opportunities.... This fact, combined with differences in both the skills demanded and the system of rewards in the United States, means that occupational status cannot be transferred from the homeland to the new land. A high proportion of Korean immigrants were thus forced to turn to small retail businesses. (1981, pp. 102–103)

Many of the businesses that Korean immigrants establish are geared to Korean ethnic tastes and cultural needs. For example, Korean immigrants are often unwilling to change their diet and will buy most of their food from Korean-owned food stores. Other shops import books, gifts, magazines, and other items from South Korea. Numerous Korean travel agencies satisfy the immigrants' strong desire to visit their homeland. But Korean small businesses are not limited to serving the needs of the immigrant population. A growing number cater to all racial and ethnic groups and supply typically American goods and services (Min, 2005).

In summing up the situation of Korean immigrants in New York, Kim points out that "until they completely master the American language, education, and culture, Koreans will be forced to rely on one another" (p. 319). For this reason, they are likely to maintain their own community within the city for at least two generations. In time, however, their commitment to education and a better life will cause them to become more fully assimilated into American society, and their ties to the Korean community will be weakened.

The case of the Koreans echoes that of many other immigrant groups, both past and present, as well as that of other groups, such as homosexuals and artists. The details may differ somewhat, but the tendency to establish ethnic or cultural enclaves in large cities is universal. In a well-known study titled *The Urban Villagers*, Herbert Gans (1984/1962) described an Italian neighborhood in Boston. This community was a long-established one with many of the characteristics that are typical of ethnic communities everywhere. At the time Gans did his research, however, the neighborhood was being destroyed by an urban renewal project. Its residents were forced to find new homes in other parts of the city, to join new churches, attend new schools, seek out new places to buy food and clothes. For some, the upheaval was so great that it produced depression and anomie (Fried, 1963).

It seems reasonable to conclude that the effects of urban life on communities and individuals are more complex than Durkheim, Simmel, and Wirth suggested. Certainly, social disorganization occurs in cities, but so does social *reorganization*. Communities that are uprooted by urban renewal eventually may be re-formed in other parts of the city or its suburbs. The new community may be less homogeneous than the old one, but it is a community nonetheless (Marwell, 2007).

The Suburbs Like city dwellers, suburban dwellers have been said to lack the close attachments that are thought to characterize rural communities. In fact, before the late 1960s, most social scientists had a rather dismal view of suburban life. Suburbs had grown rapidly after World War II as large numbers of middle-class Americans left the central cities in search of a less crowded, more pleasant lifestyle. Many suburban dwellers were corporate employees, and in the 1950s and early 1960s, they became the subject of a widely held stereotype. The suburbs, it was said, were "breeding a new set of Americans, as mass-produced as the houses they lived in, driven into a never-ending round of group activity ruled by the strictest conformity. Suburbanites were incapable of real friendships; they were bored and lonely, alienated, atomized, and depersonalized" (Gans, 1967, pp. xxvii–xxviii).

In *The Levittowners*, a classic study of a new suburban community in Pennsylvania, Gans (1967, 1976) challenged this stereotype. He found that the residents of Levittown lost no time in forming attachments to one another. At first they

associated only with their neighbors, but before long, they formed more extensive associations based on shared interests and concerns. Moreover, far from being bored and isolated, most of the Levittowners were satisfied with their lives and felt that Levittown was a good place to live. Gans concluded that "new towns are merely old social structures on new land" (1967, p. vii). In other words, the contrasts between central-city and suburban life are generally exaggerated. Other researchers have reached similar conclusions. They note that as suburban communities age and newcomers buy homes from people whose children are grown, these neighborhoods go through ethnic, racial, and generational changes that are not so different from the patterns occurring inside the cities (Jackson, 1985).

Some comparative research has focused on women born in suburban communities after the 1950s who never lived in other types of communities (Stimson, 1980; Wekerle, 1980). For example, Sylvia Fava (1985) found that younger women born in suburban communities do not think of suburban life as stifling or as limiting their opportunities. They may prefer to live in older, more diverse suburban communities within easy commuting distance of the jobs and nightlife of the central cities, but they normally do not feel that their suburban lifestyle is unsatisfactory. Indeed, as conditions continue to deteriorate in some inner-city communities—with increasing homelessness, exposure to drug markets, and violence—both older and newer suburban areas become more attractive to people seeking to escape the dangers they perceive in urban life (Lofland, 1998).

For the past few decades, the phenomenon of "white flight" from older inner-city communities to newer and presumably safer suburban communities has been a subject of concern among urban sociologists and policy makers (Wilson, 1987). Recent data show, however, that African Americans and other minority groups are also fleeing the central cities at an accelerating rate. Demographer William Frey notes that "minority suburbanization took off in the 1980s both as the black middle class came into its own and as more assimilated Latinos and Asians translated their moves up the socioeconomic ladder into a suburban life style" (quoted in De Witt, 1994, p. A1). This outward migration threatens to leave central-city communities with even greater concentrations of poor people. In cities with large numbers of immigrants, this discouraging trend is countered by increasing racial, ethnic, and class diversity (Huang, 1994), but in other cities, the trend toward greater concentration of poor people inside the city and more affluent people in the suburbs is a growing problem. Recent research by the Urban Institute on concentrated poverty in different urban areas of the United States finds that while poor people are still concentrated in central cities, there has been an appreciable decrease in high-poverty tracts in the nation's 100 largest cities, and an increase in such tracts in suburban areas of metropolitan regions. Similarly, concentrated poverty, measured by poverty rates of 40 percent or more in a given neighborhood (or census tract), has decreased among African Americans and whites but increased among Latinos and in poor neighborhoods of mixed ethnic and racial composition (Pettit & Kingsley, 2003). (See Figure 21.10.)

Private Communities The growth of private communities throughout the United States is another indication of the widespread fear of urban life (Garreau, 1991; Low, 2003). Recent data indicate that about 4 million Americans live in private communities where public through-traffic is prohibited. For those who can afford to live there, these heavily patrolled and regulated communities represent a decision to place individual comfort and family safety above more social efforts to create communities. An even larger number of Americans, about 28 million, live in communities represented by private community associations. These include condominiums and cooperatives that require all residents to be members

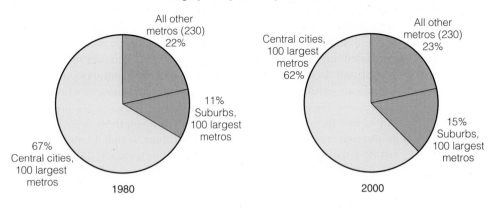

High-poverty tracts by location

1980
- All other metros (230) 22%
- 11% Suburbs, 100 largest metros
- 67% Central cities, 100 largest metros

2000
- Central cities, 100 largest metros 62%
- All other metros (230) 23%
- 15% Suburbs, 100 largest metros

Figure 21.10 High-Poverty Tracts, by Location and by Race and Ethnicity: 1980 and 2000. Source: Pettit & Kingsley, 2003.

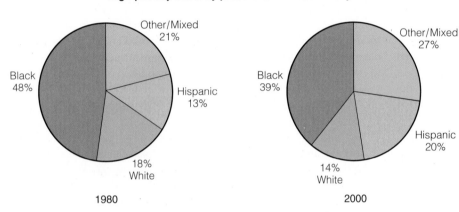

High-poverty tracts by predominant race/ethnicity

1980
- Other/Mixed 21%
- Black 48%
- Hispanic 13%
- 18% White

2000
- Other/Mixed 27%
- Black 39%
- Hispanic 20%
- 14% White

of the association, obey its rules, and pay fees to support its operations. Although they are less restrictive than private communities, these community associations also seek to control public access.

These private and quasi-private communities are the fastest-growing types of communities in the United States, and there is some concern about what this growth implies. Supporters of such communities dwell on their safety and their ability to maintain property values. Critics like Gerald Frug, an expert on local government, believe that people are trying to capture the feeling of life in a village without many other qualities of village life. "The village was open to the public," Frug observes. "The village did not have these kinds of restrictions. The village had poor people, retarded people. Somebody could hand you a leaflet. These private communities are totally devoid of random encounters" (quoted in Egan, 1995, p. A22). Frug and other critics of the trend toward private communities fear that, as people choose to live only with others like themselves, "they will become less likely to support schools, parks or roads for everyone else" (p. A22).

CITIES AND SOCIAL CHANGE

In the preceding section, we stressed the presence of a wide variety of communities and subcultures within cities. We noted that contemporary social scientists see the city as a place where many different communities coexist and thrive rather than as a place where people are isolated and do not have a feeling of belonging to a particular social group or community. At the same time, however, there is no escaping the fact that various urban communities are often in conflict. For example, ethnic and racial communities may clash in violent confrontations over such issues as busing children to achieve school integration. In this section, we examine the origins and implications of such conflicts.

Social change in an urban society is likely to be felt most deeply by city dwellers because people are most densely congregated in cities. In recent decades, this has been the case in North America as manufacturing jobs have been "exported" to Asia and Latin America. As a result of this shift, American cities are undergoing massive social change. "America's major cities are different places today from what they were in the 1960s," concludes John Kasarda (1989, p. 28), one of the nation's foremost urban sociologists. New modes of transportation, new communication technologies such as satellites and computers, and new industrial technologies (for example, automation and recycling) are transforming our cities from production and distribution centers to administrative, financial, and information centers. "In the process," writes Kasarda, "many blue-collar jobs that once constituted the economic backbone of cities and provided employment opportunities for their poorly educated residents have either vanished or moved." Many of those jobs have been replaced by "knowledge-intensive white-collar jobs with educational requirements that exclude many with substandard education" (p. 28).

Inequality and Conflict

As noted earlier, American cities and metropolitan regions have always attracted streams of migrants and immigrants. Migrants arrive from other parts of the nation—blacks from the rural South, Chicanos from the Southwest. Immigrants come from foreign lands—Haiti, Poland, China, Korea, Italy, and elsewhere. Until the 1970s, except during economic depressions or recessions, the new arrivals had no difficulty finding work in the mills and factories that produced textiles, clothing, steel, rubber, glass, cars, and trucks. And much of their education was gained while working. Formal schooling was not as crucial to job success as it is today.

Today, although there are still many manufacturing jobs, it is much more difficult to make a good living at such a job. Within the cities, where the decline in manufacturing jobs has been greatest, the number of low-status jobs, especially in restaurants and other low-paying services, has increased. So has the number of highly paid jobs in management and the professions, positions that require high levels of education. At the same time, more than 20 million legal immigrants have entered the country since 1970, usually settling in the cities. The result, often, has been fierce competition between immigrants and more established residents.

The net effect of these trends is a growing gap between the "haves" and the "have-nots," especially in large cities (Silk, 1989). The study of *social stratification*—of patterns of inequality and how they produce entire classes of people with differing opportunities to succeed in life—is at the core of sociology (see Chapter 10). Certainly the gap between those who enjoy the good things in life and those who spend their lives simply trying to survive is not unique to cities. Small towns and villages also have affluent and poor residents. But the conflicts and social problems associated with social inequality are most visible in the cities. In many poor urban neighborhoods, for example, wealthier city dwellers are buying the buildings in which the poor live, forcing them to move elsewhere. In neighborhoods where this occurs, the shops and stores that once catered to lower-income residents cannot afford the higher rents paid by stores that serve the wealthy newcomers, and are forced out of business. This process, whereby higher-income newcomers renovate poor and dilapidated neighborhoods, pushing out poor residents and merchants, is known as *gentrification.*

Status Conflict In the last section, we noted that an urban renewal project had dislocated large numbers of Italian Americans from their Boston neighborhood. Such projects have sometimes been labeled "urban removal" because they remove poor people from decaying neighborhoods and force them to find homes elsewhere. Recently urban renewal has fallen into disfavor and *redevelopment* has become popular. Redevelopment does not involve the destruction of entire neighborhoods, but it may have similar effects. For example, the redevelopment of New York City's Times Square by the Walt Disney Company and other powerful groups, both public and private, is restoring a sleazy area to its former glory as an entertainment center. The explicit goal is to make the area attractive to "decent" people and to eliminate the prostitutes, drug and pornography dealers, and petty criminals who have made Times Square a symbol of vice throughout the world (Berman, 2006; Kornblum, 1995). This is a typical instance of *status conflict,* in which different groups (for example, theater and restaurant owners versus owners of "adult" establishments) vie for territory, occupational advantages, and other benefits that will enhance their prestige in the neighborhood or community.

Some urban sociologists see the city as divided into "defended neighborhoods" or territories (Lofland, 1998; Suttles, 1972). This concept was originally developed by Park and Burgess, who viewed such neighborhoods as a type of "natural area" within the city. The defended neighborhood is a territory that a certain group of people consider to be their "turf" or base, which they are willing to defend against "invasion" by outsiders. Neighborhood defense is a common element of urban conflict, although the means used to defend neighborhoods may differ. In wealthy suburban neighborhoods, for example, it may take the form of zoning regulations that establish minimum lot sizes of an acre or more. Such regulations effectively defend the neighborhood against invasion by people who cannot afford the large lots or by developers who would like to erect apartment buildings (Perin, 1977). In less affluent neighborhoods, defense is often conducted by neighborhood improvement groups, which sometimes engage in vigilante action when they fear racial "invasion" (Hamilton, 1969). In poor neighborhoods, defense is often the province of street-corner gangs.

Gender Conflict In the nineteenth century, many social commentators were dismayed at the increasingly frequent sight of women walking alone or with other women in city streets. Women unaccompanied by men were thought to be either immoral or innocently risking their honor or their lives in a world of predatory males. As Lyn Lofland observes, "The increasing presence of 'respectable' women in urban public space served as especially vivid and provocative evidence of women's unnatural and undesirable abandonment of hearth and home" (1998, p. 128). Of course, Lofland realizes that "poor women had rarely, in the earliest cities, been barred from the streets," whereas the "spatial freedom of privileged women has been culturally and historically quite variable." What moralists were protesting against was the presence of middle-class or "respectable" women in public places.

Today, too, one can find many examples of the conflicts that women experience in urban public places. On their lunch breaks, for example, women are routinely harassed by ogling men and then accused of inviting such behavior by wearing provocative clothing. In Islamic cities, especially in more orthodox nations like Pakistan, women are prohibited from appearing in public at all; if they must be out on the streets, they are required to hide their faces and completely cover their bodies. Conflicts that result from efforts by men to "keep women in their place" thus take many forms and are a widespread form of status conflict, one that is especially visible in cities. It may be true that racial conflict is more strident and violent, but for a person with a sociological imagination, gender conflicts are easy to spot.

Racial Conflict Perhaps the most common source of intergroup conflict in American cities has been racial tension. In the belief that the presence of black residents reduces the prestige of a neighborhood and lowers the value of real estate there, whites have resisted efforts by blacks to obtain housing in "their" neighborhoods. This resistance has taken a wide variety of forms, ranging from discrimination by real estate agents to outright violence against black families moving into white neighborhoods. When blacks have succeeded in gaining a foothold in previously all-white neighborhoods, many white families have sold their homes and moved to the suburbs (Massey & Denton, 1993).

For racial and ethnic minorities, the desire to move out of segregated ghettos is based on the need to establish or maintain a residential community while pursuing opportunities to build a better life. This is among the key findings of studies of immigrant groups in South Florida. In looking at the difficulties that Haitian and Cuban refugees encounter in finding jobs in the Miami area, Alejandro Portes and Alex Stepick (1985) found that Cubans could draw on kin and ethnic networks in Miami neighborhoods in looking for jobs, but that the Haitians had no such networks to help them adapt to their new environment.

The riots that took place in 1992 in Los Angeles, Atlanta, and other major U.S. cities were a tragic example of the relationship between social change and racial violence in urban centers. Although the conditions of life had been deteriorating in central-city ghettos throughout the 1980s, it was not until televised bloodshed brought this situation to the attention of the public that significant numbers of people began to demand that the federal government take action. But this is by no means a new phenomenon in American social life. As we saw in Chapter 9, conflict between racial and other groups often raises public consciousness and contributes to major social change (Rothenberg, 1992).

Coping with Disasters in Urban Communities

Riots, epidemics, earthquakes, terrorist bombings, wars, floods, oil spills, tornadoes, and hurricanes are among the many kinds of disasters that may strike densely populated human settlements. Of course, there are important differences in the causes and outcomes of purely social disasters like riots and purely natural ones like earthquakes (see Figure 21.11). From a sociological vantage point, however, social and natural disasters have a great deal in common. Moreover, because we can expect both natural and social disasters to occur with increasing frequency on our urbanizing planet, the study of disasters is a thriving area of sociological theory and research.

Recent sociological research on disasters has focused on the longer-term consequences for the victims. In this area of research, the work of sociologist Kai Erikson is particularly noteworthy. Erikson's powerful descriptions of responses to disasters and his development of the sociological theory of trauma are having significant and positive impacts on victims' lives. In several precedent-setting cases, Erikson has convinced juries and judges that victims have suffered sociological damage, especially in the form of loss of community.

Erikson's recent work on social disasters focuses on the consequences of what he calls "a new species of trouble" that is particularly common

in urban industrial societies. The trouble may take the form of toxic spills, nuclear accidents, mercury poisoning, or many other disasters attributable to human negligence, greed, or combinations of both. The result is often an abrupt and shattering loss of community. Erikson and other sociologists who do similar research point out that disasters, especially those caused by the negligence or criminal behavior of corporations, are not distributed randomly across societies. On the contrary, they occur with far greater frequency in low-income and minority communities because dangerous facilities, dump sites, and tank farms are often located on the boundaries of those communities. This ecological pattern has been labeled "environmental racism" (Bullard, 2003).

In one such case, Erikson interviewed the residents of East Swallow Drive in Fort Collins, Colorado. This middle-income neighborhood of private homes with middle-aged or elderly residents was the site of a gasoline spill with devastating consequences. The local service station had been experiencing "inventory losses" for several years and had finally replaced three "incontinent" fuel tanks with new ones. Not long afterward, nearby residents began smelling gasoline fumes in their homes. An investigation revealed that there were large pools of gasoline in the ground under the homes along East Swallow Drive. One house near the service station was condemned, and in the months that followed, as Erikson describes it, "waves of concern began to radiate in slow-motion ripples down East Swallow Drive as people began to understand that a plume of petroleum was moving eastward with the sureness of a glacier" (1994a, pp. 99–100).

 Thinking Critically What is the difference between environmental racism and racism pure and simple? A key to answering this question is to think about racism as practiced by individuals and racism that may result from the policies of large corporations and governments that place environmentally damaging facilities near neighborhoods where poor people and minorities constitute the majority of the population.

A group of homeowners brought suit against the companies involved in the spill. As is often the case, however, the companies were able to hire teams of lawyers who could delay the suit for months. Eventually lawyers for the residents asked Erikson to prepare testimony on the sociological damage caused by the spill. He and an assistant conducted observations in the neighborhood and interviewed the residents. Later Erikson shared transcripts of the interviews with the residents and asked them for additional comments. This method, it turned out, had powerful effects. As one resident noted, "I was greatly relieved to see what other people said. . . . Seeing this report gives me a little reassurance that I wasn't feeling something that other people were not feeling" (quoted in Erikson, 1994a, p. 138).

In his report, Erikson argued that the personal distress the residents were experiencing—in the form of anxiety, sleeplessness, anger, and other emotions—was not only a sign of psychological suffering but also a symptom of sociological trauma caused by the loss of their attachments to the neighborhood. As one resident said, "Well, we feel that the value of our property has taken a nose dive. It is our only security. We don't feel that it's salable. If we can't sell it, we can't relocate, we can't move, we can't expand or change. We feel that we're sort of stuck" (quoted in Erikson, 1994a, p. 232). Erikson convinced the court not only that the residents had lost property value in terms of dollars but that they had also suffered a greater loss in a sociological sense. He noted that they had invested a great deal more than money in their homes:

> It is well understood by social and behavioral scientists who are familiar with accidents of this kind that the loss of a home can be profoundly traumatic. I wrote in my report on the Buffalo Creek flood of 1972 [in West Virginia], for example, that a home in which a person has made a real personal investment "is not simply an expression of one's taste; it is the outer edge of personality, a part of the self itself." (1994a, p. 233)

Similarly, a well-known study of the impact of a tornado in Texas reported that people who had lost their homes felt as if they had lost part of themselves (Moore, 1958). And an equally well-known study of an urban relocation project in Boston's East End was titled simply "Grieving for a Lost Home" (Fried, 1963). In the East Swallow Drive case, the residents benefited from Erikson's sociological insight: Eventually the corporations responsible for the damage had to settle with the residents and pay not only for

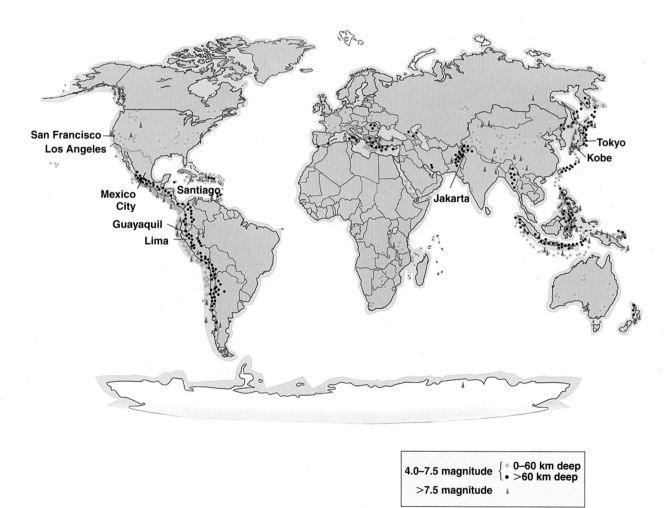

Figure 21.11 Earthquakes of Magnitude 4.0 or Greater, 1960–1989. The map shows the regions of the world where earthquake activity is greatest and where major urban disasters are likely to occur because metropoli-tan regions are located on the fault lines.

visual sociology

Women's Education and Fertility Control

Feminist sociologists and most experts on population issues agree that none of the approaches to fertility control attempted to date—contraception, abstinence, celibacy, sterilization, or delay of marriage—can have much success if women are not able to control their own reproductive lives. And they cannot do this if they remain illiterate and powerless in societies with rapidly growing populations. When given a chance to express their feelings on the matter, women and girls throughout the poor regions of the world usually agree with this crucial point. As an example, consider the outstanding visual sociological work of photographer Wendy Ewald in Vichya, in the state of Gujarat in northwestern India near the Pakistani border.

Ewald's method is to bring inexpensive cameras to the people she works with. She teaches them to take photos and to develop and print the results. Then they discuss what the photos reveal. As Ewald explains, "I gave the children assignments to expand their ideas of picture-making while staying close to what they knew. I asked them to photograph their festivals, their families, themselves, their village, and their animals" (1996, p. 14).

Tidi, the girl shown here, was ten years old when Ewald came to her village. In her reflections on her photos, note especially her desire to go to school:

I have a mommy, father, two brothers, and three sisters. One brother is married and the other is going to marry. My father laughs a lot, but I don't like him when he beats us. My small brother and cousin are at home. One younger brother and sister got smallpox right after they were born. When they died, we buried them on the roadside and marked the grave with a red cloth. I can't find it now. The sweepers must have taken it away.

I get enough food to eat—even during the drought. We went to the city then and lived in our uncle's house for six months. . . . My father did construction work. My mother helped him and washed dishes and clothes for a rich family. I don't mind washing dishes, but I'd like to go to school.

I had to leave my studies because of the housework. Anyhow, soon we have to marry. I'll be happy then. My husband will get me a scooter. I'll get on it and wander. I want sons, as many as God wants to give me—two, three, or four. I'll play with my children and scold them when they're mischievous. When I'm older I'll look after my husband's house. (Quoted in Ewald, 1996)

Tidi and her neighbors.

Wendy Ewald/Chandrakant

Hasmukh, Chandrakant, Harshad, and Dasrath learning to use the camera.

Wendy Ewald/Chandrakant

the damage to their property but for their loss of community as well.

Erikson and other sociologists who study disasters hope that as a consequence of their research—and their successes in court—there will be more safeguards against negligence and more emphasis on anticipating the risks of natural disasters. In the near term, however, Erikson fears that the balance of power is shifting toward corporations and other powerful actors, to the disadvantage of the victims or potential victims of community disasters.

Contamination and pollution of the earth and water beneath these houses, caused by the criminal dumping of toxic wastes, forced the residents to abandon their homes. Sociologist Kai Erikson has identified such disasters as a "new species of trouble" that has profound consequences for its victims.

© Kirk Condyles

SUMMARY

How do sociologists measure and predict population changes, and how does the growth of cities affect population?

- Populations change as a consequence of births, deaths, out-migration, and in-migration.

- Since World War II, the world population has been increasing at a rate of more than 1.5 percent, surpassing 6 billion in 2000.

- The *crude birthrate* is the number of births occurring during a year in a given population divided by the midyear population.

- The *crude death rate* is the number of deaths occurring during a year divided by the midyear population.

- The *rate of reproductive change* is the difference between the crude birthrate and the crude death rate for a given population.

- The *demographic transition* is a set of major changes in birth and death rates that have occurred most completely in urban industrial nations in the past 200 years. It takes place in three stages: (1) high birth and death rates, (2) declining death rates, and (3) declining birthrates. These stages are accompanied by changes in the age composition of the population.

- Life expectancy is highest in societies with advanced systems of health care and well-developed public health systems. A more sophisticated measure, disability-adjusted life expectancy (DALE), takes into account how long a person can expect to live without severe and disabling illness. On this measure, the United States ranks relatively low among wealthy nations.

- The growth of the human population in the past two centuries has begun to exert major stresses on the earth's resources of arable soil, clean air, and fresh water for drinking and irrigation, as well as on its sources of fuels.

- Urbanization is closely linked with rapid increases in population, but at the same time the nature of life in cities tends to limit the size of urban families.

- Cities grow primarily as a result of migration (which is often caused by population increases in rural areas), but new migrants do not find it easy to form families.

- Not only are increasing proportions of the world's population living in cities, but the cities themselves are also larger than ever before.

- The growth of cities in the twentieth century gave rise to the concept of the *metropolitan area,* in which a central city is surrounded by several smaller cities and suburbs that are closely related to it both socially and economically.

- The growth of cities should be distinguished from *urbanization,* which refers to the proportion of the total population that is concentrated in urban settlements. The end result of urbanization is an "urban society." Not only do cities serve as the cultural and institutional centers of such societies, but urban life has a pervasive influence on the entire society as well.

What changes can be seen in urban areas over time?

- Sociologists have devoted a great deal of study to the processes by which cities expand and to patterns of settlement within cities. An early model of urban expansion was the concentric zone model developed by Park and Burgess. In this model, a central business district is surrounded by successive zones, or rings, devoted to light manufacturing, workers' homes, higher-class apartment buildings and single-family homes, and a commuters' zone.

- This model is limited to commercial and industrial cities that formed around business centers and does not account for the rise of satellite cities and the rapid urbanization that occurs along commercial transportation corridors. The growth of satellite cities was especially rapid before World War II.

- A more current model of urbanization is known as *strip development* and describes the incorporation of smaller communities and towns into a larger metropolitan area.

- Metropolitan areas have expanded greatly since the mid-twentieth century, largely as a result of the increasing use of automobiles and the construction of a network of highways covering the entire nation. In some areas, this growth has created large multinuclear urban systems that are described by the term *megalopolis*.

- One effect of the development of such areas is decentralization, in which outlying areas become more important at the expense of the central city.

Are urban communities doomed to become anonymous, faceless places?

- Social scientists who have studied the effects of urban life have been particularly concerned with the tension between community and individualism as it relates to life in cities.

- Early studies of urban life tended to conclude that it weakens kinship ties and produces impersonal social relationships. Urban life was also thought to produce "psychic overload" and anomie.

- More recently, these conclusions have been criticized by researchers who have found that many city dwellers maintain stable, intimate relationships with kin, neighbors, and coworkers and that urban life is not necessarily stressful or anomic.

- Subcultural theory sees the city as a mosaic of social worlds or intimate social groups. Communities of all kinds can be found in cities. Those communities may be based on ethnicity, kinship, occupation, lifestyle, or similar personal attributes.

- Suburban dwellers also have been found to be far less bored and isolated than was once supposed.

In what ways can cities be described as "engines of social change?"

- Occasionally various communities within cities come into conflict. Such conflict may arise out of different class interests or the conflicting goals of different groups within the city.

- Some urban sociologists see the city as divided into "defended neighborhoods" or territories whose residents attempt to protect them from "invasion" by outsiders.

- Sociologists point out that disasters, especially those caused by the negligence or criminal behavior of corporations, are not distributed randomly across societies. On the contrary, they occur with far greater frequency in low-income and minority communities because dangerous facilities, dump sites, and tank farms are often located on the boundaries of those communities. This ecological pattern has been labeled "environmental racism."

The Kornblum Companion Website

www.cengagebrain.com

Supplement your review of this chapter by going to the companion website to take one of the tutorial quizzes, use the flash cards to master key terms, and check out the many other study aids you'll find there. You'll also find special features, such as GSS data and Web links that will put data and resources at your fingertips to help you with that special project or to do some research on your own.

Glossary

accommodation: The process by which a smaller, less powerful society is able to preserve the major features of its culture even after prolonged contact with a larger, stronger culture.

acculturation: The process by which the members of a civilization incorporate norms and values from other cultures into their own.

achieved status: A position or rank earned through the efforts of the individual.

age cohort: A set of people of about the same age who are passing through the life course together.

age grade: A set of statuses and roles based on age.

ageism: An ideology that justifies prejudice and discrimination based on age.

agencies of socialization: The groups of people, along with the interactions that occur within those groups, that influence a person's social development.

agents of socialization: Individuals who socialize others.

alienation: The feeling of being powerless to control one's own destiny; a worker's feeling of powerlessness caused by inability to control the work process.

animism: A form of religion in which all forms of life and all aspects of the earth are inhabited by gods or supernatural powers.

anomie: A state of normlessness.

anticipatory socialization: Socialization that prepares an individual for a role that he or she is likely to assume later in life.

ascribed status: A position or rank that is assigned to an individual at birth and cannot be changed.

assimilation: The process by which culturally distinct groups in a larger civilization adopt the norms, values, and language of the host civilization and are able to gain equal statuses in its groups and institutions; a pattern of intergroup relations in which a minority group is absorbed into the majority population and eventually disappears as a distinct group.

authority: Power that is considered legitimate both by those who exercise it and by those who are affected by it; power whose exercise is governed by the norms and statuses of organizations.

behaviorism: A theory stating that all behavior is learned and that this learning occurs through the process known as conditioning.

bisexuality: Sexual orientation toward members of either sex.

bureaucracy: A formal organization characterized by a clearly defined hierarchy with a commitment to rules, efficiency, and impersonality.

capitalism: A system for organizing the production of goods and services that is based on markets, private property, and the business firm or company.

caste: A social stratum into which people are born and in which they remain for life.

charisma: A special quality or "gift" that motivates people to follow a particular leader.

charismatic authority: Authority that comes to an individual through a personal calling, often claimed to be inspired by supernatural powers, and is legitimated by people's belief that the leader does indeed have God-given powers.

church: A religious organization that has strong ties to the larger society.

citizenship: The status of membership in a nation-state.

civil religion: A collection of beliefs and rituals that exists outside religious institutions.

civil society: The sphere of nongovernmental, nonbusiness social activity carried out by voluntary associations, congregations, and the like.

civilization: A cultural complex formed by the identical major cultural features of several societies.

class: A social stratum defined primarily by economic criteria such as occupation, income, and wealth.

class consciousness: A group's shared subjective awareness of its objective situation as a class.

closed question: A question that requires the respondent to choose among a predetermined set of answers.

closed society: A society in which social mobility does not exist.

closed stratification system: A stratification system in which there are rigid boundaries between social strata.

collective behavior: Nonroutine behavior that is engaged in by large numbers of people responding to a common stimulus.

community: A set of primary and secondary groups in which the individual carries out important life functions.

conditioning: The shaping of behavior through reward and punishment.

confidentiality: The promise that the information provided to a researcher by a respondent will not appear in any way that can be traced to that respondent.

conflict perspective: A sociological perspective that emphasizes the role of conflict and power in society.

control group: In an experiment, the subjects who do not experience a change in the independent variable.

controlled experiment: An experimental situation in which the researcher manipulates an independent variable in order to observe and measure changes in a dependent variable.

core state: A technologically advanced nation that has a dominant position in the world economy.

correlation: A specific relationship between two variables.

counterculture: A subculture that challenges the accepted norms and values of the larger society and establishes an alternative lifestyle.

crime: An act, or omission of an act, that is prohibited by law.

crowd: A large number of people who are gathered together in close proximity to one another.

crude birthrate: The number of births occurring during a year in a given population, divided by the midyear population.

crude death rate: The number of deaths occurring during a year in a given population, divided by the midyear population.

cult: A new religion.

cultural evolution: The process by which successful cultural adaptations are passed down from one generation to the next.

cultural lag: The time required for social institutions to adapt to a major technological change.

cultural relativity: The recognition that all cultures develop their own ways of dealing with the specific demands of their environments.

culture: All the modes of thought, behavior, and production that are handed down from one generation to the next by means of communicative interaction rather than by genetic transmission.

de facto segregation: Segregation created and maintained by unwritten norms.

de jure segregation: Segregation created by formal legal sanctions that prohibit certain groups from interacting with others or place limits on such interactions.

deference: The respect and esteem shown to an individual.

demagogue: A leader who uses personal charisma and political symbols to manipulate public opinion.

demeanor: The ways individuals present themselves to others through body language, dress, speech, and manners.

democracy: A political system in which all citizens have the right to participate in public decision making.

democratic socialism: An economic philosophy based on the belief that private property may exist at the same time that large corporations are owned by the state and run for the benefit of all citizens.

demographic transition: A set of major changes in birth and death rates that has occurred most completely in urban industrial nations in the past 200 years.

denomination: A religious organization that is on good terms with the institution from which it developed but must compete with other denominations for members.

dependent variable: The variable that a hypothesis seeks to explain.

developing nation: A nation that is undergoing a set of transformations whose effect is to increase the productivity of its people, their health, their literacy, and their ability to participate in political decision making.

deviance: Behavior that violates the norms of a particular society.

differential association: A theory that explains deviance as a learned behavior determined by the extent of a person's association with individuals who engage in such behavior.

differentiation: The processes whereby sets of social activities performed by one social institution are divided among different institutions.

discrimination: Behavior that treats people unfairly on the basis of their group membership.

diversity: A term used to refer to the heterogeneous nature of a society made up of numerous different racial, ethnic, religious, and other population groups.

double standard: The belief that women must adhere to a different and more restrictive moral code than that applied to men.

downward mobility: Movement by an individual or group to a lower social stratum.

dramaturgical approach: An approach to research on interaction in groups that is based on the recognition that much social interaction depends on the desire to impress those who may be watching.

dyad: A group consisting of two people.

education: The process by which a society transmits knowledge, values, norms, and ideologies and, in so doing, prepares young people for adult roles and adults for new roles.

educational achievement: How much the student actually learns, measured by mastery of reading, writing, and mathematical skills.

educational attainment: The number of years of school an individual has completed.

ego: According to Freud, the part of the human personality that is the individual's conception of himself or herself in relation to others.

endogamy: A norm specifying that a person brought up in a particular culture may marry within the cultural group.

endogenous force: Pressure for social change that builds within a society.

equality of opportunity: Equal opportunity to achieve desired levels of material well-being and prestige.

equality of result: Equality in the actual outcomes of people's attempts to improve their material well-being and prestige.

ethnic group: A population that has a sense of group identity based on shared ancestry and distinctive cultural patterns.

ethnic (or racial) nationalism: The belief that one's own ethnic group constitutes a distinct people whose culture is and should be separate from that of the larger society.

ethnic stratification: The ranking of ethnic groups in a social hierarchy on the basis of each group's similarity to the dominant group.

ethnocentrism: The tendency to judge other cultures as inferior to one's own.

ethnomethodology: The study of the underlying rules of behavior that guide group interaction.

exogamy: A norm specifying that a person brought up in a particular culture may marry outside the cultural group.

exogenous force: Pressure for social change that is exerted from outside a society.

experimental group: In an experiment, the subjects who are exposed to a change in the independent variable.

expulsion: The forcible removal of one population from a territory claimed by another population.

extended family: An individual's nuclear family plus the nuclear families of his or her blood relatives.

family: A group of people related by blood, marriage, or adoption.

family of orientation: The nuclear family in which a person is born and raised.

family of procreation: The nuclear family a person forms through marriage or cohabitation.

feral child: A child reared outside human society.

field experiment: An experimental situation in which the researcher observes and studies subjects in their natural setting.

folkways: Weakly sanctioned norms.

formal organization: A group that has an explicit, often written, set of norms, statuses, and roles that specify each member's relationships to the others and the conditions under which those relationships hold.

frequency distribution: A classification of data that describes how many observations fall within each category of a variable.

functionalism: A sociological perspective that focuses on the ways in which a complex pattern of social structures and arrangements contributes to social order.

gemeinschaft: A term used to refer to the close, personal relationships of small groups and communities.

gender: The culturally defined ways of acting as a male or a female that become part of an individual's personal sense of self.

gender role: A set of behaviors considered appropriate for an individual of a particular gender.

gender socialization: The ways in which we learn our gender identity and develop according to cultural norms of masculinity and femininity.

generalized other: A person's internalized conception of the expectations and attitudes held by society.

genocide: State-sponsored mass killing explicitly designed to completely exterminate a population deemed to be racially or ethnically different and threatening to the dominant population.

gerontology: The study of aging and older people.

gesellschaft: A term used to refer to the well-organized but impersonal relationships among the members of modern societies.

ghetto: A section of a city that is segregated either racially or culturally.

group: Any collection of people who interact on the basis of shared expectations regarding one another's behavior.

Hawthorne effect: The unintended effect that results from the attention given to subjects in an experimental situation.

hegemony: Undue power or influence.

hermaphrodites: Individuals whose primary sexual organs have features of both male and female organs, making it difficult to categorize the person as male or female.

heterosexuality: Sexual orientation toward members of the opposite sex.

homogamy: The tendency to marry a person who is similar to oneself in social background.

homophobia: Fear of homosexuals and same-sex attraction.

homosexuality: Sexual orientation toward members of the same sex.

horticultural society: A society whose primary means of subsistence is raising crops, which it plants and cultivates, often developing an extensive system for watering the crops.

human ecology: A sociological perspective that emphasizes the relationships among social order, social disorganization, and the distribution of populations in space and time.

hypothesis: A statement that specifies a relationship between two or more variables that can be tested through empirical observation.

id: According to Freud, the part of the human personality from which all innate drives arise.

ideas: Ways of thinking that organize human consciousness.

identification: The social process whereby an individual chooses role models and attempts to imitate their behavior.

ideologies: Systems of values and norms that the members of a society are expected to believe in and act on without question.

impression management: The strategies one uses to "set a stage" for one's own purposes.

independent variable: A variable that the researcher believes causes a change in another variable (that is, the dependent variable).

informal organization: A group whose norms and statuses are generally agreed upon but are not set down in writing.

informed consent: The right of respondents to be informed of the purpose for which the information they supply will be used and to judge the degree of personal risk involved in answering questions, even when an assurance of confidentiality has been given.

in-group: A social group to which an individual has a feeling of allegiance; usually, but not always, a primary group.

institution: A more or less stable structure of statuses and roles devoted to meeting the basic needs of people in a society.

institutional discrimination: The systematic exclusion of people from equal participation in a particular institution because of their group membership.

interactionism: A sociological perspective that views social order and social change as resulting from all the repeated interactions among individuals and groups.

interest group: An organization that attempts to influence elected and appointed officials regarding a specific issue or set of issues.

intergenerational mobility: A change in the social class of family members from one generation to the next.

internal colonialism: A theory of racial and ethnic inequality that suggests that some minorities are essentially colonial peoples within the larger society.

intragenerational mobility: A change in the social class of an individual within his or her own lifetime.

Jim Crow: The system of formal and informal segregation that existed in the United States from the late 1860s to the early 1970s.

kinship: The role relations among people who consider themselves to be related by blood, marriage, or adoption.

labeling: A theory that explains deviance as a societal reaction that brands or labels people who engage in certain behaviors as deviant.

laissez-faire capitalism: An economic philosophy based on the belief that the wealth of a nation can be measured by its capacity to produce goods and services (that is, its resources of land, labor, and machinery) and that these can be maximized by free trade.

laws: Norms that are written by specialists, collected in codes or manuals of behavior, and interpreted and applied by other specialists.

legal authority: Authority that is legitimated by people's belief in the supremacy of the law; obedience is owed not to a person but to a set of impersonal principles.

legitimacy: The ability of a society to engender and maintain the belief that the existing political institutions are the most appropriate for that society.

life chances: The opportunities an individual will have or be denied throughout life as a result of his or her social-class position.

life course: A pathway along an age-differentiated, socially created sequence of transitions.

life expectancy: The average number of years a member of a given population can expect to live.

lifetime negative experiences: Experiences that cause long-term stress, such as the death of a child or spouse.

linguistic-relativity hypothesis: The belief that language determines the possibilities for thought and action in any given culture.

lobbying: The process whereby interest groups seek to persuade legislators to vote in their favor on particular bills.

macro-level sociology: An approach to the study of society that focuses on the major structures and institutions of society.

market: An economic institution that regulates exchange behavior through the establishment of different values for particular goods and services.

mass: A large number of people who are all oriented toward a set of shared symbols or social objects.

mass public: A large population of potential spectators or participants who engage in collective behavior.

master status: A status that takes precedence over all of an individual's other statuses.

material culture: Patterns of possessing and using the products of culture.

megalopolis: A complex of cities distributed along a major axis of traffic and communication, with a total population exceeding 25 million.

mercantilism: An economic philosophy based on the belief that the wealth of a nation can be measured by its holdings of gold or other precious metals and that the state should control trade.

metropolitan area: A central city surrounded by several smaller cities and suburbs that are closely related to it both socially and economically.

micro-level sociology: An approach to the study of society that focuses on patterns of social interaction at the individual level.

middle-level sociology: An approach to the study of society that focuses on relationships between social structures and the individual.

minority group: A population that, because of its members' physical or cultural characteristics, is singled out from others in the society for differential and unequal treatment.

modernization: A term used to describe the changes that societies and individuals experience as a result of industrialization, urbanization, and the development of nation-states.

monotheistic: A term used to describe a theistic belief system centered on belief in a single all-powerful God.

mores: Strongly sanctioned norms.

multinational corporation: An economic enterprise that has headquarters in one country and conducts business activities in one or more other countries.

nationalism: The belief of a people that they have the right and the duty to constitute themselves as a nation-state.

nation-state: The largest territory within which a society's political institutions can operate without having to face challenges to their sovereignty.

natural selection: The relative success of organisms with specific genetic mutations in reproducing new generations with the new trait.

nonterritorial community: A network of relationships formed around shared goals.

normative order: The array of norms that permit a society to achieve relatively peaceful social control.

norms: Specific rules of behavior.

nuclear family: Two or more people related by blood, marriage, or adoption who share a household.

objective class: In Marxian theory, a social class that has a visible, specific relationship to the means of production.

occupational prestige: The honor or prestige attributed to specific occupations by adults in a society.

oligarchy: Rule by a few people who stay in office indefinitely rather than for limited terms.

open question: A question that does not require the respondent to choose from a predetermined set of answers; instead, the respondent may answer in his or her own words.

open society: A society in which social mobility is possible for everyone.

open stratification system: A stratification system in which the boundaries between social strata are easily crossed.

out-group: Any social group to which an individual does not have a feeling of allegiance; may be in competition or conflict with the in-group.

participant observation: A form of observation in which the researcher participates to some degree in the lives of the people being observed.

pastoral society: A society whose primary means of subsistence is herding animals and moving with them over a wide expanse of grazing land.

patriarchy: The dominance of men over women.

peer group: An interacting group of people of about the same age that has a significant influence on the norms and values of its members.

percent analysis: A mathematical operation that transforms an absolute number into a proportion as a part of 100.

peripheral area: A region that supplies basic resources and labor power to more advanced states.

plea bargaining: A process in which a person charged with a crime agrees to plead guilty to a lesser charge.

pluralist model: A theory stating that no single group controls political decisions; instead, a plurality of interest groups influence those decisions through a process of coalition building and bargaining.

pluralistic society: A society in which different ethnic and racial groups are able to maintain their own cultures and lifestyles while gaining equality in the institutions of the larger society.

political institution: A set of norms and statuses pertaining to the exercise of power and authority.

political party: An organization of people who join together to gain legitimate control of state authority.

political revolution: A set of changes in the political structures and leadership of a society.

polytheistic: A term used to describe a theistic belief system that includes numerous gods.

power: The ability to control the behavior of others, even against their will.

power elite model: A theory stating that political decisions are controlled by an elite of rich and powerful individuals even in societies with democratic political institutions.

prejudice: An attitude that prejudges a person on the basis of a real or imagined characteristic of a group to which that person belongs.

primary deviance: An act that results in the labeling of the offender as deviant.

primary group: A small group characterized by intimate, face-to-face associations.

privacy: The right of a respondent to define when and on what terms his or her actions may be revealed to the general public.

profane: A term used to describe phenomena that are not considered sacred.

profession: An occupation with a body of knowledge and a developed intellectual technique that are transmitted by a formal educational process and testing procedures.

projection: The psychological process whereby we attribute to other people behaviors and attitudes that we are unwilling to accept in ourselves.

psychosocial risk behaviors: Behaviors that are detrimental to health, such as smoking and heavy drinking.

public opinion: The values and attitudes held by mass publics.

race: An inbreeding population that develops distinctive physical characteristics that are hereditary.

racism: An ideology based on the belief that an observable, supposedly inherited trait is a mark of inferiority that justifies discriminatory treatment of people with that trait.

rate of reproductive change: The difference between the crude birthrate and the crude death rate for a given population.

recidivism: The probability that a person who has served a jail term will commit additional crimes and be jailed again.

reference group: A group that an individual uses as a frame of reference for self-evaluation and attitude formation.

relative deprivation: Deprivation as determined by comparison with others rather than by some objective measure.

religion: Any set of coherent answers to the dilemmas of human existence that makes the world meaningful; a system of beliefs and rituals that helps bind people together into a social group.

religiosity: The depth of a person's religious feelings.

resocialization: Intense, deliberate socialization designed to change major beliefs and behaviors.

rite of passage: A ceremony marking the transition to a new stage of a culturally defined life course.

ritual: A formal pattern of activity that expresses symbolically a set of shared meanings.

role: The way a society defines how an individual is to behave in a particular status.

role conflict: Conflict that occurs when a person must violate the expectations associated with one role in order to perform another role well.

role expectations: A society's expectations about how a role should be performed, together with the individual's perceptions of what is required in performing that role.

role strain: Conflict that occurs when the expectations associated with a single role are contradictory.

role taking: Trying to look at social situations from the standpoint of another person from whom one seeks a response.

sacred: A term used to describe phenomena that are regarded as extraordinary, transcendent, and outside the everyday course of events.

sample: A set of respondents selected from a specific population.

sample survey: A survey administered to a selection of respondents drawn from a specific population.

sanctions: Rewards and punishments for abiding by or violating norms.

scapegoat: A convenient target for hostility.

scientific method: The process by which theories and explanations are constructed through repeated observation and careful description.

secondary deviance: Behavior that is engaged in as a reaction to the experience of being labeled as deviant.

secondary group: A social group whose members have a shared goal or purpose but are not bound together by strong emotional ties.

sect: A religious organization that rejects the beliefs and practices of existing churches; usually formed when a group leaves the church to form a rival organization.

secularization: A process in which the dominance of religion over other institutions is reduced.

segregation: The ecological and institutional separation of races or ethnic groups.

semiperipheral area: A state or region in which industry and financial institutions are developed to some extent but that remains dependent on capital and technology provided by other states.

sex: The biological differences between males and females, including the primary sex characteristics that are present at birth.

sexism: An ideology that justifies prejudice and discrimination based on sex.

sexual dimorphism: Differences in secondary sex characteristics between males and females of a species.

sexual orientation: An enduring emotional, romantic, sexual, or affectional attraction that a person feels toward another person

sexual scripts: Ideas and fantasies about what our sexual experiences should or could be like.

sexual selection: Conscious or unconscious effort of individuals to transmit their genes to the next generation.

sexuality: The manner in which a person engages in the intimate behaviors connected with genital stimulation, orgasm, and procreation.

significant other: Any person who is important to an individual.

simple supernaturalism: A form of religion in which people may believe in a great force or spirit but do not have a well-defined concept of God or a set of rituals involving God.

slavery: The ownership of one racial, ethnic, or politically defined group by another group that has complete control over the enslaved group.

social category: A collection of individuals who are grouped together because they share a trait deemed by the observer to be socially relevant.

social change: Variations over time in the ecological ordering of populations and communities, in patterns of roles and social interactions, in the structure and functioning of institutions, and in the cultures of societies.

social conditions: The realities of the life we create together as social beings.

social control: The set of rules and understandings that control the behavior of individuals and groups in a particular culture; the ways in which a society encourages conformity to its norms and prevents deviance.

social Darwinism: The notion that people who are more successful at adapting to the environment in which they find themselves are more likely to survive and to have children who will also be successful.

social group: A set of two or more individuals who share a sense of common identity and belonging and who interact on a regular basis.

social mobility: Movement by an individual or group from one social stratum to another.

social movement: Organized collective behavior aimed at changing or reforming social institutions or the social order itself.

social revolution: A complete transformation of the social order, including the institutions of government and the system of stratification.

social stratification: The process whereby the members of a society are sorted into different statuses; a society's system for ranking people hierarchically according to such attributes as wealth, power, and prestige.

social structure: The recurring patterns of behavior that people create through their interactions, their exchange of information, and their relationships.

socialism: An economic philosophy based on the concept of public ownership of property and sharing of profits, together with the belief that economic decisions should be controlled by the workers.

socialization: The processes through which we learn to behave according to the norms of our culture.

society: A population that is organized in a cooperative manner to carry out the major functions of life.

sociobiology: The hypothesis that all human behavior is determined by genetic factors.

socioeconomic status (SES): A broad social-class ranking based on occupational status, family prestige, educational attainment, and earned income.

sociological imagination: According to C. Wright Mills, the ability to see how social conditions affect our lives.

sociology: The scientific study of human societies and human behavior in the groups that make up a society.

spatial mobility: Movement of an individual or group from one location or community to another.

state: A society's set of political structures; the set of political institutions operating in a particular society.

status: A socially defined position in a group.

status group: A category of people within a social class, defined by how much honor or prestige they receive from society in general.

status symbols: Material objects or behaviors that indicate social status or prestige.

stereotype: An inflexible image of the members of a particular group that is held without regard to whether it is true.

stigma: An attribute or quality of an individual that is deeply discrediting.

structural mobility: Movement of an individual or group from one social stratum to another caused by the elimination of an entire class as a result of changes in the means of existence.

subculture: A group of people who hold many of the values and norms of the larger culture but also hold certain beliefs, values, or norms that set them apart from that culture.

subjective class: In Marxian theory, the way members of a given social class perceive their situation as a class.

superego: According to Freud, the part of the human personality that internalizes the moral codes of adults.

technologies: The products and the norms for using them that are found in a given culture.

territorial community: A population that functions within a particular geographic area.

theism: A belief system that conceives of a god or gods as separate from humans and from other living things on the earth.

theoretical perspective: A set of interrelated theories that offer explanations for important aspects of social behavior.

theory: A set of interrelated concepts that seeks to explain the causes of an observable phenomenon.

total institution: A setting in which people undergoing resocialization are isolated from the larger society under the control of a specialized staff.

totalitarian regime: A regime that accepts no limits to its power and seeks to exert its rule at all levels of society.

traditional authority: Authority that is hereditary and is legitimated by traditional values, particularly people's idea of the sacred.

transsexuals: People who feel strongly that the sexual organs they were born with do not conform to their deep-seated sense of what their sex should be.

triad: A group consisting of three people.

upward mobility: Movement by an individual or group to a higher social stratum.

urbanization: A process in which an increasing proportion of a total population becomes concentrated in urban settlements.

values: The ideas that support or justify norms.

variable: A characteristic of an individual, group, or society that can vary from one case to another.

visual sociology: Qualitative observation techniques that rely on the use of photography, video, or film to supplement or further document social behavior that can be observed.

voluntary association: A formal organization whose members pursue shared interests and arrive at decisions through some sort of democratic process.

welfare capitalism: An economic philosophy in which markets determine what goods will be produced and how, but the government regulates economic competition.

References

Abbott, A. D. (1988). *The system of professions: An essay on the division of expert labor.* Chicago: University of Chicago Press.

Abbott, A. D. (1993). The sociology of work and occupations. *Annual Review of Sociology, 19,* 187–209.

Abbott, A. D. (2001). *Chaos of disciplines.* Chicago: University of Chicago Press.

Abbott, A. (2009). The last thirty years. *Contemporary Sociology, 38,* 507–512.

Abu-Lugod, J. (1996, January). Personal communication.

Ackerman, S. (2010, March 24). Gates will relax "Don't Ask, Don't Tell" enforcement this week. *The Washington Independent.*

Addams, J. (1895). *Hull-House maps and papers.* New York: Crowell.

Adler, M. (1997). *Drawing down the moon: Witches, druids, goddess-worshippers, and other pagans in America today.* New York: Viking.

Adler, P., & Adler, P. (Eds.) (2006). *Constructions of deviance: Social power, context, and interaction,* 5th ed. Belmond, CA: Thomson/Wadsworth.

Adult Protective Services (2006). Adult Protective Services cases. www.elderabusecenter.org.

Aganbegyan, A. (1989). *Perestroika 1989.* New York: Scribner.

Ahlburg, D. A., & De Vita, C. J. (1992, August). New realities of the American family. *Population Bulletin* (Population Reference Bureau).

Ahmed, Z. S. (2010). Fighting for the rule of law: Civil resistance and the lawyers' movement in Pakistan. *Democratization,* June, 492–513.

Airton, L. (2009). From sexuality (gender) to gender (sexuality): The aims of anti-homophobia education. *Sex Education, 9,* 129–139.

Alam, M. (1988). *Studies on brain drain.* Monticello, IL: Vance Bibliographies.

Alan Guttmacher Institute (2001). *Teenage sexual and reproductive behavior in developed countries, 2001.* New York: Alan Guttmacher Institute.

Alarid, L. F., Burton, Jr., V. S., & Cullen, F. T. (2000). Gender and crime among felony offenders: Assessing the generality of social control and differential association theories. *Journal of Research in Crime and Delinquency, 37,* 171–199.

Alarkon, W. (2009, October 7). Appropriations deal blow to border fence. The Hill. http://thehill.com.

Alba, R. D. (2009). *Blurring the color line: The new chance for a more integrated America.* Cambridge, MA: Harvard University Press.

Albright, W. F. (2006). *Archaeology and the religion of Israel.* Louisville, KY: Westminster John Knox Press.

Allen, F. L. (1931). *Only yesterday: An informal history of the 1920s.* New York: Harper.

Allen, G. (1997). Abolishing parole saves lives and property. *Corrections Today, 59,* 22.

Allen, J. L. (Ed.) (2007). *Environment 06/07.* Dubuque, IA: McGraw-Hill Contemporary Learning Series.

Altman, D. (2001). *Global sex.* Chicago: University of Chicago Press.

Alvarez, L. (2009, August 15). G.I. Jane breaks the combat barrier. *New York Times.*

Amato, P. R., & Sobolewski, J. M. (2001). The effects of divorce and marital discord on adult children's psychological well-being. *American Sociological Review, 66,* 900–921.

Ammerman, N. T. (1990). *Baptist battles: Social change and religious conflict in the Southern Baptist Convention.* New Brunswick, NJ: Rutgers University Press.

Ammerman, N. T. (1997). Organized religion in a voluntaristic society. *Sociology of Religion, 58,* 203–216.

Ammerman, N. T. (2000, March 15). New life for denominationalism. *Christian Century,* pp. 302–307.

Ammerman, N. T. (2005). *Pillars of faith: American congregations and their partners.* Berkeley: University of California Press.

Ammerman, N. T. (2007). *Everyday religion: Observing modern religious lives.* New York: Oxford University Press.

An-Na'im, A. (2002). *Islamic family law in a changing world: A global resource book.* London: Zed Books.

Annie E. Casey Foundation. Annual. Kids count: A pocket guide on America's youth. Washington, DC: Population Reference Bureau.

AP (Associated Press) (2002, January 3). A nation challenged: American Muslims. *New York Times,* p. A14.

Applebome, P. (1998, May 10). No room for children in a world of little adults. *New York Times,* sec. 4, pp. 1, 3.

Applied Research Center. (2009, May). Race and recession. http://arc.org/downloads/2009_race_recession_0909.pdf.

Archer, R. (2010). Seymour Martin Lipset and political sociology. *British Journal of Sociology, 61,* 43–52.

Archibald, K. (1947). *Wartime shipyard.* Berkeley: University of California Press.

Archibold, R. C. (2010a, April 23). Arizona enacts stringent law on immigration. *New York Times.*

Archibold, R. C. (2010b). Preemption, not profiling, in challenge to Arizona. *New York Times,* July 8, p. 15.

Ariès, P. (1962). *Centuries of childhood.* New York: Vintage.

Arnot, M. (2002). Making the difference to sociology of education: Reflections on family–school and gender relations. *Discourse: Studies in the Cultural Politics of Education, 23,* 347–355.

Aron, R. (1955). *The century of total war.* Boston: Beacon.

Aronowitz, S. (1996). Losing and winning in a conservative age. *Social Policy, 27,* 10–21.

Aronowitz, S. (1998). *From the ashes of the old.* Boston: Houghton Mifflin.

Aronowitz, S. (2003). *How class works.* New Haven, CT: Yale University Press.

Aronowitz, S., & DiFazio, W. (1994). *The jobless future: Sci-tech and the dogma of work.* Minneapolis: University of Minnesota Press.

Aronowitz, S., & Giroux, H. A. (1985). *Education under siege.* South Hadley, MA: Bergin and Garvey.

Arrow, K., Bowles, S., & Durlauf, S. (Eds.) (2000). *Meritocracy and economic inequality.* Princeton, NJ: Princeton University Press.

Asch, S. E. (1966). Effects of group pressure upon the modification and distortion of judgments. In H. Proshansky & B. Seidenberg (Eds.), *Basic studies in social psychology.* Fort Worth, IN: Holt, Rinehart and Winston.

Ash, T. G. (1999). Helena's kitchen. *New Yorker,* February 15, pp. 32–39.

Ashmawi, M. S. (1998). *Against Islamic extremism: The writings of Muhammad Sa'id al-'Ashmawi.* Gainesville: University Press of Florida.

Aspaklaria, S. (1985, September 12). A divorced father, a child, and a summer visit together. *New York Times,* p. C1.

Babbie, E. (2003). The practice of social research, 11th ed. Belmont, CA: Wadsworth.

Baber, Z. (2001). Modernization theory and the cold war. *Journal of Contemporary Asia, 31,* 71.

Bacci, M. (1997). *A concise history of world population,* 2nd ed. Malden, MA: Blackwell.

Bakalian, A., & Bozorgmehr, M. (2010). *Backlash 9/11: Middle Eastern and Muslim Americans respond.* Berkeley: University of California Press.

Bakke, E. W. (1933). *The unemployed man.* London: Nisbet.

Balandier, A. (1971/1890). *The delight makers.* Orlando: Harcourt Brace Jovanovich.

Balch, E. G. (1910). *Our Slavic fellow citizens.* New York: Charities Publication Committee.

Baldassare, M. (1986). *Trouble in paradise: The suburban transformation in America.* New York: Columbia University Press.

Baldassare, M. (2003). How the West is taxed. *New York Times,* June 27, p. A27.

Bales, R. F., & Slater, P. E. (1955). Role differentiation in small decision-making groups. In T. Parsons & R. F. Bales (Eds.), *Family, socialization, and interaction process.* New York: Free Press.

Ballantine, J. H. (1997). *The sociology of education: A systematic analysis,* 4th ed. Upper Saddle River, NJ: Prentice Hall.

Baller, R. D., & Richardson, K. K. (2002). Social integration, imitation, and the geographic patterning of suicide. *American Sociological Review, 67.*

Baltimore, D. (2001). Our genome unveiled. *Nature, 409,* 814–816.

Barash, D. (2002). *Peace and conflict studies.* Thousand Oaks, CA: Sage.

Barash, D. P. (2008). *Natural selections: Selfish altruists, honest liars, and other realities of evolution.* New York: Bellevue Literary Press.

Barboza, D. (2000). Rampant obesity, a debilitating reality for the urban poor. *New York Times,* December 26, p. F5.

Barker, M. (2002). The day of reckoning. *London Financial Times,* January 21, p. 1.

Barnet, R. J. (1994). Lords of the global economy: Stateless corporations. *The Nation,* December 19, pp. 754–758.

Barrett, D. V. (1996). *Sects, "cults," and alternative religions: A world survey and sourcebook.* London: Blandford.

Barrow, L., & Rouse, C. E. (2006). The economic value of education by race and ethnicity. *Economic Perspectives, 30,* 14–27.

Bartlett, B. (2009). Health care: Costs and reform. *Forbes,* July 3.

Bashi, V. (1991). Mentoring of at-risk students. *Focus* (University of Wisconsin, Institute for Research on Poverty), *13,* 26–32.

Bassuk, E. L. (1984, July). The homelessness problem. *Scientific American,* pp. 40–45.

Bates, A. (1999, February 15). The siege at Waco: Deadly inferno. *Booklist,* pp. 1050–1051.

Batson, C. D., Batson, J. G., & Todd, R. M. (1995). Empathy and the collective good: Caring for one or the others in a social dilemma. *Journal of Personality and Social Psychology, 68,* 619–631.

Bhattacharyya, G. (2002). *Sexuality and society: An introduction.* New York: Routledge.

Battaglia, L., & Zecchin, F. (1989). *Chroniques siciliennes.* Paris: Centre Nationale de la Photographie.

Battle, J. (2003, August). Unpublished manuscript.

Baum, S. (1987). *Financial aid to low-income college students: Its history and prospects* (Institute for Research on Poverty Discussion Paper No. 846-87). Madison: University of Wisconsin.

Bayer, R. (1991). The great drug policy debate. *Millbank Quarterly, 69,* 341–364.

Beall, J., Guha-Khasnobis, B., & Kanbur, R. (2010). Introduction: African development in an urban world: Beyond the tipping point. *Urban Forum, 21,* 187–204.

Bearak, B. (1997, April 28). Eyes on glory. *New York Times,* p. A1.

Bearman, P. S., & Brückner, H. (2001). Promising the future: Virginity pledges and first intercourse. *American Journal of Sociology, 106.*

Beauregard, R. A. (2009). Urban population loss in historical perspective: United States, 1820–2000. *Environment and Planning, 41,* 514–528.

Becker, G. S. (1997). Why every married couple should sign a contract. *Business Week,* December 29, p. 30.

Becker, H. S. (1961). *Boys in white: Student culture in medical school.* Chicago: University of Chicago Press.

Becker, H. S. (1963). *The outsiders: Studies in the sociology of deviance.* New York: Free Press.

Beebe, S. A. (1997). *Communication in small groups: Principles and practices,* 5th ed. New York: Longman.

Beer, F. A. (1981). *Peace against war: The ecology of international violence.* San Francisco: Freeman.

Bell, D. (1962). Crime as an American way of life. In D. Bell (Ed.), *The end of ideology.* New York: Free Press.

Bell, D. (1973). *The coming of post industrial society: A venture in social forecasting.* New York: Basic Books.

Bell, D. (1999/1973). *The coming of post industrial society: A venture in social forecasting.* New York: Basic Books.

Bellah, R. N. (1970). *Beyond belief.* New York: Harper.

Bellah, R. N., Madison, R., Sullivan, W. M., Swidler, A., & Tipton, S. M. (1985). *Habits of the heart: Individualism and commitment in American life.* Berkeley: University of California Press.

Belsky, J. (1988). The "effects" of infant day care reconsidered. *Early Childhood Research Quarterly, 3,* 235–272.

Belsky, J., & Steinberg, L. D. (1991). *Infancy, childhood & adolescence: Development in context.* New York: McGraw-Hill.

Bem, S. L. (1993). *The lenses of gender: Transforming the debate on sexual inequality.* New Haven, CT: Yale University Press.

Benavot, A., et al. (1991). Knowledge for the masses: World models and national curricula, 1920–1986. *American Sociological Review, 56,* 85–91.

Bendix, R. (1969). *Nation-building and citizenship.* Garden City, NY: Doubleday.

Bendix, R. (1974). *Work and authority in industry.* Berkeley: University of California Press.

Bendix, R., & Lipset, S. M. (1966). *Class, status, and power: Social stratification in comparative perspective,* 2nd ed. New York: Free Press.

Beneckson, R. E., Dunn, M., Gonzalez, A., & Marichal, C. (2000). Racial and ethnic disparities in Florida welfare reform: A study of income levels for black, Hispanic and white post-welfare recipients. Washington, DC: The Finance Project.

Bennett, M. K. (1954). *The world's food.* New York: Harper.

Benokraitis, N. V. (2007). *Marriages and families: Changes, choices, and constraints,* 6th ed. Upper Saddle River, NJ: Prentice Hall.

Bensman, D., & Lynch, R. (1987). *Rusted dreams: Hard times in a steel community.* New York: McGraw-Hill.

Benson, B. L. (2009). Unintended consequences. *Stanford Law and Policy Review, 20:* 293–357.

Bentham, J. (1789). *An introduction to the principles of morals and legislation.* London: T. Payne.

Berger, A. A. (2000). *Ads, fads, and consumer culture: Advertising's impact on American character and society.* Oxford, England: Rowman & Littlefield.

Berger, B. (1961). The myth of suburbia. *Journal of Social Issues, 17,* 38–48.

Berger, P. (1963). *Invitation to sociology: A humanistic perspective.* New York: Doubleday Anchor.

Bergmann, B. R., & Hartmann, H. (1995). A program to help working parents. *The Nation,* May 1, pp. 592ff.

Berlin, I. (1998, May 14). My intellectual path. *New York Review of Books,* pp. 53–60.

Berman, M. (2006). A Times Square for the new millennium. *Dissent,* Winter, pp. 73–79.

Bernard, J. (1964). *Academic women.* University Park: Pennsylvania State University Press.

Bernard, J. (1973). *The sociology of community.* Glenview, IL: Scott, Foresman.

Bernard, J. (1982). *The future of marriage,* 2nd ed. New Haven, CT: Yale University Press.

Bernstein, E. (2001). The meaning of the purchase: Desire, demand, and the commerce of sex. *Ethnography, 2,* 375–406.

Bernstein, E. (Ed.) (2004). *Regulating sex: The politics of intimacy and identity.* Routledge.

Berrueta-Clement, J. R., Schweinhart, L. J., Barnett, W. S., Epstein, A. S., & Weikart, D. P. (1984). *Changed lives: The effects of the Perry preschool program on youths through age 19.* Ypsilanti, MI: High/Scope Press.

Betts, J. R. (1998). The two-legged stool: The neglected role of educational standards in improving America's public schools. *Federal Reserve Bank of New York Economic Policy Review,* March, pp. 97–117.

Bianchi, S. M., & Casper, L. M. (2000). American families. *Population Bulletin,* vol. 55, no. 4. Washington, DC: Population Reference Bureau.

Bianchi, S. M., & Spain, D. (1986). *American women in transition.* New York: Russell Sage.

Bierstedt, R. (1963). *The social order.* New York: McGraw-Hill.

Binstock, R. K., & George, L. K. (Eds.) (2006). *The handbook of aging and the social sciences,* 6th ed. San Diego: Academic Press.

Bittman, M. (1993). Australians' changing use of time, 1974–1987. *Social Indicators Research, 30,* 91–109.

Blakemore, J., & Hill, C. (2008). The child gender socialization scale: A measure to compare traditional and feminist parents. *Sex Roles, 58,* 192–207.

Blau, J. R. (1993). *Social contracts and economic markets.* New York: Plenum.

Blau, P. (1964). *Exchange and power in social life.* New York: Wiley.

Blau, P., & Duncan, O. D. (1967). *The American occupational structure.* New York: Wiley.

Blau, P. M., & Scott, R. W. (1962). *Formal organizations: A comparative approach.* San Francisco: Chandler.

Blauner, R. (1964). *Alienation and freedom.* Chicago: University of Chicago Press.

Blauner, R. (1969). Internal colonialism and ghetto revolt. *Social Problems, 16,* 393–408.

Blauner, R. (1972). *Racial oppression in America.* New York: Harper.

Blauner, R. (Ed.) (1989). *Black lives, white lives: Three decades of race relations in the United States.* Berkeley: University of California Press.

Blinder, D., & Crossette, B. (1993). As ethnic wars multiply, U.S. strives for a policy. *New York Times,* February 7, p. A1.

Bloch, M. (1964). *Feudal society* (Vol. 1). Chicago: University of Chicago Press.

Block, F. (1990). *Post industrial possibilities: A critique of economic discourse.* Berkeley: University of California Press.

Bloom, B. S. (1976). *Human characteristics and school learning.* New York: McGraw-Hill.

Bloom, S. W. (2002). *The word as scalpel: A history of medical sociology.* New York: Oxford University Press.

Bluestone, B. (1992). *Negotiating the future: A labor perspective on American business.* New York: Basic Books.

Blum, D. L. (2002). *Love at Goon Park: Harry Harlow and the science of affection.* Cambridge, MA: Perseus.

Blumenfeld, W. J. (Ed.) (1992). *Homophobia.* Boston: Beacon.

Blumer, H. (1969a). Elementary collective groupings. In A. M. Lee (Ed.), *Principles of sociology,* 3rd ed. New York: Barnes & Noble.

Blumer, H. (1969b). *Symbolic interactionism.* Englewood Cliffs, NJ: Prentice Hall.

Blumer, H. (1978). Elementary collective behavior. In L. E. Genevie (Ed.), *Collective behavior and social movements.* Itasca, IL: Peacock.

Bobrick, B. (1997). *Angel in the whirlwind: The triumph of the American Revolution.* New York: Simon & Schuster.

Bodenheimer, T. (2008). The political divide in health care: A liberal perspective. *Kennedy Institute of Ethics Journal,* June.

Bodenhorn, H. (2002). Review of *Banking panics of the gilded age,* by Elmus Wicker. *Journal of Interdisciplinary History, 32,* 497–498.

Boese, W. (1998). U.S. retains top spot in latest UN conventional arms register. *Arms Control Today, 28*, 27.

Boggs, V. W., & Meyersohn, R. (1988). The profile of a Bronx salsero: Salsa's still alive! *Journal of Popular Music and Society, 11*, 7–14.

Bogue, D. J. (1969). *Principles of demography.* New York: Wiley.

Bogue, D. J. (1985). *The population of the United States: Historical trends and future projections.* New York: Free Press.

Bohannon, L., & Bohannon, P. (1953). *The Tiv of central Nigeria.* London: International African Institute.

Bohannon, L., & Bohannon, P. (1962). *Markets in Africa.* Evanston, IL: Northwestern University Press.

Bombardieri, M. (2005, January 19). Harvard women's group rips Summers. *Boston Globe.*

Bonner, R. (1994, August 25). Hutu and Tutsi mill the rice and set an example. *New York Times,* p. A1.

Bonnett, A. W. (1981). *Institutional adaptation of West Indian immigrants to America.* Lanham, MD: University Press of America.

Booth, A., & Edwards, J. N. (1985, February). Age at marriage and marital instability. *Journal of Marriage and the Family,* pp. 67–75.

Borgerhoff Mulder, M. (2009). Serial monogamy as polygyny or polyandry? *Human Nature, 10,* 130–150.

Bose, C. E., & Rossi, P. H. (1983). Gender and jobs. *American Sociological Review, 48,* 316–330.

Boserup, E. (1965). *The conditions of agricultural growth.* London: Allen and Unwin.

Bosk, C. (1979). *Forgive and remember: Managing medical failure.* Chicago: University of Chicago Press.

Bosworth, B., & Burtless, G. (Eds.) (1998). *Aging societies: The global dimension.* Washington, DC: Brookings Institution.

Bott, E. (1977). Urban families: Conjugal roles and social networks. In S. Leinhardt (Ed.), *Social networks: A developing paradigm.* Orlando: Academic Press.

Bottomore, T. (Ed.) (1973). *Karl Marx.* Englewood Cliffs, NJ: Prentice Hall.

Bourdieu, P. (1993). *Sociology in question.* London: Sage.

Bower, B. (2002, March 16). Heads up. *Science News,* pp. 1–3.

Bowker, J. W. (1997). *World religions.* New York: DK Publishers.

Bowles, S., & Gintis, H. (2000). Schooling in capitalist America revisited. *Sociology of Education, 75,* 1–18.

Bowman, K. (2009). How Americans think about wealth. *Forbes,* March 23.

Boxer, D. (2002). Nagging: The familial conflict arena. *Journal of Pragmatics, 34,* pp. 49–61.

Bracey, J. H., Meier, A., & Rudwick, E. (1970). *Black nationalism in America.* Indianapolis: Bobbs-Merrill.

Bradley, H. (1998). A new gender(ed) order? Researching and rethinking women's work. *Sociology, 32,* 869–874.

Braidwood, R. S. (1967). *Prehistoric man,* 7th ed. Glenview, IL: Morrow.

Brass, T. (2009). Capitalist unfree labour: A contradiction? *Critical Sociology, 35,* 743–765.

Braudel, F. (1976/1949). *The Mediterranean and the Mediterranean world in the age of Philip II.* New York: Harper.

Braudel, F. (1984). *The perspective of the world: Vol. 3. Civilizations and capitalism: 15th–18th century.* New York: Harper.

Braverman, H. (1974). *Labor and monopoly capital.* New York: Monthly Review Press.

Bravo, A., Cass, Y., & Tranter, D. (2008). Good food in family day care: Improving nutrition and food safety in family day care. *Nutrition and Dietetics, 65,* 47–55.

Breese, G. (1966). *Urbanization in newly developing countries.* Englewood Cliffs, NJ: Prentice Hall.

Breiner, P. (1996). *Max Weber and democratic politics.* Ithaca, NY: Cornell University Press.

Brody, D. (1960). *Steelworkers in America: The non-union era.* Cambridge, MA: Harvard University Press.

Brody, J. (1992, August 11). To predict divorce, ask 125 questions. *New York Times,* pp. C1, C9.

Brokaw, T. (1994). Learning lessons for life in football. *New York Times,* January 2, sec. 8, p. 11.

Bronfenbrenner, K. (2000, December). Raw power. *Multinational Monitor,* pp. 24–29.

Bronfenbrenner, K. (2003, May). Declining unionization, rising inequality. *Multinational Monitor,* pp. 21–24.

Bronfenbrenner, U. (1981). The ecology of human development: Experiments by nature and design. Cambridge, MA: Harvard University Press.

Brooke, J. (1997, April 28). Former cult member warns of continuing cult loyalties. *New York Times,* p. A16.

Brotherton, D., & Salazar-Atias, C. (2003). Amor de reina! The pushes and pulls of group membership among the Latin Queens. In L. Kontos, D. Brotherton, & L. Barrios (Eds.), *Gangs and society: Alternative perspectives.* New York: Columbia University Press.

Broussard, C. A., & Joseph, A. L. (1998). Tracking: A form of educational neglect? *Social Work in Education, 20,* 110–120.

Brown, C. (1966). *Manchild in the promised land.* New York: Macmillan.

Brown, D. (1970). *Bury my heart at Wounded Knee.* New York: Washington Square Press.

Brown, K. M. (1991). *Mama Lola: A Vodou priestess in Brooklyn.* Berkeley: University of California Press.

Brown, K. M. (1992, April 15). Writing about "the other." *Chronicle of Higher Education,* p. A56.

Brown, L. R. (1995). Facing food security. *World Watch,* December 10.

Brown, L. R., Flavin, C., & French, H. F. (1997). *State of the world 1998: A Worldwatch Institute report on progress toward a sustainable society.* New York: Norton.

Browne, K. (2010, October 3). Women in combat. *Newsweek.*

Browne, M. W. (1987). Relics of Carthage show brutality amid the good life. *New York Times,* September 1, pp. C1, C10.

Brownfield, D., Sorenson, A. M., & Thompson, K. M. (2001). Gang membership, race, and social class: A test of the group hazard and master status hypotheses. *Deviant Behavior, 22,* 73–89.

Brownmiller, S. (1975). *Against our will.* New York: Simon & Schuster.

Brundtland, G. H. (1989). How to secure our common future. *Scientific American,* September, p. 190.

Bucholtz, M. (2002). Youth and cultural practice. *Annual Review of Anthropology, 31,* 525–552.

Bullard, R. D. (2003, January–February). Environmental justice for all. *The New Crisis.*

Bulmer, M. (1986). *The Chicago school of sociology.* Chicago: University of Chicago Press.

Bumpass, L. L., & Raley, R. K. (1995). Redefining single-parent families: Co-habitation and changing family reality. *Demography, 32,* 97–110.

Bumpass, L. L., Raley, R. K., & Sweet, J. A. (1995). The changing character of stepfamilies: Implications of cohabitation and nonmarital childbearing. *Demography, 32,* 425–437.

Bumpass, L. L., & Sweet, J. A. (1989). Children's experience in single-parent families: Implications of cohabitation and marital transitions. *Family Planning Perspectives, 21,* 256–260.

Burawoy, M. (1980). *Manufacturing consent.* Chicago: University of Chicago Press.

Burawoy, M. (1996). A classic of its time. *Contemporary Sociology, 25,* 296–300.

Burdman, P. (2003, April 24). Investigating below the surface. *Black Issues in Higher Education,* pp. 30–33.

Burgess, E. W. (1925). The growth of the city: An introduction to a research project. In R. E. Park & E. W. Burgess (Eds.), *The city.* Chicago: University of Chicago Press.

Burris, S. (2002). Introduction: Merging law, human rights, and social epidemiology. *Journal of Law, Medicine, and Ethics, 30,* 498–509.

Burstein, R. (1991). Legal mobilization as a social movement tactic: The struggle for equal employment opportunity. *American Journal of Sociology, 96,* 1201–1225.

Burum, I., & Margalit, A. (2002, January 17). Occidentalism. *New York Review of Books,* pp. 4–8.

Bury, J. B. (1932). *The idea of progress: An inquiry into its origin and growth.* New York: Dover.

Butler, R. (1989). Dispelling ageism: The cross-cutting intervention. *Annals of the American Academy of Political and Social Science, 503,* 138–148.

Byrd-Bredbenner, C., & Murray, J. (2003). A comparison of the anthropometric measurements of idealized female body images in media directed to men, women, and mixed gender audiences. *Topics in Clinical Nutrition, 18,* 117–129.

Cacioppo, J. T., Fowler, J. H., & Cristakis, N. A. (2009). Alone in the crowd: The structure and spread of loneliness in a large social network. *Journal of Personality and Social Psychology, 97,* 977–991.

Caccamo, R. (2000). *Back to Middletown: Three generations of sociological reflections.* Stanford, CA: Stanford University Press.

Cahill, S. E. (1992). The sociology of childhood at and in an uncertain age. *Contemporary Sociology, 21,* 669–672.

Cahill, T. (2002, February 3). The one true faith: Is it tolerance? *New York Times,* sec. 4, pp. 1, 4.

Cahill, L. S., Garvey, J., & Kennedy, T. F. (2006). *Sexuality and the U.S. Catholic Church: Crisis and renewal.* Crossroad.

Cain, L. D. (1964). Life course and social structure. In R. F. Faris (Ed.), *Handbook of modern sociology.* Chicago: Rand McNally.

Caldwell, J., & Caldwell, P. (1988). Women's position and child mortality and morbidity in LDCs. In *Conference on women's position and demographic change in the course of development.* Liège: International Union for the Scientific Study of Population.

Califano, J. A. (1998, February 21). A punishment-only prison policy. *America*, pp. 3–4.

Cameron, W. B. (1966). *Modern social movements: A sociological outline.* New York: Random House.

Campbell, B. (1984). *Wigan Pier revisited: Poverty and politics in the eighties.* London: Virago.

Cancian, F. M. (1994). *Romantic longings: Love in America, 1830–1980.*

Cancian, M., et al. (2003, Summer). Income and program participation among early TANF recipients: The evidence from New Jersey, Washington, and Wisconsin. *Focus* (Institute for Research on Poverty), pp. 1–10.

Cant, S., & Sharma, U. (1998). Reflexivity, ethnography and the professions (complementary medicine): Watching you watching me watching you (and writing about both of us). *Sociological Review, 46*, 244–264.

Cantril, H., with Gaudet, H., & Herzog, J. (1982/1940). *The invasion from Mars.* Princeton, NJ: Princeton University Press.

Caplow, T., Bahr, H. M., Chadwick, B. A., Hill, R., & Williamson, M. H. (1983). *Middletown families: Fifty years of change and continuity.* New York: Bantam.

Capra, F. (2000). The challenge of the 21st century. *Tikkun*, February.

Carnegie Endowment for International Peace (2003). www.ceip.org.

Carpenter, L., & Bartley, M. (1994). Fat, female, and poor. *Lancet, 344*, 1715–1716.

Carter, L. J. (1975). *The Florida experience.* Baltimore: Johns Hopkins University Press.

Castells, M., Flecha, R., Freire, P., Giroux, H. A., Macedo, D., & Willis, P. (1998). *Critical education in the new information age.* Totowa, NJ: Rowman and Littlefield.

CBS Poll (2006, August 11–13). *Americans' views on the threat of terrorism.* New York: CBS.

Cecco, J. P., & Parker, D. A. (1995). The biology of homosexuality: Sexual orientation or sexual preference? *Journal of Homosexuality, 28*, 1–27.

Cerulo, K. A. (2008). Social relations, core values, and the polyphony of the American experience. *Sociological Forum, 23*, 351–362.

Chaddock, G. R. (1999, October 27). Teach for America turns 10—and thrives. *Christian Science Monitor*, p. 10.

Chakravarty, S. (2006). Cinema and http://transnationalism. homepage.newschool.edu.

Chalk, F., & Jonassohn, K. (1990). *The history and sociology of genocide.* New Haven, CT: Yale University Press.

Chambliss, D. F. (1996). *Beyond caring: Hospitals, nurses, and the social organization of ethics.* Chicago: University of Chicago Press.

Chambliss, W. J. (1973, December). The Saints and the Roughnecks. *Society*, pp. 23–31.

Chan, A., & Unger, J. (2009). A Chinese state enterprise under the reforms: What model of capitalism? *China Journal*, July, pp. 1–26.

Chang, T. F. H. (2003). A structural model of race, gender, class, and attitudes toward labor unions. *Social Science Journal, 40*, 189–200.

Chenault, C. (1996). *Maasai.* Unpublished manuscript, Calhoun Community College, Decatur, AL.

Cherlin, A. J. (1981). *Marriage, divorce, remarriage.* Cambridge, MA: Harvard University Press.

Cherlin, A. J. (1996). *Public and private families: An introduction.* New York: McGraw-Hill.

Cherlin, A. J. (2005, Fall). American marriage in the early twenty-first century. *The Future of Children*, pp. 33–55.

Chernin, K. (1981). *The obsession: Reflections on the tyranny of slenderness.* New York: Harper.

Chernow, R. (1998). *Titan: The life of J. D. Rockefeller Sr.* New York: Random House.

Chirot, D. (1986). *Social change in the modern era.* San Diego: Harcourt Brace Jovanovich.

Chirot, D. (1994a). *How societies change.* Thousand Oaks, CA: Pine Forge.

Chirot, D. (1994b). *Modern tyrants: The power and prevalence of evil in our age.* New York: Free Press.

Chodorow, N. (2002). Mother and child: Feminist psychology. *Feminism and Psychology, 12*, 11–17.

Chodorow, N. (2004). Psychoanalysis and women. *Annual of Psychoanalysis, 32*, 101–129.

Chomsky, N. (1965). *Aspects of the theory of syntax.* Cambridge, MA: MIT Press.

Chwast, J. (1965). Value conflicts in law enforcement. *Crime and Delinquency, 2*, 151–161.

Clark-Miller, J., & Murdock, J. (2005). Order on the edge: Remedial work in a right-wing political discussion group. In C. Morrill, D. A. Snow, & C. H. White (Eds.), *Together alone: Personal relationships in public spaces.* Berkeley: University of California Press.

Clarke, L. H. (2002). Older women's perceptions of ideal body weights: The tensions between health and appearance motivations for weight loss. *Ageing and Society, 22*, 751–773.

Clausen, J. A. (1968). *Socialization and society.* Boston: Little, Brown.

Clay, G. (1980). *Close-up: How to read the American city.* Chicago: University of Chicago Press.

Clay, G. (1994). *Real places: An unconventional guide to America's landscape.* Chicago: University of Chicago Press.

Clement, K., Tatum, K. M., Kruse, M. J., & Kunselman, J. C. (2009). Exploring agency policing models and response to domestic violence. *Policing, 32*, 92–107.

Cloud, R. N., Rowan, N., Wulff, D., & Golder, S. (2007). Posttreatment 12-step program affiliation and dropout: Theoretical model and qualitative exploration. *Journal of Social Work Practice in the Addictions, 7*, 49–74.

Cloward, R., & Ohlin, L. (1960). *Delinquency and opportunity: A theory of delinquent gangs.* New York: Free Press.

Clydesdale, T. T. (1997). Family behaviors among early U.S. baby boomers: Exploring the effects of religion and income change, 1965–1982. *Social Forces, 76*, 605–635.

Coates, J. F. (2002, September–October). What's next? Foreseeable terrorist acts. *Futurist*, p. 23.

Cockerham, W. C. (2004). *Medical sociology*, 9th ed. Upper Saddle River, NJ: Prentice Hall.

Cohen, B. E., & Burt, M. R. (1990).The homeless: Chemical dependency and mental health problems. *Social Work Research and Abstracts, 26*, 8–17.

Cohen, J. E. (1998, October 8). How many people can the earth support? *New York Review of Books*, pp. 29–31.

Cohen, N. (1961). *The pursuit of the millennium.* New York: Harper.

Cohen, R. (2000). Third party time. *New York Times Magazine*, October 22, pp. 44–45.

Coleman, J. S. (1961). *The adolescent society.* New York: Free Press.

Coleman, J. S. (1976). Liberty and equality in school desegregation. *School Policy, 6*, 9–13.

Coleman, J. S. (1994). Social capital, human capital, and investment in youth. In A. C. Petersen & T. J. Mortimer, *Youth unemployment and society.* Cambridge, England: Cambridge University Press.

Collins, E. G. C., & Scott, P. (Eds.) (1978). Everyone who makes it has a mentor. *Harvard Business Review, 56*, 89–101.

Comer, J. P. (1997). *Waiting for a Miracle: Why schools can't solve our problems and how we can.* New York: Dutton.

Comte, A. (1971/1854). The positive philosophy. In M. Truzzi (Ed.), *Sociology: The classic statements.* New York: Random House.

Concerned Women for Justice. (2006). www. charityblossom.org/nonprofit/concerned-women-for-justice.

Condon, S. (2009). Recession may have lasting impact on kids. CBS News, June 3.

Conger, D., Ge, X., & Elder, G. H., Jr. (1994). Economic stress, coercive family process, and developmental problems of adolescence. *Child Development, 65*, 541–561.

Conley, D. (1999). *Being black, living in the red: Race, wealth, and social policy in America.* Berkeley: University of California Press.

Conley, D. (2003, February 15). The cost of slavery. *New York Times*, p. A25.

Connell, E., & Hunt, A. (2006). Sexual ideology and sexual physiology in the discourses of sex advice literature. *Canadian Journal of Human Sexuality, 15*, 23–45.

Connell, R. W. (2005). *Masculinities.* Berkeley: University of California Press.

Connell, R. W., & Messerschmidt, J. (2005). Hegemonic masculinity: Rethinking the concept. *Gender and Society, 19*, 829–859.

Constantine, N., & Huberman, B. (2000, August). *European approaches to adolescent sexual health and behavior.* Florence Crittendon Roundtable, San Francisco, California.

Conversi, D. (2010). Nations, states, and violence. *Review of Politics, 72*, 337–40.

Cookson, P. W. (1997). New kid on the block? A closer look at America's private schools. *Brookings Review, 15*, 22–25.

Cookson, P. W., & Persell, C. H. (1985). *Preparing for power: America's elite boarding schools.* New York: Basic Books.

Cooley, C. H. (1909). *Social organization: A study of the large mind.* New York: Scribner.

Cooley, C. H. (1956/1902). *Human nature and the social order.* New York: Free Press.

Corsaro, W. (2005). *The sociology of childhood*, 2nd ed. Thousand Oaks, CA: Pine Forge.

Coser, L. A. (1966). *The functions of social conflict.* New York: Free Press.

Cowie, J. (1998). The terror of the machine: Technology, work, gender and ecology on the U.S.–Mexico border. *Labor History, 39*, 502.

Cramer, E., & Boyd, J. (1995). The tenure track and the parent track: A road guide. *Wilson Library Bulletin, 69*, 41–42.

Cranz, G. (1982). *The politics of park design: A history of urban parks in America.* Cambridge, MA: MIT Press.

Creel, M. W. (1988). *"A peculiar people": Slave religion and community-culture among the Gullahs.* New York: New York University Press.

Cremin, L. A. (1980). *American education: The national experience 1783–1876.* New York: Harper.

Cressey, D. R. (1971/1953). *Other people's money: A study in the social psychology of embezzlement.* Belmont, CA: Wadsworth.

Cronon, W. (1983). *Changes in the land: Indians, colonists, and the ecology of New England.* New York: Hill and Wang.

Crosley, N. (2002). *Making sense of social movements.* London: Open University Press.

Crowder, M. (1966). *A short history of Nigeria,* rev. ed. New York: Praeger.

Cuber, J. F., & Harroff, P. B. (1980). Five types of marriage. In A. Skolnick & J. Skolnick (Eds.), *Family in transition.* Boston: Little, Brown.

Cumming, E., & Henry, W. (1971). *Growing old: The process of disengagement.* New York: Basic Books.

Cummings, J. (1983, November 20). Breakup of black family imperils gains of decades. *New York Times,* p. 56.

Curtin, P. D. (1969). *The Atlantic slave trade.* Madison: University of Wisconsin Press.

Curtiss, S. (1977). *Genie: A psycholinguistic study of a modern-day "wild child."* New York: Academic Press.

Dahl, R. (1961). *Who governs?* New Haven, CT: Yale University Press.

Dahrendorf, R. (1990). *The modern social conflict: An essay on the politics of liberty.* Berkeley: University of California Press.

Dahrendorf, R. (1997). *After 1989: Morale, revolution, and civil society.* New York: St. Martin's Press.

Daniel, K. (1996). The marriage premium. In Tommasi, M., & Tommasi, K. (Eds.), *The new economics of human behavior.* Cambridge, England: Cambridge University Press.

Dankert, C. E., Mann, F. C., & Northrup, H. R. (Eds.) (1965). *Hours of work.* New York: Harper.

Danziger, S., & Gottschalk, P. (Eds.) (1993). *Uneven tides: Rising inequality in America.* New York: Russell Sage.

Darling-Hammond, L. (1997). What matters most: 21st-century teaching. *Education Digest,* November, pp. 4–10.

Dasgupta, P. (2000). Population and resources: An exploration of reproductive and environmental externalities. *Population and Development Review, 26,* 643–690.

Dasgupta, S. (Ed.) (2004). *The changing face of globalization.* Thousand Oaks, CA: Sage.

Davidson, D. (1973). The furious passage of the black graduate student. In J. Ladner (Ed.), *The death of white sociology.* New York: Vintage.

Davies, J. C. (1962). Toward a theory of revolution. *American Sociological Review, 27,* 5–19.

Davis, K. (1937). The sociology of prostitution. *American Sociological Review, 2,* 746–755.

Davis, K. (1939). Illegitimacy and the social structures. *American Journal of Sociology, 45,* 215–233.

Davis, K. (1955). The origin and growth of urbanization in the world. *American Journal of Sociology, 60,* 429–437.

Davis, K. (1965). *Human society.* New York: Macmillan.

Davis, K. (1968). The urbanization of the human population. In S. F. Fava (Ed.), *Urbanism in world perspective: A reader.* New York: Crowell.

Davis, K. (1991). Population and resources: Fact and interpretation. In K. Davis and M. S. Bernstam (Eds.), *Resources, environment, population: Present knowledge, future options.* New York: Oxford University Press.

Davis, K., & Moore, W. E. (1945). Some principles of stratification. *American Sociological Review, 10,* 242–249.

Davis, K. C. (1995, September 3). Ethnic cleansing didn't start in Bosnia. *New York Times,* sec. 4, pp. 1, 6.

Davis, L. (1992). *Natural disasters: The black plague.* New York: Facts on File.

Davis, M. (1998). *Ecology of fear: Los Angeles and the imagination of disaster.* New York: Metropolitan Books.

Davis, N. J. (1975). *Deviance: Perspectives and issues in the field.* Dubuque, IA: Brown.

Davis, T., Smith, W., & Marsden, P. V. (2007). *General social surveys, 1972–2006: Cumulative codebook.* Chicago: NORC.

Dawson, T. L. (2002). New tools, new insights: Kohlberg's moral judgment stages revisited. *International Journal of Behavioral Development, 26,* 154–167.

Deaux, K., & Wrightsman, L. S. (1988). *Social psychology,* 5th ed. Pacific Grove, CA: Brooks/Cole.

DeCoker, G. (2010). Beyond the rhetoric of national standards. *Education Week,* May 19, p. 31.

Defina, M. P., & Wetherbee, L. (1997, October). Advocacy and law enforcement: Partners against domestic violence. *FBI Law Enforcement Bulletin,* pp. 22–26.

Deierlein, K. (1994). Ideology and holistic alternatives. In W. Kornblum & C. D. Smith (Eds.), *The healing experience: Readings on the social context of health care.* Englewood Cliffs, NJ: Prentice Hall.

De Jong, G. F., & Tran, Q. G. (2001). Warm welcome, cool welcome: Mapping receptivity toward immigrants in the U.S. *Population Today, 29,* 1, 4–5.

Denis, J., & Derbyshire, I. (1996). *Political systems of the world.* New York: St. Martin's Press.

Denno, D. W. (1990). *Biology and violence: From birth to adulthood.* New York: Cambridge University Press.

de Rougemont, D. (1983). *Love in the western world.* Princeton, NJ: Princeton University Press.

Deshpande, A. (2000). Recasting economic inequality. *Review of Social Economy, 58,* 381–399.

de Tocqueville, A. (1955/1856). *The old regime and the French revolution* (S. Gilbert, Trans.). Garden City, NY: Doubleday.

de Tocqueville, A. (1956/1840). *Democracy in America.* New York: Vintage.

de Tocqueville, A. (1980/1835). *On democracy, revolution, and society* (J. Stone & S. Mennell, Trans.). Chicago: University of Chicago Press.

De Visscher, S., & Bouverne-De Bie, M. (2008). Children's presence in the neighborhood: A social-pedagogical perspective. *Children and Society, 22:* 470–481.

DeVoe, J. F., Peter, K., Noonan, M., Snyder, T. D., & Baum, K. (2005). *Indicators of school crime and safety: 2005.* Washington, DC: U.S. Department of Justice, Bureau of Justice Statistics.

Dewey, R. (1948). Charles Horton Cooley: Pioneer in psychosociology. In H. F. Barnes (Ed.), *An introduction to the history of sociology.* Chicago: University of Chicago Press.

De Witt, K. (1994). Wave of suburban growth is being fed by minorities. *New York Times,* August 15, pp. A1, B6.

De Zwart, F. (2000). The logic of affirmative action: Caste, class and quotas in India. *Acta Sociologica, 43,* 235.

Diaz Soto, L. (1997). *Language, culture, and power: Bilingual families and the struggle for quality education.* Albany: State University of New York Press.

Diaz Soto, L., & Malewski, E. (2001, Spring). A light shines in steel town: A response to He, Phillion, and Roberge's review of *Language, Culture, Power: Bilingual Families and the Struggle for Quality Education. Curriculum Inquiry,* pp. 51–57.

Djilas, M. (1982). *The new class: An analysis of the communist system.* Orlando: Harcourt Brace Jovanovich.

Dobbin, F. (Ed.) (2004). *The sociology of the economy.* New York: Russell Sage.

Dobbins, J., McGinn, J. G., Crane, K., Jones, S. G., Lal, R., Rathmell, A., Swanger, R., & Timilsina, A. (2003). *America's role in nation-building: From Germany to Iraq.* Santa Monica, CA: RAND.

Dodgson, R. A. (1987). *The European past: Social evolution and spatial order.* London: Macmillan.

Dollard, J. (1937). *Caste and class in a southern town.* New Haven, CT: Yale University Press.

Dollard, J., Miller, N., & Doob, L. (1939). *Frustration and aggression.* New Haven, CT: Yale University Press.

Domhoff, G. W. (1983). *Who rules America now?* New York: Simon & Schuster.

Domhoff, G. W. (2010). *Who rules America? Challenges to corporate and class dominance,* 6th ed. New York: McGraw-Hill.

Dornbusch, S. (1955). The military as an assimilating institution. *Social Forces, 33,* 316–321.

Dowd, M. (1985, November 17). Youth, art, hype: A different bohemia. *New York Times Magazine,* pp. 26ff.

Drake, S. C., & Cayton, H. (1970/1945). *Black metropolis: Vol. 1. A study of Negro life in a northern city,* rev. ed. Orlando: Harcourt Brace Jovanovich.

Dreifus, C. (1998, April 14). She talks to apes, and, according to her, they talk back. *New York Times,* p. F4.

Drucker, P. F. (1992). The new society of organizations. *Harvard Business Review, 70,* 95–105.

Du Bois, W. E. B. (1967/1899). *The Philadelphia Negro: A social study.* New York: Schocken.

Dudley, W. (Ed.) (2002). *Religion in America: Opposing viewpoints.* San Diego: Greenhaven Press.

Duffy, M., Dickerson, J. F., Thomas, C. B., Tumulty, K., & Weisskopf, M. (2002). Enron spoils the party. *Time,* February 4, pp. 16–23.

Dugger, C. W. (2001). Relying on hard and soft sells, India pushes sterilization. *New York Times,* June 22, p. A1.

Duncan, G. J., & Brooks-Gunn, J. (Eds.) (1997). *Consequences of growing up poor.* New York: Russell Sage.

Duneier, M. (1992). *Slim's table: Race, respectability, and masculinity.* Chicago: University of Chicago Press.

Dunn, C. W. (Ed.) (2009). *The future of religion in American politics.* Lexington: University Press of Kentucky.

Durkheim, E. (1964/1893). *The division of labor in society,* 2nd ed. New York: Free Press.

Durkheim, E. (1951/1897). *Suicide, a study in sociology.* New York: Free Press.

Duster, T. (2001, September 14). Buried alive: The concept of race in science. *Chronicle of Higher Education,* pp. 1–5.

Dworkin, S. H., & O'Sullivan, L. (2005). Actual versus desired initiation patterns among a sample of college men: Tapping disjunctures within traditional male sexual scripts. *Journal of Sex Research, 42,* 150–158.

Dyer, J. (1996, November-December). Ground zero. *Utne Reader,* pp. 80ff.

Dymski, G. A. (2000). Illegal-seizure and market-disadvantage approaches to restitution: A comparison of the Japanese American and African American cases. *Review of Black Political Economy, 27,* 49–80.

Ebbinghaus, B., & Visser, J. (2000). *Trade unions in Western Europe since 1945.* London: Macmillan.

Eckholm, E. (1985, June 25). Kanzi the chimp: A life in science. *New York Times,* pp. C1, C3.

Eckholm, E. (1992, October 6). Learning if infants are hurt when mothers go to work. *New York Times,* pp. A1, A21.

Edleson, J. I., & Tan, N. T. (1993). Conflict and family violence: The tale of two families. In P. G. Boss, W. J. Doherty, R. LaRossa, W. R. Schumm, & S. K. Steinmetz (Eds.), *Sourcebook of family theories and methods: A contextual approach.* New York: Plenum.

EEOC (U.S. Equal Employment Opportunity Commission) (1993). *Guidelines on discrimination because of sex.* Washington, DC: U.S. Government Printing Office.

Egan, T. (1995). Many seek security in private communities. *New York Times,* September 3, pp. A1, A22.

Ehrlich, P. R., & Ehrlich, A. H. (2004). *One with Nineveh: Politics, consumption, and the human future.* Washington, DC: Island Press.

Eibl-Eibesfeldt, I. (1989). *Human ethology.* Hawthorne, NY: Aldine.

Eisley, L. (1961). *Darwin's century.* Garden City, NY: Doubleday Anchor.

Eisley, L. (1970). *The invisible pyramid.* New York: Scribner.

Elder, G. H. (1981). History and the life course. In D. Berteaux (Ed.), *Biography and society: The life history approach to the social sciences.* Newbury Park, CA: Sage.

Eldridge, C. C. (1978). *Victorian imperialism.* Atlantic Highlands, NJ: Humanities Press.

Elias, N. (1978/1939). The civilizing process. In N. Elias (Ed.), *The development of manners.* New York: Urizen.

Elkin, F., & Handel, G. (1989). *The child and society: The process of socialization,* 5th ed. New York: Random House.

Ellul, J. (1964). *The technological society.* New York: Vintage.

Ellwood, D. (1988). *Poor support: Poverty in the American family.* New York: Basic Books.

Elshtain, J. B. (1997, May 5). Heaven can wait. *New Republic,* p. 23.

Elster, J. (1997). Accounting for tastes. *University of Chicago Law Review, 64,* 749–764.

Elster, J. (2000). *Ulysses unbound: Studies in rationality, precommitment, and constraints.* New York: Cambridge University Press.

Engeln-Maddox, R. (2006). Buying a beauty standard or dreaming of a new life? Expectations associated with media ideals. *Psychology of Women Quarterly, 30,* 258–266.

Enloe, C. (2000). The surprised feminist. *Signs, 25,* 1023.

Epstein, C. (1995, Fall). Affirmative action. *Dissent,* pp. 463–465.

Epstein, C. F. (1985). Ideal roles and real roles. *Research in Social Stratification and Mobility, 4,* 29–51.

Epstein, C. F. (1988). *Deceptive distinctions: Sex, gender, and the social order.* New Haven, CT: Yale University Press.

Erikson, E. (1963). *Childhood and society.* New York: Norton.

Erikson, E. (Ed.) (1965). *The challenge of youth.* Garden City, NY: Doubleday Anchor.

Erikson, K. T. (1962). Notes on the sociology of deviance. *Social Problems, 9,* 307–314.

Erikson, K. T. (1966). *Wayward puritans: A study in the sociology of deviance.* New York: Wiley.

Erikson, K. T. (1994a). *A new species of trouble.* New York: Norton.

Eshleman, J. R. (2003). *The family,* 10th ed. Boston: Allyn & Bacon.

Espin, O. O. (Ed.) (2009). *Building bridges, doing justice: Constructing a Latino/a ecumenical theology.* Maryknoll, NY: Orbis.

Esposito, J. L. (2003). *Religion and immigration: Christian, Jewish, and Muslim experiences in the United States.* AltaMira Press.

Ewald, W. (1996). *I dreamed I had a girl in my pocket.* New York: Norton.

Exum, K. J., & Messina, L. M. (Eds.) (2004). *The car and its future.* New York: H. W. Wilson.

Ezorsky, G. (1991). *Affirmative action.* Ithaca, NY: Cornell University Press.

Faison, S. (1997). Chinese happily break the "one child" rule. *New York Times,* August 17, pp. 1, 10.

Fallows, D. (1985). *A mother's work.* Boston: Houghton Mifflin.

Falsafi, S. (2010). Civil society and democracy in Japan, Iran, Iraq and beyond. *Vanderbilt Journal of Transnational Law, 43,* 357–435.

Farley, R. (1996). *The new American reality.* New York: Russell Sage.

Farley, R., & Allen, W. R. (1987). *The color line and the quality of life in America.* New York: Russell Sage.

Farrington, K., & Chertok, E. (1993). Social conflict theories of the family. In P. G. Boss, W. J. Doherty, R. LaRossa, W. R. Schumm, & S. K. Steinmetz (Eds.), *Sourcebook of family theories and methods: A contextual approach.* New York: Plenum.

Fattah, H. M. (2006, June 26). First time out, Kuwaiti women become a political force. *New York Times,* p. A3.

Fava, S. (1956). Suburbanism as a way of life. *American Sociological Review, 21,* 34–38.

Fava, S. (1985). Residential preferences in the suburban era: A new look? *Sociological Focus, 18,* 109–117.

FBI (Federal Bureau of Investigation) Annual. *Crime in the United States* (Uniform Crime Reports). Washington, DC: U.S. Department of Justice.

Feagin, J. R. (1991, November 27). Blacks still face the malevolent reality of white racism. *Chronicle of Higher Education,* p. A44.

Feiler, A., & Webster, A. (1999). Teacher predictions of young children's literacy success or failure. *Assessment in education: Principles, policy, and practice, 6,* 341–355.

Feldman, N. (2004). *What we owe Iraq: War and the ethics of nation building.* Princeton, NJ: Princeton University Press.

Fenton, C., Brooks, F., Spencer, N. H., & Morgan, A. (2010). Sustaining a positive body image in adolescence: An assets-based analysis. *Health and Social Care in the Community, 18,* 189–198.

Fernandes, F. (1968). The weight of the past. In J. H. Franklin (Ed.), *Color and race.* Boston: Beacon.

Feuer, M. J. (2009). Danger: Bell curve ahead. *Issues in Science and Technology, 25,* 91–93.

Fine, G. A. (1987). *With the boys: Little League baseball and preadolescent culture.* Chicago: University of Chicago Press.

Fine, M. (2002). Civil lessons. UC Los Angeles: UCLA's Institute for Democracy, Education, and Access. http://escholarship.org/uc/item/2ct0p7rw.

Finke, R., & Stark, R. (1992). *The churching of America 1776–1992.* New Brunswick, NJ: Rutgers University Press.

Finke, R., & Stark, R. (2005). *The churching of America, 1776–2005: Winners and losers in our religious economy.* New Brunswick, NJ: Rutgers University Press.

Fischer, C. S. (1987/1976). *The urban experience.* Orlando: Harcourt Brace Jovanovich.

Fischer, C. S. (1982). *To dwell among friends: Personal networks in town and city.* Chicago: University of Chicago Press.

Fischer, C. S., & Hout, M. (2007). *Century of difference: How America changed in the last one hundred years.* New York: Russell Sage.

Fischer, P., Kastenmüller, A, & Greitemeyer, T. (2010). Media violence and the self: The impact of personalized gaming characters in aggressive video games on aggressive behavior. *Journal of Experimental Social Psychology, 46,* 192–195.

Fishman, R. (1987). *Bourgeois utopias: The rise and fall of suburbia.* New York: Basic Books.

Fiske, M., & Chiriboga, D. A. (1990). *Change and continuity in adult life.* San Francisco: Jossey-Bass.

Fitzgerald, J. (2006). Labor unions and job advancement in the service sector economy. *Law and Society,* 2006 Annual Meeting, p. 1.

Fitzgerald, L., Swan, S., & Magley, V. J. (1997). But was it really sexual harassment? Legal, behavioral, and psychological definitions of the workplace victimization of women. In W. O'Donohue (Ed.), *Sexual harassment: Theory, research, and therapy.* Boston: Allyn & Bacon.

Fivaz-DePeursinge, E., Lopes, F., Python, M., & Favez, N. (2009). Coparenting and toddlers' interactive styles in family coalitions. *Family Process, 4,* 500–516.

Fleisher, M. S. (2005). Fieldwork research and social network analysis: Different methods creating complementary perspectives. *Journal of Contemporary Criminal Justice, 21:* 120–134.

Flink, J. J. (1975). *The car culture.* Cambridge, MA: MIT Press.

Flink, J. J. (1988). *The automobile age.* Cambridge, MA: MIT Press.

Flores, J. (2009). The diaspora strikes back: Caribeño tales of learning and turning. New York: Routledge.

Food and Agriculture Organization (2004). *The state of world fisheries and aquaculture.* www.fao.org/docrep/007/y5600e/y5600e04.htm.

Foreign Policy (2005, July–August). *The failed states index.*

Foreign Policy. (2010, June 21). Failed states index—interactive map and rankings. www.foreignpolicy.com/articles.

Foster, L., Martin, J. W., Chidester, D., & Ammerman, N. T. (1998). Forum: Interpreting Waco. *Religion and American Culture, 8,* 1–30.

Foucault, M. (1973). *Madness and civilization: A history of insanity in the Age of Reason.* New York: Vintage.

Foucault, M. (1977). *Discipline and punishment.* London: Allen Lane.

Foucault, M. (1984). *The Foucault reader*. New York: Pantheon.

Foucault, M. (2001). Power, (ed. J. D. Faubion, trans. R. Hurley et al.). London: Allen Lane.

Frank, A. G. (1966). The development of underdevelopment. *Monthly Review, 18*, 3–17.

Frank, R. H. (1988). *Passions within reason: The strategic role of the emotions*. New York: Norton.

Frazier, E. F. (1957/1939). *The Negro family in the United States*, rev. ed. New York: Macmillan.

Frazier, E. F. (1966). *The Negro church in America*. New York: Schocken.

Freeman, J. (1973). The origins of the women's liberation movement. In J. Huber (Ed.), *Changing women in a changing society*. Chicago: University of Chicago Press.

Freeman, R. (1994). *Working under different rules*. New York: Russell Sage.

Freeman, R. B. (2000). *What workers want*. Ithica, NY: ILR Press.

Freeman, R. B., & Rogers, J. (2002, posted June 24). A proposal to American labor. *Nation*.

French, H. R. (2003, July 25). Japan's neglected resource: Female workers. *New York Times*, p. A3.

Freud, S. (1905/1991). *Three essays on the theory of sexuality*. New York: Penguin.

Fried, M. (1963). Grieving for a lost home. In L. J. Duhl (Ed.), *The urban condition*. New York: Basic Books.

Friedan, B. (1963). *The feminine mystique*. New York: Dell.

Friedman, M. (1962). *Capitalism and freedom*. Chicago: University of Chicago Press.

Friend, T. (2000, September–October). Chimp culture. *International Wildlife*, pp. 28–35.

Fromm, E. (1961). *Marx's concept of man*. New York: Ungar.

Fukuyama, F. (Ed.) (2004). *Nation-building: Beyond Afghanistan and Iraq*. Baltimore, MD: Johns Hopkins University Press.

Fullan, M. (1993). *Change forces: Probing the depths of educational reform*. New York: Falmer.

Fuller, B., & Elmore, R. F. (1996). *Who chooses? Who loses? Culture, institutions, and the unequal effects of school choice*. New York: Teachers College Press.

Furedi, F. (1997). *Population and development: A critical introduction*. New York: St. Martin's Press.

Furstenberg, F. E. (2000). Teenagers; adolescence; adulthood. *Journal of Marriage and the Family, 62*, 896–1,000.

Furstenberg, F. E., & Cherlin, A. J. (1991). *Divided families: What happens to children when parents part*. Cambridge, MA: Harvard University Press.

Fyfe, C. (1976). The dynamics of African dispersal. In M. L. Kilson & R. I. Rothberg (Eds.), *The African diaspora*. Cambridge, MA: Harvard University Press.

Gabriel, T. (1995, June 12). A new generation seems ready to give bisexuality a place in the spectrum, *New York Times*, p. A12.

Gagnon, J. H. (2004). *An interpretation of desire: Essays in the study of sexuality*. Chicago: University of Chicago Press.

Gaines, D. (1998). *Teenage wasteland*, 2nd ed. Chicago: University of Chicago Press.

Galbraith, J. K. (1995). Blame history, not the liberals. *New York Times*, September 19, p. A21.

Gale, E. A. M. (2003). Is there really an epidemic of type 2 diabetes? *Lancet, 362*, 503–504.

Gallup Organization. (1997, March). *The Gallup Poll Monthly*, pp. 8–13.

Gallup Organization. (2008). www.gallup.com/poll/105802/Economic-Anxiety-Surges-Past-Year.aspx.

Galston, W. A. (2008). How big government got its groove back. *American Prospect*, June, pp. 23–26.

Galtung, J. (1985). War. In A. Kuper & J. Kuper (Eds.), *The social science encyclopedia*. London: Routledge and Kegan Paul.

Gamson, W. A. (1992). *Talking politics*. Cambridge, England: Cambridge University Press.

Ganong, L. H., & Coleman, M. (1999). Family obligations following divorce and remarriage. Mahwah, NJ: Erlbaum.

Gans, H. (1984/1962). *The urban villagers*. New York: Free Press.

Gans, H. (1976/1967). *The Levittowners: Way of life and politics in a new suburban community*. New York: Pantheon.

Gans, H. (1967). *The Levittowners*. New York: Knopf.

Gans, H. (1985). The uses of poverty: The poor pay all. In W. Feigelman (Ed.), *Sociology: Full circle*, 4th ed. Fort Worth: Holt, Rinehart and Winston.

Gans, H. (1995). *The war against the poor*. New York: Basic Books.

Garcia, A. (1996). Moral reasoning in interactional context: Strategic uses of care and justice arguments in mediation hearings. *Sociological Inquiry, 66*, 197–214.

Gardner, H. (1983). *Frames of mind: The theory of multiple intelligences*. New York: Basic Books.

Gardner, H. (2002). Performance standards. *Organizational Behavior, 15*, 88–90.

Garfinkel, H. (1991). Respecification. In G. Button (Ed.), *Ethnomethodology and the human sciences*. New York: Cambridge University Press.

Garfinkel, H. (1967). *Studies in ethnomethodology*. Englewood Cliffs, NJ: Prentice Hall.

Garfinkel, H. (2000). Studies in ethnomethodology. In W. Sharrock & M. J. Lynch (Eds.), *Harold Garfinkel*, 3rd ed. Newbury Park, CA: Sage.

Garfinkel, I., & McLanahan, S. S. (1986). *Single mothers and their children: A new American dilemma*. Washington, DC: Urban Institute.

Garmezy, N. (1993). Children in poverty: Resilience despite risk. *Psychiatry, 56*, 127–136.

Garner, R., & Tenuto, J. (1997). *Social movement theory and research: An annotated bibliographical guide*. Lanham, MD: Scarecrow Press.

Garreau, J. (1991). *Edge city: Life on the new frontier*. Garden City, NY: Doubleday Anchor.

Garrett, M. (1997). The effects of infant child care on infant-mother attachment security: Results of the NICHD Study of Early Child Care. *Child Development, 68*, 860–879.

Gbadamosi, O., & Adewoye, O. I. (2010). The rule of law as a catalyst for sustainable democracy in Nigeria. *Commonwealth Law Bulletin, 36*, 343–355.

Geertz, C. (1973). The growth of culture and the evolution of mind. In C. Geertz (Ed.), *The interpretation of culture*. New York: Basic Books.

Gelbard, A., Haub, C. & Kent, M. M. (1999). World population beyond six billion. *Population Bulletin, 54*, No. 1.

Genevie, L. E. (Ed.) (1978). *Collective behavior and social movements*. Itasca, IL: Peacock.

Geoghegan, T. (2007, September). What worker rights can do. *American Prospect*, p. 44.

Geoghegan, T. (1991). *Which side are you on?* New York: Farrar Straus Giroux.

Gerbner, G. (1990). *Violence profile*. Philadelphia: Annenberg School of Communications.

Gerbner, G. (2001). *Television and New Media, 2*, 35.

Gerson, K. (2009). Changing lives, resistant institutions: A new generation negotiates gender, work, and family change. *Sociological Forum, 24*, 735–753.

Gerth, H., & Mills, C. W. (1958). *From Max Weber: Essays in sociology*. New York: Oxford University Press.

Geschwender, J. A. (1977). *Class, race and worker insurgency*. New York: Cambridge University Press.

Gibbs, J. P. (1994). *A theory about control*. Boulder, CO: Westview.

Gilder, G. (1982). *Wealth and poverty*. New York: Basic Books.

Gilder, G. (2006). Picking up the ax. *New Leader*, January–April, pp. 32–34.

Gilligan, C. (1982). *In a different voice: Psychological theory and women's development*. Cambridge, MA: Harvard University Press.

Gilligan, C. (1987). Adolescent development reconsidered. In C. E. Irwin, Jr. (Ed.), *New directions for child development: No. 37. Adolescent social behavior and health*. San Francisco: Jossey-Bass.

Gilligan, C., Ward, J. V., Taylor, J. M., & Bardige, B. (Eds.) (1998). *Mapping the moral domain*. Cambridge, MA: Harvard University Press.

Gioioso, R. N. (2010). Life in the megalopolis: Mexico City and São Paulo. *Journal of Regional Science, 50*, 661–663.

Gist, N. P. (1968). Urbanism in India. In S. F. Fava (Ed.), *Urbanism in world perspective: A reader*. New York: Crowell.

Gitlin, T. (1993). *The sixties*. New York: Bantam.

Giuffre, K. (2009). Graffiti lives: Beyond the tag in New York's urban underground. *Contemporary Sociology, 38*: 527–552.

Glazer, N. (1975). *Affirmative discrimination: Ethnic inequality and public policy*. New York: Basic Books.

Glazer, N., & Moynihan, D. P. (1970). *Beyond the melting pot: The Negroes, Puerto Ricans, Jews, Italians, and Irish of New York City*, 2nd rev. ed. Cambridge, MA: MIT Press.

Glick, P. C., & Parke, R., Jr. (1965). New approaches in studying the life cycle of the family. *Demography, 2*, 187–202.

Glueck, S., & Glueck, E. T. (1950). *Unraveling juvenile delinquency*. New York: Commonwealth Fund.

Goffman, E. (1958). Deference and demeanor. *American Anthropologist, 58*, 488–489.

Goffman, E. (1961). *Asylums*. Garden City, NY: Doubleday.

Goffman, E. (1963). *Stigma: Notes on the management of spoiled identity*. Englewood Cliffs, NJ: Prentice Hall.

Goffman, E. (1965). *Interaction ritual: Essays on face-to-face behavior*. Garden City, NY: Doubleday.

Goffman, E. (1972). Territories of the self. In E. Goffman (Ed.), *Relations in public*. New York: Harper.

Goldberg, J. (2010, May). What kind of socialist is Barack Obama? *Commentary*.

Golden, L. (2010). A purpose for every time? The timing and length of the work week and implications for worker well-being. *Connecticut Law Review, 42*, 1181–1201.

Goldhamer, H., & Shils, E. (1939). Types of power and status. *American Journal of Sociology, 45,* 171–182.

Goldstein, Dana. (2009). Pink-collar blues. *American Prospect, 20* (5), 19–21.

Goleman, D. (1985, March 19). Dislike of own body found common among women. *New York Times*, pp. C1, C5.

Gonzalez-Mena, J. (1998). *The child in the family and the community.* Upper Saddle River, NJ: Merrill/Prentice Hall.

Goodall, J. (1994). Digging up the roots: Relationships with the Gombe chimpanzees. *Orion, 13,* 21–21.

Goodall, J. V. L. (1968). A preliminary report on expressive movements and communications in Gombe Stream chimpanzees. In P. Jay (Ed.), *Primates: Studies in adaptation and variability.* New York: Holt, Rinehart and Winston.

Goode, Eric (1994). *Deviant behavior,* 4th ed. Englewood Cliffs, NJ: Prentice Hall.

Goode, Erica (2001, July 17). A rainbow of differences in gays' children. *New York Times,* p. 1.

Goode, W. J. (1964). *The family.* Englewood Cliffs, NJ: Prentice Hall.

Goodman-Draper, J. (1995). *Health care's forgotten majority: Nurses and their frayed white collars.* Westport, CT: Auburn House.

Goodnough, A. (2002). More applicants answer the call for teaching jobs. *New York Times,* February 11, p. A1.

Goodstein, L. (2003, July 19). Homosexuality issue threatens to break Anglicanism in two. *New York Times,* p. A1.

Gordon, M. (1964). *Assimilation in American life.* New York: Oxford University Press.

Gottmann, J. (1978). Megalopolitan systems around the world. In L. S. Bourne & J. W. Simmons (Eds.), *Systems of cities: Readings on structure, growth, and policy.* New York: Oxford University Press.

Gould, S. J. (1981). *The mismeasure of man.* New York: Norton.

Gould, S. J. (1995, February). Ghosts of bell curves past. *Natural History,* pp. 12–19.

Gould, S. J. (1996). *Full house: The spread of excellence from Plato to Darwin.* New York: Harmony Books.

Gramling, R., & Krogman, N. (1997). Communities, policy and chronic technological disasters. *Current Sociology, 45,* 41–57.

Gramsci, A. (1971). *Selections from prison notebooks.* London: Routledge and Kegan Paul.

Gramsci, A. (1992/1965). *Prison notebooks.* New York: Columbia University Press.

Gramsci, A. (1995). *Further selections from the prison notebooks.* Minneapolis: University of Minnesota Press.

Granovetter, M. (1995/1974). *Getting a job: A study of contacts and careers.* Chicago: University of Chicago Press.

Grasmuck, S. (2005). *Protecting home: Class, race, and masculinity in boys' baseball.* New Brunswick, NJ: Rutgers University Press.

Graves, J. L., Jr. (2001). *The emperor's new clothes: Biological theories of race at the millennium.* New Brunswick, NJ: Rutgers University Press.

Gray, J. (1992). *Men are from Mars, women are from Venus.* New York: HarperCollins.

Greeley, A. M. (2004). *Priests: A calling in crisis.* Chicago: University of Chicago Press.

Greeley, A., Michael, R. T., & Smith, T. (1990). Americans and their sexual partners. *Society, 27,* 36–42.

Greenhouse, L. (2003, June 27). Justices, 6–3 legalize gay sexual conduct. *New York Times,* pp. 1, 19.

Greenhouse, S. (2003, July 20). Hiring for looks can get legally ugly for stores. *New York Times.*

Greenspan, S. I. (2004). *The first idea: How symbols, language, and intelligence evolved from our primate ancestors to modern humans.* Cambridge, MA: Da Capo Press.

Greer, G. (2009). To have or not to have: The critical importance of reproductive rights to the paradox of population policies in the 21st century. *International Journal of Gynecology and Obstetrics,* 106, 148–150.

Grodin, D., & Lindlof, T. R. (Eds.) (1996). *Constructing the self in a mediated world.* Thousand Oaks, CA: Sage.

Groesz, L. M., Levine, M. P., & Murnen, S. K. (2002). The effect of experimental presentation of thin media images on body satisfaction: A meta-analytic review. *International Journal of Eating Disorders, 31,* 1–16.

Gross, J. (1992, March 29). Collapse of inner-city families creates America's new orphans. *New York Times,* pp. 1, 20.

Gugler, J. (Ed.) (1997). *Cities in the developing world: Issues, theory, and policy.* New York: Oxford University Press.

Gurr, T. R. (2002). Attaining peace in divided societies: Five principles of emerging doctrine. *International Journal on World Peace, 19,* 27–51.

Gursky, D. (1998). Class size does matter. *Education Digest, 64,* 15–19.

Gusfield, J. (1966). *Symbolic crusade: Status politics and the American temperance movement.* Urbana: University of Illinois Press.

Gusfield, J. (1981). *The culture of public problems: Drinking, driving and the symbolic order.* Chicago: University of Chicago Press.

Gusfield, J. (1996). *Contested meanings: The construction of alcohol problems.* Madison: University of Wisconsin Press.

Gushulak, B., Weekers, J., & MacPherson, D. W. (2009). Migrants in a globalized world—health threats, risks and challenges: An evidence-based framework. *Emerging Health Threats, 2,* 42.

Gutman, H. (1976). *The black family in slavery.* New York: Pantheon.

Hacker, A. (1995, Fall). Affirmative action. *Dissent,* pp. 465–467.

Hacker, A. (1997). *Money: Who has how much and why.* New York: Simon & Schuster.

Hackett, D. G. (Ed.) (1995). *Religion and American culture: A reader.* New York: Routledge.

Hadaway, C. K., & Marler, P. L. (1998, May 6). Did you really go to church this week? Behind the poll data. *The Christian Century,* pp. 472–476.

Hagestad, G. O., & Neugarten, B. L. (1985). Age and the life course. In R. H. Binstock & E. Shanas (Eds.), *Handbook of aging and the social sciences* (2nd ed.). New York: Van Nostrand Reinhold.

Halbwachs, M. (1960). *Population and society: Introduction to social morphology.* Glencoe, IL: Free Press.

Hale, L. (2010). Respecting autonomy in population policy: An argument for international family planning programs. *Public Health Ethics, 3,* 157–166.

Halevy, E. (1955). *The growth of philosophical radicalism* (trans. M. Morris). Boston: Beacon.

Hall, H. I., Ruiguang, S., Rhodes, P., et al. (2008). Estimation of HIV incidence in the United States. *JAMA,* 300, 520–529.

Hall, J. R. (1987). *Gone from the promised land: Jonestown in American cultural history.* New Brunswick, NJ: Transaction.

Halle, D. (1984). *America's working man: Work, home and politics among blue-collar property owners.* Chicago: University of Chicago Press.

Halsey, M. (1946). *Color blind.* New York: Simon & Schuster.

Halstead, T. (2003, January). The American paradox. *Atlantic Monthly.*

Hamilton, C. V. (1969). The politics of race relations. In C. U. Daley (Ed.), *The minority report.* New York: Pantheon.

Hammond, J. (1998). *Fighting to learn: Popular education and guerilla war in El Salvador.* New Brunswick, NJ: Rutgers University Press.

Handel, G. (2001). Socialization. In L. Balter (Ed.), *Parenthood in America: An encyclopedia.* Santa Barbara, CA: ABC-CLIO.

Handlin, O. (1992). The newcomers. In P. S. Rotherberg (Ed.), *Race, class and gender in the United States.* New York: St. Martin's Press.

Hannah, L. (2006). Leviathans: Multinational corporations and the new global history. *Business History Review, 80,* 194–195.

Harbison, F. H. (1973). *Human resources as the wealth of nations.* New York: Oxford University Press.

Harden, B. (2001, August 12). 2-parent families rise after change in welfare laws. *New York Times,* p. 1.

Harding, S. (1998). Women, science, and society. *Science, 281,* 1.

Hardt, M., & Negri, A. (2001). *Empire.* Cambridge, MA: Harvard University Press.

Hare, A. P. (1992). *Groups, teams, and social interaction.* Westport, CT: Praeger.

Hare, A. P., Blumberg, H. H., Davis, M. F., & Kent, V. (1994). *Small group research: A handbook.* Norwood, NJ: Ablex.

Harlow, H. F. (1986). *From learning to love: The selected papers of H. F. Harlow.* New York: Praeger.

Harlow, H. F. (2008). The monkey as a psychological subject. *Integrative psychological and behavioral science,* 42: 336–347.

Harper, C. L. (1993). *Exploring social change,* 2nd ed. Englewood Cliffs, NJ: Prentice Hall.

Harper, D. A. (2006). *Good company.* Chicago: University of Chicago Press.

Harrington, M. (1973). *Socialism.* New York: Bantam.

Harrington, M. (1987). *The next life: The history of a future.* New York: Holt.

Harris, M. (1980). *Culture, people, nature: An introduction to general anthropology.* New York: Harper.

Harrison, B. B., Tilly, C., and Bluestone, B. (1986, March–April). Wage inequality takes a great U-turn. *Challenge,* pp. 26–32.

Hartmann, H. (2009). Women, the recession, and the stimulus package. *Dissent,* pp. 42–47.

Hassan, F. (2001, May 31-June 6). *Al-Ahram Weekly Online.* http://171.66.122.53/cgi/content/full/290/5489/55.

Hasws, K. A. (1998). *Carried to the wall: American memory and the Vietnam Veterans Memorial.* Berkeley, University of California Press.

Haugen, D. M., & Box, M. J. (Eds.) (2006). *Homosexuality.* San Diego, CA: Greenhaven Press.

Hauser, P. M. (1957). The changing population pattern of the modern city. In P. K. Hatt & A. J. Reiss (Eds.), *Cities and society.* New York: Free Press.

Hawkins, G. (1976). *The prison: Policy and practice.* Chicago: University of Chicago Press.

Hazler, R. J. (2000). When victims turn aggressors: Factors in the development of deadly school violence. *Professional School Counseling, 4,* 105–112.

Hearn, J. (2010). Reflecting on men and social policy: Contemporary critical debates and implications for social policy. *Critical Social Policy, 30,* 165–188.

Hechter, M. (1974). *Internal colonialism.* Berkeley: University of California Press.

Hechter, M. (1987). *Principles of group solidarity.* Berkeley: University of California Press.

Hechter, M. (2000). *Containing nationalism.* New York: Oxford University Press.

Hechter, M., & Kanazawa, S. (1996, March 29). Sociological rational choice theory. In Elster, J. (Ed.), Doing our level best (rational-choice theory—a symposium). *Times Literary Supplement,* pp. 12–14.

Heilbroner, R. L. (1995). Putting economics in its place. *Social Research, 62,* 883–898.

Heitzeg, N. (1996). *Deviance: Rulemakers and rulebreakers.* Minneapolis: West.

Helmreich, W. B. (1982). *The things they say behind your back.* Garden City, NY: Doubleday.

Henry, D. O. (1989). *From foraging to agriculture.* Philadelphia: University of Pennsylvania Press.

Henslin, J., & Briggs, M. (1971). Dramaturgical desexualization: The sociology of the vaginal examination. In J. Henslin (Ed.), *Studies in the sociology of sex.* New York: Appleton-Century-Crofts.

Herek, G. M. (2002, Spring). Gender gaps in public opinion about lesbians and gay men. *Public Opinion Quarterly,* pp. 40–66.

Herrnstein, R., & Murray, C. (1994). *The bell curve.* New York: Free Press.

Herzog, D. (2008). *The future of American politics.* New York: Basic.

Hickins, M. (1998). Reconcilable differences. *Management Review, 87,* 54–59.

Hicks, D., & Beaudry, M. C. (Eds.) (2006). *The Cambridge companion to historical archaeology.* Cambridge University Press.

Higgins, T. (2003). State of emergency. *Commonweal,* February 14.

Hill, J. (1978). Apes and language. *Annual Review of Anthropology, 7,* 89–112.

Himes, C. L. (2001). Elderly Americans. *Population Bulletin, 56,* no. 4.

Hinton, W. (1966). *Fanshen: A documentary of revolution in a Chinese village.* New York: Vintage.

Hobsbawm, E. J. (1989). *The age of empire, 1875–1914.* Vintage.

Hochschild, A. (1989). *The second shift: Working parents and the revolution at home.* New York: Viking.

Hochschild, A. (1997). *The time bind: When work becomes home and home becomes work.* New York: Henry Holt.

Hodge, R. W., & Treiman, D. J. (1968). Class identification in the United States. *American Journal of Sociology, 73,* 535–547.

Hoecker-Drysdale, S. (1992). *Harriet Martineau: First woman sociologist.* New York: Berg Publishers.

Hoffman, L. M. (1989). *The politics of knowledge: Activist movements in medicine and planning.* Albany: State University of New York Press.

Hogan, M. C. (2010). Maternal mortality for 181 countries, 1980—2008: a systematic analysis of progress towards Millennium Development Goal 5. *The Lancet, 375*(9726): 1609–1623.

Holden, C. (1996). New populations of old people. *Science, 273,* 46–48.

Hollingshead, A. (1949). *Elmtown's youth.* New York: Wiley.

Holmes, S. A. (1998, July 1). Birth rate falls to 40-year low among unwed black women. *New York Times,* pp. A1, A16.

Holt, T. C. (1980). Afro-Americans. In *Harvard encyclopedia of American ethnic groups.* Cambridge, MA: Belknap.

Homans, G. (1950). *The human group.* Orlando: Harcourt Brace Jovanovich.

Homans, G. (1961). *Social behavior: Its elementary forms.* Orlando: Harcourt Brace Jovanovich.

Hopkins, T. K., & Wallerstein, I. (Eds.) (1996). *The age of transition: Trajectory of the world-system 1945–2025.* London: Zed Books.

Horgan, J. (1997, May). Seeking a better way to die. *Scientific American,* pp. 100–105.

Horowitz, M. (1979). The jurisprudence of *Brown* and the dilemmas of liberalism. *Harvard Civil Rights–Civil Liberties Review, 14,* 599–610.

Horowitz, R. (1985). *Honor and the American dream: Culture and identity in a Chicago neighborhood.* New Brunswick, NJ: Rutgers University Press.

Horowitz, T. (1999). Run, Rudolph, run: How the fugitive becomes a folk hero. *New Yorker,* March 15, pp. 46–52.

House, J. S., Lepkowski, J. M., Kinny, A. M., Mero, R. P., Kessler, A., & Herzog, R. (1994). The social stratification of aging and health. *Journal of Health and Social Behavior, 35,* 213–234.

Howe, I. (1983). *1984 revisited: Totalitarianism in our century.* New York: Harper.

Huang, C. (1994). *Immigration and the underclass.* Unpublished doctoral dissertation, City University of New York Graduate School.

Huggins, R., & Johnston, A. (2009). The economic and innovation contribution of universities: A regional perspective. *Environment and Planning C: Government and Policy,* pp. 1088–1106.

Hughes, E. (1945). The dilemmas and contradictions of status. *American Journal of Sociology, 50,* 353–359.

Hughes, E. (1959). The study of occupations. In R. K. Merton, L. Broom, & L. S. Cottrell, Jr. (Eds.), *Sociology today: Problems and prospects.* New York: Basic Books.

Human Rights Watch, The Sentencing Project. (1998). Felony disenfranchisement: overview. www.hrw.org/reports98/vote/usvot980.htm.

Humes, E. (2006). *Over here: How the G.I. Bill transformed the American Dream.* Harcourt.

Hummin, D. (1990). *Commonplaces: Community ideology and identity in American cities.* Albany: State University of New York Press.

Humphreys, L. (1970, 1975). *Tearoom trade: Impersonal sex in public places.* Hawthorne, NY: Aldine.

Hunsberger, B. (1995). Religion and prejudice: The role of religious fundamentalism, quest, and right-wing authoritarianism. *Journal of Social Issues, 51,* 113–130.

Hunt, G., & Joe-Laidler, K. (2001). Situations of violence in the lives of girl gang members. *Health Care for Women International, 22,* 1.

Hunt, M. (1985). *Profiles of social research: The scientific study of human interactions.* New York: Russell Sage.

Hunt, M. (1999, Summer). The biological roots of religion. *Free Inquiry,* pp. 30–33.

Hunter, F. (1953). *Community power structure: A study of decision makers.* Chapel Hill: University of North Carolina Press.

Hunter-Gault, C. (2010). The third man. *The New Yorker,* July 5, pp. 24–29.

Hussein, M. (1994, December). Behind the veil of fundamentalism. *UNESCO Courier,* pp. 25–28.

Hyde, J. S. (2006). *Understanding human sexuality,* 9th ed. Boston: McGraw-Hill.

Iannaccone, L., Stark, R., & Finke, R. (1998). Rationality and the "religious mind." *Economic Inquiry, 36,* 373–390.

Ianni, F. A. J. (1998). New Mafia: Black, Hispanic and Italian styles. *Society, 35,* 115–129.

Iglesia, R. (1990). *Cronistas e historiadores de la conquista de Mexico.* Mexico City: Biblioteca de la Ciudad de Mexico.

Inter-Parliamentary Union. (2010). Women in parliament in 2009: The year in perspective. www.ipu.org.

International Association for the Evaluation of Educational Achievement (2003). *Trends in international mathematics and science study (TIMSS).*

Irvine, J. M. (2003). Introduction to "Sexual scripts: Origins, influences and changes." *Qualitative Sociology, 26:* 489–490.

Ishitani, T. T., & McKitrick, S. A. (2010). After transfer: The engagement of community college students at a four-year collegiate institution. *Community College Journal of Research and Practice, 34,* 576–594.

Ishiyama, J., & Quinn, J. J. (2006). African phoenix? Explaining the electoral performance of the formerly dominant parties in Africa. *Party Politics, 12,* 317–40.

Jackman, M. R., & Jackman, R. W. (1983). *Class awareness in the United States.* Berkeley: University of California Press.

Jackson, B. (1972). *In the life: Versions of the criminal experience.* New York: NAL.

Jackson, K. (1985). *The crab grass frontier: The suburbanization of the United States.* New York: Oxford University Press.

Jackson-Smith, D. B., & Jensen, E. (2009). Finding farms: Comparing indicators of farming dependence and agricultural importance in the United States. *Rural Sociology, 74,* 37–55.

Jacobs, J. A., & Gerson, K. (2005). *The time divide: Work, family, and gender inequality.* Cambridge, MA: Harvard University Press.

Jacobson, J. L. (1992). Improving women's reproductive health. In L. R. Brown (Ed.), *State of the world.* New York: Norton.

Jacobson, L. (2001, July 11). Looking to France. *Education Week,* www.edweek.org/ew.

Jacobson-Hardy, M. (1995). The changing landscape of labor: Workers and workplace. *Labor's Heritage, 6,* 40–57.

Jahoda, M. (1982). *Employment and unemployment: A social-psychological analysis.* London: Cambridge University Press.

Jahoda, M., Lazarsfeld, P., & Zeisel, H. (1971). *Marienthal: A study of an unemployed community.* Hawthorne, NY: Aldine.

Janowitz, M. (1960). *The professional soldier: A social and political portrait.* New York: Free Press.

Janowitz, M. (1968). Political sociology. In D. Sills (Ed.), *The international encyclopedia of the social sciences.* New York: Free Press.

Janowitz, M. (1978). *The last half century: Societal change and politics in America.* Chicago: University of Chicago Press.

Janowitz, M., & Shils, E. A. (1948). The cohesion and disintegration of the Wehrmacht in World War II. *Public Opinion Quarterly, 12,* 280–315.

Jargowsky, P. A. (1997). *Poverty and place: Ghettos, barrios, and the American city.* New York: Russell Sage.

Jaynes, G. D., & Williams, R. M., Jr. (1989). *A common destiny: Blacks and American society.* Washington, DC: National Academy Press.

Jencks, C. (1993). *Rethinking social policy: Race, poverty and the underclass.* Cambridge, MA: Harvard University Press.

Jencks, C. (1994). *The homeless.* Cambridge, MA: Harvard University Press.

Jencks, C., & Phillips, M. (1999). The black-white test score gap: How to reduce it. *Current,* January, pp. 9–16.

Jennings, F. (1975). *The invasion of America: Indians, colonialism and the cant of conquest.* New York: Norton.

Johnson, P. (1987). *A history of the Jews.* New York: Harper.

Johnson, S. M. (2008, May 21). Study finds Teach For America teachers stay in the classroom past initial commitment. Harvard Graduate School of Education. www.gse.harvard.edu/news_events/ features/2008/05/21_project.php.

Johnston, D. K. (1988). Adolescents' solutions to dilemmas in fables: Two moral orientations—two problem solving strategies. In C. Gilligan, J. V. Ward, J. M. Taylor, & B. Bardige (Eds.), *Mapping the moral domain.* Cambridge, MA: Harvard University Press.

Jones, C. (2002, November 12). The AIDS memorial quilt. *Advocate,* p. 99.

Jones-Correa, M., & Ajinkya, J. (2007). Immigration, gender and political socialization. Conference Papers, Annual Meeting, American Political Science Association.

Kagan, J. (2000, October). Adult personality and early experiences. *Harvard Mental Health Letter,* pp. 4–5.

Kagan, J., Pals, J. L., McCrae, R. R., Paunonen, S. V., Saarni, C., Shiner, R. L., Tellegen, A., Masten, A. S., & Whitbourne, S. K. (2001). Commentaries on "Personality development across the life course: The argument for change and continuity" and "Issues in the study of personality development." *Psychological Inquiry, 12,* 16–23.

Kagan, R. (2003). Of paradise and power. New York: Knopf.

Kagan, S. L., & Neuman, M. J. (1998). Lessons from three decades of transition research. *Elementary School Journal, 98,* 365–380.

Kagay, M. R., & Elder, J. (1992, August 9). Numbers are no problem for pollsters. Words are. *New York Times,* sec. 4, p. 5.

Kahl, J. A. (1965). *The American class structure.* Fort Worth: Holt, Rinehart and Winston.

Kahneman, D., Knetsch, J., & Thaler, R. (1986). Fairness and the assumptions of economics. *Journal of Business, 59,* S285–S300.

Kaiser Family Foundation (2001). Inside-OUT: A report on the experiences of lesbians, gays and bisexuals in America and the public's view on issues and policies related to sexual orientation. www.kff.org.

Kammerman, S. B. (1995). *Starting right: How America neglects its youngest children and what we can do about it.* New York: Oxford University Press.

Kammerman, S., & Kahn, A. (1981). *Child care, family benefits, and working parents: A study in comparative policy.* New York: Columbia University Press.

Kammerman, S., & Kahn, A. J. (1993, Fall). What Europe does for single parent families. *The Public Interest,* pp. 70–86.

Kanter, R. M. (1977). *Men and women of the corporation.* New York: Basic Books.

Kanter, R. M., & Stein, B. A. (Eds.) (1980). *Life in organizations: Workplaces as people experience them.* New York: Basic Books.

Kaplan, D. E. (1998, April 13). Yakuza Inc.: U.S. investors snapping up bad loans from Japanese banks will collide with organized crime. *U.S. News & World Report,* pp. 40–41.

Kaplan, M. F., & Martin, A. F. (1999). Effects of differential status of group members on process and outcome of deliberations. *Group Processes and Intergroup Relations, 2,* 347–364.

Kaplan, R. (2001). *The coming anarchy.* New York: Vintage.

Karras, T. (2003, March). You'll never believe what I heard! *Good Housekeeping.*

Kasarda, J. D. (1988). *Metropolis era.* Newbury Park, CA: Sage.

Kasarda, J. D. (1989). Urban industrial transition and the underclass. *Annals of the American Academy of Political and Social Science, 501,* 26–47.

Kasinitz, P., Mollenkopf, J., & Waters, M. (2008). Inheriting the city: The children of immigrants come of age. Cambridge, MA: Harvard University Press.

Katz, E. (1957). The two step flow of communication: An up-to-date report on an hypothesis. *Public Opinion Quarterly, 21,* 61–78.

Kazemipur, A., & Rezaei, A. (2003). Religious life under theocracy: The case of Iran. *Journal for the Scientific Study of Religion, 42,* 347–361.

Kealy, T. (2008). Trust and transparency are frowned upon while degree inflation soars. *New York Times, Higher Education Supplement,* October 16, pp. 26–27.

Keith, J. (1982). *Old people, new lives: Community creation in a retirement residence* (2nd ed.). Chicago: University of Chicago Press.

Keller, B. (Ed.) (2005). *Class matters.* Henry Holt, Times Books.

Kelley, T. (1998, February 3). Charting a course to ethical profits. *New York Times,* sec. 3, p. 1.

Kemper, T. D. (1990). *Social structure and testosterone: Explorations of the socio-bio-social chain.* New Brunswick, NJ: Rutgers University Press.

Keniston, K. (1965). Social change and youth in America. In E. H. Erikson (Ed.), *The challenge of youth.* Garden City, NY: Doubleday Anchor.

Kennedy, R. J. R. (1944). Single or triple melting pot? *American Journal of Sociology, 49,* 331–339.

Kenyon, P. (2006). Sexual selection and human reproduction. Plymouth, England: University of Plymouth, Department of Psychology, SALMON (Study and Learning Materials On-line).

Kessler, S. J., & McKenna, W. (1978). *Gender: An ethnomethodological approach.* Chicago: University of Chicago Press.

Killian, L. (1952). Group membership in disaster. *American Journal of Sociology, 57,* 309–314.

Kilson, M. (1995, Fall). Affirmative action. *Dissent,* pp. 469–470.

Kim, I. (1981). *New urban immigrants: The Korean community in New York.* Princeton, NJ: Princeton University Press.

Kimmel, M. S. (2006). *Manhood in America: A cultural history,* 2nd ed. New York: Oxford University Press.

Kimmel, M. S., Hearn, J., & Connell, R. W. (Eds.). (2005). Handbook of studies on men and masculinities. Thousand Oaks, CA: Sage Publications.

King, S. (1994, May). Diamonds are forever. *Life,* p. 26.

Kingston, P. W., Hubbard, R., Lapp, B., Schroeder, P., & Wilson, J. (2003). Why education matters. *Sociology of Education, 76,* 53–70.

Kinsey, A. C., Pomeroy, W. B., & Martin, C. E. (1948). *Sexual behavior in the human male.* Philadelphia: Saunders.

Kinsey, A. C., Pomeroy, W. B., & Martin, C. E. (1953). *Sexual behavior in the human female.* Philadelphia: Saunders.

Kitano, H. H. I. (1980). *Race relations.* Englewood Cliffs, NJ: Prentice Hall.

Klandermans, B. (1997). *The social psychology of protest.* Cambridge, MA: Blackwell.

Klass, P. (1988, October 30). Wells, Welles and the Martians. *New York Times Book Review,* pp. 1, 48–49.

Kleinberg, O. (1935). *Race differences.* New York: Harper.

Kleniewsky, N. (2006). *Cities, change, and conflict: A political economy of urban life,* 3rd ed. Belmont, CA: Thomson/ Wadsworth.

Klinenberg, E. (2010). *Alone in America.* New York: Penguin.

Kluger, J. (2001, May 21). Can gays switch sides? *Time,* p. 62.

Knecht, A., Snijders, T. A. B., Baerveldt, C., Steglich, C. E. G., & Raub, W. (2010). Friendship and delinquency: Selection and influence processes in early adolescence. *Social Development, 19,* 494–514.

Koberg, C. S., Chappell, D., & Ringer, R. C. (1997). Correlates and consequences of protégé mentoring in a large hospital. *Annual Review of Sociology, 23,* 191–215.

Kohlberg, L., & Gilligan, C. (1971, Fall). The adolescent as a philosopher: The discovery of self in a postconventional world. *Daedalus,* pp. 1051–1086.

Kohlberg, L., Levine, C., & Hewer, A. (1983). *Moral stages: A current formulation and a response to critics.* Basel, NY: Karger.

Kolbert, E. (2003). Pox populi. *New Yorker*, September 1, pp. 25–26.

Koretz, G. (2003, April 21). Demographic time bombs. *Business Week*, pp. 17–18.

Koretz, G. (2003). Equality? Not on death row. *Business Week*, June 30, pp. 6–7.

Kornblum, W. (1974). *Blue collar community*. Chicago: University of Chicago Press.

Kornblum, W. (1995). Times Square as a field research site. In R. P. McNamara, *Sex, scams, and street life: The sociology of New York City's Times Square*. Westport, CT: Greenwood.

Kornblum, W., & Julian, J. (2007). *Social problems*, 10th ed. Englewood Cliffs, NJ: Prentice Hall.

Kornblum, W., & Lawler, K. (1999). *Handbook of park user research*. Washington, DC: Urban Institute and Lila Wallace Readers Digest Foundation.

Kornhauser, W. (1952). The Negro union official: A study of sponsorship and control. *American Journal of Sociology, 57*, 443–452.

Kosmin, B. (1991). *Research report: The national survey of religious identification*. New York: Center for Jewish Studies, City University of New York Graduate Center.

Kotlowitz, A. (2004). *Never a city so real: A walk in Chicago*. Crown Journeys.

Kozol, J. (1995). *Amazing grace: The lives of children and the conscience of a nation*. New York: Crown.

Kremer, P. (1998, February 4). Sociologists rediscover the links between suicide and economic crisis. *Le Monde*, p. 8.

Kriegel, L. (1993). From the catbird seat: Football, baseball, and language. *Sewanee Review, 101*, 213–225.

Kroc, R. (1977). *Grinding it out: The making of McDonald's*. Chicago: Henry Regnery.

Kroeger, B. (2004, March 20). When a dissertation makes a difference. *New York Times*.

Kronus, C. L. (1977). Mobilizing voluntary associations into a social movement: The case of environmental quality. *Sociological Quarterly, 18*, 267–283.

Kübler-Ross, E. (1969). *On death and dying*. New York: Macmillan.

Kunen, J. S. (1995, July 10). Teaching prisoners a lesson. *New Yorker*, pp. 34–39.

Kunkel, D., Eyal, K., Donnerstein, E., Farrar, K. M., Biely, E., & Rideout, V. (2007). Sexual socialization messages on entertainment television: Comparing content trends 1997–2002. *Media Psychology, 9*, 595–622.

Kurtz, L. (1995). *Gods in the global village*. Thousand Oaks, CA: Pine Forge.

Kutner, B., Wilkins, C., & Yarrow, P. R. (1952). Verbal attitudes and overt behavior involving racial prejudice. *Journal of Abnormal and Social Psychology, 47*, 649–652.

Kwamena-Poh, M., Tosh, J., Waller, R., & Tidy, M. (1982). *African history in maps*. Burnt Hill, England: Longman.

Lachs, M. S., & Boyer, P. (2003, August). When rivalry won't die. *Prevention*, p. 3.

Ladner, J. (Ed.) (1973). *The death of white sociology*. New York: Vintage.

Lai, C. S. L., Fisher, S. E., Hurst, J. A., Vargha-Khadem, F., & Monaco, A. P. (2001). A forkhead-domain gene is mutated in a severe speech and language disorder. *Nature, 413*, 519–523.

Lamanna, M. A., & Riedmann, A. (2009). *Marriages and families: Making choices in a diverse society*, 10th ed. Belmont, CA: Thomson.

Lampman, J. (2001, June 14). Homosexuality remains focus of denominational controversy. *Christian Science Monitor*, p. 14.

Lampman, J. (2002, January 10). Muslims in America. *Christian Science Monitor*, p. 15.

Land, K. C. (1989). Review of *Predicting Recidivism Using Survival Models*, by Peter Schmidt and Ann Dryden Witte. *Contemporary Sociology, 18*, 245–246.

Land, K. (2010, June 8). *Child and youth well-being index*. Foundation for Child Development. www.fcd-us.org.

Lansford, J. E., Killeya-Jones, L. A., Miller, S., & Costanzo, P. R. (2009). Early adolescents' social standing in peer groups: Behavioral correlates of stability and change. *Journal of Youth and Adolescence, 38*, 1,084–1,095.

LaPiere, R. (1934). Attitudes vs. actions. *Social Forces, 13*, 230–237.

Lash, S. (1992). *Modernity and identity*. Cambridge, MA: Blackwell.

Laslett, P. (1972). Introduction. In P. Laslett & R. Wall (Eds.), *Household and family in past time*. Cambridge, England: Cambridge University Press.

Laslett, P. (1983). *The world we have lost* (3rd ed.). London: Methuen.

Lasswell, H. D. (1936). *Politics: Who gets what, when and how*. New York: McGraw-Hill.

Lasswell, H. D. (1941). The garrison state. *American Journal of Sociology, 46*, 455–468.

Latané, B., & Darley, J. (1970). *The unresponsive bystander: Why doesn't he help?* New York: Meredith.

Laumann, E. O., Gagnon, J. H., Michaels, R. T., & Michaels, S. (1994). *The social organization of sexuality: Sexual practices in the United States*. Chicago: University of Chicago Press.

Laumann, E. O., Paik, A., Glasser, D. B., Kang, J-H., Wang, T., Levinson, B., Moreira, E. D., Jr., Nicolosi, A., & Gingell, C. (2006). A cross-national study of subjective sexual well-being among older women and men: Findings from the global study of sexual attitudes and behaviors. *Archives of Sexual Behavior, 35*, 145–161.

Lawton, J. (1984). *The all American war game*. Oxford, England: Blackwell.

Lazreg, M. (1994). *The eloquence of silence: Algerian women in question*. New York: Routledge.

Lazreg, M. (2009). *Questioning the veil: Open letters to Muslim women*. Princeton, NJ: Princeton University Press.

LeBon, G. (1947/1896). *The crowd*. London: Ernest Bonn.

Lebovics, H. (2006). *Imperialism and the corruption of democracies*. Durham, NC: Duke University Press.

LeMasters, E. E. (1975). *Blue collar aristocrats*. Madison: University of Wisconsin Press.

Lemme, G. H. (2006). *Development in adulthood*, 4th ed. Boston: Allyn & Bacon.

Lengermann, P. M., & Niebrugge-Brantley, J. (1998). *The women founders*. New York: McGraw-Hill.

Lenski, G., & Lenski, J. (1982). *Human societies*, 4th ed. New York: McGraw-Hill.

Lerman, H. (1996). *Pigeonholing women's misery*. New York: Basic Books.

Lesthaeghe, R. (2010). The unfolding story of the second demographic transition. *Population and Development Review, 36*, 211–251.

Lever, J. (1976). Sex differences in the games children play. *Social Problems, 23*, 478–487.

Lever, J. (1978). Sex differences in the complexity of children's play and games. *American Sociological Review, 43*, 471–483.

Levi, P. (1989). *The drowned and the saved*. New York: Vintage International.

Levy, S. (1997, April 7). Blaming the Web. *Newsweek*, pp. 46–47.

Lewin, T. (1989, November 14). Aging parents: Women's burden grows. *New York Times*, pp. A1, B12.

Lewin, T. (2010). Community colleges cutting back on open access. *New York Times*, June 24, p. 15.

Lewis, J. S. (2005). The long sexual revolution: English women, sex, and contraception, 1800–1975. *Journal of Social History, 38*, 1134–1136.

Lewis, M., & Feiring, C. (1982). Some American families at dinner. In L. M. Laosa & I. E. Sigel (Eds.), *Families as learning environments for children*. New York: Plenum.

Lewontin, R. C. (1995, April 20). The social organization of sexuality. *New York Review of Books*, pp. 24–29.

Lewontin, R. C. (2003). The truth about DNA. *New York Review of Books*, 39–42.

Lieberson, S. (1980). *A piece of the pie: Blacks and white immigrants since 1880*. Berkeley: University of California Press.

Liebow, E. (1967). *Tally's corner: A study of Negro streetcorner men*. Boston: Little, Brown.

Lim, L. L. (Ed.) (1998). *The sex sector: The economic and social bases of prostitution in Southeast Asia*. Geneva, Switzerland: International Labor Organization.

Lin, M. (1995, November 1). Interview by Terri Gross on *Fresh Air*, National Public Radio.

Linton, R. (1936). *The study of man*. New York: Appleton.

Lippa, R. A. (2002). *Gender, nature, and nurture*. Mahwah, NJ: Erlbaum.

Lipset, S. M. (1979). *The first new nation*. New York: Norton.

Lipset, S. M. (1981). *Political man*. Baltimore: Johns Hopkins University Press.

Lipset, S. M. (1994). The social prerequisites of democracy revisited. *American Sociological Review, 59*, 1–22.

Lipset, S. M. (1995). America today: Malaise and resiliency. *Current*, December, pp. 3–11.

Lipset, S. M. (1998). One nation after all. *Wilson Quarterly, 22*, 100–102.

Livernash, R., & Rodenburg, E. (1998). Population change, resources, and the environment. *Population Bulletin, 53*, no. 1.

Lofland, J. F. (1981). Collective behavior: The elementary forms. In N. Rosenberg & R. H. Turner (Eds.), *Social psychology: Sociological perspectives*. New York: Basic Books.

Lofland, L. (1998). *The public realm: Exploring the city's quintessential social territory*. Hawthorne, NY: Aldine.

Lombroso, C. (1911). *Crime: Its cause and remedies*. Boston: Little, Brown.

London, B. (1992). School-enrollment rates and trends, gender, and fertility: A cross-national analysis. *Sociology of Education, 65*, 305–318.

Loomis, S., & Rodriguez, J. (2009). Institutional change and higher education. *Higher Education, 58*, 475–489.

Lorber, J. (1994). *Paradoxes of gender*. New Haven, CT: Yale University Press.

Lorber, J. (2006). Shifting paradigms and challenging categories. *Social Problems, 53*, 448–453.

Low, S. A. (2003). *Behind the gates: Life, security, and the pursuit of happiness in fortress America*. New York: Routledge.

Lowry, E. H. (1993). *Freedom and community: The ethics of interdependence*. Albany: State University of New York Press.

Lubienski, C. (2000). Whither the common good? A critique of home schooling. *Peabody Journal of Education, 75,* 207–232.

Luce, I. (1964). *Letters from the Peace Corps*. Washington, DC: Robert B. Luce.

Lutfey, K., & Mortimer, J. T. (2006). Development and socialization through the adult life course. In J. Delamater (Ed.). *Handbook of social psychology*. New York: Springer.

Lye, K. (1995). *The complete atlas of the world*. Austin: Raintree Steck-Vaughn.

Lynch, K. (1990). *The hidden curriculum*. London: Falmer Press.

Lynd, R. S., & Lynd, H. M. (1929). *Middletown: A study in American culture*. Orlando: Harcourt Brace Jovanovich.

Lynd, R. S., & Lynd, H. M. (1937). *Middletown in transition: A study in cultural conflicts*. Orlando: Harcourt Brace Jovanovich.

Machiavelli, N. (1950/1513). *The prince* and *The discourses*. New York: Modern Library.

Mack, P. (2001). Women and gender in early modern England. *Journal of Modern History, 73,* 379.

MacLeod, J. (1995/1987). *Ain't no makin' it: Leveled aspirations in a low-income neighborhood*. Boulder, CO: Westview.

MacLeod, C. (2010). Factory workers in China show more backbone. *USA Today,* June 17.

MacWhinney, B. (1998). Models of the emergence of language. *Annual Review of Psychology, 49,* 199–227.

Majundar, R. C. (Ed.) (1951). *The history and culture of the Indian people*. London: Allen and Unwin.

Maksel, R. (2002, March). Migrant Madonna. *Smithsonian,* pp. 15–16.

Malinowski, B. (1927). *Sex and repression in savage society*. London: Harcourt Brace.

Malson, L. (1972). *Wolf children and the problem of human nature* (trans. E. Fawcett, P. Aryton, & J. White). New York: Monthly Review Press.

Malthus, T. (1927–1928/1798). *An essay on population*. New York: Dutton.

Mannheim, K. (1941). *Man and society in an age of reconstruction*. Orlando: Harcourt Brace Jovanovich.

Mansnerus, L. (1992). Should tracking be derailed? *Education Life, New York Times,* November 1, pp. 14–16.

Marks, J. (1994, December). Black, white, other. *Natural History,* pp. 32–35.

Marks, P. M. (1998). *In a barren land: American Indian dispossession and survival*. New York: Morrow.

Marmor, T., Oberlander, J., & White, J. (2009). The Obama administration's options for health care cost control: Hope versus reality. *Annals of Internal Medicine, 150,* 485–489.

Marshall, T. H. (1964). *Class, citizenship and social development*. Garden City, NY: Doubleday.

Martinson, R. (1972, April 29). Planning for public safety. *New Republic,* pp. 21–23.

Marwell, G., & Ames, R. E. (1985). Experiments on the provision of public goods II: Provision points, stakes, experience, and the free-rider problem. *American Journal of Sociology, 90,* 926–937.

Marwell, N. P. (2004). Privatizing the welfare state: Nonprofit community-based organizations as political actors. *American Sociological Review, 69,* 265–291.

Marwell, N. P. (2007). *Bargaining for Brooklyn*. Chicago: University of Chicago Press.

Marx, K. (1961/1844). *Economic and philosophical manuscripts of 1844*. Moscow: Foreign Languages Publishing House.

Marx, K. (1962/1867). *Capital: A critique of political economy*. Moscow: Foreign Languages Publishing House.

Marx, K. (1963/1869). *The eighteenth Brumaire of Louis Bonaparte*. New York: International Publishers.

Marx, K., & Engels, F. (1955). Wage labor and capital. In *Selected works in two volumes*. Moscow: Foreign Languages Publishing House.

Marx, K., & Engels, F. (1969/1848). *The communist manifesto*. New York: Penguin.

Massey, D. S., & Denton, N. A. (1993). *American apartheid: Segregation and the making of the underclass*. Cambridge, MA: Harvard University Press.

Massey, D. S., Durand, J., & Malone, N. J. (2003). *Beyond smoke and mirrors: Mexican immigration in an era of economic integration*. New York: Russell Sage.

Massey, D. S., & Fischer, M. J. (2000). How segregation concentrates poverty. *Ethnic and Racial Studies, 23,* 670–691.

Mauss, M. (1966/1925). *The gift*. New York: Free Press.

Mawson, A. R. (1989). Review of *Homicide,* by Martin Daly and Margo Wilson. *Contemporary Sociology, 18,* 238–240.

Mayfield, L. (1984). *Teenage pregnancy*. Doctoral dissertation, City University of New York.

Maynard, M. (2002). Studying age, "race" and gender: Translating a research proposal into a project. *International Journal of Social Research Methodology, 5,* 31–40.

Mayo, E. (1945). *The social problems of an industrial civilization*. Boston: Harvard University, Graduate School of Business Administration.

McAdam, D. (1992). Gender as a mediator of the activist experience: The case of Freedom Summer. *American Journal of Sociology, 97,* 1211–1240.

McAdam, D., McCarthy, J. D., & Zald, N. (1988). Social movements. In N. J. Smelser (Ed.), *The handbook of sociology*. Newbury Park, CA: Sage.

McAdam, D., McCarthy, J. D., & Zald, M. N. (Eds.) (1996). *Comparative perspectives on social movements: Political opportunities, mobilizing structures, and cultural framings*. New York: Cambridge University Press.

McAdam, D., & Snow, D. S. (1997). *Social movements: Readings on their emergence, mobilization, and dynamics*. Los Angeles: Roxbury.

McAdam, D., Tarrow, S., & Tilly, C. (2001). *Dynamics of contention*. Cambridge University Press.

McCord, W. (1991). *The dawn of the Pacific century: Implications for three worlds of development*. New Brunswick, NJ: Transaction.

McCorkel, J. A. (2003). Embodied surveillance and the gendering of punishment. *Journal of Contemporary Ethnography, 32,* 41–76.

McDowell, S. A., & Ray, B. D. (2000). The home education movement in context, practice, and theory: Editors' introduction. *Peabody Journal of Education, 75,* 1–7.

McFadden, R. B. (1998). Review of *Waiting for a miracle,* by James P. Comer. *Black Issues in Higher Education, 15,* 30.

McGinn, T. A. J. (2004). *The economy of prostitution in the Roman world: A study of social history and the brothel*. Ann Arbor: University of Michigan Press.

McGoldrick, M., & Carter, E. A. (1982). *The family life cycle in normal family processes*. London: Guilford.

McGuire, M. B. (1987). *Religion: The social context,* 2nd ed. Belmont, CA: Wadsworth.

McIntyre, M. (1995). Altruism, collective action, and rationality: The case of LeChambon. *Polity, 27,* 537–557.

McLanahan, S. (2010). How are the children of single mothers faring? Evidence from the fragile families study. Population Reference Bureau. http://discuss.prb.org.

McManus, P. A., & DiPrete, T. A. (2001). Losers and winners: The financial consequences of separation and divorce for men. *American Sociological Review, 66,* 246–268.

McNamara, R. P. (1994). *Crime displacement: The other side of prevention*. East Rockway, NY: Cummings and Hathaway.

McNeill, W. (1963). *The rise of the West: A history of the human community*. Chicago: University of Chicago Press.

McNeill, W. H. (1982). *The pursuit of power: Technology, armed force, and society since A.D. 1000*. Chicago: University of Chicago Press.

McPhillips, K. (2005, September). Global violence: Some thoughts on hope and change. *Feminist Theology,* pp. 25–34.

Mead, G. H. (1971/1934). *Mind, self and society*. Chicago: University of Chicago Press.

Mead, M. (1950). *Sex and temperament in three primitive societies*. New York: NAL Mentor.

Mead, M. (1971). Comment. In J. Tanner & B. Inhelder (Eds.), *Discussions on child development*. New York: International Universities Press.

Medalie, J. H., & Cole-Kelly, K. (2002). The clinical importance of defining family. *American Family Physician, 65,* 1277.

Meier, D. (1995). *The power of their ideas: Lessons from a small school in Harlem*. Boston: Beacon.

Meier, D. (2008, March). Schooling for democracy or for the workplace? *Phi Delta Kappan,* pp. 507–510.

Melton, J. G. (1989). *The encyclopedia of American religions* (3rd ed.). Detroit: Gale Research.

Merton, R. K. (1938). Social structure and anomie. *American Sociological Review, 3,* 672–682.

Merton, R. K. (1948). Discrimination and the American creed. In R. M. MacIver (Ed.), *Discrimination and national welfare*. New York: Institute for Religious and Social Studies; dist. by Harper.

Merton, R. K. (1968). *Social theory and social structure,* 3rd ed. New York: Free Press.

Merton, R. K. (1998). Personal communication.

Merton, R. K., & Kitt, A. (1950). Contributions to the theory of reference group behavior. In R. K. Merton & P. Lazarsfeld (Eds.), *Continuities in social research*. New York: Free Press.

Metropolitan Washington Council of Governments. (2009). www.mwcog.org.

Meyer, D. S., & Tarrow, S. (Eds.) (1998). *Contentious politics for a new century*. Lanham, MD., and Oxford, England: Rowman & Littlefield.

Meyer, J. W., Ramirez, F. O., & Soysal, Y. N. (1992). World expansion of mass education, 1870–1980. *Sociology of Education, 65,* 128–149.

Meyersohn, R., & Katz, E. (2004). Notes on a natural history of fads. In J. D. Peters, & P. Simonson (Eds.), *Mass communication and American social thought: classic texts, 1919–1968.* Oxford, U.K.: Rowman and Littlefield.

Milgram, S. (1970). The experience of living in cities. *Science, 167,* 1461–1468.

Milgram, S. (1974). *Obedience to authority: An experimental view.* New York: Harper.

Milkman, R., & Voss, K. (Eds.) (2004). *Rebuilding labor: Organizing and organizers in the new union movement.* Ithaca, NY: Cornell University Press.

Miller, D. C., & Form, W. H. (1964). *Industrial sociology,* 2nd ed. New York: Harper.

Miller, N. (1989). *In search of gay America: Women and men in a time of change.* New York: Atlantic Monthly Press.

Miller, W. B. (1958). Lower class culture as a generating milieu of gang delinquency. *Journal of Social Issues, 14,* 5–19.

Mills, C. W. (1951). *White collar.* New York: Oxford University Press.

Mills, C. W. (1956). *The power elite.* New York: Oxford University Press.

Mills, C. W. (1959). *The sociological imagination.* New York: Oxford University Press.

Min, P. G. (2005). Bitter fruit: The politics of black-Korean conflict in New York City. *Amerasia Journal, 31,* 187–190.

Mintz, S. (1985). *Sweetness and power: The place of sugar in modern history.* New York: Viking.

Minuchin, S. (1974). *Families and family therapy.* Cambridge, MA: Harvard University Press.

Mishel, L., Bernstein, J., & Boushey, H. (2003). *State of working America 2002/2003.* Ithaca, NY: Cornell University Press.

Mishra, P. (2006, July 6). The myth of the New India. *New York Times,* p. A21.

Mixner, D., & Bailey, D. (2002). *Brave journeys: Profiles in gay and lesbian courage.* London: Random House International.

Mock, B. (2010). *Immigration backlash: Hate crimes against Latinos flourish.* Southern Poverty Law Center. www.splcenter.org.

Modell, J., & Hareven, T. K. (1973). Urbanization and the malleable household: An examination of boarding and lodging in American families. *Journal of Marriage and the Family, 35,* 466–479.

Moen, P. (2001). The career quandary. *PRB Reports on America,* vol. 2, no. 1. Washington, DC: Population Reference Bureau.

Money, J., & Tucker, P. (1975). *Sexual signatures.* Boston: Little, Brown.

Monnier, M. (1993). *Mapping it out: Expository cartography for the humanities and social sciences.* Chicago: University of Chicago Press.

Montagna, P. D. (1977). *Occupations and society.* New York: Wiley.

Moody, H. R. (1998). *Aging: Contexts and controversies,* 2nd ed. Thousand Oaks, CA: Pine Forge.

Moore, H. E. (1958). *Tornados over Texas.* Austin: University of Texas Press.

Moore, J. W. (1991). *Going down to the barrio: Homeboys and homegirls in change.* Philadelphia: Temple University Press.

Moore, J., & Pachon, H. (1985). *Hispanics in the United States.* Englewood Cliffs, NJ: Prentice Hall.

Moore, R., & Muller, J. (2002). The growth of knowledge and the discursive gap. *British Journal of Sociology of Education, 23,* 627–637.

Morales, E. (1986). Coca and the cocaine economy and social change in the Andes of Peru. *Economic Development and Cultural Change, 35,* 143–161.

Morales, E. (1989). *Cocaine: White gold rush in Peru.* Tucson: University of Arizona Press.

Morgan, D. L. (1998). Introduction: The aging of the baby boom. *Generations, 22,* 5–10.

Morgan, R. (Ed.) (1970). *Sisterhood is powerful: An anthology of writings from the women's liberation movement.* New York: Random House.

Mosher, W. D., Chandra, A., & Jones, J. (2005). *Sexual behavior and selected health measures: Men and women 15–44 years of age, United States, 2002.* U.S. Centers for Disease Control and Prevention.

Mosher, W. D., Martinez, G. M., Chandra, A., Abma, J. C., & Wilson, S. J. (2004). *Use of contraception and use of family planning services in the United States: 1982–2002.* U.S. Centers for Disease Control and Prevention.

Moskos, C. A. (1988). *Call to civic service.* New York: Free Press.

Moskos, C. (2006). Should the all volunteer force be replaced by universal mandatory national service? *Congressional Digest,* September, pp. 220–222.

Moynihan, D. P. (1988, Spring). Our poorest citizens. *Focus* (Institute for Research on Poverty), p. 5.

Mulkey, L. M. (1993). *Sociology of education: Theoretical and empirical investigations.* Fort Worth: Harcourt Brace.

Murdock, G. (1949). *Social structure.* New York: Macmillan.

Murdock, G. (1983). *Outline of world cultures,* 6th ed. New Haven, CT: Human Relations Area Files.

Murphy, K. (2010). Feminism and political history. *Australian Journal of Politics and History, 56,* 21–37.

Murray, C. (2000). Press release. New York: World Health Organization, June 4.

Murray, C. (2008a). *Real education: Four simple truths for bringing America's schools back to reality.* Crown Forum.

Murray, C. (2008b). Should the Obama generation drop out? *New York Times,* December 28, p. 9.

Murray, C. A. (1988). *In pursuit: Of happiness and good government.* New York: Simon & Schuster.

Myrdal, G. (1944). *An American dilemma.* New York: Harper.

Myrdal, J. (1965). *Report from a Chinese village.* New York: Random House.

Nagel, T. (1994, May 12). Freud's permanent revolution. *New York Review of Books,* pp. 34–39.

Naipaul, V. S. (1998). *Beyond belief: Islamic excursions among the converted peoples.* New York: Random House.

Naisbitt, J., & Aburdene, P. (1990). *Megatrends 2000.* New York: Morrow.

Napier, A. Y., & Whitaker, C. (1980). *The family crucible.* New York: Bantam.

National Center for Children in Poverty (2001). http://cpmcnet.columbia.edu/dept/nccp/ycpf.html.

National Center on Addiction and Substance Abuse (2005). *The importance of family dinners II.* New York: Columbia University.

National Center for Education Statistics (2006). *The condition of education 2006.* Washington, D.C.: U.S. Department of Education.

National Center for Health Statistics, U.S. Public Health Service. Annual. *Vital Statistics of the United States.*

National Center for Health Statistics (2002). *Teenagers in the United States: Sexual activity, contraceptive use, and childbearing.* www.cdc.org.

National Center for Health Statistics. (2000, March 28). National vital statistics report. www.cdc.gov.

National Commission on America's Urban Families (1993). *Families first.* Washington, DC: U.S. Government Printing Office.

National Opinion Research Center, University of Chicago (NORC). Annual. *General social survey, cumulative codebook.* Chicago: University of Chicago Press.

Neihardt, J. G. (1959/1932). *Black Elk speaks: Being the life story of a holy man of the Oglala Sioux.* New York: Washington Square Press.

Neurology (2002). Mental activity may delay Alzheimer's. *Neurology, 59,* 1910–1914.

Neugarten, B. L. (1996). *The meanings of age.* Chicago: University of Chicago Press.

Newcomb, T. (1958). Attitude development as a function of reference groups: The Bennington study. In E. Maccoby, T. M. Newcomb, & E. L. Hartley (Eds.), *Readings in social psychology,* 3rd ed. Fort Worth, IN: Holt, Rinehart and Winston.

Newman, K. S. (1988). *Falling from grace.* New York: Free Press.

Newman, K. S. (2004). *Rampage: The social roots of school shootings.* New York: Basic Books.

Newton, M. (2003). *Savage girls and wild boys: A history of feral children.* Faber and Faber.

NHPCO (National Hospice and Palliative Care Organization). (2007, November). NHPCO facts and figures: Hospice care in America. Alexandria, VA: National Hospice and Palliative Care Organization.

Niebuhr, G. (1995, October 3). With every wave of newcomers, a church more diverse. *New York Times,* p. B6.

Niebuhr, G. (2001). After the attacks: Finding fault; U.S. "secular" groups set tone for terror attacks, Falwell says. *New York Times,* September 14, p. A18.

Niebuhr, H. R. (1929). *The social sources of denominationalism.* New York: Meridian.

Ning, P. (1995). *Red in tooth and claw: Twenty-six years in communist Chinese prisons.* New York: Grove.

Nisbet, R. A. (1969). *Social change and history.* New York: Oxford University Press.

Nisbet, R. A. (1970). *The social bond.* New York: Knopf.

Nord, M., Andrews, M., & Carlson, S. (2009). *Household food security in the United States, 2008.* Economic Research Report No. ERR-83.

Norman, M. (1997). From Carol Gilligan's chair. *New York Times Magazine,* November 9, p. 50.

Nurge, D. (2003). Liberating yet limiting: The paradox of female gang membership. In L. Kontos, D. Brotherton, & L. Barrios (Eds.), *Gangs and society: Alternative perspectives.* New York: Columbia University Press.

Nussbaum, M. (1999). *Sex and social justice.* New York: Oxford University Press.

NYT Correspondents. (2005). *Class matters.* New York: Times Books.

Obermeyer, C. M. (1992). Islam, women, and politics: The demography of Arab countries. *Population and Development Review, 18,* 33–60.

Oberschall, A. (1996). *Social movements.* New Brunswick, NJ: Transaction.

Odaka, K. (1998). Portraits of the Japanese workplace: Labor movements, workers, and managers. *Journal of Comparative Economics, 26,* 825–829.

Office of Technology Assessment, U.S. Congress (1986). *Technology, public policy, and the changing structure of agriculture: Vol. 2. Background papers: Part D. Rural communities.* Washington, DC.

Office on Violence Against Women, U.S. Department of Justice (2006). *Myths and facts about sexual violence.* www.usdoj.gov/ovw.

Ofosu-Amaah, V. (1998). Declines in fertility levels evident in Africa, notes UN Population Fund. *UN Chronicle, 35,* 20–21.

Ogbu, J. U. (1994, Winter). Racial stratification and education in the United States: Why inequality persists. *Teachers College Record,* pp. 264–298.

Ogburn, W. F. (1942). Inventions, population and history. In American Council of Learned Societies, *Studies in the history of culture.* Freeport, NY: Books for Libraries Press.

Ogburn, W. F. (1957). Cultural lag as theory. *Sociology and Social Research, 41,* 167–174.

Ogburn, W. F., & Nimkoff, M. F. (1934). *The family.* Boston: Houghton Mifflin.

O'Hare, W. P., & Lamb, V. L. (2009, March). Ranking states on improvement in child well-being since 2000. A KIDS COUNT Working Paper. Baltimore: Annie E. Casey Foundation.

O'Hear, A. (Ed.) (2005). *Philosophy, biology, and life.* New York: Cambridge University Press.

Oldenburg, R. (1997). *The great good place.* New York: Marlowe.

O'Leary, J. (2006). Developing countries "cannot reverse brain drain," ACU told. *New York Times, Higher Education Supplement,* April 21, p. 11.

Olinger, D., & Emery, E. (2008). Waging internal war: The Army's suicide rate is at a high, and kin say that multiple deployments are to blame. *Denver Post,* August 27.

Oliver, M. L., & Shapiro, T. M. (1990). Wealth of a nation: A reassessment of asset inequality in America shows at least one third of households are asset-poor. *American Journal of Economics and Sociology, 49,* 129–150.

Olson, M. (1965). *The logic of collective action.* Cambridge, MA: Harvard University Press.

O'Neil, J. (1997). Building schools as communities: A conversation with James Comer. *Educational Leadership, 54,* 6–11.

O'Neill, B., & Balk, D. (2001). World population futures. *Population Bulletin, 56,* no. 3.

Opie, I., & Opie, P. (1969). *Children's games in street and playground.* Oxford, England: Oxford University Press.

Oppenheimer, M. (1999). We made it, so what? *Monthly Review,* September, pp. 5–6.

Oppenheimer, V. K. (1994). Women's rising employment and the future of the family in industrial societies. *Population and Development Review, 20,* 293–342.

Orfield, G., & Eaton, S. E. (1996). *Dismantling desegregation: The quiet reversal of Brown v. Board of Education.* New York: New Press.

Orfield, G., & Wald, J. (2001). High stakes tests attached to high school graduation lead to increased drop-out rates, particularly for poor and minority students. *In Motion,* April 29, pp. 1–2.

Oriard, M. (2009). How the 60s changed big-time college football. *Chronicle of Higher Education,* October 30, pp. B13–B15.

Ornish, D. (1998). *Love and survival: The scientific basis for the healing power of intimacy.* New York: HarperCollins.

Orwell, G. (1950). *Shooting an elephant and other essays.* New York: Harcourt Brace Jovanovich.

Pager, D. (2003). The mark of a criminal record. *American Journal of Sociology, 108,* 937–975.

Pager, D., Bonikowski, B., & Western, B. (2009). Discrimination in a low-wage labor market. *American Sociological Review, 74,* 777–799.

Pal, A. (2010). Times Square bomber a complete disgrace. *Lexington Leader-Herald,* June 23.

Palmer, J. (1995, July-August). Windows of opportunity (the motion picture industry in other countries). *UNESCO Courier,* pp. 29–31.

Park, R. E. (1914). Racial assimilation in secondary groups. *Publications of the American Sociological Society, 8,* 66–72.

Park, R. E. (1967/1925). The city: Suggestions for the investigation of human behavior in the urban environment. In R. E. Park & E. W. Burgess (Eds.), *The city.* Chicago: University of Chicago Press.

Park, R. E. (1967/1926). The urban community as a spatial pattern and a moral order. In R. H. Turner (Ed.), *Robert E. Park on social control and collective behavior.* Chicago: University of Chicago Press.

Park, R. E., & Burgess, E. W. (1921). *Introduction to the science of sociology,* rev. ed. Chicago: University of Chicago Press.

Parsons, T. (1940). An analytic approach to the theory of social stratification. *American Journal of Sociology, 45,* 841–862.

Parsons, T. (1951). *The social system.* New York: Free Press.

Parsons, T. (1960). *Structure and process in modern societies.* New York: Free Press.

Parsons, T. (1966). *Societies: Evolutionary and comparative.* Englewood Cliffs, NJ: Prentice Hall.

Parsons, T. (1968). The problem of polarization along the axis of color. In J. H. Franklin (Ed.), *Color and race.* Boston: Beacon.

Parsons, T. (1991). The integration of economic and sociological theory. *Sociological Inquiry, 61,* 10–60.

Parsons, T., & Bales, R. F. (1955). *Family, socialization, and interaction process.* New York: Free Press.

Partin, M., & Palloni, A. (1995, Spring). Accounting for the recent increases in low birth weight among African Americans. *Focus* (Institute for Research on Poverty), pp. 33–38.

Patterson, O. (1982). *Slavery and social death.* Cambridge, MA: Harvard University Press.

Patterson, O. (1991). *Freedom: Vol. 1. Freedom in the making of Western culture.* New York: Basic Books.

Patterson, O. (1999). *Rituals of blood: Consequences of slavery in two American centuries.* Washington, DC: Civitas/Counterpoint.

Pavlov, I. (1927). *Conditioned reflexes: An investigation of the physiological activity of the cerebral cortex* (trans. end ed. G. V. Anrep). London: Oxford University Press.

Pear, R. (1993, July 8). Big health gap, tied to income, is found in U.S. *New York Times,* pp. A1, B10.

Pear, R. (2010). Health plans must provide some tests at no cost. *New York Times,* July 14, p. 16.

Pedersen, E., & Faucher, T. A., with Eaton, W. W. (1978). A new perspective on the effects of first-grade teachers on children's subsequent adult status. *Harvard Educational Review, 48,* 1–31.

Peek, C. W., Zsembik, B. A., & Cloward, R. T. (1997). The changing caregiving networks of older adults. *Research on Aging, 19,* 333–361.

Peiss, K. (Ed.) (2002). *Major problems in the history of American sexuality.* Boston: Houghton Mifflin.

Peplau, L. A., Hill, C. T., & Rubin, Z. (1993). Sex role attitudes in dating and marriage: A 15-year follow-up of the Boston couples study. *Journal of Social Issues, 49,* 3–25.

Perin, C. (1977). *Everything in its place: Social order and land use in America.* Princeton, NJ: Princeton University Press.

Perls, T. T., & Silver, M. H. (1999). *Living to 100.* New York: Basic Books.

Perutz, M. F. (1992). The fifth freedom. *New York Review of Books, 39,* 5–7.

Peters, A. (1990). *Peters atlas of the world.* New York: HarperCollins.

Peterson, P. E. (2003, July–August). Schools after twenty years. *Current,* pp. 7–9.

Pettigrew, T. F. (1980). Prejudice. In *Harvard encyclopedia of American ethnic groups.* Cambridge, MA: Belknap.

Pettit, K. L. S., & Kingsley, G. T. (2003, May 19). *Concentrated poverty: A change in course.* www.urban.org.

Pew Research Center. (2010, February 10). Millennials, the generation next. Washington, DC: Pew Research Center for the People and the Press.

Peyton, L. (2005). High/Scope supporting the child, the family, the community: A report of the proceedings of the High/Scope Ireland Third Annual Conference, 12th October 2004. *Child Care in Practice, 11,* 433–456.

Pfeiffer, E., Verwoerdt, A., & Davis, G. (1972). Sexual behavior in middle life. *American Journal of Psychiatry, 128,* 1262–1267.

Phillips, T. (1998). The end of the rope, part three: The capital punishment revival. *Contemporary Review, 272,* 181–186.

Piaget, J., & Inhelder, B. (1969). *The psychology of the child.* New York: Basic Books.

Pierson, P., & Skocpol, T. (Eds.). (2007). *The transformation of American politics: Activist government and the rise of conservatism.* Princeton, NJ: Princeton University Press.

Pifer, A., & Bronte, L. (Eds.) (1986). *Our aging society: Paradox and promise.* New York: Carnegie Corporation of New York.

Pinker, S. (2002). *The blank slate.* New York: Viking.

Pittman, J. F. (1993). Functionalism may be down, but it surely is not out. In P. G. Boss, W. J. Doherty, R. LaRossa, W. R. Schumm, & S. K. Steinmetz (Eds.), *Sourcebook of family theories and methods: A contextual approach.* New York: Plenum.

Piven, F. F. (1997). *The breaking of the American social contract.* New York: New Press.

Platt, A. M. (1977). *The child savers,* 2nd ed. Chicago: University of Chicago Press.

Ploch, D. R., & Hastings, D. W. (1995). Some church; some don't. *Journal for the Scientific Study of Religion, 34,* 507–516.

Polanyi, K. (1944). *The great transformation.* Boston: Beacon.

Polsby, N. (1980). *Community power and political theory,* 2nd ed. New Haven, CT: Yale University Press.

Population Reference Bureau. (1998). Parents whose children need day care often face daily crisis. *Population Today, 26,* 1–2.

Population Reference Bureau. (2001, December). *Population Bulletin #4.* Washington, DC: Government Printing Office.

Population Reference Bureau. (2002). *2002 world population data sheet.* Washington, DC: U.S. Government Printing Office.

Population Reference Bureau. (2010). Analysis of 2007 American Community Survey. Washington, DC: Population Reference Bureau.

Portes, A., & Rumbaut, R. G. (1996). *Immigrant America: A portrait.* Berkeley: University of California Press.

Portes, A., & Stepick, A. (1985). Unwelcome immigrants. *American Sociological Review, 50,* 493–514.

Pozo, J. C., & Pozo, L. M. (2007). Paradoxes of the market wars. *Science and Society, 71,* 322–347.

Press, J. E., & Townsley, E. (1998). Wives' and husbands' housework reporting: Gender, class, and social desirability. *Gender and Society, 12,* 188–219.

Pressman, J. L., & Wildavsky, A. (1984). *Implementation,* 3rd ed. Berkeley: University of California Press.

Preston, J. (1998, March 6). Mexico's overtures to the Zapatistas bring tensions in Chiapas to a new boiling point. *New York Times,* p. A8.

Preston, S. H. (1984). Children and the elderly: Divergent paths for America's dependents. *Demography, 21,* 435–457.

Prüss-Üstün, A., Bos, R., Gore, F., & Bartram, J. (2008). *Safer water, better health.* World Health Organization.

Pryce, D. (2006, May 8). Combating modern-day slavery. *Human Events,* p. 20

Putnam, R. D. (2000). *Bowling alone: The collapse and revival of American community.* New York: Simon and Schuster.

Pyle, R. E. (1996). *Persistence and change in the Protestant establishment.* Westport, CT: Praeger.

Quinn, B. A. (2000). The paradox of complaining: Law, humor, and harassment in the everyday work world. *Law and Social Inquiry, 25,* 1151–1185.

Quinney, R. (1980). *Class, state, and crime.* White Plains, NY: Longman.

Quinney, R. (2001). *Bearing witness to crime and social justice.* Albany: State University of New York Press.

Radelet, M. L., & Bedau, H. A. (1992). *In spite of innocence: Erroneous convictions in capital cases.* Boston: Northeastern University Press.

Radick, G. (2007). *The simian tongue: The long debate about animal language.* Chicago: University of Chicago Press.

Rajan, R. S. (2001, April). Globalization. *ASEAN Economic Bulletin,* pp. 1–11.

Randall, V. (1998). *Political change and underdevelopment: A critical introduction to Third World politics,* 2nd ed. Durham, NC: Duke University Press.

Rapoport, R., et al. (2002). *Beyond work-family balance: Advancing gender equity and workplace performance.* San Francisco: Jossey-Bass.

Rawles, J. (2001). *Justice as fairness.* Cambridge, MA: Belknap Press.

Ray, B. D. (2000). Homeschooling—state versus family. *Peabody Journal of Education, 75,* 71–106.

Reardon, C. (2002, Winter). Islam and the modern world. *Ford Foundation Report,* pp. 18–23.

Redfield, R. (1947). The folk society. *American Journal of Sociology, 52,* 293–308.

Reese, S. (1998, August). Religious spirit. *American Demographics,* p. 62.

Reich, R. B. (1992). Training a skilled work force. *Dissent,* Winter, pp. 42–46.

Reichel, P. (Ed.) (2005). *Handbook of transnational crime and justice.* Thousand Oaks, CA: Sage.

Reid, S. T. (1993). *Criminal justice.* New York: Macmillan.

Reid, T. R. (2009). *The healing of America: A global quest for better, cheaper, and fairer health care.* New York: Penguin.

Reidy, D. A. (2001). Pluralism, liberal democracy, and compulsory education: Accommodation and assimilation. *Journal of Social Philosophy, 32,* 585–609.

Reiman, J. (2002). *The rich get richer and the poor get prison,* 7th ed. Boston: Allyn & Bacon.

Reinhardt, U. W. (2008). The true cost of health care. *America,* September 8, pp. 10–13.

Rejwan, N. (1998). *Arabs face the modern world: Religious, cultural, and political responses to the West.* Gainesville: University of Florida Press.

Renzetti, C. M. (1999). Editor's introduction. *Violence Against Women, 5,* 227–229.

Renzetti, C. M., & Bergen, R. K. (Eds.) (2005). *Violence against women.* Lanham, MD: Rowman & Littlefield.

Richards, E. (1989). *The knife and gun club: Scenes from an emergency room.* New York: Atlantic Monthly Press.

Ricks, T. E. (2009). *The gamble: General David Petraeus and the American military adventure in Iraq, 2006–2008.* New York: Penguin.

Riesman, D. (1957). The suburban dislocation. *Annals of the American Academy of Political and Social Science, 314,* 123–146.

Riis, J. A. (1890). *How the other half lives: Studies among the tenements of New York.* New York: Scribner.

Riley, M. W., Foner, A., & Waring, J. (1988). The sociology of age. In N. E. Smelser (Ed.), *The handbook of sociology.* Newbury Park, CA: Sage.

Riley, M. W., Kahn, R. L., & Foner, A. (1994). *Age and structural lag.* New York: Wiley.

Riley, M. W., & Riley, J. W., Jr. (1989). The lives of older people and changing social roles. *Annals of the American Academy of Political and Social Science, 503,* 14–28.

Rist, R. C. (1973). *The urban school: A factory for failure.* Cambridge, MA: MIT Press.

Roberts, A. E., Koch, J. R., & Johnson, D. P. (2001). Religious reference groups and the persistence of normative behavior: An empirical test. *Sociological Spectrum, 21,* 81–96.

Robertson, R. (2009, October 4). Housing battle reveals post-Katrina tensions. *New York Times,* p. A16.

Robertson, R. (1985). The sacred and the world system. In P. Hammond (Ed.), *The sacred in a secular age.* Berkeley: University of California Press.

Robinson, J. (1988). *The rhythm of everyday life.* Boulder, CO: Westview.

Robinson, J. P., & Godbey, G. (1997). *Time for life: The surprising ways Americans use their time.* University Park: Pennsylvania State University Press.

Robinson, P. (1994). *Freud and his critics.* Berkeley: University of California Press.

Robinson, R. (2003, February). Reparations—More than just a check. *Black Collegian,* pp. 96–98.

Rock, P. (1985). Symbolic interactionism. In A. Kuper & J. Kuper (Eds.), *The social science encyclopedia.* London: Routledge and Kegan Paul.

Rodgers, K. (2009). When do opportunities become trade-offs for social movement organizations? Assessing media impact in the global human rights movement. *Canadian Journal of Sociology, 34,* 1087–1114.

Rodriguez, C. E. (1992). The Puerto Rican community in the South Bronx: Contradictory views from within and without. In P. S. Rotherberg (Ed.), *Race, class and gender in the United States.* New York: St. Martin's Press.

Roer-Strier, D., & Rosenthal, M. K. (2001). Socialization in changing cultural contexts: A search for images of the "adaptive adult." *Social Work, 46,* 215–228.

Rosario, V. (2001, September–October). The "gay gene" is born. *Gay and Lesbian Review,* pp. 13–15.

Rose, M. (1995). *Possible lives: The promise of public education.* Boston: Houghton Mifflin.

Rosen, E. I. (1987). *Bitter choices: Blue-collar women in and out of work.* Chicago: University of Chicago Press.

Rosenberg, M., & Turner, R. H. (Eds). (1981). *Social Psychology: Sociological Perspectives.*

Rosenhan, D. L. (1973). On being sane in insane places. *Science, 179,* 250–258.

Rosow, I. (1965). Forms and functions of adult socialization. *Social Forces, 44,* 38–55.

Ross, H. L. (1963). *Perspectives on the social order.* New York: McGraw-Hill.

Rossi, A. (1964). Equality between the sexes: An immodest proposal. In R. J. Lifton (Ed.), *The woman in America.* Boston: Beacon.

Rossi, A. (1980). Aging and parenthood in the middle years. In P. B. Baltes & O. G. Brim, Jr. (Eds.), *Life-span development and behavior* (Vol. 3). Orlando: Academic Press.

Rossi, A. (1994). *Sexuality across the life course.* Chicago: University of Chicago Press.

Rossi, A. (Ed.) (2001). *Caring and doing for others: Social responsibility in the domains of family, work, and community.* Chicago: University of Chicago Press.

Roszak, T. (1969). *The making of a counterculture.* Garden City, NY: Doubleday.

Rothenberg, P. S. (Ed.) (1992). *Race, class and gender in the United States,* 2nd ed. New York: St. Martin's Press.

Rubin, Z. (1980). The love research. In A. Skolnick & J. H. Skolnick (Eds.), *Family in transition.* Boston: Little, Brown.

Rubington, E., & Weinberg, M. S. (1999). *Deviance: The interactionist perspective,* 7th ed. Boston: Allyn & Bacon.

Ruggles, S. (1997). The rise of divorce and separation in the United States, 1880–1990. *Demography, 34,* 455–467.

Ruskin, C. (1988). *The quilt: Stories from the NAMES Project.* New York: Pocket Books.

Ryan, A. (1992). Twenty-first century limited. *New York Review of Books, 39,* 20–24.

Rybczynski, W. (1987). *Home: A short history of an idea.* New York: Penguin.

Rymer, R. (1994). *Genie: A scientific tragedy.* New York: Harper.

Sabel, C. F. (1995). Bootstrapping reform: Rebuilding firms, the welfare state, and unions. *Politics & Society, 23,* 5–49.

Sadik, N. (1995). Decisions for development: Women, empowerment and reproductive health. In *The state of the world population 1995.* New York: United Nations Population Fund.

Sadker, M., & Sadker, D. M. (2005). *Teachers, schools, and society,* 7th ed. Boston: McGraw-Hill.

Sadovnik, A. R., & Semel, S. F. (2010). Education and inequality: Historical and sociological approaches to schooling and social stratification. *Paedegogica Historica, 46,* pp. 1–13.

Sagarin, E. (1975). *Deviants and deviance: A study of disvalued people and behavior.* New York: Praeger.

Sahagun, L., & Stammer, L. B. (1998, February 20). Entire Promise Keepers' staff to be laid off. *Los Angeles Times,* p. A1.

Sahlins, M. D. (1960, September). The origin of society. *Scientific American,* pp. 76–87.

Salo, M., & Siebold, G. L. (2008). Variables impacting peer group cohesion in the Finnish conscript service. *Journal of political and Military Sociology, 36,* 1–18.

Salgado, S. (1990). *An uncertain grace.* New York: Aperture.

Salisbury, R. F. (1962). *From stone to steel.* Parkville, Australia: Melbourne University Press.

Samuelson, P. A. (1998). How Foundations came to be ("Foundations of Economic Analysis"). *Journal of Economic Literature, 36,* 1375–1387.

Sanchez-Jankowski, M. S. (1991). *Islands in the street: Gangs and American urban society.* Berkeley: University of California Press.

Sandefur, G. D., & Tienda, M. (Eds.) (1988). *Divided opportunities: Minorities, poverty, and social policy.* New York: Plenum.

Sander, T. H., & Putnam, R. D. (2010). Still bowling alone? The post-9/11 split. *Journal of Democracy, 21.*

Sanders, T. (2006). Female sex workers as health educators with men who buy sex: Utilising narratives of reationalisations. *Social Science and Medicine, 62,* 2 434–2444.

Sanlo, R. L. (Ed.) (2005). *Gender identity and sexual orientation: Research, policy, and personal perspectives.* San Francisco: Jossey-Bass.

Sapir, D. (1997, October). Women in the front line. *The UNESCO Courier,* pp. 27–29.

Sapolsky, R. (1998, April). How the other half heals: Links between health and socioeconomic status. *Discover,* pp. 46–72.

Sassen, S. (1991). *Global city.* Princeton, NJ: Princeton University Press.

Sassen, S. (1998). *Globalization and its discontents: Essays on the new mobility of people and money.* New York: Free Press.

Satir, V. (1972). *Peoplemaking.* Palo Alto, CA: Science and Behavior Books.

Satya, Resnick, S. A., & Wolff, R. D. (2008). State capitalism versus communism: What happened in the USSR and the PRC? *Critical Sociology, 34,* 539–556.

Savage-Rumbaugh, E. S., & Shanker, S. G. (1998). *Ape language and the human mind.* New York: Oxford University Press.

Sawhill, I., & Morton, J. E. (2008). *Economic mobility: Is the American dream alive and well?* Pew Trust, Economic Mobility Project.

Sbraga, T. P., & O'Donohue, W. (2000). Sexual harassment. *Annual Review of Sex Research, 11,* 258–286.

Scanzoni, J. (1995). *Contemporary families and relationships: Reinventing responsibility.* New York: McGraw-Hill.

Scarf, M. (1995). *Intimate worlds: Life inside the family.* New York: Random House.

Schaller, G. B. (1964). *The year of the gorilla.* Chicago: University of Chicago Press.

Schemo, D. J. (2003). Education secretary defends school system he once led. *New York Times,* July 26, p. A9.

Scheper-Hughes, N., & Devine, J. (2003, February). Priestly celibacy and child sexual abuse. *Sexualities,* pp. 15–40.

Schlosser, E. (2001). *Fast food nation: The dark side of the all-American meal.* Boston: Houghton Mifflin.

Schmitt, E. (1993). The top soldier is torn between 2 loyalties. *New York Times,* February 6, p. A1.

Schmitt, E. (2001, April 4). Analysis of census finds segregation along with diversity. *New York Times,* p. A15.

Schoenherr, R. A. (2002). *Goodbye father: The celibate male priesthood and the future of the Catholic Church.* New York: Oxford University Press.

Schooler, C., & Miller, J. (1985). Work for the household: Its nature and consequences for husbands and wives. *American Journal of Sociology, 90,* 97–124.

Schorr, L. B. (1988). *Within our reach: Breaking the cycle of disadvantage.* Garden City, NY: Doubleday.

Schuman, H., & Scott, J. (1989). Generations and collective memories. *American Sociological Review, 54,* 359–381.

Schumpeter, J. A. (1950). *Capitalism, socialism, and democracy* (3rd ed.). New York: Harper.

Schur, E. M. (1984). *Labeling women deviant: Gender, stigma, and social control.* New York: Random House.

Schwartz, B. (1976). Images of suburbia: Some revisionist commentary and conclusions. In B. Schwartz (Ed.), *The changing face of the suburbs.* Chicago: University of Chicago Press.

Schweber, N. (2010, May 2). Queens vigil protests Arizona immigration law. *New York Times.*

Scott, T. G. (1998). *The United States of suburbia.* Amherst, NY: Prometheus.

Scott, J. (2005). Life at the top in America isn't just better, it's longer. In Keller, B. (Ed.), *Class matters.* Henry Holt, Times Books.

Scruggs, L., & Lange, P. (2002). Where have all the members gone? Globalization, institutions, and union density. *Journal of Politics, 64,* 22–39.

Scull, A. T. (1988). Deviance and social control. In N. J. Smelser (Ed.), *The handbook of sociology.* Newbury Park, CA: Sage.

Seligson, M. A., & Passe-Smith, J. T. (1993). *Development and underdevelopment: The political economy of inequality.* Boulder, CO: Renner.

Selizer, V. (1985). *Pricing the priceless child: The changing social value of children.* New York: Free Press.

Selznick, P. (1952). *The organizational weapon: A study of Bolshevik strategy and tactics.* New York: McGraw-Hill.

Selznick, P. (1966). *TVA and the grass roots: A study in the sociology of formal organization.* New York: Harper.

Sen, A. (1992). *Inequality reexamined.* New York: Russell Sage.

Sen, A. (1993, May). The economics of life and death. *Scientific American,* pp. 40–47.

Sen, A. (1997). Maximization and the act of choice. *Econometrica, 65,* 745–779.

Sen, A. (2000). *Development as freedom.* Anchor.

Sen, A. K. (2001). *Development as freedom.* New York: Oxford University Press.

Sen, A. (2009). *The idea of justice.* Cambridge, MA: Belknap.

Sennett, R., & Cobb, J. (1972). *The hidden injuries of class.* New York: Random House.

Sentencing Project. (2010). Report submitted to Inter-American Commission on Human Rights. Washington, DC: The Sentencing Project.

Settersten, R. A., Jr., & Owens, T. J. (Eds.) (2002). *Advances in life course research, Vol. 7: New frontiers in socialization.* Oxford, UK: JAI Press.

Shapiro, I., & Greenstein, R. (2001). *The widening income gulf.* Washington, DC: Center on Budget and Policy Priorities.

Shaw, C. R. (1929). *Delinquency areas: A study of the geographic distribution of school truants, juvenile delinquents, and adult offenders in Chicago.* Chicago: University of Chicago Press.

Shibutani, T., & Kwan, K. M. (1965). *Ethnic stratification.* New York: Macmillan.

Shils, E. (1970). Tradition, ecology and institution in the history of sociology. *Daedalus, 99,* 760–825.

Shils, E. (1985). Sociology. In A. Kuper & J. Kuper (Eds.), *The social science encyclopedia.* London: Routledge and Kegan Paul.

Shorris, E. (1992). *Latinos: A biography of the people.* New York: Norton.

Shotland, R. L. (1985). When bystanders just stand by. *Psychology Today, 19,* 50–55.

Shuck, P. (2001, Winter). Affirmative action. *Brookings Review,* pp. 24–27.

Siegel, A. (1992, March). Felice Schwartz tries to set the record straight. *Working Woman,* p. 17.

Silberman, C. (1980). *Criminal violence, criminal justice.* New York: Random House.

Silk, L. (1989). Rich and poor: The gap widens. *New York Times,* May 12, p. D2.

Simmel, G. (1904). The sociology of conflict. *American Journal of Sociology, 9,* 490ff.

Simmons, J. L. (1985). The nature of deviant subcultures. In E. Rubington & M. S. Weinberg (Eds.), *Deviance: The interactionist perspective.* New York: Macmillan.

Simon, J. (1996). *The ultimate resource.* Princeton, NJ: Princeton University Press.

Simon, W., & Gagnon, J. H. (1969). On psychosexual development. In D. A. Goslin, Ed., *Handbook of socialization theory and research.* Chicago: Rand McNally.

Simon, W., & Gagnon, J. H. (2003). Sexual scripts: Origins, influences and changes. *Qualitative Sociology, 26,* 491–497.

Simpson, C. E., & Yinger, J. M. (1953). *Racial and cultural minorities: An analysis of prejudice and discrimination.* New York: Harper.

Simpson, J. H. (1983). Moral issues and status politics. In R. C. Liebman & R. Wuthnow (Eds.), *The new Christian right: Mobilization and legitimation.* Hawthorne, NY: Aldine.

Singer, M. (1999). The population surprise. *Atlantic Monthly,* August, pp. 22–24.

Singer, P. A. (2006). *Children at war.* Berkeley: University of California Press.

Sizer, T. (1995). What's wrong with standard tests. *Education Life, New York Times,* January 8, p. 58.

Sjoberg, G. (1968). The preindustrial city. In S. Fava (Ed.), *Urbanism in world perspective: A reader.* New York: Crowell.

Skinner, B. F. (1976). *About behaviorism.* Vintage.

Skocpol, T. (1979). *States and social revolutions.* New York: Cambridge University Press.

Skolnick, J. H., & Currie, E. (2004). Crisis in American institutions. Boston: Allyn & Bacon.

Skousen, M. (2001). *The making of modern economics: The lives and ideas of the great thinkers.* Armonk, NY: M. E. Sharpe.

Slevin, P. (2010, May 2). Obama defends priorities. *Washington Post.*

Smelser, N. J. (1962). *Theory of collective behavior.* New York: Free Press.

Smelser, N. J. (1966). The modernization of social relations. In M. Weiner (Ed.), *Modernization.* New York: Basic Books.

Smelser, N. J., & Swedberg, R. (Eds.) (2005). *The handbook of economic sociology,* 2nd ed. Princeton, NJ: Princeton University Press.

Smil, V. (2001). *Feeding the world.* Cambridge, MA: M.I.T. Press.

Smith, A. (1965/1776). *The wealth of nations.* Modern Library.

Smith, C., Denton, M. L., Faris, R., & Regnerus, M. (2002). Mapping American adolescent religious participation. *Journal for the Scientific Study of Religion, 41,* 597–612.

Smith, C. D. (1994). *The absentee American: Repatriates' perspectives on America and its place in the contemporary world.* Putnam Valley, NY: Aletheia.

Smith, C. D. (Ed.) (1996). *Strangers at home: Essays on the effects of living overseas and coming "home" to a strange land.* Putnam Valley, NY: Aletheia.

Smith, D. (1989). *Promises* (Letter from Madurai). Oberlin, OH: Oberlin Shansi Memorial Association.

Smith, D. M., & Gates, G. J. (2001, August). Gay and lesbian families in the United States: Same-sex unmarried partner households, a preliminary analysis of 2000 census data. Washington, DC: D.C. Human Rights Campaign.

Smith, R. A. (1968). Los Angeles, prototype of supercity. In S. F. Fava (Ed.), *Urbanism in world perspective: A reader.* New York: Crowell.

Smith, T.W. (2007, April 17). Job satisfaction in the United States. NORC/University of Chicago.

Snipp, C. M. (1991). *American Indians: The first of this land.* New York: Russell Sage.

Snow, D. A., & Benford, R. D. (1992). Master frames and cycles of protest. In A. D. Morris & C. M. Mueller (Eds.), *Frontiers in social movement theory.* New Haven, CT: Yale University Press.

Sorokin, P. (1937). *Social and cultural dynamics: Vol. 3. Fluctuation of social relationships, war, and revolution.* New York: American Book.

Sowell, T. (1972). *Black education: Myths and tragedies.* New York: McKay.

Spencer, H. (1874). *The study of sociology.* New York: Appleton.

Spohn, W. (2010). Political sociology: Between civilizations and modernities. *European Journal of Social Theory, 13,* 49–66.

Stack, C. (1974). *All our kin.* New York: Harper.

Staggenborg, S. (1998). *Gender, family, and social movements.* Thousand Oaks, CA: Pine Forge.

Stark, R. (2005). *The rise of Mormonism.* New York: Columbia University Press.

Stark, R., with Bader, C. (2008). *What Americans really believe: New findings from the Baylor surveys of religion.* Waco, TX: Baylor University Press.

Starr, P. (1982). *The social transformation of American medicine.* New York: Basic Books.

Starr, P. (1995, Winter). What happened to healthcare reform? *American Prospect,* pp. 20–31.

Starr, P. (2010). Last chance for health reform. *American Prospect,* April, p. 3.

Statistical Abstract of the United States. Annual. Washington, DC: U.S. Bureau of the Census.

Stehr, N., & Adolf, M. (2010). Consumption between market and morals: A socio-cultural consideration of moralized markets. *European Journal of Social Theory, 13,* 213–228.

Stein, J. (1994, December). Space and place. *Art in America,* pp. 66–71.

Steltzer, U. (1982). *Inuit: The north in transition.* Chicago: University of Chicago Press.

Step Family Foundation. (2010). www.healthofchildren.com/S/Stepfamilies.html.

Stephen, A. (2003, August 25). More junk food, ever-larger portions and more hours spent in front of the television have made the United States the fattest nation in the world. Is it too late to reverse the trend? *New Statesman,* p. 12.

Stevens, W. K. (1988, December 20). Life in the Stone Age: New findings point to complex societies. *New York Times,* pp. C1, C15.

Stevenson, H. W. (1992). Learning from Asian schools. *Scientific American,* December, pp. 70–76.

Stevenson, H. W. (1998). A study of three cultures: Germany, Japan, and the United States. *Phi Delta Kappan,* March, pp. 524–530.

Stevenson, R. W. (1999). Texaco is said to set payment over sex bias. *New York Times,* January 6, p. C1.

Steward, J. H. (1955). *The theory of culture change: The methodology of multilinear evolution.* Urbana: University of Illinois Press.

Stewart, C. T., Jr. (2002). Inequality of wealth and income in a technologically advanced society. *Journal of Social Political and Economic Studies, 27,* 495–512.

Stewart, G. R. (1968). *Names in the land.* Boston: Houghton Mifflin.

Stimson, C. (1980). Women and the American city. *Signs, 5* (Suppl.).

Stolberg, C. G. (1997, November 23). Gay culture weighs sense and sexuality. *New York Times,* sec. 4, pp. 1, 6.

Stouffer, S. A., Suchman, E. A., DeVinney, L. C., Star, S. A., & Williams, R. A., Jr. (1949). *Studies in social psychology in World War II: Vol. 1. The American soldier: Adjustment during army life.* Princeton, NJ: Princeton University Press.

Sudman, S., & Bradburn, N. (1982). *Asking questions: A practical guide to questionnaire design.* San Francisco: Jossey-Bass.

Sudnow, D. (1967). *Passing on: The social organization of dying.* Englewood Cliffs, NJ: Prentice Hall.

Sullivan, K. (1998, November-December). Defining democracy down. *American Prospect,* pp. 91–96.

Sumner, W. G. (1940/1907). *Folkways.* Boston: Ginn.

Suro, R. (1992, May 26). For women, varied reasons for single motherhood. *New York Times,* p. A12.

Surowiecki, J. (2010). The financial page: Greater fools. *New Yorker,* July 5, p. 23.

Sutherland, E. H. (1940). White collar criminality. *American Sociological Review, 5,* 1–12.

Suttles, G. (1967). *The social order of the slum.* Chicago: University of Chicago Press.

Suttles, G. (1972). *The social construction of communities.* Chicago: University of Chicago Press.

Sutton-Smith, B., & Rosenberg, B. G. (1961). Sixty years of historical change in the game preferences of American children. *Journal of American folklore, 74,* 17–46.

Swarns, R. L. (2006). In Georgia, newest immigrants unsettle an old sense of place. *New York Times,* August 4, pp. A1, A15.

Swedberg, R. (1994). *Explanations in economic sociology.* New York: Russell Sage.

Sweet, S. A., & Meiksins, P. (2008). *Changing contours of work: Jobs and opportunities in the new economy.* Thousand Oaks, CA: Pine Forge Press.

Swift, D. C. (1998). *Religion and the American experience: A social and cultural history.* Armonk, NY: Sharpe.

Swindell, B., Allen, J., & Graham-Silverman, A. (2003). "Family friendly" bill to foster comp time flexibility gets laid off House schedule. *CQ Weekly,* June 7, p. 1.

Szelenyi, I. (1983). *Urban inequalities under state socialism.* New York: Oxford University Press.

Tabb, W. (1986). *Churches in struggle: Liberation theologies and social change.* New York: Monthly Review Press.

Taeuber, K. E., & Taeuber, A. F. (1965). *Negroes in cities.* Hawthorne, NY: Aldine.

Tagliabue, J. (1998). Communists, remodeled, keep trying in Germany. *New York Times,* August 19, p. A5.

Talbot, M. (1998, May 24). Attachment theory: The ultimate experiment. *New York Times Magazine,* pp. 24–30.

Tannen, D. (1990). *You just don't understand: Women and men in conversation.* New York: Morrow.

Tannen, D. (2001). *I only say this because I love you.* New York: Random House.

Tarde, G. (1903). *The laws of imitation.* New York: Holt.

Tarrow, S. (2008). Charles Tilly and the practice of contentious politics. *Social Movement Studies, 7,* 225–246.

Tarrow, S. G. (1994). *Power in movement: Social movements, collective action, and politics.* Cambridge, England: Cambridge University Press.

Tawney, R. A. (1966/1932). *Land and labor in China.* Boston: Beacon.

Taylor, J. M., Gilligan, C., & Sullivan, A. M. (1995). *Between voice and silence: Women and girls, race and relationship.* Cambridge, MA: Harvard University Press.

Teachman, J. (2007). Military service and educational attainment in the all-volunteer era. *Sociology of Education, 80,* 359–374.

Teitelbaum, M., & Weiner, M. (1995). *Threatened peoples, threatened borders: World migration and U.S. policy.* New York: Norton.

Tewksbury, R., & Taylor, J. M. (1996). The consequences of eliminating Pell Grant eligibility for students in post-secondary correctional education programs. *Federal Probation, 60,* 60–63.

Thio, A. (1998). *Deviant behavior,* 5th ed. New York: Longman.

Thomas, L. (1979). *The medusa and the snail.* New York: Bantam.

Thomas, W. I. (1971/1921). *Old world traits transplanted.* Montclair, NJ: Patterson.

Thomas, W. L. (1956). *Man's role in changing the face of the earth.* Chicago: University of Chicago Press.

Thompson, L. B. (2002). Defense budget battle: Did Rumsfeld get it right? *San Diego Union-Tribune,* February 17, p. G-1.

Thompson, P. M., et al. (2001). Genetic influences on brain structure. *Nature Neuroscience, 4,* 1253–1258.

Thorne, B. (1993). *Gender play: Girls and boys in school.* New Brunswick, NJ: Rutgers University Press.

Thorne, B. (2000). A telling time for women's studies. *Signs: Journal of Women in Culture and Society, 25,* 1183–1187.

Thorne, B. (2006). How can feminist sociology sustain its critical edge? *Social Problems, 53,* 473–478.

Thornton, A., & Young-DeMarco, L. (2001, November 1). Four decades of trends in attitudes toward family issues in the United States: The 1960s through the 1990s. *Journal of Marriage and Family.*

Thornton, R. (1987). *American Indian holocaust and survival: A population history since 1492.* Norman: University of Oklahoma Press.

Tienda, M., & Singer, A. (1995). Wage mobility of undocumented workers in the United States. *International Migration Review, 29,* 112–138.

Tienda, M., & Wilson, F. D. (1992). Migration and the earnings of Hispanic men. *American Sociological Review, 57,* 661–678.

Tilly, C. (1993). *European revolutions, 1492–1992.* Oxford, England: Blackwell.

Tobach, E., & Rosoff, B. (Eds.). (1994). *Challenging racism and sexism: Alternatives to genetic explanations.* New York: Feminist Press at the City University of New York.

Toffler, A. (1970). *Future shock.* New York: Bantam.

Torres, C. A., & Mitchell, T. R. (Eds.) (1998). *Sociology of education: Emerging perspectives.* Albany: State University of New York Press.

Touraine, A. (1998). Social transformations of the twentieth century. *International Social Science Journal, 50,* 165–172.

Toynbee, A. J. (1972). *A study of history.* New York: Oxford University Press.

Traven, B. (1966). Assembly line. In *The night visitor, and other stories.* New York: Hill and Wang.

Truzzi, M. (1971). *Sociology: The classic statements.* New York: McGraw-Hill.

Tuch, S. A., & Martin, J. K. (1997). *Racial attitudes in the 1990s: Continuity and change.* Westport, CT: Praeger.

Tumin, M. (1967/1984). *Social stratification: The forms and functions of inequality,* 2nd ed. Prentice Hall.

Turner, F. J. (1920/1893). *The frontier in American history.* Fort Worth, IN: Holt, Rinehart and Winston.

Turner, R. H. (1974). The theme of contemporary social movements. In R. E. L. Faris (Ed.), *Handbook of modern sociology.* Chicago: Rand McNally.

Uchitelle, L. (2006). When talk of guns and butter includes lives lost. *New York Times,* January 15, p. BU3.

Uhlenberg, P. (1980). Death and the family. *Journal of Family History, 5,* 313–320.

United Nations. Annual. *State of the World* report.

UNDP (United Nations Development Programme). Various years. *Human Development Report.* New York: Oxford University Press.

United Nations Development Programme. 2008. Millennium Goals Report. Geneva, Switzerland: United Nations.

UNICEF. Annual. The state of the world's children 2008. Geneva, Switzerland: United Nations.

United Nations. (1995, March). *Adoption of the declaration and programme of action of the World Summit for Social Development.* Geneva, Switzerland: United Nations.

United Nations. (2006). Millennium Development Goals. http://econ.lse. ac.uk/?-tbesley/papers/jep10.pdf.

United Nations Department of Public Information. (2002, March). DP/2264.

UNFPA (United Nations Population Fund). Annual. State of the world population. www.unfpa.org/swp.

United Nations Population Fund. (2005). *State of world population 2005: The promise of equality: Gender equity, reproductive health and the millennium development goals.* New York: United Nations Population Fund.

U.S. Bureau of Labor Statistics. (2010). "Education Pays." *Education Projections.* www.bls.gov/emp/ep_chart_001.htm.

U.S. Census Bureau. (2010). *Census of population: 2010.*

U.S. Census Bureau. (2005). *Annual social and economic supplement.*

U.S. Census Bureau. (2006). *Housing and household economic statistics.*

U.S. Census Bureau. Annual. *Statistical Abstract.* www.census.gov/compendia/statab/.

U.S. Census Bureau. (Annual). *Current Population Survey, Annual Social and Economic Supplement.*

U.S. Department of Agriculture. (2010). www.ers.usda.gov/statefacts/us.htm.

U.S. Department of Commerce (2001). *An aging world.*

U.S. Department of Education (1999). *How old are America's public schools?*

U.S. Department of Education (2006). *Rates of victimization in U.S. schools.*

U.S. Department of Labor. (2009). News release, August 2009. www.bls.gov/news.release/youth.

University of Chicago (2006). Gender equality leads to better sex lives among people 40 and over. Press release, www.news.uchicago.edu/releases.

Valdez, A. (2007). *Mexican American girls and gang violence: Beyond risk.* Palgrave Macmillan.

Valle, I., & Weiss, E. (2010). Participation in the figured world of graffiti. *Teaching and Teacher Education, 26:* 128–135.

Van Gennep, A. (1960/1908). *The rites of passage.* Chicago: University of Chicago Press.

Van Gestel, S., & Van Broeckhoven, C. (2003). Genetics of personality: Are we making progress? *Molecular Psychiatry, 8,* 840–852.

Vanneman, R., & Cannon, L. W. (1987). *The American perception of class.* Philadelphia: Temple University Press.

Vass, W. K. (1979). *The Bantu speaking heritage of the United States.* Los Angeles: Center for Afro-American Studies, University of California.

Ventura, S. J., et al. (2001). Births to teenagers in the United States, 1940–2000. *National Vital Statistics Reports, 49,* 10.

Vest, J. (2001, December). Fourth-generation warfare. *Atlantic Monthly,* pp. 49–50.

Vilas, C. M. (1993). The hour of civil society. *NACLA Report on the Americas, 27,* 38–43.

Viljoen, F. C. (2010). The World Health Organization's water safety plan is much more than just an integrated drinking water quality management plan. *Science and Technology, 61,* 173–179.

Villarosa, L. (2002, January 1). A conversation with Joseph Graves. *New York Times,* p. F5.

Vining, D. R., Jr. (1985). The growth of core regions in the Third World. *Scientific American,* April, pp. 42–49.

Vogel, E. (1979). *Japan as number one: Lessons for America.* Cambridge, MA: Harvard University Press.

Vogel, M.E. (2008, June). The social origins of plea bargaining: An approach to the empirical study of discretionary leniency. *Journal of Law and Society,* Supp.

Volti, R. (2008). *An introduction to the sociology of work and occupations.* Los Angeles, CA: Pine Forge Press.

Vondra, J. (1996, April). Resolving conflicts over values. *Educational Leadership,* pp. 76–79.

Wade, N. (2001, November 5). Study finds genetic link between intelligence and size of some regions of the brain. *New York Times,* p. A15.

Wallerstein, I. (1974). *The modern world system: Capitalist agriculture and the origins of the European world-economy in the sixteenth century.* Orlando: Academic Press.

Wallerstein, I. (1999). The heritage of sociology, the promise of social science. Presidential address, XIVth World Congress of Sociology, Montreal, July 26, 1998. *Current Sociology, 47,* 1–43.

Wallerstein, I. M. (2004). *World-systems analysis: An introduction.* Durham, NC: Duke University Press.

Wallerstein, J., & Blakeslee, S. (1989). *Second chances: Men, women and children a decade after divorce.* New York: Ticknor & Fields.

Wallerstein, J., Lewis, J., & Blakeslee, S. (2000). *The unexpected legacy of divorce.* New York: Hyperion.

Wallerstein, M., & Moene, K. O. (2003). Does the logic of collective action explain the logic of corporatism? *Journal of Theoretical Politics, 15,* 271–297.

Wallis, J. (2009). For the healing of the nation: 46 million reasons for health-care reform. *Sojourner,* November.

Walsh, E. J., & Warland, R. H. (1983). Social movement involvement in the wake of a nuclear accident: Activists and free riders in the TMI area. *American Sociological Review, 48,* 764–780.

Walton, J. (1992). *Western times and water wars: State, culture, and rebellion in California.* Berkeley: University of California Press.

Waltzer, M. (1980). Pluralism. In S. Thernstrom, A. Orlov, & O. Handlin (Eds.), *Harvard encyclopedia of American ethnic groups.* Cambridge, MA: Belknap.

Wang, W., & Morin, R. (2009, November 24). *Home for the holidays . . . and every other day recession brings many young adults back to the nest.* Pew Research Center.

Wang, C.-Y., Haskell, W. L., Farrell, S. W., LaMonte, M. J., Blair, S. N., Curtin, L. R., Hughes, J. P., & Burt, V. L. (2010). Cardiorespiratory fitness levels among U.S. adults 20–49 years of age: Findings from the 1999–2004 National Health and Nutrition Examination Survey. *American Journal of Epidemiology, 171,* 426–435.

Warhus, M. (1997). *Another America: Native American maps and the history of our land.* New York: St. Martin's Press.

Warner, R. (2008). Turned on its head? Norms, freedom, and acceptable terms in Internet contracting. *Tulane Journal of Technology and Intellectual Property, 11: 1–34.*

Warner, R. S. (1998). Work in progress toward a new paradigm for the sociological study of religion in the United States. *American Journal of Sociology, 98,* 1044–1093.

Warner, W. L., Meeker, M., & Calls, K. (1949). *Social class in America: A manual of procedure for the measurement of social status.* Chicago: Science Research Associates.

Warner, W. L., & Lunt, Paul S. (1941). *The Social Life of a Modern Community.* New Haven, CT: Yale University Press.

Watson, J. B. (1930). *Behaviorism.* New York: Norton.

Wayne, L. (2001). For trade protesters, "slower, sadder songs." *New York Times,* October 28, sec. 3, p. 1.

Weaver, M. A. (1994, September 12). A fugitive from justice. *New Yorker,* pp. 46–60.

Weber, M. (1922). *Gesammelte aufsatze zur Religions—soziologie.* Tubingen, Germany: Mohr.

Weber, M. (1947). *The theory of social and economic organization* (A. M. Henderson & T. Parsons, Trans.). New York: Free Press.

Weber, M. (1958/1913). World religion. In H. Gerth & C. W. Mills (Trans. & Eds.), *From Max Weber: Essays in sociology.* New York: Oxford University Press.

Weber, M. (1958/1922). Economy and society. In H. Gerth & C. W. Mills (Trans. & Eds.), *From Max Weber: Essays in sociology.* New York: Oxford University Press.

Weber, M. (1962/1921). *The city.* New York: Collier.

Weber, M. (1963/1922). *The sociology of religion* (E. Fischoff, Trans.). Boston: Beacon.

Weber, M. (1968). The concept of citizenship. In S. N. Eisenstadt (Ed.), *Max Weber on charisma and institution building.* Chicago: University of Chicago Press.

Weber, M. (1974/1904). *The Protestant ethic and the spirit of capitalism* (T. Parsons, Trans.). New York: Scribner.

Weinberg, H. (1994). Marital reconciliation in the United States: Which couples are successful? *Journal of Marriage and the Family, 56,* 80–88.

Wekerle, G. (Ed.) (1980). *New space for women.* Boulder, CO: Westview.

Wells, M. J. (1996). *Strawberry fields: Politics, class, and work in California agriculture.* Ithaca, NY: Cornell University Press.

West, C. (1992). Learning to talk of race. *New York Times Magazine,* February 8, pp. 24–25.

West, C. (1996). *Race matters.* New York: Vintage.

West, W. (1998). The U.S. armed forces: Ready to fight—nicely? *Insight on the News,* March 9, p. 48.

Western, B. (2007, July). The prison boom and the decline of American citizenship. *Society,* pp. 30–36.

Westin, A. (1967). *Privacy and freedom.* New York: Atheneum.

Wheeler, S. R. (1998, April 10). Future showing promise: Donors revitalize religious group. *Denver Post,* pp. A1, A5.

Whitchurch, G. G., & Constantine, L. L. (1993). Systems theory. In P. G. Boss, W. J. Doherty, R. LaRossa, W. R. Schumm, & S. K. Steinmetz (Eds.), *Sourcebook of family theories and methods: A contextual approach.* New York: Plenum.

White, H. (1970). *Chains of opportunity.* Cambridge, MA: Harvard University Press.

Whiting, B. B., & Edwards, C. P. (1988). *Children of different worlds.* Cambridge, MA: Harvard University Press.

Whittle, J. (2003, May). War of the worldviews. *Church and State,* pp. 8–9.

WHO (World Health Organization) (2000). WHO global database on child growth and malnutrition, 2000. www.who.int.

Whorf, B. L. (1961). The relation of habitual thought and behavior to language. In J. B. Carroll (Ed.), *Language, thought, and reality: Selected writings of Benjamin Lee Whorf.* Cambridge, MA: MIT Press.

Whyte, W. F. (1943). *Street corner society.* Chicago: University of Chicago Press.

Whyte, W. F. (1948). Human relations in the restaurant industry. McGraw-Hill.

Wiepking, S., & Maas, I. (2005). Gender differences in poverty: A cross-national study. *European Sociological Review, 21,* 187–200.

Wilgoren, J. (2003, April 13). Where race matters. *The New York Times,* sec. 4A, p. 20.

Wilkinson, A. (1989, July 24). Sugarcane. *New Yorker,* pp. 56ff.

Williams, T. (1989). *The cocaine kids.* Reading, MA: Addison-Wesley.

Williams, T. (1992). *Crack house.* Reading, MA: Addison-Wesley.

Williams, T., & Kornblum, W. (1994). *The uptown kids: Struggle and hope in the projects.* New York: Putnam.

Williams, T., & Mousa, Y. (2010). Baghdad bombing streak stokes fear of new round of sectarian violence. *New York Times,* April 6.

Willis, P. (1983). Cultural production and theories of reproduction. In L. Barton & S. Walker (Eds.), *Race, class and education.* London: Croom-Helm.

Willmott, P., & Young, M. (1971). *Family and class in a London suburb.* London: New American Library.

Wills, G. (2006, June 8). Mousiness. *New York Review of Books,* pp. 14–19.

Wilson, B. L. (1993). *Mandating academic excellence: High school responses to state curriculum reform.* New York: Teachers College Press.

Wilson, C. R. (1987). The death of Bear Bryant: Myth and ritual in the modern South. *South Atlantic Quarterly, 86,* 282–295.

Wilson, E. O. (1975). *Sociobiology.* Cambridge, MA: Belknap.

Wilson, E. O. (1979). *On human nature.* New York: Bantam.

Wilson, E. O. (1998). *Consilience: The unity of knowledge.* New York: Knopf.

Wilson, J. Q. (1977). *Thinking about crime.* New York: Vintage.

Wilson, W. J. (1984). The urban underclass. In L. W. Dunbar (Ed.), *The minority report.* New York: Pantheon.

Wilson, W. J. (1998). Inner-city dislocations. *Society, 35,* 270–278.

Winch, R. F. (1958). *Mate selection.* New York: Harper.

Winik, M. F. (1996, November 13). Sharing a legacy of rescue. *The Christian Century,* pp. 1,112–1,116.

Wirth, L. (1945). The problem of minority groups. In R. Linton (Ed.), *The science of man in the world crisis.* New York: Columbia University Press.

Wirth, L. (1968/1938). Urbanism as a way of life. In S. F. Fava (Ed.), *Urbanism in world perspective: A reader.* New York: Crowell.

Witt, S. D. (1997). Parental influence on children's socialization to gender roles. *Adolescence, 32,* 253–360.

Wittfogel, K. (1957). *Oriental despotism: A comparative study of total power.* New Haven, CT: Yale University Press.

Wixen, B. N. (1979). Children of the rich. In J. D. Call, J. D. Noshpitz, R. L. Cohen, & I. N. Berlin (Eds.), *Basic handbook of child psychiatry.* New York: Basic Books.

Wolf, A. (1998). *One nation after all.* New York: Viking.

Wolf, E. R. (1984a, November 4). The perspective of the world. *New York Times Book Review,* pp. 13–14.

Wolf, E. R. (1984b, November 4). Unifying the vision. *New York Times Book Review,* p. 11.

Wolf, G. (2007). *Religion.* Detroit: Greenhaven Press.

Wolf, N. (1991). *The beauty myth.* New York: Morrow.

Wolf, P. J., Howell, W. G., & Peterson, P. E. (2000, February). *School choice in Washington, DC: An evaluation after one year.* Cambridge, MA: Harvard University, Program on Education Policy and Governance.

Wolfe, A. (1998). The black-white test score gap. *New York Times Book Review,* October 25, p. 15.

Wolfe, T. (1980). *The right stuff.* New York: Bantam.

Wolfgang, M. E., & Riedel, M. (1973). Race, judicial discretion, and the death penalty. *Annals of the American Academy of Political and Social Science, 407,* 119–133.

Women's Health Weekly (2003, May 15). White women three times more likely to commit suicide than black women, p. 23.

World Bank (2005). *World development report 2005: A better investment climate for everyone.* Washington, DC: World Bank.

World Justice Project. (2010). Rule of law index. www.worldjusticeproject.org.

Worldwatch Institute. (2009, November). Executive summary of the report *Global Environmental Change: The Threat to Human Health.*

Wren, D. J. (1997). Adolescent females' "voice" changes can signal difficulties for teachers and administrators. *Adolescence, 32,* 463–470.

Wright, E. O. (1979). *Class structure and economic determination.* Orlando: Academic Press.

Wright, E. O. (1989). *The debate on classes.* New York: Verso.

Wright, E. O. (1997). *Class counts: Comparative studies in class analysis.* New York: Cambridge University Press.

Wright, E. O., Costello, C., Hachen, D., & Sprague, J. (1982). The American class structure. *American Sociological Review, 47,* 702–726.

Wrigley, E. A. (1969). *Population and history.* New York: McGraw-Hill.

Wrigley, J. (1995). *Other people's children.* New York: Basic Books.

Wrong, D. H. (1961). The oversocialized conception of man in modern sociology. *American Sociological Review, 24,* 772–782.

Wrong, D. H. (1999). The oversocialized conception of man. New Brunswick, NJ: Transaction.

Wu, H., & Wakeman, C. (1995). *Bitter winds: A memoir of my years in China's gulag.* New York: Wiley.

Wuthnow, R. (1988). Sociology of religion. In N. J. Smelser (Ed.), *The handbook of sociology.* Newbury Park, CA: Sage.

Wuthnow, R. (1994). *Sharing the journey: Support groups and America's new quest for community.* New York: Free Press.

Yinger, J. (1995). *Closed doors, opportunities lost: The continuing costs of housing segregation.* New York: Russell Sage.

Young, J. (2002). Crime and social exclusion. In *Oxford handbook of criminology.*

Zakaria, F. (2003). *The future of freedom: Illiberal democracy at home and abroad.* New York: Norton.

Zangwill, I. (1909). *The melting pot.* New York: Macmillan.

Zhao, X-Y, & Zhang, J. (2008). The influence of folk law on the modernization of rule of law in China. *Journal of China Lawyer and Jurist, 4,* 23–25.

Zhou, X. (2005). The institutional logic of occupational prestige rankings: Reconceptualization and reanalysis. *American Journal of Sociology, 111,* 20–140.

Zill, N. (1995, Spring). Back to the future: Improving child indicators by remembering their origins. *Focus* (Institute for Research on Poverty), pp. 17–24.

Zilliox, R. (2010). Quoted in MSF (Medecins sans Frontieres). wwwdoctorswithout borders.org/news.

Zolberg, A. (2006). *A nation by design: Immigration policy in the fashioning of America.* Cambridge, MA: Harvard University Press/Russell Sage.

Zonana, H. (1997). The civil commitment of sex offenders. *Science, 278,* 1248–1249.

Zurcher, L. A., & Snow, D. A. (1981). Collective behavior and social movements. In M. Rosenberg & R. Turner (Eds.), *Social psychology: Sociological perspectives.* New York: Basic Books.

Zweigenhaft, R., & Domhoff, G. W. (1998). *Diversity in the power elite: Have women and minorities reached the top?* New Haven, CT: Yale University Press.

Credits

Chapter 1

Page 1, Ariel Skelley/Blend Images/Getty Images; Page 6, Michael Dwyer/Stock Boston Inc.; Page 8, Karl Marx Museum Trier/Alfredo Dagli Orti/The Art Archive/Picture Desk; Page 8, Bettmann/Corbis; Page 9, Max Weber (1864-920) c.1896-97 (b/w photo), German Photographer, (19th century)/Private Collection/Archives Charmet/The Bridgeman Art Library; Page 9, Spencer Arnold/ Hulton Archive/Getty Images; Page 9, R. Gates/Hulton Archive/Getty Images; Page 9, Manjula Giri; Page 11, Library of Congress Prints and Photographs Division [LC-USZ62-75205]; Page 13, Library of Congress Prints and Photographs Division [LC-USZ62-123822]; Page 15, FR Sport Photography/Alamy; Page 16, Bill Lai/The Image Works; Page 17, Agostino Pacciani/Anzenberger/Redux; Page 18, Crandall/The Image Works; Page 18, Zbigniew Bzdak/The Image Works; Page 19, ISSOUF SANOGO/AFP/Getty Images

Chapter 2

Page 22, Jim West/Alamy; Page 23, AP Wide World Photos; Page 25, UPI/Bettman/Corbis; Page 25, AP Photos; Page 30, Courtesy of Douglas Harper/Paradigm Publishers; Page 30, William Kornblum; Page 41, AP Photos; Page 41, AP Photos

Chapter 3

Page 44, Ansres Prerz Moreno/Picture Alliance/Photoshot; Page 46, Peabody Museum of Archaeology and Ethnology; Page 49, Photofest; Page 49, Michael Newman/PhotoEdit; Page 51, guatebrian/Alamy; Page 51, David Sutherland/Alamy; Page 51, John Cancalosi/Alamy; Page 51, Chenault Studios; Page, 53, AP Photo; Page 53, AP Photos; Page 54, Underwood & Underwood/Bettmann/Corbis; Page 54, Ron C. Angle/Getty Images News/Getty Images; Page 57, Everett Collection; Page 60, Bettmann/Corbis; Page 61, Richard B. Levine; Page 63, Hans Georg Roth/Documentary/Corbis; Page 65, Gianni Dagli Orti/The Art Archive/Alamy; Page 67, Margaret Washington/Cornell University; Page 69, The Art Archive/Egyptian Museum Cairo/Alfredo Dagli Orti /Picture Desk

Chapter 4

Page 72, Martin Rose/FIFA/Getty Images; Page 73, Eugene Richards/Reportage Division/Getty Images; Page 79, Eugene Richards/Reportage Division/Getty Images; Page 79, Eugene Richards/Reportage Division/Getty Images; Page 79, Eugene Richards/Reportage Division/Getty Images; Page 82, Picture Alliance/Photoshot; Page 86, John Neubauer/PhotoEdit; Page 86, Nicolas Sapieha/Art Resource, NY; Page 90, Nicholas Roberts/Reuters /Landov; Page 90, Photo by Romana Vystaova; Page 90, Rick Friedman/WpN/Photoshot

Chapter 5

Page 96, Photosindia/Getty Images; Page 97, AP Photo/Ramon Espinosa; Page 98, AP Photos; Page 102, S. Curtiss, Genie: A Psycholinguistic Study of a Modern Day "Wild Child" (Academic Press, 1977). Used by permission; Page 103, Nina Leen/Time Life Pictures/Getty Images; Page 107, Rich Puchalsky; Page 111, Erik Isakson/Fancy/Photolibrary; Page 113, The New York Public Library, Photography Collection, Miriam and Ira D. Wallach Division of Art, Prints and Photographs; Page 114, Spencer Grant/PhotoEdit; Page 116, AP Photos; Page 116, Courtesy of the Ludin Family; Page 117, Kuttig - People/Alamy; Page 119, AP Photos; Page 119, Eric Miller/Getty Images

Chapter 6

Page 125, Paul Chesley/Getty Images; Page 126, Columbia University Press; Page 128, Marc Goff/Shutterstock.com; Page 129, Ken Heyman/Woodfin Camp & Associates; Page 136, By kind permission of the Trustees of the Wallace Collection, London/Art Resource, NY; Page 139, Patrick Durand/Sygma/Corbis; Page 141, Zoran Karapancev/Shutterstock.com; Page 142, Lee Foster/Alamy; Page 148, Bettmann/Corbis; Page 148, AP Photos; Page 150, Ken Heyman/Woodfin Camp & Associates; Page 150, Juerco Boerner/Picture Press/Photolibrary; Page 150, Luiz C. Marigo/Peter Arnold Images/PhotoLibrary

Chapter 7

Page 153, Jason Stitt/ Shutterstock.com; Page 159, 2010 Her Majesty Queen Elizabeth II / The Bridgeman Art Library; Page 160, kzenon/iStockphoto.com; Page 162, Imagno/Hulton Archive /Getty Images; Page 163, Wallace Kirkland/Time Life Pictures/Getty Images; Page 166, Bob Krist/Documentary/Corbis; Page 166, John Nordell/Index Stock Imagery/Photolibrary; Page 171, Pat Behnke/Alamy; Page 172, Leonard Freed/Magnum Photos; Page 173, AP Photos/Ron Haviv/VII; Page 175, AP Photo/Gustavo Ferrari; Page 176, Richard Levine/Alamy; Page 176, Kayte Deioma/PhotoEdit

Chapter 8

Page 179, Dale Mitchell/Shutterstock.com; Page 182, Essex Institute, Salem, Massachusetts/Peabody Essex Museum; Page 183, Erich Schlegel/Dallas Morning News/Corbis; Page 187, UPI Photo/Monika Graff/Newscom; Page 188, Rafiqur Rahman/Reuters NewMedia Inc./Corbis; Page 189, Roger Viollet Collection/Getty Images; Page 189, Bettmann/Corbis; Page 189, Alex Segre/Alamy; Page 189, A. Ramey/PhotoEdit; Page 193, Beau Lark/Fancy/Corbis; Page 201, The Granger Collection, New York; Page 201, AP Photos; Page 205, Margaret Morton author/photographer The Tunnel (Yale University Press, 1995); Page 205, Margaret Morton author/photographer The Tunnel (Yale University Press, 1995); Page 205, Margaret Morton author/photographer The Tunnel (Yale University Press, 1995)

Chapter 9

Page 208, Jim West/Photolibrary; Page 212, Collection of Palm Springs Art Museum; Page 212, AP Photo/Jack Plunkett; Page 212, Adam Osterman/Getty Images; Page 217, TANNEN MAURY/EPA /Landov; Page 220, Steven Rubin/The Image Works; Page 221, Bettmann/Corbis; Page 222, Culver Pictures, Inc; Page 227, 1987 Matt Herron/Take Stock; Page 227, 1987 Matt Herron/Take Stock; Page 227, 1987 Matt Herron/Take Stock

Chapter 10

Page 230, Mike Goldwater/Alamy; Page 237, Catherine Karnow/Corbis; Page 238, Anne-Marie Palmer/Alamy; Page 241, Xinhua /Landov; Page 243, Bibliothèque nationale de France; Page 243, Chang W. Lee/The New York Times/Redux; Page 247, AP Photos; Page 247, ITAR-TASS Photo Agency/Alamy; Page 255, The Four Horsemen of the Apocalypse, 1887 (oil on canvas), Vasnetsov, Victor Mikhailovich (1848-1926)/Museum of Religion and Atheism, St. Petersburg, Russia/The Bridgeman Art Library; Page 255, Sebastiao Salgado/Contact Press Images; Page 255, Sebastiao Salgado/Contact Press Images; Page 255, AP Photo/Ramon Espinosa

Chapter 11

Page 258, George Robinson/Alamy; Page 259, Jose More; Page 262, Jeff Greenberg/PhotoEdit; Page 273, Bettman/Corbis; Page 273, ROD LAMKEY JR/AFP/Getty Images; Page 275, Barbara Norfleet; Page 278, Elliot Erwitt/

Magnum Photos; Page 283, Nathan Benn/ Woodfin Camp & Associates; Page 284, Frances Roberts/Alamy; Page 286, Michael Jacobson-Hardy; Page 286, Michael Jacobson-Hardy; Page 286, Michael Jacobson-Hardy

Chapter 12
Page 289, ullstein bild/The Image Works; Page 290, AP Photo/David Bookstaver; Page 293, AP Photo/Jae C. Hong; Page 294, Bettmann/ Corbis; Page 294, Richard Avery/Stock Boston Inc.; Page 299, Newscom; Page 300, Gibson, James F./Library of Congress[LC-DIG-cwpb-01005]; Page 304, Sarah Hadley/Alamy; Page 307, Robert Brenner/PhotoEdit; Page 309, Courtesy Devah Pager; Page 310, Chicago Historical Society; Page 310, Peter Turnley/Corbis; Page 315, Jim Noelker/The Image Works; Page 316, William Johnson/Stock Boston Inc.; Page 316, Steve Skjold/PhotoEdit; Page 317, Dan Dry/Dan Dry & Associates

Chapter 13
Page 320, David Bunting/Alamy; Page 321, Chenault Studios/Decature, AL; Page 323, Paul Chesley/Stone/Getty Images; Page 324, KHALED DESOUKI/AFP/Getty Images; Page 325, The Kobal Collection/Picture Desk; Page 326, Dorthea Lange/The Library of Congress[LC-DIG-fsa-8b29516]; Page 326, Bill Ganzel, www.ganzelgroup.com; Page 330, Musee Conde, Chantilly, France/Giraudon/The Bridgeman Art Library; Page 333, Catherine Ursillo/Photo Researchers, Inc.; Page 334, AP Photos; Page 336, Hulton Archive/Archive Photos/Getty Images; Page 336, Easterling Studios; Page 337, The State Hermitage Museum, St. Petersburg/Photo by Vladimir Terebenin, Leonard Kheifets, Yuri Molodkovets; Page 338, FPG/Getty Images; Page 341, Chenault Studios/Decatur, AL; Page 341, Chenault Studios/ Decatur, AL; Page 341, Chenault Studios/Decatur, AL; Page 341, Chenault Studios/Decatur, AL

Chapter 14
Page 345, Keith Brofsky/Brand X Pictures/Jupiter Images; Page 346, ©1974, 2010 The Imogen Cunningham Trust www.ImogenCunningham. com; Page 352, Jim West/Alamy; Page 354, Farrel Grehan/Photo Researchers, Inc.; Page 355, AP Photos; Page 356, Huntington Library/ SuperStock; Page 356, Dave Sartin Photography; Page 356, David Deas/DK Stock/Getty Images; Page 361, Hulton Archive/Archive Photos/Getty Images; Page 364, UN Photo/Grunzweig; Page 366, ©1975, 2010 The Imogen Cunningham Trust www.ImogenCunningham.com; Page 366, ©1976, 2010 The Imogen Cunningham

Trust www.ImogenCunningham.com; Page 366, ©1973, 2010 The Imogen Cunningham Trust www.ImogenCunningham.com; Page 366, ©1968, 2010 The Imogen Cunningham Trust www.ImogenCunningham.com

Chapter 15
Page 368, Queerstock, Inc./Alamy; Page 369, Eugene Richards/Reportage Division/Getty Images; Page 369, Eugene Richards/Reportage Division/Getty Images; Page 372, Paul Conklin/ PhotoEdit; Page 372, Stephanie Maze/Woodfin Camp & Associates; Page 378, Newscom; Page 389, Erich Lessing/Art Resource, NY; Page 393, Kevin Rivoli; Page 396, Phillip Jarrell/Digital Vision/Getty Images; Page 396, Frank Fournier/ Contact Press Images; Page 396, Dan Habib Photography

Chapter 16
Page 399, Bill Pugliano/Getty Images; Page 401, AP Photo/Khalil Hamra; Page 404, David Moore/ Black Star; Page 404, Robert Goldwitz/Photo Researchers, Inc.; Page 405, Reuters/Corbis; Page 408, Edmundo Morales; Page 409, Collection of the Museum of American Folk Art, New York. Gift of Mr. and Mrs. Edwin C. Braman, 1983. 14.4; Page 409, Collection of Sandra S. Jaffe/ Preservation Hall; Page 412, Susan Kornblum; Page 414, National Anthropological Archives; Page 415, Bettmann/Corbis; Page 416, Bob Daemmrich/ Stock Boston Inc.; Page 417, Ross Marino/Corbis; Page 422, AP Photo/Alessandra Tarantino; Page 424, Tony Savino/The Image Works; Page 426, AP Photos; Page 426, Steven Rubin/The Image Works; Page 426, Steven Rubin/The Image Works

Chapter 17
Page 429, David Smith/Alamy; Page 430, Courtesy of Jonas Chartock; Page 433, George Holton/Photo Researchers, Inc.; Page 450, Bettmann/Corbis; Page 450, Mark Lewis; Page 454, Courtesy of Allison Serafin; Page 454, AP Photos/Brandon Hoffman; Page 454, Courtesy of Orin Gutlerner

Chapter 18
Page 457, vario images GmbH & Co.KG/Alamy; Page 462, The Granger Collection, New York; Page 464, adrian arbib/Alamy; Page 465, Edmundo Morales; Page 467, Simon Marcus/ PhotoLibrary; Page 468, Library of Congress Prints and Photographs Division; Page 469, China Newsphoto /Reuters /Landov; Page 484, Gary Conner/PhotoEdit; Page 484, The Print Collector/Alamy; Page 484, Billy E. Barnes/ PhotoEdit

Chapter 19
Page 487, Jim Arbogast/Digital Vision/Jupiter Images; Page 488, SHAH MARAI/AFP/Getty Images; Page 491, The Granger Collection, New York; Page 494, Fuse/Jupiter Images; Page 496, David R. Frazier Photolibrary, Inc./Alamy; Page 499, Pete Winkel; Page 500, AP Photo/ Paul Sakuma; Page 506, M. Wuerker; Page 509, AP Photo; Page 509, Bettmann/Corbis; Page 509, /AP Photo/Ed Andrieski; Page 510, AP Photo/Charles Dharapak; Page 512, AP Photo/ Jerome Delay; Page 512, Dmitri Kessel/Time Life Pictures/Getty Images; Page 512, Stephanie Sinclair/Corbis

Chapter 20
Page 516, Rachel Epstein/PhotoEdit; Page 517, © Francois Dumont/Doctors Without Borders/ Médecins Sans Frontières (MSF); Page 517, © Remy Zilliox/Doctors Without Borders/ Médecins Sans Frontières (MSF); Page 519, DESMOND KWANDE/AFP/Getty Images; Page 526, Stedelijke Musea Brugge, Memlingmuseum, Sint-Janshospitaal, Bruges; Page 529, AP PHOTO/CP,Adrian Wyld

Chapter 21
Page 532, Brasil2/iStockphoto.com; Page 535, Corbis; Page 544, AP Photo/Lenny Ignelzi; Page 545, MARWAN IBRAHIM/AFP/Getty Images; Page 545, North Wind/North Wind Picture Archives; Page 547, Peter Menzel/Stock Boston Inc.; Page 548, Bettmann/Corbis; Page 548, Kirk Condyles; Page 553, Dennis Stock/Magnum Photos; Page 553, Joe Carini/The Image Works; Page 553, J. B. Diederich/Contact Press Images; Page 560, Kirk Condyles; Page 560, Wendy Ewald; Page 560, Wendy Ewald

Chapter 22
Page 563, Boris Breuer/Getty Images; Page 568, AP Photo/Jerome Delay; Page 569, Nik Wheeler/ Corbis; Page 572, Brandt/Hulton Archive/Getty Images; Page 573, Getty Images; Page 573, George Hall/Corbis; Page 576, Michael Grecco/ Stock Boston Inc.; Page 581, MOHSIN RAZA/ Reuters /Landov; Page 583, Réunion des Musées Nationaux/Art Resource, NY; Page 588, Betty Press/Woodfin Camp & Associates; Page 588, Kyaman goldsmiths, Anna Village, Cote d'Ivoire, Photograph by Eliot Elisofon, 1972, EEPA EECL 6951, Eliot Elisofon Photographic Archives, National Museum of African Art, Smithsonian Institution; Page 589, Courtesy of United Nations/ DPI; Page 589, "A Harrist Church in the Blokosso section of Abidjan, Cote d'Ivoire. Photography by Albert Votaw, 1969, T 2 IVC 1 AV 69, Eliot Elisofon Photographic Archives; National Museum of African Art, Smithsonian Institution

Name Index

Subject Index

military institutions and, 439
social change and, 431–433
social class and, 272–273
social mobility and, 445–447
Educational achievement, 443
Educational attainment, 439–440
barriers to, 440–443
definition of, 263, 439
inequality in, 262–263
Edwards, John N., 393
Ego, 100
Egypt, Ancient, art of, 69
Eisenhower, Dwight D., 498, 513
Elder abuse, 363
Elderly, 346
among hunter-gatherers, 81–82
death and dying, 364–365
global perspective on, 350–351
in hospital emergency rooms, 139
life expectancy and, 351–353
living alone, 381
in population growth rates, 538
in poverty, 360
poverty among, 279
social life of, 361
stereotypes of, 363
theories of aging, 363–364
Election polls, 33, 34
Elections. *See also* Voting
of 1948, 509
of 1960, 507
of 1964, 226–227
of 1972, 497
of 2000, 500
of 2008, 223, 293
of 2010, 227
campaign financing for, 275, 506, 508
in Iraq, 512
Electorate, 497
Electronic media, 467
Elias, Norbert, 63–64
Elites, 234
compensation for, 250–251
in power elite model, 503–505
social networks of, 133–134
Ellwood, David, 278
Emergency rooms (ERs), 73–74, 79, 139
Emotional intelligence, 110–111
Emotional reciprocity theory, 391
Emotions, in collective behavior, 213–214
Empirical investigation, 10
Employment
discrimination against job applicants with criminal records, 309
fairness in, 148–149
finding jobs, 482–483
gender inequality in, 331–333
health insurance tied to, 522
job ceilings in, 268
mentoring in, 474–475
occupational prestige and, 263–265
in postindustrial society, 269
sexual harassment at, 335–336
socialization and, 118

unemployment and, 472–473
of women, 340–342, 374–375
work environment, 286
of working class, 277
Empowerment, 322
Endogamy, 389
Endogenous forces, 565
Energy industry, 506
Engels, Friedrich, 193, 246, 587
England
British Empire of, 65
class system in, 234
Industrial Revolution in, 86
English (language), 147
Enlightenment, Age of, 7
Enron (firm), 506
Entitlements, 273
Environment
definition of, 5
impact on population of, 539–541
in nature versus nurture dispute, 98
socialization and, 111–118
Environmental issues
BP oil spill, 506
climate change as, 583–584
in Nigeria, 93
Environmental movement, 223
Environmental racism, 558–559
Epidemiology, 527–529
Episcopalian (Anglican) Church, 422–423
Epstein, Cynthia, 331
Equal Employment Opportunity
Commission, U.S. (EEOC), 336, 338
Equality
conflict between liberty and, 50
education for, 447–448
sexual, 164
types of, 282–284
Equality of opportunity, 282
Equality of result, 282
Erikson, Erik, 120–121, 355
Erikson, Kai, 558–560
Estates (feudal France), 242
Ethical issues
life-and-death issues, 24
in Milgram's obedience studies, 146
in social research, 35–36
Ethiopia, 231
Ethnic groups, 293–296. *See also* Race
definition of, 68, 293
in postindustrial society, 582–583
Ethnic inequality, 311–314
Ethnicity, 290–291
Ethnic nationalism, 308–310
Ethnic stratification, 304
Ethnocentrism, 61–62
Ethnomethodology, 139–140
Europe, economy of, 227
Europeans, 45
Evolution
culture and, 55–59
of humans, 80–81
models on social change based in, 585–586

Scopes trial on, 423–424
sexuality and, 157–158
Ewald, Wendy, 560
Exchange theory, 15
Executions. *See* Capital punishment
Exogamy, 389
Exogenous forces, 565
Experimental groups, 31
Experiments, 31–33
Exploration, 85–86
Expressive social movements, 215
Expulsion (of immigrants), 297–298
Extended families, 373
violence in, 387

Face work, 108
Fads, 211–213, 216
Failed states, 501–502, 566
Fairness principle, 137–138
Faisal (king, Iraq), 512
False consciousness, 478
Families. *See also* Kinship; Marriage
African American, 315–316
as agent of socialization, 113–114
among African Americans,
394–395
definition of, 372
economy and, 374–378
farm families, 281–282
in Freudian theory, 100–101
gay and lesbian, 392–393
gender roles and, 581–582
homelessness among, 280
as institution, 370–372
interactionism perspective on,
13–14
kinship diagrams of, 76
life cycles of, 379–382
parenting in, 378–379
sociological perspectives on,
382–387
varieties of structures of, 373–374
violence in, 387–388
Family and Medical Leave Act
(U.S., 1993), 90
Family life cycle, 379–380
Family of orientation, 373
Family of procreation, 373
Family role relations, 370
Family values, 384–385
Famine, 255
Farmers, 281–282
Fashions, 213
Fast food industry, 226, 528–529
Fathers, 369, 384
Fava, Sylvia, 555
Feagin, Joe R., 317
Felony disenfranchisement, 202–204, 309
Female circumcision, 321
Females. *See* Women
Feminism, 325
on sexuality, 156, 174–175
Wicca and, 420
women's movement, 338–340

Michelangelo, 409
Micro-level sociology, 5–6
    social change at, 565–566
    social stratification at, 240–241
Middle classes, 275–277
    as structural prerequisites for
      democracy, 503
    in suburbs, 555
Middle East
    accommodation in, 68–69
    hunter-gatherers in, 81
    suicide bombings in, 41
    wars in, 570
Middle-level sociology, 6
Middletown and *Middletown in Transition*
    (Lynd and Lynd), 266–267
Migration. *See also* Immigration and
    immigrants
    global health and, 519
    rural to urban, 544, 557
    to small towns and cities, 542
Milgram, Stanley, 145–146, 553
Military institutions, 510
    economic role of, 510–511
    education and, 439
    gay men and women in, 116, 182
    political power of, 498
    roles and statuses in, 78
    social change and, 511–513
    socialization of, 511
    spending on, 572–573
Mill, John Stuart, 466
Millenarian movements, 215–216
Millennials, 349–350, 352
Miller, Joanne, 582
Miller, Walter, 194
Mills, C. Wright, 249, 275, 503–505
Minerals Management Services
    (MMS), 506
Minority groups, 291, 294–295. *See also*
    African Americans; Hispanic Americans
    assimilation of, 302–306
    definition of, 294
    obesity and, 529
    poverty among elderly in, 360
Miss America Pageant, 333, 334
Mississippi, 294
Modernity, 417–418, 574–575
    antimodernist movements and,
      576–577
    postmodern critiques of, 575–576
Modernization, 225
    definition of, 573
    indicators of, 578–579
    social change and, 573–578
Modes, 114
Mohammed (prophet), 64, 408, 426
Monkeys, 103
Monogamy, 161
Monotheism, 407–408
Moral development
    religion and, 115–116
    socialization and, 99
    theories of, 109–111

Morales, Edmundo, 465
Mores, 52–53
Morgan, Gertrude, 409
Mormon Church, 413
Morris, Jan (James), 160
Morris, William, 484
Moskos, Charles, 439
Mother Mosque of America, 420
Mothers, 115, 369
Moynihan, Daniel Patrick, 306, 358
Multilinear models, 586
Multinational corporations, 463
Muncie (Indiana), 266–269
Murders
    capital punishment for, 200–201
    by firearms, 198
    genocide, 296
    school shootings, 188
    worldwide rates of, 199
Murray, Charles, 103–104, 292
Muslims, 64. *See also* Islam
    post-9/11 attacks on, 400–401
    in U.S., 420
Mutations, 56
Myrdal, Gunnar, 309

Naidu, Chandrababu, 533
NAMES Project Quilt, 227
Napier, Augustus, 383, 386
Narcotics
    prison sentences for, 202
    world market for, 465
Nasrin, Taslima, 187
National Commission on Excellence in
    Education, 443
National identities, 92
Nationalism, 308–310, 493–494
National Opinion Research Center
    (NORC), 33, 270–271, 278
    on religion in U.S., 415
National service, 439
National socialist regimes, 501
Nation-states, 75. *See also* Governments
    current crisis in, 494–496
    definition of, 494
    modernization and, 574
    societies and, 92–94
    state religions of, 402–403
Native Americans, 45
    accommodation and resistance by, 69
    Columbus's contacts with, 65
    European norms and, 52
    genocide against, 296–298
    inequality among, 232
    non-Native social movement
      supporting, 215
    as pariah group, 266
    Trail of Tears of, 298
    tribalism among, 310
    war's impact on, 569
    in wars with U.S., 573
    Wounded Knee and, 414–415
Natural disasters, 558
Natural experiments, 32

Natural selection, 56
Nature versus nurture, in socialization,
    98–105
Neighborhoods, 131
Nesbet, Robert, 585
Networks, 132–134
    for finding jobs, 482–483
Newburyport (Massachusetts), 267
Newman, Katherine, 188
New Orleans (Louisiana), 3
Newspapers, 507–508
New York City, Times Square in, 557
Niebuhr, H. Richard, 413
Nigeria, 92–94
    military in, 510
    tribal societies within, 93, 461
Nisbet, Robert A., 403–404
Nixon, Richard M., 507
No Child Left Behind Act, 442, 455
No-fault divorce, 391
Nonterritorial communities, 131
Normative order, 52
Norms, 48–49
    deviance as violation of, 181
    to reinforce stratification, 241
    social control and, 52–53, 183
    socialization to, 98
    technology governed by, 52
    typology of, 54–55
Nuclear families, 372–373

Obama, Barack
    economic philosophy of, 470
    on educational reform, 452
    election of, 293
    in election of 2008, 223
    on gay men and women in
      military, 116
    on health care, 521, 523–524
    on immigration, 209
    military controlled by, 510
    on political climate, 228
    on Teach for America, 431
Obama, Michelle, 293
Obedience, 145–146
Obesity, 334
    social epidemiology of, 527–528
Objective classes, 246–247, 270
Observations, 29–31
Occupational prestige, 263–265
Occupations. *See also* Employment
    inequality and, 263–265
    in pink-collar ghettos, 340–341
    professions, 480–481
    of working class, 277
Ogburn, William Fielding, 370, 575
Ogoni people (Nigeria), 93
Old people. *See* Elderly
Oligarchies, 498
Oliver, Melvin, 295
Olmstead, Frederick Law, 548
Open classrooms, 449
Open questions, 34
Open societies, 83